清华计算机图书 译丛

Machine Vision Algorithms and Applications
Second Edition

机器视觉算法与应用
（第2版）

[德] 卡斯特恩·斯蒂格（Carsten Steger）
[德] 马克乌斯·乌尔里克（Markus Ulrich） 著
[德] 克里斯琴·威德曼（Christian Wiedemann）

杨少荣 段德山 张勇 彭潇 偏召华 译

清华大学出版社
北京

Carsten Steger, Markus Ulrich, Christian Wiedemann

Machine Vision Algorithms and Applications, Second Edition

EISBN: 978-3-527-41365-2

Copyright © 2018 by Wiley-VCH

All Rights Reserved. This translation published under license.

Simplified Chinese translation edition is published and distributed exclusively by Tsinghua University Press under the authorization by McGraw-Hill Education (Asia) Co., within the territory of the People's Republic of China only, excluding Hong Kong, Macao SAR and Taiwan. Unauthorized export of this edition is a violation of the Copyright Act. Violation of this Law is subject to Civil and Criminal Penalties.

本书中文简体字翻译版由德国 Wiley-VCH 公司授权清华大学出版社在中华人民共和国境内(不包括中国香港、澳门特别行政区和中国台湾)独家出版发行。未经许可之出口，视为违反著作权法，将受法律之制裁。未经出版者预先书面许可，不得以任何方式复制或抄袭本书的任何部分。

北京市版权局著作权合同登记号　图字 01-2018-6459 号

本书封面贴有清华大学出版社防伪标签，无标签者不得销售。

版权所有，侵权必究。举报：010-62782989，beiqinquan@tup.tsinghua.edu.cn。

图书在版编目(CIP)数据

机器视觉算法与应用/(德)卡斯特恩·斯蒂格(Carsten Steger)，(德)马克乌斯·乌尔里克(Markus Ulrich)，(德)克里斯琴·威德曼(Christian Wiedemann)著；杨少荣等译.—2 版.—北京：清华大学出版社，2019（2023.12重印）

(清华计算机图书译丛)

书名原文: Machine Vision Algorithms and Applications, 2e

ISBN 978-7-302-51905-8

Ⅰ.①机… Ⅱ.①卡… ②马… ③克… ④杨… Ⅲ.①计算机视觉–计算方法–教材 Ⅳ.①TP302.7

中国版本图书馆 CIP 数据核字(2018)第 288568 号

责任编辑：龙启铭
封面设计：常雪影
责任校对：梁　毅
责任印制：沈　露

出版发行：清华大学出版社
　　　　网　　址：https://www.tup.com.cn，https://www.wqxuetang.com
　　　　地　　址：北京清华大学学研大厦 A 座　　邮　　编：100084
　　　　社　总　机：010-83470000　　邮　　购：010-62786544
　　　　投稿与读者服务：010-62776969，c-service@tup.tsinghua.edu.cn
　　　　质量反馈：010-62772015，zhiliang@tup.tsinghua.edu.cn
印 装 者：三河市铭诚印务有限公司
经　　销：全国新华书店
开　　本：185mm×260mm　　印　张：49.75　　字　数：1118 千字
版　　次：2008 年 11 月第 1 版　　2019 年 7 月第 2 版　　印　次：2023年12月第8次印刷
定　　价：128.00 元

产品编号：080742-01

List of Abbreviations   缩略词

ADC analog-to-digital converter.
AOI area of interest.
API application programming interface.
APS active pixel sensor.
BCS base coordinate system.
BGA ball grid array.
BRDF bidirectional reflectance distribution function.
CAD computer-aided design.
CCD charge-coupled device.
CCIR Comité consultatif international pour la radio.
CCS camera coordinate system.
CD compact disk.
CFA color filter array.
CLP Camera Link Protocol.
CMOS complementary metal-oxide semiconductor.
CNN convolutional neural network.
CPU central processing unit.
CWM continuous-wave-modulated.
DCS distributed control system.
DFT discrete Fourier transform.
DHCP Dynamic Host Configuration Protocol.
DLP digital light processing.
DMA direct memory access.
DMD digital micromirror device.
DN digital number.
DPS digital pixel sensor.
DR dynamic range.
DSNU dark signal nonuniformity.
DSP digital signal processor.

EIA Electronic Industries Alliance.
EM expectation maximization.
EMVA European Machine Vision Association.
FFT fast Fourier transform.
FPGA field-programmable gate array.
GenApi Generic application programming interface for configuring cameras.
GenCP Generic Control Protocol.
GenICam Generic Interface for Cameras.
GenTL Generic Transport Layer.
GHT generalized Hough transform.
GMM Gaussian mixture model.
GPIO general-purpose input/output.
GPU graphics processing unit.
GUI graphical user interface.
GVCP GigE Vision Control Protocol.
GVSP GigE Vision Streaming Protocol.
HTTP Hypertext Transfer Protocol.
HVS human visual system.
I/O input/output.
IC integrated circuit.
ICP iterative closest point.
ICS image coordinate system.
IDE integrated development environment.
IEEE Institute of Electrical and Electronics Engineers.
IP Internet Protocol.
IPCS image plane coordinate system.
IPv4 Internet Protocol, version.
IPv6 Internet Protocol, version.
IR infrared.
IRLS iteratively reweighted least-squares.

ISO International Organization for Standardization.
kNN k nearest-neighbor.
LCD liquid-crystal display.
LCOS liquid crystal on silicon.
LED light-emitting diode.
LLA Link-Local Address.
LUT lookup table.
LVDS low-voltage differential signaling.
MCS model coordinate system.
MLP multilayer perceptron.
NCC normalized cross-correlation.
NN nearest-neighbor.
NTSC National Television System Committee.
OCR optical character recognition.
PAL phase alternating line.
PC personal computer.
PCB printed circuit board.
PFNC pixel format naming convention.
PLC programmable logic controller.
PLL phase-locked loop.
PM pulse-modulated.
PRNU photoresponse nonuniformity.
PTP Precision Time Protocol.
RANSAC random sample consensus.
ReLU rectified linear unit.
ROI region of interest.
SAD sum of absolute gray value differences.
SCARA Selective Compliant Arm for Robot Assembly.
SED mean squared edge distance.
SFNC standard features naming convention.
SGD stochastic gradient descent.
SLR single-lens reflex.
SNR signal-to-noise ratio.
SSD sum of squared gray value differences.
SVD singular value decomposition.
SVM support vector machine.
TCP Transmission Control Protocol.
TCS tool coordinate system.
TOF time-of-flight.
U3VCP USB3 Vision Control Protocol.
U3VSP USB3 Vision Streaming Protocol.
UDP User Datagram Protocol.
USB Universal Serial Bus.
UV ultraviolet.
WCS world coordinate system.
WWW World Wide Web.
XML extensible markup language.

Preface to the Second Edition 第 2 版前言

It has been almost exactly ten years since the first edition of this book was published. Many things that we stated in the preface to the first edition of this book have remained constant. Increasing automation has continued to provide the machine vision industry with above-average growth rates. Computers have continued to become more powerful and have opened up new application areas.

On the other hand, many things have changed in the decade since the first edition was published. Efforts to standardize camera–computer interfaces have increased significantly, leading to several new and highly relevant standards. MVTec has participated in the development of many of these standards. Furthermore, sensors that acquire 3D data have become readily available in the machine vision industry. Consequently, 3D machine vision algorithms play an increasingly important role in machine vision applications, especially in the field of robotics. Machine learning (classification) is another technology that has become increasingly important.

The second edition of this book has been extended to reflect these changes. In Chapter 2, we have added a discussion of the latest camera–computer interface and image acquisition standards. Furthermore, we have included a discussion of 3D image acquisition devices. Since many of these sensors use Scheimpflug optics, we have also added a discussion of this important principle. In Chapter 3, we have extended the description of the algorithms that are used in 3D image acquisition

本书第 1 版出版差不多已经十年了。在第 1 版前言中讲的许多事情没有发生变化。日益更新的自动化技术给机器视觉工业持续带来高于平均水平的增长速度，计算机变得更加强大并且开辟了一些新的应用领域。

另外一方面，自从本书第 1 版出版以来的十年中很多事情发生了变化。对标准化摄像机-计算机接口所做的努力已经显著增加，产生了许多新的且高度相关的标准。MVTec 参与了许多新标准的开发。此外，机器视觉工业领域中采集三维数据的传感器已经很容易得到，因此，三维机器视觉算法在机器视觉应用中扮演着越来越重要的角色，尤其是在机器人领域。另外一个越来越重要的技术是机器学习（分类）。

在本书第 2 版中扩展的内容已经体现了这些变化。在第 2 章中，我们讨论了最新的摄像机-计算机接口和图像采集标准。而且，包含了三维图像采集设备的讨论。由于这些传感器许多都用到了沙姆光学，我们也增加了对这一重要原理的讨论。在第 3 章中，扩展了三维图像采集设备用来进行三维重构的算法，此外，介绍了使用沙姆光学的摄像机模型和标定算法。为

devices to perform the 3D reconstruction. Furthermore, we describe camera models and calibration algorithms for cameras that use Scheimpflug optics. The growing importance of 3D processing is reflected by new sections on hand–eye calibration and 3D object recognition. Furthermore, the section on classification has been extended by algorithms that have become increasingly important (in particular, novelty detection and convolutional neural networks). In Chapter 4, we have added two new application examples that show how the 3D algorithms can be used to solve typical 3D applications. Overall, the book has grown by more than 35%.

The applications we present in this book are based on the machine vision software HALCON, developed by MVTec Software GmbH. To make it possible to also publish an electronic version of this book, we have changed the way by which HALCON licenses can be obtained. MVTec now provides the HALCON Student Edition for selected universities and academic research institutes. Please contact your lecturer or local distributor to find out whether you are entitled to participate in this program. Note that the student version of HALCON 8.0 is no longer available. To download the applications discussed in Chapter 4, please visit www.machine-vision-book.com.

The first edition of this book has been used extensively in the lectures "Image understanding I: Machine vision algorithms" given by Carsten Steger at the Department of Informatics of the Technical University of Munich, "Industrial Photogrammetry" given by Markus Ulrich at the Department of Civil, Geo, and Environmental

了体现日益增长的三维图像处理的重要性，增加了新的章节：手眼标定和三维物体识别。而且，分类章节对算法进行了扩展，这些算法也变得越来越重要（尤其是，异常（新奇）检测和卷积神经网络）。在第 4 章中，增加了两个实际案例来介绍三维算法如何用于解决典型的三维应用问题。总之，本书第 2 版内容增加了 35% 以上。

本书中的应用基于德国 MVTec Software GmbH 公司开发的 HALCON 软件。为了出版本书的电子版，我们更改了 HALCON 许可的获取方式。MVTec 对挑选出的一些大学和学术研究机构提供 HALCON 学生版，请联系你的讲师或本地代理商来了解是否有权参与这个计划（是否在授权名单中）。注意 HALCON 8.0 学生版已经不再可用。要下载第 4 章中的应用案例，请访问 www.mahine-vision-book.cn。

本书第 1 版内容已经广泛用于以下讲座中，Carsten Steger 在慕尼黑科技大学信息系所开设的讲座"图像理解 I：机器视觉算法"，Markus Ulrich 在慕尼黑科技大学土木、地球和环境工程系开设的讲座"工业摄影测量法"，以及 Markus Ulrich 在卡尔

Engineering of the Technical University of Munich, and "Industrielle Bildverarbeitung und Machine Vision" given by Markus Ulrich at the Institute of Photogrammetry and Remote Sensing of the Karlsruhe Institute of Technology. We have integrated the feedback we have received from the students into this edition of the book. A substantial part of the new material is based on the lecture "Image understanding II: Robot vision" given by Carsten Steger since 2011 at the Department of Informatics of the Technical University of Munich.

We would like to express our gratitude to several of our colleagues who have helped us in the writing of the second edition of this book. Jean-Marc Nivet provided the images in Figures 3.129–3.131 and proof-read Sections 2.5 and 3.10. Julian Beitzel supported us by preparing the pick and place example described in Section 4.4. We are also grateful to the following colleagues for proof-reading various sections of this book: Thomas Hopfner (Section 2.4), Christoph Zierl (Section 2.4), Andreas Hofhauser (Section 3.12.1), Bertram Drost (Section 3.12.3), Tobias Bötger (Section 3.13), Patrick Follmann (Sections 3.13 and 3.15.3.4), and David Sattlegger (Section 3.15.3.4). Finally, we would like to thank Martin Preuß and Stefanie Volk of Wiley-VCH who were responsible for the production of this edition of the book.

We invite you to send us suggestions on how to improve this book. You can reach us at authors@machine-vision-book.com.

München, July 2017

斯鲁厄理工学院摄影测量与遥感研究所开设的讲座"工业图像处理和机器视觉"。我们把收到的学生反馈也整合到本书第 2 版中了，另外新资料的很大一部分基于 Carsten Steger 自从 2011 年以来在慕尼黑科技大学信息系所开设的讲座"图像理解 II：机器人视觉"。

在此我们想感谢在撰写本书第 2 版过程中提供帮助的几位同事，Jean-Marc Nivet 提供了图 3.129-3.131 中的图像并且校对了 2.5 节和 3.10 节。Julia Beitzel 帮助准备了 4.14 节机器人取放案例。我们也感谢以下同事对本书各个章节所做的校对工作：Thomas Hopfner (2.4 节), Christoph Zierl (2.4 节), Andreas Hofhauser (3.12.1 节), Bertram Drost (3.12.3 节), Tobias Böttger(3.13 节), Patrick Follmann (3.13 节和 3.15.3.4 节), 和 David Sattlegger（3.15.3.4 节）。最后，我们由衷感谢 Wiley-VCH 出版社负责本书第 2 版出版工作的 Martin Preuß 和 Stefanie Volk。

欢迎大家就如何完善本书提供宝贵意见，我们的联系方式是 authors@machine-vision-book.com。

Carsten Steger, Markus Ulrich, Christian Wiedemann

Preface to the First Edition 第 1 版前言

The machine vision industry has enjoyed a growth rate well above the industry average for many years. Machine vision systems currently form an integral part of many machines and production lines. Furthermore, machine vision systems are continuously deployed in new application fields, in part because computers get faster all the time and thus enable applications to be solved that were out of reach just a few years ago.

Despite its importance, there are few books that describe in sufficient detail the technology that is important for machine vision. While there are numerous books on image processing and computer vision, very few of them describe the hardware components that are used in machine vision systems to acquire images (illuminations, lenses, cameras, and camera–computer interfaces). Furthermore, these books often only describe the theory, but not its use in real-world applications. Machine vision books, on the other hand, often do not describe the relevant theory in sufficient detail. Therefore, we feel that a book that provides a thorough theoretical foundation of all the machine vision components and machine vision algorithms, and that gives non-trivial practical examples of how they can be used in real applications, is highly overdue.

The applications we present in this book are based on the machine vision software HALCON, developed by MVTec Software GmbH. To enable you to get a hands-on experience with the machine vision algorithms and applications that we

由于计算机的运算速度逐年增长，机器视觉在许多新的领域不断得到应用，而在几年前这些应用还无法实现。机器视觉多年来的增长速度均高于工业平均增长速度，目前机器视觉已成为许多机器和生产线的一部分。

目前市面上缺少详细介绍机器视觉技术的书籍，尽管这类书籍非常重要。已有的大量书籍介绍了图像处理及计算机视觉，但书中对于机器视觉中获取图像的硬件部分，如照明、镜头、摄像机及摄像机与计算机的接口却少有介绍，这些书籍更多的是介绍机器视觉的理论，而不是如何在现实中应用。另一方面，机器视觉的书籍对于机器视觉的相关理论又没有足够详细的介绍。因此，我们觉得一本充分介绍机器视觉硬件各个部分的理论基础及算法、同时提供如何在实际中应用的典型案例的书是非常必要的。

本书中的应用基于德国 MVTec Software GmbH 公司研发的 HALCON 软件。为使读者更好地掌握书中所讲机器视觉算法及应用，书中含有免费下载学生版 HALCON 软件及

discuss, this book contains a registration code that enables you to download, free of charge, a student version of HALCON as well as all the applications we discuss. For details, please visit www.machine-vision-book.com.

应用案例的注册码。更多详细信息请访问 www.machine-vision-book.cn。

While the focus of this book is on machine vision applications, we would like to emphasize that the principles we will present can also be used in other application fields, e.g., photogrammetry or medical image processing.

本书虽然重点讨论机器视觉，但书中所述原理同样可以用于如照相测量、医学图像处理等其他应用领域。

We have tried to make this book accessible to students as well as practitioners (OEMs, system integrators, and end-users) of machine vision. The text requires only a small amount of mathematical background. We assume that the reader has a basic knowledge of linear algebra (in particular, linear transformations between vector spaces expressed in matrix algebra), calculus (in particular, sums and differentiation and integration of one- and two-dimensional functions), Boolean algebra, and set theory.

本书既适合学生，同时也适合于 OEM 厂商、系统集成商及最终用户这样的机器视觉从业者。本书只要求读者稍有数学知识背景，对于线性代数和微积分有所了解，特别是了解以矩阵表示的矢量空间线性变换和一维、二维函数和、差分及积分。

This book is based on a lecture and lab course entitled "Machine vision algorithms" that Carsten Steger has given annually since 2001 at the Department of Informatics of the Technical University of Munich. Parts of the material have also been used by Markus Ulrich in a lecture entitled "Close-range photogrammetry" given annually since 2005 at the Institute of Photogrammetry and Cartography of the Technical University of Munich. These lectures typically draw an audience from various disciplines, e.g., computer science, photogrammetry, mechanical engineering, mathematics, and physics, which serves to emphasize the interdisciplinary nature of machine vision.

本书主要基于 Carsten Steger 先生自 1999 年以来每年为慕尼黑科技大学信息系所作的题为"机器视觉算法"的讲座及实验课程。部分材料来源于 Markus Ulrich 先生自 2005 年每年在慕尼黑科技大学测绘研究所所作的题为"近距离照相测量"的讲座。这些讲座的听众既有来自计算机科学、照相测量、机械工程，也有物理、数学等学科，充分体现了机器视觉的多学科交叉的本质。

We would like to express our gratitude to several of our colleagues who have helped us in the writing of this book. Wolfgang Eckstein, Juan Pablo de la Cruz Gutiérez, and Jens Heyder designed or wrote several of the application examples in Chapter 4. Many thanks also go to Gerhard Blahusch, Alexa Zierl, and Christoph Zierl for proofreading the manuscript. Finally, we would like to express our gratitude to Andreas Thoβ and Ulrike Werner of Wiley-VCH for having the confidence that we would be able to write this book during the time HALCON 8.0 was completed.

We invite you to send us suggestions on how to improve this book. You can reach us at authors@machine-vision-book.com.

在此我们要感谢 Wolfgang Eckstein, Juan Pablo de la Cruz Gutiérrez 及 Jens Heyder 设计或撰写了第 4 章部分应用案例。感谢 Gerhard Blahusch, Alexa Zierl 及 Christoph Zierl 校对原稿。最后我们衷心感谢 Wiley-VCH 出版社的 Andreas Thoβ 和 Ulrike Werner，是他们使我们在 HALOCN 8.0 研制过程中有信心完成本书。

欢迎大家就如何完善本书提出宝贵意见。我们的联系方式是 authors@machine-vision-book.com。

München, May 2007

Carsten Steger, Markus Ulrich, Christian Wiedemann

目 录

缩略词 I

第 2 版前言 III

第 1 版前言 VII

1 Introduction 简介 1

2 Image Acquisition 图像采集 6
   2.1 Illumination 照明 6
      2.1.1 Electromagnetic Radiation 电磁辐射 6
      2.1.2 Types of Light Sources 光源类型 9
      2.1.3 Interaction of Light and Matter 光与被测物间的相互作用 12
      2.1.4 Using the Spectral Composition of the Illumination 利用照明的光谱 14
      2.1.5 Using the Directional Properties of the Illumination 利用照明的方向性 18
   2.2 Lenses 镜头 25
      2.2.1 Pinhole Cameras 针孔摄像机 26
      2.2.2 Gaussian Optics 高斯光学 27
      2.2.3 Depth of Field 景深 37
      2.2.4 Telecentric Lenses 远心镜头 42
      2.2.5 Tilt Lenses and the Scheimpflug Principle 倾斜镜头和沙姆定律 48
      2.2.6 Lens Aberrations 镜头的像差 53
   2.3 Cameras 摄像机 61
      2.3.1 CCD Sensors CCD 传感器 62
      2.3.2 CMOS Sensors CMOS 传感器 69
      2.3.3 Color Cameras 彩色摄像机 72
      2.3.4 Sensor Sizes 传感器尺寸 75
      2.3.5 Camera Performance 摄像机性能 77
   2.4 Camera–Computer Interfaces 摄像机-计算机接口 84
      2.4.1 Analog Video Signals 模拟视频信号 85
      2.4.2 Digital Video Signals 数字视频信号 92
      2.4.3 Generic Interfaces 通用接口 116
      2.4.4 Image Acquisition Modes 图像采集模式 131
   2.5 3D Image Acquisition Devices 三维图像采集设备 134

2.5.1 Stereo Sensors 立体视觉传感器 135
2.5.2 Sheet of Light Sensors 片光（激光三角测量）传感器 139
2.5.3 Structured Light Sensors 结构光传感器 142
2.5.4 Time-of-Flight Cameras 飞行时间摄像机 151

3 Machine Vision Algorithms 机器视觉算法 157
3.1 Fundamental Data Structures 基本数据结构 157
3.1.1 Images 图像 158
3.1.2 Regions 区域 160
3.1.3 Subpixel-Precise Contours 亚像素精度轮廓 164
3.2 Image Enhancement 图像增强 165
3.2.1 Gray Value Transformations 灰度值变换 165
3.2.2 Radiometric Calibration 辐射标定 170
3.2.3 Image Smoothing 图像平滑 181
3.2.4 Fourier Transform 傅里叶变换 198
3.3 Geometric Transformations 几何变换 205
3.3.1 Affine Transformations 仿射变换 206
3.3.2 Image Transformations 图像变换 209
3.3.3 Projective Image Transformations 投影图像变换 216
3.3.4 Polar Transformations 极坐标变换 218
3.4 Image Segmentation 图像分割 220
3.4.1 Thresholding 阈值分割 220
3.4.2 Extraction of Connected Components 提取连通区域 233
3.4.3 Subpixel-Precise Thresholding 亚像素精度阈值分割 237
3.5 Feature Extraction 特征提取 240
3.5.1 Region Features 区域特征 241
3.5.2 Gray Value Features 灰度值特征 248
3.5.3 Contour Features 轮廓特征 254
3.6 Morphology 形态学 256
3.6.1 Region Morphology 区域形态学 257
3.6.2 Gray Value Morphology 灰度值形态学 282
3.7 Edge Extraction 边缘提取 288
3.7.1 Definition of Edges 边缘定义 289
3.7.2 1D Edge Extraction 一维边缘提取 295
3.7.3 2D Edge Extraction 二维边缘提取 305
3.7.4 Accuracy and Precision of Edges 边缘的准确度和精确度 317

- 3.8 Segmentation and Fitting of Geometric Primitives 几何基元的分割和拟合 328
  - 3.8.1 Fitting Lines 直线拟合 329
  - 3.8.2 Fitting Circles 圆拟合 336
  - 3.8.3 Fitting Ellipses 椭圆拟合 338
  - 3.8.4 Segmentation of Contours 轮廓分割 341
- 3.9 Camera Calibration 摄像机标定 347
  - 3.9.1 Camera Models for Area Scan Cameras with Regular Lenses
    普通镜头与面阵摄像机组成的摄像机模型 349
  - 3.9.2 Camera Models for Area Scan Cameras with Tilt Lenses
    倾斜镜头和面阵摄像机组成的摄像机模型 357
  - 3.9.3 Camera Model for Line Scan Cameras 线阵摄像机的摄像机模型 363
  - 3.9.4 Calibration Process 标定过程 370
  - 3.9.5 World Coordinates from Single Images 从单幅图像中提取世界坐标 380
  - 3.9.6 Accuracy of the Camera Parameters 摄像机参数的准确度 386
- 3.10 3D Reconstruction 三维重构 390
  - 3.10.1 Stereo Reconstruction 立体重构 390
  - 3.10.2 Sheet of Light Reconstruction 激光三角测量法（片光）重建 412
  - 3.10.3 Structured Light Reconstruction 结构光重建 416
- 3.11 Template Matching 模板匹配 424
  - 3.11.1 Gray-Value-Based Template Matching 基于灰度值的模板匹配 426
  - 3.11.2 Matching Using Image Pyramids 使用图形金字塔进行匹配 434
  - 3.11.3 Subpixel-Accurate Gray-Value-Based Matching 基于灰度值的亚像素精度匹配 441
  - 3.11.4 Template Matching with Rotations and Scalings 带旋转与缩放的模板匹配 441
  - 3.11.5 Robust Template Matching 可靠的模板匹配算法 443
- 3.12 3D Object Recognition 三维物体识别 476
  - 3.12.1 Deformable Matching 变形匹配 478
  - 3.12.2 Shape-Based 3D Matching 基于形状的三维匹配 493
  - 3.12.3 Surface-Based 3D Matching 基于表面的三维匹配 510
- 3.13 Hand–Eye Calibration 手眼标定 526
  - 3.13.1 Introduction 前言 527
  - 3.13.2 Problem Definition 问题定义 529
  - 3.13.3 Dual Quaternions and Screw Theory 对偶四元数和螺旋理论 533
  - 3.13.4 Linear Hand–Eye Calibration 线性手眼标定 540
  - 3.13.5 Nonlinear Hand–Eye Calibration 非线性手眼标定 545
  - 3.13.6 Hand–Eye Calibration of SCARA Robots SCARA 机器人手眼标定 547
- 3.14 Optical Character Recognition 光学字符识别 (OCR) 551
  - 3.14.1 Character Segmentation 字符分割 552

3.14.2　Feature Extraction　特征提取 555
3.15　Classification　分类 560
 3.15.1　Decision Theory　决策理论 560
 3.15.2　Classifiers Based on Estimating Class Probabilities
    基于估计概率的分类器 566
 3.15.3　Classifiers Based on Constructing Separating Hypersurfaces
    基于构造分离超曲面的分类器 573
 3.15.4　Example of Using Classifiers for OCR　使用分类器用于 OCR 的例子 606

4　Machine Vision Applications　机器视觉应用 608
4.1　Wafer Dicing　半导体晶片切割 608
 4.1.1　Determining the Width and Height of the Dies　确定芯片的宽度和高度 609
 4.1.2　Determining the Position of the Dies　确定芯片的位置 612
 4.1.3　exercises　练习 616
4.2　Reading of Serial Numbers　序列号读取 617
 4.2.1　Rectifying the Image Using a Polar Transformation
   使用极坐标变换对图像进行校正 618
 4.2.2　Segmenting the Characters　字符分割 622
 4.2.3　Reading the Characters　读取字符 624
 4.2.4　exercises　练习 625
4.3　Inspection of Saw Blades　锯片检测 626
 4.3.1　Extracting the Saw Blade Contour　提取锯片的轮廓 627
 4.3.2　Extracting the Teeth of the Saw Blade　提取锯片上的锯齿 628
 4.3.3　Measuring the Angles of the Teeth of the Saw Blade　测量锯片锯齿的角度 630
 4.3.4　exercises　练习 632
4.4　Print Inspection　印刷检测 632
 4.4.1　Creating the Model of the Correct Print on the Relay
   创建继电器上正确印刷信息的模型 633
 4.4.2　Creating the Model to Align the Relays　创建一个用于对齐继电器的模型 635
 4.4.3　Performing the Print Inspection　印刷检测 636
 4.4.4　exercises　练习 637
4.5　Inspection of Ball Grid Arrays　BGA 封装检查 638
 4.5.1　Finding Balls with Shape Defects　找出有形状缺陷的焊锡球 639
 4.5.2　Constructing a Geometric Model of a Correct BGA
   构造一个正确的 BGA 几何模型 642
 4.5.3　Finding Missing and Extraneous Balls　检测缺失或多余的焊锡球 644
 4.5.4　Finding Displaced Balls　检测位置错误的焊锡球 647

	4.5.5	exercises 练习 649

4.6 Surface Inspection 表面检测 649
- 4.6.1 Segmenting the Doorknob 分割门把手 651
- 4.6.2 Finding the Surface to Inspect 找到需要检测的平面 652
- 4.6.3 Detecting Defects 缺陷检测 657
- 4.6.4 exercises 练习 660

4.7 Measurement of Spark Plugs 火花塞测量 660
- 4.7.1 Calibrating the Camera 标定摄像机 662
- 4.7.2 Determining the Position of the Spark Plug 确定火花塞的位置 664
- 4.7.3 Performing the Measurement 测量 666
- 4.7.4 exercises 练习 669

4.8 Molding Flash Detection 模制品披峰检测 669
- 4.8.1 Molding Flash Detection Using Region Morphology
  区域形态学方法检测模制品毛边 671
- 4.8.2 Molding Flash Detection with Subpixel-Precise Contours
  使用亚像素精度轮廓检测模制品毛边 675
- 4.8.3 exercises 练习 679

4.9 Inspection of Punched Sheets 冲孔板检查 679
- 4.9.1 Extracting the Boundaries of the Punched Sheets 提取冲孔板的边界 ... 681
- 4.9.2 Performing the Inspection 边缘检测 683
- 4.9.3 exercises 练习 685

4.10 3D Plane Reconstruction with Stereo
使用双目立体视觉系统进行三维平面重构 685
- 4.10.1 Calibrating the Stereo Setup 标定立体视觉系统 686
- 4.10.2 Performing the 3D Reconstruction and Inspection 进行三维重构及检测 ... 688
- 4.10.3 exercises 练习 695

4.11 Pose Verification of Resistors 电阻姿态检验 695
- 4.11.1 Creating Models of the Resistors 创建电阻模型 696
- 4.11.2 Verifying the Pose and Type of the Resistors 检测电阻的位姿和类型 ... 700
- 4.11.3 exercises 练习 703

4.12 Classification of Non-Woven Fabrics 非织造布分类 704
- 4.12.1 Training the Classifier 训练分类器 704
- 4.12.2 Performing the Texture Classification 进行纹理分类 708
- 4.12.3 exercises 练习 710

4.13 Surface Comparison 表面比对 711
- 4.13.1 Creating the Reference Model 创建参考模型 711
- 4.13.2 Reconstructing and Aligning Objects 重构和对齐物体 714

  4.13.3 Comparing Objects and Classifying Errors　对比物体并且对错误进行分类........715

  4.13.4 exercises　练习 722

 4.14 3D Pick-and-Place　三维取放 722

  4.14.1 Performing the Hand–Eye Calibration　手眼标定 723

  4.14.2 Defining the Grasping Point　定义抓取点 728

  4.14.3 Picking and Placing Objects　取放物体 731

  4.14.4 exercises　练习 733

References 参考文献 735

Index 索引 751

1 Introduction

1. 简介

Machine vision is one of the key technologies in manufacturing because of increasing demands on the documentation of quality and the traceability of products. It is concerned with engineering systems, such as machines or production lines, that can perform quality inspections in order to remove defective products from production or that control machines in other ways, e.g., by guiding a robot during the assembly of a product.

由于对产品质量记录及可追溯性文档的需求越来越多，机器视觉已成为生产过程中关键技术之一。在机器或生产线上，机器视觉可以检测产品质量以便将不合格产品剔除，或者指导机器人完成组装工作，因此，机器视觉与整个系统密切相关。

Some of the common tasks that must be solved in machine vision systems are as follows (Fraunhofer Allianz Vision, 2003):

下面举几个常见的、必须有机器视觉系统参与的任务（Fraunhofer Allianz Vision，2003）。

- Object identification is used to discern different kinds of objects, e.g., to control the flow of material or to decide which inspections to perform. This can be based on special identification symbols, e.g., character strings or bar codes, or on specific characteristics of the objects themselves, such as their shape.

- 目标识别：用来甄别不同的被测物体。比如物流控制或者根据不同目标进行不同的检测。识别可以基于特殊的识别特征，比如字符串、条码或被测物体的形状等特性。

- Position detection is used, for example, to control a robot that assembles a product by mounting the components of the product at the correct positions, such as in a pick-and-place machine that places electronic components onto a printed circuit board (PCB). Position detection can be performed in two or three dimensions, depending on the requirements of the application.

- 位置检测：用来控制机器人在组装生产线上将产品的组件放置到正确位置。如贴片机就是将元器件放置到印刷电路板（PCB）上的正确位置。根据不同应用，位置检测可以是二维或三维的。

- Completeness checking is typically performed after a certain stage of the assembly of a pro-

- 完整性检测：通常用于产品装配进行到一定阶段后。比如当元器件

duct has been completed, e.g., after the components have been placed onto a PCB, to ensure that the product has been assembled correctly, i.e., that the right components are in the right place.

- Shape and dimensional inspection is used to check the geometric parameters of a product to ensure that they lie within the required tolerances. This can be used during the production process but also after a product has been in use for some time to ensure that the product still meets the requirements despite wear and tear.

- Surface inspection is used to check the surface of a finished product for imperfections such as scratches, indentations, protrusions, etc.

Figure 1.1 displays an example of a typical machine vision system. The object (1) is transported mechanically, e.g., on a conveyor belt. In machine vision applications, we would often like to image the object in a defined position. This requires mechanical handling of the object and often also a trigger that triggers the image acquisition, e.g., a photoelectric sensor (4). The object is illuminated by a suitably chosen or specially designed illumination (3). Often, screens (not shown) are used to prevent ambient light from falling onto the object and thereby lowering the image quality. The object is imaged with a camera (2) that uses a lens that has been suitably selected or specially designed for the application. The camera delivers the image to a computer (5) through a camera–computer interface (6), e.g., a frame grabber. The device driver of the camera–computer interface as-

安放于印刷电路板后要通过检测确保产品装配是正确的，也就是说正确的元器件被安放在正确的位置。

- 形状和尺寸检测：用于检测产品的几何参数来保障其在允许的公差范围。这种检测可用于生产过程中；也可以用于产品使用一段时间之后，通过检测来确认产品经磨损后是否仍然满足要求。

- 表面检测：用于检查完成的产品是否存在缺陷，如是否有划痕，是否凹凸不平等。

图1.1为典型的机器视觉系统的例子。被测物（1）在传送带上运动，在机器视觉应用中，通常在相对固定的位置采集被测物的图像。这就要求有相应的机械部分，同时需要外触发信号来触发采集。（4）就是一种产生触发信号的光电传感器。被测物体需要合适的标准或定制光源（3）照明。通常情况下会加上遮光隔板（本例未画出）以防止环境光落到被测物体上降低图像质量。被测物的图像通过摄像机（2）及针对本应用选择或定制的合适的镜头采集得到。摄像机通过与计算机的接口（6）如图像采集卡将采集到的图像传至计算机（5），接口设备驱动程序将图像（7）放置计算机内存。如果图像采集是通过图像卡，照明可能由图像卡的闪光灯控制信号控

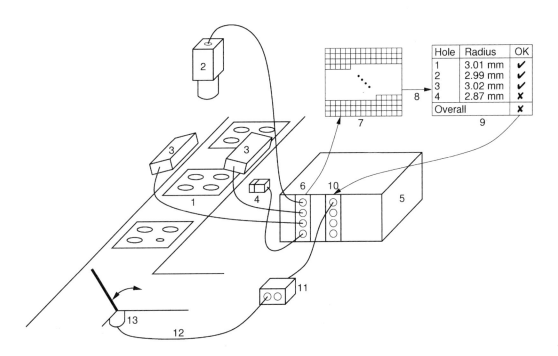

图 1.1 典型机器视觉系统组成。被测物（1）的图像由摄像机（2）获取。（3）为照明，（4）为触发图像采集的光电传感器。计算机（5）通过摄像机-计算机接口（6）获取图像，本例中接口为图像采集卡。光电传感器与图像采集卡相连接，图像采集卡触发闪光灯。驱动软件控制获取图像（7）并将图像放置计算机内存。机器视觉软件（8）检测被测物并返回检测结果（9）。通过数字 I/O（10）检测结果与 PLC（11）通信。PLC 通过现场总线接口（12）控制执行机构（13）。执行机构如电机驱动分流器将不合格被测物从生产线上剔除

sembles the image (7) in the memory of the computer. If the image is acquired through a frame grabber, the illumination may be controlled by the frame grabber, e.g., through strobe signals. If the camera–computer interface is not a frame grabber but a standard interface, such as IEEE 1394, USB, or Ethernet, the trigger will typically be connected to the camera and illumination directly or through a programmable logic controller (PLC). The computer can be a standard industrial PC or a specially designed computer that is directly built into the camera. The latter configuration is often called a smart camera. The computer may use a standard processor, a digital signal processor (DSP), a field-programmable gate array (FPGA),

制。如果摄像机与计算机的接口不是图像采集卡，而是像 IEEE 1394, USB 或网络等标准接口，外触发信号通常接至摄像机和照明光源，或通过可编程逻辑控制器 PLC 完成。计算机可以是标准的工业 PC 或直接做在摄像机内部的定制计算机，后一种方式通常被称作智能摄像机。计算机可以使用标准处理器、数字信号处理器（DSP）、现场可编程门阵列（FPGA）或以上几个部分合用。机器视觉软件（8）检测被测物并给出检测结果（9）。检测结果与可编程控制器（PLC）或分布式控制系统（DCS）等控制器（11）通信。通常情况下，这种通信由数字

or a combination of the above. The machine vision software (8) inspects the objects and returns an evaluation of the objects (9). The result of the evaluation is communicated to a controller (11), e.g., a PLC or a distributed control system (DCS). Often, this communication is performed by digital input/output (I/O) interfaces (10). The PLC, in turn, typically controls an actuator (13) through a communication interface (12), e.g., a fieldbus or serial interface. The actuator, e.g., an electric motor, then moves a diverter that is used to remove defective objects from the production line.

As can be seen from the large number of components involved, machine vision is inherently multidisciplinary. A team that develops a machine vision system will require expertise in mechanical engineering, electrical engineering, optical engineering, and software engineering.

To maintain the focus of this book, we have made a conscious decision to focus on the aspects of a machine vision system that are pertinent to the system until the relevant information has been extracted from the image. Therefore, we will forgo a discussion of the communication components of a machine vision system that are used after the machine vision software has determined its evaluation. For more information on these aspects, please consult Caro (2003); Berge (2004); Mahalik (2003):

In this book, we will try to give you a solid background on everything that is required to extract the relevant information from images in a machine vision system. We include the information that we wish someone had taught us when we started working in the field. In particular,

I/O 接口（10）完成。而 PLC 一般是通过通信接口（12）如现场总线或串口控制执行机构（13）。执行机构如电机则控制分流器将有问题的被测物从生产线上剔除。

从机器视觉系统包含这么多部件可以看出，机器视觉的确是多学科交叉的技术。开发机器视觉系统的团队需要机械工程、电子工程、光学工程及软件工程多方面的经验。

为了突出重点，本书不考虑机器视觉软件检测出结果之后的通信等部件，仅讨论至从图像中得到相关信息为止。详细系统可参考文献（Caro，2003；Berge，2004；Mahalik，2003）。

本书将介绍机器视觉系统从图像中得到相关信息的各个环节的背景知识。当我们开始进入这一领域时我们所需要的各种信息都包含在内。特别是我们讲到了与不同应用密切相关的硬件的一些特性，这些知识是我们必

we mention several idiosyncrasies of the hardware components that are highly relevant in applications, which we had to learn the hard way.

The hardware components that are required to obtain high-quality images are described in Chapter 2: illumination, lenses, cameras, and camera–computer interfaces. We hope that, after reading this chapter, you will be able to make informed decisions about which components and setups to use in your application.

Chapter 3 discusses the most important algorithms that are commonly used in machine vision applications. It is our goal to provide you with a solid theoretical foundation that will help you in designing and developing a solution for your particular machine vision task.

To emphasize the engineering aspect of machine vision, Chapter 4 contains a wealth of examples and exercises that show how the machine vision algorithms discussed in Chapter 3 can be combined in non-trivial ways to solve typical machine vision applications.

须掌握的。

第 2 章介绍了为得到高质量图像所需的硬件，包括照明、镜头、摄像机及摄像机与计算机接口。我们希望通过阅读本章节，读者可以学会在自己的应用中如何选择合适的部件及如何安装使用。

第 3 章论述了机器视觉应用中常用的重要算法，目的是使读者学到足够的理论知识，帮助读者完成特定机器视觉任务解决方案的设计和研发。

为强调机器视觉的工程应用，第 4 章以大量的实例及练习向读者展示如何将在第 3 章所讲各种机器视觉算法结合起来，解决实际的机器视觉应用问题。

2 Image Acquisition

In this chapter, we will take a look at the hardware components that are involved in obtaining an image of the scene we want to analyze with the algorithms presented in Chapter 3. Illumination makes the essential features of an object visible. Lenses produce a sharp image on the sensor. The sensor converts the image into a video signal. Finally, camera–computer interfaces (frame grabbers, bus systems like USB, or network interfaces like Ethernet) accept the video signal and convert it into an image in the computer's memory.

2.1 Illumination

The goal of illumination in machine vision is to make the important features of the object visible and to suppress undesired features of the object. To do so, we must consider how the light interacts with the object. One important aspect is the spectral composition of the light and the object. We can use, for example, monochromatic light on colored objects to enhance the contrast of the desired object features. Furthermore, the direction from which we illuminate the object can be used to enhance the visibility of features. We will examine these aspects in this section.

2.1.1 Electromagnetic Radiation

Light is electromagnetic radiation of a certain range of wavelengths, as shown in Table 2.1. The range of wavelengths visible for humans is 380–780 nm. Electromagnetic radiation with

2. 图像采集

本章将讲述为了得到被测物图像而需要的硬件部件,只有得到图像才可以使用第 3 章的算法进行分析。照明使得被测物的基本特征可见,镜头使得在传感器上得到清晰的图像,传感器将图像转换为视频信号。最后,摄像机与计算机的接口接收视频信号并将其放置到计算机内存。接口可能是图像采集卡、USB,也可能是 Ethernet 网络接口。

2.1 照明

机器视觉中照明的目的是使被测物的重要特征显现,而抑制不需要的特征。为达到此目的,我们需要考虑光源与被测物间的相互作用。其中一个重要的因素就是光源和被测物的光谱组成。我们可以用单色光照射彩色物体以增强被测物相应特征的对比度。照明的角度可以用于增强某些特征。本节将介绍上述这些内容。

2.1.1 电磁辐射

如表 2.1 所示,光是一定波长范围内的电磁辐射。人眼可视的波长范围为 380~780 nm。比此波长短的电磁辐射称作紫外线(UV)。更短的电磁

表 2.1 与光学和光子学有关的电磁波谱。红外辐射和紫外辐射的范围名称对应 ISO 20473:2007。可见光颜色名称参考 Lee（2005）。

光谱范围	名称	缩写	波长 λ
紫外线	极短紫外	–	1~100 nm
	真空紫外		100~190 nm
	深紫外	UV-C	190~280 nm
	中紫外	UV-B	280~315 nm
	近紫外	UV-A	315~380 nm
可见光	蓝紫色		380~430 nm
	蓝色		430~480 nm
	绿蓝色		480~490 nm
	蓝绿色		490~510 nm
	绿色		510~530 nm
	黄绿色		530~570 nm
	黄色		570~580 nm
	橙色		580~600 nm
	红色		600~720 nm
	紫红色		720~780 nm
红外线	近红外	IR-A	780~1.4 μm
		IR-B	1.4~3 μm
	中波红外		3~50 μm
	远红外	IR-C	50 μm~1 mm

shorter wavelengths is called ultraviolet (UV) radiation. Electromagnetic radiation with even shorter wavelengths consists of X-rays and gamma rays. Electromagnetic radiation with longer wavelengths than the visible range is called infrared (IR) radiation. Electromagnetic radiation with even longer wavelengths consists of microwaves and radio waves.

Monochromatic light is characterized by its wavelength λ. If light is composed of a range of wavelengths, it is often compared to the spectrum of light emitted by a black body. A black body

辐射为 X 射线和伽马射线。比可见光波长长的电磁辐射称作红外线（IR）。比红外更长的波长的电磁辐射为微波和无线电波。

单色光以其波长 λ 表征。对于由多个波长组成的光，则通常将其与黑体辐射的光谱相比较。黑体可以吸收所有落到其表面的电磁辐射，因此可

is an object that absorbs all electromagnetic radiation that falls onto it and thus serves as an ideal source of purely thermal radiation. Therefore, the light spectrum of a black body is directly related to its temperature. The spectral radiance of a black body is given by Planck's law (Planck, 1901; Wyszecki and Stiles,1982）:

以看作理想的纯热辐射源，所以黑体光谱与其温度直接相关。黑体光谱辐射符合普朗克定律（Planck，1901；Wyszecki and Stiles，1982）:

$$I(\lambda, T) = \frac{2hc^2}{\lambda^5} \frac{1}{e^{hc/(\lambda kT)} - 1} \tag{2.1}$$

Here, $c = 2.99792458 \times 10^8 \,\mathrm{m\,s^{-1}}$ is the speed of light, $h = 6.6260693 \times 10^{-34}\,\mathrm{J\,s}$ is the Planck constant, and $k = 1.3806505 \times 10^{-23}\,\mathrm{J\,K^{-1}}$ is the Boltzmann constant. The spectral radiance is the energy radiated per unit wavelength by an infinitesimal patch of the black body into an infinitesimal solid angle of space. Hence, its unit is $\mathrm{W\,sr^{-1}\,m^{-2}\,nm^{-1}}$.

Figure 2.1 displays the spectral radiance for different temperatures T. It can be seen that black bodies at 300 K radiate primarily in the middle and far IR range. This is the radiation range that is perceived as heat. Therefore, this range of wavelengths is also called thermal IR. The radiation of an object at 1000 K just starts to enter the visible range. This is the red glow that can be seen first when objects are heated. For $T = 3000\,\mathrm{K}$, the spectrum is that of an incandescent lamp (see Section 2.1.2). Note that it has a strong red component. The spectrum for $T = 6500\,\mathrm{K}$ is used to represent average daylight. It defines the spectral composition of white light. The spectrum for $T = 10000\,\mathrm{K}$ produces light with a strong blue component.

其中，$c = 2.99792458 \times 10^8 \mathrm{m/s}$，为光速；$h = 6.6260693 \times 10^{-34}\mathrm{J \cdot s}$，为普朗克常数；$k = 1.3806505 \times 10^{-23}\mathrm{J/K}$，为玻尔兹曼常数。光谱辐射即为单位面积的黑体在单位立体角内、单位波长内辐射出的能量。因此，其单位为瓦特每球面平方米每纳米（$\mathrm{W \cdot sr^{-1} \cdot m^{-2} \cdot nm^{-1}}$）。

图 2.1 显示了不同温度（T）下的光谱辐射。从中可以看出黑体温度在 300 K 时的辐射主要在中红外和远红外，此辐射范围就是我们感觉到的热。因此这段波长也称作热红外。1000 K 的物体辐射开始进入可见光范围，这就是当物体被加热我们首先所看到的红辉。$T = 3000$ K 是白炽灯的谱线（见 2.1.2 节）。注意谱线中含有很强的红的成分。$T = 6500$ K 用来表示日光光谱即白光的光谱。$T=10000$ K 为蓝光成分很强的光的谱线。

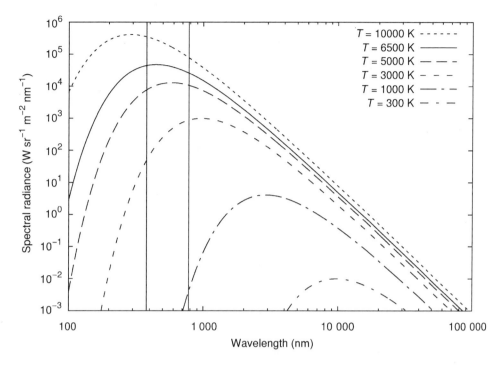

图 2.1 黑体在不同温度下的辐射光谱。两条垂直线内为可见光部分

Because of the correspondence of the spectra with the temperature of the black body, the spectra also define so-called correlated color temperatures (CIE 15:2004).

光谱与黑体温度相关，又称为色温（CIE 15:2004）。

2.1.2 Types of Light Sources

Before we take a look at how to use light in machine vision, we will discuss the types of light sources that are commonly used in machine vision.

Incandescent lamps create light by sending an electrical current through a thin filament, typically made of tungsten. The current heats the filament and causes it to emit thermal radiation. The filament is contained in a glass envelope that contains either a vacuum or a halogen gas, such as iodine or bromine, which prevents oxidation of the filament. Filling the envelope with a halogen

2.1.2 光源类型

在讨论机器视觉中如何使用光源之前，首先看看在机器视觉中常用的光源有哪些。

白炽灯通过在细细的灯丝中传输电流产生光，通常情况下，灯丝是用钨制成的。电流加热灯丝使其产生热辐射。灯丝的温度非常高，其辐射在电磁辐射谱线的可见光范围内。灯丝在真空或充有卤素气体的密闭玻璃灯泡中，常见的卤素气体为碘或溴，以防止灯丝氧化。充满卤素气体比起真

gas has the advantage that the lifetime of the lamp is increased significantly compared to using a vacuum. The advantage of incandescent lamps is that they are relatively bright and create a continuous spectrum with a correlated color temperature of 3000–3400 K. Furthermore, they can be operated with low voltage. One of their disadvantages is that they produce a large amount of heat: only about 5% of the power is converted to light; the rest is emitted as heat. Other disadvantages are short lifetimes and the inability to use them as flashes. Furthermore, they age quickly, i.e., their brightness decreases significantly over time.

Xenon lamps consist of a sealed glass envelope filled with xenon gas, which is ionized by electricity, producing a very bright white light with a correlated color temperature of 5500–12 000 K. They are commonly divided into continuous-output short- and long-arc lamps as well as flash lamps. Xenon lamps can produce extremely bright flashes at a rate of more than 200 flashes per second. Each flash can be extremely short, e.g., 1–20 µs for short-arc lamps. One of their disadvantages is that they require a sophisticated and expensive power supply. Furthermore, they exhibit aging after several million flashes.

Like xenon lamps, fluorescent lamps are gas-discharge lamps that use electricity to excite mercury vapor in a noble gas, e.g., argon or neon, causing UV radiation to be emitted. This UV radiation causes a phosphor salt coated onto the inside of the tube that contains the gas to fluoresce, producing visible light. Different coatings can be chosen, resulting in different spectral distributions of the visible light with correlated color temperatures of 3000–6000 K. Fluorescent lamps

空可使灯泡的寿命大大延长。白炽灯的优点是相对较亮，而且可以产生相关联的色温为 3000~3400 K 的连续光谱。还有一个优点是白炽灯可以工作在低电压。缺点是发热严重：仅有 5% 左右的能量转换为光，其他都以热的形式散发了。另一个的缺点是寿命短，而且不能用作闪光灯。此外，白炽灯老化快，随着时间的推移，亮度迅速下降。

氙灯是在密闭的玻璃灯泡中充上氙气，氙气被电离产生色温在 5500~12 000 K 的非常亮的白光。常被分为连续发光的短弧灯、长弧灯以及闪光灯。氙灯可做成每秒 200 多次的非常亮的闪光灯。对于短弧灯，每次亮的时间可以短至 1~20µm。氙灯的缺点是供电复杂且昂贵。此外，在几百万次闪光后会出现老化。

与氙灯类似，荧光灯也是一类气体放电光源，通过电流激发在如氩、氖等惰性环境气体中的水银蒸气，产生紫外光辐射。这些紫外光辐射使得封装惰性气体的管壁上的磷盐涂层发荧光，产生可见光。使用不同的涂层，可以产生 3000~6000 K 色温的可见光。荧光灯由交流电供电，因此产生与供电相同频率的闪烁。对于机器视觉应用来说，为了避免图像明暗的变

are driven by alternating current. This results in a flickering of the lamp with the same frequency as the current. For machine vision, high-frequency alternating currents of 22 kHz or more must be used to avoid spurious brightness changes in the images. The main advantages of fluorescent lamps are that they are inexpensive and can illuminate large areas. Some of their disadvantages are a short lifetime, rapid aging, and an uneven spectral distribution with sharp peaks for certain frequencies. Furthermore, they cannot be used as flashes.

A light-emitting diode (LED) is a semiconductor device that produces narrow-spectrum (i.e., quasi-monochromatic) light through electroluminescence: the diode emits light in response to an electric current that passes through it. The color of the emitted light depends on the composition and condition of the semiconductor material used. The possible range of colors comprises IR, visible, and near UV radiation. White LEDs can also be produced: they internally emit blue light, which is converted to white light by a coating with a yellow phosphor on the semiconductor. One advantage of LEDs is their longevity: lifetimes larger than 100 000 hours are not uncommon. Furthermore, they can be used as flashes with fast reaction times and almost no aging. Since they use direct current, their brightness can be controlled easily. In addition, they use comparatively little power and produce little heat. The main disadvantage of LEDs is that their performance depends on the ambient temperature of the environment in which they operate. The higher the ambient temperature, the lower the performance of the LED and the shorter its lifetime. However,

化，需要使用不低于 22 kHz 的供电频率。荧光灯的优点是价格便宜，照明面积大。缺点是寿命短、老化快，光谱分布不均匀，在有些频率下有尖峰。还不能用作闪光灯。

发光二极管（LED）是一种通过电致发光的半导体，能产生类似单色光的非常窄的光谱的光。其发光亮度与通过二极管的电流相关。发出的光的颜色取决于所用半导体材料的成分。可以制作成红外、可见光及近紫外。也可做成白光 LED。实际上白光 LED 内部产生的光是蓝色的，通过在半导体上加上黄磷涂层将蓝光转换为白光。LED 光源的一大优点是寿命长，寿命超过 10 万小时非常普遍。另外 LED 可用作闪光灯，响应速度很快，几乎没有老化现象。由于 LED 采用直流供电，因此，亮度非常容易控制。另外 LED 光源功耗小，发热小。主要缺点是 LED 的性能与环境温度有关。环境温度越高，LED 的性能越差，寿命越短。由于 LED 有如此多的优点，目前是机器视觉中应用最多的一种光源。

since LEDs have so many practical advantages, they are currently the primary illumination technology used in machine vision applications.

2.1.3 Interaction of Light and Matter

Light can interact with objects in various ways, as shown in Figure 2.2.

2.1.3 光与被测物间的相互作用

如图 2.2 所示，光与被测物有多种相互作用方式。

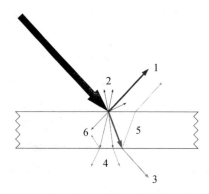

图 2.2　光与被测物相互作用
落到物体的光用黑箭头表示。（1）镜面反射；（2）漫反射；（3）定向透射；
（4）漫透射；（5）背反射；（6）吸收

Reflection occurs at the interfaces between different media. The microstructure of the object (essentially the roughness of its surface) determines how much of the light is reflected diffusely and how much specularly. Diffuse reflection scatters the reflected light more or less evenly in all directions. For specular reflection, the incoming and reflected light ray lie in a single plane. Furthermore, their angles with respect to the surface normal are identical. Hence, the macrostructure (shape) of the object determines the direction into which the light is reflected specularly. In practice, however, specular reflection is never perfect (as it is for a mirror). Instead, specular reflection causes a lobe of intense reflection for certain viewing angles, depending on the angle of the incident light,

反射发生在不同介质的分界面上。被测物表面的粗糙程度等微细结构，决定了光线有多少发生漫反射；多少发生镜面反射。漫反射在各个方向上散射的光线基本是均匀的。对于镜面反射，入射光与反射光在同一个平面，而且它们与反射面法线的夹角是相等的。因此，被测物的形状等宏观结构决定了镜面反射的方向。然而在现实中镜面反射几乎不可能像镜子一样是理想的。相反，镜面反射在一定角度产生较强的波瓣形反射，如图 2.3 所示。入射光的角度决定是否有波瓣形反射，物体表面的微细结构决定波瓣的宽度。

as shown in Figure 2.3. The width of the side lobe is determined by the microstructure of the surface.

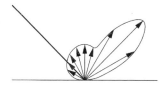

图 2.3 漫反射和镜面反射共同产生的光线分布

Reflection at metal and dielectric surfaces, e.g., glass or plastics, causes light to become partially polarized. Polarization occurs for diffuse as well as specular reflection. In practice, however, polarization caused by specular reflection dominates.

The fraction of light reflected by the surface is given by the bidirectional reflectance distribution function (BRDF). The BRDF is a function of the direction of the incoming light, the viewing direction, and the wavelength of the light. If the BRDF is integrated over both directions, the reflectivity of the surface is obtained. It depends only on the wavelength.

Note that reflection also occurs at the interface between two transparent media. This may cause backside reflections, which can lead to double images.

Transmission occurs when the light rays pass through the object. Here, the light rays are refracted, i.e., they change their direction at the interface between the different media. This is discussed in more detail in Section 2.2.2.1. Depending on the internal and surface structure of the object, the transmission can be direct or diffuse. The fraction of light that passes through the

发生在金属或绝缘材料（如玻璃或者塑料）表面的反射使光线产生部分偏振。偏振的产生源于漫反射和镜面反射，但实际上主要是由镜面反射决定的。

物体表面反射光的多少由双向反射率分布函数（BRDF）表示。双向反射率分布函数是光线入射方向、观察方向及波长的函数。将双向反射分布函数在两个方向上积分，即可得到表面的反射率，表面反射率仅取决于波长。

在两个透明的介质分界面也会有反射，这样产生了背反射，从而产生重影。

光线通过物体产生透射。当光线到达不同介质的分界面，光的传播方向发生改变产生折射。见 2.2.2.1 节。物体的内部和表面结构决定透射为漫透射或定向透射。透过物体的光线多少称作透射率。同反射率一样，透射率也决定于光线的波长。

object is called its transmittance. Like reflectivity, it depends, among other factors, on the wavelength of the light.

Finally, absorption occurs if the incident light is converted into heat within the object. All light that is neither reflected nor transmitted is absorbed. If we denote the light that falls onto the object by I, the reflected light by R, the transmitted light by T, and the absorbed light by A, the law of conservation of energy dictates that $I = R + T + A$. In general, dark objects absorb a significant amount of light.

All of the above quantities except specular reflection depend on the wavelength of the light that falls onto the object. Wavelength-dependent diffuse reflection and absorption give opaque objects their characteristic color. Likewise, wavelength-dependent transmission gives transparent objects their characteristic color.

Finally, it should be noted that real objects are often more complex than the simple model described above. For example, an object may consist of several layers of different materials. The top layer may be transparent to some wavelengths and reflect others. Further layers may reflect parts of the light that has passed through the layers above. Therefore, establishing a suitable illumination for real objects often requires a significant amount of experimentation.

2.1.4 Using the Spectral Composition of the Illumination

As mentioned in the previous section, colored objects reflect certain portions of the light spectrum,

入射光除反射和透射外剩下的光线都被吸收，吸收就是入射光在物体内部被转换成热。如果我们假定落到一个物体上的光为 I，反射光为 R，透射光为 T，吸收为 A，根据能量守恒定律 $I = R + T + A$。通常黑色的物体能吸收大量的光。

除镜面反射外，上述各个物理量均取决于投射到物体的光的波长。不透明物体特有的颜色就是由与波长相关的漫反射及吸收决定的。而透明物体的颜色是与波长相关的透射决定的。

实际的物体要比上述简单模型复杂得多。比如有的物体是由几层不同的材料组成的，表层对于一定波长的光透明，而反射其他波长的光，下一层又可能反射部分从上一层透过的光。因此，为实物确定一个合适的光源常常需要大量的实验。

2.1.4 利用照明的光谱

如上节所述，彩色物体反射了一部分光谱，其他部分被吸收。我们可

while absorbing other portions. This can often be used to enhance the visibility of certain features by employing an illumination source that uses a range of wavelengths that is reflected by the objects that should be visible and is absorbed by the objects that should be suppressed. For example, if a red object on a green background is to be enhanced, red illumination can be used. The red object will appear bright, while the green object will appear dark.

Figure 2.4 illustrates this principle by showing a printed circuit board (PCB) illuminated with white, red, green, and blue light. While white light produces a relatively good average contrast, the contrast of certain features can be enhanced significantly by using colored light. For example, the contrast of the large component in the lower left part of the image is significantly better under red illumination because the component itself is light orange, while the print on the component is dark orange. Therefore, the component itself can be segmented more easily with red illumination. Note, however, that red illumination reduces the contrast of the print on the component, which is significantly better under green illumination. Red illumination also enhances the contrast of the copper-colored plated through holes. Blue illumi-

以利用这一特点来增强我们需要的特征。比如使用合适的照明光源使其光谱范围正好是希望看到的波长范围被物体反射，不希望看到的波长范围被物体吸收。举个例子，如果绿色背景上面的红色被测物需要增强，就可以使用红色照明，这时红色物体会更加明亮，同时绿色物体会变得暗淡。

图 2.4 是印刷电路板（PCB 板）在白色、红色、绿色和蓝色照明下得到的不同效果，据此可以演示上述原理。使用白光可以得到相对较好的平均对比度，而使用其他颜色的光可以显著提高一些特征的对比度。比如左下角大芯片由于本身是浅橘黄色而上面的印刷字符是深橘黄色，芯片的对比度在红光照明下大大提高。因此，在红色照明下，这个元件非常容易被分割出来。请注意，此时元件上的印刷字符在红色照明下对比度是下降的，而在绿色照明下，印刷字符的对比度显著提高。红色照明同时改善了铜色的过孔的对比度。此外，蓝色照明使得位于五个电阻中间的那个浅蓝色电阻的对比度非常好。

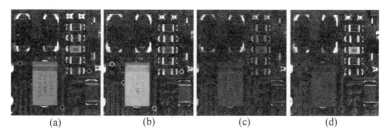

图 2.4 在不同光照明下的 PCB 板
（a）白光；（b）红光；（c）绿光；（d）蓝光照明下的 PCB 板

nation, on the other hand, maximizes the contrast of the light blue resistor in the center of the row of five small components.

Since charge-coupled device (CCD) and complementary metal-oxide semiconductor (CMOS) sensors are sensitive to IR radiation (see Section 2.3.3.4), IR radiation can often also be used to enhance the visibility of certain features, as shown in Figure 2.5. For example, the tracks can be made visible easily with IR radiation since the matt green solder resist that covers the tracks and makes them hard to detect in visible light is transparent to IR radiation.

由于 CCD 和 CMOS 传感器对于红外光比较敏感（见 2.3.3.4 节），我们也常用红外光来增强某些特征，如图 2.5 所示，在红外光下，PCB 走线非常容易看到，由于走线上覆盖有无光泽的绿色阻焊剂，在可见光下很难看到，在红外照明下走线就清晰了。

图 2.5　红外照明下的 PCB 板

All of the above effects can also be achieved through the use of white light and color filters. However, since a lot of efficiency is wasted if white light is created only to be filtered out later, it is almost always preferable to use colored illumination from the start.

以上效果也可通过白光加滤镜得到。然而，如果白光这样被过滤，其发光效率将大大降低，所以通常情况下从开始就使用彩色照明。

Nevertheless, there are filters that are useful for machine vision. As we have seen previously, CCD and CMOS sensors are sensitive to IR radiation. Therefore, an IR cut filter is often useful to avoid unexpected brightness or color changes in the image. On the other hand, if the object is illuminated with IR radiation, an IR pass filter, i.e., a filter that suppresses the visible spectrum and lets only IR radiation pass, is often helpful.

然而，滤镜在机器视觉中的用途还是很多的。上面我们谈到 CCD 和 CMOS 对于红外敏感，因此，常常需要加上红外截止滤光片来避免图像过亮以及图像颜色变化。反过来，如果被测物是用红外照明的，使用红外透过滤光镜抑制可见光部分而仅让红外光通过将非常有助于得到好的图像。

Another useful filter is a polarizing filter. As mentioned in Section 2.1.3, light becomes partially polarized through reflection at metal and dielectric surfaces. To suppress this light, we can mount a polarizing filter in front of the camera and turn it in such a way that the polarized light is suppressed. Since unpolarized light is only partially polarized through reflection, an even better suppression of specular reflections can be achieved if the light that falls onto the object is already polarized. This principle is shown in Figure 2.6. The polarizing filters in front of the illumination and camera are called polarizer and analyzer, respectively.

另外一种非常有用的滤光片就是偏振片。在 2.1.3 节讲到光线在金属和绝缘体表面反射时光线会产生部分偏振。我们可以在摄像机前面加上偏振滤镜并调整方向来抑制偏振光。由于非偏振光反射后只有部分偏振，所以更好地抑制某些反射的方法是先使光线成为偏振光，然后再落到物体表面。图 2.6 说明了此方法。加在照明和摄像机前的滤光片分别称作起偏镜和检偏镜。

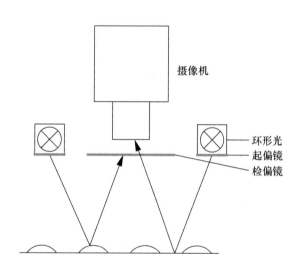

图 2.6　起偏镜和检偏镜原理

安装在光源前的滤光片称作起偏镜，加在摄像机前的滤光片称作检偏镜，如果将检偏镜与起偏镜呈 90°，反射造成的偏振光就被检偏镜抑制了

The effect of using polarizing filters is shown in Figure 2.7. Figure 2.7(a) shows a PCB illuminated with a directed ring light. This causes specular reflections on the board, solder, and metallic parts of the components. Using a polarizer and analyzer almost completely suppresses the specular reflections, as shown in Figure 2.7(b).

图 2.7 显示了使用偏振滤镜的作用。图 2.7（a）为直接环形光照明下的 PCB 板，可看到电路板、焊点及其他金属元件的镜面反射，而图 2.7（b）为使用了起偏镜和检偏镜的 PCB 板图像，镜面反射几乎全部被抑制。

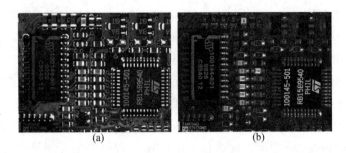

图 2.7　（a）直接环形光照明下的 PCB 板。可看到电路板、焊点及其他金属元件的镜面反射；（b）使用起偏镜和检偏镜后 PCB 板图像，此时镜面反射几乎全部被抑制

2.1.5 Using the Directional Properties of the Illumination

2.1.5 利用照明的方向性

While the spectral composition of light can often be used advantageously, in machine vision most often the directional properties of the illumination are used to enhance the visibility of the essential features.

在有效地利用照明中光谱成分的同时，照明的方向性通常也可以用在机器视觉中来增强被测物的必要特征。

By directional properties, we mean two different effects. On the one hand, the light source may be diffuse or directed. In the first case, the light source emits the light more or less evenly in all directions. In the second case, the light source emits the light in a very narrow range of directions. In the limiting case, the light source emits only parallel light rays in a single direction. This is called telecentric illumination. Telecentric illumination uses the same principles as telecentric lenses (see Section 2.2.4).

谈到方向性，有两种效果。一方面，光源可以是漫射或直射的。漫射时，光在各个方向的强度几乎是一样的；直射时，光源发出的光集中在非常窄的空间范围内。在特定情况下，光源仅发出单向平行光，称作平行光照明。平行光照明与远心镜头（见 2.2.4 节）的原理是相同的。

On the other hand, the placement of the light source with respect to the object and camera is important. Here, we can discern different aspects. If the light source is on the same side of the object as the camera, we speak of front light. This is also often referred to as incident light. However, since incident light often means the light that falls onto

另一方面，可以从几个方面看出光源与摄像机和被测物的相对位置也是非常重要的。如果光源与摄像机位于被测物的同一侧，我们称作正面光，通常也叫入射光。由于入射光一般指投射到物体上的光线，所以在本书中都使用术语正面光。如果光源与摄像

an object, we will use the term front light throughout this book. If the light source is on the opposite side of the object to the camera, we speak of back light. This is sometimes also called transmitted light, especially if images of transparent objects are acquired. If the light source is placed at an angle to the object so that most of the light is reflected to the camera, we speak of bright-field illumination. Finally, if the light is placed in such a way that most of the light is reflected away from the camera, and only light of certain parts of the object is reflected to the camera, we speak of dark-field illumination.

All of the above criteria are more or less orthogonal to each other and can be combined in various ways. We will discuss the most commonly used combinations below.

2.1.5.1 Diffuse Bright-Field Front Light Illumination

A diffuse bright-field front light illumination can be built in several ways, as shown in Figure 2.8. LED panels or ring lights with a diffuser in front of the lights have the advantage that they are easy to construct. The light distribution, however, is not perfectly homogeneous, unless the light panel or ring is much larger than the object to be imaged. Furthermore, no light comes from the direction in which the camera looks through the illumination. Coaxial diffuse lights are regular diffuse light sources for which the light is reflected onto the object through a semi-transparent mirror, through which the camera also acquires the images. Since there is no hole through which the camera must look, the light distribution is more uniform. Howe-

机位于被测物两侧，此时的光称作背光，特别是当被测物是透明物体时称作透射光。如果光源与被测物成一定角度，使得绝大部分光反射到摄像机，我们称作明场照明；如果光源位置使得大部分的光没有反射到摄像机，仅仅将照射到被测物体的特定部分的光反射到摄像机，我们称此种照明为暗场照明。

上述各种划分标准基本互相独立，可以有多种组合方式。我们下面将讲述常见的组合。

2.1.5.1 明场漫射正面照明

如图 2.8 所示，可以有多种产生明场漫射正面照明的方式。在 LED 平板或环形光前端加上漫射板的方式具有易于构造的优点。缺点是除非 LED 平板或环形光比被测物大很多，否则很难得到均匀的照明。而且，摄像机的轴向没有光。同轴漫射光通常是通过半透半反镜将光反射到被测物上，通过同一面镜子，摄像机同样采集到图像。由于摄像机无须通过观测孔来采集图像，这种方法使得照明较为均匀。然而，半透半反镜可能产生鬼像。半球形照明可以得到非常均匀的照明。得到半球形照明可以有两种方法：一是在半球形照明光源前加上

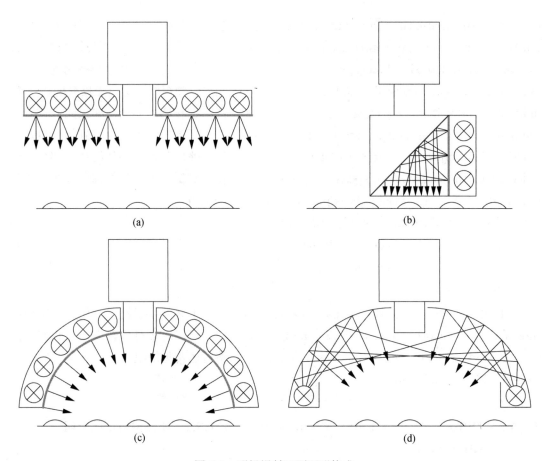

图 2.8 明场漫射正面照明构成

(a) 前端安装了漫射板的 LED 平板或环形光；(b) 在光源前安装有漫射板和 45° 半透半反镜的同轴漫射光；(c) 安装有漫射板的半球光源；(d) 光源是 LED 环形光、由半球表面作为漫射板的半球照明

ver, the semi-transparent mirror can cause ghost images. Dome lights try to achieve a perfectly homogeneous light distribution either by using a dome-shaped light panel with a diffuser in front of the lights or by mounting a ring light into a diffusely reflecting dome. In the latter case, multiple internal reflections cause the light to show no preferred direction. Like the LED panels or ring lights, the dome lights have the slight disadvantage that no light comes from the direction of the hole through which the camera looks at the object.

漫射板；另外的方法是在漫反射半球中安装 LED 环形光。在后一种方法中，多次内部漫反射使得光线没有固定方向。同 LED 平板或环形光一样，半球形照明的缺点是摄像机对着物体的方向没有光。

Diffuse bright-field front light illumination is typically used to prevent shadows and to reduce or prevent specular reflections. It can also be used to look through transparent covers of objects, e.g., the transparent plastic on blister packs (see Figure 2.9).

明场漫射正面照明方式常用于防止产生阴影，并用于减少或防止镜面反射。也可以用于透过被测物体的透明包装，比如图 2.9 中硬质衬垫包装上的透明塑料。

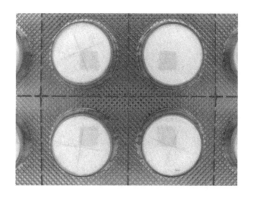

图 2.9　在如图 2.8（d）所示的半球照明下的硬质泡沫塑料衬垫包装图像

2.1.5.2　Directed Bright-Field Front Light Illumination

Directed bright-field front light illumination comes in two basic varieties, as shown in Figure 2.10. One way to construct it is to use tilted ring lights. This is typically used to create shadows in cavities or around the objects of interest. One disadvantage of this type of illumination is that the light distribution is typically uneven. A second method is to use a coaxial telecentric illumination to image specular objects. Here, the principle is that object parts that are parallel to the image plane reflect the light to the camera, while all other object parts reflect the light away from the camera. This requires a telecentric lens. Furthermore, the object must be aligned very well to ensure that the light is reflected to the camera. If this is not ensured mechanically, the object may appear completely dark.

2.1.5.2　明场直接正面照明

如图 2.10 所示，构造明场直接正面照明有两种方法。一种是使用倾斜的环形光，常用于使孔洞或感兴趣区域产生阴影，这种照明方式的缺点是光线分布不均匀。另外一种方法是使用同轴平行光，常用于采集会产生镜面反射的物体的图像。这种方法的原理是平行于像平面的被测物表面反射到摄像机，而物体的其他表面将光反射到远离摄像机的其他方向，此时需要远心镜头。此外，必须准确调整被测物的位置以确保需要的光线反射到摄像机中，如果机械安装无法保证这点，被测物的图像可能是全黑的。

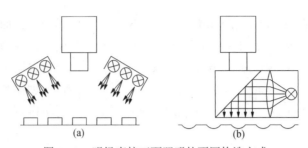

图 2.10　明场直接正面照明的不同构造方式
（a）聚焦的环形光；（b）含 45° 角的半透半反镜的同轴远心平行照明

2.1.5.3 Directed Dark-Field Front Light Illumination

Directed dark-field front light illumination is typically constructed as an LED ring light, as shown in Figure 2.11. The ring light is mounted at a very small angle to the object's surface. This is used to highlight indentations and protrusions of the object. Hence, the visibility of structures like scratches, texture, or engraved characters can be enhanced. An example of the last two applications is shown in Figure 2.12.

2.1.5.3 暗场直接正面照明

暗场直接正面照明通常由 LED 环形光构成，如图 2.11 所示，环形光与物体表面呈非常小的角度，这样可以突出被测物的缺口及凸起，所以像划痕、纹理或雕刻文字等被增强，看得更加清晰。图 2.12 是最后两类应用的示例。

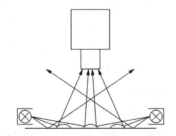

图 2.11　暗场直接正面照明（这种照明通常由 LED 环形光构成）

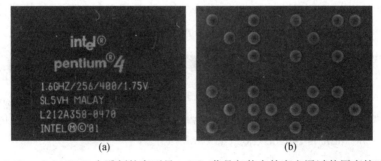

图 2.12　（a）CPU 上雕刻的序列号；（b）药品包装上的盲文通过使用直接暗场正面照明得到增强。可以看到 CPU 上的字符和划痕均得到增强

2.1.5.4 Diffuse Bright-Field Back Light Illumination

Diffuse bright-field back light illumination consists of a light source, often made from LED panels or fluorescent lights, and a diffuser in front of the lights. The light source is positioned behind the object, as shown in Figure 2.13(a). Back light illumination only shows the silhouette of opaque objects. Hence, it can be used whenever the essential information can be derived from the object's contours. For transparent objects, back light illumination can be used in some cases to make the inner parts of the objects visible because it avoids the reflections that would be caused by front light illumination. One drawback of diffuse bright-field back light illumination is shown in Figure 2.13(b). Because the illumination is diffuse, for objects with a large depth, some parts of the object that lie on the camera side of the object can be illuminated. Therefore, diffuse back light illumination is typically only used for objects that have a small depth.

2.1.5.4 明场漫射背光照明

明场漫射背光照明通常在光源前面安有漫射板，光源常采用 LED 平板或荧光灯。如图 2.13（a）所示，光源安放在被测物体的后面。背光方式只显示不透明物体的轮廓，所以这种照明方式用于被测物需要的信息可以从其轮廓得到的场合。对于透明物体，背光可以用于检测被测物的内部部件，这样可以避免使用正面照明造成的反射。明场漫射背光照明的缺陷如图 2.13（b）所示，由于是漫反射，对于有一定高度的被测物，其在摄像机一侧的某些部分也可能会被照亮。因此，漫射背光照明主要用于厚度不大的被测物。

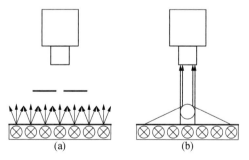

图 2.13 （a）明场漫射背光照明。这种照明通常由光源前装有漫射板的 LED 平板灯或荧光灯组成；（b）对于厚度较大的被测物，其在摄像机一侧的部分可能产生反射

Figure 2.14 displays two kinds of objects that are usually illuminated with a diffuse bright-field

图 2.14 为平板金属工件和灯泡中的灯丝，是常用明场漫射背光照明

back light illumination: a flat metal workpiece and a filament in a light bulb.

的两类被测物。

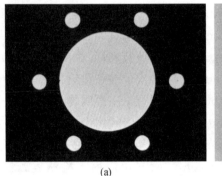 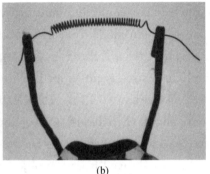

图 2.14　使用明场漫射背光照明的：（a）金属工件；（b）灯泡中的灯丝

2.1.5.5　Telecentric Bright-Field Back Light Illumination

Directed bright-field back light illumination is usually constructed using telecentric illumination, as shown in Figure 2.15, to prevent the problem that the camera side of the object is also illuminated. For a perspective lens, the image of the telecentric light would be a small spot in the image because the light rays are parallel. Therefore, telecentric back light illumination must use telecentric lenses. The illumination must be carefully aligned with the lens. This type of illumination produces very sharp edges at the silhouette of the object. Furthermore, because telecentric lenses are used,

2.1.5.5　明场平行光背光照明

为了解决上述的摄像机一侧的被测物某些部分也被照亮的问题，可以使用平行照明构造的直接明场背光照明，如图 2.15 所示。对于透视镜头，由于照明光源是平行光，光线在图像中是一个小点。因此，平行背光照明需要使用远心镜头配合，而且照明与镜头位置需要仔细调整。这种照明会使被测物轮廓非常锐利，此外，由于使用远心镜头，图像也没有透射变形，所以这种照明常用于测量应用。

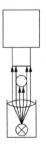

图 2.15　明场平行背光照明

there are no perspective distortions in the image. Therefore, this kind of illumination is frequently used in measurement applications.

Figure 2.16 compares the effects of using diffuse and telecentric bright-field back light illuminations on part of a spark plug. In both cases, a telecentric lens was used. Figure 2.16(a) clearly shows the reflections on the camera side of the object that occur with the diffuse illumination. These reflections do not occur for the telecentric illumination, as can be seen from Figure 2.16(b).

图 2.16 比较了对于火花塞使用明场漫射背光照明和明场平行背光照明的不同效果。二者均使用了远心镜头。图 2.16（a）可以清楚地看到使用漫射背光照明时被测物在摄像机一侧部分的反射，而使用平行照明就不会出现反射，如图 2.16（b）所示。

(a) (b)

图 2.16 （a）使用明场漫射背光照明的火花塞；(b）使用明场平行背光照明的火花塞。注意由于使用漫射照明时摄像机一侧的被测物部分所产生的反射

2.2 Lenses

A lens is an optical device through which light is focused in order to form an image inside a camera, in our case on a digital sensor. The purpose of the lens is to create a sharp image in which fine details can be resolved. In this section, we will take a look at the image geometry created by different kinds of lenses as well as the major lens aberrations that may cause the image quality to be less than perfect, and hence may influence the accuracy of some of the algorithms described in Chapter 3.

2.2 镜头

镜头是一种光学设备，用于聚集光线在摄像机内部成像，本书中则是指在数字传感器上成像。镜头的作用是产生锐利的图像，以得到被测物的细节。本节将讨论使用不同镜头产生不同的成像几何，同时将讲述镜头的主要像差，像差会影响图像质量，就可能会影响第 3 章所讲述的一些算法的精度。

2.2.1 Pinhole Cameras

If we neglect the wave nature of light, we can treat light as rays that propagate in straight lines in a homogeneous medium. Consequently, a model for the image created by a camera is given by the pinhole camera, as shown in Figure 2.17. Here, the object on the left side is imaged on the image plane on the right side. The image plane is one of the faces of a box in which a pinhole has been made on the opposite side to the image plane. The pinhole acts as the projection center. The pinhole camera produces an upside-down image of the object.

2.2.1 针孔摄像机

如果忽略光的波的特性，我们可以将光看作在同类介质中直线传播的光线。图 2.17 表示了针孔摄像机成像的模型。左端物体在右边像平面上成像。像平面相当于一个方盒子的一个面，在这个面的对面是针孔所在的面，针孔相当于投影的中心。针孔摄像机所成的像为物体的倒像。

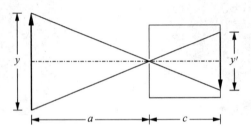

图 2.17 针孔摄像机。被测物在像平面成像，像平面与针孔形成一个盒子状，针孔相当于投影中心

From the similar triangles on the left and right sides of the projection center, it is clear that the height y' of the image of the object is given by

$$y' = y\frac{c}{a} \qquad (2.2)$$

where y is the height of the object, a is the distance of the object to the projection center, and c is the distance of the image plane to the projection center. The distance c is called the camera constant or the principal distance. Equation (2.2) shows that, if we increase the principal distance c, the size y' of the image of the object also increases. On the other hand, if we increase the object distance a, then y' decreases.

从投影中心左右两侧的相似三角形可以得到像的高度 y'

其中 y 为物体高度，a 为物体到投影中心的距离，c 为像平面到投影中心的距离。c 被称作摄像机常数或主距。从等式（2.2）可以看出，增加主距 c，像高 y' 也会增加，反过来，如果增加物距 a，y' 就会减小。

2.2.2 Gaussian Optics

The simple model of a pinhole camera does not model real lenses sufficiently well. Because of the small pinhole, very little light will actually pass to the image plane, and we would have to use very long exposure times to obtain an adequate image. Therefore, real cameras use lenses to collect light in the image. Lenses are typically formed from a piece of shaped glass or plastic. Depending on their shape, lenses can cause the light rays to converge or to diverge.

2.2.2.1 Refraction

Lenses are based on the principle of refraction. A light ray traveling in a certain medium will travel at a speed v, slower than the speed of light in vacuum c. The ratio $n = c/v$ is called the refractive index of the medium. The refractive index of air at the standard conditions for temperature and pressure is 1.0002926, which is typically approximated by 1. Different kinds of glasses have refractive indices roughly between 1.48 and 1.62.

When a light ray meets the border between two media with different refractive indices n_1 and n_2 at an angle of α_1 with respect to the normal to the boundary, it is split into a reflected ray and a refracted ray. For the discussion of lenses, we are interested only in the refracted ray. It will travel through the second medium at an angle of α_2 to the normal, as shown in Figure 2.18. The relation of the two angles is given by the law of refraction (Born and Wolf, 1999; Lenhardt, 2017):

2.2.2 高斯光学

针孔摄像机这种简单模型并不能足够好地模拟真实的镜头，由于针孔太小，只有极少量的光线能够通过小孔到达像平面，因此必须采用非常长的曝光时间以得到亮度足够的图像。因此真正的摄像机使用镜头收集光线。镜头通常由一定形状的玻璃或塑料构成。玻璃或塑料的形状决定了镜头可能使光线发散或汇聚。

2.2.2.1 折射

镜头是基于折射原理构造而成的。光线在一定介质中的传播速度 v 小于在真空中的传播速度 c，其比值 $n = c/v$ 称作此介质的折射率。在常温常压下，空气的折射率为 1.0002926，接近于 1。不同的玻璃的折射率大致在 1.48~1.62 之间。

假设第一种介质折射系数为 n_1，第二种介质折射系数为 n_2，当光线以入射角 α_1 到达介质一与介质二分界面时，光线将分成折射光与反射光，其中入射角 α_1 是入射光线与分界面法线的夹角。对于将要讲述的镜头，我们只关注折射光。如图 2.18 所示，折射光以出射角 α_2 传输通过第二种介质，其中出射角 α_2 是出射光线与分界面法线的夹角。这两个角度之间的关系可以用折射定律表示（Born and Wolf，1999；Lenhardt，2017）：

$$n_1 \sin \alpha_1 = n_2 \sin \alpha_2 \qquad (2.3)$$

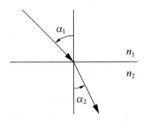

图 2.18 折射原理：自上而下的入射光在不同介质的分界处发生折射，两种介质的折射率分别为 n_1 和 n_2

The refractive index n actually depends on the wavelength λ of the light: $n = n(\lambda)$. Therefore, if white light, which is a mixture of different wavelengths, is refracted, it is split up into different colors. This effect is called dispersion.

As can be seen from Eq. (2.3), the law of refraction is nonlinear. Therefore, it is obvious that imaging through a lens is, in contrast to the pinhole model, a nonlinear process. In particular, this means that in general light rays emanating from a point (a so-called homocentric pencil) will not converge to a single point after passing through a lens. For small angles α to the normal of the surface, however, we may replace $\sin \alpha$ with α (Lenhardt, 2017). With this paraxial approximation, the law of refraction becomes linear again:

折射率 n 实际上决定于波长：$n = n(\lambda)$。白光是由多种不同波长的光组成，因此当白光折射时会散成多种颜色，这种效果称作色散。

从等式（2.3）可以看出折射定律是非线性的。显然，与针孔模型不同，镜头成像是非线性过程。也就是说同心光束通过镜头后将不能完全汇聚在一点。当入射角 α 很小时我们可以用 α 代替 $\sin\alpha$(Lenhardt, 2017)，通过这个近轴近似，我们可以得到线性的折射定律：

$$n_1 \alpha_1 = n_2 \alpha_2 \qquad (2.4)$$

The paraxial approximation leads to Gaussian optics, in which a homocentric pencil will converge again to a single point after passing through a lens that consists of spherical surfaces. Hence, Gaussian optics is the ideal for all optical systems. All deviations from Gaussian optics are called aberra-

根据近轴近似可以得到高斯光学，在高斯光学中同心光束通过由球面透镜构成的镜头后又汇聚到一点。高斯光学是理想化的光学系统，所有与高斯光学的背离均称作像差。光学系统设计的目标就是使得镜头的结构

tions. The goal of lens design is to construct a lens for which Gaussian optics hold for angles α that are large enough that they are useful in practice.

2.2.2.2 Thick Lens Model

Let us now consider what happens with light rays that pass through a lens. For our purposes, a lens can be considered as two adjacent refracting centered spherical surfaces with a homogeneous medium between them. Furthermore, we will assume that the medium outside the lens is identical on both sides of the lens. Lenses have a finite thickness. Hence, the model we are going to discuss, shown in Figure 2.19, is called the thick lens model (Born and Wolf, 1999; Lenhardt, 2017). For the discussion, we use the conventions and notation of DIN 1335:2003-12. Light rays travel from left to right. All horizontal distances are measured in the direction of the light. Consequently, all distances in front of the lens are negative. Furthermore, all upward distances are positive, while all downward distances are negative.

在满足高斯光学基础上使入射角足够大，以满足实际应用。

2.2.2.2 厚透镜模型

现在看看光线通过一个镜头将会发生什么。为了这一目的，可以将镜头看作是由两个球心位于同一直线的折射球面组成，两个球面之间为一种均匀介质。镜头外两侧介质也是相同的，镜头具有一定厚度。如图 2.19 所示的模型我们称为厚透镜模型（Born and Wolf，1999；Lenhardt，2017），我们将要讨论的就是这样一种模型。注意光线是从左向右传播的，所有水平间距均按光的方向测量，因此所有在镜头前的水平间距为负。而且，所有向上的间距为正，向下的间距为负。

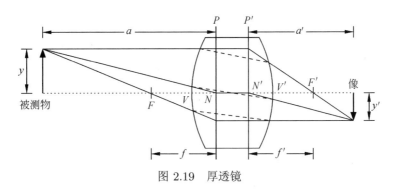

图 2.19 厚透镜

The object in front of the lens is projected to its image behind the lens. The lens has two focal points F and F', at which rays parallel to the optical axis that enter the lens from the opposite

位于镜头前方的物体在镜头后成像。镜头有两个焦点 F 和 F'，在镜头一侧的平行于光轴的光线经过镜头后汇聚到另一侧的对应焦点。主平面

side of the respective focal point converge. The principal planes P and P' are given by the intersection of parallel rays that enter the lens from one side with the corresponding rays that converge to the focal point on the opposite side. The focal points F and F' have a distance of f and f' from P and P', respectively. Since the medium is the same on both sides of the lens, we have $f = -f'$, and f' is the focal length of the lens. The object is at a distance of a from P (the object distance), while its image is at a distance of a' from P' (the image distance). The optical axis, shown as a dotted line in Figure 2.19, is the axis of symmetry of the two spherical surfaces of the lens. The surface vertices V and V' are given by the intersection of the lens surfaces with the optical axis. The nodal points N and N' have a special property that is described below. Since the medium is the same on both sides of the lens, the nodal points are given by the intersection of the principal planes with the optical axis. If the media were different, the nodal points would not lie in the principal planes.

With the above definitions, the laws of imaging with a thick lens can be stated as follows:

- A ray parallel to the optical axis before entering the lens passes through F'.
- A ray that passes through F leaves the lens parallel to the optical axis.
- A ray that passes through N passes through N' and does not change its angle with the optical axis.

As can be seen from Figure 2.19, all three rays converge at a single point. Since the imaging geometry is completely defined by F, F', N, and

P 和 P' 可以由镜头一侧入射的平行光线与另一侧过焦点的对应光线的交点得到，该平面与光轴垂直。相应的焦点 F 和 F' 与主平面 P 和 P' 的距离为 f 和 f'。由于镜头两侧的介质相同，因此 $f = -f'$，f' 为镜头焦距。物体到主平面 P 的距离为 a（物距），而像到主平面 P' 的距离为 a'（像距）。图 2.19 中虚点线表示的是光轴，为镜头两个折射球面的旋转对称轴。折射球面与光轴的交点为顶点 V 和 V'。节点 N 和 N' 的特点是当镜头两边介质相同，节点 N 和 N' 为主平面与光轴的交点。如果介质不同，节点就不在主平面上。

在上述定义下，厚透镜成像法则如下：

- 镜头前平行于光轴的光线过 F'。
- 过 F 点的光线通过镜头后平行于光轴。
- 过 N 点的光线也会过 N' 点，并且通过镜头之前与通过镜头之后与光轴夹角不变。

从图 2.19 可以看出，3 条光线聚于一点，由于像的几何尺寸完全取决于 F、F'、N 和 N'，这 4 个点称作

N', they are called the cardinal elements of the lens. Note that, for all object points that lie in a plane parallel to P and P', the corresponding image points also lie in a plane parallel to P and P'. This plane is called the image plane.

As for the pinhole camera, we can use similar triangles to determine the essential relationships between the object and its image. For example, it is easy to see that $y/a = y'/a'$. Consequently, similar to Eq. (2.2) we have

$$y' = y\frac{a'}{a} \qquad (2.5)$$

By introducing the magnification factor $\beta = y'/y$, we also see that $\beta = a'/a$. By using the two similar triangles above and below the optical axis on each side of the lens having F and F' as one of its vertices and the optical axis as one of its sides, we obtain (using the sign conventions mentioned above): $y'/y = f/(f-a)$ and $y'/y = (f'-a')/f'$. Hence, with $f = -f'$ we have

$$\frac{1}{a'} - \frac{1}{a} = \frac{1}{f'} \qquad (2.6)$$

This equation is very interesting. It tells us where the light rays will intersect, i.e., where the image will be in focus, if the object distance a is varied. For example, if the object is brought closer to the lens, i.e., the absolute value of a is decreased, the image distance a' must be increased. Likewise, if the object distance is increased, the image distance must be decreased. Therefore, focusing corresponds to changing the image distance. It is interesting to look at the limiting cases. If the object moves to infinity, all

镜头的基本要素。注意，对于平行于主平面 P 和 P' 的物面上的所有物点，其对应的像点也会在平行于 P 和 P' 的平面上，这个平面叫做像平面。

与针孔摄像机一样，我们同样可以利用相似三角形来确定物像之间的基本关系。可以看出 $y/a = y'/a'$，类似等式（2.2）我们得到

定义放大系数为 $\beta = y'/y$，可以得到 $\beta = a'/a$，利用光轴上下两侧的相似三角形，可以得出 $y'/y = f/(f-a)$ 及 $y'/y = (f'-a')/f'$，这两个三角形分别位于镜头两侧，并且光轴是它们的其中一条边，F 和 F' 是其中的一个顶点，同时正负符号根据前面所提到的符号定义。因此，当 $f = -f'$，可以推出

这个等式非常有用，从中可以推出当物距 a 变化时，通过镜头的光线将相交于何处，即物体将于何处成像。例如物体靠近镜头，也就是 a 的绝对值变小，像距 a' 就会变大；同理，如果物距变大，像距就会变小。所以，聚焦过程就相当于改变像距的过程。其极限情况非常有意思：如果物距无穷远，所有的光线都会成为平行光，此时 $a' = f'$。从另一方面讲，如果把被测物置于 F，像平面将在无穷远处。

the light rays are parallel, and consequently $a' = f'$. On the other hand, if we move the object to F, the image plane would have to be at infinity. Bringing the object even closer to the lens, we see that the rays will diverge on the image side. In Eq. (2.6), the sign of a' will change. Consequently, the image will be a virtual image on the object side of the lens, as shown in Figure 2.20. This is the principle that is used in a magnifying glass.

如果继续把物体向镜头移动使其位于 F 之内,将看到光线在成像端发散,等式 (2.6) 中 a' 的正负号发生变化,其像为在物体同一侧的虚像,如图 2.20 所示,这就是放大镜的主要原理。

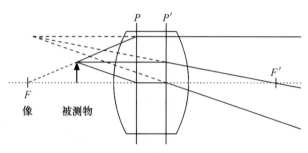

图 2.20　物距比焦距还小时成虚像

From $\beta = y'/y = f/(f-a) = f'/(f'+a)$, we can also see that, for constant object distance a, the magnification β increases if the focal length f' increases.

Real lens systems are more complex than the thick lens we have discussed so far. To minimize aberrations, they typically consist of multiple lenses that are centered on their common optical axis. An example of a real lens is shown in Figure 2.21. Despite its complexity, a system of lenses can still be regarded as a thick lens, and can consequently be described by its cardinal elements. Figure 2.21 shows the position of the focal points F and F' and the nodal points N and N' (and thus the position of the principal planes). Note that for this lens the object side focal point F lies within the second lens of the system. Also note that N' lies in front of N.

从 $\beta = y'/y = f/(f-a) = f'/(f'+a)$ 可以得出,对于相同物距 a,随着焦距 f' 的增加,放大倍率 β 也会增加。

实际的镜头系统远比已经讨论过的厚透镜要复杂得多。为了减少像差,通常镜头由多个球心位于同一光轴上的光学镜片组成。图 2.21 是一个真实的镜头的例子。尽管真实的镜头更加复杂,一个镜头系统仍可以看作是一个厚透镜,因此也可以用它的主要元素来描述。图 2.21 表示了焦点 F 和 F',节点 N 和 N' 及主平面的位置。请注意,在这个镜头中,物方焦点 F 位于第二个镜片内部。而且 N' 在 N 的前面。

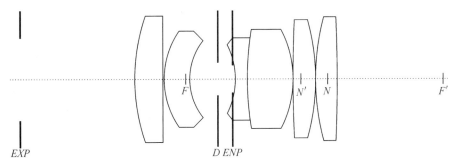

图 2.21 镜头主要要素。D 为光阑；ENP 为入瞳；EXP 为出瞳

2.2.2.3 Aperture Stop and Pupils

Real lenses have a finite extent. Furthermore, to control the amount of light that falls on the image plane, lenses typically have a diaphragm (also called an iris) within the lens system. The size of the opening of the diaphragm can typically be adjusted with a ring on the lens barrel. The diaphragm of the lens in Figure 2.21 is denoted by D. In addition, other elements of the lens system, such as the lens barrel, may limit the amount of light that falls on the image plane. Collectively, these elements are called stops. The stop that limits the amount of light the most is called the aperture stop of the lens system (ISO 517:2008). Note that the aperture stop is not necessarily the smallest stop in the lens system because lenses in front or behind the stop may magnify or shrink the apparent size of the stop as the light rays traverse the lens. Consequently, a relatively large stop may be the aperture stop for the lens system.

Based on the aperture stop, we can define two important virtual apertures in the lens system (Born and Wolf, 1999; ISO 517:2008). The entrance pupil defines the area at the entrance of the lens system that can accept light. The entrance pupil is the (typically virtual) image of

2.2.2.3 孔径光阑和瞳孔

真实的镜头有一定孔径大小限制。为了控制可以到达像平面光线的多少，镜头系统中一般都设计有可变光阑。在镜头筒上有个环可以用来调整光阑大小。在图 2.21 中以 D 来表示系统的光阑。镜头的其他组成部件，比如镜筒也会限制到达像平面光线的总量。这些因素统称为光阑，其中最大程度限制通光量的光阑称作镜头的孔径光阑（ISO 517: 2008）。需要注意的是并不是最小的光阑即为镜头的孔径光阑，因为在光穿过镜头时，光阑前后的镜片可能放大或缩小光阑的实际尺寸。因此，镜头中相对较大的光阑也可能成为镜头系统的孔径光阑。

基于孔径光阑，定义镜头系统中两个重要的虚拟光阑（Born and Wolf，1999；ISO 517:2008）：入瞳与出瞳。入瞳决定镜头入口可以接收光线的面积。入瞳是孔径光阑被其前面的光学系统在物方所成的像，通常为

the aperture stop in the optics that come before it. Rays that pass through the entrance pupil are able to enter the lens system and pass through it to the exit. Similarly, the exit pupil is the (typically virtual) image of the aperture stop in the optics that follow it. Only rays that pass through this virtual aperture can exit the system. The entrance and exit pupils of the lens system in Figure 2.21 are denoted by ENP and EXP, respectively.

We can single out a very important light ray from the pencil that passes through the lens system: the principal or chief ray. It passes through the center of the aperture stop. Virtually, it also passes through the center of the entrance and exit pupils. Figure 2.22 shows the actual path of the principal ray as a thick solid line. The virtual path of the principal ray through the centers of the entrance pupil Q and the exit pupil Q' is shown as a thick dashed line. In this particular lens, the actual path of the principal ray passes very close to Q. Figure 2.22 also shows the path of the rays that touch the edges of the entrance and exit pupils as thin lines. As for the principal ray, the actual paths are shown as solid lines, while the virtual paths to the edges of the pupils are shown as dashed lines. These rays determine the light

虚像。入瞳是物面上所有各点发出的光束的共同入口，通过入瞳的光线可以进入到镜头系统中，在镜头中传播并通过镜头；同样，出瞳是孔径光阑被其后面的光学系统在像方所成的像，通常也为虚像。只有能通过出瞳的光线才能通过整个光学系统。在图 2.21 中入瞳和出瞳以 ENP 和 EXP 表示。

可以从同心光束中挑选出一条通过镜头系统的重要光线——主光线。主光线过孔径光阑中心，其在物方和像方的对应光线或光线延长线也分别过入瞳和出瞳的中心，在图 2.22 中主光线的实际光路以粗实线表示，过入瞳中心 Q 和出瞳中心 Q' 的主光线的延长虚拟光路以粗虚线表示。在这个镜头中，主光线真实的传播路径非常靠近 Q 点。图 2.22 也画出了过入瞳和出瞳边缘的光线，以细线表示，与主光线一样，实际光路以实线表示，而过入瞳和出瞳边缘的延长虚拟光路则以虚线表示。这些光线形成了进、出镜头系统的光锥，从而决定了能够到达像平面的光通量。

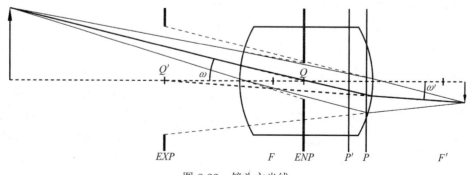

图 2.22 镜头主光线

cone that enters and exits the lens system, and consequently the amount of light that falls onto the image plane.

Another important characteristic of the lens is the pupil magnification factor

$$\beta_p = \frac{d_{\text{EXP}}}{d_{\text{ENP}}} \qquad (2.7)$$

i.e., the ratio of the diameters of the exit and entrance pupils. It can be shown that it also relates the object and image side field angles to each other (Lenhardt, 2017):

$$\beta_p = \frac{\tan \omega}{\tan \omega'} \qquad (2.8)$$

Note that ω' in general differs from ω, as shown in Figure 2.22. The angles are identical only if $\beta_p = 1$. Equation (2.7) gives us a simple method to check whether a lens has different image and object side field angles: we simply need to look at the lens from the front and the back. If the sizes of the entrance and exit pupils differ, the field angles differ.

2.2.2.4 Relation of the Pinhole Model to Gaussian Optics

With the above, we can now discuss how the pinhole model is related to Gaussian optics. We can see that the light rays in the pinhole model correspond to the principal rays in Gaussian optics. In the pinhole model, there is a single projection center, whereas in Gaussian optics there are two projection centers: one in the entrance pupil for the object side rays and one in the exit pupil for the image side rays. Furthermore, for Gaussian optics, in general $\omega' \neq \omega$, whereas $\omega' = \omega$ for the

镜头的另外一个重要参数是光瞳放大率

也就是出射光瞳的直径与入射光瞳直径之比。从（Lenhardt，2017）中可以看到，β_p 的大小也与物方和像方视场角相关。

需要注意通常情况下 ω' 与 ω 不相等，如图 2.22 所示，仅在 $\beta_p = 1$ 时两者才一致。等式（2.7）提供了一个简单方法来检查一个镜头是否有不同的像方和物方视场角：我们只需要从前方和后方观察镜头，如果入瞳和出瞳尺寸不同，则视场角也不同。

2.2.2.4 针孔模型和高斯光学的关系

在以上基础上来讨论针孔模型与高斯光学的关系。可以看出针孔模型中的光线相当于高斯光学中的主光线。在针孔模型中仅有一个投影中心，而在高斯光学中有两个投影中心，一个是针对物方光束，位于入射光瞳，另一个是针对像方光束，位于出射光瞳。此外，在通常情况下，高斯光学中 $\omega' \neq \omega$，而在针孔摄像机中 $\omega' = \omega$。为了使二者一致，我们必须保证在物

pinhole camera. To reconcile these differences, we must ensure that the object and image side field angles are identical. In particular, the object side field angle ω must remain the same since it is determined by the geometry of the objects in the scene. Furthermore, we must create a single projection center. Because ω must remain constant, this must be done by (virtually) moving the projection center in the exit pupil to the projection center in the entrance pupil. As shown in Figure 2.23, to perform these modifications, we must virtually shift the image plane to a point at a distance c from the projection center Q in the entrance pupil, while keeping the image size constant. As described in Section 2.2.1, c is called the camera constant or the principal distance. This creates a new image side field angle ω''. From $\tan \omega = \tan \omega''$, we obtain $y/a = y'/c$, i.e.,

方和像方的视场角一样，特别是物方视场角 ω 必须保持不变，因为它仅取决于被测物的几何尺寸。此外要建立一个单一投影中心系统，由于 ω 必须保持不变，因此只能通过将出瞳上的投影中心移动到入瞳上的投影中心来实现这一目的，如图 2.23 所示，同时我们还必须将像平面虚拟地移到与入瞳投影中心 Q 的距离为 c 的点上，而保持像的大小不变。如 2.2.1 节所述，c 称为摄像机常数或主距。这时就有了新的像方视场角 ω''，根据 $\tan \omega = \tan \omega''$，可以得到 $y/a = y'/c$，也就是

$$c = \frac{y'}{y}a = \beta a = f'\left(1 - \frac{\beta}{\beta_p}\right) \qquad (2.9)$$

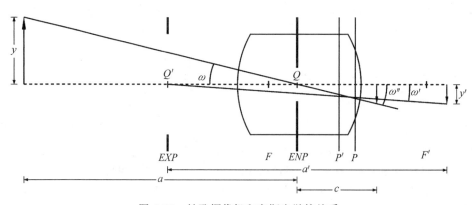

图 2.23　针孔摄像机和高斯光学的关系

The last equation is derived in Lenhardt (2017). The principal distance c is the quantity that can be determined through camera calibration (see Section 3.9). Note that it can differ substantially from the focal length f' (in our example

最后一个等式在文献（Lenhardt，2017）中有推导。3.9 节中论述了如何通过摄像机标定得到主距 c。需要注意的是主距与焦距 f' 可能不同。在我们的例子中相差 10% 左右。c 取决

by almost 10%). Also note that c depends on the object distance a. Because of Eq. (2.6), c also depends on the image distance a'. Consequently, if the camera is focused to a different depth it must be recalibrated.

2.2.3 Depth of Field

Up to now, we have focused our discussion on the case in which all light rays converge again to a single point. As shown by Eq. (2.6), for a certain object distance a, the image plane (IP) must lie at the corresponding image distance a'. Hence, only objects lying in a plane parallel to the image plane, i.e., perpendicular to the optical axis, will be in focus. Objects not lying in this focusing plane (FP) will appear blurred. For objects lying farther from the camera, e.g., the object at the distance a_f in Figure 2.24, the light rays will intersect in front of IP at an image distance of a'_f. Likewise, the light rays of objects lying closer to the camera, e.g., at the distance a_n, will intersect behind IP at an image distance of a'_n. In both cases, the points of the object will be imaged as a circle of confusion having a diameter of d'_f and d'_n, respectively.

于物距 a。从等式 (2.6) 可以看出 c 同时取决于像距 a'。因此，如果摄像机聚焦于另外一个平面必须重新标定。

2.2.3 景深

到目前为止，讨论都是基于所有的光线聚于一点。如等式（2.6）所示，对于物距 a，像平面一定在像距为 a' 的像平面（IP）上。因此只有位于垂直于光轴且与像平面平行的平面上的被测物才能聚焦。不在这个对焦面（FP）上的物所成的像将是模糊的。对于到摄像机的距离比对焦面更远的，物距为 a_f 的物面上的物体，如图 2.24 中所示，将成像于 IP 之前，像距为 a'_f。同样，对于到摄像机的距离比对焦面近的，物距为 a_n 的物面上的物体，将成像于 IP 之后，像距为 a'_n。在这两种情况下，物点在像平面 IP 上都将成一弥散圆斑，直径分别为 d'_f 和 d'_n。

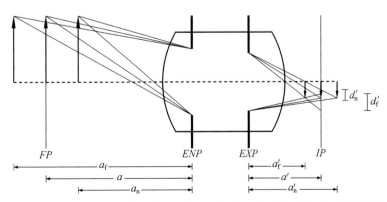

图 2.24 镜头景深。对焦距离为 a 的被测物可以很好地聚焦。比对焦距离远或近的物体都将成模糊像

Often, we want to acquire sharp images even though the object does not lie entirely within a plane parallel to the image plane. Since the sensors that are used to acquire the image have a finite pixel size, the image of a point will still appear in focus if the size of the circle of confusion is in the same order as the size of the pixel (between 1.25 µm and 10 µm for current sensors).

As we can see from Figure 2.24, the size of the blur circle depends on the extent of the ray pencil, and hence on the diameter of the entrance pupil. In particular, if the diaphragm is made smaller, and consequently the entrance pupil becomes smaller, the diameter of the circle of confusion will become smaller. Let us denote the permissible diameter of the circle of confusion by d'. If we set $d' = d'_f = d'_n$, we can calculate the image distances a_f and a_n of the far and near planes that will result in a diameter of the circle of confusion of d' (see Lenhardt, 2017):

通常情况下希望即使被测物所在平面不完全与像平面平行，也能够得到锐利的图像。由于用于捕获图像的传感器每个像素尺寸有一定大小，因此如果弥散圆斑的尺寸与像素尺寸差不多，则可以认为是聚焦的。目前传感器的像素尺寸为 1.25~10 µm。

从图 2.24 可以看出，弥散圆的大小取决于同心光束的大小，即取决于入瞳的直径。如果我们把可变光阑调小，入瞳也相应变小，则弥散圆的直径将会变小。假定可以被接受的弥散圆直径为 d'，如果设 $d' = d'_f = d'_n$，我们可以计算出产生弥散圆直径为 d' 的 a_f 和 a_n（Lenhardt，2017）：

$$a_{f,n} = \frac{f'^2 a}{f'^2 \pm F d'(a + f'/\beta_p)} \tag{2.10}$$

In terms of the magnification β, Eq. (2.10) can also be written as (see Lenhardt, 2017):

等式（2.10）中的放大率 β 也可以写为（Lenhardt，2017）

$$a_{f,n} = \frac{(1 - \beta/\beta_p) F d'}{\beta^2 \pm F^2 d'^2 / f'^2} \tag{2.11}$$

Here, F denotes the f-number of the lens, which is given by

这里 F 定义为镜头的 f 值

$$F = f'/d_{\text{ENP}} \tag{2.12}$$

The f-number can typically be adjusted with a ring on the lens barrel. The f-numbers are usually specified using the standard series of f-number markings as powers of $\sqrt{2}$, namely: $f/1$, $f/1.4$,

通常 f 值可以用镜头筒上的环来调节。F 值通常如（ISO 517：2008）标准标注，是以 $\sqrt{2}$ 的幂表示：$f/1$, $f/1.4$, $f/2$, $f/2.8$, $f/4$, $f/5.6$, $f/8$,

$f/2, f/2.8, f/4, f/5.6, f/8, f/11, f/16, f/22$, etc. (ISO 517:2008). A ratio of $\sqrt{2}$ is chosen because the energy E that falls onto the sensor is proportional to the exposure time t and proportional to the area A of the entrance pupil, i.e., inversely proportional to the square of the f-number:

$f/11, f/16, f/22$ 等。按 $\sqrt{2}$ 的幂表示是由于到达传感器的能量与时间 t 和入瞳面积 A 成正比,也就是与 f 数的平方成反比:

$$E \sim tA \sim \frac{t}{F^2} \tag{2.13}$$

Hence, increasing the f-number by one step, e.g., from $f/4$ to $f/5.6$, halves the image brightness.

这样增大一级 f 数,比如从 $f/4$ 到 $f/5.6$,可使图像的亮度减少一半。

From Eq. (2.10), it follows that the depth of field Δa is given by

从等式(2.10)得到景深 Δa,

$$\Delta a = \frac{2f'^2 aFd'(a + f'/\beta_{\rm p})}{f'^4 - F^2 d'^2 (a + f'/\beta_{\rm p})^2} \tag{2.14}$$

If we assume that a is large compared to $f'/\beta_{\rm p}$, we can substitute $a + f'/\beta_{\rm p}$ by a. If we additionally assume that f'^4 is large with respect to the rest of the denominator, we obtain

如果假设 a 比 $f'/\beta_{\rm p}$ 大,我们可以用 a 代替 $a + f'/\beta_{\rm p}$,如果还假设 f'^4 比分母中其他项都大,可以得到

$$\Delta a \approx \frac{2a^2 Fd'}{f'^2} \tag{2.15}$$

Hence, we can see that the depth of field Δa is proportional to $1/f'^2$. Consequently, a reduction of f' by a factor of 2 will increase Δa by a factor of 4. On the other hand, increasing the f-number by a factor of 2 will lead to an increase of Δa by a factor of 2. Note that this means that the exposure time will need to be increased by a factor of 4 to obtain the same image brightness.

从上式可以看出景深 Δa 与 $1/f'^2$ 成正比。f' 减少一半,景深将增加 4 倍。另一方面,f 值增加一倍会使景深也增加一倍。这也意味着曝光时间需要是原来的 4 倍才能得到同样亮度的图像。

From Eq. (2.11), the depth of field is given by

根据等式(2.11),景深为

$$\Delta a = \frac{2(1 - \beta/\beta_{\rm p})Fd'}{\beta^2 + F^2 d'^2/f'^2} \tag{2.16}$$

If we assume that β is much larger than $F^2 d'^2/f'^2$ and introduce the effective f-number $F_{\rm e} = F(1 - \beta/\beta_{\rm p})$, we obtain

如果我们假设 β 比 $F^2 d'^2/f'^2$ 更大,并且引入有效的 f 值 $F_{\rm e} = F(1 - \beta/\beta_{\rm p})$,可以得到

$$\Delta a \approx \frac{2F_e d'}{\beta^2} \tag{2.17}$$

Thus, the depth of field Δa is proportional to $1/\beta^2$, i.e., it becomes much smaller as the magnification of the lens increases.

Figure 2.25 displays two images of a depth-of-field target taken with $F = 2$ and $F = 16$ with a lens with $f' = 12.5\,\mathrm{mm}$. The depth-of-field target is a set of scales and lines mounted at an angle of 45° on a metal block. As predicted from Eq. (2.15), the image with $F = 2$ has a much smaller depth of field than the image with $F = 16$.

这样，景深 Δa 和 $1/\beta^2$ 成正比，也就是说，当镜头放大率增加时，景深变得更小。

图 2.25 是使用 $f' = 12.5\mathrm{mm}$ 的镜头在 $F = 2$ 和 $F = 16$ 时对一个具有一定视场深度的目标所采集到的两幅图像。被测物是成 45° 角的金属尺子上的刻度及数字。如等式（2.15）所预料的，$F = 2$ 时的景深比 $F = 16$ 时的景深要小得多。

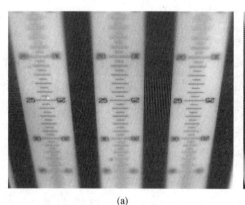 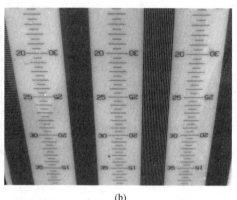

(a)   (b)

图 2.25 （a）$F=2$ 时的图像；（b）$F=16$ 时的图像。两幅图都采用 $f' = 12.5\mathrm{mm}$。注意（a）的景深小，（b）的景深大。同时请注意图像中的相应畸变

It should be noted that it is generally impossible to increase the depth of field arbitrarily. If the aperture stop becomes very small, the wave nature of light will cause the light to be diffracted at the aperture stop. Because of diffraction, the image of a point in the focusing plane will be a smeared-out point in the image plane. This will limit the sharpness of the image. Born and Wolf (1999) and Lenhardt (2017) show that, for a circular aperture stop with an effective f-number of F_e, the image of a point will be given by the Airy disk

需要注意的是不能任意加大景深，因为由于光的波动特性，如果使用非常小的孔径光阑，光线将在光阑处发生衍射，这将会使对焦平面上的物点在像面上所成的像为条纹圆环斑，从而影响图像的清晰度。从文献（Born and Wolf, 1999）和（Lenhardt, 2017）可以看出，对于有效 f 数为 F_e 的圆形孔径光阑，一个物点的像为一艾里衍射圆斑。

$$I = \left(\frac{2J_1(v)}{v}\right)^2 \tag{2.18}$$

where $J_1(v)$ is the Bessel function of the first kind and $v = \pi\sqrt{x'^2 + y'^2}/(\lambda F_e)$. Figure 2.26 shows the two-dimensional (2D) image of the Airy disk as well as a one-dimensional (1D) cross-section. The radius of the first minimum of Eq. (2.18) is given by $r' = 1.21967 \lambda F_e$. This is usually taken as the radius of the diffraction disk. Note that r' increases as the effective f-number increases, i.e., as the aperture stop becomes smaller, and that r' increases as the wavelength λ increases. For $\lambda = 500$ nm (green light) and $F_e = 8$, r' will already be approximately 5 μm, i.e., the diameter of the diffraction disk will be in the order of 1–8 pixels. We should keep in mind that the Airy disk will occur for a perfectly focused system. Points lying in front of or behind the focusing plane will create even more complicated brightness distributions that have a radius larger than r' (Born and Wolf, 1999; Lenhardt, 2017).

其中 $J_1(v)$ 是一阶贝塞尔函数，$v = \pi\sqrt{x'^2 + y'^2}/(\lambda F_e)$。图 2.26 为艾里圆斑的二维图像及一维横截面。等式（2.18）的第一个极小值半径为 $r' = 1.21967 \lambda F_e$，通常被看作是衍射斑的半径。注意，当有效 f 值变大，即孔径光阑变小时，r' 则会增大，波长 λ 增大，r' 也会增大。对于 $\lambda=500$ nm 的绿光，当 $F_e = 8$ 时 r' 已经大约为 5μm 了。也就是说衍射斑直径已经达到 1~8 个像素了。即使在非常好的聚焦情况下，艾里斑也会出现。位于对焦平面前后的物点将会产生更加复杂的亮度分布，其像斑半径大于 r'（Born and Wolf，1999；Lenhardt，2017）。

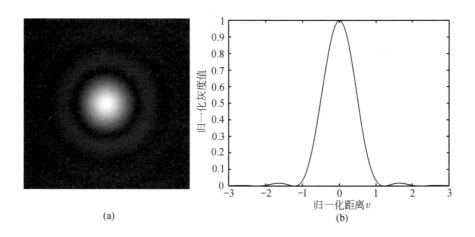

图 2.26 衍射会使一个点的像成为艾里斑
（a）艾里斑二维图像，为使外围环可视，使用了 $\gamma = 0.4$ 的 gamma 函数查找表；（b）一维艾里斑截面图

2.2.4 Telecentric Lenses

The lens systems we have considered so far all perform a perspective projection of the world. The projection is perspective both in object and in image space. These lenses are called entocentric lenses. Because of the perspective projection, objects that are closer to the lens produce a larger image, which is obvious from Eqs. (2.2) and (2.9). Consequently, the image of objects not lying in a plane parallel to the image plane will exhibit perspective distortions. In many measurement applications, however, it is highly desirable to have an imaging system that performs a parallel projection in object space because this eliminates the perspective distortions and removes occlusions of objects that occur because of the perspective distortions.

2.2.4.1 Object-Side Telecentric Lenses

From a conceptual point of view, a parallel projection in object space can be achieved by placing an infinitely small pinhole aperture stop at the image-side focal point F' of a lens system. From the laws of imaging with a thick lens in Section 2.2.2.2, we can deduce that the aperture stop only lets light rays parallel to the optical axis on the object side pass, as shown in Figure 2.27. Consequently, the lens must be at least as large as the object we would like to image.

Like the pinhole camera, this construction is impracticable because too little light would reach the sensor. Therefore, the aperture stop must have a finite extent, as shown in Figure 2.28. Here, for simplicity the principal planes have been

2.2.4 远心镜头

至今讨论的镜头系统都是实际物体的透视投影。投影在物方空间和像方空间都是透视的，这种镜头叫做近心镜头（普通光学镜头）。从等式(2.2)和等式（2.9）明显可以看出距离镜头越近，物体所成的像就越大。因此，与像平面不平行的被测物体所成的像将会变形。然而，在许多测量应用中，非常需要在物方空间产生平行投影的成像系统，以消除透视变形以及由于透视变形产生的被测物的遮挡。

2.2.4.1 物方远心镜头

从概念上讲，可以通过在镜头系统像方焦点 F' 处安装无限小的针孔孔径光阑来实现物方空间平行投影。根据 2.2.2.2 节的厚透镜成像定律，可以推断出这个孔径光阑仅允许物方平行于光轴的光线通过，如图 2.27 所示。因此镜头至少要与被测物体一样大。

如同针孔摄像机一样，这种结构能够到达传感器的光线太少，没有实用意义。因此，孔径光阑必须有一定的大小，如图 2.28 所示。为了简化起见，主平面画在了一起，也就是

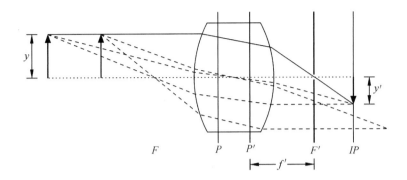

图 2.27 远心镜头原理
在像方焦点 F' 处放置针孔孔径光阑，仅允许物方平行于主轴的光线通过

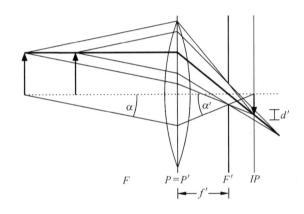

图 2.28 物方远心镜头
从以粗实线表示的主光线可以看出，物距不同的被测物在同一位置成像。
与一般普通镜头一样，不在对焦平面上的物体将会产生弥散圆斑

drawn at the same position ($P = P'$). As can be seen from the principal ray, objects at different object distances are imaged at the same position. As for regular lenses, objects not in the focusing plane will create a circle of confusion.

If we recall that the entrance pupil is the virtual image of the aperture stop in the optics that come before it, and that the aperture stop is located at the image-side focal point F', we can see that the entrance pupil is located at $-\infty$ and has infinite extent. Since the center of the entrance

$P = P'$，从主光线可以看出，不同物距的被测物成像在相同位置。与一般普通镜头一样，不在对焦平面上的物体将会产生弥散圆斑。

入瞳是孔径光阑被其前面的光学系统在物方所成的虚像，现在孔径光阑位于像方焦点 F' 处，则可以推出入瞳位于物方无穷远，其大小为无穷大。由于入瞳中心扮演投影中心的角色，而投影中心处于非常远处，因此

pupil acts as the projection center, the projection center is infinitely far away, giving rise to the name telecentric perspective for this parallel projection. In particular, since the projection center is at infinity on the object side, such a lens system is called an object-side telecentric lens. Note that the exit pupil of the lens in Figure 2.28 is the aperture stop because no optical elements are behind it. This is a simplified model for object-side telecentric lenses. In real lenses, there typically are optical elements behind the aperture stop. Nevertheless, the exit pupil will still be at a finite location. Furthermore, note that an object-side telecentric lens still performs a perspective projection in image space. This is immaterial if the image plane is perpendicular to the optical axis.

In contrast to the simple pinhole telecentric lens discussed above, an object-side telecentric lens must be larger than the object by an amount that takes the size of the aperture stop into account, as shown by the rays that touch the edge of the aperture stop in Figure 2.28.

The depth of field of an object-side telecentric lens is given by (see Lenhardt, 2017)

这种平行投射方式被称为远心，特别是当投影中心位于物方无穷远处时，镜头系统被称为物方远心镜头。注意，图 2.28 所示镜头的出瞳即为孔径光阑，因为在孔径光阑后面没有其他光学元件。这是物方远心镜头的一个简化模型，对于实际的镜头，在孔径光阑后面通常会有光学元件。然而，出瞳仍然位于有限的位置。而且，需要注意的是物方远心镜头在像方空间仍然遵循透视投影。当然，如果像平面垂直于光轴，这些就不重要了。

与上面论述的简单针孔远心镜头不同，如图 2.28 有到达孔径光阑边缘的光线，因此需要考虑到孔径光阑的大小，物方远心镜头必须要比被测物体大。

物方远心镜头的景深为（Lenhardt, 2017）

$$\Delta a = -\frac{d'}{\beta \sin \alpha} = \frac{d'}{\beta^2 \sin \alpha'} \tag{2.19}$$

As before, d' is the permissible diameter of the circle of confusion (see Section 2.2.3), β is the magnification of the lens, given by $\beta = -\sin \alpha/\sin \alpha'$, and $A = \sin \alpha$ is the numerical aperture of the lens. It is related to the f-number by $F = 1/(2A)$. Likewise, $\sin \alpha'$ is the image-side numerical aperture.

d' 为可接受的弥散圆的直径（见 2.2.3 节），β 为镜头的放大倍率，$\beta = -\sin \alpha/\sin \alpha'$，$A = \sin \alpha$ 为镜头的数值孔径，与 f 数的关系为 $F = 1/(2A)$，同样，$\sin \alpha'$ 为像方数值孔径。

2.2.4.2 Bilateral Telecentric Lenses

Another kind of telecentric lens can be constructed by positioning a second lens system behind the aperture stop of an object-side telecentric lens in such a way that the image-side focal point F_1' of the first lens system coincides with the object-side focal point F_2 of the second lens system (see Figure 2.29). From the laws of imaging with a thick lens in Section 2.2.2.2, we can deduce that now the light rays on the image side of the second lens system also will be parallel to the optical axis. This construction will move the exit pupil to ∞. Therefore, these lenses are called bilateral telecentric lenses.

2.2.4.2 双远心镜头

另外一种远心镜头是在物方远心镜头孔径光阑后面再加上第二个镜头系统，使第一个镜头的像方焦点 F_1' 与第二个镜头的物方焦点 F_2 重合，如图 2.29 所示。根据 2.2.2.2 节的厚透镜成像定律可以得出第二个镜头的像方主光线也将平行于光轴。这种结构将出瞳也移到了无穷远，因此也称作双远心镜头。

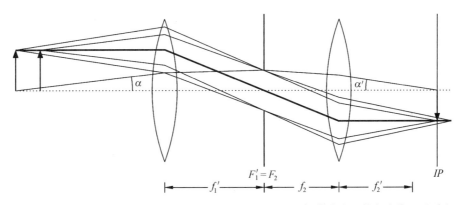

图 2.29 双远心镜头。孔径光阑后面放置镜头，使 $F_1' = F_2$。从以粗实线表示的主光线可以看出，物距不同的被测物在同一位置成像。与一般普通镜头相似，不在对焦平面上的物体将会产生弥散圆斑

The magnification of a bilateral telecentric lens is given by $\beta = -f_2'/f_1'$ (Lenhardt, 2017). Therefore, the magnification is independent of the object position and of the position of the image plane, in contrast to object-side telecentric lenses, where the magnification is only constant for a fixed image plane.

双远心镜头的放大倍率可表示为 $\beta = -f_2'/f_1'$（Lenhardt，2017）。因此放大倍率与被测物的位置及像平面的位置无关。而在物方远心镜头中，对应一个固定的像平面，放大倍率是一个常数。

The depth of field of a bilateral telecentric lens is given by (see Lenhardt, 2017)

双远心镜头的景深可表示如下（Lenhardt，2017）

$$\Delta a = -\frac{d'}{\beta \sin \alpha} = \frac{d'}{\beta^2 \sin \alpha'} \qquad (2.20)$$

Figure 2.30 shows an image of a depth-of-field target taken with an object-side telecentric lens with $\beta = 0.17$ and $F = 5.6$. Note that, in contrast to the perspective lens used in Figure 2.25, there are no perspective distortions in the image.

图 2.30 为 $\beta = 0.17$、$F = 5.6$ 时物方远心镜头对一个具有一定视场深度的目标所采集的图像。与图 2.25 中使用的镜头形成对比,这张图中没有透视畸变。

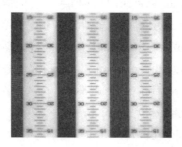

图 2.30　使用 $F = 5.6$、$\beta = 0.17$ 远心镜头所采集的图像。注意图像中没有透视畸变

2.2.4.3　Image-Side Telecentric Lenses

There is a further type of telecentric lens. If the aperture stop is placed at the object-side focal point F of a lens system, the laws of imaging with a thick lens in Section 2.2.2.2 imply that the principal ray is parallel to the optical axis in image space, as shown in Figure 2.31. The entrance pupil is located at a finite location, while the exit pupil is at ∞. Therefore, these lenses are called image-side telecentric lenses. The projection in object space is perspective, as for entocentric lenses. Note that the lens behind the aperture stop in a bilateral telecentric lens essentially constitutes an image-side telecentric lens.

Image-side telecentric lenses have the advantage that the light rays impinge on the sensor perpendicularly. This helps to avoid pixel vignetting, i.e., the effect that solid-state sensors are less sensitive to light the more the angle of the light ray deviates from being perpendicular to the sensor.

2.2.4.3　像方远心镜头

还有一种远心镜头类型。如果在某个镜头系统中物方焦点 F 处放置孔径光阑,根据 2.2.2.2 节的厚透镜成像定律,可以推断出在像方空间主光线平行于光轴,如图 2.31 所示。入瞳位于有限的位置,而出瞳在无穷远处。因此,这种镜头叫做像方远心镜头。和普通近心镜头一样,在物方空间是透视投影。需要注意的是,在双远心镜头孔径光阑后面的镜头部分本质上就是一个像方远心镜头。

像方远心镜头的优点是光线会垂直进入传感器,这有助于避免像素光晕现象,即对于固态传感器,在垂直于传感器方向光线的偏离角度越大越不敏感。

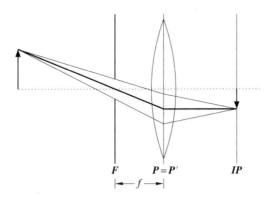

图 2.31 像方远心镜头，主光线用粗实线表示，在像方空间中平行于主光轴

2.2.4.4 Projection Characteristics of Lenses

We conclude this section by listing the projection characteristics of the different lens types we have discussed so far in Table 2.2. If the image plane is perpendicular to the optical axis, the projection characteristics in image space become immaterial from a geometric point of view since in this configuration a perspective and a parallel projection in image space have the same effect. Note that, in this case, the parallel projection becomes an orthographic projection. Therefore, if the image plane is perpendicular to the optical axis, we will simply speak of a perspective or a telecentric lens, with the understanding that in this case the terms refer to the object-side projection characteristics.

2.2.4.4 镜头的投影特征

通过列举到目前为止我们所讨论过的表 2.2 中不同类型镜头的投影特征，在本节我们可以得出结论。如果像平面垂直于光轴，从几何学的观点来看，在像方空间的投影特征变得不重要了，因为对于这种结构，透视投影和平行投影在像方空间的效果是相同的。注意，在这种情况下，平行投影变成了正交投影。因此，如果像平面垂直于光轴，我们可以简单地说是一个透视或者远心镜头，可以这样理解，这种情况下术语可以参考物方投影特征。

表 2.2 不同类型的镜头及其投影特征

镜头类型	物方投影	像方投影
近心（普通镜头）	透视	透视
像方远心	透视	平行
物方远心	平行	透视
双远心	平行	平行

2.2.5 Tilt Lenses and the Scheimpflug Principle

The discussion of the depth of field in Section 2.2.3 has assumed that the image plane is perpendicular to the optical axis. Equation (2.11) shows that the depth of field in this case is a region in space that is bounded by two planes that are parallel to the image plane. Furthermore, Eq. (2.17) shows that the depth of field is inversely proportional to the square of the magnification of the lens. Consequently, the larger the magnification of the camera, the smaller the depth of field.

The small depth of field at high magnifications becomes problematic whenever it is necessary to image objects in focus that lie in or close to a plane that is not parallel to the image plane. With regular lenses, this is only possible by reducing the size of the aperture stop, i.e., by increasing the f-number of the lens. However, as discussed in Section 2.2.3, there is a limit to this approach for two reasons. First, if the aperture stop is made too small, the image will appear blurred because of diffraction. Second, a small aperture stop causes less light to reach the sensor. Consequently, a high-powered illumination is required to achieve reasonable exposure times, especially when images of moving objects must be acquired.

There are several practical applications in which a plane in object space that is not parallel to the image plane must be imaged in focus (Steger, 2017). One example is stereo reconstruction, where typically the cameras are used in a converging setup (see Sections 2.5.1 and 3.10.1). As shown in Figure 2.32, this setup causes the volume

2.2.5 倾斜镜头和沙姆定律

在 2.2.3 节讨论景深时假设像平面垂直于光轴。等式（2.11）表明在这种情况下景深是一个空间范围，这个范围限制在和像平面平行的两个平面内。而且，等式（2.17）表明景深和镜头放大率的平方成反比。因此，摄像机的放大率越大，景深越小。

每当必须对那些位于或者接近于某个和像平面不平行的平面上的物体清晰成像时，高放大率下的浅景深就会变得有问题。对于普通镜头来说，唯一的可能就是减小孔径光阑的尺寸，也就是说，通过增加镜头的 f 值。然而，如 2.2.3 节讨论，这种方法有局限性。首先，如果孔径光阑做得太小，因为衍射，图像会出现模糊；其次，孔径光阑太小会导致到达传感器的光线太少。因此，需要一个高能量的光源来获得比较理想的曝光时间，尤其当需要对移动物体成像时。

在许多实际应用中，需要对物体空间中的某个平面清晰成像，而这个平面和像平面不平行（Steger, 2017）。其中一个应用就是立体重构，摄像机通常在一个汇聚装置中使用（见 2.5.1 节和 3.10.1 节）。如图 2.32 所示，这套装置在物体空间中会形成一个无

in object space for which both cameras produce a sharp image to be a rhomboid-shaped infinite prism. This problem is typically ignored at small magnifications because the common depth of field is large enough. For large magnifications (e.g., larger than 0.1), however, the volume is small enough to cause significant defocus.

限长的菱形棱柱体，在这个棱柱体范围内两个摄像机都能清晰成像。这个问题在低放大率下一般可以忽略，因为在这种情况下通常景深足够大。然而，对于较大的放大率（例如，大于0.1），这个范围即使很小也足够引起严重的失焦。

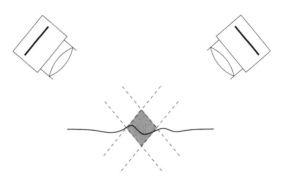

图 2.32　一个使用汇聚摄像机的常规立体视觉装置。像平面用粗实线表示，两个摄像机的景深用虚线表示，共同的景深用灰色菱形表示。为了能够清晰地表示共同的景深，两个摄像机的角度被夸大。待重构的表面用粗实线表示

Another application where a tilted object plane must be imaged in focus is sheet of light 3D reconstruction (see Sections 2.5.2 and 3.10.2. Here, a laser projects a line onto objects in the world, and a camera acquires images of the laser line. The projection of the laser line forms a plane in space that is not perpendicular to the optical axis of the camera. Different object distances cause different displacements of the laser line in the image, which allow a 3D reconstruction of the scene. To obtain maximum accuracy, it is necessary that the laser line is in focus for all 3D depths that must be reconstructed, i.e., the entire 3D laser plane emitted by the projector should ideally be in focus.

需要对倾斜物体平面清晰成像的另外一个应用就是基于激光三角测量法（片光）的三维重构（见 2.5.2 节和 3.10.2 节）。在这个应用中，激光器在实际空间中的物体上投影一条激光线，摄像机采集激光线图像。激光线的投影在空间中会形成一个平面，这个平面和摄像机光轴是不垂直的。不同的物体距离在图像中会产生不同的激光线位移，因此可以实现场景的三维重构。为了获得最高的精度，激光线必须在所有重建的三维深度内都能够清晰成像，也就是说，投影仪投射的完整三维激光平面应该能够理想聚焦。

One more application where it is important to image a plane in focus that is tilted with respect to the image plane is structured light 3D reconstruction (see Sections 2.5.3 and 3.10.3). Here, a 2D projector replaces one of the cameras of a stereo camera setup. Consequently, this application is geometrically equivalent to the stereo camera setup described above.

The Scheimpflug principle states that an arbitrary plane in object space can be imaged in focus by tilting the lens with respect to the image plane (Steger, 2017). If we assume a thin lens, i.e., a lens for which the two principal planes coincide in a single plane, the Scheimpflug principle states the following: the focusing plane (the plane in object space that is in focus), the principal plane, and the image plane must all meet in a single line, called the Scheimpflug line. For thick lenses, the condition must be modified as follows: the Scheimpflug line is split into two lines, one in each principal plane of the lens, that are conjugate to each other, i.e., have the same distance and orientation with respect to the principal points of the lens (see Figure 2.33). The angles of the focusing and image planes with respect to the principal planes can be derived from Eq. (2.6) and are given by (see Steger, 2017)

$$\tan \tau' = \frac{f}{a-f} \tan \tau \qquad (2.21)$$

where τ is the angle of the focusing plane with respect to the object-side principal plane, τ' is the angle of the image plane with respect to the image-side principal plane, and a is the distance of the intersection point of the optical axis with the focusing plane from the object-side principal point.

结构光三维重构（见 2.5.3 节和 3.10.3 节）也是一个需要对倾斜物体平面清晰成像的应用。这里，二维投影仪代替了立体摄像机中的一台摄像机。因此，这个应用在几何结构上来说，和上面描述的立体摄像机是相同的。

沙姆定律表明通过相对于像平面来倾斜镜头，在物体空间中的任意一个平面都可以清晰成像（Steger, 2017）。如果假定一个薄透镜，即透镜的两个主平面重合，则沙姆定律描述如下：焦平面（在物体空间中能够清晰成像的平面），主平面和像平面都必须汇聚到一条线上，叫做沙姆线。对于厚透镜来说，情况有些变化：沙姆线分成两条直线，分别在镜头的主平面内，彼此相交，即相对于镜头的主点有相同的距离和方向（见图 2.33）。焦平面和像平面相对于主平面的角度可以通过等式（2.6）得到，等式如下（Steger，2017）

其中 τ 是焦平面相对于物方主平面的角度，τ' 是像平面相对于像方主平面的角度，a 是光轴和焦平面的交点相对于物方主点的距离。

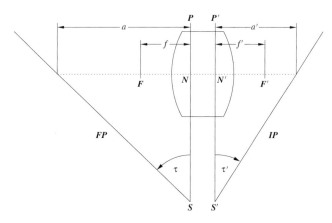

图 2.33 沙姆定律，S 和 S' 分别是物方和像方沙姆线

If a plane that is parallel to the image plane is drawn through the object-side principal point, and this plane is intersected with the object-side focal plane, a straight line is obtained (see Figure 2.34). This construction can also be performed analogously on the image side of the lens. These lines are called hinge lines (Evens, 2008; Merklinger, 2010; Steger, 2017). The object-side hinge line H has an important geometric significance: if the image is refocused by changing the distance of the image plane with respect to the image-side principal point, the focusing plane will rotate around the object-side hinge line if the tilt of the image plane remains fixed (Evens, 2008; Merklinger, 2010; Steger, 2017). Furthermore, the depth of field is an infinite wedge-shaped region that has the hinge line as its edge (Evens, 2008; Merklinger, 2010; Steger, 2017). This can also be seen from Figure 2.34. If we interpret the image planes IP_1 and IP_2 as the limits of the depth of focus, the limits of the depth of field are given by FP_1 and FP_2. Note that positioning the image plane parallel to the principal planes moves the hinge and Scheimpflug lines to infinity, and produces the regular depth-of-field geometry.

如果平行于像平面的某个平面通过物方主点，并且这个平面和物方焦平面相交，则可以得到一条直线（见图 2.34）。这种结构也可以在镜头的像方模拟完成。这些直线称为铰合线（Evens, 2008; Merklinger, 2010; Steger, 2017）。物方铰合线 H 有重要的几何意义：如果通过改变像平面相对于像方主点的距离来重新聚焦图像，则焦平面会沿着物方铰合线旋转，前提是像平面的倾斜程度保持不变（Evens, 2008; Merklinger, 2010; Steger, 2017）。而且，景深会是一个以铰合线为边界的无限大的楔形区域（Evens, 2008; Merklinger, 2010; Steger, 2017），从图 2.34 也可以看出，如果我们把像平面 IP_1 和 IP_2 作为景深的范围，则景深的范围表示为 FP_1 和 FP_2。注意，如果放置像平面平行于主平面则铰合线和沙姆线会移动到无穷远，此时就是常规的景深几何结构。

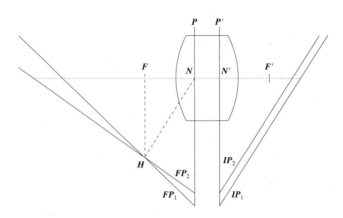

图 2.34 相对于主平面 P' 移动像平面从 IP_1 到 IP_2 重新聚焦，则焦平面会沿铰合线 H 旋转从 FP_1 到 FP_2

With the Scheimpflug principle, it is obvious how to solve the focusing problems in the applications discussed earlier (Steger, 2017). For example, for stereo cameras, the image planes must be tilted as shown in Figure 2.35. A similar principle holds for structured light systems. For sheet of light systems, the focusing plane in Figure 2.35 corresponds to the laser plane and there is only one camera.

利用沙姆定律，则非常清楚如何解决之前讨论过的应用中的聚焦问题（Steger，2017）。例如，对于立体摄像机，则像平面必须倾斜，如图 2.35 所示。对于结构光系统来说，遵循相同的原则。对于激光三角测量系统（片光），图 2.35 所示的焦平面相当于激光平面并且只有一个摄像机。

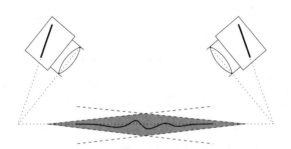

图 2.35 使用汇聚摄像机的立体视觉装置，按照沙姆定律像平面倾斜。像平面用粗实线表示，两个摄像机的景深用从铰合线延伸出的虚线表示。共同的景深区域用灰色菱形表示，为了能够清晰地表示共同的景深，两个摄像机的角度被夸大。待重构的表面用粗实线表示

To construct a camera with a tilted lens, there are several possibilities (Steger, 2017). One popular option is to construct a special camera housing with the desired tilt angle to which the lens can be attached. This is typically done for specialized

构造带有倾角镜头的摄像机，有多种可能性（Steger，2017）。一种比较流行的方式是，构造一个带有期望倾角的特殊接口摄像机，镜头可以安装到接口上。通常用于特定的应用，

applications for which the effort of constructing the housing is justified, e.g., specialized sheet of light or structured light sensors. In these applications, the lens is typically tilted around the vertical or horizontal axis of the image, as required by the application. Another option is to use lenses that have been designed specifically to be tilted in an arbitrary direction. In the machine vision industry, these lenses are typically called Scheimpflug lenses or Scheimpflug optics. They are available as perspective or telecentric lenses. In the consumer SLR camera market, these lenses are typically called tilt/shift lenses or perspective correction lenses (although, technically, perspective correction only requires that the lens can be shifted). Since the ability to tilt the lens is its essential feature, we will call these lenses tilt lenses in this book.

2.2.6 Lens Aberrations

Up to now, we have assumed Gaussian optics, where a homocentric pencil will converge to a single point after passing through a lens. For real lenses, this is generally not the case. In this section, we will discuss the primary aberrations that can occur (Born and Wolf, 1999; Mahajan, 1998).

2.2.6.1 Spherical Aberration

Spherical aberration, shown in Figure 2.36(a), occurs because light rays that lie far from the optical axis will not intersect at the same point as light rays that lie close to the optical axis. This happens because of the increased refraction that occurs at the edge of a spherical lens. No matter where we place the image plane, there will always

对于这些应用花精力构造接口是合理的，例如，定制的激光三角测量或者结构光传感器等。在这些应用中，根据实际需要，镜头通常沿着图像的横轴或者纵轴倾斜。另外一种方式是使用经过特殊设计的可以在任意方向倾斜的镜头。在机器视觉工业应用中，这些镜头通常叫做沙姆镜头或者沙姆光学。透视镜头和远心镜头两种形式都有。对于消费级 SLR 摄像机市场，这些镜头通常叫做移轴镜头或者透视校正镜头（尽管技术上来讲，透视校正仅需要镜头可以移动）。由于倾斜镜头的能力是它最基本的特征，所以在本书中称为倾斜镜头。

2.2.6 镜头的像差

到目前为止，一直假设在高斯光学中同心光束通过镜头后汇聚于一点。在实际中通常不是这种情况。本节将讨论主要的像差（Born and Wolf，1999；Mahajan，1998）。

2.2.6.1 球差

图 2.36（a）表示了球差。球差的产生是由于远离光轴的光线与近轴光线并不交于同一点，这是由于球面镜头边缘的折射增大造成的。无论像平面在哪，都会有一个弥散圆，只能使这个圆尽量小。减小球差的一个办法是使用较大的 f 值，也就是使用较小

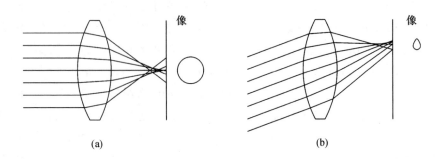

图 2.36 （a）球差。近轴光线与远轴光线通过镜头后不交于同一点；（b）彗差。与光轴成一定角度的光线通过镜头后不交于一点

be a circle of confusion. The best we can do is to make the circle of confusion as small as possible. The spherical aberration can be reduced by setting the aperture stop to a larger f-number, i.e., by making the diaphragm smaller because this prevents the light rays far from the optical axis from passing through the lens. However, as described at the end of Section 2.2.3, diffraction will limit how small we can make the aperture stop. An alternative is to employ lenses that use non-spherical surfaces instead of spherical surfaces. Such lenses are called aspherical lenses.

2.2.6.2 Coma

A similar aberration is shown in Figure 2.36(b). Here, rays that pass through the lens at an angle to the optical axis do not intersect at the same point. In this case, the image is not a circle, but a comet-shaped figure. Therefore, this aberration is called coma. Note that, in contrast to spherical aberration, coma is asymmetric. This may cause feature extraction algorithms like edge extraction to return wrong positions. Coma can be reduced by setting the aperture stop to a larger f-number.

的光圈可以阻止远离光轴的光线通过镜头。然而如在 2.2.3 节尾所述，由于衍射的存在，无法使孔径光阑无限小。另外一种办法就是在镜头中使用非球面来代替球面，这样的镜头称为非球面镜头。

2.2.6.2 彗差

图 2.36（b）表示了另外一种类似的像差。与光轴成一定角度的光线通过镜头后不聚于一点。在这种情况下，像不是圆形的而是类似彗星的形状。因此这种像差被称作彗差。与球差不一样，彗差是非对称的。这种像差会使如边缘提取这类的特征提取算法得到错误的位置。可通过使用大的 f 值来减小彗差。

2.2.6.3 Astigmatism

Figure 2.37 shows a different kind of aberration. Here, light rays in the tangential plane, defined by the object point and the optical axis, and the sagittal plane, perpendicular to the tangential plane, do not intersect at points, but in lines perpendicular to the respective plane. Because of this property, this aberration is called astigmatism. The two lines are called the tangential image I_T and a sagittal image I_S. Between them, a circle of least confusion occurs. Astigmatism can be reduced by setting the aperture stop to a larger f-number or by careful design of the lens surfaces.

2.2.6.3 像散

图 2.37 表示另外一种像差。轴外物点和光轴所定义的平面称为子午平面，而过光轴且与子午平面相互垂直的平面称为弧矢平面。轴外物点发出的在子午平面和弧矢平面上的光线通过镜头后不交于一点，而是聚焦成两条短线，分别垂直于子午平面和弧矢平面。基于这一特性，这种像差称为像散，这两条短线分别称为子午焦线 I_T 和弧矢焦线 I_S。在这两者之间的某一位置则会得到最小的弥散圆斑。可以通过使用大 f 值的孔径光阑或者通过仔细的镜头设计来减小像散。

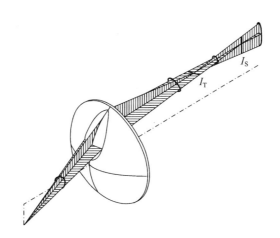

图 2.37　像差。子午方向和弧矢方向的光线不聚焦于同一点，产生子午焦线 I_T 和弧矢焦线 I_S 两条正交的线

2.2.6.4 Curvature of Field

A closely related aberration is shown in Figure 2.38(a). Above, we have seen that the tangential and sagittal images do not necessarily lie at the same image distance. In fact, the surfaces in which the tangential and sagittal images lie (the

2.2.6.4 场曲

图 2.38（a）表示了另外一种与像散关系非常密切的像差。子午像与弧矢像不一定在同一像平面上。实际上，子午像和弧矢像所在的面（称为子午焦面和弧矢焦面）甚至不是平面，这

tangential and sagittal focal surfaces) may not even be planar. This aberration is called curvature of field. It leads to the fact that it is impossible to bring the entire image into focus. Figure 2.38(b) shows an image taken with a lens that exhibits significant curvature of field: the center of the image is in focus, while the borders of the image are severely defocused. Like the other aberrations, curvature of field can be reduced by setting the aperture stop to a larger f-number or by carefully designing the lens surfaces.

种像差称为场曲。场曲导致了不可能使整个图像完全聚焦。图 2.38（b）是由一个具有较大场曲像差的镜头所采集的图像，图像中心聚焦清晰，而边缘则散焦很严重。与其他像差一样，也可以通过使用大 f 值的孔径光阑或者通过仔细的镜头设计来减小场曲。

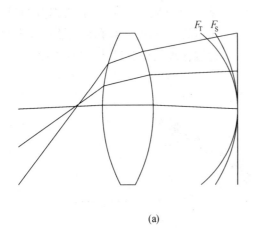 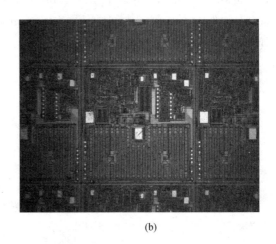

(a) (b)

图 2.38　(a) 场曲，子午图像与弧矢图像位于两个弯曲的焦面 F_T 和 F_S 上；(b) 由一个具有较大场曲像差的镜头所采集的图像，图像中心聚焦清晰，而边缘散焦很严重

2.2.6.5　Distortion

In addition to the above aberrations, which, apart from coma, mainly cause problems with focusing the image, lens aberrations may cause the image to be distorted, leading to the fact that straight lines not passing through the optical axis will no longer be imaged as straight lines. This is shown in Figure 2.39. Because of the characteristic shapes of the images of rectangles, two kinds

2.2.6.5　畸变

以上像差除彗差外都会影响图像聚焦。除此之外，镜头像差还会造成图像变形，也就是会使不经过光轴的直线通过镜头后成的像不再是一条直线，见图 2.39。根据一个矩形成像后的特征可以定义两种畸变：枕形畸变和桶形畸变。注意通过光轴的直线不产生畸变。而且，以光轴为圆心的

of distortion can be identified: pincushion and barrel distortion. Note that distortion will not affect lines through the optical axis. Furthermore, circles with the optical axis as center will produce larger or smaller circles. Therefore, these distortions are called radial distortions. In addition, if the elements of the lens are not centered properly, decentering distortions may also occur. Note that there is an ISO standard ISO 9039:2008 that defines how distortions are to be measured and reported by lens manufacturers.

圆所成的像是或大或小的圆。因此，这样的畸变称为径向畸变。此外，如果镜头的各个光学元件的中心线不在一条直线上，则会产生偏心畸变。注意，ISO 标准（ISO 9039:2008）定义了镜头厂商如何测量和报告畸变。

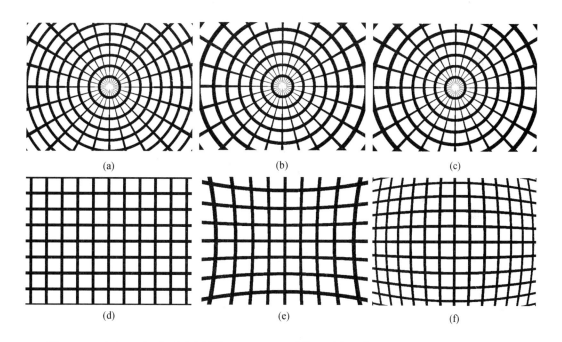

图 2.39　（a）、（d）为无畸变图像；（b）、（e）为枕形畸变图像；（c）、（f）为桶形畸变图像

2.2.6.6　Chromatic Aberration

All of the above aberrations already occur for monochromatic light. If the object is illuminated with light that contains multiple wavelengths, e.g., white light, chromatic aberrations may occur, which lead to the fact that light rays having different wavelengths do not intersect at a single

2.2.6.6　色差

上述所有像差都是单色光像差。如果被测物被白光等多波长光照明，还会产生色差。如图 2.40 所示，不同波长的光线不聚在同一点。对于彩色摄像机，色差会在物体边缘产生彩条，而对于黑白摄像机，色差会造成

point, as shown in Figure 2.40. With color cameras, chromatic aberrations cause colored fringes at the edges of objects. With black-and-white cameras, chromatic aberrations lead to blurring. Furthermore, because they are not symmetric, they can cause position errors in feature extraction algorithms like edge extraction. Chromatic aberrations can be reduced by setting the aperture stop to a larger f-number. Furthermore, it is possible to design lenses in such a way that two different wavelengths will be focused at the same point. Such lenses are called achromatic lenses. It is even possible to design lenses in which three specific wavelengths will be focused at the same point. Such lenses are called apochromatic lenses. Nevertheless, the remaining wavelengths still do not intersect at the same point, leading to residual chromatic aberrations. Note that there is an ISO standard ISO 15795:2002 that defines how chromatic aberrations are to be measured and reported by lens manufacturers.

模糊。而且，由于色差是非对称的，它会使边缘提取等特征提取算法产生位置错误。色差可以通过采用大 f 值的孔径光阑来减小。而且，可以通过镜头设计使两种不同波长的光线聚于一点。这种镜头称作消色差镜头。甚至可以通过镜头设计使三种指定波长的光线聚于同一点，这种镜头称作复消色差镜头。然而，其他波长的光线仍然不能聚于同一点，导致了剩余色差。注意，ISO 标准（ISO 15795:2002）定义了镜头厂商如何测量和报告畸变。

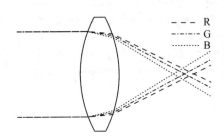

图 2.40 色差；不同波长的光不聚于同一点

2.2.6.7 Edge-Spread Function

It is generally impossible to design a lens system that is free from all of the above aberrations, as well as higher-order aberrations that we have not discussed (Born and Wolf, 1999). Therefore, the lens designer must always make a trade-off

2.2.6.7 边缘扩散函数

通常情况下不可能设计出完美的镜头系统消除上述所有像差，以及没有谈到的高阶像差（Born and Wolf, 1999）。因此，镜头设计者需要在各种像差之间权衡。尽管剩余像差可能已

between the different aberrations. While the residual aberrations may be small, they will often be noticeable in highly accurate feature extraction algorithms. This happens because some of the aberrations create asymmetric gray value profiles, which will influence all subpixel algorithms. Aberrations cause ideal step edges to become blurred. This blur is called the edge-spread function. Figure 2.41 shows an example of an edge-spread function of an ideal step edge that would be produced by a perfect diffraction-limited system as well as edge-spread functions that were measured with a real lens for an edge through the optical axis and an off-axis edge lying on a circle located at a distance of 5 mm from the optical axis (i.e., the edge-spread function for the off-axis edge is perpendicular to the circle). Note that the ideal edge-spread function and the on-axis edge-spread function are symmetric, and consequently cause no problems. The off-axis edge-spread function, however, is asymmetric, which causes subpixel-accurate edge detection algorithms to return incorrect positions, as discussed in Section 3.7.4.

经很小了，但是在高精度特征提取算法中仍然很明显，这是由于像差会造成不对称灰度轮廓，影响所有的亚像素算法。图 2.41 分别为完美衍射极限镜头所得到的理想阶跃边缘剖面图、使用实际镜头对轴上阶跃边缘和对距离光轴 5mm 的圆上的边缘扩散函数（也就是轴外阶跃的边缘扩散函数与圆正交）所得到的边缘扩散函数。注意衍射极限镜头产生的理想边缘扩散函数和轴上边缘扩散函数是轴对称的，因此不会产生问题。而轴外边缘扩散函数是非轴对称的，将会使亚像素精度的边缘检测算法得出错误的位置，见 3.7.4 节。

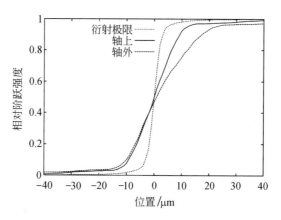

图 2.41 理想衍射极限镜头所生成的理想边缘扩散函数以及使用实际镜头对轴上阶跃边缘和对距离光轴 5mm 的圆上的轴外阶跃边缘（也就是轴外阶跃的边缘剖面与圆正交）所产生的边缘剖面图。注意轴外边缘扩散函数是非轴对称的（Lenhardt，2017）

2.2.6.8 Vignetting

We conclude this section by discussing an effect that is not considered an aberration per se, but nevertheless influences the image quality. As can be seen from Figure 2.38(b), for some lenses there can be a significant drop in image brightness toward the border of the image. This effect is called vignetting. It can have several causes. First of all, the lens may have been designed in such a way that for a small f-number the diaphragm is no longer the aperture stop for light rays that form a large angle with the optical axis. Instead, as shown in Figure 2.42, one of the lenses or the lens barrel becomes the aperture stop for a part of the light rays. Consequently, the effective entrance pupil for the off-axis rays will be smaller than for rays that lie on the optical axis. Thus, the amount of light that reaches the sensor is smaller for off-axis points. This kind of vignetting can be removed by increasing the f-number so that the diaphragm becomes the aperture stop for all light rays.

2.2.6.8 渐晕

本节最后来讨论一种效果，这种效果本身不是像差，但仍然影响图像质量。从图 2.38（b）可以看出对于有些镜头，其采集的图像边缘亮度有很大下降。这种效果称作渐晕。造成渐晕的原因有多种。首先，镜头在小 f 值时，对于与光轴成大角度的光线而言，可变光阑不再是系统的孔径光阑，孔径光阑变成了某一个镜片或者镜筒，如图 2.42 所示。对于轴外点发出的光线，其有效入瞳要比轴上点的有效孔径小，因此，从轴外点到达传感器的光通量也就小了。对于这种光晕，可以增加 f 值来消除，这样对于所有的光线可变光阑都是孔径光阑。

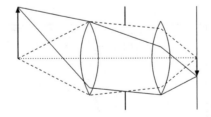

图 2.42 对于部分轴外点光线，可变光阑已不再是孔径光阑从而产生了渐晕

The second cause of vignetting is the natural light fall-off for off-axis rays. By looking at Figure 2.22, we can see that light rays that leave the exit pupil form an image-side field angle ω' with the optical axis. This means that for off-axis points with $\omega' \neq 0$, the exit pupil will no

第二个造成光晕的原因是轴外点光线强度降低。从图 2.22 可看出，轴外点光线从出瞳出来与光轴成一个像方视场角 ω'，这就意味着当 $\omega' \neq 0$ 时轴外点的出瞳不是圆而是比圆小一个系数为 $\cos\omega'$ 的椭圆，因此，轴外

longer be a circle, but an ellipse that is smaller than the circle by a factor of $\cos \omega'$. Consequently, off-axis image points receive less light than axial image points. Furthermore, the length of the light ray to the image plane will be longer than for axial rays by a factor of $1/\cos \omega'$. By the inverse square law of illumination (Born and Wolf, 1999), the energy received at the image plane is inversely proportional to the square of the length of the ray from the light source to the image point. Hence, the light received by off-axis points will be reduced by another factor of $(\cos \omega')^2$. Finally, the light rays fall on the image plane at an angle of ω'. This further reduces the amount of light by a factor of $\cos \omega'$. In total, we can see that the amount of light for off-axis points is reduced by a factor of $(\cos \omega')^4$. A more thorough analysis shows that, for large aperture stops (f-numbers larger than about 5), the light fall-off is slightly less than $(\cos \omega')^4$ (Mahajan, 1998). This kind of vignetting cannot be reduced by increasing the f-number.

2.3 Cameras

The camera's purpose is to create an image from the light focused in the image plane by the lens. The most important component of the camera is a digital sensor. In this section, we will discuss the two main sensor technologies: CCD and CMOS. They differ primarily in their readout architecture, i.e., in the manner in which the image is read out from the chip (El Gamal and Eltoukhy, 2005; Holst and Lomheim, 2011). Furthermore, we will discuss how to characterize the performance of a camera. Finally, since we do not discuss this in this chapter, it should be mentioned that the camera contains the necessary electronics to create a video signal that can be transmitted to a computer as well as electronics that permit the image acquisition to be controlled externally by a trigger.

还会提到摄像机中包含的能够产生向计算机传输的视频信号的电路，也包含了在外部触发信号控制下的图像采集电路，但这部分内容不作详细讨论。

2.3.1 CCD Sensors

2.3.1.1 Line Sensors

To describe the architecture of CCD sensors, we start by examining the simplest case: a linear sensor, as shown in Figure 2.43. The CCD sensor consists of a line of light-sensitive photodetectors. Typically they are photogates or photodiodes (Holst and Lomheim, 2011). We will not describe the physics involved in photodetection. For our purposes, it suffices to think of a photodetector as a device that converts photons into electrons and corresponding positively charged holes and collects them to form a charge. Each type of photodetector has a maximum charge that can be stored, depending, among other factors, on its size. The charge is accumulated during the exposure of the photodetector to light. To read out the charges, they are moved through transfer gates to serial readout registers (one for each photodetector). The serial readout registers are also light-sensitive, and thus must be covered with a metal shield to prevent the image from receiving additional exposure during the time it takes to perform the readout. The readout is performed by shifting the charges to the charge conversion unit, which converts the charge into a voltage, and by amplifying the resulting voltage (Holst and Lomheim, 2011). The transfer gates and serial readout

2.3.1 CCD 传感器

2.3.1.1 线阵传感器

从图 2.43 所示的线阵摄像机这种最简单的情况来描述 CCD 传感器的结构。CCD 传感器由一行光线敏感的光电探测器组成，光电探测器一般为光栅晶体管或光电二极管（Holst and Lomheim, 2011）。我们不涉及光电探测器所含的物理问题，我们仅把光电探测器看作能将光子转为电子并将电子转为电流的设备。每种光电探测器都有最多可以存储电子数量的限制，常取决于光电探测器的大小。曝光时光电探测器累积电荷，通过转移门电路，电荷被移至串行读出寄存器从而读出。每个光电探测器对应一个读出寄存器。串行读出寄存器也是光敏的，必须由金属护罩遮挡以避免读出期间其接收到其他光子。读出的过程是将电荷转移到电荷转换单元，转换单元将电荷转换为电压，并将电压放大（Holst and Lomheim, 2011）。转移门电路及串行读出电路是电子耦合设备。每一个 CCD 由许多门组成，这些门在一定方向上传输电荷。电荷传输过程的细节可参看文献（Holst and Lomheim, 2011）。电荷转换为电压并放大后，就可以转换为模拟或数字视

registers are CCDs. Each CCD consists of several gates that are used to transport the charge in a given direction. Details of the charge transportation process can be found, for example, in Holst and Lomheim (2011). After the charge has been converted into a voltage and has been amplified, it can be converted into an analog or digital video signal. In the latter case, the voltage would be converted into a digital number (DN) through an analog-to-digital converter (ADC).

频信号。对于数字视频信号，是由模拟电压通过模数转换器（ADC）转换为数字（DN）电压的。

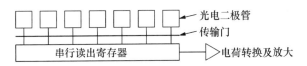

图 2.43 线阵 CCD 传感器。光在光电探测器中转换为电荷，转移至串行读出寄存器并通过电荷转换器和放大器读出。对于线性传感器光电探测器常为光电二极管

A line sensor as such would create an image that is one pixel high, which would not be very useful in practice. Therefore, typically many lines are assembled into a 2D image. Of course, for this image to contain any useful content, the sensor must move with respect to the object that is to be imaged. In one scenario, the line sensor is mounted above the moving object, e.g., a conveyor belt. In a second scenario, the object remains stationary while the camera is moved across the object, e.g., to image a PCB. This is the principle that is also used in flatbed scanners. In this respect, a flatbed scanner is a line sensor with integrated illumination.

线阵传感器只能生成高度为一个像素高度的图像，在实际中用途有限。因此常通过多行组成二维图像。为得到有效图像，线阵传感器必须相对被测物体运动。一种方法是将传感器安置在运动的被测物（如传送带）上方；第二种办法是被测物不动而传感器相对被测物运动，如 PCB 成像。平板扫描仪的原理也是同样的。平板扫描仪是由一个传感器和一个集成光源组成的。

To acquire images with a line sensor, the line sensor itself must be parallel to the plane in which the objects lie and must be perpendicular to the movement direction to ensure that rectangular pixels are obtained. Furthermore, the frequency with which the lines are acquired must match the

使用线阵传感器采集图像时传感器本身必须与被测物平面平行并与运动方向垂直以保障得到矩形像素。同时根据线阵传感器的分辨率，线采集频率必须与摄像机、被测物间相对运动速度匹配以得到方形像素。假定运

speed of relative movement of the camera and the object and the resolution of the sensor to ensure that square pixels result. It must be assumed that the speed is constant to ensure that all the pixels have the same size. If this is not the case, an encoder must be used, which essentially is a device that triggers the acquisition of each image line. It is driven, for example, by the stepper of the motor that causes the relative movement. Since perfect alignment of the sensor to the object and movement direction is relatively difficult to achieve, in some applications the camera must be calibrated using the methods described in Section 3.9 to ensure high measurement accuracy.

The line readout rates of line sensors vary between 10 and 200 kHz. This obviously limits the exposure time of each line. Consequently, line scan applications require very bright illumination. Furthermore, the diaphragm of the lens must typically be set to a relatively small f-number, which can severely limit the depth of field. Therefore, line scan applications often pose significant challenges for obtaining a suitable setup.

2.3.1.2 Full Frame Array Sensors

We now turn our attention to area sensors. The logical extension of the principle of the line sensor is the full frame sensor, shown in Figure 2.44. Here, the light is converted into charge in the photodetectors and is shifted row by row into the serial readout registers, from where it is converted into a video signal in the same manner as in the line sensor.

During the readout, the photodetectors are still exposed to light, and thus continue to accumulate charge. Because the upper pixels are

动速度是恒定的，这样就可以保证所有像素采集到的图像具有一致性。如果运动速度是变化的，就需要编码器来触发传感器采集每行图像。相对运动可以由步进电机驱动来产生。由于很难做到使传感器非常好地与运动方向匹配，在有些应用中必须利用 3.9 节中的摄像机标定方法来确保测量精度。

线阵传感器的线读出速度在 10~200 kHz 之间，显然会限制每行的曝光时间，因此线扫描应用要求非常强的照明。同时镜头的光圈通常要在较小 f 值，从而严重地限制了景深。所以线扫描应用系统中参数的设定是很有挑战性的。

2.3.1.2 全帧转移面阵传感器

现在来看看面阵传感器。图 2.44 表示了线阵传感器扩展为全帧转移面阵传感器的基本原理。这里光在光电探测器中转换为电荷，电荷按行的顺序转移到串行读出电路寄存器，然后与线阵传感器的方式一样转换为视频信号。

在读出过程中，光电传感器还在曝光，仍有电荷在积累。由于上面的像素要经过下面的像素移位移出，因

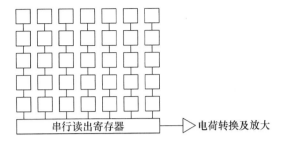

图 2.44　全帧转移型 CCD 传感器。光在光电探测器中转换为电荷，一行一行地转移到串行读出寄存器，然后进行电荷转换和放大读出

shifted through all the lower pixels, they accumulate information from the entire scene, and consequently appear smeared. To avoid the smear, a mechanical shutter or a strobe light must be used. This is also the biggest disadvantage of the full frame sensor. Its biggest advantage is that the fill factor (the ratio of the light-sensitive area of a pixel to its total area) can reach 100%, maximizing the sensitivity of the pixels to light and minimizing aliasing.

此像素积累的全部场景信息就会发生拖影现象。为了避免拖影，必须加上机械快门或利用闪光灯，这是全帧转移面阵传感器的最大缺点。其最大的优点是填充因子（填充因子是像素光敏感区域与整个靶面之比）可达 100%。100% 填充因子使得像素的光灵敏度最大化并使图像失真最小化。

2.3.1.3　Frame Transfer Sensors

To reduce the smearing problem in the full frame sensor, the frame transfer sensor, shown in Figure 2.45, uses an additional sensor that is covered by a metal light shield, and can thus be used as a storage area. In this sensor type, the image is created in the light-sensitive sensor and then transferred into the shielded storage array, from which it can be read out at leisure. Since the transfer between the two sensors is quick (usually less than 500 µs (Holst and Lomheim, 2011)), smear is significantly reduced.

The biggest advantage of the frame transfer sensor is that it can have fill factors of 100%. Furthermore, no mechanical shutter or strobe light needs to be used. Nevertheless, a residual smear

2.3.1.3　帧转移传感器

为了解决全帧转移传感器的拖影问题，全帧转移传感器加上另外的传感器用于存储，在这个传感器上覆盖有金属光屏蔽层，构成帧转移传感器，见图 2.45。对于这种类型的传感器，图像产生于光敏感传感器，然后转移至有光屏蔽的存储阵列，在空闲时从存储阵列中读出。由于两个传感器间转移速度很快，通常小于 500 µs（Holst and Lomheim, 2011），因此拖影可以大大地减少。

帧转移型传感器的最大优点是其填充因子可达 100%，而且不需要机械快门或闪光灯。然而在两个传感器间传输数据的短暂时间内图像还是在

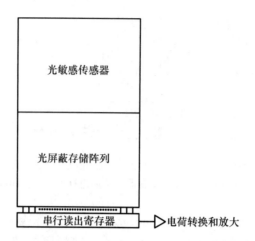

图 2.45 帧转移型 CCD 传感器。光在光敏感传感器中转为电荷,快速转移至屏蔽存储阵列,从存储阵列中一行行读出

may remain because the image is still exposed during the short time required to transfer the charges to the second sensor. The biggest disadvantage of the frame transfer sensor is its high cost, since it essentially consists of two sensors.

Because of the above characteristics (high sensitivity and smearing), full frame and frame transfer sensors are most often used in scientific applications for which the exposure time is long compared to the readout time, e.g., in astronomical applications.

曝光,因而还是有残留的拖影存在。帧转移型传感器的缺点是其通常由两个传感器组成,因此成本高。

由于高灵敏度和拖影等特征,全帧转移型传感器和帧转移型传感器通常用于像天文等曝光时间比读出时间长的科学研究等应用领域。

2.3.1.4 Interline Transfer Sensors

The final type of CCD sensor, shown in Figure 2.46, is the interline transfer sensor. In addition to the photodetectors, which in most cases are photodiodes, there are vertical transfer registers that are covered by an opaque metal shield. After the image has been exposed, the accumulated charges are shifted through transfer gates (not shown in Figure 2.46) to the vertical transfer registers. This can typically be done in less than 1 µs (Holst and Lomheim, 2011). To create the video signal, the

2.3.1.4 行间转移传感器

最后一种 CCD 传感器是图 2.46 中的行间转移传感器。除光电探测器外(通常情况下为光电二极管),这种传感器还有一个带有不透明的金属屏蔽层的垂直移位寄存器。图像曝光后,累积到的电荷通过传输门电路(图 2.46 中没有显示)转移到垂直移位寄存器。这一过程通常在小于 1µs(Holst and Lomheim, 2011)内完成。电荷通过垂直移位寄存器移至串行读

charges are then shifted through the vertical transfer registers into the serial readout registers, and read out from there.

出寄存器然后读出形成视频信号。

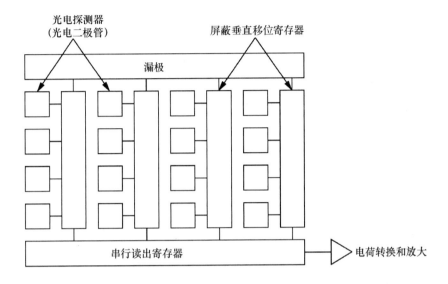

图 2.46　行间转移型 CCD 传感器。光在光敏感传感器中转换为电荷，快速地传输至屏蔽垂直移位寄存器，按行转移至串行读出寄存器并读出

Because of the quick transfer from the photodiodes into the shielded vertical transfer registers, there is no smear in the image, and consequently no mechanical shutter or strobe light needs to be used. The biggest disadvantage of the interline transfer sensor is that the transfer registers take up space on the sensor, causing fill factors that may be as low as 20%. Consequently, aliasing effects may increase. To increase the fill factor, microlenses are typically located in front of the sensor to focus the light onto the light-sensitive photodiodes, as shown in Figure 2.47. Nevertheless, fill factors of 100% typically cannot be achieved.

One problem with CCD sensors is an effect called blooming: when the charge capacity of a photodetector is exhausted, the charge spills over into the adjacent photodetectors. Thus, bright

由于从光电二极管传输至屏蔽垂直移位寄存器的速度很快，因此图像没有拖影，所以不需要机械快门和闪光灯。行间转移型传感器的最大缺点是由于其移位寄存器需要在传感器上占用空间，所以其填充因子可能低至 20%。图像失真会增加。为了增大填充因子，常利用在传感器上加上微镜头来使光聚焦至光敏光电二极管，见图 2.47。然而即使这样也不可能使其填充因子达到 100%。

CCD 传感器的一个问题是其高光溢出效应。也就是当积累的电荷超过光电探测器的容量时，电荷将会溢出到相邻的光电探测器中，因此图像

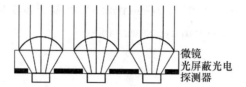

图 2.47 行间转移型传感器常利用微镜来增大填充因子。微镜使光线在光敏感光电二极管上聚焦

areas in the image are significantly enlarged. To prevent this problem, anti-bloom drains can be built into the sensor (Holst and Lomheim, 2011). The drains form an electrostatic potential barrier that causes extraneous charge from the photodetectors to flow into the drain. The drains can be located either next to the pixels on the surface of the sensor (lateral overflow drains), e.g., on the opposite side of the vertical transport registers, or can be buried in the substrate of the device (vertical overflow drains). Figure 2.46 displays a vertical overflow drain, which must be imagined to lie underneath the transfer registers.

An interesting side-effect of building anti-bloom drains into the sensor is that they can be used to create an electronic shutter for the camera. By setting the drain potential to zero, the photodetectors discharge. Afterwards, the potential can be set to a high value during the exposure time to accumulate the charge until it is read out. The anti-bloom drains also facilitate the construction of sensors that can immediately acquire images after receiving a trigger signal. Here, the entire sensor is reset immediately after receiving the trigger signal. Then, the image is exposed and read out as usual. This operation mode is called asynchronous reset.

中亮的区域就显著放大了。为了防止这个问题，可在传感器上增加溢流沟道（Holst and Lomheim, 2011）。加在沟道的电势差使得光电探测器中多余的电荷通过沟道流向衬底。溢流沟道可位于传感器平面中每个像素的侧边（侧溢流沟道），也可埋于设备的底部（垂直溢流沟道）。侧溢流沟道常位于垂直移位寄存器的相反一侧。图2.46 是垂直溢流沟道，其一定在移位寄存器下面。

在传感器上增加溢流沟道带来的一个有趣的效果是其可以用来作为摄像机的电子快门。将沟道的电位置为0，光电探测器不再充电，然后可以将沟道的电位在曝光时间内置为高，即可以积累电荷直至读出。溢流沟道还使传感器可以在接收到外触发信号后立刻开始采集图像，也就是接收到外触发信号后整个传感器可以立刻复位，图像开始曝光然后正常读出。这种操作模式称作异步复位。

2.3.1.5 Readout Modes

We conclude this section by describing a readout mode that is often implemented in analog CCD

2.3.1.5 读出模式

本节的最后讲一下模拟 CCD 摄像机常用的读出模式。2.4.1 节中描述

cameras because the image is transmitted with one of the analog video standards described in Section 2.4.1. The analog video standards require an image to be transmitted as two fields, one containing the odd lines of the image and one containing the even lines. This readout mode is called interlaced scan. Because of the readout architecture of CCD sensors, this means that the image will have to be exposed twice. After the first exposure, the odd rows are shifted into the transfer registers, while after the second exposure, the even rows are shifted and read out. If the object moves between the two exposures, its image will appear serrated, as shown in Figure 2.48(a).

The mode of reading out the rows of the CCD sensor sequentially is called progressive scan. From Figure 2.48(b), it is clear that this mode is essential for capturing correct images of moving objects.

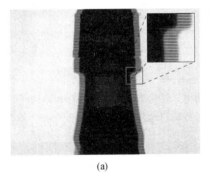
(a)

(b)

图 2.48　(a) 隔行扫描摄像机与 (b) 逐行扫描摄像机采集运动物体图像的比较。可以看出逐行扫描是采集运动物体正确图像所必需的

2.3.2 CMOS Sensors

2.3.2.1 Sensor Architecture

CMOS sensors, shown in Figure 2.49, typically use photodiodes for photodetection (El Gamal and Eltoukhy, 2005; Yadid-Pecht and Etienne-Cummings, 2004; Holst and Lomheimm, 2011).

In contrast to CCD sensors, the charge of the photodiodes is not transported sequentially to a readout register. Instead, each row of the CMOS sensor can be selected directly for readout through the row and column select circuits. In this respect, a CMOS sensor acts like a random access memory. Furthermore, as shown in Figure 2.49, each pixel has its own amplifier. Hence, this type of sensor is also called an active pixel sensor (APS). CMOS sensors typically produce a digital video signal. Therefore, the pixels of each image row are converted in parallel to DNs through a set of ADCs.

2004；Holst and Lomheimm，2011）。与 CCD 传感器不同，光电二极管中的电荷不是顺序地转移到读出寄存器，CMOS 传感器的每一行都可以通过行和列选择电路直接选择并读出。这方面，CMOS 传感器可以当作随机存取存储器。如图 2.49 所示 CMOS 每个像素都有一个自己的独立放大器。这种类型传感器也称作主动像素传感器（APS）。CMOS 传感器常用数字视频作输出。因此，图像每行中的像素通过模数转换器阵列并行地转化为数字信号。

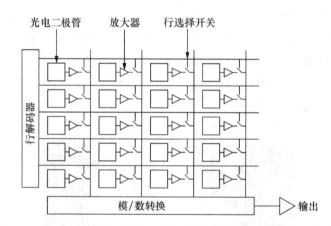

图 2.49　CMOS 传感器。光电二极管将光转换为电荷，通过行和列选择电路 CMOS 传感器的每行可以选择直接读出

Since the amplifier and row and column select circuits typically use a significant amount of the area of each pixel, CMOS sensors, like interline transfer CCD sensors, have low fill factors and therefore normally use microlenses to increase the fill factor and to reduce aliasing (see Figure 2.47).

The random access behavior of CMOS sensors facilitates an easy readout of rectangular areas of interest (AOIs) from the image. This gives them a significant advantage over CCD sensors in some applications since it enables much higher

因为放大器及行、列选择电路常会用到每个像素大部分面积，因此与行间转移型 CCD 传感器一样，CMOS 传感器的填充因子很低。因此常使用微镜来增加填充因子和减少图像失真，见图 2.47。

CMOS 传感器的随机读取特性使其很容易实现图像的矩形感兴趣区域（AOI）读出方式。与 CCD 传感器相比，对于有些应用这点有很大优势，在较小的 AOI 时可以得到更高的帧

frame rates for small AOIs. While AOIs can also be implemented for CCDs, their readout architecture requires that all rows above and below the AOI are transferred and then discarded. Since discarding rows is faster than reading them out, this results in a speed increase. However, typically no speed increase can be obtained by making the AOI smaller horizontally since the charges must be transferred through the charge conversion unit. Another big advantage is the parallel analog-to-digital conversion that is possible in CMOS sensors. This can give CMOS sensors a speed advantage even if AOIs are not used. It is even possible to integrate the ADCs into each pixel to further increase the readout speed (El Gamal and Eltoukhy, 2005). Such sensors are also called digital pixel sensors (DPSs).

Although Figure 2.49 shows an area sensor, the same principle can also be applied to construct a line sensor. Here, the main advantage of using a CMOS sensor is also the readout speed.

2.3.2.2 Rolling and Global Shutters

Since each row in a CMOS sensor can be read out individually, the simplest strategy to acquire an image is to expose each line individually and to read it out. Of course, exposure and readout can be overlapped for consecutive image lines. This is called a rolling shutter. Obviously, this readout strategy creates a significant time lag between the acquisition of the first and last image lines. As shown in Figure 2.50(a), this causes sizeable distortion when moving objects are acquired. For these kinds of applications, sensors with a global shutter must be used (Holst and Lomheim, 2011; Wäny and Israel, 2003). This requires a separate storage area for each pixel, and thus further lowers the fill factor. As shown in Figure 2.50(b), the global shutter results in a correct image for moving objects.

图 2.50（b）对于运动物体全局曝光可以得到正确的图像。

(a) (b)

图 2.50 对于运动物体使用（a）行曝光和（b）全局曝光采集图像的比较。使用行曝光会使被测物有明显的变形

Because of their architecture, it is easy to support asynchronous reset for triggered acquisition in CMOS cameras.

由于 CMOS 的结构使其很容易支持异步复位外触发采集。

2.3.3 Color Cameras

2.3.3.1 Spectral Response of Monochrome Cameras

CCD and CMOS sensors are sensitive to light with wavelengths ranging from near UV (200 nm) through the visible range (380–780 nm) into the near IR (1100 nm). A sensor responds to the incoming light with its spectral response function. The gray value produced by the sensor is obtained by multiplying the spectral distribution of the incoming light by the spectral response of the sensor and then integrating over the range of wavelengths for which the sensor is sensitive.

Figure 2.51 displays the spectral responses of typical CCD and CMOS sensors and the human visual system (HVS) under photopic conditions

2.3.3 彩色摄像机

2.3.3.1 黑白摄像机的光谱响应

CCD 和 CMOS 传感器对于近紫外 200 nm 至可见光 380~780 nm 直至近红外 1100 nm 波长范围都有响应。每个传感器都是按其光谱响应函数对于入射光作出响应。传感器产生的灰度是传感器所能感应的所有波长范围内入射光的积累后按传感器光谱响应的结果。

图 2.51 是日光下典型 CCD、CMOS 和人眼的光谱响应曲线。传感器的光谱响应范围要比人眼范围广

(daylight viewing). Note that the spectral response of each sensor is much wider than that of the HVS. This can be used advantageously in some applications by illuminating objects with IR strobe lights and using an IR pass filter that suppresses the visible spectrum and only lets IR radiation pass to the sensor. Since the HVS does not perceive the IR radiation, it may be possible to use the strobe without a screen. On the other hand, despite the fact that the sensors are also sensitive to UV radiation, often no special filter is required because lenses are usually made of glass, which blocks UV radiation.

许多。在有些应用中可以利用红外闪光灯照明，在传感器上使用红外通过滤镜使可见光得到抑制，仅使红外光到达传感器。由于人眼对于红外光没有响应，使用红外闪光灯可以不要屏蔽。另外，尽管传感器对紫外也有响应，但是由于通常情况下镜头是玻璃制作的，阻止紫外光，因此通常不需要特殊滤光片滤掉紫外，当需要紫外响应时，需要特殊的镜头。

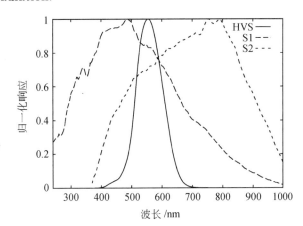

图 2.51 典型 CCD、CMOS 传感器和人眼在日光下的光谱响应曲线（S1 和 S2）。横坐标单位为纳米。由于做过归一化，所以最大响应为 1

2.3.3.2 Single-Chip Cameras

Since CCD and CMOS sensors respond to all frequencies in the visible spectrum, they are unable to produce color images. To construct a color camera, a color filter array (CFA) can be placed in front of the sensor, which allows only light of a specific range of wavelengths to pass to each photodetector. Since this kind of camera only uses one chip to obtain the color information, it is called a single-chip camera.

2.3.3.2 单芯片摄像机

由于 CCD 和 CMOS 传感器对于整个可见光波段全部有响应，所以无法产生彩色图像。为了产生彩色图像，需要在传感器前面加上彩色滤镜阵列（CFA）使得一定范围的光到达每个光电探测器。由于这种传感器仅使用一个芯片得到彩色信息，所以称作单芯片摄像机。

Figure 2.52 displays the most commonly used Bayer CFA (Bayer, 1976). Here, the CFA consists of three kinds of filters, one for each primary color that the HVS can perceive (red, green, and blue). Note that green is sampled at twice the frequency of the other colors because the HVS is most sensitive to colors in the green range of the visible spectrum. Also note that the colors are subsampled by factors of 2 (green) and 4 (red and blue). Since this can lead to severe aliasing problems, often an optical anti-aliasing filter is placed before the sensor (Holst and Lomheim, 2011). Furthermore, to obtain each color with full resolution, the missing samples must be reconstructed through a process called demosaicking. The simplest method to do this is to use methods such as bilinear or bicubic interpolation. This can, however, cause significant color artifacts, such as colored fringes. Methods for demosaicking that cause fewer color artifacts have been proposed, for example, by Alleysson *et al.* (2005) and Hirakawa and Parks (2005).

图 2.52 表示了最常见的 Bayer 滤镜阵列（Bayer，1976）。这种滤镜阵列由三种滤镜组成，每种滤镜可以透过人眼敏感的三基色红、绿、蓝中的一种。由于人眼对绿色最为敏感，所以滤镜阵列中绿色采样频率是其他两种的两倍。值得注意的是由于绿色采样是 1/2，红、蓝是 1/4，这就导致了严重的图像失真。通常在传感器前加上抗图像失真滤光片（Holst and Lomheim，2011）。单芯片彩色摄像机传感器前加有 Bayer 滤镜阵列，使得一定波长范围的光到达每个光电传感器。为了得到传感器全分辨率下的彩色图像，少采样的部分需要通过称作颜色插值的处理来重建。颜色重建的最简单方法是双线性或双三次插值。这种方法会产生彩条等明显的人为颜色缺陷。产生很少的人为颜色缺陷的颜色插值方法已经提出，例如，文献（Alleysson *et al.*，2005）和（Hirakawa and Parks，2005）。

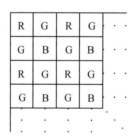

图 2.52 单芯片彩色摄像机传感器前加有彩色滤镜阵列使得一定波长范围内的光到达每个光电探测器，这个图展示了 Bayer 彩色滤镜阵列

2.3.3.3 Three-Chip Cameras

A second method to construct a color camera is shown in Figure 2.53. Here, the light beam coming from the lens is split into three beams by a

2.3.3.3 三芯片摄像机

图 2.53 是构造彩色摄像机的第二种方法。通过镜头的光线被分光器或棱镜分为三束光然后到达三个传感器。

beam splitter or prism and sent to three sensors that have different color filters placed in front of them (Holst and Lomheim, 2011). Hence, these cameras are called three-chip cameras. This construction obviously prevents the aliasing problems of single-chip cameras. However, since three sensors must be used and carefully aligned, three-chip cameras are much more expensive than single-chip cameras.

每个传感器前有一个不同的滤光片（Holst and Lomheim, 2011）。这种摄像机称作三芯片摄像机。这种结构很显然可以克服单芯片摄像机的图像失真问题。然而由于必须使用三个传感器，而且三个传感器需要很仔细地调整位置，因此三芯片摄像机比单芯片摄像机贵许多。

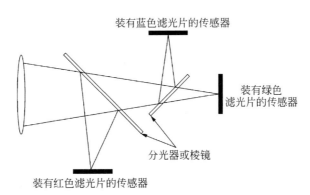

图 2.53　在三芯片彩色摄像机中，来自镜头的光线被分光器或棱镜分成三束光后到达具有不同滤光片的三个传感器上

2.3.3.4　Spectral Response of Color Cameras

Figure 2.54 shows the spectral response of a typical color sensor. Note that the sensor is still sensitive to near IR radiation. Since this may cause unexpected colors in the image, an IR cut filter should be used.

2.3.3.4　彩色摄像机的光谱响应

图 2.54 表示了典型彩色 CCD 传感器的光谱响应曲线，可见其在近红外是敏感的，这会使图像产生不希望的颜色，因此必须加上红外滤光片。

2.3.4　Sensor Sizes

CCD and CMOS sensors are manufactured in various sizes, typically specified in inches. The most common sensor sizes and their typical widths, heights, and diagonals are shown in Table 2.3. Note that the size classification is a remnant of

2.3.4　传感器尺寸

CCD 和 CMOS 有多种生产尺寸，常以英寸表示。最常见的传感器的长、宽及对角线列于表 2.3。这种分类是延续了当电视机使用摄像管时的分类方法。这一尺寸是摄像管的外接圆直

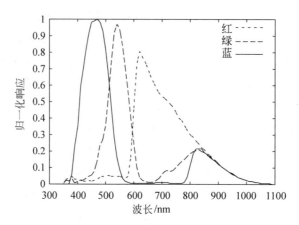

图 2.54 归一化后典型 CCD 传感器光谱响应，在近红外是有响应的。
响应曲线已作了归一化处理使得最大响应为 1

the days when video camera tubes were used in television. The size defined the outer diameter of the video tube. The usable area of the image plane of the tubes was approximately two-thirds of their diameter. Consequently, the diagonals of the sensors in Table 2.3 are roughly two-thirds of the respective sensor sizes. An easier rule to remember is that the width of the sensors is approximately half the sensor size.

径大小。摄像管有效的像平面大约是这一尺寸的 2/3，因此表 2.3 中传感器对角线的尺寸大约是传感器标称尺寸的 2/3。有个简单的方法可以记住这些数据，就是传感器的宽度大约是传感器标称尺寸的一半。

表 2.3 典型传感器尺寸及 640×480 分辨率时对应的像素间距

尺寸/inch	宽/mm	高/mm	对角线/mm	间距/μm
1	12.8	9.6	16	20
2/3	8.8	6.6	11	13.8
1/2	6.4	4.8	8	10
1/3	4.8	3.6	6	7.5
1/4	3.2	2.4	4	5

It is imperative to use a lens that has been designed for a sensor size that is at least as large as the sensor that is actually used. If this is not done, the outer parts of the sensor will receive no light. For example, a lens designed for a 1/2 inch chip cannot be used with a 2/3 inch chip.

Table 2.3 also displays typical pixel pitches for an image size of 640 × 480 pixels. If the sensor

为传感器选择镜头时必须使镜头的设计不小于传感器实际使用大小。如果不这样，传感器外围就没有光线到达，举个例子，1/2″ 镜头就不可以用于 2/3″ 的传感器。

表 2.3 还表示了对于 640 × 480 图像大小时像素间距。当传感器分辨

has a higher resolution, the pixel pitches decrease by the corresponding factor, e.g., by a factor of 2 for an image size of 1280 × 960.

Sensors with a very high resolution fall outside the above classification scheme because they are often larger than a 1 inch sensor (otherwise, the pixels would become too small). For these sensors, a classification scheme from analog photography is sometimes used, e.g., "35 mm" for sensors that have a size of around 36 × 24 mm (the traditional analog camera format) or "medium format" for sensors that are even larger (in analogy to traditional analog medium format cameras). For example, "645" is sometimes used for sensors that have a size of around 6 × 4.5 cm ("645" is one of several film formats for medium format cameras).

CCD and CMOS area sensors are manufactured in different resolutions, from 640 × 480 to 8856 × 5280. The resolution is often specified in megapixels. For line sensors, the resolution ranges from 2048 pixels to 16 384 pixels. As a general rule, the maximum frame and line rates decrease as the sensors get larger.

2.3.5 Camera Performance

When selecting a camera for a particular application, it is essential that we are able to compare the performance characteristics of different cameras. For machine vision applications, the most interesting questions are related to noise and spatial nonuniformities of the camera. Therefore, in this section we take a closer look at these two characteristics. Our presentation follows the EMVA standard 1288 (EMVA, 2016a). There is also an ISO standard ISO 15739:2013 for noise measure-

率提高，像素间距就会相应减小。比如当图像大小为 1280 × 960 时，像素间距减小一半。

更高分辨率的传感器不在上面的分类表里，因为靶面尺寸经常大于 1 英寸（否则像素太小）。对于这些传感器，模拟摄像机的分类方式是，对于靶面尺寸是 36 mm × 24 mm（传统的模拟摄像机格式）左右的传感器有时用"35 mm"表示，对于更大靶面尺寸（传统的模拟中画幅摄像机）的传感器用"中画幅"表示。例如，"645"有时用来表示靶面尺寸是 6 cm × 4.5 cm（对于中画幅摄像机，"645"是许多菲林格式中的一种格式）左右的传感器。

CCD 和 CMOS 传感器可产生出不同分辨率，从 640 × 480 到 8856 × 5280。分辨率经常用百万像素表示。对于线扫摄像机，分辨率从 512 到 16 384 像素。在一般情况下，传感器分辨率越高帧率和行频就会越低。

2.3.5 摄像机性能

为了某特定应用选择摄像机时，能够比较不同摄像机的性能特点是有必要的。对于机器视觉来讲，最值得注意的问题与摄像机的噪声和空间不均匀性有关。本节将详细讨论摄像机的这两个特性。我们的描述会遵循 EMVA 标准 1288（EMVA，2016a）。除此之外也有用于噪声测量的 ISO 标准（ISO 15739:2013）。然而，这一标准适合于具有非线性灰度响应的家

ment. However, it is tailored to consumer cameras, which generally have nonlinear gray value responses. Therefore, we will not consider this standard here.

2.3.5.1 Noise

We begin by looking at how a gray value in an image is produced. During the exposure time, a number μ_p of photons fall on the pixel area (including the light-insensitive area) and create a number μ_e of electrons and holes (for CCD sensors) or discharge the reset charge, i.e., annihilate μ_e electron–hole pairs (for CMOS sensors). The ratio of electrons created or annihilated per photon is called the total quantum efficiency η. The total quantum efficiency depends on the wavelength λ of the radiation: $\eta = \eta(\lambda)$. To simplify the discussion, we will ignore this dependency in this section. The electron–hole pairs form a charge that is converted into a voltage, which is amplified and digitized by an ADC, resulting in the digital gray value y that is related to the number of electrons by the overall system gain K.

Note that this process assumes that the camera produces digital data directly. For an analog camera, we would consider the system consisting of camera and frame grabber.

Different sources of noise modify the gray value that is obtained. First of all, the photons arrive at the sensor not at regular time intervals, but according to a stochastic process that can be modeled by the Poisson distribution. This random fluctuation of the number of photons is called photon noise. Photon noise is characterized by the variance σ_p^2 of the number of photons. For the Poisson distribution, σ_p^2 is identical to the mean

2.3.5.1 噪声

首先来看看一幅图像的灰度值是如何产生的。在曝光期间，有 μ_p 个光子落到传感器的区域内（包括光线不敏感区域）对于 CCD 来讲产生 μ_e 个电子和空穴，对于 CMOS 来讲消灭 μ_e 个电子-空穴对。每个光子产生或消灭的电子的比率称作量子效率 η。总的量子响应取决于辐射波长 λ：$\eta = \eta(\lambda)$，为了讨论简化，本节中我们会忽略这个因素。电子-空穴对形成电荷然后转换为电压，电压经过放大并通过模数转换数字化产生灰度值 y，y 与整个系统的增益 K 有关。

这一过程是假设摄像机直接产生数字信号。对于模拟摄像机，可以认为整个系统由摄像机和图像采集卡组成。

有不同的噪声使得真正得到的灰度值有所改变。首先，光子并不是等时间间隔到达传感器，而是按泊松分布模型的随机分布到达。这种光子的随时间不一致性称作光子噪声。光子噪声使用光子数的平方 σ_p^2 来表示。对于泊松分布 σ_p^2 与到达像素的光子的均值 μ_p 一致。这也意味着光线具有一致的信噪比 $\text{SNR}_p = \mu_p/\sigma_p = \sqrt{\mu_p}$。

number μ_p of photons arriving at the pixel. Note that this means that light has an inherent signal-to-noise ratio (SNR) $\mathrm{SNR}_\mathrm{p} = \mu_\mathrm{p}/\sigma_\mathrm{p} = \sqrt{\mu_\mathrm{p}}$. Consequently, using more light typically results in a better image quality.

During the exposure, each photon creates an electron with a certain probability η that depends on the optical fill factor of the sensor (including the increase in fill factor caused by the microlenses), the material of the sensor (its quantum efficiency), and the wavelength of the light. The product of the optical fill factor and the quantum efficiency of the sensor material is called the total quantum efficiency. Hence, the number of electrons $\mu_\mathrm{e} = \eta\mu_\mathrm{p}$ is also Poisson distributed with $\mu_\mathrm{e} = \eta\mu_\mathrm{p}$ and $\sigma_\mathrm{e}^2 = \mu_\mathrm{e} = \eta\mu_\mathrm{p} = \eta\sigma_\mathrm{p}^2$.

During the readout of the charge from the pixel, different effects in the electronics create random fluctuations of the resulting voltages (Holst and Lomheim, 2011). Reset noise occurs because a pixel's charge may not be completely reset during the readout. It can be eliminated through correlated double sampling, i.e., subtracting the voltage before the readout from the voltage after the readout. Dark current noise occurs because thermal excitation also causes electron–hole pairs to be created. This is only a problem for very long exposure times, and can usually be ignored in machine vision applications. If it is an issue, the camera must be cooled. Furthermore, the amplifier creates amplifier noise (Holst and Lomheim, 2011). All these noise sources are typically combined into a single noise value called the noise floor or dark noise, because it is present even if the sensor receives no light. Dark noise can be modeled by a Gaussian distribution with mean μ_d and vari-

光线强就可以得到好的图像。

曝光期间，每个光子产生电子的概率为 η，η 取决于传感器的填充因子（包括由于使用微镜而提高的部分）、传感器材料（量子效率）及光线的波长。填充因子与传感器材料的量子效率相乘称作总的量子效率。因此电子数 $\mu_\mathrm{e} = \eta\mu_\mathrm{p}$ 也是泊松分布的，其中 $\mu_\mathrm{e} = \eta\mu_\mathrm{p}$，$\sigma_\mathrm{e}^2 = \mu_\mathrm{e} = \eta\mu_\mathrm{p} = \eta\sigma_\mathrm{p}^2$。

从像素中读出电荷的过程中，电路中多种影响也会造成读出电压的随机浮动（Holst and Lomheim，2011）。复位噪声是在读出时像素电荷不能完全复位造成的。通过双采样可以去除这一噪声，也就是在读出电压中减去读出后像素中的电压。暗电流噪声是由于热也会激发生成电子-空穴。这一问题通常在长时间曝光时比较严重。在机器视觉应用中常可以忽略不计，当这一噪声成为问题时，摄像机必须制冷。放大器会产生放大噪声（Holst and Lomheim，2011）。所有这些噪声即使没有任何光线也会有，这些噪声合在一起构成本底噪声或称作暗噪声。暗噪声可以用均值为 μ_d、方差为 σ_d^2 的高斯分布模型来表示。最后，电压通过模数转换器转换为数字，会引入量化噪声，当 $\mu_\mathrm{q} = 0$，$\sigma_\mathrm{q}^2 = 1/12$ 时是均匀分布的。

ance σ_d^2. Finally, the voltage is converted to a DN by the ADC. This introduces quantization noise, which is uniformly distributed with mean $\mu_q = 0$ and variance $\sigma_q^2 = 1/12$.

The process of converting the charge into a DN can be modeled by a conversion factor K called the overall system gain. Its inverse $1/K$ can be interpreted as the number of electrons required to produce unit change in the gray value. With this, the mean gray value is given by

将电荷转换为数字的过程可用转换因子为 K 的模型表示，K 称作整个系统的增益，其倒数 $1/K$ 可以理解为电子数目，这个电子数目可以产生灰度值单位变化。基于此，平均灰度用以下等式表示

$$\mu_y = K(\mu_e + \mu_d) = K\mu_e + \mu_{y.dark} \tag{2.22}$$

If the noise sources are assumed to be stochastically independent, we also have

假设噪声是随机分布的，可以得到

$$\sigma_y^2 = K^2(\sigma_e^2 + \sigma_d^2) + \sigma_q^2 = K^2\sigma_d^2 + \sigma_q^2 + K(\mu_y - \mu_{y.dark}) \tag{2.23}$$

Note that the above noise sources constitute temporal noise because they can be averaged out over time (for temporal averaging, see Section 3.2.3).

注意，以上噪声源称作随机噪声，可以通过长时间平均去掉（有关随机噪声平均见 3.2.3 节）。

2.3.5.2 Signal-to-Noise Ratio

A very important characteristic that can be derived from the above quantities is the SNR at a particular illumination level μ_p:

2.3.5.2 信噪比

从上述噪声可以得出一个重要的特性信噪比，在特定的光照水平 μ_p 下：

$$\text{SNR}(\mu_p) = \frac{\mu_y - \mu_{y.dark}}{\sigma_y} = \frac{\eta\mu_p}{\sqrt{\sigma_d^2 + \sigma_q^2/K^2 + \eta\mu_p}} \tag{2.24}$$

For low light levels ($\eta\mu_p \ll \sigma_d^2 + \sigma_q^2/K^2$), we have

对于低光照水平（$\eta\mu_p \ll \sigma_d^2 + \sigma_q^2/K^2$），得出

$$\text{SNR}(\mu_p) \approx \frac{\eta\mu_p}{\sqrt{\sigma_d^2 + \sigma_q^2/K^2}} \tag{2.25}$$

while for high light levels ($\eta\mu_p \gg \sigma_d^2 + \sigma_q^2/K^2$), we have

对于高亮度情况（$\eta\mu_p \gg \sigma_d^2 + \sigma_q^2/K^2$），得出

$$\text{SNR}(\mu_p) \approx \sqrt{\eta\mu_p} \tag{2.26}$$

This implies that the slope of all SNR curves changes from a linear increase at low light levels to a square root increase at high light levels. To characterize the performance of a real sensor, its SNR curve can be compared to that of an ideal sensor with perfect quantum efficiency ($\eta = 1$), no dark noise ($\sigma_\text{d} = 0$), and no quantization noise ($\sigma_\text{q}/K = 0$). Consquently, the SNR of an ideal sensor is given by $\text{SNR}(\mu_\text{p}) = \sqrt{\mu_\text{p}}$.

The SNR is often specified in decibels (dB): $\text{SNR}_\text{dB} = 20 \log \text{SNR}$, where $\log x$ is the base-10 logarithm. It also can be specified as the number of significant bits: $\text{SNR}_\text{bit} = \lg \text{SNR}$, where $\lg x = \log_2 x$ is the base-2 logarithm.

2.3.5.3 Dynamic Range

Another interesting performance metric is the dynamic range (DR) of the sensor. In simplified terms, it compares the output of the sensor at an illumination close to the sensor's capacity to its output at an illumination at which the image content can just be discriminated from the noise. The saturation illumination $\mu_\text{p.sat}$ is defined not by the maximum gray level the camera supports, but by a gray level such that the noise in Eq. (2.23) can still be represented without clipping (EMVA, 2016a). From it, the saturation capacity can be computed: $\mu_\text{e.sat} = \eta \mu_\text{p.sat}$. Furthermore, the illumination $\mu_\text{p.min}$ for which the SNR (2.24) is 1 is usually taken as the smallest detectable amount of light and is called the absolute sensitivity threshold. It can be approximated by (see EMVA (2016a) for details of the derivation of this formula):

这意味着所有 SNR 曲线的斜率从低光照下的线性增加变化到高亮度下的平方根增加。为了描绘一个真实传感器的性能特点,其 SNR 曲线可以和理想传感器的 SNR 做对比,理想传感器 SNR 具有完美的量子效率($\eta = 1$),没有暗噪声($\sigma_\text{d} = 0$)并且没有量化噪声($\sigma_\text{q}/K = 0$)。因此,理想传感器的 SNR 为 $\text{SNR}(\mu_\text{p}) = \sqrt{\mu_\text{p}}$。

SNR 常以分贝表示 (dB):$\text{SNR}_\text{dB} = 20 \log \text{SNR}$,$\log x$ 是以 10 为底的对数。也可以用有效位数表示:$\text{SNR}_\text{bit} = \lg \text{SNR}$,$\lg x = \log_2 x$ 是以 2 为底的对数。

2.3.5.3 动态范围

另外一个值得注意的性能指标是传感器的动态范围(DR)。简单地说,动态范围就是在接近传感器饱和能力照明下的传感器输出比上图像恰好能够和噪声相区分的照明条件下的传感器输出。饱和照明 $\mu_\text{p.sat}$ 并不是定义为摄像机所支持的最大灰度值,而是等式(2.23)中噪声仍然能够无裁剪表示时的灰度值(EMVA,2016a)。基于此,饱和量可以计算为:$\mu_\text{e.sat} = \eta \mu_\text{p.sat}$。而且,对于 SNR 等式(2.24)为 1 的照明 $\mu_\text{p.min}$ 通常作为可以检测到的最小光照量,称为绝对灵敏度阈值。近似可以表示为(公式的推导细节参见文献(EMVA,2016a))

$$\mu_\text{p.min} \approx \frac{1}{\eta}\left(\sqrt{\sigma_\text{d}^2 + \sigma_\text{q}^2/K^2} + \frac{1}{2}\right) = \frac{1}{\eta}\left(\frac{\sigma_\text{y.dark}}{K} + \frac{1}{2}\right) \qquad (2.27)$$

From the above quantities, we can derive the DR of the camera (see EMVA, 2016a):

$$DR = \frac{\mu_{p.sat}}{\mu_{p.min}} \qquad (2.28)$$

Like the SNR, the DR can be specified in dB or in bits.

2.3.5.4 Nonuniformities

The noise sources we have examined so far are assumed to be identical for each pixel. Manufacturing tolerances, however, cause two other effects that produce gray value changes. In contrast to temporal noise, these gray value fluctuations cannot be averaged out over time. Since they look like noise, they are sometimes called spatial noise or pattern noise (Holst and Lomheim, 2011). However, they are actually systematic errors. Consequently, we will call them nonuniformities (EMVA, 2016a).

One effect is that the dark signal may not be identical for each pixel. This is called dark signal nonuniformity (DSNU). The second effect is that the responsivity of the pixels to light may be different for each pixel. This is called photoresponse nonuniformity (PRNU). CMOS sensors often exhibit significantly larger nonuniformities than CCD sensors because each pixel has its own amplifier, which may have a different gain and offset.

To characterize the nonuniformities, two illumination levels are used: one image y_{dark} with no illumination, i.e., with the sensor completely dark, and one image y_{50} corresponding to 50% of the sensor saturation level, where the image must be illuminated evenly across the entire image. From these images, the mean gray values across the entire image are computed (see Eq. (3.67) in Section 3.5.2.1):

整个图像上，基于这些图像，整个图像的平均灰度值可以通过以下等式计算（见 3.5.2.1 节的等式（3.67））

$$\mu_{\text{y.dark}} = \frac{1}{wh} \sum_{r=0}^{h-1} \sum_{c=0}^{w-1} y_{\text{dark};r,c} \tag{2.29}$$

$$\mu_{\text{y.50}} = \frac{1}{wh} \sum_{r=0}^{h-1} \sum_{c=0}^{w-1} y_{50;r,c} \tag{2.30}$$

where w and h denote the width and height of the image, respectively. Then, the spatial variances of the two images are computed (see Eq. (3.68)):

其中 w 和 h 表示图像宽高。那么，两幅图像的空间差异可以用以下等式计算（见等式（3.68））

$$s_{\text{y.dark}}^2 = \frac{1}{wh-1} \sum_{r=0}^{h-1} \sum_{c=0}^{w-1} (y_{\text{dark};r,c} - \mu_{\text{y.dark}})^2 \tag{2.31}$$

$$s_{\text{y.50}}^2 = \frac{1}{wh-1} \sum_{r=0}^{h-1} \sum_{c=0}^{w-1} (y_{50;r,c} - \mu_{\text{y.50}})^2 \tag{2.32}$$

With these quantities, the DSNU and PRNU are defined in EMVA (2016a) as follows:

基于以上内容，在 EMVA 中 DSNU 和 PRNU 定义如下：

$$\text{DSNU}_{1288} = s_{\text{y.dark}}/K \tag{2.33}$$

$$\text{PRNU}_{1288} = \frac{\sqrt{s_{\text{y.50}}^2 - s_{\text{y.dark}}^2}}{\mu_{\text{y.50}} - \mu_{\text{y.dark}}} \tag{2.34}$$

Note that the DSNU is specified in units of the number of electrons. It can be converted to gray levels by multiplying it by the overall system gain K. In contrast, the PRNU is specified as a percentage relative to the mean gray level.

注意，DSNU 的单位用电子数量的单位表示。通过乘以总的系统增益 K，也可以转换为灰度阶。与之对比，PRNU 表示为相对于平均灰度值的百分比。

In addition to these performance characteristics, the EMVA standard 1288 describes two other criteria by which the nonuniformities can be characterized. One measure charac-

除了以上性能特点之外，EMVA 标准 1288 描述了其他两个标准，可以描述不均匀特性，一个是不均匀性的测量特征周期变化，另外一个是特

terizes periodic variations of the nonuniformities, while the other characterizes defect pixels. The interested reader is referred to EMVA (2016a) for details.

The EMVA standard 1288 requires that the sensitivity, linearity, and nonuniformity parameters of the camera are determined using a monochromatic light source. Furthermore, if the quantum efficiency $\eta(\lambda)$ is used to characterize the camera (which is not mandated by the standard), it must be measured over the entire range of wavelengths to which the sensor is sensitive. Detailed instructions about how the above performance parameters are to be measured are specified in EMVA (2016a).

2.4 Camera–Computer Interfaces

As described in the previous section, the camera acquires an image and produces either an analog or a digital video signal. In this section, we will take a closer look at how the image is transmitted to the computer and how it is reconstructed as a matrix of gray or color values. We start by examining how the image can be transmitted via an analog signal. This requires a special interface card to be present on the computer, which is conventionally called a frame grabber. We then describe the different means by which the image can be transmitted in digital form—digital video signals that require a frame grabber or some kind of standard interface, e.g., IEEE 1394, USB, or Gigabit Ethernet. Finally, we discuss the different acquisition modes with which images are typically acquired.

征缺陷像素。感兴趣的读者可以参见文献（EMVA，2016a）。

EMVA 标准 1288 要求摄像机的灵敏度，线性度和不均匀性等参数可以用单色光源来决定。而且，如果量子效率 $\eta(\lambda)$ 用来描述摄像机的特性（不包含在标准中），必须测量传感器所能感应的完整波长范围。关于以上性能参数如何测量，详细说明请参见文献（EMVA，2016a）。

2.4 摄像机-计算机接口

上节谈到摄像机捕获图像然后输出模拟或数字视频信号。本节将讨论图像是如何传到计算机中，又是如何重建成为灰度或彩色矩阵图像。首先看看模拟信号如何传递到计算机中的。这一过程需要在计算机中安装一块通常称作图像采集卡的专用接口卡。然后讲述数字图像的不同传输方法—数字视频信号也需要图像采集卡或者某种标准接口卡，例如 IEEE 1394 卡、USB 卡和千兆网卡。最后讨论图像采集的不同模式。

2.4.1 Analog Video Signals

Analog video standards have been defined since the early 1940s. Because of the long experience with these standards, they were the dominant technology in the machine vision industry for a long time. However, as we will discuss in this section, analog video transmission can cause problems that deteriorate the image quality, which can lower the measurement accuracy or precision. Since digital video transmission prevents these problems, the use of analog video transmission has declined in recent years to the point where it is almost exclusively used to continue to operate existing machine vision systems. Newly developed machine vision systems typically use digital video transmission.

2.4.1.1 Analog Video Standards

While several analog video standards for television are defined in ITU-R BT.470-6 (1998), only four of them are important for machine vision. Table 2.4 displays their characteristics. EIA-170 and CCIR are black-and-white video standards, while NTSC and PAL are color video standards. The primary difference between them is that EIA-170 and NTSC have a frame rate of 30 Hz with 525 lines per image, while CCIR and PAL have a frame rate of 25 Hz with 625 lines per image. In all four standards, it takes roughly the same time to transmit each line. Of the 525 and 625 lines, nominally 40 and 50 lines, respectively, are used for synchronization, e.g., to indicate the start of a new frame. Furthermore, 10.9 µs (EIA-170 and NTSC) and 12 µs (CCIR and PAL) of each line are also used for synchronization. This usually

表 2.4 模拟视频标准

标准	类型	帧率 f/s	行数	行周期 /μs	行频 /s^{-1}	像素周期 /ns	图像大小 /像素×像素
EIA-170	B/W	30.00	525	63.49	15 750	82.2	640 × 480
CCIR	B/W	25.00	625	64.00	15 625	67.7	768 × 576
NTSC	Color	29.97	525	63.56	15 734	82.3	640 × 480
PAL	Color	25.00	625	64.00	15 625	67.7	768 × 576

results in an image size of 640 × 480 (EIA-170 and NTSC) and 768 × 576 (CCIR and PAL). From these characteristics, it can be seen that a pixel must be sampled roughly every 82 ns or 68 ns, respectively.

Figure 2.55 displays an overview of the EIA-170 video signal. The CCIR signal is very similar. We will describe the differences below.

640 像素×480 像素,而 CCIR 和 PAL 为 768 像素 × 576 像素。从以上特征可以看出,每个像素的采样时间大约为 82 ns 和 68 ns。

图 2.55 是 EIA-170 视频信号概况。CCIR 与其非常类似,我们会在下文中描述二者的区别。

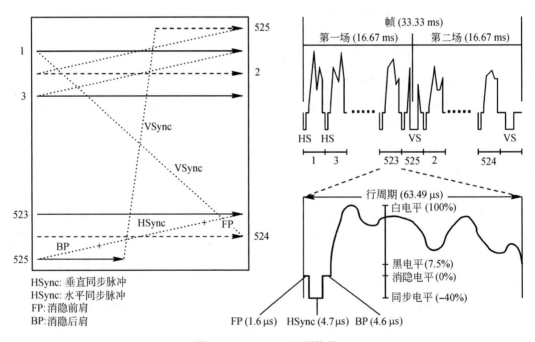

图 2.55 EIA-170 视频信号

As already mentioned in Section 2.3.1.5, an image (a frame) is transmitted interlaced as two fields. The first field consists of the even-numbered lines and the second field of the odd-

在 2.3.1.5 节已经提到过一幅图像是以两场隔行传输的。对于 EIA-170 第一场包括所有偶数行,第二场包括所有奇数行。对于 CCIR 顺序为

numbered lines for EIA-170 (for CCIR, the order is odd lines, then even lines). Each line consists of a horizontal blanking interval that contains the horizontal synchronization information and the active line period that contains the actual image signal of the line. The horizontal blanking interval consists of a front porch, where the signal is set to the blanking level (0%), a horizontal synchronization pulse, where the signal is set to the synchronizing level (−40% for EIA-170, −43% for CCIR), and a back porch, where again the signal is set to the blanking level. For CCIR, the length of the front and back porches is 1.5 µs and 5.8 µs, respectively. The purpose of the front porch was to allow voltage levels to stabilize in older television sets, preventing interference between lines. The purpose of the horizontal synchronization pulse is to indicate the start of the valid signal of each line. The purpose of the back porch was to allow the slow electronics in early television sets time to respond to the synchronization pulse and to prepare for the active line period. During the active line period, the signal varies between the white and black levels. For CCIR, the black level is 0%, while for EIA-170 it is 7.5%.

Each field starts with a series of vertical synchronization pulses. In Figure 2.55, for simplicity, they are drawn as single pulses. In reality, they consist of a series of three different kinds of pulses, each of which spans multiple lines during the vertical blanking interval. The vertical blanking interval lasts for 20 lines in EIA-170 and for 25 lines in CCIR.

The vertical blanking interval was originally needed to allow the beam to return from bottom to top in cathode-ray-tube television sets because

奇数行然后偶数行。每行都由包含水平同步信息的水平消隐间隔和包含实际图像信号的行有效间隔组成。水平消隐间隔由消隐前肩、水平同步脉冲以及消隐后肩组成。消隐前肩视频信号被置为消隐电平；水平同步脉冲时视频信号被置于同步电平（EIA-170 为 −40%，CCIR 为 −43%）。消隐后肩视频信号又被置为消隐电平。对于 CCIR 消隐前肩和消隐后肩为 1.5µs 和 5.8µs。消隐前肩的作用是使老式电视机的视频信号电平稳定下来，以避免行间串扰。水平同步脉冲用于表示每行有效信号的开始。消隐后肩的作用是使早期的电视定时器中较慢的电子器件有时间响应同步脉冲并为有效信号做好准备。在有效行周期，信号在黑、白电平之间变化。对于 CCIR 黑电平为 0%，而 EIA-170 为 7.5%。

每一场都是以几个垂直同步脉冲序列开始的。为了简化图 2.55 画了一个脉冲。实际上垂直同步脉冲由三种不同的脉冲序列组成，每种脉冲占垂直消隐周期的几行。EIA-170 的垂直消隐周期为 20 行，CCIR 为 25 行。

垂直消隐周期原用于使阴极射线管电视的扫描电子束从底部回到顶部，因为使电子束垂直偏转的磁化线

of the inductive inertia of the magnetic coils that deflect the electron beam vertically. The magnetic field, and hence the position of the spot on the screen, cannot change instantly. Likewise, the horizontal blanking interval allowed the beam to return from right to left.

Color can be transmitted in three different ways. First of all, the color information (called chrominance) can be added to the standard video signal, which carries the luminance (brightness) information, using quadrature amplitude modulation. Chrominance is encoded using two signals that are 90° out of phase, known as I (in-phase) and Q (quadrature) signals. To enable the receiver to demodulate the chrominance signals, the back porch contains a reference signal, known as the color burst. This encoding is called composite video. It has the advantage that a color signal can be decoded by a black-and-white receiver simply by ignoring the chrominance signal. Furthermore, the signal can be transmitted over a cable with a single wire. A disadvantage of this encoding is that the luminance signal must be low-pass filtered to prevent crosstalk between high-frequency luminance information and the color information. In S-Video (separate video), also called Y/C, the two signals are transmitted separately in two wires. Therefore, low-pass filtering is unnecessary, and the image quality is higher. Finally, color information can also be transmitted directly as an RGB signal in three wires. This results in an even better image quality. In RGB video, the synchronization signals are transmitted either in the green channel or through a separate wire.

While interlaced video can fool the human eye into perceiving a flicker-free image (which is

圈具有感应惯性，磁场和屏幕上的点的位置不能立刻改变。同理水平消隐使得电子束从右回到左。

彩色视频信号有三种不同的传输方式。第一种可以将称作色度的颜色信息采用正交调制的方式调制到含有亮度信息的标准视频信号中。色度信号以相差 90° 相位的 I 信号和 Q 信号编码。为了在接收端可以解出色度信号，在消隐后肩中含有称作彩色色同步的参考信号。这种编码形式称作复合视频。这种视频信号的优点是彩色信号可以被黑白接收器解码，只需要忽略色度信号就可以了，另外这种信号可以通过单线传输。这种编码的缺点是亮度信号必须经过低通滤波器，以抑制亮色串扰。亮、色分离信号 S-Video 又称作 Y/C 信号，亮度与色度通过两线传输。因此不需要低通滤波器，图像质量有所改进。最后，颜色信息可以通过三线传输方式直接传输红、绿、蓝信号，这样得到的图像更好。对于红、绿、蓝视频，同步信号可以与绿通道一起传输或使用单独的电缆。

尽管电视使用隔行扫描视频，而且隔行信号可以使人眼看到无闪烁图

the reason why it is used in television), we have already seen in Section 2.3.1.5 that it causes severe artifacts for moving objects since the image must be exposed twice. For machine vision, images should be acquired in progressive scan mode. The above video standards can be extended easily to handle this transmission mode. Furthermore, the video standards can also be extended for image sizes larger than those specified in the standards.

2.4.1.2 Analog Frame Grabbers

To reconstruct the image in the computer from the video signal, a frame grabber card is required. A frame grabber consists of components that separate the synchronization signals from the video signal in order to create a pixel clock signal (see below) that is used to control an ADC that samples the video signal. Furthermore, analog frame grabbers typically also contain an amplifier for the video signal. It is often possible to multiplex video signals from different cameras, i.e., to switch the acquisition between different cameras, if the frame grabber has multiple connectors but only a single ADC. If the frame grabber has as many ADCs as connectors, it is also possible to acquire images from several cameras simultaneously. This is interesting, for example, for stereo reconstruction or for applications where an object must be inspected from multiple sides at the same time. After sampling the video signal, the frame grabber transfers the image to the computer's main memory through direct memory access (DMA), i.e., without using up valuable processing cycles of the CPU.

During the image acquisition, the frame grabber must reconstruct the pixel clock with which

像，但是在 2.3.1.5 节我们看到对于运动物体由于两次曝光会造成严重的伪图像。对于机器视觉，需要逐行扫描模式。以上视频标准很容易扩充为逐行采集信号，这些标准也可以扩充到比上述标准更高的图像尺寸。

2.4.1.2 模拟图像采集卡

计算机需要图像采集卡以便从视频信号中重建图像。图像采集卡包含同步分离部分可以从视频信号中分离出同步信号用来产生像素时钟信号并用来控制采样视频信号的模数转换器。模拟图像采集卡常含有视频信号放大器。图像卡可以采集多路视频；如果图像卡仅有一路模数转换器，但是有多路输入插座，图像卡就可以在与其连接的多只摄像机间切换。如果图像卡有多路模数转换器，就可能同时采集多路信号，对于立体视觉或被测物需要在多角度同时观测时非常有用。图像采集卡采样后通过 DMA 将图像传到计算机内存，不需占用 CPU 资源。

图像卡在采样期间需要重建像素时钟以对齐摄像机中的像素，这一信

the camera has stored its pixels in the video signal, since this is not explicitly encoded in the video signal. This is typically performed by phase-locked loop (PLL) circuitry in the frame grabber.

Reconstructing the pixel clock can create two problems. First of all, the frequencies of the pixel clock in the camera and the frame grabber may not match exactly. If this is the case, the pixels will be sampled at different rates. Consequently, the aspect ratio of the pixels will change. For example, square pixels on the camera may no longer be square in the image. Although this is not desirable, it can be corrected by camera calibration (see Section 3.9).

The second problem is that the PLL may not be able to reconstruct the start of the active line period with perfect accuracy. As shown in Figure 2.56, this causes the sampling of each line to be offset by a time Δt. This effect is called line jitter or pixel jitter. Depending on the behavior of the PLL, the offset may be random or systematic. From Table 2.4, we can see that $\Delta t = \pm 7$ ns will already cause an error of approximately ± 0.1 pixels. In areas of constant or slowly changing gray values, this usually causes no problems. However, for pixels that contain edges, the gray value may change by as much as 10% of the amplitude of the edge. If the line jitter is truly random, i.e., averages out for each line across multiple images of the same scene, the precision with which the edges can be extracted will decrease while the accuracy does not change, as explained in Section 3.7.4. If the line jitter is systematic, i.e., does not average out for each line across multiple images of the same scene, the accuracy of the edge positions will decrease, while the precision is unchanged.

号没有直接编码在视频信号中。通常在图像卡中使用锁相环电路（PLL）来完成这一工作。

像素时钟重建会造成两个问题。首先，图像卡的时钟频率与摄像机的时钟频率不会完全一致，因此，采样频率就会不同。这就会造成像素横、纵比率的改变。比如摄像机中的方像素在图像中不再是方像素。这个问题可以通过摄像机标定来解决，见3.9节。

第二个问题是锁相环电路不能够精确重建有效行周期的开始。从图2.56可以看出每行采样有Δt的时间偏移，称作列抖动或像素抖动。锁相环电路决定了这一偏移是随机的还是系统的。从表2.4我们可以看到当$\Delta t = \pm 7$ns时造成的抖动约为± 0.1个像素。灰度没有变化或者变化不大时没有问题，但是对于含有边缘的像素可能会造成约10%的灰度变化。如果列抖动是随机的，可以通过多幅图像叠加来消除，这时精度虽然降低但是不会影响准确度，见3.7.4节。如果列抖动是系统的，不可以通过多幅图像叠加来消除，这时准确度降低但是精确度不变。

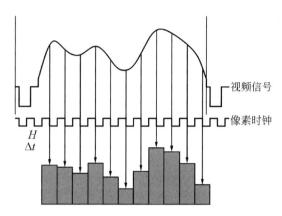

图 2.56　如果图像采集卡的像素时钟不能与有效行周期完全一致就会产生列抖动。每行采样就会有 Δt 的随机或系统时间差

Line jitter can be detected easily by acquiring multiple images of the same object, computing the temporal average of the images (see Section 3.2.3.1), and subtracting the temporal average image from each image. Figures 2.57(a) and (c) display two images from a sequence of 20 images that show horizontal and vertical edges. Figures 2.57(b) and (d) show the result of subtracting the temporal average image of the 20 images from the images in Figures 2.57(a) and (c). Note that, as predicted above, there are substantial gray value differences at the vertical edges, while there are no gray value differences in the homogeneous areas and at the horizontal edges. This clearly shows that the lines are offset by line jitter. At first glance, the effect of line jitter is increased noise at the vertical edges. However, if you look closely you will see that the gray value differences look very much like a sine wave in the vertical direction. Although this is a systematic error in this particular image, it averages out over multiple images (the sine wave will have a different phase). Hence, for this frame grabber the precision of the edges will decrease significantly.

通过采集同一物体多幅图像叠加求出平均（见 3.2.3.1 节），然后减去每幅图像就可以看出列抖动。对于含有水平边缘和垂直边缘的图像，每种各采集 20 幅，图 2.57（a）和图 2.57（c）是其中的一幅。图 2.57（b）和 2.57（d）是减去这 20 幅图像的均值后的图像。如我们所预料，垂直边缘有灰度的变化，而均匀的地方及水平边缘没有灰度的变化，清晰地显示了列抖动造成的行偏移。乍一看列抖动增加了垂直边缘的噪声。然而仔细看看可以发现在垂直方向上灰度的变化非常像正弦波。尽管是系统误差，由于正弦波有不同的相位，通过多幅图像平均可以去除。但图像采集卡采集到的边缘的精度有很大下降。

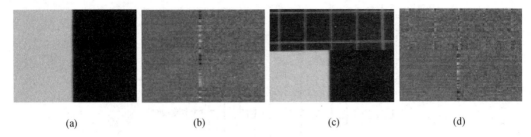

图 2.57 (a) 垂直边缘和 (c) 垂直和水平边缘。(b)，(d) 由于列抖动造成的（a）和（b）灰度值起伏。为了便于观察，灰度值的起伏放大了 5 倍。灰度变化仅在垂直边缘出现表明实际上列抖动导致了每行的偏移

To prevent line jitter and non-square pixels, the pixel clock of the camera can be fed into the frame grabber (if the camera outputs its pixel clock and the frame grabber can use this signal). However, in these cases it is typically better to transmit the video signal digitally.

2.4.2 Digital Video Signals

In contrast to analog video signals, in which the synchronization information is embedded into the signal, digital video signals make this information explicit. For Camera Link (see Section 2.4.2.1), which is based on a parallel transmission of the digital video data, the frame valid signal, which replaces the vertical synchronization signal, is asserted for the duration of a frame (see Figure 2.58). Similarly, the line valid signal, which replaces the horizontal synchronization signal, is asserted for the duration of a line. The pixel clock is transmitted explicitly in the packets. All other digital video standards that we will discuss in this section are packet-based and explicitly encode the beginning and end of a transmitted frame. Furthermore, the pixel data is transmitted explicitly. Therefore, digital video transmission prevents all the problems inherent in analog video that were

如果摄像机输出其像素时钟，而且图像采集卡可以使用这个信号，就可以将摄像机的像素时钟反馈到图像采集卡来避免列抖动和非方像素。然而在这种情况下最好的办法是直接传输数字信号。

2.4.2 数字视频信号

模拟视频的同步信息是嵌入在信号中的，与其相反，数字视频的同步信息是非隐性的。CameraLink（见2.4.2.1 节）是基于并行传输的数字视频信号，其表示一帧持续时间的帧有效信号取代了垂直同步信号（见图2.58）。同样，表示一行持续时间的行有效信号取代了水平同步信号。像素时钟在包中是非隐性传输的。本节所讨论的其他数字视频标准都是基于包传输，每帧的开始及结束都是显性的。数字视频传输避免了上节末讲述的所有模拟视频固有的问题。对于数字摄像机，图像像素的宽高比与摄像机的宽高比相同。而且，不产生列抖动。为了生成数字视频信号，摄像机会将传感器输出的电压进行模数转换，然后将产生的数值串行或并行地传输到图

described at the end of the previous section. For digital cameras, the aspect ratio of the pixels of the image is identical to the aspect ratio of the pixels on the camera. Furthermore, there is no line jitter. To create the digital video signal, the camera performs an analog-to-digital conversion of the voltage of the sensor and transmits the resulting DN to the computer.

像采集卡。

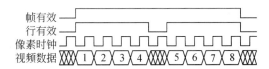

图 2.58　典型数字视频信号。与模拟信号不同，数字视频的同步信息通过帧有效、行有效和像素时钟信号直接表达

Apart from better image quality, digital cameras offer many other advantages. For example, digital video signals offer much higher resolutions and frame rates than analog video signals. Furthermore, analog cameras typically are controlled by setting DIP switches in the camera and configuring the frame grabber appropriately. In contrast, digital cameras can be controlled directly in software. Finally, digital cameras are typically much smaller and require less power than analog cameras.

数字摄像机除了图像质量更好之外，还有很多其他好处，例如比模拟摄像机更高的分辨率和更高的帧率，另外模拟摄像机一般是通过设置摄像机上面的拨码开关来控制摄像机和采集卡配合工作的，而数字摄像机则可以在软件中直接控制。数字摄像机一般比模拟摄像机体积小很多而且功耗降低。

2.4.2.1　Camera Link

Until 2000, the machine vision industry lacked a standard even for the physical connector between the camera and the frame grabber, not to mention the lack of a standard digital video format. Camera manufacturers used a plethora of different connectors, making cable production for frame grabber manufacturers very burdensome and the cables extremely expensive. In October 2000, the connector problem was addressed through

2.4.2.1　Camera Link

直到 2000 年，机器视觉工业缺少一个标准，甚至连摄像机与图像采集卡之间的物理连接接口标准都没有，更别提数字视频格式标准了。摄像机生产厂商使用多种不同的连接器，使得为图像采集卡厂商生产线缆的工作非常繁重，同时线缆也极其的昂贵。在 2000 年 10 月 Camera Link 规范的推出解决了连接器问题，它定义了

the introduction of the Camera Link specification, which defines a 26-pin MDR connector as the standard connector. Later versions of the standard also define smaller 26-pin SDR and HDR connectors as well as an even smaller 14-pin HDR connector. The current version of the Camera Link specification is 2.0 (AIA, 2012b).

The Camera Link specification defines not only the connector, but also the physical means by which digital video is transmitted. The basic technology is low-voltage differential signaling (LVDS). LVDS transmits two different voltages, which are compared at the receiver. This difference in voltage between the two wires is used to encode the information. Hence, two wires are required to transmit a single signal. One advantage of LVDS is that the transmission is very robust to disturbances since they affect both wires equally and consequently are eliminated by the calculation of the difference at the receiver. Another advantage of LVDS is that very high transmission speeds can be achieved.

Camera Link is based on Channel Link, a solution that was developed for transmitting video signals to flat panel displays and then extended for general-purpose data transmission. Camera Link consists of a driver and receiver pair. The driver is a chip or an FPGA on the camera that accepts 28 single-ended data signals and a single-ended clock signal and serializes the data 7:1, i.e., the 28 data signals are transmitted serially over four wire pairs with LVDS. The clock is transmitted over a fifth wire pair. The receiver is a similar chip or an FPGA on the frame grabber that accepts the four data signals and the clock signal, and reconstructs the original 28 data signals from them. Camera

26 针 MDR 连接器作为标准连接器。后续版本又添加了 SDR 和 HDR 小 26 针以及更小的 14 针 HDR。目前 Camera Link 规范为 2.0（AIA，2012b）。

Camera Link 规范不仅定义了连接器还定义了数字信号的物理传输方式。基本技术是低电压差分信号技术 LVDS。LVDS 传输两个不同的电压，在接收端进行比较。两条线上的电压差异值用于编码信息。因此对于一个信号需要两根电缆。LVDS 的一个优点是抗干扰能力强，因为干扰信号会同样作用于两根线缆上，在接收端计算差值时就被消除了。LVDS 的最大优点是传输速度可以非常高。

Camera Link 基于 Channel Link。Channel Link 最初是将视频信号传输到平板显示器的解决方案，后来用于通用数据传输。Camera Link 由驱动器和接收器对组成。驱动器是摄像机上的一个芯片或者 FPGA，可以接收 28 个单向的数据信号和一个单向的时钟信号，将数据按 7:1 串行输出。也就是将 28 个数据信号用 4 对信号线按 LVDS 传输。时钟信号通过第 5 对信号线传输。接收器是在图像采集卡上的类似的芯片或者 FPGA，接收 4 组信号数据和时钟信号，还原成 28 个数据信号。Camera Link 规

Link specifies that four of the 28 data signals are used for so-called enable signals (frame valid, line valid, and data valid, as well as a signal reserved for future use). Hence, 24 data bits can be transmitted with a single chip. The chips can run at up to 85 MHz, resulting in a maximum data rate of 255 MB s^{-1} (megabytes per second).

Camera Link also specifies that four additional LVDS pairs are reserved for general-purpose camera control signals, e.g., triggering. How these signals are used is up to the camera manufacturer. Finally, two LVDS pairs are used for serial communication to and from the camera, e.g., to configure the camera. Camera Link also specifies a serial communications application programming interface (API) that describes how to transmit control data to and from the camera, but leaves the protocol for configuring the camera up to the manufacturer. Optionally, i.e., if the frame grabber manufacturer supports it, GenICam (discussed in Section 2.4.3.1) can be used for this purpose through the GenICam CLProtocol (EMVA, 2011). Using GenICam has the advantage that a Camera Link camera can be configured in a generic manner that is independent of the actual serial protocol used by the camera manufacturer.

The above configuration is called the base configuration. Camera Link also defines one additional configuration for a lower data rate and three additional configurations for even higher data rates. The lite configuration uses the smaller 14-pin connector and supports data rates of up to 100 MB s^{-1}. The configurations with higher data rates require two connectors. If a second driver–receiver chip pair is added, 24 additional data bits and four enable signals can be transmitted, result-

定了 28 个数据中有 4 个用于所谓的使能信号（帧有效、行有效、数据有效及一个备用信号）。因此，还有 24 位数据可以使用同一片芯片传输，芯片传输速度可达 85 MHz，因此数据传输率可达 255 MB/s。

Camera Link 规范还定义了 4 对 LVDS 用于一般摄像机控制，例如外触发等。如何使用这些信号取决于摄像机生产厂商。另外还有 2 对 LVDS 信号用于设置摄像机，与摄像机做双向通信。Camera Link 还规范了摄像机如何双向发送控制信号的串行通信应用编程接口（API），这个标准将摄像机设置协议留给了摄像机厂商。如图像采集卡生产厂商支持 GenICam（在 2.4.3.1 节中介绍），可以通过 GenICam CLProtocol（EMVA，2011）来实现摄像机设置。使用 GenICam 的优势是可以不使用摄像机厂商的串口协议而使用通用的协议设置摄像机。

以上配置称作 Base 配置。Camera Link 规范还为低速数据传输定义了一个配置以及为更高数据量定义了另外三种配置。Lite 配置使用 14 针小连接器，最高支持 100 MB/s。更高配置需要第二个连接器。如果加上第二个驱动-接收芯片组对，就可以传输另外 24 位数据和 4 个有效信号，使得最大传输速率可达 510 MB/s，称作 Medium 配置。如果加上第三个

ing in a maximum data rate of 510 MB s^{-1}. This is called the medium configuration. If a third driver–receiver chip pair is added, 16 additional data bits and four enable signals can be transmitted, resulting in a maximum data rate of 680 MB s^{-1}. This is called the full configuration. Finally, by using all available lines and by using some of the lines that are normally used to carry enable signals for data transmission, a configuration that is able to transmit 80 bits of data simultaneously is defined, leading to a data rate of 850 MB s^{-1}. This is called the 80 bit configuration.

The Camera Link standard optionally allows the camera to be powered over the camera link cable. The maximum cable length is 10 m for the lite, base, and medium configurations, 5 m for the full configuration, and 4 m for the 80 bit configuration.

驱动-接收芯片组对，就可以传输另外 16 位数据和 4 个有效信号，使得最大传输速率可达 680 MB/s，称作 Full 配置。还有，如果把一些通常用于传输使能信号的线也用于信号传输，使用所有可用线路，可以同时传输 80 位数据，数据传输速率达到 850 MB/s，称作 80 bit 配置。

Camera Link 标准可以通过 Camera Link 电缆供电。对于 Lite、Base、Medium 最长支持 10 m 电缆。Full 配置可以 5 m，而 80 bit 最长 4 m。

2.4.2.2 Camera Link HS

The Camera Link HS specification was released in 2012 (AIA, 2012a) to support data rates that are even higher than those supported by the Camera Link specification (see Section 2.4.2.1). Despite the similar name, Camera Link HS uses technology that is completely different from that used in Camera Link. Whereas Camera Link uses a parallel transmission format, Camera Link HS uses a packet-based protocol that is based on technology used in network or data storage equipment. Like Camera Link, Camera Link HS requires a frame grabber.

The Camera Link HS specification defines three types of cables (connectors) that can be used to attach cameras to frame grabbers. One type of cable (C2) is a copper cable, while two other types

2.4.2.2 Camera Link HS

Camera Link HS 规范发布于 2012 年（AIA，2012a），用于支持比 Camera Link（见 2.4.2.1 节）协议更高的数据传输。除了名字类似，其实 Camera Link HS 使用的技术与 Camera Link 完全不同。Camera Link 是并行传输模式，而 Camera Link HS 使用的是用于网络和存储设备的基于包的协议。和 Camera Link 一样，Camera Link HS 也需要采集卡。

Camera Link HS 定义了三种摄像机和采集卡连接的电缆（连接器）。一种（C2）是铜的，另外两种（F1 和 F2）是光缆。C2 电缆可达 15 m，是基

of cable (F1 and F2) are fiber-optic cables. The C2 cable, which can be up to to 15 m long, is based on the InfiniBand networking technology and uses an SFF-8470 connector. It has up to eight data lanes, one of which is used for camera control. Each data lane supports a data rate of $300\,\text{MB}\,\text{s}^{-1}$, leading to a maximum data rate of $2100\,\text{MB}\,\text{s}^{-1}$. The fiber-optic cables F1 and F2 are based on the SFP and SFP+ network transceiver technologies, respectively. Their maximum length is at least 300 m. The maximum data rate of F1 cables is $300\,\text{MB}\,\text{s}^{-1}$, that of F2 cables is $1200\,\text{MB}\,\text{s}^{-1}$. To increase the data rates further, Camera Link HS allows up to eight cables per camera, leading to maximum data rates of $16\,800\,\text{MB}\,\text{s}^{-1}$ (C2), $2400\,\text{MB}\,\text{s}^{-1}$ (F1), and $9600\,\text{MB}\,\text{s}^{-1}$ (F2). The data can be transmitted to more than one frame grabber on different computers, which facilitates parallel processing of image data. Neither cable type supports powering the camera over the cable.

To control the camera, each cable type provides a dedicated uplink channel from the frame grabber to the camera at $300\,\text{MB}\,\text{s}^{-1}$ (C2, F1) or $1200\,\text{MB}\,\text{s}^{-1}$ (F2). The downlink channel from the camera to the frame grabber is shared with the image data.

At the physical and data link layers, Camera Link HS defines two different protocols: the so-called M- and X-protocols. The M-protocol is used with C2 and F1 connectors, the X-protocol with F2 connectors. Both protocols provide error correction (either directly in the X-protocol or through error detection and a resend mechanism in the M-protocol). In addition to the connector type and the protocol, a further option is the num-

于 InfiniBand 网络技术，使用 SFF-8470 连接器，最多可以有 8 个数据通道，其中一个用于摄像机控制。每个通道支持 300 MB/s 数据，因此 C2 最高速率可到 2100 MB/s。F1 和 F2 分别基于 SFP 和 SFP+ 网络收发技术，长度可达 300 m 以上。F1 速率为 300 MB/s，F2 速率为 1200 MB/s，Camera Link HS 每个摄像机最多支持 8 根电缆，因此 C2、F1、F2 速率可达 16 800MB/s、2400 MB/s 和 9600 MB/s。数据可以被传输到不同的采集卡或者不同计算机上进行并行处理，三种都不支持通过电缆供电。

每种电缆都有专门用于控制摄像机的从采集卡到摄像机的上行链路，C2 和 F1 是 300 MB/s，F2 是 1200 MB/s。从摄像机到采集下行链路是与图像数据共享的。

Camera Link HS 在物理和数据链路层定义了 M 协议和 X 协议两个不同的协议。M 协议和 C2 与 F1 连接器一起使用，X 协议和 F2 连接器一起使用。两个协议都提供纠错功能，在 X 协议中是直接纠错，在 M 协议中是发现错误重传机制。除了连接器和协议，另外的选项是每根电缆上的数据通道数量，比如 C2 电缆支持

ber of data lanes per cable (for example, the C2 cable supports between 1 and 7 data lanes). Furthermore, the number of connectors can also be chosen. Cameras and frame grabbers with different connectors or different protocols are incompatible. Using a frame grabber with fewer connectors or fewer lanes per connector than the camera is compatible, but will prevent the peak data rate from being achieved. The user who wishes to connect a camera to a frame grabber must ensure that the two devices are compatible. For this purpose, Camera Link HS defines a naming convention that should be used to identify a product's capabilities. For example, "C2,7M1" designates a camera or frame grabber with a C2 connector that uses the M-protocol with up to 7 data lanes and a single command channel. In contrast, "Quad-F1,1M1" designates a device with four F1 connectors that use the M-protocol with one lane each and a single command channel each.

At the protocol layer, Camera Link HS defines a message-based protocol that uses several different message types with different priorities. Higher priority messages can interrupt lower priority messages to ensure low-latency performance. We will discuss the relevant message types in the order of their priority, highest priority first. Pulse messages are used to trigger and control the acquisition of images. Acknowledge messages are used, for example, to request a resending of image data if a transmission error has been detected. General-purpose input/output (GPIO) messages can be used to control I/O channels on the camera or frame grabber, e.g., to control strobe lights. Camera Link HS supports 16 bidirectional GPIO signals. Video messages are used to transmit video

1~7个数据通道。另外连接器的数量是可选的,摄像机和采集卡如果连接器不同或者协议不同是不兼容的。采集卡的连接器数量或者通道数量少于摄像机的是可以的,就是会影响能达到的最高速率。用户在使用摄像机和采集卡时需要确定二者是兼容的。为此 Camera Link HS 规定了命名约定。比如一个使用 C2 连接器、最多 7 个通道 M 协议及一个命令通道的摄像机或者采集卡叫做 "C2,7M1"。而一个 4 个 F1 连接器、每个连接器一个通道 M 协议、一个命令通道的设备命名为 "Quad-F1,1M1"。

Camera Link HS 在协议层规定了消息基协议,对于不同优先级使用几种不同的消息类型。为保证延迟小,高优先级的消息可以打断低优先级的消息。我们会按照优先级高低来讨论相应的信息类型。脉冲信息触发和控制图像采集。确认信息用于在发现传输错误时请求重传。GPIO 信息可用于控制摄像机或者采集卡上面的 I/O 通道,比如控制闪光灯。Camera Link HS 支持 16 个 GPIO 信号。视频消息用于从摄像机向采集卡传输视频数据。控制信息通过内存读、写改变寄存器内容来控制摄像机。Camera Link HS 是通过 GenICam 和 GenICam GenCP 实现

data from the camera to the frame grabber. The control of the camera is through register access (memory reads and writes) using command messages. Camera Link HS uses the GenICam and GenICam GenCP standards for this purpose (see Section 2.4.3.1). This means that the camera must provide an extensible markup language (XML) file that describes the access to the registers. The names of the registers must follow the GenICam standard features naming convention (SFNC) and the XML file must be compatible with GenICam GenApi.

The Camera Link HS specification also defines the image data that can be transmitted. The pixel data includes monochrome, raw, Bayer, BGR (but not RGB), and individual BGR channels. BGR images can be transmitted interleaved or planar (i.e., as separate single-channel images). The supported bit depths for each format are 8, 10, 12, 14, and 16 bits per pixel (channel). Bit depths that are not multiples of 8 are transmitted in a packed fashion to maximize the data throughput. Camera Link HS supports the transmission of rectangular regions of interest of an image.

2.4.2.3 CoaXPress

CoaXPress (often abbreviated as CXP) is another approach to achieve higher data rates than those possible with Camera Link. The first version of the standard was released in 2010. The current version is 1.1.1 (JIIA CXP-001-2015). One of the design goals of CoaXPress was to use standard coaxial (coax) cables, which are also used for analog video transmission, for high-speed digital video transmission. This means that existing cabling can be reused if an analog machine vision

的（见 2.4.3.1 节）。摄像机需要提供可扩展标记语言文件（XML）来使用寄存器，寄存器的命名必须遵从 GenICam 标准特征命名规范（SFNC），XML 文件必须兼容 GenICam GenApi。

Camera Link HS 还定义了包括黑白、Raw、Bayer、BGR（不是 RGB）以及 BGR 独立通道的图像数据传输。BGR 图像可以通过交错或者独立的一个通道单独传输。像素数据深度可以是 8、10、12 和 16 位。当数据深度不是 8 的倍数时是以打包的形式传输的，这样可以使数据吞吐量最大化。Camera Link HS 支持矩形感兴趣区域。

2.4.2.3 CoaXPress

CoaXPress（简写为 CXP）是比 Camera Link 数据量高的另外一种协议。第一版于 2010 年发布，目前是 1.1.1 版本（JIIA CXP-001-2015）。这个协议的目标之一就是使用用于传输模拟视频的同轴电缆来进行高速数据传输。这样当把模拟机器视觉系统升级到数字视觉系统时，原电缆可以继续使用，如果是新的系统，同轴电缆的优点是非常便宜。CoaXPress 定义

system is upgraded to digital video. If a new machine vision system is being built, one advantage is that coax cables are very cheap. CoaXPress specifies that the widely used BNC connectors and the smaller DIN 1.0/2.3 connectors can be used. Multiple DIN connectors can also be combined into a multiway connector. Like Camera Link and Camera Link HS, CoaXPress requires a frame grabber.

CoaXPress can transmit data at rates of $1.25\,\text{Gb}\,\text{s}^{-1}$ (CXP-1), $2.5\,\text{Gb}\,\text{s}^{-1}$ (CXP-2), $3.125\,\text{Gb}\,\text{s}^{-1}$ (CXP-3), $5.0\,\text{Gb}\,\text{s}^{-1}$ (CXP-5), and $6.25\,\text{Gb}\,\text{s}^{-1}$ (CXP-6). Because of the encoding that is used on the physical medium, this translates to data rates of roughly 120–$600\,\text{MB}\,\text{s}^{-1}$ on a single cable. To further increase the throughput, more than one coax cable can be used. CoaXPress does not limit the number of cables. For example, if four cables are used at CXP-6 speed, a data rate of $2400\,\text{MB}\,\text{s}^{-1}$ can be achieved. Unlike Camera Link HS, CoaXPress currently does not support sending the data to multiple frame grabbers.

The maximum cable length depends on the speed at which it is operated. With CXP-1, the maximum length is more than $200\,\text{m}$, while for CXP-6 it is about $25\,\text{m}$. The camera can be powered over the coax cable.

To control the camera, CoaXPress uses a dedicated uplink channel from the frame grabber to the camera at $20.8\,\text{Mb}\,\text{s}^{-1}$. This channel uses part of the bandwidth of the coax cable. Optionally, the bandwidth of the uplink channel can be increased to $6.25\,\text{Gb}\,\text{s}^{-1}$ with a dedicated coax cable. Such a high-speed uplink channel might be required, for example, for line scan cameras for which each line is triggered individually. The downlink channel from the camera to the frame grabber is shared with the image data.

可以使用常用的 BNC 连接器或者更小的 DIN 1.0/2.3 连接器。多个 DIN 可以组合成多路连接器。与 Camera Link 和 Camera Link HS 一样，CoaXPress 也需要采集卡。

CoaXPress 支持以下几种数据传输速率：1.25 Gb/s（CXP-1）、2.5 Gb/s（CXP-2）、3.125 Gb/s（CXP-3）、5.0 Gb/s（CXP-5）和 6.25 Gb/s（CXP-6）。因为解码是使用物理介质的，所以单根电缆的数据传输速率大致为 120~600 MB/s。为了进一步提高带宽，可以使用更多的线缆。CoaXPress 并没有限制电缆的数量。例如，如果 CXP-6 使用 4 根线缆，则传输速率可达 2400 MB/s。和 Camera Link HS、CoaXPress 不同的是，目前不支持发送数据到多个采集卡。

CoaXPress 传输电缆长度取决于速度高低。对于 CXP-1，最长电缆可以达到 200 m，而 CXP-6 大约 25 m。可以通过同轴电缆供电。

CoaXPress 利用从采集卡到摄像机的一个独立上行通道控制摄像机，这个通道共享同轴电缆的带宽，为 20.8 Mb/s；也可以使用单独的同轴电缆控制摄像机，这时可到 6.25 Gb/s。这种高速的上行通道对于每行都需要触发的线阵摄像机是必要的。从摄像机到采集卡的下行通道用于图像数据。

To transmit data, CoaXPress defines a packet-based protocol that is designed to be immune to single bit errors. Furthermore, the protocol facilitates error detection. Unlike Camera Link HS, CoaXPress does not define a resend mechanism. Therefore, if a transmission error is detected, an error will be returned to the application.

The protocol defines a set of logical channels that carry specific types of data, such as stream data (e.g., images), I/O (e.g., real time triggers), and device control. For this purpose, specific packet types called trigger, stream, and control are defined. Trigger packets have the highest priority, followed by I/O acknowledgment packets (which are used to acknowledge trigger packets). All other packet types have the lowest priority. Higher priority packets are inserted into lower priority packets (effectively interrupting the lower priority packets). This ensures that triggering the camera has low latency, even over the relatively slow $20.8\,\mathrm{Mb\,s^{-1}}$ control channel. Unlike Camera Link HS, CoaXPress currently has no packet type for GPIO.

Video is sent through stream data packets. The supported pixel types follow the GenICam pixel format naming convention (PFNC); see Section 2.4.3.1. However, the bit-level encoding of the pixel formats and their packing over the CoaXPress link is CoaXPress-specific, i.e., different from the encoding defined by the GenICam PFNC. Pixel data is packed as tightly as possible to maximize throughput. CoaXPress supports rectangular as well as arbitrarily shaped regions of interest.

为了传输数据，CoaXPress 定义了基于包的协议，这种设计是为了避免受单一位错误的影响，而且协议便于发现错误。CoaXPress 和 Camera Link HS 不同，并没有重传机制，因此如果发现传输错误，这个错误就会返回给应用程序。

CoaXPress 协议定义了一组逻辑通道，其中包含特定数据类型，例如，流数据（图像）、I/O（实时触发信号）和设备控制，为此，定义了特定数据包类型：触发、流和控制。触发包的优先级最高，然后是 I/O 应答包（用于应答收到触发信号）。其他包优先级较低。高优先级包可以打断低优先级包插入到低优先级包中以保证即使在相对低速的控制通道 20.8 Mb/s 时也可以做到低延时触发摄像机。CoaXPress 目前没有 GPIO 包，这点与 Camera Link HS 不同。

视频信号通过流数据包传输支持的像素类型符合 GenICam 像素格式命名规范（PFNC）。见 2.4.3.1 节。像素格式的位级编码和基于 CoaXPress link 的打包与 GenICam PFNC 不同，是 CoaXPress 定义的。像素数据以最紧凑方式打包这样可以达到最高的速率。CoaXPress 支持矩形和任意形状感兴趣区域。

The camera is controlled through control data packets (which are sent over the control channel). Control packets are sent by the frame grabber and are acknowledged by the camera. The control of the camera is through register access (memory reads and writes); GenICam is used for this purpose (see Section 2.4.3.1). The camera must provide an XML file that describes the access to the registers. The names of the registers must follow the GenICam SFNC, and the XML file must be compatible with GenICam GenApi. Furthermore, camera control and image acquisition are performed through GenICam GenTL (see Section 2.4.3.2).

2.4.2.4　IEEE 1394

IEEE 1394, also known as FireWire, is a standard for a high-speed serial bus system. The original standard IEEE Std 1394-1995 was released in 1995. It defines data rates of 98.304, 196.608, and 393.216 Mb s^{-1}, i.e., 12.288, 24.576, and 49.152 MB s^{-1}. Annex J of the standard stipulates that these data rates should be referred to as 100, 200, and 400 Mb s^{-1}. A six-pin connector is used. The data is transmitted over a cable with two twisted-pair wires used for signals and one wire each for power and ground. Hence, low-power devices can be used without having to use an external power source. A latched version of the connector is also defined, which is important in industrial applications to prevent accidental unplugging of the cable. IEEE Std 1394a-2000 adds various clarifications to the standard and defines a four-pin connector that does not include power. IEEE Std 1394b-2002 defines data rates of 800 and 1600 Mb s^{-1} as well as the architecture for

摄像机是通过控制通道的控制数据包来控制的。控制包由采集卡发出，摄像机应答。摄像机控制通过寄存器访问也就是内存读、写完成。GenICam 就是用于这个目的的，见 2.4.3.1 节。摄像机需要提供与 GenICam GenApi 兼容的 XML 文件来描述如何使用寄存器。这些寄存器的命名需要满足 GenICam 标准特征命名规范（SFNC）。摄像机的控制和图像获取是通过 GenICam GenTL（见 2.4.3.2 节）实现的。

2.4.2.4　IEEE 1394

IEEE 1394 又称作火线，是高速串行总线标准。最初的标准 IEEE Std 1394–1995 颁布于 1995 年。规定了数据量为 98.304 Mb/s、196.608 Mb/s 和 393.216 Mb/s，也就是 12.288 MB/s、24.576 MB/s 和 49.152MB/s。附录 J 标准规定这些数据速率称为 100、200 和 400 Mb/s。使用 6 针连接器。数据传输时使用 2 对双绞线传输信号、一根电缆用于电源、另外一根电缆用于地线。因此，对于低功耗产品可以直接使用电缆提供的电源而不需要额外电源。标准还定义了带锁固的连接器，这对于工业应用非常重要，锁固的电缆防止了电缆线意外脱开。IEEE Std. 1394a–2000 增加了其他分类，包括不含电源的 4 针连接方式 IEEE 1Std 1394b–2002 定义了 800 Mb/s 和 1600 Mb/s 速率以及 3200 Mb/s 结构（实际上是 98.304 Mb/s 的倍数）。

$3200\,\text{Mb}\,\text{s}^{-1}$ (these are again really multiples of $98.304\,\text{Mb}\,\text{s}^{-1}$). A nine-pin connector as well as a fiber optic connector and cable are defined. The latest version of the standard is IEEE Std 1394-2008. It consolidates the earlier versions of the standard. The use of IEEE 1394 cameras is declining since USB and Ethernet are currently the dominant technologies for standard digital interfaces.

A standard that describes a protocol to transmit digital video for consumer applications over IEEE 1394 was released as early as 1998 (IEC 61883-1:1998; IEC 61883-1:2008; IEC 61883-2:1998; IEC 61883-2:2004; IEC 61883-3:1998; IEC 61883-3:2004; IEC 61883-4:1998; IEC 61883-4:2004; IEC 61883-5:1998; IEC 61883-5:2004; IEC 61883-8:2008). This led to the widespread adoption of IEEE 1394 in devices such as camcorders. Since this standard did not address the requirements for machine vision, a specification for the use of IEEE 1394 for industrial cameras was developed by the 1394 Trade Association Instrumentation and Industrial Control Working Group, Digital Camera Sub Working Group (1394 Trade Association, 2008). This standard is commonly referred to as IIDC.

IIDC defines various standard video formats, which define the resolution, frame rate, and pixel data that can be transmitted. The standard resolutions range from 160×120 up to 1600×1200. The frame rates range from 1.875 frames per second up to 240 frames per second. The pixel data includes monochrome (8 and 16 bits per pixel), RGB (8 and 16 bits per channel, i.e., 24 and 48 bits per pixel), YUV in various chrominance compression ratios (4:1:1, 12 bits per pixel; 4:2:2, 16

同时也对 9 针连接器和光缆接口以及电缆作了规定。最新版本 IEEE Std. 1394-2008 整理了以前的版本。目前标准数字接口主要是 USB 和网络接口，IEEE 1394 摄像机使用呈下降趋势。

早在 1998 年就颁布了基于 IEEE 1394 总线的消费级应用的数字视频信号传输协议（IEC 61883-1:1998；IEC 61883-1:2008；IEC 61883-2:1998；IEC 61883-2:2004；IEC 61883:1998；IEC 61883-3:2004；IEC 61883-4:1998；IEC 61883-4:2004；IEC 61883-5:1998；IEC 61883-5:2004；IEC 61883-8:2008）。这导致了便携式摄像机等 IEEE 1394 设备的广泛使用。由于此标准不是针对机器视觉应用的，1394 贸易协会仪器和工业控制小组数字摄像机分组（1394 贸易协会 2008）对 IEEE 1394 针对工业摄像机进行了修订，这一修订后的标准常称作 IIDC。

IIDC 定义了多种视频输出格式，包括分辨率、帧率以及传输的像素数据格式。标准中的分辨率从 160×120 到 1600×1200。帧率从 1.875 帧/秒到 240 帧/秒。像素数据格式包括黑白（每像素 8 位和 16 位），RGB(每通道 8 位和 16 位，也就是每个像素 24 位和 48 位)，不同色度压缩比率的 YUV（4:1:1，每像素 12 位；4:2:2，每像素 16 位；4:4:4，每像素 24 位）以及

bits per pixel; 4:4:4, 24 bits per pixel), as well as raw Bayer images (8 and 16 bits per pixel). Not all combinations of resolution, frame rate, and pixel data are supported. In addition to the standard formats, arbitrary resolutions, including rectangular areas of interest, can be used through a special video format called Format 7.

Furthermore, IIDC standardizes the means to configure and control the settings of the camera, e.g., exposure (called shutter in IIDC), diaphragm (called iris), gain, trigger, trigger delay, and even control of pan-tilt cameras. Cameras are controlled through reading and writing of registers in the camera that correspond to the implemented features. The register layout is fixed and defined by the IIDC standard.

IEEE 1394 defines two modes of data transfer: asynchronous and isochronous. Asynchronous transfer uses data acknowledge packets to ensure that a data packet is received correctly. If necessary, the data packet is sent again. Consequently, asynchronous mode cannot guarantee that a certain bandwidth is available for the signal, and hence it is not used for transmitting digital video signals.

Isochronous data transfer, on the other hand, guarantees a desired bandwidth, but does not ensure that the data is received correctly. In isochronous mode, each device requests a certain bandwidth. One of the devices on the bus, typically the computer to which the IEEE 1394 devices are connected, acts as the cycle master. The cycle master sends a cycle start request every 125 µs. During each cycle, it is guaranteed that each device that has requested a certain isochronous bandwidth can transmit a single

原始 Bayer 图像（每像素 8 位和 16 位）。以上每种分辨率、帧率和数据格式并不是可以随意组合。除标准格式外，任意分辨率以及矩形感兴趣区域在 Format 7 中做了定义。

另外，IIDC 还标准化了设置和控制摄像机方法，比如曝光（IIDC 中称为快门）、光圈（称作 iris）、增益、外触发、外触发延时及摄像机的上、下、左、右旋转等。IIDC 定义和规范了摄像机内部用于控制摄像机的寄存器，通过读、写相应的寄存器来控制摄像机。

IEEE 1394 定义了异步传输和等时传输两种数据传输模式。异步传输使用数据应答包技术来保障每包数据都能正确接收。如果需要，整包数据可以重传。但是异步传输模式不能为数据保证一个固定的传输带宽，因此不用于传输数字视频信号。

等时传输正相反，可以保证所需的传输带宽，但是不能保证接收到的数据都是正确无误的。等时传输模式时，每个设备都申请一定的带宽。总线上一个设备，通常是 IEEE 1394 设备连接的计算机，充当周期控制器，周期控制器每 125µs 发出一个周期开始请求。在每个周期中，保障每个申请固定带宽的设备传输一个数据包。大约总带宽的 80% 用于传输等时数据，其他带宽用于异步数据传输和控

packet. Approximately 80% of the bandwidth can be used for isochronous data; the rest is used for asynchronous and control data. The maximum isochronous packet payload per cycle is 1024 bytes for $100\,\mathrm{Mb\,s^{-1}}$, and is proportional to the bus speed (i.e., it is 4096 bytes for $400\,\mathrm{Mb\,s^{-1}}$). An example of two isochronous devices on the bus is shown in Figure 2.59.

制信号的传输。对于 100 Mb/s，每个周期最大的有效等时数据包长为 1024 字节，并与总线速度成正比。也就是对于 400 Mb/s，等时数据包长最大为 4096 字节。图 2.59 为一个总线上接两台等时传输设备的例子。

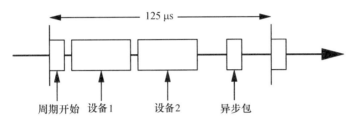

图 2.59　IEEE 1394 的一个数据传输周期。当周期控制器发出周期开始请求时一个周期开始。每一个同步传输设备可以发送一定大小的数据包，剩余周期用于异步数据包传输

IIDC uses asynchronous mode for transmitting control data to and from the camera. The digital video data is sent in isochronous mode by the camera. Since cameras are forced to send their data within the limits of the maximum isochronous packet payload, the maximum data rate that a camera can use is $31.25\,\mathrm{MB\,s^{-1}}$ for a bus speed of $400\,\mathrm{Mb\,s^{-1}}$. Hence, the maximum frame rate supported by a $400\,\mathrm{Mb\,s^{-1}}$ bus is 106 frames per second for a 640×480 image. If images are acquired from multiple cameras at the same time, the maximum frame rate drops accordingly.

IIDC 使用异步模式来传输与摄像机间往来的控制信号。摄像机使用等时传输来发送数字视频信号。由于摄像机受限于最大有效等时数据包长，对于 400 Mb/s 的总线速度，摄像机能传输的最大数据率为 31.25 MB/s。因此，对于 640 像素×480 像素的图像，400 Mb/s 的总线可以支持的最高帧率为 106 帧/秒。如果同时连接多个摄像机采集图像，帧率会相应下降。

2.4.2.5　USB 2.0

The Universal Serial Bus (USB) was originally intended to replace the various serial and parallel ports that were available at the time the USB 1.0 specification was released in 1996. It was designed to support relatively low transfer speeds $\mathrm{MB\,s^{-1}}$

2.4.2.5　USB 2.0

1996 年发布 USB1.0 规范时，通用串行总线（USB）的初衷是取代各种串口和并口。因此，它被设计成支持较低的传输速率：1.5 Mb/s 和 12 Mb/s，也就是 0.1875 MB/s 和

of $1.5\,\mathrm{Mb\,s^{-1}}$ and $12\,\mathrm{Mb\,s^{-1}}$, i.e., 0.1875 and $1.5\,\mathrm{MB\,s^{-1}}$. While this allowed accessing peripheral devices such as keyboards, mice, and mass storage devices at reasonable speeds, accessing scanners was relatively slow, and acquiring images from video devices such as webcams offered only small image sizes and low frame rates. The USB 2.0 specification (USB Implementers Forum, 2000), which was adopted later as the international standard IEC 62680-2-1:2015, supports transfer speeds of up to $480\,\mathrm{Mb\,s^{-1}}$ ($60\,\mathrm{MB\,s^{-1}}$). Of the nominal $480\,\mathrm{Mb\,s^{-1}}$, only $280\,\mathrm{Mb\,s^{-1}}$ ($35\,\mathrm{MB\,s^{-1}}$) are effectively usable due to bus access constraints. The potential throughput and the fact that instead of a frame grabber a widely available communication technology can be used makes USB 2.0 very attractive for machine vision.

USB 2.0 uses four-pin connectors that use one pair of wires for signal transmission and one wire each for power and ground. Hence, low-power devices can be used without having to use an external power source. There is no standard for a connector that can be latched or otherwise fixed to prevent accidental unplugging of the cable. The maximum USB 2.0 cable length is 5 m. However, this can be extended to 30 m by using up to five USB hubs.

While there is a USB video class specification (USB Implementers Forum, 2012), its intention is to standardize access to webcams, digital camcorders, analog video converters, analog and digital television tuners, and still-image cameras that support video streaming. It does not cover the requirements for machine vision. Consequently, manufacturers of USB 2.0 machine vision cam-

1.5 MB/s。此规范访问外围设备，如键盘，鼠标和大容量存储设备的速度一般，访问扫描仪的速度较慢，而对于像网络摄像机这种视频设备只能采集低分辨率和低帧率的图像。USB 2.0（USB Implements Forum, 2000）后来成为国际标准，支持的传输速率可高达 480 Mb/s（60 MB/s），这个 480 Mb/s 只是理论上的，由于总线的限制，实际上只有 280 Mb/s（35 MB/s）可有效利用。潜在的数据吞吐率以及替代图像采集卡有大量可以直接使用的通信技术，使得 USB 2.0 对机器视觉有很大的吸引力。

USB 使用 4 针连接器，一对电缆用于信号传输，另外两根电缆分别用于电源和地线。因此，对于低功耗产品可以不外接电源。没有锁紧插头或其他固定防止电缆意外脱落的规范。USB 2.0 电缆长度为 5 m，可通过使用最多 5 个 USB 中继器达到 30 m。

USB 中有视频类规范（USB Implementers Forum, 2012），目的在于规范网络摄像机、数字便携式摄像机、模拟视频转换器、模拟和数字电视调谐器和支持视频流的静态图像摄像机。但此规范并不包括机器视觉的要求。因此，机器视觉的 USB 摄像机厂商通常使用自己的传输协议和设备

eras typically use proprietary transfer protocols and their own device drivers to transfer images.

USB 2.0 is a polled bus. Each bus has a host controller, typically the USB device in the computer to which the devices are attached, that initiates all data transfers. When a device is attached to the bus, it requests a certain bandwidth from the host controller. The host controller periodically polls all attached devices. The devices can respond with data that they want to send or can indicate that they have no data to transfer. The USB architecture defines four types of data transfers. Control transfers are used to configure devices when they are first attached to the bus. Bulk data transfers typically consist of larger amounts of data, such as data used by printers or scanners. Bulk data transfers use the bandwidth that is left over by the other three transfer types. Interrupt data transfers are limited-latency transfers that typically consist of event notifications, such as characters entered on a keyboard or mouse coordinates. For these three transfer types, data delivery is lossless. Finally, isochronous data transfers can use a pre-negotiated amount of USB bandwidth with a pre-negotiated delivery latency. To achieve the bandwidth requirements, packets can be delivered corrupted or can be lost. No error correction through retries is performed. Video data is typically transferred using isochronous or bulk data transfers.

USB 2.0 divides time into chunks of 125 µs, called microframes. Up to 20% of each microframe is reserved for control transfers. On the other hand, up to 80% of a microframe can be allocated for periodic (isochronous and interrupt) transfers. The remaining bandwidth is used for bulk transfers.

驱动来传输图像。

USB 是轮询总线。每个总线都有一个主控制器发起所有的数据传输，通常为与设备连接的计算机端的 USB 设备。当一个设备接到总线上，就会从主控制器申请一定的带宽。主控制器周期性查询所有设备。连接的设备可以响应需要传输的数据或表示没有数据需要传输。USB 体系结构定义了 4 种数据传输类型。控制传输用于设备首次接到总线时的配置。批量传输通常用于打印机、传真机等大量数据传输，批量传输带宽为其他 3 种传输类型剩余的带宽。中断传输为有限延时传输，通常是事件通知，比如键盘和鼠标的输入。对于以上 3 种传输模式，数据不会有丢失。等时传输可以使用预先商定的 USB 带宽和预先商定的传输延时。为了达到要求的带宽，数据传输包可以有传输错误，也可以被丢掉，而且不会重传数据修正错误。视频数据常用等时传输或批量传输方式。

USB 2.0 将时间分成 125µs 的块，称作微帧。每个微帧有 20% 留作控制传输，最多 80% 可分配作等时和中断周期传输，剩余带宽用于批量传输。

For isochronous data transfers, each end point can request up to three 1024-byte packets of bandwidth per microframe. (An end point is defined by the USB specification as "a uniquely addressable portion of a USB device that is the source or sink of information in a communication flow between the host and device.") Hence, the maximum data rate for an isochronous end point is $192\,\mathrm{Mb\,s^{-1}}$ ($24\,\mathrm{MB\,s^{-1}}$). Consequently, the maximum frame rate a 640×480 camera could support over a single end point with USB 2.0 is 78 frames per second. To work around this limitation, the camera could define multiple end points. Alternatively, if a guaranteed data rate is not important, the camera could try to send more data or the entire data using bulk transfers. Since bulk transfers are not limited to a certain number of packets per microframe, this can result in a higher data rate if there are no other devices that have requested bandwidth on the bus.

对于等时传输,每个端点可以申请每微帧 3 个 1024 字节包长的带宽。在 USB 规范中,每个端点定义为 USB 设备唯一可寻址部分,是主机与设备间通信的信息源或接收源。等时传输一个端点的最大速率为 $192\,\mathrm{Mb/s}$($24\,\mathrm{MB/s}$),因此,对于 640×480 分辨率的摄像机一个 USB 2.0 端点支持的最大帧率为 78 帧/s。为了绕过这一限制,摄像机可以定义多个端点。如果有保障的数据传输率不重要,则摄像机可以使用批量传输,尝试发送更多数据或者整块数据。由于批量传输不受每个微帧特定数量数据包的限制,当总线上没有其他设备申请带宽时,批量传输可以得到较高的数据传输率。

2.4.2.6 USB3 Vision

The USB 3.0 specification was first released in 2008. It adds a new transfer rate, called SuperSpeed, with a nominal data rate of $5\,\mathrm{Gb\,s^{-1}}$ ($625\,\mathrm{MB\,s^{-1}}$). Due to the encoding in the physical layer (called "Gen 1 encoding"), the data rate effectively available for applications is at most $4\,\mathrm{Gb\,s^{-1}}$ ($500\,\mathrm{MB\,s^{-1}}$). The maximum data rate that can be achieved in practice is approximately $450\,\mathrm{MB\,s^{-1}}$. In 2013, the USB 3.1 specification (USB Implementers Forum, 2013) replaced the USB 3.0 specification. It defined a new data rate, called SuperSpeedPlus (often written as SuperSpeed+) with a nominal data rate of $10\,\mathrm{Gb\,s^{-1}}$ ($1250\,\mathrm{MB\,s^{-1}}$). Due to a more efficient

2.4.2.6 USB3 Vision

最早发布于 2008 年的 USB 3.0 增加了被称作"SuperSpeed"的、理论值达 $5\,\mathrm{Gb/s}$($625\,\mathrm{MB/s}$)新的传输速度。由于物理层编码(称作"Gen 1 编码")效率,应用中最快可达 $4\,\mathrm{Gb/s}$($500\,\mathrm{MB/s}$)。实际上最快大约是 $450\,\mathrm{MB/s}$。在 2013 年 USB 3.1 规范(USB Implementers Forum, 2013)取代了 USB 3.0。USB 3.1 增加了被称作"SuperSpeedPlus"(常写作 SuperSpeed+)的新的传输速度,其理论速率为 $10\,\mathrm{Gb/s}$($1250\,\mathrm{MB/s}$)。USB 3.1 提高了物理层编码(称为"Gen 2 编码")效率,应用的有效速率最高达

encoding in the physical layer (called "Gen 2 encoding"), the data rate effectively available for applications is at most $9.6\,\mathrm{Gb\,s^{-1}}(1200\,\mathrm{MB\,s^{-1}})$. The maximum data rate that can be achieved in practice is approximately $1000\,\mathrm{MB\,s^{-1}}$. In contrast to USB 2.0, which is half duplex, USB 3.1 is full duplex, i.e., it offers its data rates simultaneously from the host to the device and vice versa. Furthermore, USB 3.1 no longer uses a polled protocol: devices can asynchronously request service from the host. USB 3.1 devices can be powered over the cable. USB 3.0 and 3.1 cables are very cheap. The maximum specified cable length is around 5 m. Cables with lengths of up to 3 m are readily available. However, there are extenders available on the market that increase the maximum cable length to 100 m through multi-mode fiber optic cables.

All of the above features and the fact that no frame grabber is required make USB 3.1 very attractive for machine vision. Consequently, many camera manufacturers offer USB 3.1 cameras. In many cases, the cameras use the same principles that are also used in USB 2.0 cameras (see Section 2.4.2.5), i.e., they use proprietary protocols to control the camera and stream image data. For this reason, a specification for machine vision cameras, called USB3 Vision (often abbreviated as U3V), was released in 2013. The current version of the standard is 1.0.1 (AIA, 2015). It defines how the camera can be configured and controlled as well as how video data is transmitted. Furthermore, since the USB 3.1 specification does not define a connector that can be locked, the USB3 Vision specification also defines extensions of standard USB 3.1 connectors with locking screws that prevent the camera from being inadvertently unplugged.

9.6 Gb/s（1200 MB/s），实际应用中可以达到的最大的速率大致是 1000 MB/s。USB 2.0 是半双工，USB 3.1 与之不同，是全双工，即从主机到设备可双向同时传输。而且 USB 3.1 不再使用轮询协议，设备可以异步向主机申请。USB 3.1 可以通过电缆供电。USB 3.0 和 USB 3.1 的电缆长度可达 5 m 而且都很便宜。市面上 3 m 以内的电缆很方便可以买到。通过使用多模光纤电缆，最长电缆长度可以达到 100 m。

由于以上特点以及不需要采集卡，USB 3.1 对于机器视觉很有吸引力，因此有很多厂商可以提供 USB 3.1 摄像机。和 USB 2.0 一样 USB 3.1 摄像机也是使用专用协议控制摄像机和图像数据流。为此 2013 年发布了针对机器视觉摄像机的规范 USB3 Vision（常被简写为 U3V）。目前的版本是 1.0.1（AIA, 2015）。U3V 定义了如何设置、控制摄像机以及如何传输图像数据。USB 3.1 没有定义带锁的连接器，为此 USB3 Vision 定义了可以用螺钉固定的连接器以防止摄像机被意外拔掉。

While USB 3.1 is compatible with USB 2.0 at slower speeds through a separate USB 2.0 bus, if data is transmitted at SuperSpeed or SuperSpeedPlus, a completely different physical layer is used. At the protocol layer, however, USB 3.1 is quite similar to USB 2.0. There are still control transfers, bulk data transfers, interrupt data transfers, and isochronous data transfers (see Section 2.4.2.5). The first three types of transfers are guaranteed to be delivered without errors. Since this leads to the fact that higher layer protocols do not need to be concerned about data integrity and, therefore, can be simpler, USB3 Vision exclusively uses bulk data transfers.

On top of the above layers, USB3 Vision defines protocols for identifying cameras on the USB bus, for controlling the device, and for streaming data (video). The protocol for identifying cameras basically specifies how cameras must present themselves in the standard USB device enumeration mechanism. For example, USB3 Vision cameras are assigned to the "miscellaneous" device class with special subclasses that indicate the different protocol endpoints.

To configure and control the camera, the USB3 Vision Control Protocol (U3VCP) is defined. The configuration of the camera is performed through GenICam (see Section 2.4.3.1). The camera must provide an XML file that describes the access to the corresponding camera registers, which must be compatible with GenApi. The names of the registers should follow the GenICam SFNC, and the XML file must be compatible with GenICam GenApi. The control of the camera is performed through GenICam GenCP (see Section 2.4.3.1). The U3VCP defines how to map GenCP control messages to USB bulk transfers.

USB 3.1 在使用 SuperSpeed 和 SuperSpeed+ 传输数据时使用与 USB 2.0 完全不同的物理层，在低速时使用独立的 USB 2.0 总线与 USB 2.0 兼容。USB 3.1 协议层与 USB 2.0 非常类似，同样有控制传输、批量传输、中断传输和等时传输（见 2.4.2.5 节）。前 3 种传输方式可以保障没有错误。高层协议不必考虑数据的完整性，为了简单，USB3 Vision 全部使用批量传输。

在上述层之上，USB3 Vision 定义了用于在 USB 总线上发现摄像机、控制摄像机和传输视频数据的规范。规范中主要指明摄像机在 USB 标准设备枚举机制中如何体现。比如 USB3 Vision 摄像机划归为"其他"设备，并用特殊子类表示不同的协议端点。

USB3 Vision 控制协议（U3VCP）用于设置和控制摄像机。摄像机需要提供兼容 GenApi 来描述如何读、写摄像机相应寄存器的 XML 文件。寄存器命名需要遵从 GenICam SFNC，XML 文件必须兼容 GenICam GenApi。通过 GenICam GenCP 控制摄像机（见 2.4.3.1 节）。U3VCP 确定了如何将 GenCP 的控制信息映射到 USB 的批量传输中。

To transfer images, the USB3 Vision Streaming Protocol (U3VSP) is defined. To transfer an image, a so-called leader is transmitted in a single bulk transfer. The leader contains information about the type of information that is transferred, for example, the pixel format and position of the rectangular region of interest of the image to be transmitted. Then, the image data is transmitted in as many bulk transfers as necessary. Finally, a so-called trailer is transmitted in a single bulk transfer. The trailer can be used, for example, to indicate the actual size of the image that was transmitted, which might be useful for line-scan camera applications in which the number of lines to capture is controlled by triggering and therefore is not known in advance. U3VSP is quite flexible with respect to the type of image data that can be transferred. It supports regular image data as well as so-called chunk data. With chunk data, it is possible to transfer image data as well as additional data, e.g., metadata about the image (such as the exposure time), the region of interest and pixel format of the chunk, or data that was extracted from the image. If chunk data is supported, the chunks should be defined in the XML description file of the camera. The description of the pixel types and their data format must follow the GenICam PFNC (see Section 2.4.3.1).

USB3 Vision 传输流协议（U3VSP）是为图像传输制定的。图像传输时先有单独一个称作头的块传输，头中包含像素格式、被传输的矩形感兴趣区域位置等信息。然后传输的是图像数据，按照需要有不同块数。最后是一个单独称作尾的块传输。尾信息可以用来表示图像实际大小，这个信息对于线阵摄像机应用比较有意义，因为线阵摄像机靠外触发控制采集行数，事先无法知道。U3VSP 灵活支持多种数据传输，既支持一般图像数据也支持被称作数据块数据的传输。对于数据块数据，在传输图像数据的同时也传输图像的元数据（比如曝光时间）、感兴趣区域和像素格式或者从图像中提取出的数据。如支持数据块传输，则块需要在摄像机的 XML 文件中确定。像素类型和数据格式描述必须满足 GenICam PFNC（见 2.4.3.1 节）。

2.4.2.7 GigE Vision

Ethernet was invented in the 1970s (Metcalfe and Boggs, 1976) as a physical layer for local area networks. The original experimental Ethernet provided a data rate of $2.94\,\text{Mb}\,\text{s}^{-1}$. Ethernet was first standardized in 1985 with a data rate of $10\,\text{Mb}\,\text{s}^{-1}$. The current version of the Ethernet

2.4.2.7 GigE Vision

20 世纪 70 年代发明了作为局域网物理层的以太网（Metealife and Boggs,1976)。最初实验性的以太网可以提供 $2.94\,\text{Mb/s}$ 的速率。1985 年以太网第一次被标准化为 $10\,\text{Mb/s}$。目前以太网标准 IEEE Std. 802.3-2015

standard, IEEE Std 802.3-2015, defines data rates from $1\,\mathrm{Mb\,s^{-1}}$ up to $100\,\mathrm{Gb\,s^{-1}}$. Currently, Ethernet at data rates of $1\,\mathrm{Gb\,s^{-1}}$ (Gigabit Ethernet) or higher is in widespread use. These high data rates and the fact that no frame grabber is required make Ethernet very attractive for machine vision. Another attractive feature of Ethernet is that it uses widely available, and therefore cheap, cables and connectors. Cable lengths of up to 100 m are supported via standard CAT-5e, CAT-6a, and CAT-7 copper cables. These cables optionally allow power to be supplied to the camera. Using fiber optic cables, cable lengths of up to 5000 m are possible. However, in this case the camera cannot be powered via the cable.

Ethernet comprises the physical and data link layers of the TCP/IP model (or Internet reference model) for communications and computer network protocol design. As such, it is not used to directly transfer application data. On top of these two layers, there is a network layer, which provides the functional and procedural means of transferring variable-length data sequences from a source to a destination via one or more networks. This is implemented using the Internet Protocol (IP). There are two IP versions: version 4 (IPv4; Postel, 1981a) and version 6 (IPv6; Deering and Hinden, 1998). One additional layer, the transport layer, provides transparent transfer of data between applications, e.g., by segmenting and merging packets as required by the lower layers. The Transmission Control Protocol (TCP) (Postel, 1981b) and User Datagram Protocol (UDP) (Postel, 1980) are probably the best-known protocols since they form the basis of almost all Internet networking software. TCP provides connection-oriented, reliable transport, i.e.,

定义了数据率从 $1\,\mathrm{Mb/s}$ 至 $100\,\mathrm{Gb/s}$。$1\,\mathrm{Gb/s}$（千兆以太网）和更高速以太网目前被广泛使用。高数据率、不需要采集卡使得以太网对于机器视觉很有吸引力，另外由于以太网被广泛使用，因此电缆和连接器非常便宜。标准超五类、六类和七类铜线可以支持到 100 m，使用铜线可以同时供电。如使用光缆，可使电缆长度达 5000 m，但是使用光缆不能提供供电。

以太网由用于通信和计算机网络协议设计的 TCP/IP 模型（或因特网参考模型）的物理层和数据链路层构成。因此不直接传输应用数据。在这两层之上还有网络层用于提供一个或者多个网络把可变长度数据流从源传输到目标。使用因特网协议（IP）来完成。目前 IP 协议有版本 4（IPv4；Postel，1981a）和版本 6（IPv6；Deering and Hinden，1998）两个版本。还有一层传输层，提供应用间数据的透明传输。比如应下一层的要求将数据包分段或合并。传输控制协议（TCP）和自带寻址信息的用户数据报协议（UDP）是广为人知的协议，这两个协议是组成几乎所有因特网软件的基础。TCP 提供面向连接、可靠的传输。数据以发送时同样的顺序完全到达。UDP 正相反，提供非连接、不可靠传输。也就是说数据报可能丢失或重复，也可能以与发送时不一样的顺序到达。UDP 与 TCP 相比

the data will arrive complete and in the same order in which it was sent. UDP, on the other hand, provides connectionless, unreliable transport, i.e., datagrams may be lost or duplicated, or may arrive in a different order from that in which they were sent. UDP requires less overhead than TCP. The final layer in the TCP/IP model is the application layer. This is the layer from which an application actually sends and receives its data. Probably the best-known application-layer protocol is the Hypertext Transfer Protocol (HTTP) (Fielding et al., 1999), which forms the basis of the World Wide Web (WWW).

An application-layer protocol for machine vision cameras was standardized in 2006 under the name GigE Vision (often abbreviated as GEV). The current version of the standard is 2.0.03 (AIA, 2013). Although its name refers to Gigabit Ethernet, the standard explicitly states that it can be applied to lower or higher Ethernet speeds.

While IEEE 1394 and USB are plug-and-play buses, i.e., devices announce their presence on the bus and have a standardized manner to describe themselves, things are not quite as simple for Ethernet. The first hurdle that a GigE Vision camera must overcome when it is connected to the Ethernet is that it must obtain an IP address (GigE Vision currently only supports IPv4). This can be done through the Dynamic Host Configuration Protocol (DHCP) (Alexander and Droms, 1997; Droms, 1997) or through the dynamic configuration of Link-Local Addresses (LLAs) (Cheshire et al., 2005). DHCP requires that the camera's Ethernet MAC address is entered into the DHCP server. Another method of assigning a valid IP address to a camera is called ForceIP. It allows

需要较少的管理。TCP/IP 模型的最后一层是应用层。这层是应用程序实际发送和接收数据层。可能知名度最高的应用层协议是 World Wide Web（WWW）构成基础的超文本传输协议（HTTP）（Fielding et al.，1999）。

2006 年机器视觉摄像机应用层协议得到标准化，称作 GigE Vision（简称 GEV）目前版本是 2.0.03（AIA，2013）。尽管从名字上看是千兆以太网，但是标准明确指出此标准可以用于更低或更高的以太网速度。

IEEE 1394 和 USB 都是即插即用总线，也就是设备可以自动在总线上宣布存在，并且有标准方式来描述设备本身，但是对于以太网就没有这么简单了。第一个障碍就是当摄像机连接到网络时，必须获取 IP 地址（GigE Vision 目前仅支持 IPv4）。GigE Vision 摄像机通过动态主机配置协议（DHCP）（Droms，1997；Alexander and Droms，1997）或链路本地地址动态配置（LLA）（Cheshire et al.,2005）来获取 IP 地址。DHCP 要求摄像机的 MAC 地址输入到 DHCP 服务器。另外一种给摄像机赋予有效 IP 地址的方法叫做强制 IP, 这种方法通过在应用程序中发送特殊包给

changing the IP address of an idle camera from the application by sending a special packet. A final option is to store a fixed IP address in the camera.

A machine vision application can inquire which cameras are connected to the Ethernet through a process called device enumeration. To do so, it must send a special UDP broadcast message and collect the responses by the cameras.

To control the camera, GigE Vision specifies an application-layer protocol called GigE Vision Control Protocol (GVCP). It is based on UDP. Since UDP is unreliable, explicit reliability and error recovery mechanisms are defined in GVCP. For example, the host can request that control messages are explicitly acknowledged by the camera. If no acknowledgment is received, it can retransmit the control message. GVCP establishes a control channel over the connectionless UDP protocol. The control channel is used to control the camera by writing and reading of registers and memory locations in the camera. Each camera must describe its capabilities via an XML file according to the GenICam standard (see Section 2.4.3.1). GigE Vision requires that the register names must adhere to the GenICam SFNC.

GVCP also can be used to create a message channel from the camera to the host. The message channel can be used to transmit events from the camera to the application, e.g., an event when a trigger has been received or when the acquisition of an image has started or finished.

Furthermore, GVCP can be used to create from 1 to 512 stream channels to the camera. The stream channels are used to transfer the actual image data. For this purpose, GigE Vision defines

闲置的摄像机改变 IP 地址。还有一种可能就是在摄像机中存储一个 IP 地址。

机器视觉应用可以通过称作设备枚举的过程查询到与以太网连接的摄像机。在这个过程中向以太网发出特殊的 UDP 广播信息，然后收集摄像机的响应。

GigE Vision 定义了称作 GVCP（GigE Vision 控制协议）的应用层协议来控制摄像机。GVCP 是基于 UDP 的，由于 UDP 不可靠，在 GVCP 中定义了可靠性和错误恢复机制。举个例子，主机可以要求摄像机对于控制信号明确应答，如果收不到应答，就重新发控制信号。GVCP 对非连接 UDP 协议建立了控制通道。控制通道通过读、写摄像机的寄存器和存储器来控制摄像机。每个摄像机必须按照 GenICam 标准（见 2.4.3.1 节）通过 XML 文件来描述其性能。

GVCP 可以用来生成一个从摄像机到主机的信息通道。信息通道可以为摄像机至应用程序传递事件，例如触发接收后或者一幅图像采集开始或完毕时的一个事件。

GVCP 还可以为摄像机产生 1~512 个流通道。流通道用于传递实际的图像数据。为此目的，GigE Vision 定义了称作 GVSP（GigE Vision 流协

a special application-layer protocol called GigE Vision Streaming Protocol (GVSP). It is based on UDP and by default does not use any reliability and error recovery mechanisms. This is done to maximize the data rate that can be transmitted. However, applications can optionally request the camera to resend lost packets.

Images are transferred by sending a data leader packet, multiple data payload packets, and a data trailer packet. The data leader packet contains information about the type of data being transferred, for example, the pixel format and position of the rectangular region of interest of the image to be transmitted (if the payload constitutes an image). Then, the data is transmitted in as many data payload packages as necessary. Finally, a data trailer packet is transmitted. It can be used, for example, to indicate the actual size of the image that was transmitted, which might be useful for line-scan camera applications in which the number of lines to capture is controlled by triggering and therefore is not known in advance. GVSP is quite flexible with respect to the type of data that can be transferred. It supports regular image data as well as so-called chunk data. With chunk data, it is possible to transfer image data as well as additional data, e.g., metadata about the image (such as the exposure time), the region of interest and pixel format of the chunk, or data that was extracted from the image. If chunk data is supported, the layout of the chunks should be defined in the XML description file of the camera.

GVSP also defines the image data that can be transmitted. The specification is effectively a superset of the GenICam PFNC (see Section 2.4.3.1) that defines a few extra pixel formats for reasons of backward compatibility.

议) 的特殊应用层协议。此协议基于 UDP，缺省不包含可靠性和错误恢复机制。这样可以使可传输的数据量最大化。然而，应用程序可以要求摄像机重传丢掉的数据包。

图像传输时发送一个数据头包、若干数据载荷包和一个数据尾包。数据头包含有像素格式、需要传输的矩形感兴趣区域图像位置等。然后是图像数据需要的若干个数据包，最后传输的是数据尾包，图像尾可以用于表明图像传输的实际大小，这对于线阵摄像机的应用有意义，因为线阵摄像机采集的行数是靠外触发控制的，事先无法知道。GVSP 在可以传输的数据类型方面很灵活。既可以支持一般的图像数据也可以支持数据块数据。对于数据块数据，在传输图像数据的同时也传输图像的元数据（比如曝光时间）、感兴趣区域和像素格式或者从图像中提取出的数据。如支持数据块传输，块需要在摄像机的 XML 文件中确定。

GVSP 还定义了可以传输的图像数据。这个规范是 GenICam PFNC （见 2.4.3.1 节）的超集，为了可以向下兼容定义了一些其他的像素格式。

One interesting aspect of the GigE Vision standard is that it provides a mechanism to trigger multiple cameras at the same time through a special GVCP message. This can be convenient because it avoids having to connect each camera to a separate trigger I/O cable. To be able to trigger the cameras simultaneously with low jitter, the clocks of the cameras must be synchronized using the Precision Time Protocol (PTP) (IEEE Std 1588-2008). PTP is an optional feature in GigE Vision, so not all cameras support it.

Another interesting feature is that GVSP packets can be sent to multiple hosts simultaneously through an IP feature called multicasting. This could, for example, be used to facilitate parallel processing on multiple computers.

The data rates that can be achieved with Ethernet vary based on which Ethernet type is used. A data rate of around $920\,\mathrm{Mb\,s^{-1}}$ ($115\,\mathrm{MB\,s^{-1}}$) can be achieved for Gigabit Ethernet. For 10 Gigabit Ethernet, the data rate increases accordingly to around $1100\,\mathrm{MB\,s^{-1}}$. Currently, the highest data rate supported by GigE Vision cameras available on the market is 10 Gigabit Ethernet. The GigE Vision standard specifies that a camera may have up to four network interfaces. If the computer to which the camera is attached also has four network interfaces, the data rate can be increased by a factor of up to 4.

2.4.3 Generic Interfaces

2.4.3.1 GenICam

From Sections 2.4.1 and 2.4.2, it can be seen that there are many different technologies to acquire images from cameras. As discussed above, there

GigE Vision 标准一个有趣的方面是其通过 GVCP 信息提供同时触发多只摄像机的机制。这样可以不必每只摄像机都和一个单独的触发电缆相连。摄像机时钟必须使用精准时间协议（PTP）（IEEE Std. 1588-2008）同步，才能减小同时触发多只摄像机时的抖动。因为 PTP 是 GigE Vision 中可选项。所以不是所有摄像机都支持。

GVSP 包通过 IP 称作多点传送的特征同时传输到多个主机，可以用于多台计算机的并行处理。

以太网可以达到的速率取决于以太网的类型。对于千兆网大约可达 $920\,\mathrm{Mb/s}$（$115\,\mathrm{MB/s}$）。万兆网的速率增加到 $1100\,\mathrm{MB/s}$。目前市场上最高速率的 GigE Vision 摄像机是万兆网。GigE Vision 标准规定每台摄像机最多可以有 4 个网络接口，所以如果摄像机连接的计算机也有 4 个网络接口，则速率可以增加 3 倍。

2.4.3 通用接口

2.4.3.1 GemICam

从 2.4.1 节和 2.4.2 节描述的内容，可以看到有多种不同的技术可以从摄像机获取图像。正如上面所讨论的，

are standards for the physical and protocol layers for many of the technologies. Unfortunately, what is missing is a generic software API layer that facilitates accessing arbitrary cameras. In practice, this means that new software to acquire images must be written whenever the acquisition technology is changed, e.g., if a different type of frame grabber is used or sometimes even if different camera models are used. The latter problem occurs, for example, if USB 2.0 cameras from different manufacturers are used. These cameras will invariably have different software APIs. This problem is especially grave for software libraries, like HALCON, that provide a generic image acquisition interface. Because of the plethora of software APIs for different image acquisition technologies, more than 50 separate image acquisition interfaces have been developed for HALCON in the 20 years since it was first released. Each image acquisition interface is a software library that makes the functionality of a particular image acquisition technology available through HALCON's generic interface.

HALCON has provided a generic image acquisition interface since 1996 with the aim of making it easy for HALCON users to switch from one camera or image acquisition technology to another with little programming effort. However, some problems still remain. For example, without standardization, different camera manufacturers frequently name features with the same semantics differently (for example, the amount of time with which the sensor is exposed to light is sometimes called exposure and sometimes shutter). If the application is switched to a different image acquisition technology, this alone requires a change in

对于多种技术有一些对应的物理层和协议层标准。遗憾的是没有一个通用的软件 API 层使得可以很容易访问任意摄像机。也就是说如果换用不同的采集卡或者摄像机型号不同，也就是说如果采集图像的技术是不同的，就需要重新编写采集软件。举例来说，不同厂商的 USB 2.0 摄像机就需要不同的采集软件。这些摄像机总是使用不同的 API。对于像 HALCON 这种提供通用图像采集接口的软件库这个问题就比较严重。从 HALCON 发布以来的 20 年间，HALCON 开发了 50 多种不同的图像采集接口，以应对不同图像采集技术所对应的众多软件 API。每个图像采集接口都是一个通过 HALOCN 可以支持的特定图像采集技术的软件库。

从 1996 年开始，HALCON 提供通用采集接口的目的就是使用户在更换摄像机或者图像采集技术时基本不用重新编程。可还是存在不少问题。不同的摄像机厂商对于同一个语义有不同叫法，比如摄像机传感器暴露在光线下的时间有时叫做曝光，有时叫做快门。因此如果应用程序更换不同采集技术，尽管使用通用采集接口还是需要改变代码。而且当图像采集技术和摄像机有了新的发展，如果应用中想使用新的特性，必须重新编写代码才行。对于 HALCON 来说，必须

the code, even if a generic image acquisition interface is used. Furthermore, features of image acquisition technologies and cameras might evolve over time. If these features are to be used in an application, often completely new code to make use of this feature must be written. For HALCON's image acquisition interfaces, this meant that a new release of the respective image acquisition interface was necessary whenever a new feature had to be exposed to the user.

These problems were also recognized industry-wide at the time the GigE Vision standard was first developed. A solution was first proposed in 2006 with the GenICam standard. The GenICam standard provides a generic programming interface for configuring and controlling cameras that is independent of the underlying communication technology (called the "transport layer" in GenICam). The GenICam standard has evolved over time. It now consists of the GenICam standard, version 2.1.1 (EMVA, 2016e), the GenICam Standard Features Naming Convention, version 2.3 (GenICam SFNC; EMVA, 2016d), the GenICam Pixel Format Naming Convention, version 2.1 (GenICam PFNC; EMVA, 2016c), the GenICam Generic Control Protocol, version 1.2 (GenICam GenCP; EMVA, 2016b), and a GenICam reference implementation, called GenApi, version 3.0.2 (EMVA, 2017a). As mentioned in Section 2.4.2.1, there is also a specification that describes how the serial communication offered by Camera Link cameras can be integrated into the GenICam framework (GenICam CLProtocol, version 1.1; EMVA, 2011).

GenICam and GenApi only define the means to configure and control the camera. They pro-

发布新的图像采集接口才能让用户用到新的特性。

GigE Vision 首次发布时就意识到这个问题。因此于 2006 年使用 GenICam 标准首次提出解决方案。GenICam 标准提供了独立于底层通信技术（在 GenICam 中称作"透明传输"）的设置和控制摄像机的通用编程接口。GenICam 也在不断进步。目前已经包含了 GenICam 标准版本 2.1.1（EMVA，2016e）、GenICam 标准特性命名规范版本 2.3（GenICam SFNC；EMVA，2016d），GenICam 像素格式命名规范版本 2.1（GenICam PFNC；EMVA 2016e）、GenICam 通用控制协议版本 1.2（GenICam GenCP；EMVA 2016b）和称作 GenApi 的 GenICam 参考实现版本 3.0.2（EMVA，2017a）。在 2.4.2.1 节中也谈到了协议也定义了如何将 Camera Link 摄像机中的串口通信集成到 GenICam 架构中（GenICam CL Protocal，版本 1.1；EMVA，2011）。

GenICam 和 GenApi 并没有提供图像数据和控制信号的传输机制，

vide no mechanism to transport data to and from the camera, neither for control data nor for image data. Consequently, GenApi users must implement the transport layer themselves. For this reason, a standard for the transport layer was published in 2008. This standard is called GenICam GenTL. It will be discussed in Section 2.4.3.2.

The first problem the GenICam standard addresses is how an application can find out what features a camera provides and how to configure and control them. For this purpose, the GenICam standard requires that a camera must provide a file that contains a description of every feature the camera provides and how each feature maps to control registers on the camera. This file must be provided as an XML file (Bray et al., 2008). For example, if the camera's exposure time can be controlled, the camera's XML file will contain an entry that describes the name of the feature (`ExposureTime`), which data type this feature has (the exposure time is a floating-point value), which control register this feature maps to (e.g., `0xacdc`), the length of the register (e.g., 4 bytes), that the access to the register is read/write, and further information that will be discussed below. From this information, the application or software library can infer that to change the exposure time, it must write a floating point value into the register with the address `0xacdc`. To obtain the current exposure time, it can read the value of this register.

The advantage of the above mechanism is that there is an abstraction layer between the actual register layout of the camera and the software layer, which uses the name of a feature to control the camera. Consequently, the register layout of

而是仅仅规定了设置和控制摄像机的方法。所以 GenApi 用户必须自己在传输层实现。传输层规范被称作 GenICam GenTL，发布于 2008 年，将在 2.4.3.2 节中讨论。

GenICam 标准解决的第一件事就是如何发现摄像机的特性并设置和控制这台摄像机。为此 GenICam 标准要求摄像机必须提供描述摄像机每个特征的文件并给出每个特征与摄像机上控制寄存器间的映射关系。这个文件以 XML 文件（Bray et al., 2008）形式提供。举个例子：如果一台摄像机的曝光时间可控，这台摄像机 XML 文件的一个条目会包含摄像机特性的名字（ExposureTime）、这个特性的数据类型（曝光时间是浮点数）、这个特征映射到的控制寄存器（比如 0xacdc）、寄存器长度（比如 4 个字节）、寄存器可读可写等，更多信息在后面章节会讨论。有了上述信息，应用程序或者软件库可以通过向地址为 0xacdc 的寄存器写入一个浮点数来改变曝光时间，通过读取这个寄存器的值可以知道当前的曝光时间。

以上机制的好处是在摄像机的实际寄存器布局和软件层之间有了一个抽象层，这个抽象层使用摄像机某个特性的名字就可以控制摄像机，这样 2 台摄像机可能在寄存器布局上是不

two cameras may differ, but the way in which cameras' features are controlled is generic.

The XML file is typically provided by the camera. It can also be provided in the file system of the host that controls the camera or through an HTTP access. One advantage of this mechanism is that applications that use the GenICam standard no longer need to be reprogrammed if new features are made available in a camera. The camera manufacturer only needs to provide an updated XML file if new features become available, e.g., through an upgrade of the camera's firmware. The most significant advantage, however, is that an application or software library can configure and control cameras from different manufacturers (and using different transport layer technologies, provided a transport layer implementation is available) through a single software API.

In addition to the basic information discussed above, the XML file also provides information that can be used to construct a generic graphical user interface (GUI) to control the camera. For example, in addition to the name of a feature, a tool tip can be specified. This information can be used to display a tool tip in a GUI to provide the user with a short documentation of the feature. Furthermore, each feature is assigned to a feature category. This can be used to group features in the GUI into feature sets that are thematically related. A feature also has a certain visibility, which can be "beginner" "expert" "guru" and "invisible". This can be used in the GUI to make certain features available based on the user's experience level. As mentioned above, each feature also has an interface type. This allows maping of particular data types to certain GUI elements.

同的，但是摄像机特性的控制方式是通用的。

XML 文件通常是摄像机提供的。也可以由控制摄像机的主机文件系统或者通过 HTTP 提供。这种机制的好处是如果摄像机添加了新特性，则使用 GenICam 标准的应用程序不需要重新编程，摄像机厂商通过升级摄像机固件来更新 XML 文件就可以了。然而，最大的好处在于一个应用或者软件包可以统一使用一个软件 API 来设置和控制不同厂商的使用不同传输层技术的摄像机。

XML 文件除了上述基本信息外，还提供了用于构建通用图形界面 GUI 的信息来控制摄像机。比如在 GUI 中可以提供特性名称和特性的工具提示，给用户提供每个特性的简短文字说明，这些说明会在工具提示中显示出来。另外还可以给特性分类，在 GUI 中把主题相关的特性放在一组。每个特性都有一个特定的名字，可以是"新手""行家""专家"和"不可见"，这样在 GUI 中可以针对不同水平的用户开放不同的特性。每种特性都有类型，因此特定的数据类型可以映射到对应的 GUI 元素比如整数和浮点数可以对应滚动条，字符串对应编辑栏，枚举对应下拉框，布尔值对应复选框，命令类型对应命令按键等。GenICam 标准提供了给不同特性之

For example, integers and floating-point values can be mapped to sliders, strings to edit boxes, enumerations to drop-down boxes, Boolean values to check boxes, commands to command buttons, etc. To support GUI elements like sliders, the GenICam standard provides mechanisms to model dependencies between features. For example, the XML file might model that the camera has a feature called `Gain`, whose range depends on two other features that provide the minimum and maximum of the feature's range, e.g., `GainMin` and `GainMax`. This can be used to adjust a slider's range. The GenICam standard also allows modeling much more complex dependencies by allowing expressions between different features to be described in an XML file. For example, value ranges may depend on multiple features (e.g., the width of the chosen AOI might depend on the physical width of the sensor and on the column coordinate of the upper left corner of the AOI) or the access mode of certain features can be set to "not available" depending on the values of other features (implying that they can be grayed out in the GUI).

Figure 2.60 shows how GenICam can be used in practice to construct a generic GUI. HALCON's generic image acquisition interface supports a mechanism for all image acquisition interfaces to describe themselves, in particular, their features, feature types, feature ranges, etc. It maps the GenICam mechanism discussed above to this generic mechanism. This is used in HDevelop, HALCON's integrated development environment (IDE), for example, to construct a generic image acquisition assistant that can be used to configure arbitrary image acquisition devices. Figure 2.60 shows HDevelop's image acquisition assistant

间建立联系的机制，这样就可以支持像滚动条这类的 GUI 元素了。比如，摄像机称作增益（Gain）的特性，其范围可以被 XML 文件对应到另外两个特性，这两个特性就是 GainMin 和 GainMax，分别是增益的最小和最大值，用于控制滚动条的范围。GenICam 标准通过允许在 XML 文件中使用不同特性间的描述，可以对更复杂的相关性建模，比如，一个值的范围可能和多个特性相关（像感兴趣区域的宽度取决于多个特性，包括图像传感器的物理宽度、感兴趣区域左上角的坐标）或者有些特性是否可用有时取决于其他特性（意味着在 GUI 上是灰色不可用的）。

图 2.60 显示了应用程序如何利用 GenICam 构建通用 GUI。HALCON 通用图像采集接口提供各种采集设备描述其特性、特性类型、特性范围等参数的机制，把上述 GenICam 机制映射到通用机制。在 HALCON 集成开发环境（IDE）中，HDevelop 构建了用于控制各种图像采集设备的通用图像采集助手。图 2.60 显示了用于设置一个 GigE Vision 摄像机的图像采集助手。在 GenICam 内部 GenApi 用于设置 GUI。

being used to configure a GigE Vision camera. Internally, the GenICam standard, in particular, GenApi, is used to configure the GUI.

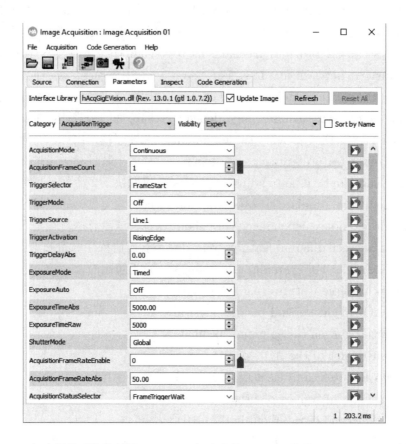

图 2.60　HDevelop 中图像采集是利用 GenICam 标准中的 GenApi 来设置 GigE Vision 摄像机的。如文中所述，摄像机的特性对应 GUI 的元素比如滚动条、下拉框等。摄像机特性通过 GUI 的分类分组，用户水平可以通过 GUI 的可见性来选择

To prevent different camera manufacturers from naming features with identical semantics differently, the GenICam SFNC defines the names of more than 500 standard camera features. Furthermore, it groups the features into more than 20 categories and defines the visibility (the user's experience level) for each feature. The 20 categories are very comprehensive and include device control, image format control, acquisition control, digital I/O control, event control, transfer con-

GenICam SFNC 定义了超过 500 种摄像机特性的名字以避免不同的摄像机厂商对于完全相同的特性使用不同的名字。这些特性又被分成 20 多类，并确定了每个特性适合的用户水平。这 20 多个类非常复杂，包括设备控制、图像格式控制、采集控制、数字 I/O 控制、事件控制、传输控制、三维扫描控制等。如果一个摄像机是 GenICam 兼容的，GenICam 要求摄

trol, and 3D scan control, to name just a few. The GenICam standard requires that if a camera implements a feature and the camera is to be GenICam-compliant, the feature must follow the GenICam SFNC.

In addition to the feature names, the GenICam SFNC defines the semantics of the features. For example, the image format control category contains a precise description of how a rectangular AOI can be defined, what parameters are used to control it, and how they affect the AOI. As a further example, the acquisition control category describes the semantics of different image acquisition modes (e.g., continuous, single-frame, multi-frame, software-triggered, hardware-triggered, etc.). As a final example, the 3D scan control defines in detail how to configure and control 3D image acquisition devices. For more information on 3D image acquisition, refer to Section 2.5.

One remaining issue is the definition of the pixel formats that a camera supports. The GenICam SFNC defines various features that allow specification of the pixel format of an image. For example, there is a feature `PixelFormat` in the category `ImageFormatControl` and a feature `ChunkPixelFormat` in the category `ChunkDataControl`. The standardization of the pixel formats is addressed by the GenICam PFNC. It specifies a general scheme as to how pixel formats are named and in what bit layout the pixel data is delivered to the host, i.e., the transport layer might use a different low-level bit encoding as long as the data is delivered to the host in the bit encoding specified by the PFNC.

像机的特性必须符合 GenICam SFNC GenICam 才能实现此特性。

GenICam SFNC 除了特性的名字以外还定义了特性的语义。比如图像格式控制分类包含了如何定义长方形感兴趣区域精准的描述。采集控制分类描述了不同采集模式的语义，如连续采集、单帧、多帧、软触发、硬触发等。三维扫描控制定义了如何设置和控制三维采集设备。三维采集设备详细信息请见 2.5 节。

摄像机支持的像素格式也是由 GenICam SFNC 定义的，GenICam 定义了多种特性来规定图像的像素格式。比如在 ImageFormatControl 分类中有 PixelFormat 特性，在 ChunkDataControl 分类中有 ChunkPixelFormat 特性。像素格式标准化是通过像素格式命名规范（GemICam PFNC）实现的，这个规范规定了一些通用的方案，例如像素格式如何命名、像素数据以何种位布局传输到主机的方案等，也就是说当像素数据按照像素格式命名规范（PFNC）的位编码传输到主机时，在传输层可以使用不同的低层的位编码。

The PFNC covers 2D images as well as 3D image data. The pixel format names are generally constructed from five fields that are merged into one string: component (and, optionally, component location), number of bits, data type (optional), packing (optional), and an interface-specific string (optional). The component string can be, for example, `Mono` (for monochromatic images), `RGB` (for RGB color images), `Coord3D_ABC` (for 3D data), `BayerRG` (for raw Bayer images with the first two pixels in the top left corner being R and G pixels), etc. The number of bits can be 1, 2, 4, 5, 6, 8, 10, 12, 14, and 16 for integer data types and 32 and 64 for floating-point data types. The data type string can be left empty or can be set to `u` to indicate unsigned data. It can also be set to `s`, indicating signed data, and to `f`, indicating floating point data. For example, `Mono8` indicates an unsigned 8-bit monochrome image, `Mono8s` a signed 8-bit monochrome image, and `Coord3D_ABC32f` indicates a 3D coordinate image with 32-bit floating-point components. The packing string can be empty, implying that data must be padded with 0 to the next byte boundary, it can be set to `p` to indicate that the data is packed with no bit left in between components, etc. The packing string can optionally be followed by a number of bits to pack the data into. For example, `Mono10p` indicates a 10-bit monochrome image with components packed tightly, whereas `RGB10p32` indicates that three 10-bit RGB components are packed into 32 bits of data (implying that the remaining 2 bits are padded with 0). With packed data, a different number of bits can be specified for each component. For example, `RGB565p` indicates an RGB color image with 5 bits for the red and blue channels and 6 bits

像素格式命名规范（PFNC）覆盖二维和三维图像。像素格式命名通常由 5 部分组成一个字符串，包含元素（以及可选的元素位置）、位数、数据类型（可选）、打包（可选），以及接口说明（可选）。元素字符串可以是 Mono（对于黑白摄像机）、RGB（对 RGB 彩色图像）、Coord3D_ABC（对三维数据）、BayerRG（对于左上角前两个像素是 RG 的 Bayer 原始数据）等。位数对于整数数据类型可以是 1、2、4、5、6、8、10、12、14 和 16，对于浮点数据类型可以是 32 和 64。数据类型字符串可以为空，也可以设为 u 代表无符号数据，s 代表有符号数据，f 代表浮点数据。比如 Mono8 表示无符号 8 位黑白图像，Mono8s 代表有符号 8 位黑白图像，Coord3D_ABC32f 代表 32 位浮点三维图像。打包字符串可以为空，表示必须与下一个数据间填补 0，设置为 p 时表示各个成分间打包是没有数据留下来。打包字符串后面可以跟着位数，代表打包到多少位，比如 Mono10p，代表 10 位图像数据完全打包，而 RGB10p32 代表 3 个 10 位的 RGB 数据打包到 32 位数据中（也就是说有两位用 0 补充）。对于打包数据每个成分也可以指定为不同位数，比如 RGB565p 代表一个 5 位红和蓝、6 位绿的 RGB 彩色图像被打包为 16 位。对于打包数据，如果需要，由主机负责解压缩以便处理。可选的接口说明设置为 Planar 代表图像每个通道都是独立传输的，比如 RGB10_Planar 代表 RGB 彩色图像分别以 10 位红、绿、蓝三幅独立图像传输。

for the green channel packed into 16 bits. For packed data, it is the responsibility of the host to unpack the data into a representation that is easier to process, if required. The optional interface-specific string can be set to `Planar`, for example. This indicates that each component of a multi-channel image is transmitted as a single-channel image. For example, `RGB10_Planar` indicates that an RGB color image is transmitted in three separate images that contain the 10-bit red, green, and blue channels of the image, respectively.

There are two more parts of the GenICam standard that we will discuss briefly. The GenICam GenCP standard defines a generic packet-based control protocol that can be used to exchange messages between the host and the camera. In addition, it defines the corresponding packet layout. It has been developed based on the GigE Vision GVCP (see Section 2.4.2.7) and is very similar to it. Unfortunately, there are subtle differences that make the GVCP and GenICam GenCP control packets incompatible. For example, register addresses are 32 bits wide in GVCP and 64 bits wide in GenICam GenCP. Like GVCP, GenICam GenCP defines a protocol to read and write registers on the camera and a simple command and acknowledge mechanism that can be used to resend packets in the event that they were lost. In addition, a protocol for sending events from the camera to the host over a separate message channel is defined.

Finally, the GenICam CLProtocol module describes how the GenICam standard can be used based on the Camera Link serial communication mechanism. It requires that an interface software library must be provided by the camera manufac-

接下来简要地讨论一下 GenICam 标准的另外两部分内容。GenICam GenCP 标准规范了基于通用包的控制协议用于主机和摄像机间信息交换。另外还定义了对应的包布局。GenICam GenCP 是基于 GigE Vision GVCP（见 2.4.2.7 节）发展而来，与其非常类似。遗憾的是，由于细微差距，使得 GVCP 和 GenICam GenCP 不兼容。比如 GVCP 寄存器地址是 32 位，而 GenICam GenCP 是 64 位。和 GVCP 一样，GenICam GenCP 定义了摄像机寄存器读写协议，以及简单的命令和应答机制用于数据丢失时数据包重传。另外还定义了从摄像机至主机单独通道的事件传输协议。

GenICam CL Protocol 描述了 GenICam 如何基于 Camera Link 串行通信机制使用。要求摄像机厂商提供与 GenICam 标准兼容的（也就是和 GenApi 兼容）基于虚拟寄存器可

turer that provides a virtual register-based access to the camera's features that is compatible with the GenICam standard (and, consequently, with GenApi). It is the responsibility of this interface library to map the virtual register-based access to the actual commands that are exchanged with the camera over the serial link.

As already mentioned in Section 2.4.2, the GenICam standard has been adopted to different degrees in various image acquisition standards. For Camera Link, the GenICam CLProtocol module is optional. For Camera Link HS, CoaXPress, and USB3 Vision, the GenICam standard (GenApi), the GenICam SFNC, and GenICam GenCP are mandatory. For GigE Vision, the GenICam standard (GenApi) and the GenICam SFNC are mandatory. Note that the GenICam PFNC is mandatory if the GenICam SFNC is mandatory.

2.4.3.2 GenICam GenTL

As discussed in the previous section, the GenICam standard and GenApi only define a mechanism and API to configure and control cameras. They offer no solution for transporting data to and from the camera. Users of the GenICam standard and GenApi must implement the transport layer themselves.

This problem was addressed in 2008 with the release of the GenICam GenTL standard. The current version of the standard is 1.5 (EMVA, 2015). Like GenICam, the GenICam GenTL standard has a companion standard, the GenICam GenTL SFNC, currently at version 1.1.1 (EMVA, 2017b), that standardizes the names and semantics of features that are used to control the GenICam GenTL interface itself (i.e., not the camera).

The GenICam GenTL standard defines a software API that offers a transport layer abstraction that is independent of the actual underlying transport layer (such as Camera Link, Camera Link HS, CoaXPress, USB3 Vision, GigE Vision, or even a transport layer that is not based on any standard). Therefore, cameras can be accessed and images can be acquired in a manner that is independent of the actual underlying transport layer technology. This has the advantage that applications that use GenICam GenTL can easily switch from one camera model to another, no matter which transport layer technology the cameras use.

To implement GenICam GenTL, a so-called GenTL producer must be provided, either by the camera manufacturer or by an independent software vendor. The GenTL producer is a software library that implements the GenICam GenTL specification. It is implemented on top of the actual transport layer. The GenTL producer enables an application or software library (collectively called GenTL consumer in the GenICam GenTL standard) to enumerate cameras, access camera registers, stream image data from a camera to the host, and deliver asynchronous events from the camera to the host. GenICam GenTL is mandatory for CoaXPress (see Section 2.4.2.3).

GenICam GenTL uses a layered architecture of modules to provide its functionality. The system module provides an abstraction for different transport layer interfaces that are present in a host. For example, the host could have a Camera Link HS interface and a GigE vision interface. The system module can be used, for example, to enumerate all available interfaces and to instantiate a particular interface.

GenICam GenTL 标准定义了不依赖于实际的传输层（像 Camera Link，Camera Link HS，CoaXPress，USB3 Vision，GigE Vision 或者非标传输层）的抽象传输层软件 API。这样可以不依赖于实际的传输层技术就可以访问摄像机并且采集图像。这样做的好处就是摄像机无论使用哪一个传输层技术，使用 GenICam GenTL 的应用可以很容易从一款摄像机型号切换到另外型号。

为了实现 GenICam GenTL，需要由摄像机厂商或者独立软件厂家提供一个 GenTL producer（实现者）。GenTL producer（实现者）是在实际的传输层之上实现 GenICam GenTL 的软件库函数。GenTL producer（实现者）使得应用或者软件库（在 GenICam GenTL 标准中统称为 GenTL 用户）可以枚举摄像机、使用摄像机寄存器、将摄像机中图像数据流向主机、把异步事件从摄像机传输到主机。对于 CoaXPress（见 2.4.2.3 节）GenICam GenTL 是强制性的。

GenICam GenTL 的功能是利用模块分层体系结构实现的。系统模块在主机中为不同传输层接口提供抽象概念。比如主机可以具有 Camera Link 接口和 GigE Vision 接口。系统模块可以枚举所有可用的接口和列举一个特定接口。

The interface module provides an abstraction of a particular physical interface that is available on the host. For example, for a Camera Link HS interface, this would be a particular frame grabber, while for a GigE Vision interface, this would be a particular network interface. The GenICam GenTL standard requires that an interface represents a single transport layer technology. The interface module can be used, for example, to enumerate all devices that are available on an interface (i.e., all cameras that are connected to a particular interface) and to instantiate a particular device.

The device module provides a proxy for a remote device. This is an abstraction of a particular device, typically a camera. It enables the control and configuration of device-specific parameters that cannot be performed on the remote device (an example is discussed below). It also can be used to enable communication with the device, to enumerate all available data streams, and to instantiate a particular data stream. (As discussed in the previous sections, many transport layer standards allow a camera to provide multiple, logically independent data streams from the camera to the host.)

The data stream module provides an abstraction for a particular data stream from the camera to the host. It provides the actual acquisition engine and maintains the internal buffer pool.

Finally, the buffer module provides an abstraction for a particular buffer into which the data from the data stream is delivered. To enable streaming of data, at least one buffer must be registered with a data stream module instance.

接口模块提供主机端可用的每种物理接口的抽象概念.比如对于 Camera Link HS 接口是个特定的接口卡,对于 GigE Vision 接口是特定网口。GenICam GenTL 标准要求一种接口代表一个传输层技术。接口模块可以枚举接口上面所有设备和列举一个特定设备。

设备模块提供远程设备的替代,这是某种设备特别是摄像机的抽象概念,使得可以控制和设置针对设备的参数,本来是无法在远程设备上实现的(下面会举例说明)。还可以用于和设备通信、枚举所有可用数据流、列举某个数据流。(前面讨论过,许多传输层标准允许摄像机提供给主机多个逻辑上不相关的数据流。)

数据流模块提供了从摄像机到主机某个具体数据流的抽象概念,提供实际的采集工具并保证内部缓存寄存器。

缓存模块提供将数据流传输后存放的特定缓存的抽象概念。为了数据流动,至少有一个缓存需要注册为数据流模块实例。

GenICam GenTL provides signaling from the camera to the GenTL consumer, i.e., camera events are transported from the camera and are forwarded to the GenTL consumer. Furthermore, all of the GenICam GenTL modules provide their own signaling capabilities to the GenTL consumer. For example, the GenTL consumer can be configured to receive an event whenever a buffer has been filled with data, e.g., when an image has been acquired.

Additionally, all modules provide means for the GenTL consumer to configure the module. Note that the modules are different from the devices. The configuration of a module is an operation that is performed on the host, while the configuration of a device is an operation that requires communication with the device through the transport layer. For example, the data stream module provides a feature `StreamBufferHandlingMode` that allows configuring in what order the buffers are filled and returned to the GenTL consumer. As another example, the device module has a parameter `LinkCommandTimeout` that specifies a timeout for the control channel communication. As a final example, the buffer module offers a parameter `BufferPixelFormat` that defines the pixel format of the image buffer on the host. Note that `BufferPixelFormat` may differ from `PixelFormat` on the camera, which means that the device driver or device (e.g., the frame grabber) will have to convert the data in this case.

The configuration of the GenTL producer is performed through a similar mechanism to the configuration of a device. The GenTL producer provides a set of virtual registers that are described through an XML file that conforms to the

GenICam GenTL 提供从摄像机到 GenTL 用户的信号，也就是摄像机的事件从摄像机传送到 GenTL 用户。而且 GenICam GenTL 所有模块均能提供信号给 GenTL 用户。比如可以在一幅图像采集后也就是等缓存填满数据时 GenTL 用户设置接收到一个事件。

另外所有的模块都给 GenTL 用户提供设置模块的方法。请注意，模块和设备是不同的，设置模块在主机端实现，而设置设备是需要通过传输层通信到设备端。比如数据流模块中的 `StreamBufferHandlingMode` 设置了缓存填充的顺序并返回给 GenTL 用户。设备模块中的 `LinkCommandTimeout` 参数用于确定控制通道通信的超时。缓存模块中的 `BufferPixelFormat` 参数定义了主机中图像缓存的像素格式。请注意 `BufferPixelFormat` 与摄像机的 `PixelFormat` 可能不一样，也就是说在这种情况下设备驱动或者设备本身（比如图像采集卡）需要数据转换。

设置 GenTL 发生器与设置设备的机制类似。GenTL 发生器通过 XML 文件提供符合 GenICam 标准的虚拟寄存器。这些虚拟寄存器必须遵从 GenICam GenTL SFNC。XML

GenICam standard. The virtual registers must follow the GenICam GenTL SFNC. The XML file can be parsed, e.g., with the GenApi library. Access to the GenTL producer's configuration is provided through a virtual port for each module, which allows performing of read and write operations on the virtual registers. To configure the device (camera) itself, the device module provides a port to the camera through which the actual registers of the camera can be read or written via a communication through the actual transport layer. The available register names and addresses on the camera can be obtained through the GenICam standard, typically through the GenApi library. With GenICam GenTL, an application would typically perform the following steps to acquire images from a camera:

- Initialize the GenTL producer.
- Optionally, enumerate all available transport layers.
- Obtain a handle to the desired transport layer (e.g., GigE Vision).
- Optionally, enumerate all available interfaces.
- Obtain a handle to a specific interface (e.g., a particular network interface).
- Optionally, enumerate all cameras that are available through the interface.
- Obtain a handle to a specific camera.
- Open a port to the device for configuration and control.
- Read out the camera's XML file through the port and configure the camera.
- Optionally, enumerate all available data streams of the camera.
- Open a specific data stream from the camera.
- Allocate and queue buffers for the image acquisition.

文件可以被 GenApi 库解析。对应每个模块有虚拟口，这样可以读、写对应的虚拟寄存器以实现对 GenTL 发生器的设置。设备模块给摄像机提供一个口来通过实际的传输层通信读、写摄像机实际寄存器以设置设备（摄像机）。摄像机可用的寄存器名字和地址可通过 GenICam 标准通常是 GenApi 库来获取。借助 GenICam GenTL，一个应用通常经过以下步骤从摄像机获取图像：

- 初始化 GenTL producer（实现者）。
- 枚举所有可用的传输层（可选）。
- 获取需要的传输层（比如 GigE Vision）的句柄。
- 枚举所有可用的接口（可选）。
- 获取特定接口的句柄（比如特定网络接口）。
- 枚举接口上面所有可用的摄像机（可选）。
- 获取某个摄像机的句柄。
- 打开端口来设置和控制设备。
- 通过端口读出摄像机的 XML 文件设置摄像机。
- 枚举摄像机所有可用的数据流。
- 为图像采集分配和排列缓存。

- Start the acquisition.
- Repeat:
  - Wait for events that signal that a buffer has been acquired.
  - De-queue and process the buffer.
  - Re-queue the buffer that has been processed.
- Stop the acquisition.
- Close all open handles.
- Deinitialize the GenTL producer.

2.4.4 Image Acquisition Modes

We conclude this section by looking at the different timing modes with which images can be acquired. We start with probably the simplest mode of all, shown in Figure 2.61. Here, the camera is free-running, i.e., it delivers its images with a fixed frame rate that is not influenced by any external events. To acquire an image, the application issues a command to start the acquisition to the device driver. Since the camera is free-running, the transport layer must wait for the start of the frame. The transport layer can then start to reconstruct the image i that is transmitted by the camera. The camera has exposed the image dur-

- 开始采集。
- 重复：
  - 等待缓存采集的事件信号。
  - 出列并处理缓存。
  - 重新排队已经处理过的缓存。
- 停止采集。
- 关闭所有句柄。
- GenTL producer（实现者）去初始化。

2.4.4 图像采集模式

本节最后讨论一下图像采集不同的模式，首先看一下图 2.61 这种最简单的模式。摄像机处于自由采集模式也就是摄像机不受外触发信号影响而按固定帧率传送图像。应用程序发出指令给设备驱动开始采集一幅图像。由于摄像机处于连续采集模式，传输层必须等待帧开始，然后传输层开始重构摄像机传输的第 i 幅图像，在上一个帧周期中图像曝光。驱动在内存中将图像构成，应用程序就可以处理图像了。第 i 帧图像处理完毕后，采集周期再次开始。因为图像采集和处理

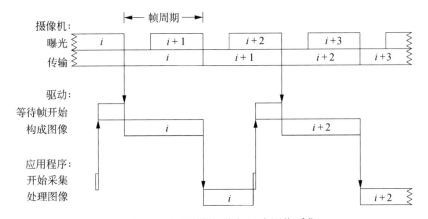

图 2.61 自由采集摄像机同步图像采集

ing the previous frame cycle. After the image has been created in memory by the driver, the application can process the image. After the processing of image i has finished, the acquisition cycle starts again. Since the acquisition and processing must wait for the image to be transferred, we call this acquisition mode synchronous acquisition.

As we can see from Figure 2.61, synchronous acquisition has the big disadvantage that the application spends most of its time waiting for the image acquisition and image transfer to complete. In the best case, when processing the image takes less than the frame period, only every second image can be processed.

Note that synchronous acquisition could also be used with asynchronous reset cameras. This would allow the driver and application to omit the waiting for the frame start. Nevertheless, this would not gain much of an advantage since the application would still have to wait for the frame to be transferred and the image to be created in memory.

All device drivers are capable of delivering the data to memory asynchronously, either through DMA or by using a separate thread. We can make use of this capability by processing the previous image during the acquisition of the current image, as shown in Figure 2.62 for a free-running camera. Here, image i is processed while image $i+1$ is acquired and transferred to memory. This mode has the big advantage that every frame can be processed if the processing time is less than the frame period.

Asynchronous grabbing can also be used with asynchronous reset cameras. However, this would

必须等待图像传送，这种采集模式称作同步采集。

从图 2.61 可以看出，同步采集模式最大缺点是应用程序有许多时间用于等待图像采集和传输的完成。在最好的情况下，也就是图像处理时间少于帧周期时，每隔一帧图像才能得到处理。

同步采集也可用于异步复位摄像机。驱动和应用不必去等待帧开始。然而这样也不会带来太多优势。因为应用程序仍然必须等待帧传输和图像在内存中重建。

所有设备驱动都可以通过 DMA 方式或者单独的线程将数据异步传递到内存。对于自由采集摄像机可在采集当前图像时处理上一幅图像。如图 2.62，第 i 帧图像在 $i+1$ 帧图像采集并传送到内存时进行处理。这种模式的优点是如果处理时间小于帧周期，每一幅图像都可以处理。

异步采集也可以用于异步复位摄像机。然而用途不大，因为在处理速

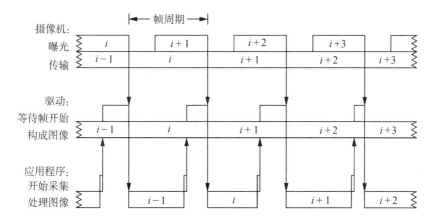

图 2.62 自由采集摄像机异步图像采集

be of little use as such since asynchronous grabbing enables the application to process every frame anyway (at least if the processing is fast enough). Asynchronous reset cameras and asynchronous grabbing are typically used if the image acquisition must be synchronized with some external event. Here, a trigger device, e.g., a proximity sensor or a photoelectric sensor, creates a trigger signal if the object of which the image should be acquired is in the correct location. This acquisition mode is shown in Figure 2.63. The process starts when the application instructs the transport layer to start the acquisition. The transport layer instructs the camera or frame grabber to wait for the trigger. At the same time, the image of the previous acquisition command is returned to the application. If necessary, the transport layer waits for the image to be created completely in memory. After receiving the trigger, the camera resets, exposes the image, and transfers it to the device driver, which reconstructs the image in memory. As in standard asynchronous acquisition mode, the application can process image i while image $i + 1$ is being acquired.

度足够快时，异步采集已经使得应用程序可以处理每帧图像了。当图像采集必须与外部事件同步时通常使用异步复位摄像机和异步采集。这时像接近开关或光电传感器等触发设备在被测物到达应该采集的正确位置时产生触发信号。图 2.63 显示了这种采集模式。当应用程序命令传输层开始采集时过程开始。传输层命令摄像机或图像卡等待触发信号。同时，上次采集的图像回到应用程序中。如果需要，传输层会等待图像在内存中创建完毕。摄像机接收到触发信号，会复位，然后开始曝光，将图像传输到设备驱动，然后至内存重构。同标准的异步采集模式一样，当 $i + 1$ 帧图像在采集时，应用程序处理第 i 帧图像。

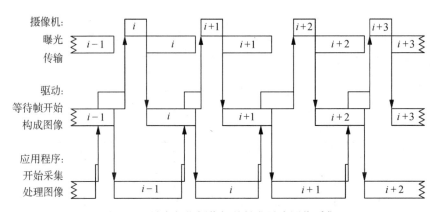

图 2.63 异步复位摄像机外触发异步图像采集

It should be noted that all of the above modes also apply to acquisition from line scan cameras since, as discussed in Section 2.3.1.1, line scan cameras typically are configured to return a 2D image. Nevertheless, here it is often essential that the acquisition does not miss a single line. Since the command to start the acquisition takes a finite amount of time, it is essential to pre-queue the acquisition commands with the driver. This mode is also sometimes necessary for area scan cameras in order to ensure that no trigger signal or frame is missed. This acquisition mode is often called queued acquisition. An alternative is to configure the camera and/or driver into a continuous acquisition mode. Here, the application no longer needs to issue the commands to start the acquisition. Instead, the camera and/or driver automatically prepare for the next acquisition, e.g., through a trigger signal, as soon as the current image has been transferred.

上述模式同样适用于线阵摄像机，2.3.1.1 节中谈到，线阵摄像机通常也是得到二维图像。但是通常情况下必须保证一行都不能丢。由于开始采集的命令需要占用些时间，必须在驱动中事先安排好采集的命令，以保证每个触发信号或每帧都不丢，这种模式有时对于面阵摄像机也是必要的，以保证所有触发信号和每帧图像都不会丢失。这种采集模式常称作排队采集。另外一种办法就是将摄像机和驱动设置为连续采集模式。这时，应用程序不再需要发出开始采集命令。取而代之的是在完成一幅图像传输后摄像机和驱动自动准备好采集下一幅图像。

2.5　3D Image Acquisition Devices

2.5　三维图像采集设备

The acquisition of the 3D shape of objects has become increasingly important in machine vision

在机器视觉应用中物体三维形状的采集变得越来越重要。在本节中，

applications. In this section, we will discuss the most common technologies that are used in the 3D image acquisition devices that are currently available in the machine vision industry. We will also briefly describe the principles by which the 3D reconstruction is performed in the different technologies. The underlying algorithms will be discussed in detail in Section 3.10.

There are two basic principles that are used for 3D reconstruction: triangulation and time-of-flight (TOF) measurement. Triangulation uses the basic mathematical theorem that the coordinates of the third point of a triangle can be computed once the coordinates of the first two points and the two angles from the side connecting the first two points to the rays to the third point are known. In contrast, TOF cameras emit radiation and measure the time until the reflection of the radiation returns to the sensor.

None of the sensors described in this section returns a full 3D reconstruction of an object. Instead, they return a distance to the object for every point in a 2D plane. Consequently, they can reconstruct at most half of an object's surface. Therefore, these sensors sometimes are called 2½D sensors to distinguish them from sensors that can reconstruct the full 3D surface of an object. This typically is done by using multiple sensors, or rotating and moving the object in front of a sensor and merging the resulting 2½D reconstructions. Despite this, we will follow the standard industry practice and call these sensors 3D sensors.

2.5.1 Stereo Sensors

Stereo sensors use the same principle as the HVS:

我们将讨论机器视觉行业主流三维图像采集设备中最常用到的技术。同时，也会简要介绍这些不同技术中三维重构的原理。底层算法会在3.10节详细讨论。

三维重构基于两种基本原理：三角测量法和飞行时间（TOF）测量法。三角测量法基于最基本的数学定理：对于一个三角形，如果知道其中两个顶点的坐标，以及这两个顶点之间的连线到第三个点之间的两个夹角，那么第三个顶点的坐标通过计算可以得到。与之相比，TOF摄像机的原理是：TOF摄像机会发出辐射，然后测量辐射反射回传感器所经过的时间。

本节中所介绍的三维传感器通常不是重构物体完整的三维形状，而是返回某个二维平面内每个点到物体之间的距离。因此这些传感器最多只能重构物体表面的一半。正因为如此，为了和那些能够重构物体完整三维形状的传感器区分，本节中的传感器有时也称为2.5维传感器。通过使用多个传感器，或者使用一个传感器然后旋转和移动物体，最后合并这些重构结果，可以获取物体完整的三维形状。尽管如此，为了遵循标准工业应用惯例，我们仍然称2.5维传感器为三维传感器。

2.5.1 立体视觉传感器

立体视觉传感器采用和人类视觉

the scene is observed with two cameras from different positions (see Figure 2.64). To be able to reconstruct moving objects correctly, it is essential that the cameras are synchronized, i.e., acquire the images at the same time, and, if CMOS cameras are used, that the cameras have a global shutter. As discussed in Section 2.2.5, it might be necessary to use a Scheimpflug configuration, i.e., to tilt the image planes, if the magnification of the lenses is large.

系统（HVS）相同的原理：两台摄像机从不同的位置观察同一场景（见图2.64）。为了能够准确地重构移动的物体，同步摄像机是非常必要的，也就是说，摄像机需要同时采集图像，另外，如果使用 CMOS 摄像机，需要支持全局快门（帧曝光）。如 2.2.5 节所讨论的，如果镜头的放大率很大，有必要使用沙姆结构，也就是说，需要倾斜成像平面。

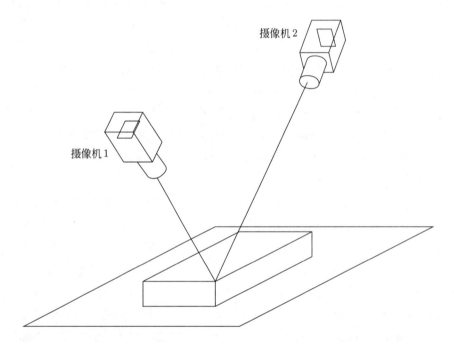

图 2.64　立体视觉传感器由两台摄像机组成，用来观察同一场景。摄像机中的传感器分别用两个矩形表示。为了重构三维空间中的一个点，需要找到两幅图像中的对应点，构建光线，然后在空间中相交。为了更清晰地表示立体视觉传感器的几何结构，图中摄像机间的夹角比较夸张。另外，两台摄像机通常会安装在一个外壳里面

The different viewpoints of the cameras lead to the fact that a point at a certain distance from the cameras is imaged at different positions in the two images, creating a parallax. Parallax is the essential information that enables the 3D reconstruction of a point in the scene.

两台不同视角的摄像机采集同一场景会产生这样的现象：距离摄像机某个距离的某个点会在两台摄像机图像上不同的位置成像，这种现象叫做视差。视差就是重构场景中这个点三维坐标的基本信息。

As discussed in Section 3.10.1.1, the cameras must be calibrated and their relative orientation must be determined to be able to perform the 3D reconstruction. The calibration is typically performed by the sensor manufacturer. The calibration data enables a stereo sensor to triangulate a 3D point. The known side of the triangle is the base of the stereo system, i.e., the line segment that connects the two projection centers (the entrance pupils) of the two cameras. Furthermore, once the interior orientation of the two cameras is known, the cameras are essentially angle measurement devices. Therefore, if a point in each of the two images is identified that corresponds to the same point in the scene, the angles of the rays to the point in the world can be computed. Thus, the 3D point can be triangulated. An equivalent way to describe the 3D reconstruction is the following: once the corresponding points have been identified in the images, their optical rays in space can be computed. The reconstructed 3D point is given by the intersection of the two optical rays.

Note that some points in a scene may be occluded from the point of view of one or both cameras. For example, in Figure 2.64, all points around the bottom edges of the rectangular cuboid are occluded in at least one camera. Obviously, the 3D position of these points cannot be reconstructed. This problem occurs for all triangulation-based sensors.

As described in Section 3.10.1.6, corresponding points are typically determined by matching small rectangular windows in one image with the other image. From the discussion in that section, it can be seen that it is essential that the object contains a sufficient amount of structure or

3.10.1.1 节将介绍在进行三维重构前，必须标定摄像机以及确定摄像机之间的相对位姿。标定过程传感器厂商通常会完成。立体视觉传感器基于这些标定数据来重构三维点云。三角形的已知边称为立体视觉系统的基线，也就是摄像机两个投影中心（入射光瞳）之间的线段。此外，一旦两个摄像机的内参已知，立体摄像机就成为一个角度测量装置。因此，如果两幅图像中的某个点能够确定是场景中的同一个点，那么在空间中到这个点的两条光线之间的角度就可以计算得到。因此，点的三维坐标能够通过三角测量法获得。可以用这种方式来描述三维重构：一旦图像中对应点能够确定，在空间中经过它们的光线也能够计算，则两条光线的交点就是所要重构的三维点。

需要注意的是在某些场景中，有些点从一个摄像机或者两个摄像机的视角来看，会存在遮挡现象。例如，图 2.64 所示，矩形立方体底边附近的所有点，至少在一个摄像机的视野里是被遮挡的。这个问题对于所有基于三角测量原理的传感器来说都会存在。

3.10.1.6 节将介绍一般通过在两幅图像中匹配小的矩形窗口来确定对应点。3.10.1.6 节中也将介绍物体必须包含足够的纹理或者结构特征，否则不能唯一地确定对应点。因为工业领域的许多被测对象缺少纹理特征，

texture. Otherwise, the corresponding points cannot be determined uniquely. Since many industrial objects are texture-less, it is advantageous to project a random texture onto the scene (Besl, 1988). The projector is typically mounted between the cameras. It is sufficient to project a single static pattern onto the scene. An alternative is to project multiple random patterns onto the scene (Zhang et al., 2003; Davis et al., 2005; Schaffer et al., 2010; Wiegmann et al., 2006); this enables a more accurate 3D reconstruction at depth discontinuities since the window size of the stereo correlation algorithms can be reduced. This helps to alleviate the problem that larger window sizes can lead to incorrect reconstructions at depth discontinuities (cf. Section 3.10.1.6). With a sufficiently large number of images, the window size can be reduced to 1×1, which can result in very accurate 3D reconstructions at depth discontinuities (Davis et al., 2005). One disadvantage of using multiple patterns is that the objects must not move. If the algorithm is unable to compensate for the object's motion, the 3D reconstruction will be incorrect (Davis et al., 2005). If the algorithm is able to compensate for the motion, the time to reconstruct the scene will typically be prohibitively long (Zhang et al., 2003).

The stereo sensors that we have described in this section are based on area scan cameras. It is also possible to perform stereo reconstruction using line scan cameras (Calow et al., 2010; Ilchev et al., 2012). Stereo line sensors can achieve a very high resolution 3D reconstruction. However, the object must be moved with respect to the sensor to perform the 3D reconstruction.

因此在场景中投射随机的纹理图案是有帮助的（Besl，1988）。投影仪一般安装在两个摄像机之间，投射单一的静态图案即可。另外一种方法是在场景中投射多个随机图案（Zhang et al., 2003；Davis et al., 2005；Schaffer et al., 2010；Wiegmann et al., 2006）；因为可以减小立体视觉算法的窗口尺寸，这种方法在深度不连续的情况下可以获得更准确的三维重构结果。在深度不连续的情况下更大的窗口尺寸会导致不正确的结果，这个问题也会得到缓解（见 3.10.1.6 节）。如果使用足够多的图像，窗口尺寸可以减小到 1×1，在深度不连续的情况下能够得到非常准确的三维重构结果（Davis et al., 2005）。使用多个图案的缺点是物体一定不能移动。如果重构算法不能补偿物体移动带来的影响，三维重构结果则会不正确（Davis et al., 2005）。如果重构算法能够补偿物体运动的影响，重构场景的时间则会非常长（Zhang et al., 2003）。

本节所介绍的立体视觉传感器是基于面阵摄像机的。当然使用线阵摄像机实现立体重构也是可能的（Calow et al., 2010；Ilchev et al., 2012）。而且立体线阵传感器能够获得分辨率非常高的三维重构结果。然而，物体和传感器必须有相对移动，才能实现三维重构。

2.5.2 Sheet of Light Sensors

Like stereo sensors, sheet of light sensors rely on triangulation to reconstruct the scene in 3D. Compared to stereo sensors, one camera is replaced by a laser projector that projects a laser plane (the sheet of light) onto the objects in the scene (Besl, 1988; Blais, 2004). A typical sheet of light setup is shown in Figure 2.65. This technology often is also called laser triangulation.

2.5.2 片光（激光三角测量）传感器

和立体传感器一样，片光（激光三角测量）传感器也是基于三角测量法重构三维场景。和立体传感器相比，片光（激光三角测量）传感器使用一台激光投影仪代替其中一台摄像机，投影仪在场景中的物体上投射一个激光平面（片光）（Besl, 1988; Blais, 2004）。典型的片光（激光三角测量）结构如图 2.65 所示。片光技术也经常被称为激光三角测量法。

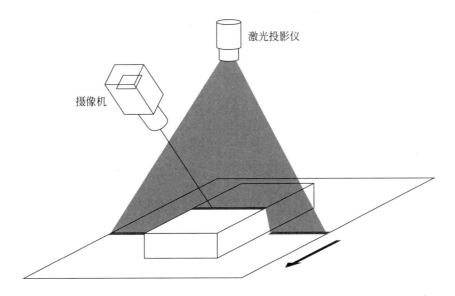

图 2.65　片光（激光三角测量）传感器由激光组成，激光在场景中的物体上投射激光平面（片光，图中灰色部分）。激光在激光平面和物体表面相交的地方形成特征线（图中粗实线部分）。摄像机在不同的视角采集这些特征线。摄像机中的传感器用一个矩形表示，物体到激光投影仪的不同距离在摄像机图像上会产生视差，视差用来重构物体的三维坐标。这个过程通过计算经过图像中激光线上某个点的光线并与三维空间中激光平面相交来实现。物体和片光之间必须有相对位移才能够重构物体形状。为了更清晰地表示片光传感器的几何结构，摄像机和激光投影仪之间的夹角比较夸张。另外，摄像机和激光投影仪通常会安装在一个外壳里面

The laser projector is constructed by sending a collimated laser beam through a cylindrical lens (Besl, 1988), a Powell lens (Powell, 1989), or a

激光投影仪的构造原理是：发射一束平行激光束，通过柱面透镜（Besl, 1988）、Powell 透镜（Powell, 1989）或

raster lens (Connolly, 2010) to fan out the laser beam. Powell and raster lenses have the advantage that they produce a more even brightness across the entire sheet of light than cylindrical lenses. Raster lenses, however, cannot be used for large working distances (Connolly, 2010).

The laser light is scattered at the points where the laser plane intersects the objects in the scene, forming characteristic lines in the scene. A camera captures the lines from a different viewpoint. As shown in Figure 2.65, to maximize the lateral resolution of the sheet of light sensor, the camera's sensor is typically mounted such that its longer side is parallel to the light plane. As discussed in Section 2.2.5, it is advantageous to use a tilt lens, i.e., to tilt the image plane of the camera, to ensure that the laser plane is in focus for the entire measurement range (Besl, 1988; Blais, 2004).

The different distances of the objects to the laser projector produce parallaxes in the camera image. If the camera sensor is aligned as described above, these cause vertical offsets of the laser line in the image. These vertical offsets are used to reconstruct the 3D coordinates of the objects within the sheet of light. This is done by computing the optical ray of a point on the laser line in the image and intersecting it with the laser plane in 3D. As described in Section 3.10.2, this requires that the camera is calibrated, i.e., that its interior orientation is known. Furthermore, the pose of the laser plane with respect to the camera must be calibrated. The calibration is typically performed by the sensor manufacturer.

In the setup in Figure 2.65, the object must be moved relative to the sheet of light to reconstruct its shape. Therefore, the relative motion

者光栅透镜（Connolly，2010）后输出激光束。Powell 透镜、光栅透镜和柱面透镜相比的优点是在整个片光上亮度更均匀，然而不能用于工作距离较长的情况（Connolly，2010）。

激光光束在激光平面和物体表面相交的地方会形成特征线。摄像机在不同的视角能够采集到特征线。如图 2.65 所示，为了能够最大化利用片光传感器的横向分辨率，摄像机传感器安装时长边和光平面保持平行。如 2.2.5 节所描述，使用倾斜镜头，也就是说倾斜摄像机的成像平面确保激光平面在整个测量范围内都能够聚焦是有利的（Besl，1988；Blasi，2004）。

物体到激光投影仪之间的不同距离在摄像机图像上会产生视差。如果摄像机传感器的安装方式如上图所示，激光线在成像时会产生垂直偏移。这些垂直偏移用来重构片光系统中物体的三维坐标。这个过程通过计算经过图像中激光线上某个点的光线并与三维空间中激光平面相交来实现。如 3.10.2 节所描述，三维重构需要标定摄像机，也就是标定内参。除此之外，激光平面与摄像机之间的相对位姿必须标定。这些标定过程通常摄像机厂商会完成。

如图 2.65 所示的结构中，物体和片光之间必须有相对位移，才能够重构物体三维形状。因此，摄像机和物

of the camera and object must be known. Typically, a linear motion is used. To ensure a constant relative motion between subsequent frames, encoders are typically used to trigger the image acquisition. As an alternative setup, the orientation of the sheet of light can be altered continuously to scan an object. This is typically done by projecting the sheet of light onto a rotating planar mirror.

Typically, not all points in the scene can be reconstructed because of shadows and occlusions. As shown in Figure 2.65, the top part of the rectangular cuboid shadows the laser plane to some extent. This leads to the fact that no laser line is visible in a small part of the scene to the right of the cuboid. Furthermore, if the laser plane is located slightly behind the cuboid, the laser line will be visible in the scene, but will be occluded by the cuboid from the point of view of the camera and therefore not visible in the camera image. Consequently, these points cannot be reconstructed.

Shadows and occlusions create a tradeoff in the design of sheet of light sensors. On the one hand, the reconstruction accuracy increases as the angle between the optical axis of the camera and the laser plane increases because this increases the resolution of the parallaxes. On the other hand, the areas that are occluded in the scene become larger as this angle increases.

Apart from the angle of triangulation, another factor that influences the reconstruction accuracy is speckle noise (Dorsch et al., 1994). Speckle noise is created by the mutual interference of different wavefronts of the coherent laser radiation that are reflected by the objects in the

体之间的相对运动必须已知，一般使用直线运动机构。为了保证相邻帧间的相对运动恒定不变，通常使用编码器来触发采集图像。另外一种方式是，通过连续不断地改变片光的方向来扫描物体，一般通过把片光投射在一个旋转平面镜上来实现。

一般情况下，因为阴影或者遮挡，并不是场景中所有的点都能够重构。如图 2.65 所示，矩形立方体的顶部在某种程度上会遮挡激光平面，这会导致立方体右侧的一小部分场景激光线是不可见的。此外，如果激光平面投射到立方体的后面，在场景中激光线是可见的，但对于摄像机视角来说，激光线被立方体所遮挡，因此无法成像到图像上。所以这些点无法重构。

在设计片光（激光三角测量）传感器时必须权衡阴影和遮挡现象。一方面，增加摄像机光轴与激光平面之间的夹角，会增加视差分辨率，重构精度会相应提高；另一方面，随着角度的增加，场景中被遮挡的面积也会增大。

三角测量法中角度是影响精度的一个重要因素，除此之外，另一个因素是斑点噪声（Dorsch et al., 1994）。斑点噪声是由场景中物体反射的相干激光辐射线的不同波阵面的相互干扰产生的。通过使用低相干辐射线的激

scene. Speckle noise can be reduced by using projectors that use radiation of low coherence (i.e., radiation that is made up of different wavelengths), e.g., LED line projectors.

光投影仪可以减少斑点噪声（也就是说辐射线由不同的波长组成），例如，LED 线投影仪。

2.5.3 Structured Light Sensors

Structured light sensors are the third category of sensors that rely on triangulation to reconstruct the scene in 3D. The setup is similar to a stereo sensor. However, one camera is replaced by a projector that projects structured patterns onto the scene, as shown in Figure 2.66. As discussed in Section 2.2.5, it might be necessary to use a Scheimpflug configuration, i.e., to tilt the image plane and the projector device, if the magnification of the lenses is large.

2.5.3 结构光传感器

结构光传感器是第三类基于三角测量原理重构三维场景的传感器，结构和立体视觉传感器类似。然而，不同的是其中一台摄像机用投影仪所替代，投影仪可以在场景中投射结构图案，如图 2.66 所示。如 2.2.5 节所述，如果镜头的放大率很大，有必要使用沙姆结构，也就是说，需要倾斜成像平面。

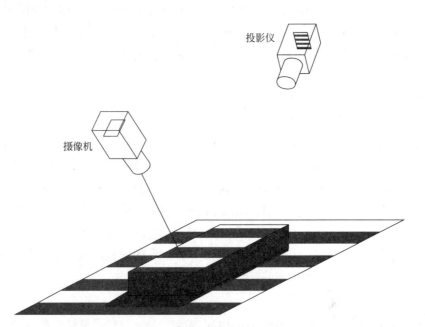

图 2.66 结构光传感器由一台投影仪和一台摄像机组成，投影仪在场景中投射多个不同频率的条纹图案，摄像机用来观察整个场景。摄像机中的传感器和投影仪中的投影器件在图中分别用矩形表示。三维空间中点的重构通过计算摄像机图像中这个点的投影仪列坐标和这个点对应的光线来计算。投影仪列坐标定义了三维空间中一个平面。这个点的三维坐标通过光线和这个平面相交来获得。为了更清晰地表示结构光传感器的几何结构，摄像机和投影仪之间的夹角比较夸张。另外，摄像机和投影仪通常会安装在一个单独的外壳里面

Many types of structured light patterns have been proposed over the years. See Bell et al. (2016); Besl (1988); Blais (2004); Geng (2011); Gorthi and Rastog (2010); Salvi et al. (2010) for reviews of some of the different possibilities. The technology that is dominant in currently available machine vision sensors is to project multiple striped patterns of different frequencies onto the scene. The stripes are designed in such a way that it is possible to identify the column coordinate of the projector that has illuminated a particular scene point easily and accurately (see Section 3.10.3.1 for details of the decoding). The interior and relative orientation of the camera and projector must be calibrated (see Section 3.10.3.2). The calibration is typically performed by the sensor manufacturer. Based on the interior orientation of the camera, the optical ray corresponding to a point in the image can be computed. Furthermore, based on the column coordinate of the projector, the interior orientation of the projector, and the relative orientation of the camera and projector, the 3D plane corresponding to the projector column can be computed. The 3D reconstruction of a point is obtained by the intersection of the optical ray and the plane.

Note that, as for stereo and sheet-of-light sensors, shadows and occlusions are typically present and prevent the reconstruction of some points in the scene. For example, in Figure 2.66, all points around the bottom edges of the rectangular cuboid are either occluded or shadowed and therefore cannot be reconstructed.

2.5.3.1 Pattern Projection

To project the patterns, the technology that is used in video projectors can be used: liquid-

多年来许多不同类型的结构光图案被采用。参见文献（Besl, 1988; Blais, 2004; Gorthi and Rastog, 2010; Salvi et al., 2010; Geng, 2011; Bell et al., 2016）可以回顾一下这些不同的结构光方案。目前在机器视觉传感器所用技术中比较占优势的是在场景中投射多个不同频率的条纹图案。这些条纹图案的设计要求是能够容易且准确地识别出照亮一个特定场景点的投影仪的列坐标（解码细节见 3.10.3.1 节）。摄像机和投影仪的内参及相对位姿必须标定（见 3.10.3.2 节），标定过程通常摄像机厂商会完成。基于摄像机内参，图像中对应某个点的光线能够计算得出。此外，基于投影仪的列坐标，投影仪的内参，摄像机与投影仪的相对位姿，以及对应投影仪某列的三维平面也能够计算得出。最后，通过光线和平面相交可以得到这个点的三维坐标。

需要注意的是，对于立体视觉传感器和结构光（原文是片光，根据上下文应该是结构光）传感器，阴影和遮挡一般都是存在的，并且会阻碍场景中某些点的重构。例如，图 2.66 所示，矩形立方体底部边缘附近的点或是被遮挡或是有阴影，因此不能被重构。

2.5.3.1 图案投影

为了投影图案，通常会用到视频投影仪中所用到的技术：液晶显示器

crystal display (LCD), liquid crystal on silicon (LCOS), and digital light processing (DLP) projectors (Proll *et al.*, 2003; Gorthi and Rastog, 2010; Bell *et al.*, 2016; Van der Jeught and Dirckx, 2016). Since DLP projectors are the dominant technology for structured light sensors, we will describe this technology in more detail (Huang *et al.*, 2003; Bell *et al.*, 2016). At the heart of a DLP projector is a digital micromirror device (DMD). A DMD consists of an array of micromirrors whose orientation can be switched between two states electronically. In one orientation, the light from a light source is reflected through the lens onto the scene (the "on" state). In the other orientation, the light is reflected into a heat sink (the "off" state). Thus, it seems that a DMD only can project binary patterns. However, the micromirrors can be switched very rapidly. Therefore, gray values can be created by switching the mirrors on and off multiple times during a certain period. The relevant period is the exposure time of the camera. The ratio of time that the micromirrors are on during the exposure to the exposure time determines the apparent gray value. Current DMD devices are able to project binary patterns at frequencies of up to 32 000 Hz and 8-bit gray scale patterns at frequencies up to 1900 Hz.

DMDs are manufactured in two configurations, as shown in Figure 2.67: a regular pixel layout in which the micromirrors are aligned with the edges of the DMD and a diamond pixel layout in which the micromirrors are rotated by 45° and every second row is shifted by half a pixel. In each case, the micromirrors rotate around their diagonal. This implies that the light source can be mounted to the left of a diamond pixel array

（LCD），硅基液晶（LCOS）和数字光处理器（DLP）投影仪（Proll *et al.*, 2003; Gorthi and Rastog, 2010; Bell *et al.*, 2016; Van der Jeught and Dirckx, 2016）等。由于在结构光传感器中 DLP 投影仪技术占主导地位，我们将详细介绍这种技术（Huang *et al.*, 2003; Bell *et al.*, 2016）。DLP 投影仪的中心位置是一个数字微镜器件（DMD），DMD 由一排微镜组成，微镜的方向可以在两种状态之间电子切换。其中一个方向，光源发出的光通过镜头反射到场景中（"开"状态），另一个方向，光被反射进散热片（"关"状态）。因此，DMD 投影仪看起来只能够投射二值图案。然而，由于微镜的状态可以快速的切换，在一个特定的时间周期中通过多次切换微镜的开关状态就可以生成灰度值，这个时间周期就是摄像机的曝光时间。在曝光周期中微镜开启的时间和曝光时间的比例决定了灰度值的大小。目前 DMD 设备能够投射频率高达 32 000 Hz 的二值图案和 1900 Hz 的灰阶图案。

DMD 有两种构造方式，如图 2.67 所示：一种是规则的像素布局，微镜和 DMD 边缘对齐；另外一种是菱形像素布局，微镜旋转 45°，每两行像素移动半个像素。两种架构，微镜都是绕对角线旋转，这意味着对于菱形像素阵列 DMD，光源能够安装在左侧，而对于规则像素阵列 DMD，光源必须安装在左上角。因此，菱形像

DMD, whereas it must be mounted in the direction of the upper left corner of a regular pixel array DMD. Therefore, the diamond pixel array layout can lead to smaller projectors. On the other hand, the rotated pixel layout must be taken into account when projecting the patterns, as discussed below.

素阵列布局可以使投影仪更小。另外，当投影图案时，必须考虑旋转的像素布局，下面会讨论。

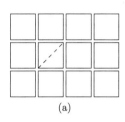
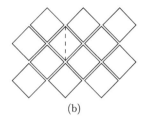

图 2.67　不同的 DMD 布局。(a) 规则的像素阵列布局；(b) 菱形像素阵列布局。微镜旋转的轴线用虚线表示。对于菱形像素布局，光源必须安装在 DMD 左侧。对于规则像素布局，光源需安装在 DMD 左上角位置

2.5.3.2　Gray Codes

One possible set of patterns that permit a very simple decoding is to project binary stripes of different frequencies onto the scene. For example, a pattern with two stripes can be projected (one dark stripe and one bright stripe, each covering one half of the projector). Then, four, eight, ⋯ stripes are projected. If n binary patterns are projected, 2^n different columns in the projector can be identified uniquely. Each pixel in the camera image can be decoded separately. For each of the n images, we only need to determine whether the pixel is bright (on) or dark (off). The decoding returns an n-bit binary code for each pixel, which directly corresponds to a certain column in the projector. One problem with using this n-bit binary code is that adjacent stripes may differ by up to $n-1$ bits. Therefore, if a decoding error of only 1 bit is made, the decoded column may differ by

2.5.3.2　格雷编码

一种使解码过程非常简单的方案是在场景中投射不同频率的二值条纹。例如，可以投射两种条纹的图案（一个暗条纹和一个亮条纹，每个条纹覆盖投影仪的一半范围）。以此类推，可以投射 4 个、8 个，甚至更多的条纹。如果投射 n 个二值图案，在投影仪中就能够唯一确定 2^n 个不同的列。摄像机图像中每个像素可以分别解码，对于 n 幅图像中的每一幅图像，我们只需确认像素是否是亮的（开）或者暗的（关）。对每一个像素进行解码可以返回一个 n 位的二进制码，直接对应投影仪中某一列。使用 n 位二进制码所带来的问题是相邻的条纹最多可以差 $n-1$ 位。因此，如果解码错了一位，解码列最多可能会相差投影仪宽度的一半。例如，如果

up to half the projector's width. For example, if the most significant bit of the binary code 0111 (7 in decimal notation) is decoded erroneously as 1, the decoded code is 1111 (15 in decimal notation).

For this reason, Gray codes are typically used to encode the projector columns (Blais, 2004; Geng, 2011). The ith code word of a Gray code is computed as $i \oplus \lfloor i/2 \rfloor$, where $\oplus$ denotes the binary XOR operation. Gray codes have the advantage that adjacent code words differ in exactly one bit, i.e., their Hamming distance is 1. Therefore, a wrong decoding, which typically occurs at the edges of the projection of a code word, has a very limited effect (Brenner *et al.*, 1998). Another effect of the Gray code is that the stripes at the finest resolution are twice as wide as in a binary code. This facilitates a correct decoding of parts of the object whose surface normal makes a large angle with the optical ray, where the projected stripes are much denser (Brenner *et al.*, 1998). A 5-bit Gray code is displayed in the upper part of Figure 2.68.

二进制码 0111（十进制的 7）的最高有效位错误地解码成 1，则解码结果是 1111（十进制的 15）。

为此，格雷码通常用来编码投影仪的列（Blais，2004；Geng，2011）。格雷码的第 i 个码字用 $i \oplus \lfloor i/2 \rfloor$ 计算，$\oplus$ 表示异或运算。格雷码的优点是相邻码字仅差一位，也就是说，汉明距离是 1。因此，错误的解码，通常发生在码字投影的边缘，影响非常有限（Brenner *et al.*, 1998）。格雷码的另外一个效果是最高分辨率的条纹宽度是二进制码的两倍，对于表面法线产生较大角度反射光线的物体部分，通常这些部位投影条纹更密集（Brenner *et al.*, 1998），格雷码有助于正确解码。图 2.68 上半部分所示为 5 位格雷码。

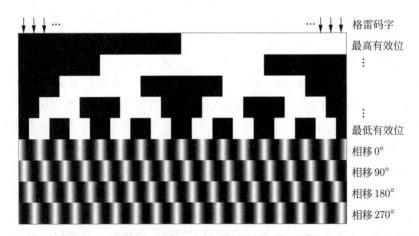

图 2.68 投影图案是一个 5 位格雷码和四个相移为 90° 倍数的正弦条纹的组合。32 个格雷码字对应的列用箭头表示，投影图案的顺序是从最高有效位到最低有效位，也就是从最粗糙的图案到最精细的图案。格雷码之后，投影相移条纹。格雷码的特点是相邻码字之间只有一位发生变化。这使得最高分辨率的条纹是两个格雷码字的宽度。条纹频率是每两个格雷码字一个周期，也就是，每个周期对应最高分辨率格雷码图案的每个条纹

For binary patterns, it is essential that the pattern is aligned with the pixels of the projector. If this is not the case, the patterns will appear jagged or will have to be interpolated to transform them to the array layout, in which case they will no longer be binary. Therefore, if a DMD with a diamond pixel layout is used, the stripes should be oriented at a 45° angle.

The following extension of Gray codes is often used in practice: in addition to the pure Gray code, two more images are acquired—one with the projector completely on and one with the projector completely off (Sansoni *et al.*, 1997). As described in Section 3.10.3.1, this makes the decoding of the stripes much easier. Furthermore, the fully illuminated image allows the surface texture, i.e., the brightness of each point in the scene, to be determined.

2.5.3.3 Fringe Projection

While Gray codes are conceptually simple, there are certain limitations in practice. On the one hand, a relatively large number of patterns is required to achieve a high spatial resolution of the projector column (Geng, 2011). On the other hand, the finer the resolution of the stripes becomes, the harder it is to decode them correctly because the optics of the projector and camera will introduce blurring. A type of pattern that is invariant to blurring is a sinusoidal pattern (Zhang, 2010): the blurring simply reduces the amplitude of the sinusoids, but leaves their frequency unchanged. Obviously, the gray value of the projection of a sinusoid is related to the projector column. However, it is relatively difficult to infer the projector column directly from a single projected

对于二进制编码图案，图案和投影仪的像素对齐是非常必要的。如果没有对齐，图案会出现锯齿，或者图案必须进行插值操作把它们转化为阵列布局，这样不再是二进制编码。因此，如果 DMD 使用菱形像素布局，条纹应该调整 45°。

在实际应用过程中，格雷码通常进行如下扩展：除了纯格雷码之外，通常会采集两幅更多的图像，一幅图像对应投影仪全开，一幅图像对应投影仪全关（Sansoni *et al.*, 1997）。如 3.10.3.1 节描述，这种扩展使条纹解码变得更容易。此外，充分照明的图像可以确定表面纹理，也就是场景中每个点的亮度。

2.5.3.3 条纹投影

尽管格雷码从概念上讲非常简单，但在实际应用中存在一定的局限性。一方面，相对大量的图案需要获得投影仪列的较高空间分辨率（Geng, 2011），另一方面，条纹的分辨率越高，正确解码会越困难，因为投影仪和摄像机的光学结构会引入模糊现象。对模糊现象有较高鲁棒性的一种图案是正弦图案（Zhang, 2010）：模糊现象仅仅会减小正弦曲线的振幅，但频率不会发生变化。很明显，正弦曲线投影的灰度值对应投影仪列。然而，从单个正弦投影曲线来推导投影仪列是比较困难的，因为物体的反射比可能是变化的，因此需要调节正弦信号。基于此，通常使用相移 $2\pi/m$ 的 m

sinusoid because the reflectance of the object may not be constant and therefore may modulate the sinusoid. For this reason, m sinusoids with a phase shift of $2\pi/m$ are typically used (Geng, 2011; Gorthi and Rastog, 2010). The 1D profile of these patterns is given by

正弦曲线（Gorthi and Rastog, 2010; Geng, 2011）。这些图案的一维轮廓用以下等式表示

$$P_k(\phi) = \frac{1}{2}\left(1 + \cos\left(\phi - \frac{2\pi k}{m}\right)\right) \qquad (2.35)$$

3D reconstruction systems that use sinusoidal patterns are called fringe projection systems (Gorthi and Rastog, 2010; Geng, 2011). Three phase shifts are sufficient to recover the phase ϕ from the camera images (Huang et al., 2003; Zhang, 2010; Geng, 2011; Bell et al., 2016). However, in practice, more than three phase shifts are frequently used. This leads to a decoding that is less sensitive to surface reflectance variations, has better immunity to ambient light, and results in increased accuracy of the 3D reconstruction (Bell et al., 2016). Often, four phase shifts of $\pi/2$ are used (Brenner et al., 1998; Sansoni et al., 1999). Four sinusoidal patterns with phase shifts of $\pi/2$ are displayed in the lower part of Figure 2.68. Let $I_k(r,c)$, $k = 0,\cdots,3$ denote the camera images obtained from projecting four patterns of Eq. (2.35) with phase shifts of $k\pi/2$. Then the phase ϕ can be decoded in a very simple manner by

使用正弦图案的三维重构系统称为条纹投影系统（Gorthi and Rastog, 2010; Geng, 2011）。三个相移足够复原摄像机图像中的相位 ϕ（Huang et al., 2003; Zhang, 2010; Geng, 2011; Bell et al., 2016）。然而，实际使用时经常会超过 3 个相移，使得解码时对表面反射变化不是很敏感，对环境光有更好的抗干扰能力，因此会增加三维重构的准确度（Bell et al., 2016）。经常会用到 4 个 $\pi/2$ 相移（Brenner et al., 1998; Sansoni et al., 1999）。如图 2.68 下半部分所示为 4 个相移 $\pi/2$ 的正弦图案。用 $I_k(r,c)$, $k = 0,\cdots,3$ 表示摄像机图像，图像从等式（2.35）中投影的 4 个图案获得，相移为 $k\pi/2$。那么相位 ϕ 的解码方式非常简单，可以用等式表示

$$\phi(r,c) = \operatorname{atan2}(I_1(r,c) - I_3(r,c), I_0(r,c) - I_2(r,c)) \qquad (2.36)$$

where atan2 denotes the two-argument arctangent function that returns its result in the full 2π range, and we assume that the result has been normalized to $[0, 2\pi)$.

其中 $atan2$ 表示双参数反正切函数，返回结果在整个 2π 范围内，我们假设结果已经归一化到 $[0, 2\pi)$。

The advantage of fringe projection methods over Gray codes is that the phase can be computed very precisely. Brenner et al. (1998) show that the precision of the phase is given by

条纹投影方法比格雷码的优势是可以非常精确地计算相位。文献（Brenner et al., 1998）提出相位精度可以用以下等式表示

$$\sigma_\phi = \frac{\sigma_g}{\sqrt{2}C} \tag{2.37}$$

where $2C$ is the contrast of the sinusoid, and σ_g is the standard deviation of the noise in the image. Note that $2C/\sigma_g$ is the SNR of the stripes. Equation (2.37) shows that the the higher the SNR, the more precisely phase can be determined. For example, for $C = 25$ and $\sigma_g = 2$, we have $\sigma_\phi = 0.05657$, i.e., 1/111 of a sinusoid cycle. To compare this to Gray codes, let us assume that two Gray code words correspond to one sinusoid cycle (since this is the highest frequency that would be supported by any binary pattern). Then, the phase of the sinusoid can be determined with a precision of 1/55.5 of the width of one Gray code word, even for this relatively low SNR.

While Eq. (2.36) is very elegant and simple, it does not allow recovery of a unique projector column since the arctangent function wraps around at 2π. Therefore, if there are multiple sinusoidal stripes, a phase unwrapping algorithm must be used to remove the discontinuities in the recovered phase (Bell et al., 2016; Gorthi and Rastog, 2010). Phase unwrapping algorithms are relatively complex and slow. Furthermore, they do not allow the system to recover the absolute phase of a scene point, which would be necessary to decode the projector column. To recover the absolute phase, there are multiple strategies. One class of algorithms makes use of the fact that the absolute phase can be recovered once the absolute phase of a single point (Zhang, 2010) or a single fringe (Wang et al., 2010) is known and therefore projects a special marker into the scene (Zhang, 2010). Another class of algorithms projects fringes at multiple frequencies that are relatively prime

其中 $2C$ 是正弦曲线的对比度，σ_g 表示图像噪声的标准偏差。注意 $2C/\sigma_g$ 表示条纹的信噪比。等式（2.37）表明信噪比越高，确定相位的精度就越高。例如，对于 $C = 25$ 和 $\sigma_g = 2$，可以得出 $\sigma_\phi = 0.05657$，也就是 1/111 个正弦周期。为了和格雷码方法对比，假设两个格雷码字对应一个正弦周期（因为这是任何二进制图案都能够支持的最高频率）。那么，正弦曲线的相位精度可以达到格雷码字宽度的 1/55.5，即便是在相对较低的对比度情况下。

尽管等式（2.36）看起来非常简单优雅，但并不能够恢复唯一的投影仪列，因为反正切函数封装在 2π。因此，如果有多个正弦条纹，必须使用相位展开算法来消除所恢复相位中的不连续点（Bell et al., 2016; Gorthi and Rastog, 2010）。相位展开算法相对复杂且较慢，而且并不能使系统恢复场景中点的绝对相位，而这对于解码投影仪列是必要的。为了能够恢复绝对相位，有多个策略。其中一类算法是利用一旦单个点（Zhang, 2010）或者单个条纹（Wang et al., 2010）的绝对相位已知那么绝对相位就可以恢复这个事实，因此在场景中投射特殊的标记（Zhang, 2010）。另一类算法是在场景中投射多种频率的相对互质的条纹（Gorthi and Rastog, 2010; Zhang, 2010; Bell et al., 2016），通常至少使用 3 种频率。如果某个频率

to each other into the scene (Gorthi and Rastog, 2010; Zhang, 2010; Bell *et al.*, 2016). At least three frequencies are typically used. The absolute phase can be recovered if one of the frequencies corresponds to one cycle per width of the projector (Wang *et al.*, 2010). The second strategy increases the number of images that must be projected considerably.

Fringe projection relies on the fact that the sinusoids actually appear as sinusoids in the image. This implies that the camera and projector of the structured light sensor must be linear. If this is not the case, the phase cannot be recovered correctly, which results in 3D reconstruction errors (Huang *et al.*, 2003; Gorthi and Rastog, 2010; Wang *et al.*, 2010; Zhang, 2010). Section 3.10.3.2 discusses how a structured light sensor can be calibrated radiometrically.

2.5.3.4 Hybrid Systems

Because of the advantages and disadvantages that were discussed above, hybrid approaches are frequently used. The ease of decoding of Gray codes is combined with the high precision of fringe projection (Sansoni *et al.*, 1997; Brenner *et al.*, 1998; Sansoni *et al.*, 1999; Chen *et al.*, 2009; Geng, 2011). First, an *n*-bit Gray code is projected. Optionally, a completely dark and a completely bright pattern are projected. Finally, four phase-shifted sinusoids are projected. An example of this hybrid approach is displayed in Figure 2.68. One cycle of the sinusoidal patterns corresponds to two Gray code words. As discussed above, in this setup the projector column can be determined with a precision of much better than 1/10 of a Gray code word, even with very low SNRs. This is the technology that we will discuss in Section 3.10.3.

对应投影仪每个宽度的某个周期，那么绝对相位就能够恢复（Wang *et al.*, 2010）。第 2 个策略通常会相当大地增加投影图像的数量。

条纹投影依赖于正弦曲线在图像上需要真实地表现为正弦曲线，这意味着结构光传感器中的摄像机和投影仪必须是线性的。如果不是这样，则相位不能正确恢复，三维重构结果也是错误的（Huang *et al.*, 2003; Gorthi and Rastog, 2010; Wang *et al.*, 2010; Zhang，2010）。3.10.3.2 节会讨论结构光传感器怎样进行辐射标定的。

2.5.3.4 混合系统

因为以上讨论的系统都存在优点和缺点，因此经常使用混合方法，格雷码的容易解码和条纹投影的高精度经常结合在一起（Sansoni *et al.*，1997; Brenner *et al.*，1998; Sansoni *et al.*，1999; Chen *et al.*，2009; Geng, 2011）。首先，投射 *n* 位格雷码，然后视情况而定，投射一个全黑和一个全亮的图案，最后投射 4 个相移正弦曲线。图 2.68 所示是一个混合方法的示例，正弦图案的一个周期对应 2 个格雷码字。如上面所讨论的，这个系统中，投影仪列的精度比一个格雷码字的 1/10 还要好，甚至在信噪比非常低的情况下。我们会在 3.10.3 节中讨论这个技术。

2.5.4 Time-of-Flight Cameras

As already mentioned, time-of-flight (TOF) cameras emit radiation and measure the time until the reflection of the radiation returns to the sensor. There are two basic technologies to achieve this. Pulse-modulated (PM) TOF cameras emit a radiation pulse and directly or indirectly measure the time it takes for the radiation pulse to travel to the objects in the scene and back to the camera. Continuous-wave-modulated (CWM) TOF cameras, on the other hand, emit amplitude-modulated radiation and measure the phase difference between the emitted and received radiation (Horaud et al., 2016). We will briefly discuss these two technologies below without going into details of the physics and electronics that are used in these sensors. The interested reader is referred to Remondino and Stoppa (2013) for details.

In the past few years, the dominant design was CWM TOF cameras (Foix et al., 2011; Fürsattel et al., 2016). Note, however, that many of the cameras that are discussed in Foix et al. (2011); Fürsattel et al. (2016) are no longer commercially available. This indicates that there seems to be a trend towards PM TOF cameras (Horaud et al., 2016).

2.5.4.1 Continuous-Wave-Modulated Time-of-Flight Cameras

CWM TOF cameras emit radiation, typically in the near IR range, that is amplitude-modulated as a sine wave of a certain frequency f_m, typically, between 15 and 30 MHz (Fürsattel et al., 2016). The radiation is reflected by the objects in the scene and received by the camera, which measures

2.5.4 飞行时间摄像机

正如上面已经提到过的，飞行时间（TOF）摄像机发射光线并测量光线反射回传感器所经过的时间。目前有两种基本技术来实现此功能。一是脉冲调制（PM）TOF摄像机发出辐射脉冲，然后直接或者间接测量辐射脉冲到达场景中物体并返回摄像机所需的时间；二是连续波调制（CWM）TOF摄像机发出调幅射线，然后测量发射射线和接收射线之间的相位差（Horaud et al., 2016）。下面我们简要讨论这两种技术，但不会详细描述这些传感器中所用到的物理学和电子学知识，感兴趣的读者可以参见文献（Remondino and Stoppa, 2013）。

在过去几年中，主流设计是CWM TOF 摄像机（Foix et al., 2011; Fürsattel et al., 2016）。然而需要注意的是，在文献（Foix et al., 2011; Fürsattel et al., 2016）中所描述的许多摄像机已经不再商用。这看起来趋势似乎在转向 PM TOF 摄像机（Horaud et al., 2016）。

2.5.4.1 连续波调制飞行时间摄像机

CWM TOF 摄像机发出的辐射，通常在近红外波长范围内，经过调幅频率 f_m 一般在 15~30 MHz 之间（Fürsattel et al., 2016）。辐射经过场景中的物体反射后被摄像机接收，在一个调制辐射周期内摄像机会测量

the amount of reflected radiation four times during a cycle of the modulated radiation (e.g., at 120 MHz for $f_m = 30$ MHz). This creates measurements m_i, $i = 0, \cdots, 3$, that are spaced at intervals of $\pi/2$ of the received wave. The phase shift ϕ between the emitted and received wave can be demodulated as follows (Foix et al., 2011; Horaud et al., 2016)

4 倍数量的反射辐射（例如，对于 $f_m = 30$ MHz，120 MHz），这会产生 m_i 个测量，其中 $i = 0, \cdots, 3$，中间间隔为所接收光波的 $\pi/2$。发射光波和接收光波之间的相移 ϕ 使用以下等式来解调（Foix et al., 2011; Horaud et al., 2016）

$$\phi = \mathrm{atan2}(m_3 - m_1, m_0 - m_2) \tag{2.38}$$

where atan2 denotes the two-argument arctangent function that returns its result in the full 2π range, and we assume that the result has been normalized to $[0, 2\pi)$. Furthermore, the four measurements can be used to calculate the scene intensity

其中 atan2 表示双参数反正切函数，返回结果在整个 2π 范围内，我们假设结果已经归一化到 $[0, 2\pi)$。因此，4 次测量可以用来计算场景灰度和振幅，可以预测测量质量。

$$b = \frac{m_0 + m_1 + m_2 + m_3}{4} \tag{2.39}$$

and an amplitude

$$a = \frac{\sqrt{(m_3 - m_1)^2 + (m_0 - m_2)^2}}{2} \tag{2.40}$$

that can be used to predict the quality of the measurements. The above measurements are performed multiple times during the integration time (exposure time) of the sensor to increase the SNR. The phase ϕ is only unique within the so-called ambiguity-free distance range, which is given by (see Foix et al., 2011)

为了提高信噪比，以上测量在传感器的积分时间（曝光时间）内会执行多次。相位 ϕ 在所谓的 ambiguity-free 距离范围内是唯一的，其中，距离用以下等式（参见 Foix et al., 2011）表示，

$$d_{\max} = \frac{c}{2f_m} \tag{2.41}$$

where c is the speed of light. For example, for $f_m = 30$ MHz, we have $d_{\max} = 4.997$ m. If objects with a distance larger than $d_{\max}$ are present in the scene, the phase will wrap around. Therefore, a phase unwrapping algorithm would have to be used, which is time-consuming and therefore almost never used. Therefore, for objects at distances greater than $d_{\max}$, a wrong object distance will be returned.

c 表示光速。例如，对于 $f_m = 30$ MHz，可以得出 $d_{\max} = 4.997$ m。如果场景中物体的距离大于 $d_{\max}$，相位会 wrap around。因此，必须使用相位展开算法，但算法非常耗时几乎很少使用。所以，对于距离大于 $d_{\max}$ 的物体，会返回错误的距离值。

If the object distance d lies within the ambiguity-free range, it can be computed as (see Foix *et al.*, 2011)

$$d = d_{\max} \frac{\phi}{2\pi} = \frac{c\phi}{4\pi f_{\mathrm{m}}} \tag{2.42}$$

CWM TOF cameras exhibit several systematic and random errors (Foix *et al.*, 2011; Fürsattel *et al.*, 2016). For example, they exhibit a relatively large temporal noise (in the range of up to a few centimeters), their distance measurements may depend on the integration time, which can cause errors between 1 and 10 centimeters, radiation scattering within the camera can cause relatively large distance errors, the distance measurements may depend on the reflectance of the objects in the scene, and there may be a systematic depth distortion (wiggling) because the emitted radiation is not exactly a sine wave.

CWM TOF cameras are capable of high frame rates of up to 30 Hz. However, one of their drawbacks is their comparatively small resolution: most of the commercially available sensors have a resolution of less than 320×240 pixels.

2.5.4.2 Pulse-Modulated Time-of-Flight Cameras

PM TOF cameras emit a radiation pulse, typically in the near IR range, and directly or indirectly measure the time of flight of the radiation pulse from the radiation source to the object and back to the sensor. Let this round-trip time be called t_{d}. Then, the distance from the camera to the object is given by (see Remondino and Stoppa, 2013)

$$d = \frac{c}{2} t_{\mathrm{d}} \tag{2.43}$$

While direct measurement of t_d is possible in principle, in PM TOF cameras the round-trip time is inferred indirectly through measurement of the radiation intensity received by the sensor (Remondino and Stoppa, 2013).

A PM TOF camera based on indirect measurement of t_d works by emitting a radiation pulse of a certain duration t_p, e.g., $t_p = 30\,\text{ns}$ (Spickermann et al., 2011). The pulse duration determines the maximum object distance (distance range) that can be measured (see Spickermann et al., 2011)

$$d_{\max} = \frac{c}{2} t_p \qquad (2.44)$$

For example, for $t_p = 30\,\text{ns}$, the maximum distance is 4.497 m. The radiation returned to the sensor is measured during three integration periods of duration t_p. The first integration period is simultaneous with the emission of the radiation pulse. The second integration period immediately follows the first integration period. These two integration periods measure the intensity of the radiation emitted by the camera that is reflected by the objects in the scene. As described below, these two measurements are the main pieces of data that are used to infer the distance of the objects in the scene. The third integration period happens a sufficient time before the radiation pulse is emitted or sufficiently long after it has been emitted. Its purpose is to correct the distance measurement for effects of ambient radiation reflected by the objects in the scene.

Let us examine the first two integration periods and assume for the moment that there is no ambient radiation reflected by the scene. During the first integration period, the radiation reflected

尽管直接测量 t_d 原则上是可行的，但对于 PM TOF 摄像机，往返时间是通过测量传感器接收到的辐射光强来间接得出的（Remondino and Stoppa，2013）。

PM TOF 摄像机间接测量 t_d，工作时会发射特定时长为 t_p 的光波脉冲，例如，$t_p = 30\,\text{ns}$（Spickermann et al.，2011）。脉冲时长决定了能够测量的最大距离（距离范围）（Spickermann et al.，2011）

例如，对于 $t_p = 30\,\text{ns}$，最大距离是 4.497 m。在时长 t_p 的 3 个积分周期内，测量返回到传感器的光波。第 1 个积分周期和发射光波脉冲是同时的，第 2 个积分周期紧跟第 1 个积分周期。这两个积分周期测量场景中物体反射的摄像机发射的光波强度。如下面所述，这两个测量结果就是用来得出场景中物体距离的主要数据。第 3 个积分周期发生在光波脉冲发射之前或者光波脉冲发射之后的足够长的一段时间内。目的是用来纠正场景中物体反射环境光对测量距离产生的影响。

让我们来看一下前两个积分周期，假设场景中目前没有反射的环境光。在第 1 个积分周期内，距离摄像机近的物体反射的光波对传感器积累

by objects that are close to the camera will make a strong contribution to the charge on the sensor. The closer the object is to the sensor, the higher the charge on the camera will be. On the other hand, during the second integration period, the radiation reflected by objects that are farther from the camera make a strong contribution to the charge on the sensor. The farther the object is to the sensor, the higher the charge on the camera will be. For example, an object at $d = 0$ will produce a charge exclusively during the first integration period and no charge during the second integration period since the entire radiation pulse returns to the sensor during the first integration period. On the other hand, an object at $d = d_{\max}$ will produce a charge exclusively during the second integration period since the pulse cannot return to the sensor during the first integration period. As a final example, an object at $d = d_{\max}/2$ will produce equal charges during the first and second integration period. This shows that the time of flight t_d can be inferred from the charges on the sensor. Let the charges of the first two integration periods be denoted by q_1 and q_2. Then, it can be shown that

$$t_d = t_p \frac{q_2}{q_1 + q_2} \qquad (2.45)$$

Note that for objects at distances greater than $d_{\max}$, Eq. (2.45) reduces to $t_d = t_p$ since $q_1 = 0$ in this case. This is the reason why objects at distances greater than $d_{\max}$ cannot be measured correctly.

Equation (2.45) assumes that there is no ambient radiation. This is rarely true in practice. The third integration period can be used to correct the depth measurement for the effects of

的电荷贡献最大，物体离传感器越近，摄像机积累的电荷越高。另外，在第 2 个积分周期内，远离摄像机的物体反射的光波对传感器上积累电荷贡献最大，物体离传感器越远，摄像机积累的电荷越高。例如，当物体距离 $d = 0$ 时，仅仅在第 1 个积分周期内产生电荷，在第 2 个积分周期内不会产生电荷，因为在第 1 个积分周期内全部辐射脉冲会返回传感器。当物体距离 $d = d_{\max}$ 时，仅在第 2 个积分周期内产生电荷，因为在第 1 个积分周期内脉冲不能返回传感器。最后，如果物体距离 $d = d_{\max}/2$，在这两个积分周期内会产生相等的电荷。以上表明飞行时间 t_d 可以通过传感器的电荷来得到。前两个积分周期的电荷分别用 q_1 和 q_2 表示，如以下等式所示

注意，当物体距离大于 $d_{\max}$ 时，等式（2.45）会得出 $t_d = t_p$，因为此时 $q_1 = 0$。这就是为什么物体距离大于 $d_{\max}$ 时无法准确测量的原因。

等式（2.45）的前提是假设没有环境光波影响，但实际很少存在这种情况。第 3 个积分周期用来纠正环境光波对深度测量产生的影响。第 3 个

ambient radiation. Let the charge created during the third integration period be denoted by q_3. Then, q_3 can be subtracted from q_1 and q_2 to cancel the ambient radiation, resulting in (see Spickermann et al., 2011)

积分周期内的电荷用 q_3 表示，然后，q_1 和 q_2 分别和 q_3 相减来抵消环境光波影响，结果如下（Spickermann et al., 2011）

$$t_\mathrm{d} = t_\mathrm{p} \frac{q_2 - q_3}{(q_1 - q_3) + (q_2 - q_3)} = t_\mathrm{p} \frac{q_2 - q_3}{q_1 + q_2 - 2q_3} \quad (2.46)$$

Therefore, the distance of a point in the scene is given by

因此，场景中某个点的距离可以用如下等式表示

$$d = \frac{c}{2} t_\mathrm{p} \frac{q_2 - q_3}{q_1 + q_2 - 2q_3} = d_\mathrm{max} \frac{q_2 - q_3}{q_1 + q_2 - 2q_3} \quad (2.47)$$

The above measurements are performed multiple times during the exposure time of the sensor to increase the SNR.

为了提高信噪比，以上测量过程在摄像机曝光时间内会执行多次。

A thorough analysis of systematic and random errors that affect PM TOF cameras currently is nonexistent. It can be assumed that some of the effects that are known to exist for CWM TOF cameras (see Foix et al., 2011; Fürsattel et al., 2016) also affect PM TOF cameras. For example, it is known that PM TOF cameras can be affected by a depth bias that depends on the reflectance of the objects in the scene (Driewer et al., 2016). The temporal noise of PM TOF cameras can be of the order of a few centimeters.

对于影响 PM TOF 摄像机的系统误差和随机误差，目前没有一个全面的分析。可以认为对于 CWM TOF 摄像机已知存在的影响也会影响 PM TOF 摄像机（Foix et al., 2011; Fürsattel et al., 2016）。例如，场景中物体反射比带来的深度偏差会影响 PM TOF 摄像机（Driewer et al., 2016）。PM TOF 时间噪声的影响是几厘米。

PM TOF cameras support high frame rates (up to 30 Hz). Furthermore, they offer higher resolutions than CWM TOF cameras. The resolutions of PM TOF cameras that are currently commercially available range from 320×240 up to 1280×1024 pixels.

PM TOF 摄像机支持高帧率（最高到 30Hz）。而且，比 CWM TOF 摄像机的分辨率也要高。目前商用 PM TOF 摄像机的分辨率范围是从 320 像素 × 240 像素到 1280 像素 × 1024 像素。

3 Machine Vision Algorithms

In the previous chapter, we examined the different hardware components that are involved in delivering an image to the computer. Each of the components plays an essential role in the machine vision process. For example, illumination is often crucial to bring out the objects we are interested in. Triggered frame grabbers and cameras are essential if the image is to be captured at the right time with the right exposure. Lenses are important for acquiring a sharp and aberration-free image. Nevertheless, none of these components can "see," i.e., extract the information we are interested in from the image. This is analogous to human vision. Without our eyes, we cannot see. Yet, even with eyes we cannot see anything without our brain. The eye is merely a sensor that delivers data to the brain for interpretation. To extend this analogy a little further, even if we are myopic we can still see—only worse. Hence, it is clear that the processing of the images delivered to the computer by the sensors is truly the core of machine vision. Consequently, in this chapter, we will discuss the most important machine vision algorithms.

3.1 Fundamental Data Structures

Before we can delve into the study of machine vision algorithms, we need to examine the fundamental data structures that are involved in machine vision applications. Therefore, in this section we will take a look at the data structures for images, regions, and subpixel-precise contours.

3.1.1 Images

An image is the basic data structure in machine vision, since this is the data that an image acquisition device typically delivers to the computer's memory. As we saw in Section 2.3, a pixel can be regarded as a sample of the energy that falls on the sensor element during the exposure, integrated over the spectral distribution of the light and the spectral response of the sensor. Depending on the camera type, the spectral response of the sensor typically will comprise the entire visible spectrum and optionally a part of the near IR spectrum. In this case, the camera will return one sample of the energy per pixel, i.e., a single-channel gray value image. RGB cameras, on the other hand, will return three samples per pixel, i.e., a three-channel image. These are the two basic types of sensors that are encountered in machine vision applications. However, cameras capable of acquiring images with tens to hundreds of spectral samples per pixel are possible (Hagen and Kudenov, 2013; Lapray et al., 2014; Eckhard, 2015). Therefore, to handle all possible applications, an image can be considered as a set of an arbitrary number of channels.

Intuitively, an image channel can simply be regarded as a two-dimensional (2D) array of numbers. This is also the data structure that is used to represent images in a programming language. Hence, the gray value at the pixel $(r,c)^\top$ can be interpreted as an entry in a matrix: $g = f_{r,c}$. In a

本数据结构。因此，本节中我们先介绍一下表示图像、区域和亚像素轮廓的数据结构。

3.1.1 图像

在机器视觉里，图像是基本的数据结构，它所包含的数据通常是由图像采集设备传送到计算机的内存中的。如2.3节所述，一个像素能被看成对能量的采样结果，此能量是在曝光过程中传感器上一个感光单元所累积得到的，它累积了在传感器光谱响应范围内的所有光能。根据摄像机的类型不同，传感器的光谱响应通常包括全部可见光谱和部分近红外光谱。正因为这样，黑白摄像机会返回每个像素所对应的一个能量采样结果，这些结果组成了一幅单通道灰度值图像。而对于RGB彩色摄像机，它将返回每个像素所对应的三个采样结果，也就是一幅三通道图像。单通道摄像机和三通道摄像机是机器视觉应用中所涉及的两类基本的图像传感器。然而，摄像机是可以采集更多通道图像的，例如每个像素对应成百上千的光谱采样（Hagen and Kudenov, 2013; Lapray et al., 2014; Eckhard, 2015）。因此，为处理所有可能的应用，图像可被视为由一组任意多的通道组成的。

很直观地，图像通道可以被简单地看作是一个二维数组，这也是程序设计语言中表示图像时所使用的数据结构。因此，在像素$(r,c)^\top$处的灰度值可以被解释为矩阵$g = f_{r,c}$中的一个元素。使用更正规的描述方式，

more formalized manner, we can regard an image channel f of width w and height h as a function from a rectangular subset $R = \{0, \cdots, h-1\} \times \{0, \cdots, w-1\}$ of the discrete 2D plane $\mathbb{Z}^2$ (i.e., $R \subset \mathbb{Z}^2$) to a real number, i.e., $f : R \mapsto \mathbb{R}$, with the gray value g at the pixel position $(r, c)^\top$ defined by $g = f(r, c)$. Likewise, a multichannel image can be regarded as a function $\boldsymbol{f} : R \mapsto \mathbb{R}^n$, where n is the number of channels.

In the above discussion, we have assumed that the gray values are given by real numbers. In almost all cases, the image acquisition device will discretize not only the image spatially but also the gray values to a fixed number of gray levels. In most cases, the gray values will be discretized to 8 bits (1 byte), i.e., the set of possible gray values will be $\mathbb{G}_8 = \{0, \cdots, 255\}$. In some cases, a higher bit depth will be used, e.g., 10, 12, or even 16 bits. Consequently, to be perfectly accurate, a single-channel image should be regarded as a function $f : R \mapsto \mathbb{G}_b$, where $\mathbb{G}_b = \{0, \cdots, 2^b - 1\}$ is the set of discrete gray values with b bits. However, in many cases this distinction is unimportant, so we will regard an image as a function to the set of real numbers.

Up to now, we have regarded an image as a function that is sampled spatially, because this is the manner in which we receive the image from an image acquisition device. For theoretical considerations, it is sometimes convenient to regard the image as a function in an infinite continuous domain, i.e., $\boldsymbol{f} : \mathbb{R}^2 \mapsto \mathbb{R}^n$. We will use this convention occasionally in this chapter. It will be obvious from the context which of the two conventions is being used.

我们可以把某个宽度为 w、高度为 h 的图像通道 f 作为一个函数，该函数表述从离散二维平面 $\mathbb{Z}^2$（也即，$R \subset \mathbb{Z}^2$）的一个矩形子集 $R = \{0, \cdots, h-1\} \times \{0, \cdots, w-1\}$ 到某一个实数的关系；$f : R \mapsto \mathbb{R}$，像素位置是 $(r, c)^\top$ 处的灰度值 g 定义为 $g = f(r, c)$。同理，一个多通道图像可被视为一个函数 $\boldsymbol{f} : R \mapsto \mathbb{R}^n$，这里的 n 表示通道的数目。

在上面的讨论中，我们已经假定了灰度值是由实数表示的。在几乎所有的情况下，图像采集设备不但在空间上把图像离散化，同时也会把灰度值离散化到某一固定的灰度级范围内。多数情况下，灰度值将被离散化为 8 位（一个字节），也就是，所有可能的灰度值所组成的集合是 $\mathbb{G}_8 = \{0, \cdots, 255\}$。在有些情况下，需要使用更高的位深，如 10, 12, 甚至是 16 位，相应地，为了更精确起见，一个单通道图像应被视为某个函数 $f : R \mapsto \mathbb{G}_b$，此处 $\mathbb{G}_b = \{0, \cdots, 2^b - 1\}$ 是位深 b 时的灰度值集合。然而，这一差别在很多情况下并不重要，所以我们将图像视为到实数集的一个函数。

到目前为止，我们已经视一幅图像为一个在空间上采样的函数，如此处理是因为我们就是以这种方式从图像采集设备获取图像的。出于理论上的考虑，有时为了方便会将图像视为无限连续域上的函数，即，$\boldsymbol{f} : \mathbb{R}^2 \mapsto \mathbb{R}^n$。我们将在本章部分地方使用此种约定，读者将可根据上下文内容明显区分出是使用此两种约定中的哪一种。

3.1.2 Regions

One of the tasks in machine vision is to identify regions in the image that have certain properties, e.g., by performing a threshold operation (see Section 3.4). Therefore, at the minimum we need a representation for an arbitrary subset of the pixels in an image. Furthermore, for morphological operations, we will see in Section 3.6.1 that it will be essential that regions can also extend beyond the image borders to avoid artifacts. Therefore, we define a region as an arbitrary subset of the discrete plane: $R \subset \mathbb{Z}^2$.

The choice of the letter R is intentionally identical to the R that is used in the previous section to denote the rectangle of the image. In many cases, it is extremely useful to restrict processing to a certain part of the image that is specified as a region of interest (ROI). In this context, we can regard an image as a function from the ROI to a set of numbers, i.e., $f : R \mapsto \mathbb{R}^n$. The ROI is sometimes also called the domain of the image because it is the domain of the image function f. We can even unify the two views: we can associate a rectangular ROI with every image that uses the full number of pixels. Therefore, from now on, we will silently assume that every image has an associated ROI, which will be denoted by R.

In Section 3.4.2, we will also see that often we will need to represent multiple objects in an image. Conceptually, this can simply be achieved by considering sets of regions.

From an abstract point of view, it is therefore simple to talk about regions in the image. It is not

3.1.2 区域

机器视觉的任务之一就是识别图像中包含某些特性的区域，比如执行一个阈值分割处理（见 3.4 节）。因此，至少我们还需要一种数据结构，它可以表示一幅图像中一个任意的像素子集。此外，对于形态学处理，在 3.6.1 节我们将会读到为避免伪像，能将区域延伸到图像边界外是重要的。故而，我们把区域定义为离散平面的一个任意子集：$R \subset \mathbb{Z}^2$。

这里选用字母 R 来表示区域是有意与前一节中用来表示矩形图像的 R 保持一致。在很多情况下，将图像处理限制在图像上某一特定的感兴趣区域（ROI）内是极其有用的。就此而论，可以视一幅图像为一个从某感兴趣区域到某一数据集的函数：$f : R \mapsto \mathbb{R}^n$。这个感兴趣区域有时也被称为图像的定义域，因为它是图像函数 f 的定义域。我们可以将这两种图像表示的方法统一：对任意一幅图像，可以用一个包含该图像所有像素点的矩形感兴趣区域来表示此幅图像。所以，从现在开始，我们默认每幅图像都有一个相关的感兴趣区域存在，这个感兴趣区域用 R 来表示。

在 3.4.2 节中我们将看到，很多时候需要描述一幅图像上的多个物体。从概念上讲，它们可以由区域的集合来简单地表示。

从抽象的观点看，讨论图像中的区域将使问题更简单。但现在我们还

immediately clear, however, how best to represent regions. Mathematically, we can describe regions as sets, as in the above definition. An equivalent definition is to use the characteristic function of the region:

$$\chi_R(r,c) = \begin{cases} 1, & (r,c) \in R \\ 0, & (r,c) \notin R \end{cases} \quad (3.1)$$

This definition immediately suggests the use of binary images to represent regions. A binary image has a gray value of 0 for points that are not included in the region and 1 (or any other number different from 0) for points that are included in the region. As an extension to this, we could represent multiple objects in the image as label images, i.e., as images in which the gray value encodes the region to which the point belongs. Typically, a label of 0 would be used to represent points that are not included in any region, while numbers > 0 would be used to represent the different regions.

The representation of regions as binary images has one obvious drawback: it needs to store (sometimes very many) points that are not included in the region. Furthermore, the representation is not particularly efficient: we need to store at least 1 bit for every point in the image. Often, the representation actually uses 1 byte per point because it is much easier to access bytes than bits. This representation is also not particularly efficient for run time purposes: to determine which points are included in the region, we need to perform a test for every point in the binary image. In addition, it is a little awkward to store regions that extend to negative coordinates as binary images, which also leads to cumbersome algorithms. Finally, the representation of multiple regions as label images leads to the fact that

不清楚最好地表示区域的方法。从数学上，我们能把区域描述成集合，如上文中的定义。另一种等价定义将使用区域的特征函数：

这个定义引入了二值图像来描述区域。一个二值图像用灰度值 0 表示不在区域内的点，用 1（或其他非 0 的数）表示被包含在区域内的点。作为此定义的延伸，我们可以将图像中的多个目标物体描绘成多个标记图像，标记图像中每个像素的灰度值表示此像素应属于哪个区域。典型地，标记 0 将被用来表示不被任何区域包含的那些点，而大于 0 的标记将被用来表示不同的区域。

以二值图像来描述区域的方法存在着一个明显的缺陷：它必须存储那些区域外的（有时是非常多的）点。此外，这种表示法的效率也不高：图像上的每个点至少需要占用一个位来保存。通常情况下，这种表示法会使用一个字节而不是一个位来表示图像上的一个点，因为访问字节比访问位更容易。这种表示法在运行时间上也不经济：为了确定哪些点在区域内，我们必须对二值图像上的所有点进行检测。此外，以二值图像来保存扩展到负坐标的区域是不易实现的，因为那需要借助复杂的算法才可以实现。最后，用标记图像表示多个区域时无法描述交叠区域，所以如果基于这些区域进行形态学处理将会引起问题。所

overlapping regions cannot be represented, which will cause problems if morphological operations are performed on the regions. Therefore, a representation that only stores the points included in a region in an efficient manner would be very useful.

Figure 3.1 shows a small example region. We first note that, either horizontally or vertically, there are extended runs in which adjacent pixels belong to the region. This is typically the case for most regions. We can use this property and store only the necessary data for each run. Since images are typically stored line by line in memory, it is better to use horizontal runs. Therefore, the minimum amount of data for each run is the row coordinate of the run and the start and end columns of the run. This method of storing a region is called a run-length representation or run-length encoding. With this representation, the example region can be stored with just 4 runs, as shown in Figure 3.1. Consequently, the region can also be regarded as the union of all of its runs:

以，如果能通过一种高效率的方式仅存储区域内所包含的点来表示区域，那么这种表示法将非常有用。

图 3.1 是区域的一个小示例。我们首先注意到沿水平或垂直方向，行程延伸所覆盖的邻近像素点属于一个区域。大多数的区域都是这种情况。我们能使用这种特征来仅保存每次行程的必要数据。由于图像数据在内存中通常是一行紧接着一行保存的，所以最好使用水平行程。因此，表示每次行程的最小量的数据包括该行程的纵坐标值、行程开始和行程结束对应横坐标值。这一保存区域的方法被称作行程表示法或行程编码。使用此种方法，本例所示的区域能用 4 个行程来保存，如图 3.1 所示。因此，区域也可以表示为该区域全部行程的一个并集：

$$R = \bigcup_{i=1}^{n} \mathbf{r}_i \qquad (3.2)$$

Here, $\mathbf{r}_i$ denotes a single run, which can also be regarded as a region. Note that the runs are sorted in lexicographic order according to their row and

此处，$\mathbf{r}_i$ 表示一个行程，也可以表示一个区域。注意，行程存储的顺序是根据其纵坐标和起始横坐标的字典序确定

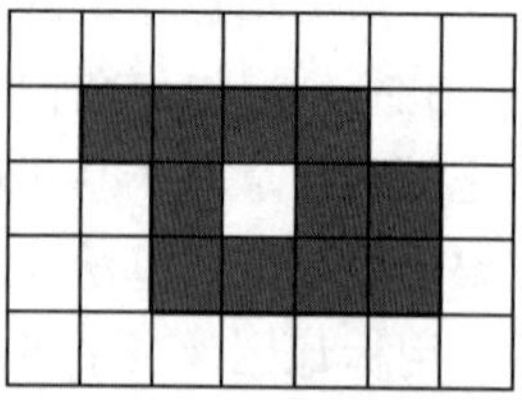

行程	行	起始列	结束列
1	1	1	4
2	2	2	2
3	2	4	5
4	3	2	5

图 3.1 一个区域的行程编码

start column coordinates. This means that there is an order of the runs $\mathbf{r}_i = (r_i, cs_i, ce_i)$ in R defined by: $\mathbf{r}_i \prec \mathbf{r}_j \Leftrightarrow r_i < r_j \vee r_i = r_j \wedge cs_i < cs_j$. This order is crucial for the execution speed of algorithms that use run-length encoded regions.

In the above example, the binary image can be stored with 35 bytes if 1 byte per pixel is used or with 5 bytes if 1 bit per pixel is used. If the coordinates of the region are stored as 2-byte integers, the region can be represented with 24 bytes in the run-length representation. This is already a saving, albeit a small one, compared to binary images stored with 1 byte per pixel, but no saving if the binary image is stored as compactly as possible with 1 bit per pixel. To get an impression of how much this representation really saves, we can note that we are roughly storing the boundary of the region in the run-length representation. On average, the number of points on the boundary of the region will be proportional to the square root of the area of the region. Therefore, we can typically expect a very significant saving from the run-length representation compared to binary images, which must at least store every pixel in the surrounding rectangle of the region. For example, a full rectangular ROI of a $w \times h$ image can be stored with h runs instead of $w \times h$ pixels in a binary image (i.e., wh or $\lceil w/8 \rceil h$ bytes, depending on whether 1 byte or 1 bit per pixel is used). Similarly, a circle with diameter d can be stored with d runs as opposed to at least $d \times d$ pixels. Thus, the run-length representation often leads to an enormous reduction in memory consumption. Furthermore, since this representation only stores the points actually contained in the region, we do not need to perform a test to see whether a point

的。也就是说，行程 $\mathbf{r}_i = (r_i, cs_i, ce_i)$ 在 R 内的存储顺序定义为：$\mathbf{r}_i \prec \mathbf{r}_j \Leftrightarrow r_i < r_j \vee r_i = r_j \wedge cs_i < cs_j$。对于任何使用了行程编码区域的算法，此排序方法对于算法的执行速度是至关重要的。

在上例中，如果每个像素占用一个字节，那么采用二值图像来描述区域要占用 35 个字节，如果每个像素占用一个位，用二值图像来描述此区域也需要 5 个字节。而采用行程编码表示此区域时，如果区域的坐标值保存在 16 位整数中，只需要 24 个字节即可。虽然与每个像素只占一个位的二值图像相比，行程编码没有节约任何存储空间。但同每像素一个字节的二值图像相比，行程编码已经节省了存储空间，尽管节省的空间有限。为加深理解行程编码如何节省存储空间，我们注意到使用行程编码时仅仅需要保存区域的边界。一般说来，区域边界上的点的数量与区域面积的平方根成比例。由于二值图像法至少需要保存区域外接矩形内所有像素点，所以同二值图像法相比，使用行程编码通常会明显减少存储空间的使用。例如，对于一个 $w \times h$ 的矩形区域，采用行程编码只需要存储 h 个行程，而二值图像法需要保存 $w \times h$ 个像素点（即，wh 或 $\lceil w/8 \rceil h$ 个字节，取决于二值图中每个像素是占一个字节还是一个位）。同理，直径是 d 的圆采用行程编码只需保存 d 个行程，而二值法需要保存至少 $d \times d$ 个像素。因此，采用行程编码通常可以显著地降低内存的使用。另外，行程编码仅存储区域内的点，所以无须检查像素是在区域

lies in the region or not. These two features can save a significant amount of execution time. Also, with this representation, it is straightforward to have regions with negative coordinates. Finally, to represent multiple regions, lists or arrays of run-length encoded regions can be used. Since in this case each region is treated separately, overlapping regions do not pose any problems.

3.1.3 Subpixel-Precise Contours

The data structures we have considered so far are pixel-precise. Often, it is important to extract subpixel-precise data from an image because the application requires an accuracy that is higher than the pixel resolution of the image. The subpixel data can, for example, be extracted with subpixel thresholding (see Section 3.4.3) or subpixel edge extraction (see Section 3.7.3). The results of these operations can be described with subpixel-precise contours. Figure 3.2 displays several example contours. As we can see, the contours can basically be represented as a polygon, i.e., an ordered set of control points $(r_i, c_i)^\top$, where the ordering defines which control points are connected to each other. Since the extraction typically is based on the pixel grid, the distance between the control points of the contour is approximately 1 pixel on average. In the computer, the contours are simply represented as arrays of floating-point row and column coordinates. From Figure 3.2, we can also see that there is a rich topology associated with the contours. For example, contours can be closed (contour 1) or open (contours 2–5). Closed contours are usually represented by having the first contour point identical to the last contour

内还是在区域外。行程编码的这两个优点可以显著减少执行的时间。并且，使用行程编码可以容易地表示含有负坐标的区域。最后一点，为表示多个区域，我们可以使用链表或数组来保存采用行程编码描述的多个区域，此时由于每个区域的信息是被独立保存和处理的，因此处理交叠区域也没有问题。

3.1.3 亚像素精度轮廓

目前为止我们讨论的数据结构都是像素精度的。通常，因为某些应用中需要达到比图像像素分辨率更高的精度，因此从图像中提取亚像素精度数据是很重要的。亚像素数据可以通过亚像素阈值分割（见 3.4.3 节）或亚像素边缘提取（见 3.7.3 节）来获得。这些处理得到的结果可以用亚像素精度轮廓来表示。图 3.2 显示了几个轮廓的例子。可以看到，轮廓基本上可以被描绘成多边形，即，一组排序后的控制点 $(r_i, c_i)^\top$ 的集合，排序是用来说明哪些控制点是彼此相连接的。由于典型的轮廓提取是基于像素网格的，所以轮廓上控制点之间的距离平均约为一个像素。在计算机里，轮廓只是用浮点数表示的横和纵坐标所构成的数组来表示的。从图 3.2 我们还可以发现轮廓有多种空间拓扑结构，比如，轮廓可以是闭的（图中轮廓 1）或是开的（图中轮廓 2~5）。闭合轮廓通常使用同一个坐标来表示轮廓上的第一个点和最后一个点，或使用一个特殊属性来表示。此外，几个轮廓能在一个接合点汇合，如轮廓 3~5。

point or by a special attribute that is stored with the contour. Furthermore, we can see that several contours can meet at a junction point, e.g., contours 3–5. It is sometimes useful to explicitly store this topological information with the contours.

与轮廓一起明确地保存这些拓扑信息有时是有用的。

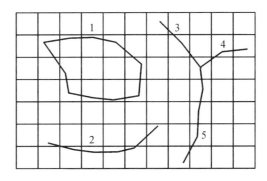

图 3.2　不同类型的亚像素精度轮廓。轮廓 1 是闭合轮廓，轮廓 2~5 均是开轮廓。轮廓 3、4 和 5 相交在接合点

3.2　Image Enhancement

In Chapter 2, we have seen that we have various means at our disposal to obtain a good image quality. The illumination, lenses, cameras, and image acquisition devices all play a crucial role here. However, although we try very hard to select the best possible hardware setup, sometimes the image quality is not sufficient. Therefore, in this section we will take a look at several common techniques for image enhancement.

3.2.1　Gray Value Transformations

Despite our best efforts in controlling the illumination, in some cases it is necessary to modify the gray values of the image. One of the reasons for this may be weak contrast. With controlled illumination, this problem usually only occurs locally. Therefore, we may only need to increase the

3.2　图像增强

在第 2 章中讲述了提高采集图像的质量的各种途径。光源、镜头、摄像机、图像采集卡等都起到了至关重要的作用。尽管我们尽力来选择最佳的硬件设置，但是有时图像还是不够好，因此，本节我们将介绍几种通用的图像增强技术。

3.2.1　灰度值变换

除控制照明光源外，某些情况下通过算法调整图像的灰度值是必要的。调整图像灰度值的一个原因是由于图像对比度太弱，通过对照明光源的调整，这个问题通常只会局部发生，所以，我们也许只需增强局部的对比

contrast locally. Another possible reason for adjusting the gray values may be that the contrast or brightness of the image has changed from the settings that were in effect when we set up our application. For example, illuminations typically age and produce a weaker contrast after some time.

A gray value transformation can be regarded as a point operation. This means that the transformed gray value $t_{r,c}$ depends only on the gray value $g_{r,c}$ in the input image at the same position: $t_{r,c} = f(g_{r,c})$. Here, $f(g)$ is a function that defines the gray value transformation to apply. Note that the domain and range of $f(g)$ typically are $\mathbb{G}_b$, i.e., they are discrete. Therefore, to increase the transformation speed, gray value transformations can be implemented as a lookup table (LUT) by storing the output gray value for each possible input gray value in a table. If we denote the LUT as f_g, we have $t_{r,c} = f_g[g_{r,c}]$, where the [ ] operator denotes table look-up.

度。调整灰度值的另一个原因是图像的对比度或者亮度同最初系统设定时相比已经发生了变化。比如，在工作一段时间后由于光源的老化而造成图像对比度变弱。

灰度值变换可被视为一种点处理。这意味着变换后的灰度值 $t_{r,c}$ 仅仅依赖于输入图像上同一位置的原始灰度值 $t_{r,c} = f(g_{r,c})$。这里，$f(g)$ 是表示进行灰度值变换的函数。注意函数 $f(g)$ 的值域范围通常是 $\mathbb{G}_b$，也就是说是离散的。为提高变换的速度，灰度值变换通常通过查找表（LUT）来进行，即将每个输入灰度值变换后得到的输出值保存在这个查找表内。如果用 f_g 表示 LUT，则 $t_{r,c} = f_g[g_{r,c}]$，此处符号 [ ] 表示的是查表的操作。

3.2.1.1 Contrast Enhancement

The most important gray value transformation is a linear gray value scaling: $f(g) = ag + b$. If $g \in \mathbb{G}_b$, we need to ensure that the output value is also in $\mathbb{G}_b$. Hence, we must clip and round the output gray value as follows:

3.2.1.1 对比度增强

最重要的灰度值变换是灰度值线性比例缩放：$f(g) = ag+b$。若 $g \in \mathbb{G}_b$，我们需要保证输出灰度值也在 $\mathbb{G}_b$ 内。所以，我们需要按等式（3.3）对输出灰度值进行裁剪和四舍五入处理。

$$f(g) = \min(\max(\lfloor ag + b + 0.5 \rfloor, 0), 2^b - 1) \qquad (3.3)$$

For $|a| > 1$, the contrast is increased, while for $|a| < 1$, the contrast is decreased. If $a < 0$, the gray values are inverted. For $b > 0$, the brightness is increased, while for $b < 0$, the brightness is decreased.

Figure 3.3(a) shows a small part of an image of a PCB. The entire image was acquired such

当 $|a| > 1$ 时，对比度增加；当 $|a| < 1$ 时，对比度则降低；当 $a < 0$ 时，灰度值反转。$b > 0$ 时，亮度值增加；$b < 0$ 时，亮度值降低。

图 3.3（a）显示的是印刷电路板图像的一个局部。整幅图像在采集时

that the full range of gray values is used. Three components are visible in the image. As we can see, the contrast of the components is not as good as it could be. Figures 3.3(b)–(e) show the effect of applying a linear gray value transformation with different values for a and b. As we can see from Figure 3.3(e), the component can be seen more clearly for $a = 2$.

已经覆盖了完整的灰度值范围。在图中可看到有三个电子元件，这三个元件的对比度没有达到最理想的效果。图 3.2（b）～（e）显示了通过设置不同的 a 值和 b 值进行灰度值线性比例缩放变换后得到的图像效果。如图 3.3（e）所示，当 $a = 2$ 时图上的元件更清晰。

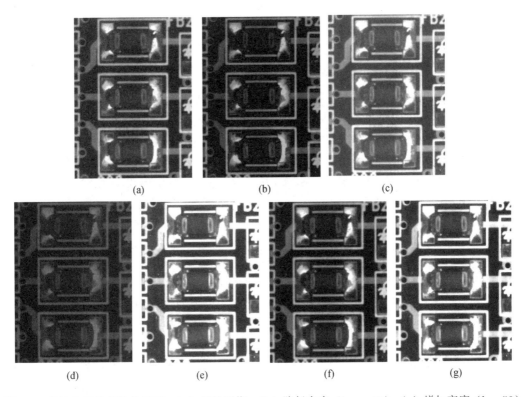

图 3.3　线性灰度值变换的例子。(a) 原始图像；(b) 降低亮度（$b = -50$）；(c) 增加亮度（$b = 50$）；(d) 降低对比度（$a = 0.5$）；(e) 增强对比度（$a = 2$）；(f) 灰度值归一化处理；(g) 鲁棒的灰度值归一化处理（$pl = 0$，$pu = 0.8$）

3.2.1.2　Contrast Normalization

The parameters of the linear gray value transformation must be selected appropriately for each application and adapted to changed illumination conditions. Since this can be quite cumbersome, ideally we would like to have a method that selects

3.2.1.2　对比度归一化

线性灰度值变换的参数必须根据不同的应用以及不同的照明进行合理地选择。由于选择合适的参数是相当麻烦的，我们希望能有一种可基于当前图像情况自动确定 a 和 b 的方法。

a and b automatically based on the conditions in the image. One obvious method to do this is to select the parameters such that the maximum range of the gray value space $\mathbb{G}_b$ is used. This can be done as follows: let $g_{\min}$ and $g_{\max}$ be the minimum and maximum gray value in the ROI under consideration. Then, the maximum range of gray values will be used if $a = (2^b-1)/(g_{\max}-g_{\min})$ and $b = -ag_{\min}$. This transformation can be thought of as a normalization of the contrast. Figure 3.3(f) shows the effect of the contrast normalization of the image in Figure 3.3(a). As we can see, the contrast is not much better than in the original image. This happens because there are specular reflections on the solder, which have the maximum gray value, and because there are very dark parts in the image with a gray value of almost 0. Hence, there is not much room to improve the contrast.

3.2.1.3 Robust Contrast Normalization

The problem with contrast normalization is that a single pixel with a very bright or dark gray value can prevent us from using the desired gray value range. To get a better understanding of this point, we can take a look at the gray value histogram of the image. The gray value histogram is defined as the frequency with which a particular gray value occurs. Let n be the number of points in the ROI under consideration and n_i be the number of pixels that have the gray value i. Then, the gray value histogram is a discrete function with domain $\mathbb{G}_b$ that has the values

$$h_i = \frac{n_i}{n} \tag{3.4}$$

In probabilistic terms, the gray value histogram can be regarded as the probability density of the

一种显而易见的方法是通过设置参数让变换后的图像灰度值覆盖 $\mathbb{G}_b$ 的最大取值范围。可按照如下方法进行操作：用 $g_{\min}$ 和 $g_{\max}$ 分别保存当前 ROI 图像中的最小灰度值和最大灰度值。然后，当 $a = (2^b-1)/(g_{\max}-g_{\min})$ 且 $b = -ag_{\min}$ 时转换后的输出灰度值可以覆盖 $\mathbb{G}_b$ 的最大的取值范围。这种变换可以被理解为灰度值的归一化处理。图 3.3（f）给出了对图 3.3（a）进行灰度归一化处理的结果。我们发现与原始图像相比，对比度并没有明显改善，这是由于图中焊锡的镜面反射造成的，这些反光部分已经达到了最高的灰度值，而又因为图像上暗的区域的灰度值也几乎达到了最小灰度值 0，所以，已经没有多少空间来进行对比度提升处理了。

3.2.1.3 可靠的对比度归一化

如果图像中存在一个非常亮或非常暗的像素值，通过上述的灰度值归一化处理就无法得到想要的灰度值范围。为了更好理解这个问题，我们来看一下图像的灰度值直方图。灰度值直方图显示每一灰度值在图像中出现的频率。用 n 表示在图像感兴趣区域中的像素点总数，n_i 表示图像中灰度值是 i 的像素点的总数，则灰度值直方图的在 $\mathbb{G}_b$ 域的离散函数为

在概率术语中，灰度值直方图表示灰度值 i 出现的概率密度。可按下式计

occurrence of gray value i. We can also compute the cumulative histogram of the image as follows:

$$c_i = \sum_{j=0}^{i} h_j \qquad (3.5)$$

This corresponds to the probability distribution of the gray values.

Figure 3.4 shows the histogram and cumulative histogram of the image in Figure 3.3(a). The specular reflections on the solder create a peak in the histogram at gray value 255. Furthermore, the smallest gray value in the image is 16. This explains why the contrast normalization did not increase the contrast significantly. Note that the dark part of the gray value range contains the most information about the components, while the bright part contains the information corresponding to the specular reflections as well as the printed rectangles on the board. Therefore, to get a more robust contrast normalization, we can simply ignore a part of the histogram that includes a fraction p_l of the darkest gray values and a fraction $1 - p_u$ of the brightest gray values. This can be done based on the cumulative histogram by selecting the smallest gray value for which $c_i \geqslant p_l$ and the largest gray value for which $c_i \leqslant p_u$. This corresponds to intersecting the cumulative histogram with the lines $p = p_l$ and $p = p_u$. Figure 3.4(b) shows two example probability thresholds superimposed on the cumulative histogram. For the example image in Figure 3.3(a), it is best to ignore only the bright gray values that correspond to the reflections and print on the board to get a robust contrast normalization. Figure 3.3(g) shows the result that is obtained with $p_l = 0$ and $p_u = 0.8$. This improves the contrast of the components significantly.

算出图像的累积直方图：

它与灰度值的概率分布相对应。

图 3.4 给出了在图 3.3（a）中所示图像的灰度直方图和累积直方图。从直方图上可以看到，图像中焊点镜面反光部分造成了在灰度值 255 处出现一个尖峰。此外，原始图像中的最小灰度值是 16。这就解释了为何对比度归一化处理没有显著地提高图像的对比度。注意，灰度值范围中暗的部分包含了与电子元件有关的大部分信息，而亮的部分包括了与焊点镜面反光部分和电路板上丝印矩形框相关的信息。因此，要得到可靠的对比度归一化处理方法，我们只要忽略直方图中所包含的一小部分最暗的灰度值以及一小部分最亮的灰度值即可，在累积直方图上通过设定 $c_i \geqslant p_l$ 来确定最小灰度值，通过设定 $c_i \leqslant p_u$ 来确定最大灰度值就可以达到这个目的。从概念上讲，就是在累积直方图上用水平线 $p = p_l$ 和 $p = p_u$ 进行相应的截取。图 3.4（b）显示的是添加了两个概率阈值 p_l 和 p_u 的累积直方图。对于图 3.3（a）的原始图像进行改进后的灰度归一化处理，最好只忽略与焊料镜面反射和电路板上丝印线相关的高亮度灰度值部分。图 3.3（g）给出了在 $p_l = 0$ 且 $p_u = 0.8$ 时进行对比度归一化处理的结果。这显著提高了图中元件的对比度。

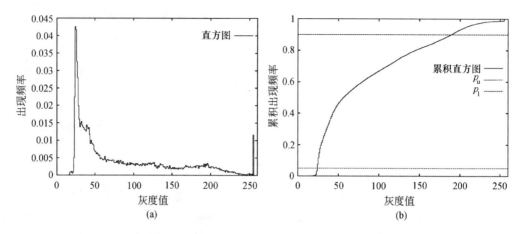

图 3.4 （a）图 3.3（a）图像的灰度值直方图；（b）添加了两个概率阈值 p_l 和 p_u 的累积直方图

Robust contrast normalization is an extremely powerful method that is used, for example, as a feature extraction method for optical character recognition (OCR), where it is used to make the OCR features invariant to illumination changes (see Section 3.14). However, it requires transforming the gray values in the image, which is computationally expensive. If we want to make an algorithm robust to illumination changes, it is often possible to adapt the parameters to the changes in the illumination. For example, if one of the thresholding approches in Section 3.4.1 is used, we simply need to adapt the thresholds.

可靠的对比度归一化处理是一种非常有效的方法，例如，在光学字符识别（见 3.14 节）中进行特征提取时，采用本方法可以使 OCR 特征提取不受照明变化的影响。但是，它需要对图像的灰度值进行变换，这需要大量的计算。如想要使一个算法在光照变化下更加鲁棒，通常更可行的方法是通过调整该算法的参数以适应光照变化，例如，如果使用 3.4.1 节中介绍的图像分割方法，我们仅需要调整阈值。

3.2.2　Radiometric Calibration

Many image processing algorithms rely on the fact that there is a linear correspondence between the energy that the sensor collects and the gray value in the image, namely $G = aE + b$, where E is the energy that falls on the sensor and G is the gray value in the image. Ideally, $b = 0$, which means that twice as much energy on the sensor leads to twice the gray value in the image. However, $b = 0$

3.2.2　辐射标定

很多图像处理算法是建立在传感器收集的能量和图像灰度值之间存在线性响应这一事实基础上的：$G = aE + b$，式中 E 是传感器累积到的能量，G 是图像上的灰度值。理想情况下，$b = 0$，这意味着传感器上两倍的能量会得到 2 倍的图像灰度值。但是，$b = 0$ 对测量的准确度不是必需的，只

is not necessary for measurement accuracy. The only requirement is that the correspondence is linear. If the correspondence is nonlinear, the accuracy of the results returned by these algorithms typically will degrade. Examples of this are the subpixel-precise threshold (see Section 3.4.3), the gray value features (see Section 3.5.2), and, most notably, subpixel-precise edge extraction (see Section 3.7, in particular Section 3.7.4). Unfortunately, sometimes the gray value correspondence is nonlinear, i.e., either the camera or the frame grabber produces a nonlinear response to the energy. If this is the case and we want to perform accurate measurements, we must determine the nonlinear response and invert it. If we apply the inverse response to the images, the resulting images will have a linear response. The process of determining the inverse response function is known as radiometric calibration.

3.2.2.1 Chart-Based Radiometric Calibration

In laboratory settings, traditionally calibrated targets are used to perform the radiometric calibration. Figure 3.5 displays examples of target types that are commonly used. Consequently, the corresponding algorithms are called chart-based. The procedure is to measure the gray values in the different patches and to compare them to the known reflectance of the patches (ISO 14524:2009). This yields a small number of measurements (e.g., 15 independent measurements in the target in Figure 3.5(a) and 12 in the target in Figure 3.5(b)), through which a function is fitted, e.g., a gamma response function that includes gain and offset, given by

需要响应是线性的就可以了。如果是非线性响应，一些算法结果的准确度将会降级，例如，亚像素精度阈值分割（见 3.4.3 节），灰度值特征提取（见 3.5.2 节），尤其对亚像素精度边缘提取影响最为显著（见 3.7 节，特别是 3.7.4 节）。然而，有时灰度值响应就是非线性的，也就是说，摄像机或者图像采集卡产生了对能量的非线性响应。在这种情况下，如果我们需要进行准确的测量就必须确定非线性响应并求其逆响应。如果对非线性响应的图像使用了逆响应，那么结果图像就是线性响应的。确定逆响应函数的过程就被称作辐射标定。

3.2.2.1 基于图表的辐射标定

在实验室配置中，传统的方法是采用经过标定的目标物（典型的标定目标物是灰阶卡，如图 3.5 所示）来进行辐射标定。因此，相应的算法被称作基于图表的标定算法，它测量不同梯度条的灰度值并将这些灰度值与这些梯度条已知的反射系数进行比较（ISO 14524:2009）。这种比较将产生一系列的独立测量（例如，使用图 3.5（a）中的标定板将要进行 15 个独立测量和图 3.5（b）中 12 个独立测量），通过这一组测量完成函数的拟合，比如，由下式给出的包含增益和偏移的一个伽马响应函数：

$$f(g) = (ag + b)^\gamma \tag{3.6}$$

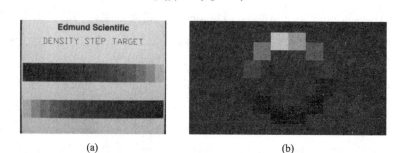

图 3.5　一个密度标定板实例，此类标定板通常被用于实验室中进行辐射标定：（a）密度阶梯标定板（使用线性响应的摄像机采集到的图像）；（b）Twelve-patch 标定板（ISO 14524:2009）（模拟使用线性响应摄像机采集到的图像）

There are several problems with this approach. First of all, it requires a very even illumination throughout the entire field of view in order to be able to determine the gray values of the patches correctly. For example, ISO 14524:2009 requires less than 2% variation of the illuminance incident on the calibration target across the entire target. While this may be achievable in laboratory settings, it is much harder to achieve in a production environment, where the calibration often must be performed. Furthermore, effects like vignetting may lead to an apparent light drop-off toward the border, which also prevents the extraction of the correct gray values. This problem is always present, independent of the environment. Another problem is that there is a great variety of target layouts, and hence it is difficult to implement a general algorithm for finding the patches on the targets and to determine their correspondence to the true reflectances. In addition, the reflectances on the targets are often specified as a linear progression in density, which is related exponentially to the reflectance. For example, the targets in Figure 3.5 both have a linear density progression,

这种方法存在一些问题。首先，它需要整个视野内有非常均匀的照明以便能够正确地检测出梯度条的灰度值。例如，ISO 14524:2009 需要整个标定板上的光照强度变化少于 2%。在实验室中这或许是可以实现的，但在生产现场环境中则很难实现，然而通常在现场环境下是必须进行此类标定的。此外，晕影效应可能导致表面光向边界分散，这也会妨碍我们提取正确的灰度值。这个问题在任何测试环境下都是存在的。另一个问题是有千变万化的标定板布局，所以很难实现一个通用的算法来寻找标定板上的梯度条并检测它们和真实反射系数之间的响应关系。此外，标定板的反射系数与密度通常是线性级数关系，密度与反射系数是指数相关的。例如，图 3.5 所示的标定板都有一个线性密度级数，即一个指数灰度值级数。这就是说拟合曲线上的样本不是均匀分布的，这将导致拟合得到的结果在曲线上样本间距较大的部分存在较低的精度。最后，摄像机响应函数的值域受

i.e., an exponential gray value progression. This means that the samples for the curve fitting are not evenly distributed, which can cause the fitted response to be less accurate in the parts of the curve that contain the samples with the larger spacing. Finally, the range of functions that can be modeled for the camera response is limited to the single function that is fitted through the data.

3.2.2.2 Chartless Radiometric Calibration

Because of the above problems, a radiometric calibration algorithm that does not require any calibration target is highly desirable. These algorithms are called chartless radiometric calibration. They are based on taking several images of the same scene with different exposures. The exposure can be varied by changing the aperture stop of the lens or by varying the exposure time of the camera. Since the aperture stop can be set less accurately than the exposure time, and since the exposure time of most industrial cameras can be controlled very accurately in software, varying the exposure time is the preferred method of acquiring images with different exposures. The advantages of this approach are that no calibration targets are required and that the images do not require an even illumination. Furthermore, the range of possible gray values can be covered with multiple images instead of a single image as required by the algorithms that use calibration targets. The only requirement on the image content is that there should be no gaps in the histograms of the different images within the gray value range that each image covers. Furthermore, with a little extra effort, even overexposed (i.e., saturated) images can be handled.

通过数据拟合得到的单一函数限制。

3.2.2.2 无图表辐射标定

因为上述这些原因，我们更需要不使用标定板的辐射标定算法。这类算法被称作"无图表辐射标定"。它们基于在不同曝光下对同一场景所拍摄的一系列图像来实现。不同的曝光既可以通过改变镜头的光圈也可通过改变摄像机曝光时间来实现。因为改变光圈比使用不同长度的曝光时间精度低，且因为大部分工业摄像机可以通过软件来精确地设定曝光时间的长度，所以设定不同长度的曝光时间以得到不同曝光条件下的一系列图像是我们的首选方法。本算法的好处是既不需要标定板，也不需要在视野内有均匀的照明。而且，所有可能的灰度值取值范围能够被多幅图像所覆盖，而使用标定板的标定算法仅能使用一幅图像所覆盖的灰度值范围。对图像内容的唯一要求就是在每幅图像的灰度值直方图中，图像灰度值所覆盖的范围内不应存在空隙。而且，稍加努力，即使过度曝光的（即饱和的）图像也可以被处理。

To derive an algorithm for chartless calibration, let us examine what two images with different exposures tell us about the response function. We know that the gray value G in the image is a nonlinear function r of the energy E that falls onto the sensor during the exposure e (Mann and Mann, 2001):

$$G = r(eE) \qquad (3.7)$$

Note that e is proportional to the exposure time and proportional to the area of the entrance pupil of the lens, i.e., proportional to $1/F^2$, where F is the f-number of the lens. As described above, in industrial applications we typically leave the aperture stop constant and vary the exposure time. Therefore, we can think of e as the exposure time.

The goal of the radiometric calibration is to determine the inverse response $q = r^{-1}$. The inverse response can be applied to an image via an LUT to achieve a linear response.

Now, let us assume that we have acquired two images with different exposures e_1 and e_2. Hence, we know that $G_1 = r(e_1 E)$ and $G_2 = r(e_2 E)$. By applying the inverse response q to both equations, we obtain $q(G_1) = e_1 E$ and $q(G_2) = e_2 E$. We can now divide the two equations to eliminate the unknown energy E, and obtain

$$\frac{q(G_1)}{q(G_2)} = \frac{e_1}{e_2} = e_{1,2} \qquad (3.8)$$

As we can see, q depends only on the gray values in the images and on the ratio $e_{1,2}$ of the exposures, but not on the exposures e_1 and e_2 themselves. Equation (3.8) is the defining equation for all chartless radiometric calibration algorithms.

为推导出无图表标定算法，让我们先看看使用不同曝光拍摄的两幅图像可以提供给我们哪些与响应函数有关的信息。我们知道图像中的灰度值 G 是传感器在曝光过程 r 中累计能量 E 的一个非线性函数 e（Mann and Manner，2001）：

$$(3.7)$$

式（3.7）中 e 与曝光时间成比例，也与镜头入射光孔的面积成比例，即与 $1/F^2$ 成比例。此处 F 是镜头的 f 值。如上所述，在工业应用中我们通常固定光圈而改变曝光时间。因此，可以将 e 看作曝光时间。

辐射标定的目的是确定逆响应 $q = r^{-1}$。通过一个查找表 LUT，逆响应能够被应用到一幅图像上以实现输出图像灰度值与传感器累积的能量之间的线性响应。

现在，假设我们用不同的曝光 e_1 和 e_2 采集了两幅图像。故而，$G_1 = r(e_1 E)$ 且 $G_2 = r(e_2 E)$。通过对这两个等式应用逆响应 q，得到 $q(G_1) = e_1 E$ 和 $q(G_2) = e_2 E$。将两个等式相除来消去未知能量 E 可以得到：

可以看到，q 仅依赖于图像的灰度值和曝光比值 $e_{1,2}$ 而不依赖于 e_1 和 e_2 的取值。式（3.8）是所有无图表辐射标定算法的定义式。

One way to determine q based on Eq. (3.8) is to discretize q in an LUT. Thus, $q_i = q(G_i)$. To derive a linear algorithm to determine q, we can take logarithms on both sides of Eq. (3.8) to obtain $\log(q_1/q_2) = \log e_{1,2}$, i.e., $\log(q_1) - \log(q_2) = \log e_{1,2}$ (Mann and Mann, 2001). If we set $Q_i = \log(q_i)$ and $E_{1,2} = \log e_{1,2}$, each pixel in the image pair yields one linear equation for the inverse response function $\boldsymbol{Q}$:

基于公式（3.8）计算 q 的一个方法是将 q 离散到一个查找表中。这样，$q_i = q(G_i)$。为推导出一个线性算法以确定 q，对等式（3.8）的两侧进行对数运算，$\log(q_1/q_2) = \log e_{1,2}$，即 $\log(q_1) - \log(q_2) = \log e_{1,2}$（Mann and Mann, 2001）。如果设定 $Q_i = \log(q_i)$ 且 $E_{1,2} = \log e_{1,2}$，对于逆响应函数 $\boldsymbol{Q}$，图像中的每个像素都可根据该像素在两幅图像中的灰度值得出一个线性方程：

$$Q_1 - Q_2 = E_{1,2} \tag{3.9}$$

Hence, we obtain a linear equation system $\boldsymbol{AQ} = \boldsymbol{E}$, where $\boldsymbol{Q}$ is a vector of the LUT for the logarithmic inverse response function, while $\boldsymbol{A}$ is a matrix with 256 columns for byte images. The matrix $\boldsymbol{A}$ and the vector $\boldsymbol{E}$ have as many rows as ther are pixels in the image, e.g., 307 200 for a 640 × 480 image. Therefore, this equation system is much too large to be solved in an acceptable time. To derive an algorithm that solves the equation system in an acceptable time, we can note that each row of the equation system has the following form:

因此，得到一个线性方程组 $\boldsymbol{AQ} = \boldsymbol{E}$，此处向量 $\boldsymbol{Q}$ 表示对数逆响应函数的查找表，而 $\boldsymbol{A}$ 是一个 256 列的矩阵（对于 8 位图像）。矩阵 $\boldsymbol{A}$ 和 $\boldsymbol{E}$ 的行数与图像中像素个数相等，如果图像大小为 640 × 480，那么矩阵 $\boldsymbol{A}$ 和向量 $\boldsymbol{E}$ 有 307 200 行。这个线性方程组过于庞大，因此不能在一个可接受的时间内计算得到结果。为推导出更快速的算法，注意到此方程组中的每个方程都存在类似如下形式：

$$(0 \cdots 0\ 1\ 0 \cdots 0\ -1\ 0 \cdots 0)\boldsymbol{Q} = E_{1,2} \tag{3.10}$$

The indices of the 1 and −1 entries in the above equation are determined by the gray values in the first and second image. Note that each pair of gray values that occurs multiple times leads to several identical rows in $\boldsymbol{A}$. Also note that $\boldsymbol{AQ} = \boldsymbol{E}$ is an overdetermined equation system, which can be solved through the normal equations $\boldsymbol{A}^\top \boldsymbol{AQ} = \boldsymbol{A}^\top \boldsymbol{E}$. This means that each row that occurs k times in $\boldsymbol{A}$ will have weight k in the normal equations. The same behavior is obtained by multiplying the row (3.10) that corresponds to the

上式中的 1 所在的位置与第一幅图像的灰度值相等，而 −1 出现的位置和第二幅图像的灰度值相等。注意多次出现的一对灰度值会在 $\boldsymbol{A}$ 中产生内容重复的行。也要注意 $\boldsymbol{AQ} = \boldsymbol{E}$ 是一个超定方程组，这个超定方程组可以通过正规方程组 $\boldsymbol{A}^\top \boldsymbol{AQ} = \boldsymbol{A}^\top \boldsymbol{E}$ 进行求解。这就意味在 $\boldsymbol{A}$ 中出现 k 次的每一行在正规方程中的权都为 k。通过将与这一对灰度值相对应的行 (3.10) 与 $\sqrt{k}$ 相乘并使该行仅在 $\boldsymbol{A}$

gray value pair by $\sqrt{k}$ and to include that row only once in A. This typically reduces the number of rows in A from several hundred thousand to a few thousand, and thus makes the solution of the equation system feasible.

The simplest method to determine k is to compute the 2D histogram of the image pair. The 2D histogram determines how often gray value i occurs in the first image while gray value j occurs in the second image at the same position. Hence, for byte images, the 2D histogram is a 256×256 image in which the column coordinate indicates the gray value in the first image, while the row coordinate indicates the gray value in the second image. It is obvious that the 2D histogram contains the required values of k. We will see examples of 2D histograms below.

Note that the discussion so far has assumed that the calibration is performed from a single image pair. It is, however, very simple to include multiple images in the calibration since additional images provide the same type of equations as in (3.10), and can thus simply be added to A. This makes it much easier to cover the entire range of gray values. Thus, we can start with a fully exposed image and successively reduce the exposure time until we reach an image in which the smallest possible gray values are assumed. We could even start with a slightly overexposed image to ensure that the highest gray values are assumed. However, in this case we must take care that the overexposed (saturated) pixels are excluded from A because they violate the defining equation (3.8). This is a very tricky problem to solve in general since some cameras exhibit a bizarre saturation behavior. Suffice it to say that for many cameras

中包含一次即可得到同样的结果。这样就显著地减少了矩阵 A 中的行数，此法可将 A 的行数从几十万行减少到几千行，这样就使对上述方程组的求解更加可行。

确定 k 的最简单的方法是计算这对图像的二维直方图。从二维直方图中可以得到第一幅图像中灰度值为 i 的那些像素在第二幅图像的同样位置上灰度值是 j 的情况所出现的频率。因此，对于 8 位图像，其二维直方图是一个 256×256 图像，图像的横坐标表示第一幅图像上的灰度值，而纵坐标表示第二幅图像上的灰度值。很显然，从二维直方图中可以得到所需的参数 k 的值。我们将在下面看到二维直方图的例子。

截止到目前所进行的讨论都是假定采用一对图像进行标定的情况。但是，使用更多的图像进行标定也是很简单的，因为新增加的图像提供的方程与等式（3.10）类型相同，因此只须将这些方程添加到矩阵 A 中即可。这就使得覆盖全部灰度值范围变得更容易。因此，可以从一幅曝光充分的图像开始，不断降低曝光时间直到采集的图像中包含可能出现的最小灰度值。我们甚至可以从一幅曝光轻微过度的图像开始以确保最高的灰度值被覆盖。但是，此时必须注意将过曝的（饱和）像素从 A 中排除，因为这些像素点已经不符合式（3.8）的原理。通常这是一个非常棘手的问题，因为一些摄像机具有怪异的饱和行为。对于大部分的摄像机而言，只要将那些灰度值最高的像素点从 A 中排除出

it is sufficient to exclude pixels with the maximum gray value from $\boldsymbol{A}$.

Despite the fact that $\boldsymbol{A}$ has many more rows than columns, the solution $\boldsymbol{Q}$ is not uniquely determined because we cannot determine the absolute value of the energy $\boldsymbol{E}$ that falls onto the sensor. Hence, the rank of $\boldsymbol{A}$ is at most 255 for byte images. To solve this problem, we could arbitrarily require $q(255) = 255$, i.e., scale the inverse response function such that the maximum gray value range is used. Since the equations are solved in a logarithmic space, it is slightly more convenient to require $q(255) = 1$ and to scale the inverse response to the full gray value range later. With this, we obtain one additional equation of the form

去就足够了。

尽管矩阵 $\boldsymbol{A}$ 有很多的行和列，但求解出的 $\boldsymbol{Q}$ 并不唯一，因为我们无法确定传感器上累积的能量 $\boldsymbol{E}$ 的绝对值。因此，对于 8 位图像，矩阵 $\boldsymbol{A}$ 的秩最大是 255。为解决此问题，我们可以直接令 $q(255) = 255$，即按比例缩放逆响应曲线以使其可覆盖最大的灰度值范围。由于该方程在对数空间中进行求解，因此可以首先令 $q(255) = 1$ 并在求出结果后再缩放逆响应函数以使其覆盖最大的灰度值范围，这样稍微方便一些。由此，我们得到一个附加的方程：

$$(0 \quad \cdots \quad 0 \quad k)\boldsymbol{Q} = 0 \qquad (3.11)$$

To enforce the constraint $q(255) = 1$, the constant k must be chosen such that Eq. (3.11) has the same weight as the sum of all other equations (3.10), i.e., $k = \sqrt{wh}$, where w and h are the width and height of the image.

Even with this normalization, we still face some practical problems. One problem is that, if the images contain very little noise, the equations in (3.10) can become decoupled, and hence do not provide a unique solution for $\boldsymbol{Q}$. Another problem is that, if the possible range of gray values is not completely covered by the images, there are no equations for the range of gray values that are not covered. Hence, the equation system will become singular. Both problems can be solved by introducing smoothness constraints for $\boldsymbol{Q}$, which couple the equations and enable an extrapolation of $\boldsymbol{Q}$ into the range of gray values that is not cov-

要强制令 $q(255) = 1$，确定常量 k 时必须使等式（3.11）的权与其他所有等式（3.10）的权的总和相等，即 $k = \sqrt{wh}$，这里 w 和 h 分别是图像的宽度和高度。

即便采用了这种归一化处理，仍然要面对很多实际的问题。一个问题是当图像中的噪声非常小时，等式（3.10）会变成解耦方程从而不能计算得到 $\boldsymbol{Q}$ 的唯一解；另一个问题是如果采集的图像不能覆盖最大的灰度值范围，就不会存在与没有覆盖到的灰度值相对应的方程，所以，方程组会变成奇异方程组。这两个问题都可以通过引入 $\boldsymbol{Q}$ 的平滑约束来解决。使用这个约束可以耦合这些方程式并可以通过外推法得到当前图像没有覆盖到的灰度值所对应的 $\boldsymbol{Q}$ 值。平滑约

ered by the images. The smoothness constraints require that the second derivative of $\boldsymbol{Q}$ should be small. Hence, for byte images, they lead to 254 equations of the form

$$(0 \; \cdots \; 0 \; s \; -2s \; s \; 0 \; \cdots \; 0)\boldsymbol{Q} = 0 \qquad (3.12)$$

The parameter s determines the amount of smoothness that is required. As for Eq. (3.11), s must be chosen such that Eq. (3.12) has the same weight as the sum of all the other equations, i.e., $s = c\sqrt{wh}$, where c is a small number. Empirically, $c = 4$ works well for a wide range of cameras.

The approach of tabulating the inverse response q has two slight drawbacks. First, if the camera has a resolution of more than 8 bits, the equation system and 2D histograms become very large. Second, the smoothness constraints lead to straight lines in the logarithmic representation of q, i.e., exponential curves in the normal representation of q in the range of gray values that is not covered by the images. Therefore, it sometimes may be preferable to model the inverse response as a polynomial (Mitsunaga and Nayar, 1999). This model also leads to linear equations for the coefficients of the polynomial. Since polynomials are not very robust in extrapolation into areas in which no constraints exist, we also must add smoothness constraints in this case by requiring that the second derivative of the polynomial is small. Because this is done in the original representation of q, the smoothness constraints will extrapolate straight lines into the gray value range that is not covered.

Let us now consider two cameras: one with a linear response and one with a strong gamma

束要求 $\boldsymbol{Q}$ 的二阶导数要小。所以，对于 8 位图像，会产生 254 个如下形式的方程

参数 s 决定所需要的平滑程度。与等式（3.11）相似，确定 s 时必须使等式（3.12）的权与其他全部的方程的权的总和相等，即 $s = c\sqrt{wh}$，c 是一个比较小的数，按照经验，对大部分摄像机来说，$c = 4$ 时效果比较好。

将逆响应 q 制成表格的方法存在两个小缺点。首先，如果摄像机的位深超过 8 位，那么方程组和二维直方图都将变得非常庞大。另外，平滑约束将会导致在图像未覆盖的灰度范围内，q 的对数表示法中使用的外推直线，将在 q 的一般表示法中对应为指数曲线。因此，有时更可取的方法是建立逆响应的多项式模型，参考文献（Mitsunaga and Nayar, 1999）。此模型会引出求解多项式系数的多个线性方程。由于多项式在外推到不存在约束的区域时鲁棒性不是非常好，所以这种情况下同样不得不加入平滑约束，这就需要多项式的二阶导数要小。因为这里 q 的平滑约束是加在 q 的原始表示法中，因此使用直线外推就可得到图像未覆盖灰度值区域内的 q。

现在有两个摄像机：一个摄像机具备线性响应，另一个摄像机则有很

response, i.e., with a small γ in Eq. (3.6), and hence with a large γ in the inverse response q. Figure 3.6 displays the 2D histograms of two images taken with each camera with an exposure ratio of 0.5. Note that in both cases, the values in the 2D histogram correspond to a line. The only difference is the slope of the line. A different slope, however, could also be caused by a different exposure ratio. Hence, we can see that it is quite important to know the exposure ratios precisely if we want to perform radiometric calibration.

强的伽马响应，也就是说在等式 (3.6) 中的 γ 值很小，因此逆响应 q 中 γ 值很大。图 3.6 中显示了每个摄像机在曝光比为 0.5 时各采集两幅图像后得到的二维直方图。注意两个二维直方图上的值都对应为一条直线，只是线的斜率有所不同。然而，一个不同的斜率也可能是由于一个不同的曝光比引起的。所以，如果想进行辐射标定，精确地知道曝光比是非常重要的。

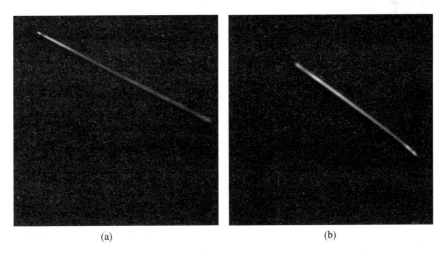

图 3.6 （a）曝光比 0.5 时使用一个线性摄像机拍摄的一对图像计算所得到的二维直方图；（b）曝光比 0.5 时采用一强伽马响应的摄像机拍摄的一对图像计算所得到的二维直方图。为更好地将结果图形化，这里使用了一个平方根查找表来显示二维直方图。注意，这两种情况中二维直方图上的值都对应为一条直线，所以不知道确切的曝光比是不能区分线性响应还是伽马响应的

To conclude this section, we give two examples of radiometric calibration. The first camera is a linear camera. Here, five images were acquired with exposure times of 32, 16, 8, 4, and 2 ms, as shown in Figure 3.7(a). The calibrated inverse response curve is shown in Figure 3.7(b). Note that the response is linear, but the camera has set a slight offset in the amplifier, which prevents very small gray values from being assumed. The second camera is a camera with a gamma

我们以两个辐射标定的示例来结束本节的内容。第一只摄像机是一个线性摄像机，标定中使用了它在曝光时间分别是 32、16、8、4 和 2 ms 时采集的 5 幅图像，如图 3.7（a）所示。经过辐射标定后得到的逆响应曲线如图 3.7（b）所示。注意该逆响应也是线性的，但由于摄像机的放大器中存在一个微小的偏移量，所以对灰度值非常小的部分有影响。第二只摄像机

response. In this case, six images were taken with exposure times of 30, 20, 10, 5, 2.5, and 1.25 ms, as shown in Figure 3.7(c). The calibrated inverse response curve is shown in Figure 3.7(d). Note the strong gamma response of the camera. The 2D histograms in Figure 3.6 were computed from the second and third brightest images in both sequences.

是一个伽马响应的摄像机。标定中使用了它在曝光时间分别为 30、20、10、5、2.5 和 1.25 ms 时采集的 6 幅图像，如图 3.7（c）所示。经过辐射标定后得到的逆响应曲线如图 3.7（d）所示。注意图中摄像机的强伽马响应。图 3.6 所示的二维直方图是通过每个图像序列中亮度排在第二和第三的图像对计算得到的。

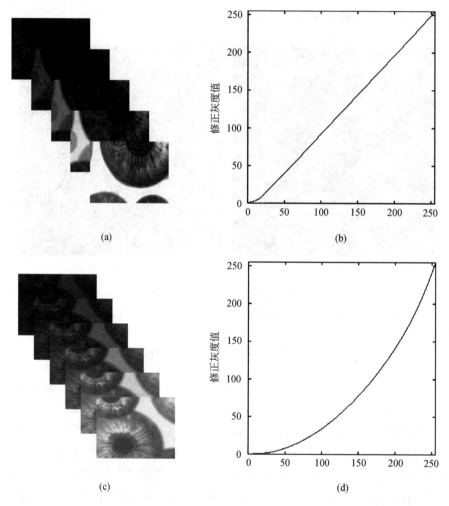

图 3.7 （a）用一线性摄像机在曝光时间分别为 32、16、8、4 和 2 ms 时采集得到的 5 幅图像；（b）标定后得到的逆响应曲线。注意响应是线性的，但由于摄像机的放大器中存在微小的偏移量，所以阻碍了非常小的灰度值的假设；（c）用一强伽马响应摄像机在曝光时间分别为 30、20、10、5、2.5 和 1.25 ms 时采集得到的 6 幅图像；（d）标定后得到的逆响应曲线，注意摄像机的强伽马响应特征

3.2.3 Image Smoothing

Every image contains some degree of noise. For the purposes of this chapter, noise can be regarded as random changes in the gray values, which occur for various reasons, e.g., because of the randomness of the photon flux. In most cases, the noise in the image will need to be suppressed by using image smoothing operators.

In a more formalized manner, noise can be regarded as a stationary stochastic process (Papoulis and Pillai, 2002). This means that the true gray value $g_{r,c}$ is disturbed by noise $n_{r,c}$ to get the observed gray value: $\hat{g}_{r,c} = g_{r,c} + n_{r,c}$. We can regard the noise $n_{r,c}$ as a random variable with mean 0 and variance σ^2 for every pixel. We can assume a mean of 0 for the noise because any mean different from 0 would constitute a systematic bias of the observed gray values, which we could not detect anyway. "Stationary" means that the noise does not depend on the position in the image, i.e., it is identically distributed for each pixel. In particular, σ^2 is assumed constant throughout the image. The last assumption is a convenient abstraction that does not necessarily hold because the variance of the noise sometimes depends on the gray values in the image. However, we will assume that the noise is always "stationary".

Figure 3.8 shows an image of an edge from a real application. The noise is clearly visible in the bright patch in Figure 3.8(a) and in the horizontal gray value profile in Figure 3.8(b). Figures 3.8(c) and (d) show the actual noise in the image. How

3.2.3 图像平滑

每幅图像都包含某种程度的噪声。为了达成本章的叙述目标，我们将噪声看作是由多种原因造成的灰度值的随机变化，比如由光子通量的随机性而产生的噪声。在大多数情况下，图像中的噪声必须通过图像平滑处理进行抑制。

在更正规的方式下，噪声被视为一种平稳随机过程（Papoulis and pillai，2002）。这表示真正的像素值 $g_{r,c}$ 被一噪声项 $n_{r,c}$ 干扰后得到的灰度值：$\hat{g}_{r,c} = g_{r,c} + n_{r,c}$。我们可以视噪声 $n_{r,c}$ 为一个针对每一像素的平均值为 0 且方差是 σ^2 的随机变量。假定噪声的平均值是 0 是因为任何非 0 的平均值将构成一个系统偏置值，该系统偏置值会作用于每个获取到的像素值上，而这是我们无法检测的。"平稳"是指噪声与图像上像素的位置无关，也就是，对于图像上的每个像素，噪声都是同样地分布。特别是方差 σ^2 被认为在整个图像上都是不变的。后一个假设，即方差不变，是为了方便起见而引入的一个抽象概念，所以不一定必须满足，因为噪声的方差有时是由图像上的灰度值来决定的。但是，我们将假设噪声总是"平稳"的。

图 3.8 是来自于实际应用中的一个边缘的图像。在图 3.8(a) 的白色部分中可清晰地看到噪声，该图的水平灰度值剖面线如图 3.8（b）所示。图 3.8（c）和（d）分别给出了该图像的

the noise has been calculated is explained below. It can be seen that there is slightly more noise in the dark patch of the image.

实际噪声图和噪声图对应的水平灰度值剖面线。如何得到的实际噪声图我们稍候再解释。可看到图像中的黑色部分包含的噪声稍多。

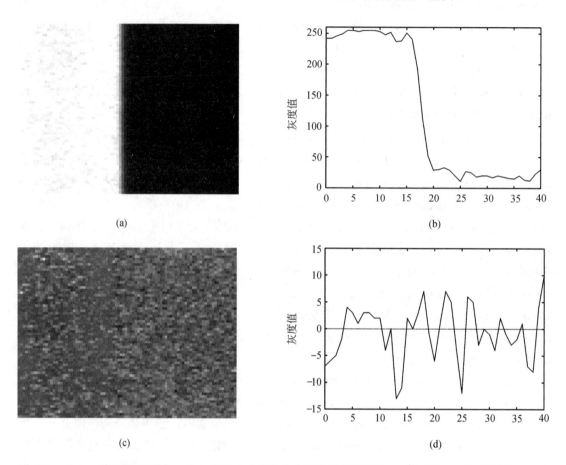

图 3.8 （a）一幅边缘的图像；（b）图像中心部分的水平灰度值剖面线；（c）放大 5 倍后的图（a）中所包含的噪声；（d）噪声的水平灰度值剖面线

3.2.3.1 Temporal Averaging

With the above discussion in mind, noise suppression can be regarded as a stochastic estimation problem, i.e., given the observed noisy gray values $\hat{g}_{r,c}$, we want to estimate the true gray values $g_{r,c}$.

An obvious method to reduce the noise is to acquire multiple images of the same scene and to

3.2.3.1 时域平均

通过以上的讨论，噪声抑制能被视为随机估计问题，也就是说，用实测到的包含有噪声的像素值 $\hat{g}_{r,c}$ 来估计像素值的真值 $g_{r,c}$。

一个最显而易见的降噪方法就是采集同一场景的多幅图像并对这些图

simply average these images. Since the images are taken at different times, we will refer to this method as temporal averaging or the temporal mean. If we acquire n images, the temporal average is given by

$$g_{r,c} = \frac{1}{n}\sum_{i=1}^{n} \hat{g}_{r,c;i} \qquad (3.13)$$

where $\hat{g}_{r,c;i}$ denotes the noisy gray value at position $(r,c)^\top$ in image i. This approach is frequently used in X-ray inspection systems, which inherently produce quite noisy images. From probability theory (Papoulis and Pillai, 2002), we know that the variance of the noise is reduced by a factor of n by this estimation: $\sigma_m^2 = \sigma^2/n$. Consequently, the standard deviation of the noise is reduced by a factor of $\sqrt{n}$. Figure 3.9 shows the result of acquiring 20 images of an edge and computing the temporal average. Compared to Figure 3.9(a), which shows one of the 20 images, the noise has been reduced by a factor of $\sqrt{20} \approx 4.5$, as can be seen from Figure 3.9(b). Since this temporally averaged image is a very good estimate for the true gray values, we can subtract it from any of the images that were used in the averaging to obtain the noise in that image. This is how the image in Figure 3.8(c) was computed.

3.2.3.2 Mean Filter

One of the drawbacks of the temporal averaging is that we have to acquire multiple images to reduce the noise. This is not very attractive if the speed of the application is important. Therefore, other means for reducing the noise are required in most cases. Ideally, we would like to use only one image to estimate the true gray value. If we turn to the theory of stochastic processes again, we see

像进行平均。由于多幅图像是在不同时间采集的，我们称此方法为时域平均或时域平均值。如果采集了 n 幅图像，时域平均值的计算方法如下：

上式中 $\hat{g}_{r,c;i}$ 代表第 i 幅图像上位置 $(r,c)^\top$ 处的灰度值。本降噪方法在 X 光检测系统中应用广泛，因为通常 X 光系统中采集的图像会有较多的噪声。从概率论中得知（Papoulis and Pillai, 2002），$\sigma_m^2 = \sigma^2/n$，所以噪声的方差降低到原来的 $1/n$。噪声的标准偏差相应地降低到原来的 $1/\sqrt{n}$。图 3.9 给出了采集 20 幅图像进行时域平均后得到的结果，图 3.9（a）是 20 幅图像中的某一幅图，比较后可以看到噪声下降到原来的 $1/(\sqrt{20} \approx 4.5)$，如图 3.9（b）所示。因为这个时域平均图像是对灰度值真值的一种出色估计，所以在用来计算平均图像的一组原始图像中，任意一幅图像减去此平均图像就可以计算得到此幅图像中所包含的噪声。图 3.8（c）就是采用此方法计算得到的。

3.2.3.2 均值滤波器

时域平均法的缺点之一就是必须采集多幅图像才能进行噪声抑制。在对速度要求很高的机器视觉应用中此方法就不那么有吸引力了。因此，在大多数情况下，需要其他的降噪方法。理想情况下，我们希望仅仅在一幅图像上就可以对灰度值真值进行估计。如果再看一下随机过程理论，我们就

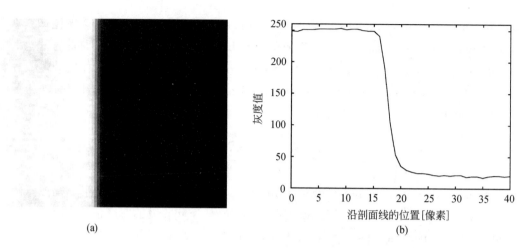

图 3.9 （a）通过对 20 幅边缘图像平均后得到的边缘图像；（b）图像中心部分的水平灰度值剖面线

that the temporal averaging can be replaced with a spatial averaging if the stochastic process, i.e., the image, is ergodic (Papoulis and Pillai, 2002). This is precisely the definition of ergodicity, and we will assume for the moment that it holds for our images. Then, the spatial average or spatial mean can be computed over a window (also called a mask) of $(2n+1) \times (2m+1)$ pixels as follows:

可以发现：如果某随机过程（即某图像）是遍历的（Papoulis and Pillai, 2002），时域平均法就可以被空间平均法所取代。这是遍历性的定义，假设此时它对图像也是适用的。这样，空间平均或者空间平均值可以通过像素数是 $(2n+1) \times (2m+1)$ 的一个窗口（也称作掩码）按如下方法计算：

$$g_{r,c} = \frac{1}{(2n+1)(2m+1)} \sum_{i=-n}^{n} \sum_{j=-m}^{m} \hat{g}_{r-i,c-j} \quad (3.14)$$

This spatial averaging operation is also called a mean filter. As in the case of temporal averaging, the noise variance is reduced by a factor that corresponds to the number of measurements that are used to calculate the average, i.e., by $(2n+1)(2m+1)$. Figure 3.10 shows the result of smoothing the image of Figure 3.8 with a 5×5 mean filter. The standard deviation of the noise is reduced by a factor of 5, which is approximately the same as the temporal averaging in Figure 3.9. However, we can see that the edge is no longer as sharp as with temporal averaging. This happens, of course, because the images are not ergodic in

此空间平均操作也被称为均值滤波器。与时域平均法类似，采用了几次平均值测量，噪声的方差就降低到原来的几分之一，也就是降低到 $1/((2n+1)(2m+1))$。图 3.10 是采用 5×5 均值滤波器对图 3.8 进行平滑操作的结果。噪声的标准偏差下降到 $1/5$，这与图 3.9 中采用时域平均法平滑后的结果几乎一致。但是，空间平均法平滑后的边缘不如时域平均法的锐利。发生这种情况是因为一般而言，图像并不是遍历的，只有在图中亮度一致的区域才是遍历的。因此，与时

general, but are only in areas of constant intensity. Therefore, in contrast to the temporal mean, the spatial mean filter blurs edges.

域平均法相比，空间均值滤波器使边缘模糊。

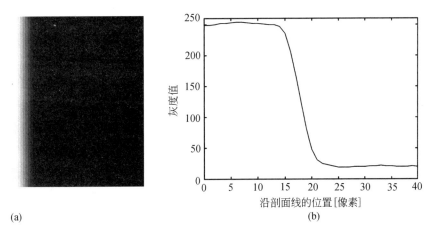

图 3.10　（a）对图 3.8（a）采用 5×5 均值滤波器平滑后的结果；（b）图像中心部分的水平灰度值剖面线

3.2.3.3　Border Treatment of Filters

In Eq. (3.14), we have ignored the fact that the image has a finite extent. Therefore, if the mask is close to the image border, it will partially stick out of the image and consequently will access undefined gray values. To solve this problem, several approaches are possible. A very simple approach is to calculate the filter only for pixels for which the mask lies completely within the image. This means that the output image is smaller than the input image, which is not very helpful if multiple filtering operations are applied in sequence. We could also define that the gray values outside the image are 0. For the mean filter, this would mean that the result of the filter would become progressively darker as the pixels get closer to the image border. This is also not desirable. Another approach would be to use the closest gray value on the image border for pixels outside the image. This approach would still create unwanted edges

3.2.3.3　滤波器边缘处理

等式（3.14）中忽略了图像在二维空间上是有限的这一事实。因此，如果掩码靠近图像的边界，掩码中就会有部分延伸到图像外而导致访问到未定义的灰度值。有几种办法可以解决这一问题。很简单的一个办法是只针对掩码能完全覆盖的图像内的像素进行滤波计算。这就意味着处理后的结果图像要比处理前的输入图像小，这样不便于对输入图像进行一系列的滤波处理。我们也可以将图像外的灰度值都定义为 0。对于均值滤波器，这意味着在滤波后的结果图像上，像素越靠近图像边界越暗，这也是不能令人满意的。另一个方法是令图像边界以外的像素灰度值等于图像边界上最邻近的像素点的灰度值。此方法也会在结果图像的边界上产生原本不希望产生的边缘。因此，通常以图像边界

at the image border. Therefore, typically the gray values are mirrored at the image border. This creates the least amount of artifacts in the result.

3.2.3.4 Runtime Complexity of Filters

As was mentioned above, noise reduction from a single image is preferable for reasons of speed. Therefore, let us take a look at the number of operations involved in the calculation of the mean filter. If the mean filter is implemented based on Eq. (3.14), the number of operations will be $(2n+1)(2m+1)$ for each pixel in the image, i.e., the calculation will have the complexity $O(whmn)$, where w and h are the width and height of the image, respectively. For $w = 640$, $h = 480$, and $m = n = 5$ (i.e., an 11×11 filter), the algorithm will perform 37 171 200 additions and 307 200 divisions. This is quite a substantial number of operations, so we should try to reduce the operation count as much as possible. One way to do this is to use the associative law of the addition of real numbers as Eq. (3.15):

3.2.3.4 滤波器运行时间复杂度

如前所述，从单幅图像上降低噪声对于对速度有一定要求的机器视觉应用是更适用的，因此，我们来考察一下与均值滤波器相关的操作次数。基于等式（3.14）进行均值滤波处理时，对图像中的每个像素需要进行 $(2n+1)(2m+1)$ 次操作，即计算的复杂度是 $O(whmn)$，此处的 w 和 h 分别代表图像的宽度和高度。对于 $w = 640$，$h = 480$，且 $m = n = 5$（即，一个 11×11 滤波器），算法将执行 37 171 200 次加法和 307 200 次除法。这个计算次数是相当可观的，所以我们应试图尽可能地降低运算次数。降低运算次数的一个办法是采用加法结合律，如等式（3.15）：

$$g_{r,c} = \frac{1}{(2n+1)(2m+1)} \sum_{i=-n}^{n} \left(\sum_{j=-m}^{m} \hat{g}_{r-i,c-j} \right) \qquad (3.15)$$

This may seem like a trivial observation, but if we look closer we can see that the term in parentheses only needs to be computed once and can be stored, e.g., in a temporary image. Effectively, this means that we are first computing the sums in the column direction of the input image, saving them in a temporary image, and then computing the sums in the row direction of the temporary image. Hence, the double sum in Eq. (3.14) of complexity $O(nm)$ is replaced by two sums of

这种改进看起来微不足道，但如果我们仔细观察就可以发现，圆括号内包含的项仅需要计算一次且可以保存下来，比如保存到某一临时图像中。这意味着我们可以先计算输入图像在纵向上的和并将此结果保存在一幅临时图像中，然后再计算临时图像在横向上的和。因此，等式（3.14）中的复杂度 $O(nm)$ 被复杂度 $O(n+m)$ 所取代。整个计算的复杂度也相应地从

total complexity $O(n + m)$. Consequently, the complexity drops from $O(whmn)$ to $O(wh(m + n))$. With the above numbers, now only 6 758 400 additions are required. The above transformation is so important that it has its own name: whenever a filter calculation allows a decomposition into separate row and column sums, the filter is called separable. It is obviously of great advantage if a filter is separable, and it is often the best speed improvement that can be achieved. In this case, however, it is not the best we can do. Let us take a look at the column sum, i.e., the part in parentheses in Eq. (3.15), and let the result of the column sum be denoted by $t_{r,c}$. Then, we have

$$t_{r,c} = \sum_{j=-m}^{m} \hat{g}_{r,c-j} = t_{r,c-1} + \hat{g}_{r,c+m} - \hat{g}_{r,c-m-1} \tag{3.16}$$

i.e., the sum at position $(r, c)^\top$ can be computed based on the already computed sum at position $(r, c-1)^\top$ with just two additions. The same also holds, of course, for the row sums. The result of this is that we need to compute the complete sum only once for the first column or row, and can then update it very efficiently. With this, the total complexity is $O(wh)$. Note that the mask size does not influence the run time in this implementation. Again, since this kind of transformation is so important, it has a special name. Whenever a filter can be implemented with this kind of updating scheme based on previously computed values, it is called a recursive filter. For the above example, the mean filter requires just 1 238 880 additions for the entire image. This is more than a factor of 30 faster for this example than the naive implementation based on Eq. (3.14). Of course, the advantage becomes even greater for larger mask sizes.

$O(whmn)$ 下降到 $O(wh(m+n))$。采用此方法后只需要进行 6 758 400 次加法。上面讨论的变换是很重要的，它有其自己的名字：只要一个滤波器在运算时允许分别在行和列上求和，那么此滤波器就被称为可分滤波器。可分滤波器的优点是显而易见的，并且通常是提升计算速度的最佳方法。但是在这里，我们还有更佳的方法。让我们查看纵向的和，也就是等式（3.15）中圆括号内的部分，用 $t_{r,c}$ 来表示纵向的和，就有：

即，在位置 $(r, c)^\top$ 处的和可以基于已经计算出来的位置 $(r, c-1)^\top$ 处的和以及两次加法得到。这个规则当然也适用于横向。应用这个规则，我们仅需要在第一行和第一列上计算完整的和，然后在这些结果上高效率地进行少量加法就可以得到需要的新的结果。这样，计算的总复杂度是 $O(wh)$。注意此时掩码窗口的尺寸不影响滤波器的执行时间。同可分滤波器一样，这种重要的变换规则也有自己的名字。只要某滤波器在运算时能采用此种机制，即在前一个计算出的值的基础上计算出新的值，这种滤波器叫做递归滤波器。这样，对于上面的例子，均值滤波只需要 1 238 880 次加法就可以完成整幅图像的处理，执行速度较基于等式（3.14）的原始算法快了 30 倍。当然，在使用大尺寸掩码窗口时其优势就更明显了。

3.2.3.5 Linear Filters

In the above discussion, we have called the process of spatial averaging a mean filter without defining what is meant by the word "filter." We can define a filter as an operation that takes a function as input and produces a function as output. Since images can be regarded as functions (see Section 3.1.1), for our purposes a filter transforms an image into another image.

The mean filter is an instance of a linear filter. Linear filters are characterized by the following property: applying a filter to a linear combination of two input images yields the same result as applying the filter to the two images and then computing the linear combination. If we denote the linear filter by h, and the two images by f and g, we have

$$h\{af(\boldsymbol{p}) + bg(\boldsymbol{p})\} = ah\{f(\boldsymbol{p})\} + bh\{g(\boldsymbol{p})\} \tag{3.17}$$

where $\boldsymbol{p} = (r,c)^\top$ denotes a point in the image and the $\{\ \}$ operator denotes the application of the filter. Linear filters can be computed by a convolution. For a one-dimensional (1D) function on a continuous domain, the convolution is given by

$$f * h = (f * h)(x) = \int_{-\infty}^{\infty} f(t)\, h(x - t)\, \mathrm{d}t \tag{3.18}$$

Here, f is the image function and the filter h is specified by another function called the convolution kernel or the filter mask. Similarly, for 2D functions we have

$$f * h = (f * h)(r,c) = \int_{-\infty}^{\infty}\int_{-\infty}^{\infty} f(u,v)\, h(r-u, c-v)\, \mathrm{d}u\, \mathrm{d}v \tag{3.19}$$

3.2.3.5 线性滤波器

在上面的讨论中，我们把空间平均处理的过程称为均值滤波，但并没有说明此处的"滤波"一词是指什么。滤波指的是一个操作，此操作采用某个函数作为输入并产生某个函数作为输出。既然图像能被看作是函数（见3.1.1节），所以，滤波可以将一幅图像变换成另一幅图像。

均值滤波器是线性滤波器中的一个例子。线性滤波器的特点如下：应用一个滤波器到两幅输入图像的一个线性组合上所产生的结果，与对这两幅图像分别应用此滤波器后将结果按同一线性规则进行组合得到的结果完全一致。如果用 h 来表示此线性滤波器，用 f 和 g 表示这两幅图像，则有：

式中 $\boldsymbol{p} = (r,c)^\top$ 表示图像上的一个点，操作符 $\{\ \}$ 表示使用滤波器。线性滤波器可以通过卷积来计算。对于在连续域上的一维函数卷积操作如下：

这里，f 表示图像函数，滤波器 h 由另一函数确定，此函数被称为滤波器掩码的卷积核。同理，对于二维函数：

For functions with discrete domains, the integrals are replaced by sums:

对于离散域上的函数，积分被求和取代：

$$f * h = \sum_{i=-\infty}^{\infty} \sum_{j=-\infty}^{\infty} f_{i,j} \, h_{r-i, c-j} \qquad (3.20)$$

The integrals and sums are formally taken over an infinite domain. Of course, to be able to compute the convolution in a finite amount of time, the filter $h_{r,c}$ must be 0 for sufficiently large r and c. For example, the mean filter is given by

在正规情况下，积分和求和要在无限域上进行。当然，为了能够在有限时间内计算卷积，滤波器 $h_{r,c}$ 在大于 r 和 c 的情况下必须为 0。例如，均值滤波器可表示为：

$$h_{r,c} = \begin{cases} \dfrac{1}{(2n+1)(2m+1)}, & |r| \leqslant n \wedge |c| \leqslant m \\ 0, & \text{其他} \end{cases} \qquad (3.21)$$

The notion of separability can be extended for arbitrary linear filters. If $h(r,c)$ can be decomposed as $h(r,c) = s(r)t(c)$ (or as $h_{r,c} = s_r t_c$), then h is called separable. As for the mean filter, we can factor out s in this case to get a more efficient implementation:

可分滤波器的概念可以被扩展到任意线性滤波器上。如果 $h(r,c)$ 能被分解为 $h(r,c) = s(r)t(c)$（或 $h_{r,c} = s_r t_c$），h 被称为可分的。此时为更有效率地实现均值滤波，可以提出因子 s：

$$\begin{aligned} f * h &= \sum_{i=-n}^{n} \sum_{j=-n}^{n} f_{i,j} \, h_{r-i, c-j} = \sum_{i=-n}^{n} \sum_{j=-n}^{n} f_{i,j} \, s_{r-i} t_{c-j} \\ &= \sum_{i=-n}^{n} s_{r-i} \left(\sum_{j=-n}^{n} f_{i,j} \, t_{c-j} \right) \end{aligned} \qquad (3.22)$$

Obviously, separable filters have the same speed advantage as the separable implementation of the mean filter. Therefore, separable filters are preferred over non-separable filters. There is also a definition for recursive linear filters, which we cannot cover in detail. The interested reader is referred to Deriche (1990). Recursive linear filters have the same speed advantage as the recursive implementation of the mean filter, i.e., the run time does not depend on the filter size. Unfortunately, many interesting filters cannot be implemented as recursive filters; usually they can only be approximated by a recursive filter.

显而易见地，可分滤波器与前面讨论的均值滤波的可分实现有着同样的速度优势。因此，相对于不可分滤波器，应该首选使用可分滤波器，对递归线性滤波器的应用也是同样道理，这里就不深入讨论了，感兴趣的读者可以参见文献（Deriche, 1990）。递归线性滤波器与均值滤波器的递归实现有着同样的速度优势，即运行时间与滤波器的尺寸无关。遗憾的是，很多常用的滤波器不能以递归滤波器的方式来实现，它们通常只能被近似为一个递归滤波器。

3.2.3.6 Frequency Response of the Mean Filter

Although the mean filter produces good results, it is not the optimum smoothing filter. To see this, we can note that noise primarily manifests itself as high-frequency fluctuations of the gray values in the image. Ideally, we would like a smoothing filter to remove these high-frequency fluctuations. To see how well the mean filter performs this task, we can examine how the mean filter responds to certain frequencies in the image. The theory of how to do this is provided by the Fourier transform (see Section 3.2.4). Figure 3.11(a) shows the frequency response of a 3×3 mean filter. In this plot, the row and column coordinates represent the frequencies as cycles per pixel. If both coordinates are 0, this corresponds to a frequency of 0 cycles per pixel, which represents the average gray value in the image. At the other extreme, row and column coordinates of ± 0.5 represent the highest possible frequencies in the image. For example, the frequencies with column coordinate 0 and row coordinate ± 0.5 correspond to a grid with alternating one-pixel-wide vertical bright and dark lines. From Figure 3.11(a), we can see that the 3×3 mean filter removes certain frequencies completely. These are the points for which the response has a value of 0. They occur for relatively high frequencies. However, we can also see that the highest frequencies are not removed completely. To illustrate this, Figure 3.11(b) shows an image with one-pixel-wide lines spaced three pixels apart. From Figure 3.11(c), we can see that this frequency is completely removed by the 3×3 mean filter: the

3.2.3.6 均值滤波器的频率响应

尽管均值滤波器提供了不错的结果，但它还不是最适宜的平滑滤波器。因为我们可以注意到，噪声主要是以图像中灰度值高频波动的方式暴露出来。理想情况下，我们希望能有一个平滑滤波器来消除这些高频波动。为了确定某均值滤波器消除高频波动的效果，我们可以检查此均值滤波器对图像中某些频率的响应情况。傅里叶变换（见 3.2.4 节）给我们提供了此操作的理论依据。图 3.11（a）显示了一个 3×3 均值滤波器的频率响应。在此坐标图中，横坐标和纵坐标表示像素周期的频率，如果横坐标和纵坐标都为 0，则对应像素周期的频率是 0，这表示的是图像中的灰度值的平均值。另外一个极端情况，横纵坐标值是 ± 0.5 时代表的是图像中可能出现的最高频率。举例说明，在横坐标为 0、纵坐标为 ± 0.5 时的频率所对应的图像是宽度均为一个像素、交替出现的黑色和白色垂线所组成的栅格。从图 3.11（a）中我们可以看到此 3×3 的均值滤波器完全消除了某些频率，即响应值为 0 的那些点，它们发生在相对高频的部分。然而，我们也能看到频率最高的部分没有被完全消除。用图解说明，图 3.11（b）显示的图像是由多条宽度为一个像素、自间距为三个像素的线组成。从图 3.11（c）中我们看到这种频率是可以用此 3×3 均值滤波器完全消除的：输出图像中只包含一种固定的灰度值。如果

output image has a constant gray value. If we change the spacing of the lines to two pixels, as in Figure 3.11(d), we can see from Figure 3.11(e) that this higher frequency is not removed completely. This is an undesirable behavior since it means that noise is not removed completely by the mean filter. Note also that the polarity of the lines has been reversed by the mean filter, which is also undesirable. This is caused by the negative parts of the frequency response. Furthermore, from Figure 3.11(a) we can see that the frequency response of the mean filter is not rotationally symmetric, i.e., it is anisotropic. This means that diagonal structures are smoothed differently than horizontal or vertical structures.

我们将线的间距修改为两个像素，如图 3.11（d）所示，那么从图 3.11（e）中我们能发现这种更高的频率是不能被完全消除的。这不是我们期望的结果，因为通过此均值滤波器并不能彻底地消除噪声。我们也注意到了滤波后线的亮暗极性发生了反转，这也不是我们想得到的结果，这种反转是由于频率响应中的负值部分造成的。更有甚者，从图 3.11（a）中我们也能看到均值滤波器的频率响应不是旋转对称的，也就是说是各向异性的。这意味着倾斜方向上的结构与水平或垂直方向上的结构在应用同一滤波器时会经历不一样的平滑处理。

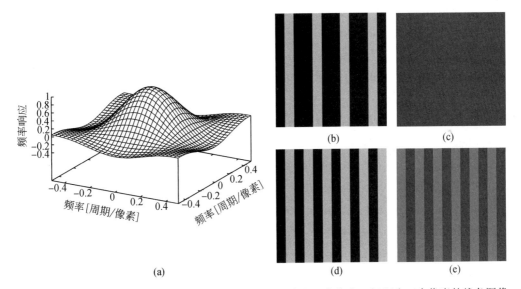

图 3.11　（a）3×3 均值滤波器的频率响应；（b）宽度为一个像素、间距为三个像素的线条图像；（c）对 (b) 所示图像应用此 3×3 均值滤波器后得到的结果，注意所有的线都被平滑掉了；（d）宽度为一个像素、间距为两个像素的线条图像；（e）对 (d) 所示图像应用此 3×3 均值滤波器后得到的结果，注意尽管与图（b）的线条相比图（d）的线条有着更高的频率，但线条并没有被彻底地平滑掉，另外线条的亮暗也反转了

3.2.3.7　Gaussian Filter

Because the mean filter has the above drawbacks, the question of which smoothing filter is optimal

3.2.3.7　高斯滤波器

由于均值滤波器有上述缺点，自然会出现这样一个问题，即采用何种

arises. One way to approach this problem is to define certain natural criteria that the smoothing filter should fulfill, and then to search for the filters that fulfill the desired criteria. The first natural criterion is that the filter should be linear. This is natural because we can imagine an image being composed of multiple objects in an additive manner. Hence, the filter output should be a linear combination of the input. Furthermore, the filter should be position-invariant, i.e., it should produce the same results no matter where an object is in the image. This is automatically fulfilled for linear filters. Also, we would like the filter to be rotation-invariant, i.e., isotropic, so that it produces the same result independent of the orientation of the objects in the image. As we saw above, the mean filter does not fulfill this criterion. We would also like to control the amount of smoothing (noise reduction) that is being performed. Therefore, the filter should have a parameter t that can be used to control the smoothing, where higher values of t indicate more smoothing. For the mean filter, this corresponds to the mask sizes m and n. We have already seen that the mean filter does not suppress all high frequencies, i.e., noise, in the image. Therefore, a criterion that describes the noise suppression of the filter in the image should be added. One such criterion is that, the larger t gets, the more local maxima in the image should be eliminated. This is a very intuitive criterion, as can be seen in Figure 3.8(a), where many local maxima due to noise can be detected. Note that, because of linearity, we only need to require maxima to be eliminated. This automatically implies that local minima are eli-

平滑滤波器最理想。一个解决办法是先定义一组自然准则，平滑滤波器必须完全满足这些准则。然后寻找可以完全满足这些准则的全部滤波器。第一个自然准则就是滤波器应该是线性的，这是因为我们能想象一幅图像是由多个物体以相加的方式组合而成，因此，滤波处理的输出应该是输入的一个线性组合。而且，滤波应是与位置无关的，也就是说无论一个物体出现在图像中的哪个位置，滤波都应该能产生同样的结果。线性滤波器是可以满足这个标准的。我们也希望滤波器是旋转对称的，也就是各向同性的，这样对图像中不同方向的物体应用滤波器后产生的结果是一致的。正如我们在上面看到的，均值滤波器是不符合这个标准的。我们也希望能控制正在执行的平滑（噪声降低）处理的程度，所以，滤波器应该有一个参数 t 来控制平滑的程度，当 t 值越高时表示平滑程度越大。对于均值滤波器而言，这是与掩码尺寸 m 和 n 相对应的。在上面我们已经看到了均值滤波器不能抑制图像中全部的高频部分，即全部的噪声，所以应该增加一个准则，此准则用来描述图像中滤波器的噪声抑制情况。一个可行的准则是：得到的 t 值越大，图像中就应该有更多的局部最大值被消除。这是相当直观的准则，如在图 3.8(a) 中看到的，由噪声造成了很多的局部最大值。请注意，由于滤波器是线性的，所以我们仅需要要求消除局部最大值，这自然意味着局部最小值也被消除掉了。

minated as well. Finally, sometimes we would like to execute the smoothing filter several times in succession. If we do this, we would also like to have a simple means to predict the result of the combined filtering. Therefore, first filtering with t and then with s should be identical to a single filter operation with $t + s$. It can be shown that, among all smoothing filters, the Gaussian filter is the only filter that fulfills all of the above criteria (Lindeberg, 1994). Other natural criteria for a smoothing filter have been proposed (Witkin, 1983; Babaud et al., 1986; Florack et al., 1992), which also single out the Gaussian filter as the optimal smoothing filter.

In one dimension, the Gaussian filter is given by

$$g_\sigma(x) = \frac{1}{\sqrt{2\pi}\,\sigma} e^{-x^2/(2\sigma^2)} \qquad (3.23)$$

This is the function that also defines the probability density of a normally distributed random variable. In two dimensions, the Gaussian filter is given by

$$g_\sigma(r,c) = \frac{1}{2\pi\sigma^2} e^{-(r^2+c^2)/(2\sigma^2)} = \frac{1}{\sqrt{2\pi}\,\sigma} e^{-r^2/(2\sigma^2)} \frac{1}{\sqrt{2\pi}\,\sigma} e^{-c^2/(2\sigma^2)}$$
$$= g_\sigma(r) g_\sigma(c) \qquad (3.24)$$

Hence, the Gaussian filter is separable. Therefore, it can be computed very efficiently. In fact, it is the only isotropic, separable smoothing filter. Unfortunately, it cannot be implemented recursively. However, some recursive approximations have been proposed (Deriche, 1993; Young and van Vliet, 1995). Figure 3.12 shows plots of 1D and 2D Gaussian filters with $\sigma = 1$. The frequency response of a Gaussian filter is also a Gaussian function, albeit with σ inverted (see

最后，我们有时希望连续地多次执行平滑滤波器。如果这样做时，我们也会需要一个简单的办法来预测组合滤波操作的结果。因此，先用 t 滤波然后用 s 滤波应该与用 $t + s$ 进行的单一滤波器处理是一样的。在所有的平滑滤波器中，高斯滤波器是唯一可以符合上述全部准则的滤波器（Lindeberg, 1994）。其他一些已经被提出的针对平滑滤波器的自然准则（Witkin, 1983; Babaud et al., 1986; Florack et al., 1992）也表明高斯滤波器是最理想的平滑滤波器。

一维高斯滤波器由下式给出：

这个函数也描述了正态分布随机变量的概率密度。二维高斯滤波器由下式给出：

因此，高斯滤波器是可分的，所以可以被非常高效率地计算出来。实际上，它是唯一的各向同性的可分平滑滤波器。不幸的是，它不能以递归的方式来执行。但是，一些递归的近似方法已经被提出了（Deriche, 1993; Young and van Vlict, 1995）。图 3.12 的坐标系中显示了当 $\sigma = 1$ 时的一维高斯滤波器和二维高斯滤波器。高斯滤波器的频率响应也是一个高斯函数，只

Eq. (3.32)). Therefore, Figure 3.12(b) also gives a qualitative impression of the frequency response of the Gaussian filter. It can be seen that the Gaussian filter suppresses high frequencies much better than the mean filter.

是 σ 变成了其自身的倒数（见等式（3.32））。所以，图 3.12（b）也是高斯滤波器的频率响应的定性描述。与均值滤波器相比，高斯滤波器更好地抑制了高频部分。

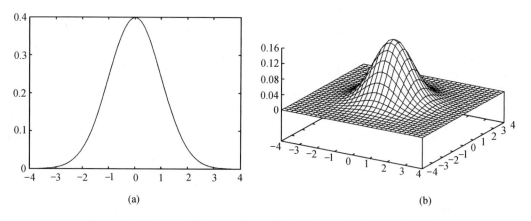

图 3.12 (a) $\sigma = 1$ 的一维高斯滤波器；(b) $\sigma = 1$ 的二维高斯滤波器

3.2.3.8 Noise Suppression by Linear Filters

3.2.3.8 线性滤波器的噪声抑制

Like the mean filter, any linear filter will change the variance of the noise in the image. It can be shown that, for a linear filter $h(r,c)$ or $h_{r,c}$, the noise variance is multiplied by the following factor (Papoulis and Pillai, 2002):

与均值滤波器类似，任何线性滤波器都将会改变图像所包含噪声的方差。对于一个线性滤波器 $h(r,c)$ 或 $h_{r,c}$，噪声的方差将乘以如下倍数（Papoulis and Pillai, 2002）：

$$\int_{-\infty}^{\infty}\int_{-\infty}^{\infty} h(r,c)^2 \, dr \, dc \quad \text{or} \quad \sum_{i=-\infty}^{\infty}\sum_{j=-\infty}^{\infty} h_{r,c}^2 \tag{3.25}$$

For a Gaussian filter, this factor is $1/(4\pi\sigma^2)$. If we compare this to a mean filter with a square mask with parameter n, we see that, to get the same noise reduction with the Gaussian filter, we need to set $\sigma = (2n+1)/(2\sqrt{\pi})$. For example, a 5×5 mean filter has the same noise reduction effect as a Gaussian filter with $\sigma \approx 1.41$.

对于一个高斯滤波器，这个倍数是 $1/(4\pi\sigma^2)$。如果与一个 $n \times n$ 的均值滤波器相比，为了让高斯滤波器具有同样的噪声抑制效果，我们必须让 $\sigma = (2n+1)/(2\sqrt{\pi})$。例如，一个 5×5 的均值滤波器与一个 $\sigma \approx 1.41$ 的高斯滤波器有着同样的噪声抑制效果。

Figure 3.13 compares the results of the Gaussian filter with those of the mean filter of an equivalent size. For small filter sizes ($\sigma = 1.41$ and 5×5), there is hardly any noticeable difference between the results. However, if larger filter sizes are used, it becomes clear that the mean filter turns the edge into a ramp, leading to a badly defined edge that is also visually quite hard to locate, whereas the Gaussian filter produces a much sharper edge. Hence, we can see that the Gaussian filter produces better results, and consequently it is usually the preferred smoothing filter if the quality of the results is the primary concern. If speed is the primary concern, then the mean filter is preferable.

图 3.13 比较了应用高斯滤波器及效果相当的均值滤波器后得到的结果。在滤波器尺寸较小时（$\sigma = 1.41$ 和 5×5），结果之间没有什么显著的差异。但是，如果使用更大的滤波器尺寸，很明显均值滤波器将边缘变成了灰度值缓慢变化的斜坡，这导致了即便用肉眼观察都很难准确地确定边缘的位置，而高斯滤波器产生了相对锐利的边缘。因此，高斯滤波器能输出更好的结果，在更关注结果的质量时，高斯滤波器是首选的平滑滤波器，而在更关注执行速度时，首选使用均值滤波器。

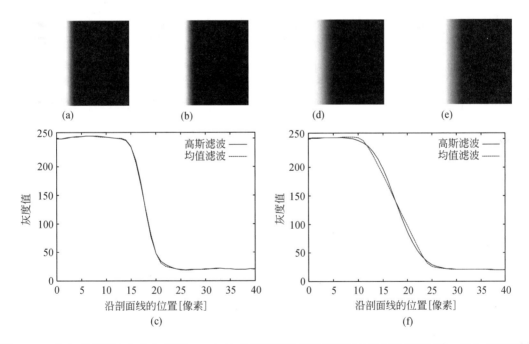

图 3.13 （a）是对图 3.8（a）使用 $\sigma = 1.41$ 的高斯滤波器进行平滑处理后的边缘图像；（b）是对同样图像使用 5×5 均值滤波器平滑后的边缘图像，它们的灰度值剖面线见（c）；注意此例中两种不同滤波器平滑处理的结果基本一致；（d）是使用 $\sigma = 3.67$ 的高斯滤波器进行平滑处理后的图像；（e）是使用 13×13 均值滤波器平滑后的图像，它们的灰度值剖面线见（f）。注意此时用均值滤波器处理后的边缘出现了灰度值缓慢变化的斜坡，我们很难准确定义边缘的位置，但高斯滤波器却可以给出更锐利的边缘

3.2.3.9 Median and Rank Filters

We close this section with a nonlinear filter that can also be used for noise suppression. The mean filter is a particular estimator for the mean value of a sample of random values. From probability theory, we know that other estimators are also possible, most notably the median of the samples. The median is defined as the value for which 50% of the values in the probability distribution of the samples are smaller and 50% are larger.

From a practical point of view, if the sample set contains n values g_i, $i = 0, \cdots, n-1$, we sort the values g_i in ascending order to get s_i, and then select the value median$(g_i) = s_{n/2}$. Hence, we can obtain a median filter by calculating the median instead of the mean inside a window around the current pixel. Let W denote the window, e.g., a $(2n+1) \times (2m+1)$ rectangle as for the mean filter. Then the median filter is given by

$$g_{r,c} = \underset{(i,j)^\top \in W}{\mathrm{median}}\, \hat{g}_{r-i,\,c-j} \qquad (3.26)$$

With sophisticated algorithms, it is possible to obtain a run time complexity (even for arbitrary mask shapes) that is comparable to that of a separable linear filter: $O(whm)$, where m is the number of horizontal boundary pixels of the mask, i.e., the pixels that are at the left or right border of a run of pixels in the mask (Huang et al., 1979; Van Droogenbroeck and Talbot, 1996). For rectangular masks, it is possible to construct an algorithm with constant run time per pixel (analogous to a recursive implementation of a linear filter) (Perreault and Hébert, 2007).

3.2.3.9 中值和排序滤波器

在本节内容结束之前，我们讨论一下同样可以用于噪声抑制的非线性滤波器。均值滤波是对随机值样本平均值的一个特殊估计。根据概率论，我们知道其他的估计方法也是可能的，比如中值就是一个特别适合的估计方法。中值被定义为这样一个值：在样本概率分布中 50% 的值要小于此值而另外 50% 要大于此值。

从实用的角度出发，如果样本包含 n 个值 g_i, $i = 0, \cdots, n-1$，我们以升序对 g_i 进行排序后得到 s_i，那么 g_i 的中值 median$(g_i) = s_{n/2}$。这样，我们通过计算当前窗口内覆盖像素的中值而不是平均值就得到一个中值滤波器。用 W 表示窗口，比如，一个 $(2n+1) \times (2m+1)$ 矩形窗口，这与使用均值滤波器时对窗口的定义是一样的。此时中值滤波器由下式给出：

通过使用优化的算法，中值滤波器的运行时间（甚至任意掩码形状）和一个可分线性滤波器的运行时间相差无几：$O(whn)$，其中 m 是掩码横向边缘像素数量，即掩码左边界或者右边界的一系列像素（Huang et al., 1979; Van Droogenbroeck and Talbot, 1996）。对于矩形掩码，构造的算法，每个像素运行时间是固定的（类似于线性滤波器的递归实现）（Perreault and Hebert, 2007）。

The properties of the median filter are quite difficult to analyze. We can note, however, that it performs no averaging of the input gray values, but simply selects one of them. This can lead to surprising results. For example, the result of applying a 3 × 3 median filter to the image in Figure 3.11(b) would be a completely black image—the median filter would remove the bright lines because they cover less than 50% of the window. This property can sometimes be used to remove objects completely from an image. On the other hand, applying a 3 × 3 median filter to the image in Figure 3.11(d) would swap the bright and dark lines. This result is as undesirable as the result of the mean filter on the same image.

On the edge image of Figure 3.8(a), the median filter produces quite good results, as can be seen from Figure 3.14. In particular, it should be noted that the median filter preserves the sharp-

虽然中值滤波器的特性很难分析，但我们还是注意到它不是对输入的灰度值进行平均，而仅仅是从这些灰度值中选择一个值。这往往能导致很多令人惊讶的结果。比如，对图3.11(b)应用一个3×3中值滤波器的结果将是一幅全黑的图像。中值滤波器会消除全部的亮线条是因为它们在窗口中小于50%。这个特性有时被用来将图像中的某些物体完全删除。另一方面，对图3.11（d）应用一个3×3中值滤波器将会把图中线条的明暗反转。这个结果与使用均值滤波器输出的结果类似，也不是我们期望得到的结果。

对图3.8（a）的边缘图像，中值滤波器给出了相当不错的处理结果，见图3.14。特别注意的是，即便使用大窗口尺寸的中值滤波器也不会降低

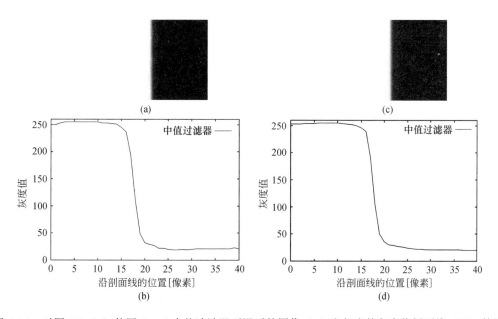

图 3.14 对图 3.8（a）使用 5×5 中值滤波器平滑后的图像（a）和相应的灰度值剖面线（b）；使用 13×13 中值滤波器平滑处理后的图像（c）和相应的灰度值剖面线（d）。注意中值滤波器在很大程度上保留了边缘的锐利程度

ness of the edge even for large filter sizes. However, it cannot be predicted if and by how much the position of the edge is changed by the median filter, which is possible for the linear filters. Furthermore, we cannot estimate how much noise is removed by the median filter, in contrast to the linear filters. Therefore, for high-accuracy measurements, the Gaussian filter should be used.

Finally, it should be mentioned that the median filter is a special case of the more general class of rank filters. Instead of selecting the median $s_{n/2}$ of the sorted gray values, the rank filter would select the sorted gray value at a particular rank r, i.e., s_r. We will see other cases of rank operators in Section 3.6.2.

3.2.4 Fourier Transform

3.2.4.1 Continuous Fourier Transform

In the previous section, we considered the frequency responses of the mean and Gaussian filters. In this section, we will take a look at the theory that is used to derive the frequency response: the Fourier transform (Brigham, 1988; Press et al., 2007). The Fourier transform of a 1D function $h(x)$ is given by

$$H(f) = \int_{-\infty}^{\infty} h(x)\, e^{2\pi i f x} \, dx \tag{3.27}$$

It transforms the function $h(x)$ from the spatial domain into the frequency domain, i.e., $h(x)$, a function of the position x, is transformed into $H(f)$, a function of the frequency f. Note that $H(f)$ is in general a complex number. Because of Eq. (3.27) and the identity $e^{ix} = \cos x + i \sin x$, we can think of $h(x)$ as being composed of sine and cosine waves of different frequencies and different amplitudes. Then $H(f)$ describes precisely which frequency occurs with which amplitude and with which phase (overlaying sine and cosine terms of the same frequency simply leads to a phase-shifted sine wave). The inverse Fourier transform from the frequency domain to the spatial domain is given by

各个频率以何种振幅和相位叠加（叠加相同频率的正弦项和余弦项即得到一个相位移动的正弦波）。从频率域到空间域的傅里叶逆变换由下式给出：

$$h(x) = \int_{-\infty}^{\infty} H(f) e^{-2\pi i f x} \, df \qquad (3.28)$$

Because the Fourier transform is invertible, it is best to think of $h(x)$ and $H(f)$ as being two different representations of the same function.

因为傅里叶变换是可逆变换，将 $h(x)$ 和 $H(f)$ 视为同一函数的两种不同表现形式。

In 2D, the Fourier transform and its inverse are given by

二维傅里叶变换和逆变换如下：

$$H(u,v) = \int_{-\infty}^{\infty} \int_{-\infty}^{\infty} h(r,c) e^{2\pi i (ur+vc)} \, dr \, dc \qquad (3.29)$$

$$h(r,c) = \int_{-\infty}^{\infty} \int_{-\infty}^{\infty} H(u,v) e^{-2\pi i (ur+vc)} \, du \, dv \qquad (3.30)$$

In image processing, $h(r,c)$ is an image, for which the position $(r,c)^\top$ is given in pixels. Consequently, the frequencies $(u,v)^\top$ are given in cycles per pixel.

在图像处理中，$h(r,c)$ 是一幅图像，图像位置 $(r,c)^\top$ 是以像素形式给出的。相应地，频率 $(u,v)^\top$ 表示的图像中每个像素的周期数。

Among the many interesting properties of the Fourier transform, probably the most interesting one is that a convolution in the spatial domain is transformed into a simple multiplication in the frequency domain: $(g * h)(r,c) \Leftrightarrow G(u,v)H(u,v)$, where the convolution is given by Eq. (3.19). Hence, a convolution can be performed by transforming the image and the filter into the frequency domain, multiplying the two results, and transforming the result back into the spatial domain.

在傅里叶变换众多令人感兴趣的特征中，最有趣的一个性质应该是在空间域的卷积被变换为在频率域的一个简单相乘：$(g * h)(r,c) \Leftrightarrow G(u,v)H(u,v)$，卷积由等式（3.19）给出。因此，通过将图像及使用的滤波器变换到频率域，将此二者在频率域的转换结果相乘后再转换回空间域就实现了卷积操作。

Note that the convolution attenuates the frequency content $G(u,v)$ of the image $g(r,c)$ by the frequency response $H(u,v)$ of the filter. This justifies the analysis of the smoothing behavior of the mean and Gaussian filters that we have performed in Sections 3.2.3.6 and 3.2.3.7. To make this analysis more precise, we can compute the Fourier transform of the mean filter in Eq. (3.21). It is given by

$$H(u,v) = \frac{1}{(2n+1)(2m+1)} \operatorname{sinc}((2n+1)u)\operatorname{sinc}((2m+1)v) \qquad (3.31)$$

where $\operatorname{sinc} x = (\sin \pi x)/(\pi x)$. See Figure 3.11 for a plot of the response of the 3×3 mean filter. Similarly, the Fourier transform of the Gaussian filter in Eq. (3.24) is given by

$$H(u,v) = e^{-2\pi^2 \sigma^2 (u^2+v^2)} \qquad (3.32)$$

Hence, the Fourier transform of the Gaussian filter is again a Gaussian function, albeit with σ inverted. Note that, in both cases, the frequency response becomes narrower if the filter size is increased. This is a relation that holds in general: $h(x/a) \Leftrightarrow |a|H(af)$.

Another interesting property of the Fourier transform is that it can be used to compute the correlation

$$g \star h = (g \star h)(r,c) = \int_{-\infty}^{\infty} \int_{-\infty}^{\infty} g(r+u, c+v) h(u,v)\, du\, dv \qquad (3.33)$$

Note that the correlation is very similar to the convolution in Eq. (3.19). The correlation is given in the frequency domain by $(g \star h)(r,c) \Leftrightarrow G(u,v)H(-u,-v)$. If $h(r,c)$ contains real numbers, which is the case for image processing, then

注意通过滤波器的频率响应 $H(u,v)$，卷积操作削弱了图像 $g(r,c)$ 所对应的频率响应 $G(u,v)$ 中的频率含量。这证明了在 3.2.3.6 节和 3.2.3.7 节中对使用均值滤波器和高斯滤波器进行平滑处理的分析是正确的。为了使这个分析更精确，我们能够计算均值滤波器等式（3.21）的傅里叶变换。变换结果如下：

其中 $\operatorname{sinc} x = (\sin \pi x)/(\pi x)$。图 3.11 给出了 3×3 均值滤波器的频率响应的图示。同理，等式（3.24）高斯滤波器的傅里叶变换结果如下：

可见，高斯滤波器的傅里叶变换还是一个高斯函数，只是 σ 变成了其自身的倒数。注意应用以上两种滤波器时如果滤波器自身的尺寸增加，那么滤波器的频率响应将变窄。一般来讲，这种对应关系总是适用的，即：$h(x/a) \Leftrightarrow |a|H(af)$。

傅里叶变换的另一个有趣特性是它可以被用于计算相关性。

注意相关与卷积等式（3.19）类似。相关在频率域内由下式给出 $(g \star h)(r,c) \Leftrightarrow G(u,v)H(-u,-v)$。如果 $h(r,c)$ 包含实数，这是图像处理常见的情况，那么 $H(-u,-v) = \overline{H(u,v)}$，

$H(-u,-v) = \overline{H(u,v)}$, where the bar denotes complex conjugation. Hence, $(g \star h)(r,c) \Leftrightarrow G(u,v)\overline{H(u,v)}$.

3.2.4.2 Discrete Fourier Transform

Up to now, we have assumed that the images are continuous. Real images are, of course, discrete. This trivial observation has profound implications for the result of the Fourier transform. As noted above, the frequency variables u and v are given in cycles per pixel. If a discrete image $h(r,c)$ is transformed, the highest possible frequency for any sine or cosine wave is $1/2$, i.e., one cycle per two pixels. The frequency $1/2$ is called the Nyquist critical frequency. Sine or cosine waves with higher frequencies look like sine or cosine waves with correspondingly lower frequencies. For example, a discrete cosine wave with frequency 0.75 looks exactly like a cosine wave with frequency 0.25, as shown in Figure 3.15. Effectively, values of $H(u,v)$ outside the square $[-0.5, 0.5] \times [-0.5, 0.5]$ are mapped to this square by repeated mirroring at the borders of the square. This effect is known as aliasing. To avoid aliasing, we must ensure that

横线表示共轭复数。因此,$(g \star h)(r,c) \Leftrightarrow G(u,v)\overline{H(u,v)}$。

3.2.4.2 离散傅里叶变换

到目前为止,我们假定图像数据是连续的。但是真正的图像数据是离散的。这个微小的发现会与傅里叶变换的结果存在着较大的牵连。如前所述,频率变量 u 和 v 是每一个像素的周期数。如果一幅离散图像 $h(r,c)$ 经历了变换,那么对任意的正弦或余弦波,可能出现的最高频率都是 $1/2$,也就是,每两个像素一个周期。频率 $1/2$ 被称为尼奎斯特临界频率。高于此频率的正弦或余弦波看起来与相应的更低频率的正弦或余弦波一样。例如,频率 0.75 的离散余弦波看起来同频率 0.25 的余弦波是完全一致的,如图 3.15 所示。$H(u,v)$ 在正方形 $[-0.5, 0.5] \times [-0.5, 0.5]$ 以外的值被映射到正方形内,这是通过以正方形边界为中心反复镜像来实现的。这就是混淆现象。为消除频率混淆,在对

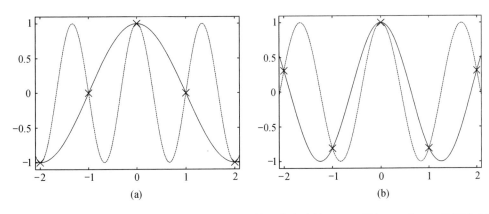

图 3.15 混淆效应的例子。(a) 频率是 0.25 和 0.75 的两个余弦波;(b) 频率 0.4 和 0.6 的两个余弦波。注意对所有函数在整数位置进行采样,即用十字叉标明的部分,采集的离散样品值是一模一样的

frequencies higher than the Nyquist critical frequency are removed before the image is sampled. During image acquisition, this can be achieved by optical low-pass filters in the camera. Aliasing, however, may also occur when an image is scaled down (see Section 3.3.2.4). Here, it is important to apply smoothing filters before the image is sampled at the lower resolution to ensure that frequencies above the Nyquist critical frequency are removed.

Real images are not only discrete, they are also only defined within a rectangle of dimension $w \times h$, where w is the image width and h is the image height. This means that the Fourier transform is no longer continuous, but can be sampled at discrete frequencies $u_k = k/h$ and $v_l = l/w$. As discussed above, sampling the Fourier transform is useful only in the Nyquist interval $-1/2 \leq u_k, v_l < 1/2$. With this, the discrete Fourier transform (DFT) is given by

图像进行采样前必须将高于尼奎斯特临界频率的高频部分消除。在图像获取过程中，这可以通过摄像机的光学低通滤镜来实现。但是，当一幅图像按比例缩小时也可能发生混淆现象（见 3.3.2.4 节）。这样，对图像进行低分辨率采样前，先使用平滑滤波器以保证消除了所有高于尼奎斯特临界频率的高频部分是很重要的。

实际的图像不但是离散的，而且是在矩形区域 $w \times h$ 内被定义的，w 是图像的宽，h 是图像的高。这意味着傅里叶变换不再是连续的，但能够在离散频率上被采样 $u_k = k/h$ 和 $v_l = l/w$。如上面讨论过的，对傅里叶变换进行采样仅在尼奎斯特间隔 $-1/2 \leq u_k, v_l < 1/2$ 时是有用的。据此，离散傅里叶变换（DFT）由下式给出：

$$H_{k,l} = H(u_k, v_l) = \sum_{r=0}^{h-1}\sum_{c=0}^{w-1} h_{r,c} e^{2\pi i(u_k r + v_l c)} = \sum_{r=0}^{h-1}\sum_{c=0}^{w-1} h_{r,c} e^{2\pi i(kr/h + lc/w)} \quad (3.34)$$

Analogously, the inverse DFT is given by

类似地，离散傅里叶逆变换如下：

$$h_{r,c} = \frac{1}{wh} \sum_{k=0}^{h-1}\sum_{l=0}^{w-1} H_{k,l} e^{-2\pi i(kr/h + lc/w)} \quad (3.35)$$

As noted above, conceptually, the frequencies u_k and v_l should be sampled from the interval $(-1/2, 1/2]$, i.e., $k = -h/2 + 1, \cdots, h/2$ and $l = -w/2 + 1, \cdots, w/2$. Since we want to represent $H_{k,l}$ as an image, the negative coordinates are a little cumbersome. It is easy to see that Eqs. (3.34) and (3.35) are periodic with period h and w. Therefore, we can map the negative frequencies to their positive counterparts,

如前所述，概念上，频率 u_k 和 v_l 应该在 $(-1/2, 1/2]$ 范围内被采样，即，$k = -h/2 + 1, \cdots, h/2$ 和 $l = -w/2 + 1, \cdots, w/2$。由于我们想用 $H_{k,l}$ 来表示一幅图像，负值坐标就有问题了。从式（3.34）和式（3.35）很容易看出它们都是周期性的，分别以 h 和 w 为周期。因此，我们能把负频率映射为对应的正频率，也就是将

i.e., $k = -h/2+1, \cdots, -1$ is mapped to $k = h/2+1, \cdots, h-1$; and likewise for l.

We noted above that for real images, $H(-u,-v) = \overline{H(u,v)}$. This property still holds for the DFT, with the appropriate change of coordinates as defined above, i.e., $H_{h-k,w-l} = \overline{H_{k,l}}$ (for $k, l > 0$). In practice, this means that we do not need to compute and store the complete Fourier transform $H_{k,l}$ because it contains redundant information. It is sufficient to compute and store one half of $H_{k,l}$, e.g., the left half. This saves a considerable amount of processing time and memory. This type of Fourier transform is called the real-valued Fourier transform.

To compute the Fourier transform from Eqs. (3.34) and (3.35), it might seem that $O((wh)^2)$ operations are required. This would prevent the Fourier transform from being useful in image processing applications. Fortunately, the Fourier transform can be computed in $O(wh \log(wh))$ operations for $w = 2^n$ and $h = 2^m$ (Press et al., 2007) as well as for arbitrary w and h (Frigo and Johnson, 2005). This fast computation algorithm for the Fourier transform is aptly called the fast Fourier transform (FFT). With self-tuning algorithms (Frigo and Johnson, 2005), the FFT can be computed in real time on standard processors.

As discussed above, the Fourier transform can be used to compute the convolution with any linear filter in the frequency domain. While this can be used to perform filtering with standard filter masks, e.g., the mean or Gaussian filter, typically there is a speed advantage only for relatively large filter masks. The real advantage of using the Fourier transform for filtering lies in the fact

$k = -h/2+1, \cdots, -1$ 映射到 $k = h/2+1, \cdots, h-1$，同理适用于 l。

在上文中我们说过对于真正的图像，$H(-u,-v) = \overline{H(u,v)}$。在使用上面定义的坐标变换后，这个特征对于离散傅里叶变换仍然是适用的，即 $H_{h-k,w-l} = \overline{H_{k,l}}(k,l>0)$。在实践中，因为变换结果中包含了冗余信息，这意味着我们不需要计算和保存全部傅里叶变换的结果 $H_{k,l}$。计算和保存 1/2 的 $H_{k,l}$ 就足够了，比如左边一半。这节省了相当大的处理时间和内存。此类傅里叶变换称作实值傅里叶变换。

从等式 (3.34) 和等式 (3.35) 计算傅里叶变换，理论上需要 $O((wh)^2)$ 次操作，这就极大降低了傅里叶变换在图像处理应用中的可用性。幸运的是，在 $w = 2^n$ 和 $h = 2^m$ 时（Press et al., 2007），或者对于任意 w 和 h（Frigo and Johnson, 2005）时，傅里叶变换可以通过 $O(wh \log(wh))$ 次操作完成计算。傅里叶变换的此种实现方法被称为快速傅里叶变换（FFT）。通过采用自校正算法（Frigo and Johnson, 2005），FFT 可以在标准处理器上实时进行。

如上所述，傅里叶变换能被用来在频率域内计算与任意线性滤波器的卷积。这点可被用来配合标准滤波器掩码实现滤波操作，比如与均值滤波器或高斯滤波器配合，但通常仅在与相对大尺寸滤波器配合使用时才有速度优势。使用傅里叶变换进行滤波处理的真正好处是：可以通过使用定制

that filters can be customized to remove specific frequencies from the image, which occur, for example, for repetitive textures.

Figure 3.16(a) shows an image of a map. The map is drawn on a highly structured paper that exhibits significant texture. The tex-

的滤波器来消除图像中某些特定的频率，例如，这些特定频率可能代表着图像中重复出现的纹理。

图 3.16（a）是地图的图像。地图被画在了结构高度有序的纸上，从纸上可以看到明显的纹理。这些纹

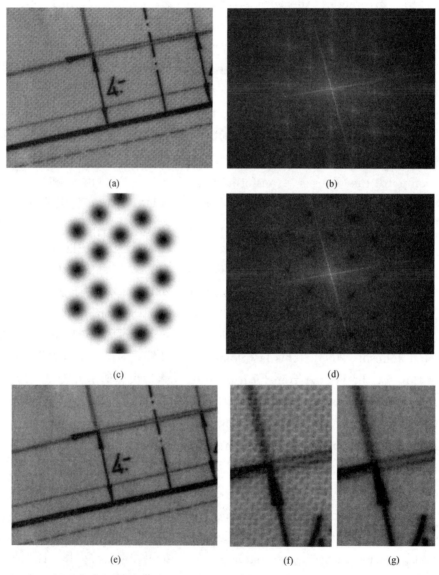

图 3.16 （a）带有纸张纹理的地图图像；（b）图（a）的傅里叶变换，由于结果的高动态范围，显示的是 $H_{u,v}^{1/16}$。注意 $H_{u,v}$ 中那些明显的峰，它们对应的是纸张的纹理；（c）用于消除与纹理对应的频率的滤波器 $G_{u,v}$；（d）$H_{u,v}G_{u,v}$ 卷积的结果；（e）图（d）的傅里叶逆变换结果；（f），（g）图（a）和图（e）的细节。注意纹理被消除了

ture makes the extraction of the data in the map difficult. Figure 3.16(b) displays the Fourier transform $H_{u,v}$. Note that the Fourier transform is cyclically shifted so that the zero frequency is in the center of the image to show the structure of the data more clearly. Hence, Figure 3.16(b) displays the frequencies u_k and v_l for $k = -h/2 + 1, \cdots, h/2$ and $l = -w/2 + 1, \cdots, w/2$. Because of the high dynamic range of the result, $H_{u,v}^{1/16}$ is displayed. It can be seen that $H_{u,v}$ contains several highly significant peaks that correspond to the characteristic frequencies of the texture. Furthermore, there are two significant orthogonal lines that correspond to the lines in the map. A filter $G_{u,v}$ that removes the characteristic frequencies of the texture is shown in Figure 3.16(c), while the result of the convolution $H_{u,v}G_{u,v}$ in the frequency domain is shown in Figure 3.16(d). The result of the inverse Fourier transform, i.e., the convolution in the spatial domain, is shown in Figure 3.16(e). Figures 3.16(f) and (g) show details of the input and result images, which show that the texture of the paper has been removed. Thus, it is now very easy to extract the map data from the image.

3.3 Geometric Transformations

In many applications, one cannot ensure that the objects to be inspected are always in the same position and orientation in the image. Therefore, the inspection algorithm must be able to cope with these position changes. Hence, one of the problems is to detect the position and orientation, also called the pose, of the objects to be examined.

理使提取地图上的数据变得困难。图 3.16（b）显示的是原图的傅里叶变换结果 $H_{u,v}$。请注意傅里叶变换被循环平移了半个周期，以便让零频率出现在图像正中心，这样可更清楚地显示数据的结构。因此，图 3.16（b）显示了在 $k = -h/2 + 1, \cdots, h/2$ 和 $l = -w/2 + 1, \cdots, w/2$ 处的频率 u_k 和 v_l。由于结果的高动态范围，所以在图中实际显示的是 $H_{u,v}^{1/16}$。能看到在 $H_{u,v}$ 中包含了一些明显的峰，这些峰对应的是纹理的特征频率。此外，两条直角交叉的直线对应的是地图上的线。图 3.16（c）显示的是用来消除纹理特征频率的滤波器 $G_{u,v}$。将 $H_{u,v}G_{u,v}$ 在频率域卷积处理的结果显示在图 3.16（d）中。傅里叶逆变换后的结果，即在空间域的卷积显示在图 3.16（e）。图 3.16（f）和图 3.16（g）给出了处理前的图像和结果图像的细节，从中可看到纸张的纹理被消除了。所以，现在从图上提取地图数据就很容易了。

3.3 几何变换

在许多应用中，并不能保证被测物在图像中总是处于同样的位置和方向。所以，检测算法必须能够应对这种位置的变化。因此，首先要解决的问题就是检测出被测物的位置和方向，即被测物的位姿。这将在本章后续几节中进行介绍。为在本节中讨论

This will be the subject of later sections of this chapter. For the purposes of this section, we assume that we know the pose already. In this case, the simplest procedure to adapt the inspection to a particular pose is to align the ROIs appropriately. For example, if we know that an object is rotated by 45°, we could simply rotate the ROI by 45° before performing the inspection. In some cases, however, the image must be transformed (aligned) to a standard pose before the inspection can be performed. For example, the segmentation in OCR is much easier if the text is either horizontal or vertical. Another example is the inspection of objects for defects based on a reference image. Here, we also need to align the image of the object to the pose in the reference image, or vice versa. Therefore, in this section we will examine different geometric image transformations that are useful in practice.

3.3.1 Affine Transformations

If the position and rotation of the objects cannot be kept constant with the mechanical setup, we need to correct the rotation and translation of the object. Sometimes the distance of the object to the camera changes, leading to an apparent change in size of the object. These transformations are part of a very useful class of transformations called affine transformations, which are transformations that can be described by the following equation:

$$\begin{pmatrix} \tilde{r} \\ \tilde{c} \end{pmatrix} = \begin{pmatrix} a_{11} & a_{12} \\ a_{21} & a_{22} \end{pmatrix} \begin{pmatrix} r \\ c \end{pmatrix} + \begin{pmatrix} t_r \\ t_c \end{pmatrix} \tag{3.36}$$

Hence, an affine transformation consists of a linear part given by a 2×2 matrix and a translation

方便，我们先假设位姿已知。此时，调整物体到检测所需位姿的最简单方法就是对 ROI 的位姿进行适当的调整。例如，如果我们知道一个物体被旋转了 45°，那么在进行对物体的检测前，我们只需将 ROI 也旋转 45° 即可。但在一些情况下，图像必须先被变换（对准）到一个标准位姿，然后进行检测。例如，进行 OCR 中的文本分割时，如果文本是水平的或是垂直，那么图像分割会更容易些。再如，当基于参考图像来检测物体的缺陷时，我们需要将物体的图像进行变换，使图中物体的位姿与参考图像中的物体位姿相同，反之亦然。所以，在本节中我们将讨论有实用价值的一些不同的几何图像变换方法。

3.3.1 仿射变换

如果在机械装置上物体的位置和旋转角度不能保持恒定，我们必须对物体进行平移和旋转角度修正。有时由于摄像机和物体间的距离发生了变化，所以导致图像中物体的尺寸发生了明显变化。这些情况下使用的变换称为仿射变换，它可以用以下等式描述：

上式中，一个仿射变换包括一个由 2×2 矩阵给定的线性部分和一个平

The above notation is a little cumbersome, however, since we always have to list the translation separately. To circumvent this, we can use a representation where we extend the coordinates with a third coordinate of 1, which enables us to write the transformation as a simple matrix multiplication:

移部分。但由于总是需要将平移部分单独列出，所以此种表述有些烦琐。为解决这个问题，在原坐标基础上引入第三个数值为 1 的坐标，这种表示方法能让我们用简单的矩阵乘法来表示仿射变换：

$$\begin{pmatrix} \tilde{r} \\ \tilde{c} \\ 1 \end{pmatrix} = \begin{pmatrix} a_{11} & a_{12} & a_{13} \\ a_{21} & a_{22} & a_{23} \\ 0 & 0 & 1 \end{pmatrix} \begin{pmatrix} r \\ c \\ 1 \end{pmatrix} \tag{3.37}$$

Note that the translation is represented by the elements a_{13} and a_{23} of the matrix $\boldsymbol{A}$. This representation with an added redundant third coordinate is called homogeneous coordinates. Similarly, the representation with two coordinates in Eq. (3.36) is called inhomogeneous coordinates. We will see the true power of the homogeneous representation below. Any affine transformation can be constructed from the following basic transformations, where the last row of the matrix has been omitted:

注意平移部分由矩阵 $\boldsymbol{A}$ 中的元素 a_{13} 和 a_{23} 表示。此种附加了多余的第三个坐标的表示法叫做齐次坐标。类似地，等式（3.36）中含两坐标的表示法叫做非齐次坐标。后面我们将看到齐次坐标表示法的真正威力。任何仿射变换都能由以下基本变换构造而来，此处基本变换矩阵的最后一行已被省略。

$$\begin{pmatrix} 1 & 0 & t_r \\ 0 & 1 & t_c \end{pmatrix} \quad \text{平移} \tag{3.38}$$

$$\begin{pmatrix} s_r & 0 & 0 \\ 0 & s_c & 0 \end{pmatrix} \quad \text{行列方向缩放} \tag{3.39}$$

$$\begin{pmatrix} \cos\alpha & -\sin\alpha & 0 \\ \sin\alpha & \cos\alpha & 0 \end{pmatrix} \quad \text{以角度 } \alpha \text{ 旋转} \tag{3.40}$$

$$\begin{pmatrix} \cos\theta & 0 & 0 \\ \sin\theta & 1 & 0 \end{pmatrix} \quad \text{以角度 } \theta \text{ 将纵轴倾斜} \tag{3.41}$$

The first three basic transformations need no further explanation. The skew (or slant) is a rotation of only one axis, in this case the row axis. It is quite useful for rectifying slanted characters in OCR.

前三个基本变换无须更多解释。倾斜是仅让一个轴旋转，此处是让纵轴旋转一个角度，这在 OCR 矫正倾斜字符的时非常有用。

3.3.1.1 Projective Transformations

An affine transformation enables us to correct almost all relevant pose variations that an object may undergo. However, sometimes affine transformations are not general enough. If the object in question is able to rotate in 3D, it will undergo a general perspective transformation, which is quite hard to correct because of the occlusions that may occur. However, if the object is planar, we can model the transformation of the object by a 2D perspective transformation, which is a special 2D projective transformation (Hartley and Zisserman, 2003; Faugeras and Luong, 2001). Projective transformations are given by

$$\begin{pmatrix} \tilde{r} \\ \tilde{c} \\ \tilde{w} \end{pmatrix} = \begin{pmatrix} h_{11} & h_{12} & h_{13} \\ h_{21} & h_{22} & h_{23} \\ h_{31} & h_{32} & h_{33} \end{pmatrix} \begin{pmatrix} r \\ c \\ w \end{pmatrix} \tag{3.42}$$

Note the similarity to the affine transformation in Eq. (3.37). The only changes that were made are that the transformation is now described by a full 3×3 matrix and that we have replaced the 1 in the third coordinate with a variable w. This representation is actually the true representation in homogeneous coordinates. It can also be used for affine transformations, which are special projective transformations. With this third coordinate, it is not obvious how we are able to obtain a transformed 2D coordinate, i.e., how to compute the corresponding inhomogeneous point. First of all, it must be noted that in homogeneous coordinates, all points $\boldsymbol{p} = (r, c, w)^\top$ are only defined up to a scale factor, i.e., the vectors $\boldsymbol{p}$ and $\lambda \boldsymbol{p}$ ($\lambda \neq 0$) represent the same 2D point (Hartley and Zisserman, 2003; Faugeras and Luong, 2001).

3.3.1.1 投影变换

仿射变换几乎能校正物体所有可能发生的与位姿相关的变化，但并不能应对所有情况。如果被讨论的物体在三维空间发生了旋转，它就经历了一个很平常的投影变换，由于可能出现阴影或遮挡，所以此投影变换是很难被修正的。但如果物体是平面的，我们能通过二维投影变换对此物体的三维变换进行模型化，这就是一个专用的二维投影变换（Hartley and Zisserman, 2003; Faugeras and Luong, 2001）。投影变换可由下式给出：

请注意上式与等式（3.37）仿射变换有相似之处。仅有的几个区别就是投影变换是用一个完整的 3×3 矩阵来描述的，并且将仿射变换里第 3 个坐标从 1 变为一个变量 w。实际上这个表示法是真正的齐次坐标表示法，它也能被用来描述仿射变换，因为仿射变换是特殊的投影变换。由于有了第 3 个坐标，我们如何能够得到一个变换后的二维坐标，即如何计算对应的非齐次点，并不是显而易见的。首先，必须注明在齐次坐标中的全部点 $\boldsymbol{p} = (r, c, w)^\top$ 只能被定义为一个比例因子，即向量 $\boldsymbol{p}$ 和 $\lambda \boldsymbol{p}$ ($\lambda \neq 0$) 代表同样的 2D 点（Hartley and Zisserman, 2003; Faugeras and Luong, 2001）。相应地，由矩阵 $\boldsymbol{H}$ 给定的投影变换也

Consequently, the projective transformation given by the matrix H is also defined only up to a scale factor, and hence has only eight independent parameters. To obtain an inhomogeneous 2D point from the homogeneous representation, we must divide the homogeneous vector by w. This requires $w \neq 0$. Such points are called finite points. Conversely, points with $w = 0$ are called points at infinity because they can be regarded as lying infinitely far away in a certain direction (Hartley and Zisserman, 2003; Faugeras and Luong, 2001).

Since a projective transformation has eight independent parameters, it can be uniquely determined from four corresponding points (Hartley and Zisserman, 2003; Faugeras and Luong, 2001). This is how projective transformations will usually be determined in machine vision applications. We will extract four points in an image, which typically represent a rectangle, and will rectify the image so that the four extracted points will be transformed to the four corners of the rectangle, i.e., to their corresponding points. Unfortunately, because of space limitations, we cannot give the details of how the transformation is computed from the point correspondences. The interested reader is referred to Hartley and Zisserman (2003) or Faugeras and Luong (2001).

3.3.2 Image Transformations

After having taken a look at how coordinates can be transformed with affine and projective transformations, we can consider how an image should be transformed. Our first idea might be to go through all the pixels in the input image, to transform their coordinates, and to set the gray value of the transformed point in the output image.

只能定义为一个比例因子，因此只能有 8 个独立参数。为从齐次表达式中计算出它对应的非齐次 2D 点，我们必须用齐次向量除以 w，这要求 $w \neq 0$。这样的点被称为有限点。反之，$w = 0$ 的点被称为无限点，这是因为它们位于某一方向上无限远处（Hartley and Zisserman，2003；Faugeras and Luong，2001）。

既然一个投影变换有 8 个独立参数，那么投影变换就可以通过 4 个相应的点被唯一地确定（Hartley and Zisserman，2003；Faugeras and Luong，2001）。这是在机器视觉应用中通常确定投影变换的方法。我们从图像上提取 4 个点，典型情况下是一个矩形，然后矫正图像使这 4 个被提取出来的点能被变换到矩形的 4 个角上，即变换到它们对应的点上。遗憾的是，由于篇幅有限，我们就不能详细介绍如何通过点的对应关系来计算变换矩阵。感兴趣的读者请参考（Hartley and Zisserman，2003）或者（Faugeras and Luong，2001）。

3.3.2 图像变换

在已经学习了如何用仿射和投影变换来实现坐标变换后，我们能够考虑如何对图像进行变换。最简单的一个想法或许是在输入图像上遍历全部像素，计算它们变换后的坐标，然后设置输出图像上这些坐标处的灰度值。不幸的是，这个简单的策略并不能成

Unfortunately, this simple strategy does not work. This can be seen by checking what happens if an image is scaled by a factor of 2: only one quarter of the pixels in the output image would be set. The correct way to transform an image is to loop through all the pixels in the output image and to calculate the position of the corresponding point in the input image. This is the simplest way to ensure that all relevant pixels in the output image are set. Fortunately, calculating the positions in the original image is simple: we only need to invert the matrix that describes the affine or projective transformation, which results again in an affine or projective transformation.

When the image coordinates are transformed from the output image to the input image, typically not all pixels in the output image transform back to coordinates that lie in the input image. This can be taken into account by computing a suitable ROI for the output image. Furthermore, we see that the resulting coordinates in the input image will typically not be integer coordinates. An example of this is given in Figure 3.17, where the input image is transformed by an affine transformation consisting of a translation, rotation, and scaling. Therefore, the gray values in the output image must be interpolated.

功，我们可以通过对一幅图像放大 2 倍时发生的事情来说明：输出图像中只有 1/4 的像素可以被设置。变换一幅图像的正确方法是在输出图像内遍历所有像素并计算其在输入图像中相对应的点的位置。这是保证能够对输出图像中所有相关像素进行设定的最简单方法。幸运的是，在原始图像中计算对应点的位置是容易的：我们仅需要先对表示仿射变换或投影变换的矩阵求逆，然后使用此矩阵进行仿射或投影变换即可。

当图像坐标被从输出图像变换到输入图像时，通常情况下不是输出图像中的所有像素都能被变换回位于输入图像内的坐标上。我们能通过为输出图像计算一个合适的 ROI 来解决此问题。而且，在输入图像中的结果坐标通常不是整数坐标，在图 3.17 中我们给出一个这样的例子，此例中对输入图像进行了一个包括平移、旋转和缩放的仿射变换。所以，输出图像内的灰度值必须是由插值得到。

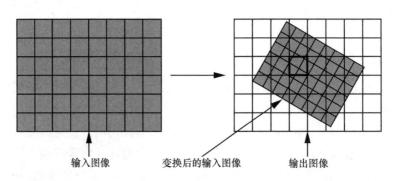

图 3.17　一幅图像的仿射变换。注意输出图像内的整数坐标变换为原始图像中的非整数坐标，所以灰度值必须由插值得到

3.3.2.1 Nearest-Neighbor Interpolation

The interpolation can be done in several ways. Figure 3.18(a) displays a pixel in the output image that has been transformed back to the input image. Note that the transformed pixel center lies on a non-integer position between four adjacent pixel centers. The simplest and fastest interpolation method is to calculate the closest of the four adjacent pixel centers, which only involves rounding the floating-point coordinates of the transformed pixel center, and to use the gray value of the closest pixel in the input image as the gray value of the pixel in the output image, as shown in Figure 3.18(b). This interpolation method is called nearest-neighbor interpolation. To see the effect of this interpolation, Figure 3.19(a) displays an image of a serial number of a bank note, where the characters are not horizontal. Figures 3.19(c)

3.3.2.1 最近邻域插值法

存在着多种插值方法。图3.18（a）显示的是输出图像中的一个像素已被变换回输入图像中，注意转换后这个像素的中心点是位于邻近的四个像素中心点之间，是一个非整数的位置。最简单快速的插值法是先对转换后像素中心的非整数坐标进行取整处理，以找到与此坐标相邻的四个像素点的中心位置中最近的一个，然后将输入图像里的这个最邻近位置的像素的灰度值视为输出图像内相应像素点的灰度值，如图3.18（b）所示。这种插值方法被称为最近邻域插值法。为了直观看到此插值算法效果，图3.19（a）中显示的是一张纸币上的序列号，序列号中的字符不是水平的。图3.19（c）和（d）显示的是采用最近邻域插值法

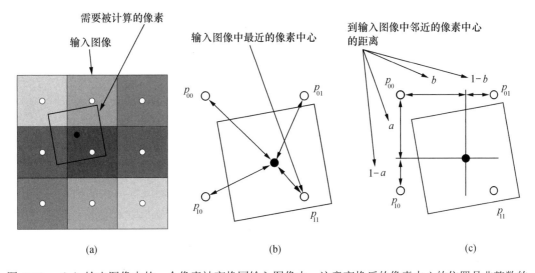

图 3.18 （a）输出图像中的一个像素被变换回输入图像中，注意变换后的像素中心的位置是非整数的，位于相邻的四个像素中心之间；（b）最近邻域插值法确定输入图像中最近的像素后在输出图像中直接使用此像素的灰度值；（c）双线性内插法先确定到四个相邻像素中心的距离，然后用距离作为它们灰度值的权重进行插值计算

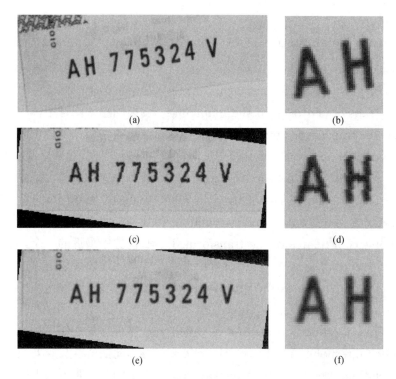

图 3.19 （a）纸币上的一个序列号图像；（b）图（a）的细节；（c）使用最近邻域插值法旋转图像以使序列号呈水平；（d）图（c）的细节，注意字符的锯齿状外观；（e）使用双线性插值法旋转图像；（f）图（e）的细节，注意字符的平滑边缘

and (d) display the result of rotating the image such that the serial number is horizontal using this interpolation. Note that because the gray value is taken from the closest pixel center in the input image, the edges of the characters have a jagged appearance, which is undesirable.

3.3.2.2 Bilinear Interpolation

The reason for the jagged appearance in the result of the nearest-neighbor interpolation is that essentially we are regarding the image as a piecewise constant function: every coordinate that falls within a rectangle of extent ± 0.5 in each direction is assigned the same gray value.

This leads to discontinuities in the result, which cause the jagged edges. This behavior is

将此序列号旋转为水平后的结果。注意由于灰度值是从最近的像素中心上取得的，所以字符的边缘出现了锯齿状外观，这个结果不是我们想得到的。

3.3.2.2 双线性插值法

最近邻域插值法的结果中会出现锯齿状外观的根本原因是我们把图像看成了一个分段常值函数：落在整数坐标 ± 0.5 的矩形区域内的每个坐标都被赋值为同一个灰度值。

这就导致了结果的不连续性，而这种结果的不连续性造成了锯齿状

especially noticeable if the image is scaled by a factor > 1. To get a better interpolation, we can use more information than the gray value of the closest pixel. From Figure 3.18(a), we can see that the transformed pixel center lies in a square of four adjacent pixel centers. Therefore, we can use the four corresponding gray values and weight them appropriately. One way to do this is to use bilinear interpolation, as shown in Figure 3.18(c). First, we compute the horizontal and vertical distances of the transformed coordinate to the adjacent pixel centers. Note that these are numbers between 0 and 1. Then, we weight the gray values according to their distances to get the bilinear interpolation:

$$\tilde{g} = b(ag_{11} + (1-a)g_{01}) + (1-b)(ag_{10} + (1-a)g_{00}) \tag{3.43}$$

Figures 3.19(e) and (f) display the result of rotating the image of Figure 3.19(a) using bilinear interpolation. Note that the edges of the characters now have a very smooth appearance. This much better result more than justifies the longer computation time (typically a factor of around 2).

3.3.2.3 Bicubic Interpolation

Bilinear interpolation works very well as long as the image is not zoomed by a large amount. However, if the image is scaled by factors of more than approximately 4, interpolation artifacts may become visible, as shown in Figure 3.20(a). These artifacts are caused by two effects. First, bilinear interpolation is continuous but not smooth. There can be sharp bends along the horizontal and vertical lines through the pixel centers of the

外观。这在对图像进行一倍以上的放大处理时尤其明显。为得到更好的插值算法，我们应在处理中使用更多的信息而不仅仅是使用最近像素的灰度值。从图 3.18（a）中可以看出，变换后像素的中心位于一个正方形中，此正方形的四个顶点就是与此像素相邻的四个像素的中心点。所以，我们可以用适当的权重配合这四个灰度值进行插值运算。这样做的方法之一就是使用双线性插值法，如图 3.18（c）所示。首先，我们分别计算转换后的坐标到四个相邻像素中心点的垂直方向和水平方向的距离。注意这些计算出的距离值在 0 到 1 之间。然后，根据距离值计算出不同灰度值所占的权重后得到双线性插值的结果：

图 3.19（e）和 3.19（f）显示的是对图 3.19（a）进行旋转时使用双线性插值处理后的结果。注意现在字符边缘的外观变得非常平滑了。这个更佳的结果需要更长的计算时间（通常是原计算时间的 2 倍左右）。

3.3.2.3 双三次插值法

在图像没有缩放很多的情况下，双线性插值法表现很好。然而，如果图像缩放因子超过 4，插值的痕迹会变得比较明显，如图 3.20（a）所示。这个结果由两个因素产生。首先，双线性插值法连续但不平滑，沿着通过原始图像像素中心的水平线和竖直线方向会有尖锐的急弯，在图像边缘部分尤其显著。此外，双线性插值在图

original image, which are especially noticeable at edges. Furthermore, bilinear interpolation leads to apparent blurring in the vicinity of edges.

像边缘附近会产生明显的模糊。

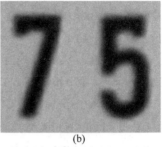

图 3.20　（a）图 3.19（a）使用双线性插值法进行旋转和缩放，缩放因子为 16 时的部分图像效果；（b）对同样的图像使用双三次插值法得到的效果

To prevent these problems, higher-order interpolation can be used. The next natural interpolation order after bilinear interpolation is bicubic interpolation (Keys, 1981). To derive a bicubic interpolation algorithm, we can first note that linear interpolation can be described as a convolution at a subpixel position with the following kernel:

为了避免这些问题，可以使用更高阶的插值方法。双线性插值方法之后另一个自然插值法是双三次插值法（Keys，1981）。为了推导双三次插值算法，首先需要清楚线性插值可以描述为一个亚像素位置的卷积，其核为：

$$l(x) = \begin{cases} 1 - |x|, & |x| \leqslant 1 \\ 0, & |x| > 1 \end{cases} \quad (3.44)$$

Here, the subpixel position x corresponds to the coordinates a and b in Figure 3.18(c). The interpolation kernel is shown in Figure 3.21(a). To obtain bilinear interpolation, we perform two linear interpolations horizontally at the row coordinates 0 and 1 using the column coordinate b and then perform a linear interpolation of the gray values thus obtained at the vertical coordinate a.

这里，亚像素位置 x 对应图 3.18（c）中的坐标 a 和 b，插值核如图 3.21（a）所示。为了得到双线性插值，首先使用列坐标 b 在行坐标 0 和 1 位置水平执行两次线性插值，然后对于在垂直坐标 a 处得到的灰度值执行一次线性插值。

For cubic interpolation, we can use the following interpolation kernel (see Keys, 1981):

对于双三次插值，我们可以使用下面的插值核（参见 Keys，1981）：

$$c(x) = \begin{cases} \frac{3}{2}|x|^3 - \frac{5}{2}|x|^2 + 1, & |x| \leqslant 1 \\ -\frac{1}{2}|x|^3 + \frac{5}{2}|x|^2 - 4|x| + 2, & 1 < |x| \leqslant 2 \\ 0, & |x| > 2 \end{cases} \quad (3.45)$$

This kernel is shown in Figure 3.21(b). To obtain bicubic interpolation, we perform four cubic interpolations horizontally at the row coordinates $-1, 0, 1$, and 2 using the column coordinate b followed by a cubic interpolation of the gray values thus obtained at the vertical coordinate a. From Eqs. (3.44) and (3.45), it is obvious that bicubic interpolation is computationally much more expensive than bilinear interpolation. Instead of 2×2 points, 4×4 points are used and the kernel requires many more arithmetic operations. Therefore, it should only be used if an image is zoomed by scale factors that are significantly larger than 1 and if the quality requirements justify the additional run time.

核如图 3.21（b）所示，为了得到双三次插值，首先使用列坐标 b 在行坐标 -1，0，1 和 2 位置水平执行四次三次线性插值，然后对于在垂直坐标 a 处得到的灰度值执行一次三次线性插值。从等式（3.44）和等式（3.45）可以看出，双三次插值比双线性插值计算成本明显高很多。使用 4×4 个点代替 2×2 个点，插值核需要更多的算术运算。因此，仅有当图像缩放因子明显大于 1，或者图像质量要求很高可以合理增加运行时间时，应该使用双三次插值。

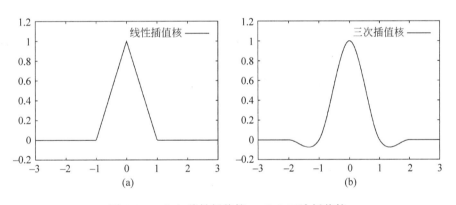

图 3.21　（a）线性插值核；（b）三次插值核

Figure 3.20 compares bilinear and bicubic interpolation. A part of the image in Figure 3.19(a) is rotated and scaled by a factor of 16. Bilinear interpolation leads to some visible artifacts, as shown in Figure 3.20(a). The non-smoothness of the bilinear interpolation and the blurred appearance of the edges are clearly visible. In contrast, as shown in Figure 3.20(b), bicubic interpolation is smooth and produces much sharper edges.

图 3.20 比较了双线性插值和双三次插值。图 3.19（a）中的图像一部分进行了旋转和缩放，缩放因子 16，如图 3.20（a）所示，双线性插值会产生明显的处理痕迹。双线性插值的不平滑特征以及边缘的模糊现象都清晰可见。作为对比，图 3.20（b）所示，双三次插值是平滑的而且边缘会更清晰。

3.3.2.4 Smoothing to Avoid Aliasing

To conclude the discussion on interpolation, we discuss the effects of scaling an image down. In bilinear interpolation, we would interpolate from the closest four pixel centers. However, if the image is scaled down, adjacent pixel centers in the output image will not necessarily be close in the input image. Imagine a larger version of the image of Figure 3.11(b) (one-pixel-wide vertical lines spaced three pixels apart) being scaled down by a factor of 4 using nearest-neighbor interpolation: we would get an image with one-pixel-wide lines that are four pixels apart. This is certainly not what we would expect. For bilinear interpolation, we would get similar unexpected results. If we scale down an image, we are subsampling it. As a consequence, we may obtain aliasing effects (see Section 3.2.4.2). An example of aliasing can be seen in Figure 3.22. The image in Figure 3.22(a) is scaled down by a factor of 3 in Figure 3.22(c) using bilinear interpolation. Note that the stroke widths of the vertical strokes of the letter H, which are equally wide in Figure 3.22(a), now appear to be quite different. This is undesirable. To improve the image transformation, the image must be smoothed before it is scaled down, e.g., using a mean or a Gaussian filter. Alternatively, the smoothing can be integrated into the gray value interpolation. Figure 3.22(e) shows the result of integrating a mean filter into the image transformation. Because of the smoothing, the strokes of the H now have the same width.

3.3.3 Projective Image Transformations

In the previous sections, we have seen the useful-

3.3.2.4 通过平滑来避免混淆现象

在结束有关插值算法的讨论前，我们再看看图像缩小的效果。在双线性插值的方案中，我们从最近的四个像素中心点来计算插值。但如果对图像进行缩小，在输出图像中相邻的像素不一定在输入图像中相邻。设想一幅如图 3.11（b）那样图案（宽度为一个像素、线间距为三个像素的垂线）的更大尺寸图像被缩小到原来的 1/4，使用最近邻域插值法：我们得到的结果图上会是宽度为一个像素、线间距为四个像素的垂线。这当然不是我们希望得到的结果。使用双线性插值法，我们也会得到类似的意外结果。如果我们缩小一幅图像时，从本质上讲我们在对其进行二次采样。因此，我们可能碰到混淆现象（见 3.2.4.2 节）。混淆现象的例子见图 3.22。用双线性插值对图 3.22（a）缩小 3 倍得到图 3.22（c）。注意图 3.22（a）中的字母 H 的垂直笔画宽度都相等，但在图 3.22（c）中就明显不同了。这是我们不想得到的结果。为改善图像变换的结果，图像必须在缩小处理前先进行平滑处理，比如，可使用均值滤波器或高斯滤波器进行平滑处理。另一种解决方法是将平滑处理整合到灰度值插值处理中。图 3.22（e）显示了在图像变换中结合了一个均值滤波器进行处理后的结果。由于进行了平滑处理，现在字符 H 的笔画宽度又相等了。

3.3.3 投影图像变换

在上例中，我们已经看到了仿射

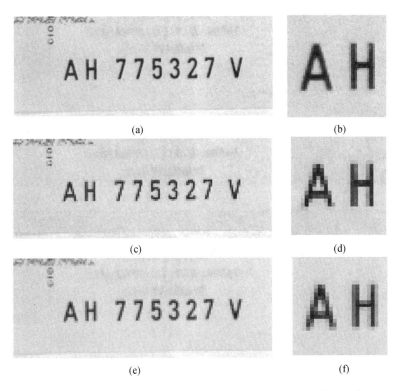

图 3.22 （a）纸币上的一个序列号图像；（b）图（a）的细节；（c）用双线性插值对图（a）缩小 3 倍；（d）图（c）的细节，注意字母 H 的垂直笔画宽度变得不同了，这是由于混淆现象造成的；（e）结合了一个平滑滤波器（此处是均值滤波器）到图像变换处理中对图像缩小后的结果；（f）图（e）的细节

ness of affine transformations for rectifying text. Sometimes, however, an affine transformation is not sufficient for this purpose. Figures 3.23(a) and (b) show two images of license plates on cars. Because the position of the camera with respect to the car could not be controlled, the images of the license plates show perspective distortions. Figures 3.23(c) and (d) show the result of applying projective transformations to the images that cut out the license plates and rectify them. Hence, the images in Figures 3.23(c) and (d) would result if we had looked at the license plates perpendicularly from in front of the car. It is now much easier to segment and read the characters on the license plates.

变换在矫正文本时的有效性。但有时这样的矫正仅使用一个仿射变换还是不够的。图 3.23（a）和图 3.23（b）给出了两幅汽车牌照图像。由于本例中摄像机相对于汽车的位置不可控，所以车牌的图像存在一些透视畸变。图 3.23（c）和图 3.23（d）是针对原图中的车牌应用投影变换并矫正后的结果。所以，图 3.23（c）和图 3.23（d）与我们从汽车正前方观看车牌时所看到的结果相似。很明显，此时更容易分割和读取车牌上的字符了。

图 3.23 （a），(b) 车牌图像；（c），(d) 用于矫正车牌透视畸变的投影变换后的结果

3.3.4 Polar Transformations

Another useful geometric transformation is the polar transformation. This transformation is typically used to rectify parts of images that show objects that are circular or that are contained in circular rings in the image. An example is shown in Figure 3.24(a): the inner part of a CD contains a ring with a bar code and some text. To read the bar code, the part of the image that contains the bar code can be rectified with the polar transformation. It converts the image into polar coordinates $(d, \phi)^\top$, i.e., into the distance d to the center of the transformation and the angle ϕ of the vector to the center of the transformation. Let the center of the transformation be given by $(m_r, m_c)^\top$. Then, the polar coordinates of a point $(r, c)^\top$ are given by

3.3.4 极坐标变换

另一个非常有用的几何变换是极坐标变换。极坐标变换通常被用来矫正图像中的圆形物体或被包含在圆环中的物体。图 3.24（a）是一个极坐标变换的例子。此例中 CD 盘在靠近圆心的部分有一个环形条码和部分文本。为了读取此条码，可以用极坐标变换对图像中包含条码的部分进行矫正。极坐标变换将图像坐标变换为极坐标 $(d, \phi)^\top$，即相对于变换中心的距离 d 和向量角度 ϕ。变换中心由 $(m_r, m_c)^\top$ 给出，那么，某点 $(r, c)^\top$ 的极坐标是：

$$d = \sqrt{(r - m_r)^2 + (c - m_c)^2}$$
$$\phi = \operatorname{atan2}(-(r - m_r), c - m_c) \tag{3.46}$$

where atan2 denotes the two-argument arctangent function that returns its result in the range $[-\pi, \pi)$. Note that the transformation of a point into polar coordinates is quite expensive to compute because of the square root and the arctangent. Fortunately, to transform an image, as for affine and projective transformations, the inverse of the polar transformation is used, which is given by

$$r = m_r - d \sin \phi$$
$$c = m_c + d \cos \phi \qquad (3.47)$$

Here, the sines and cosines can be tabulated because they only occur for a finite number of discrete values, and hence only need to be computed once. Therefore, the polar transformation of an image can be computed efficiently. Note that by restricting the ranges of d and ϕ, we can transform arbitrary circular sectors.

Figure 3.24(b) shows the result of transforming a circular ring that contains the bar code in Figure 3.24(a). Note that because of the polar transformation the bar code is straight and horizontal, and consequently can be read easily.

这里 atan2 表示两个参数的反正切函数，返回结果范围在 $[-\pi, \pi)$。注意将某个点变换到极坐标是相当耗时的运算，因为要进行开平方和反正切计算。幸运的是，与仿射变换和投影变换类似，进行变换处理时可以使用极坐标变换的逆变换，即：

这里，由于 ϕ 的值是数量有限的离散值，所以它们的正弦和余弦值能被列入表格，故而只需要计算一次。所以，一幅图像的极坐标变换能被高效率地计算出来。注意，通过限定 d 和 ϕ 的取值范围，我们能对圆的任意扇形区域进行变换。

图 3.24（b）是对图 3.24（a）中包含条码的圆环进行变换的结果。注意极坐标变换后的条码又直又平，所以很容易读取。

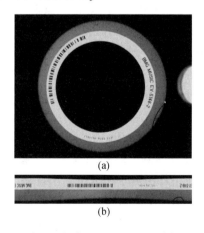

图 3.24 （a）CD 的中间部分包含一个圆形的条码；（b）包含条码的圆环的极坐标变换结果，注意变换后的条码是又平又直的

3.4 Image Segmentation

In the preceding sections, we have looked at operations that transform an image into another image. These operations do not give us information about the objects in the image. For this purpose, we need to segment the image, i.e., extract regions from the image that correspond to the objects we are interested in. More formally, segmentation is an operation that takes an image as input and returns one or more regions or subpixel-precise contours as output.

3.4.1 Thresholding

3.4.1.1 Global Thresholding

The simplest segmentation algorithm is to threshold the image. The threshold operation is defined by

$$S = \{(r,c)^\top \in R \mid g_{\min} \leqslant f_{r,c} \leqslant g_{\max}\} \tag{3.48}$$

Hence, the threshold operation selects all points in the ROI R of the image that lie within a specified range of gray values into the output region S. Since the thresholds are identical for all points within in the ROI, this operation is also called global thresholding. For reasons of brevity, we will simply call this operation thresholding. Often, $g_{\min} = 0$ or $g_{\max} = 2^b - 1$ is used. If the illumination can be kept constant, the thresholds $g_{\min}$ and $g_{\max}$ are selected when the system is set up and are never modified. Since the threshold operation is based on the gray values themselves, it can be used whenever the object to be segmented and the background have significantly different gray values.

Figures 3.25(a) and (b) show two images of integrated circuits (ICs) on a PCB with a rectangular ROI overlaid in light gray. The result of thresholding the two images with $g_{\min} = 90$ and $g_{\max} = 255$ is shown in Figures 3.25(c) and (d). Since the illumination is kept constant, the same threshold works for both images. Note also that there are some noisy pixels in the segmented regions. They can be removed, e.g., based on their area (see Section 3.5) or based on morphological operations (see Section 3.6.1).

图 3.25（a）和图 3.25（b）显示的是两幅印刷电路板（PCB）上集成电路（IC）的图像，每幅图像中都叠加显示了一个浅灰色的矩形 ROI。图 3.25（c）和图 3.25（d）是在 $g_{\min} = 90$ 和 $g_{\max} = 255$ 时对这两幅图像进行阈值分割处理后的结果。由于照明保持恒定所以同样的阈值设定对两幅图都起作用。同时注意在分割出来的区域中存在一些干扰像素。这些干扰像素能通过一些方法被消除掉，比如通过基于面积（见 3.5 节）或基于形态学（见 3.6.1 节）的方法。

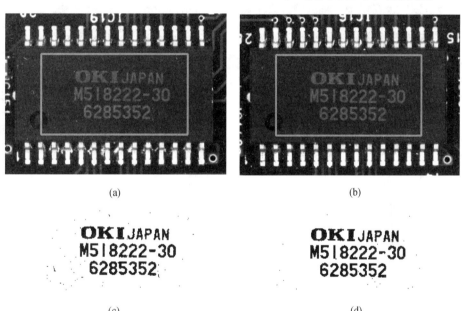

图 3.25　（a）与（b）是印刷在 IC 上字符的图像，图像上叠加显示了一个浅灰色矩形 ROI；（c）与（d）是在 $g_{\min} = 90$ 和 $g_{\max} = 255$ 时对图（a）和图（b）进行阈值分割后的结果

3.4.1.2　Automatic Threshold Selection

The constant threshold works well only as long as the gray values of the object and the background do not change. Unfortunately, this occurs less frequently than one would wish, e.g., because of changing illumination. Even if the illumination

3.4.1.2　自动阈值分割

固定阈值仅在物体的灰度值和背景的灰度值不变时效果很好。不幸的是，这种情况发生的频率比期望的要少，比如，照明变化后物体和背景的灰度值就会发生变化。即便使用的照

is kept constant, different gray value distributions on similar objects may prevent us from using a constant threshold. Figure 3.26 shows an example of this. In Figures 3.26(a) and (b), two different ICs on the same PCB are shown. Despite the identical illumination, the prints have a substantially different gray value distribution, which will not allow us to use the same threshold for both images. Nevertheless, the print and the background can be separated easily in both cases. Therefore, ideally, we would like to have a method that is able to determine the thresholds automatically.

明是恒定不变的，相似物体间不同的灰度值分布也会使固定阈值分割的结果不理想。图3.26就是一个典型的例子。图3.26（a）和图3.26（b）给出了PCB上两种不同IC的图像。尽管使用了同样的照明，印在IC上的字符本质上还是有着不同的灰度值分布，这就让我们不能对两幅图像使用同样的阈值。但不管怎样，这两幅图中的字符和背景都能被很容易地分开。因此，理想情况下希望有一种能够自动确定阈值的方法。

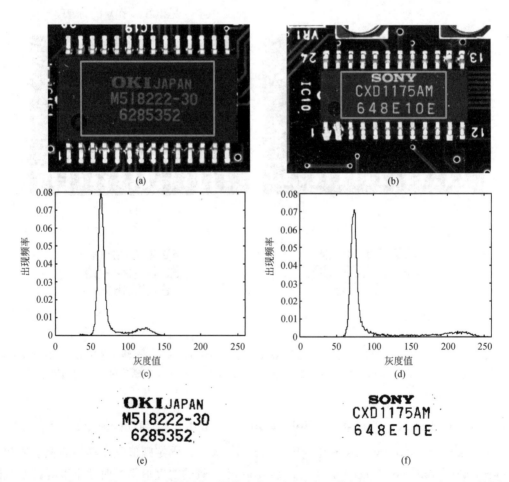

图3.26　（a）与（b）是IC上印字的图像，图像中有一浅灰色标记的矩形ROI；（c）与（d）是图（a）和图（b）在各自的ROI内的灰度值直方图；（e）与（f）通过基于灰度值直方图自动选择阈值对图（a）和图（b）进行阈值分割后的结果

The threshold can be determined based on the gray value histogram of the image. Figures 3.26(c) and (d) show the histograms of the images in Figures 3.26(a) and (b). It is obvious that there are two relevant peaks (maxima) in the histograms in both images. The one with the smaller gray value corresponds to the background, while the one with the higher gray value corresponds to the print. Intuitively, a good threshold corresponds to the minimum between the two peaks in the histogram. Unfortunately, neither the two maxima nor the minimum is well defined because of random fluctuations in the gray value histogram. Therefore, to robustly select the threshold that corresponds to the minimum, the histogram must be smoothed, e.g., by convolving it with a 1D Gaussian filter. Since it is not clear which σ to use, a good strategy is to smooth the histogram with progressively larger values of σ until two unique maxima with a unique minimum in between are obtained. The result of using this approach of selecting the threshold automatically is shown in Figures 3.26(e) and (f). As can be seen, suitable thresholds have been selected for both images.

The above approach of selecting the thresholds is not the only approach. Further approaches are described, for example, in (Haralick and Shapiro, 1992; Jain et al., 1995). All these approaches have in common that they are based on the gray value histogram of the image. One example of such an approach is to assume that the gray values in the foreground and background each have a normal (Gaussian) probability distribution, and to jointly fit two Gaussian densities to the histogram. The threshold is then defined as the gray value for which the two Gaussian densities have equal probabilities.

阈值可以基于图像的灰度直方图来确定。图3.26（c）和图3.26（d）显示的是图3.26（a）和图3.26（b）的灰度直方图。很明显，在每幅图像的直方图中都存在两个峰（最大值）。灰度值小一些的峰对应的是背景，而灰度值大些的峰对应的是IC上的字符。直观地讲，一个好的阈值应该对应着直方图中两个峰之间的最小值。不幸的是，由于灰度值直方图中的随机波动，两个峰尖的最大值和它们之间谷底的最小值都不能被很好地确定。所以，如果希望以可靠的方法选定与最小值对应的阈值，就必须先对直方图进行平滑处理，即将直方图与一个一维高斯滤波器进行卷积来完成平滑处理。为了给高斯滤波器选定合适的σ值，一个好的策略是逐渐增大σ并平滑直方图，直到可以从平滑后的直方图中得到两个唯一的最大值和它们之间的一个唯一最小值。用本法自动选择阈值进行分割处理后的结果见图3.26（e）和图3.26（f）。从图中可以看到，这两个图像都选定了合适的阈值。

上面用来选定阈值的方法不是唯一的方法。更多的方法介绍见（Haralick and Shapiro，1992；Jain et al.，1995）。所有这些方法都是大同小异的，因为它们都基于图像的灰度值直方图。比如，其中一种方法就是假设前景的灰度值和背景的灰度值有各自的正态（高斯）概率分布，然后在直方图上拟合两个高斯密度。阈值就选定在两个高斯密度概率相等的灰度值处。

Another approach is to select the threshold by maximizing a measure of separability of the gray values of the region and the background. Otsu (1979) uses the between-class variance of these two classes as the measure of separability. It is given as a function of the threshold t by

$$\sigma^2(t) = \frac{(\mu_1(g_b)\mu_0(t) - \mu_1(t))^2}{\mu_0(t)(1 - \mu_0(t))} \tag{3.49}$$

Here, $g_b = 2^b - 1$ is the maximum gray value that an image with b bits can represent and $\mu_j(k)$ denotes the moment of order j of the gray value histogram of all the gray values from 0 to k:

$$\mu_j(k) = \sum_{i=0}^{k} i^j h_i \tag{3.50}$$

where h_i is the probability of the occurrence of gray value i, given by Eq. (3.4). The optimal threshold t is determined by maximizing Eq. (3.49).

3.4.1.3 Dynamic Thresholding

While calculating the thresholds from the histogram often works extremely well, it fails whenever the assumption that there are two peaks in the histogram is violated. One such example is shown in Figure 3.27. Here, the print is so noisy that the gray values of the print are extremely spread out, and consequently there is no discernible peak for the print in the histogram. Another reason for the failure of the desired peak to appear is inhomogeneous illumination. This typically destroys the relevant peaks or moves them so that they are in the wrong location. Uneven illumination often even prevents us from using a threshold operation altogether because there are no fixed thresholds that work throughout the

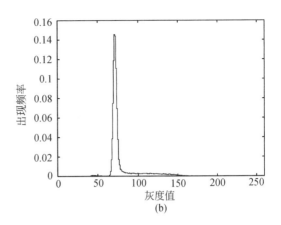

图 3.27 （a）是 IC 上字符的图像，图像中叠加显示了一浅灰色矩形 ROI；（b）图（a）ROI 区域内的灰度直方图。注意在直方图里只存在一个明显的最大值，不存在明显的最小值

entire image. Fortunately, the objects of interest often can be characterized by being locally brighter or darker than their local background. The prints on the ICs we have examined so far are a good example of this. Therefore, instead of specifying global thresholds, we would like to specify by how much a pixel must be brighter or darker than its local background. The only problem we have is how to determine the gray value of the local background. Since a smoothing operation, e.g., the mean, Gaussian, or median filter (see Section 3.2.3), calculates an average gray value in a window around the current pixel, we can simply use the filter output as an estimate of the gray value of the local background. The operation of comparing the image to its local background is called dynamic thresholding. Let the image be denoted by $f_{r,c}$ and the smoothed image be denoted by $g_{r,c}$. Then, the dynamic thresholding operation for bright objects is given by

暗。前面例子中，IC 上的字符就有这个特点。因此，我们希望指定一个像素必须比其所处的背景亮多少或暗多少，而不是指定一个全局阈值。唯一的问题就是如何确定局部区域的背景灰度值。因为使用均值、高斯或中值滤波器（见 3.2.3 节）进行平滑处理就可以计算出以当前像素为中心的窗口内的平均灰度值，所以可以使用滤波处理的输出结果作为对背景灰度值的估计。将图像与其局部背景进行比较的操作被称为动态阈值分割处理。用 $f_{r,c}$ 表示输入图像，用 $g_{r,c}$ 表示平滑后的图像，则对亮物体的动态阈值分割处理如下：

$$S = \{(r,c)^\top \in R \mid f_{r,c} - g_{r,c} \geqslant g_{\text{diff}}\} \tag{3.51}$$

while the dynamic thresholding operation for dark objects is given by

而对暗物体的动态阈值分割处理是：

$$S = \{(r,c)^\top \in R \mid f_{r,c} - g_{r,c} \leqslant -g_{\text{diff}}\} \tag{3.52}$$

Figure 3.28 gives an example of how dynamic thresholding works. In Figure 3.28(a), a small part of a print on an IC with a one-pixel-wide horizontal ROI is shown. Figure 3.28(b) displays the gray value profiles of the image and the image smoothed with a 9×9 mean filter. It can be seen that the text is substantially brighter than the local background estimated by the mean filter. Therefore, the characters can be segmented easily with a dynamic thresholding operation.

图 3.28 给出了一个例子来说明动态阈值分割的工作机制。图 3.28（a）显示的是 IC 上字符的一部分图像，图中有一个高度为一个像素的水平 ROI。图 3.28（b）显示了此 ROI 区域内原始图像的灰度值剖面线及采用 9×9 均值滤波处理后的灰度值剖面线。以滤波后的结果作为局部背景，可以看到文字还是比背景要明亮得多。因此，使用动态阈值分割处理可以很容易地把字符分割出来。

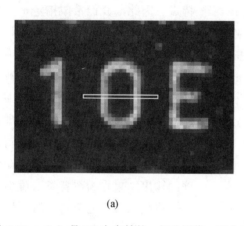
(a)

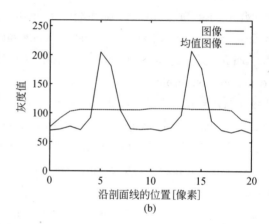
(b)

图 3.28 （a）是 IC 上字符的一部分图像，图中有一个高度为一个像素的水平 ROI；（b）ROI 中原始图像的灰度值剖面线及用 9×9 均值滤波器平滑处理后的图像的灰度值剖面线。注意字符较由均值滤波器估算出的局部背景要亮得多

In the dynamic thresholding operation, the size of the smoothing filter determines the size of the objects that can be segmented. If the filter size is too small, the local background will not be estimated well in the center of the objects. As a rule of thumb, the diameter of the mean filter must be larger than the diameter of the objects to be recognized. The same holds for the median filter, and an analogous relation exists for the Gaussian filter. Furthermore, in general, if larger filter sizes

在动态阈值分割处理中，平滑滤波器的尺寸决定了能被分割出来的物体的尺寸。如果滤波器尺寸太小，那么在物体的中心估计出的局部背景将不理想。凭经验，均值滤波器的宽度必须大于被识别物体的宽度。这一规则同样适用于中值滤波器。对高斯滤波器也存在类似的关系。此外，总的说来如果均值滤波器或高斯滤波器的尺寸越大，那么滤波后的结果越能更

are chosen for the mean and Gaussian filters, the filter output will be more representative of the local background. For example, for light objects the filter output will become darker within the light objects. For the median filter, this is not true since it will completely eliminate the objects if the filter mask is larger than the diameter of the objects. Hence, the gray values will be representative of the local background if the filter is sufficiently large. If the gray values in the smoothed image are more representative of the local background, we can typically select a larger threshold g_{diff}, and hence can suppress noise in the segmentation better. However, the filter mask cannot be chosen arbitrarily large because neighboring objects might adversely influence the filter output. Finally, it should be noted that the dynamic thresholding operation returns a segmentation result not only for objects that are brighter or darker than their local background but also at the bright or dark region around edges.

Figure 3.29(a) again shows the image of Figure 3.27(a), which could not be segmented with an automatic threshold. In Figure 3.29(b), the result of segmenting the image with a dynamic thresholding operation with $g_{\text{diff}} = 5$ is shown. The local background was obtained with a 31×31 mean filter. Note that the difficult print is segmented very well with dynamic thresholding.

As described so far, the dynamic thresholding operation can be used to compare an image with its local background, which is obtained by smoothing the image. With a slight modification, the dynamic thresholding operation can also be used to detect errors in an object, e.g., for print inspection. Here, the image $g_{r,c}$ is an image of the

好地代表局部背景。例如，亮物体滤波后会变得暗些。对于中值滤波就不仅仅是变暗了，因为当中值滤波器的半径大于被提取物体宽度时，滤波操作将会在结果图中完全消除此物体。因此，当滤波器足够大时，滤波后的灰度值就能代表局部背景。如果平滑后图像的灰度值能够更好地代表局部背景，就能选定一个更大的阈值 g_{diff}，这样能够在分割中更好地抑制噪声。但滤波器的掩码窗口也不能无限大，因为相邻近的物体可能对滤波结果产生不利的影响。最后应该注意的是动态阈值分割的结果不仅包括比局部背景更亮或更暗的物体，也包括亮区域或暗区域的边缘。

图 3.29（a）与图 3.27（a）完全一致，此图不能够应用自动阈值分割处理。图 3.29（b）是在 $g_{\text{diff}} = 5$ 时采用动态阈值分割处理后的结果。局部背景是通过一个 31×31 的均值滤波处理得到的。注意很难提取的字符通过动态阈值分割被很好地提取出来。

讨论到目前为止，我们已经知道了动态阈值分割处理能被用来将图像与图像的局部背景进行对比，局部背景是由图像平滑处理得到的。稍作调整，动态阈值分割处理就可以被用来检测某一物体上的缺陷，比如，检测印刷缺陷。此时，图像 $g_{r,c}$ 代表理想

(a) (b)

图 3.29 （a）是印刷在 IC 上字符的图像，图像中叠加显示了一浅灰色矩形 ROI；（b）对图 (a) 在 $g_{\text{diff}} = 5$ 且采用一个 31×31 均值滤波器的动态阈值分割处理后的结果

ideal object, i.e., the object without errors; $g_{r,c}$ is called the reference image. To detect deviations from the ideal object, we can simply look for too bright or too dark pixels in the image $f_{r,c}$ by using Eq. (3.51) or Eq. (3.52). Often, we are not interested in whether the pixels are too bright or too dark, but simply in whether they deviate too much from the reference image, i.e., the union of Eqs. (3.51) and (3.52), which is given by

物体，即无缺陷物体的图像。$g_{r,c}$ 被称为参考图像。为检测出同理想物体的偏差，仅需要使用等式（3.51）或等式（3.52）找到图像 $f_{r,c}$ 中太亮或太暗的那些像素即可。通常我们不感兴趣这些像素是太亮了还是太暗了，我们仅想知道它们与参考图像相比是否偏差太大，也就是说，可将等式（3.51）与等式（3.52）合并为下式：

$$S = \{(r,c)^\top \in R \mid |f_{r,c} - g_{r,c}| > g_{\text{abs}}\} \quad (3.53)$$

Note that this pixel-by-pixel comparison requires that the image $f_{r,c}$ of the object to check and the reference image $g_{r,c}$ are aligned very accurately to avoid spurious gray value differences that would be interpreted as errors. This can be ensured either by the mechanical setup or by finding the pose of the object in the current image, e.g., using template matching (see Section 3.11), and then transforming the image to the pose of the object in the ideal image (see Section 3.3).

这种像素一一对应的比较处理是有前提的：待测物的图像 $f_{r,c}$ 和参考图像 $g_{r,c}$ 必须精确地对准，只有这样才能避免将虚假的灰度值差错误地解释为缺陷。此前提可以通过设置机械装置或确定当前图像中物体的位姿来保证。例如，使用模板匹配（见 3.11 节）确定位姿，然后将图像变换成与参考图像中的物体位姿一致的图像（见 3.3 节）。

3.4.1.4 Variation Model

This kind of dynamic thresholding operation is

3.4.1.4 变差模型

此类动态阈值分割处理对物体形

very strict on the shape of the objects. For example, if the size of the object increases by half a pixel and the gray value difference between the object and the background is 200, the gray value difference between the current image and the model image will be 100 at the object's edges. This is a significant gray value difference, which would surely be larger than any reasonable g_{abs}. In real applications, however, small variations of the object's shape typically should be tolerated. On the other hand, small gray value changes in areas where the object's shape does not change should still be recognized as an error. To achieve this behavior, we can introduce a thresholding operation that takes the expected gray value variations in the image into account. Let us denote the permissible variations in the image by $v_{r,c}$. Ideally, we would like to segment the pixels that differ from the reference image by more than the permissible variations:

状的要求很严格。例如，如果物体尺寸增加半个像素且物体与背景的灰度值差异是 200 时，那么当前图像与参考图像在物体边缘部分的灰度值差异将会是 100。这是相当显著的灰度值差异了，肯定要超过任何合理的 g_{abs} 值。但在现实应用中，算法应该能够容忍物体形状有轻微的抖动。另一方面，即便物体形状没有改变但物体所处的这部分图像区域的灰度值发生小的变化时同样会导致错误的结果。为了改善算法，在进行阈值分割处理时，我们要考虑图像中预期的灰度值偏差。用 $v_{r,c}$ 表示灰度值的容许偏差，理想情况下，我们将分割出与参考图像的偏差大于容许偏差的那些像素：

$$S = \{(r,c)^\top \in R \mid |f_{r,c} - g_{r,c}| > v_{r,c}\} \tag{3.54}$$

The permissible variations can be determined by learning them from a set of training images. For example, if we use n training images of objects with permissible variations, the standard deviation of the gray values of each pixel can be used to derive $v_{r,c}$. If we use n images to define the variations of the ideal object, we might as well use the mean of each pixel to define the reference image $g_{r,c}$ to reduce noise. Of course, the n training images must be aligned with sufficient accuracy. The mean and standard deviation of the n training images are given by

容许偏差可以从一组训练图像中学习得到。例如，如果存在 n 幅在容许偏差范围内的训练图像，则可使用每个灰度值的标准偏差来推导 $v_{r,c}$。如果用 n 幅图像确定理想物体的容许偏差，那么为了降低噪声，在定义参考图像 $g_{r,c}$ 时也应该使用每个像素的平均值。当然，n 幅图像必须充分地对齐。n 幅训练图像的平均值和标准偏差如下：

$$m_{r,c} = \frac{1}{n}\sum_{i=1}^{n} g_{r,c;i} \tag{3.55}$$

$$s_{r,c} = \sqrt{\frac{1}{n}\sum_{i=1}^{n}(g_{r,c;i} - m_{r,c})^2} \qquad (3.56)$$

The images $m_{r,c}$ and $s_{r,c}$ model the reference image and the allowed variations of the reference image. Hence, we can call this approach a variation model. Note that $m_{r,c}$ is identical to the temporal average of Eq. (3.13). To define $v_{r,c}$, ideally we could simply set $v_{r,c}$ to a small multiple c of the standard deviation, i.e., $v_{r,c} = cs_{r,c}$, where, for example, $c = 3$. Unfortunately, this approach does not work well if the variations in the training images are extremely small, e.g., because the noise in the training images is significantly smaller than in the test images, or because parts of the object are near the saturation limit of the camera. In these cases, it is useful to introduce an absolute threshold a for the variation images, which is used whenever the variations in the training images are very small: $v_{r,c} = \max(a, cs_{r,c})$. As a further generalization, it is sometimes useful to have different thresholds for too bright and too dark pixels. With this, the variation threshold is no longer symmetric with respect to $m_{r,c}$, and we need to introduce two threshold images for the too bright and too dark pixels, respectively. If we denote the threshold images by $u_{r,c}$ and $l_{r,c}$, the absolute thresholds by a and b, and the factors for the standard deviations by c and d, the variation model segmentation is given by

图像 $m_{r,c}$ 可用来模型化参考图像，$s_{r,c}$ 则可来模型化参考图像的容许偏差。所以，可以称此方法为变差模型法。注意 $m_{r,c}$ 等同于等式（3.13）的时域平均值。理想情况下仅通过令标准偏差与一个小倍数 c 相乘就可得到 $v_{r,c}$，如 $v_{r,c} = cs_{r,c}$, $c = 3$。不幸的是，如果一组训练图像的偏差非常小时，比如由于训练图像中的噪声很明显地小于用于检测的图像时，或者由于物体的一部分接近摄像机的饱和极限时，此方法效果不好。在这些情况下，为偏差图像引入一个绝对阈值 a 是很有用处的，任何时候当训练图像的偏差太小时，就使用此绝对阈值：$v_{r,c} = \max(a, cs_{r,c})$。进一步概括，对太暗的或太亮的像素使用不同的阈值在有些情况下是很有用的。这样，偏差阈值相对于 $m_{r,c}$ 就不再是对称的，我们需要为太亮的和太暗的像素引入两幅阈值图像。如果用 $u_{r,c}$ 和 $l_{r,c}$ 表示阈值图像，a 和 b 表示绝对阈值，c 和 d 表示与标准偏差相乘的因数，则采用变差模型进行的图像分割由等式（3.57）给出：

$$S = \{(r,c)^\top \in R \mid f_{r,c} < l_{r,c} \vee f_{r,c} > u_{r,c}\} \qquad (3.57)$$

where 其中

$$u_{r,c} = m_{r,c} + \max(a, cs_{r,c})$$
$$l_{r,c} = m_{r,c} - \max(b, ds_{r,c}) \qquad (3.58)$$

Figures 3.30(a) and (b) display two images of a sequence of 15 showing a print on the clip of a pen. All images are aligned such that the MVTec logo is in the center of the image. Note that the letter V in the MVTec logo moves with respect to the rest of the logo and that the corners of the letters may change their shape slightly. This hap-

图 3.30(a) 和图 3.30(b) 是一组 15 幅图像中的两幅，图中显示的是笔夹上的印刷图案。所有图像都经过对准处理以使 MVTec 标识出现在图像的正中间，注意 MVTec 标识中的字母 V 相对于标识中的其他部分有一些移动，字母的棱角处形状可能稍微有了

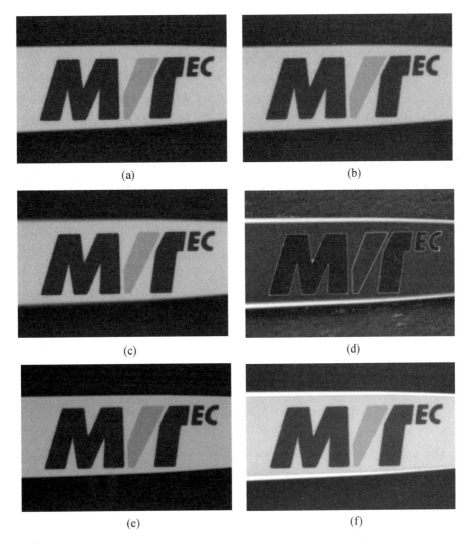

图 3.30 （a）与（b）是 15 幅图像中的两幅，图中显示的是笔夹上的印刷图案。注意 MVTec 标识中的字母 V 相对于标识的其他部分会有稍许移动；（c）在 15 幅图像基础上用变差模型法计算出的参考图像 $m_{r,c}$；（d）标准偏差图像 $s_{r,c}$，为得到更好的显示效果，图中显示的是 $s_{r,c}^{1/4}$；（e）与（f）分别是最小阈值图像 $l_{r,c}$ 和最大阈值图像 $u_{r,c}$，这是在 $a = b = 20$ 和 $c = d = 3$ 的条件下计算出的

pens because of the pad printing technology used to print the logo. The two colors of the logo are printed with two different pads, which can move with respect to each other. Furthermore, the size of the letters may vary because of slightly different pressures with which the pads are pressed onto the clip. To ensure that the logo has been printed correctly, the variation model can be used to determine the mean and variation images shown in Figures 3.30(c) and (d), and from them the threshold images shown in Figures 3.30(e) and (f). Note that the variation is large at the letter V of the logo because this letter may move with respect to the rest of the logo. Also note the large variation at the edges of the clip, which occurs because the logo's position varies on the clip.

Figure 3.31(a) shows a logo with errors in the letters T (small hole) and C (too little ink). From Figure 3.31(b), it can be seen that the two errors can be detected reliably. Figure 3.31(c) shows a different kind of error: the letter V has moved too high and to the right. This kind of error can also be detected easily, as shown in Figure 3.31(d).

As described so far, the variation model requires n training images to construct the reference and variation images. In some applications, however, it is possible to acquire only a single reference image. In these cases, there are two options to create the variation model. The first option is to create artificial variations of the model, e.g., by creating translated versions of the reference image. Another option can be derived by noting that the variations are necessarily large at the edges of the object if we allow small size and position tolerances. This can be clearly seen in Figure 3.30(d). Consequently, in the absence of training images

一点改变，这都是由于用于印刷标识的移印技术造成的。标识中的两种颜色要用两个移印胶头印刷，这两个移印胶头彼此间可相对移动。此外，由于每次移印胶头压在笔夹上的压力略有不同，这造成了印刷出的字母尺寸也可能不一致。为保证能正确地印刷标识，由变差模型法确定的均值图像见图 3.30（c），标准偏差图像见图 3.30（d），在这两幅图基础上得到的两幅阈值图像见图 3.30（e）和图 3.30（f）。注意标识中字母 V 上的偏差大是因为 V 相对于标识的其他部分可能会移动。注意在笔夹边缘上的变差大是因为标识在笔夹上的印刷位置是不一致的。

图 3.31（a）显示的标识中字母 T（小孔）和 C（缺墨）有缺陷。由图 3.31（b）可看到这两个缺陷都被可靠地检查出了。图 3.31（c）是另一种不同缺陷：字母 V 位置向右上移动了。从图 3.31（d）看到这类缺陷也被轻松地检查出来了。

讨论到目前为止，变差模型需要 n 幅训练图像来构造参考图像和偏差图像。但在一些应用中只可能采集一幅参考图像。这样的情况下，建立变差模型有两个方法：一个方法是人为创建一些偏差，如将参考图像进行平移；另一个方法是，当允许尺寸和位置存在小的公差时，那么在物体边缘处就存在着相应的最大偏差，由此即可推导出变差模型。这能从图 3.30（d）中很清楚地看出。因此，在不存在能计算出物体真正偏差的训练图像时，可以使用 3.7.3 节中介绍的任意一种

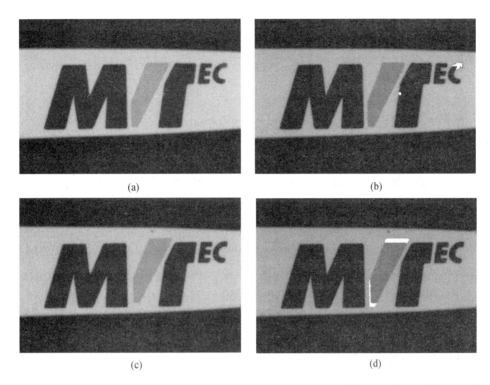

图 3.31 （a）标识中字母 T（小孔）和 C（缺墨）存在缺陷；（b）用图 3.27 的变差模型进行分割后，缺陷用白色标出；（c）标识中的字母 V 向右上移动；（d）分割后得到的缺陷

that show the real variations of the object, a reasonable approximation for $s_{r,c}$ is given by computing the edge amplitude image of the reference image using one of the edge filters described in Section 3.7.3.

边缘滤波器计算参考图像的边缘幅度图像，用此幅度图像作为 $s_{r,c}$ 的合理近似值。

3.4.2 Extraction of Connected Components

The segmentation algorithms in the previous section return one region as the segmentation result (recall the definitions in Eqs. (3.48)–(3.52)). Typically, the segmented region contains multiple objects that should be returned individually. For example, in the examples in Figures 3.25–3.29, we are interested in obtaining each character as a separate region. Typically, the objects we are

3.4.2 提取连通区域

前一节中的分割算法返回一个区域作为分割的结果（见等式（3.48）～等式（3.52））。通常情况下，分割后得到的区域中所包含的多个物体在返回结果中应该是彼此独立的。例如，在图 3.25～图 3.29 中，我们所感兴趣的是将每个字符作为一个独立的区域来获取。通常情况下，我们所感兴趣

interested in are characterized by forming a connected set of pixels. Hence, to obtain the individual regions we must compute the connected components of the segmented region.

To be able to compute the connected components, we must define when two pixels should be considered connected. On a rectangular pixel grid, there are only two natural options to define the connectivity. The first possibility is to define two pixels as being connected if they have an edge in common, i.e., if the pixel is directly above, below, left, or right of the current pixel, as shown in Figure 3.32(a). Since each pixel has four connected pixels, this definition is called 4-connectivity or 4-neighborhood. Alternatively, the definition can be extended to also include the diagonally adjacent pixels, as shown in Figure 3.32(b). This definition is called 8-connectivity or 8-neighborhood.

的物体是由一些相互连通的像素集合而成。所以，为获得每一个区域，必须计算出分割后所得到的区域内包含的所有连通区域（分支）。

为了能够计算出连通区域，必须定义何时两个像素应被视为彼此连通。在一个矩形像素网格上，对于连通性有两种自然的定义。两个像素连通的第一种定义是这两个像素有共同的边缘，也就是说，一个像素在另一个像素的上方、下方、左侧或右侧，如图 3.32（a）。由于每个像素有四个相连通的像素，此定义被称为 4 连通或 4 邻域。第二种定义是第一种定义的扩展，将对角线上的相邻像素也包括进来，如图 3.32（b）。此定义被称为 8 连通或 8 邻域。

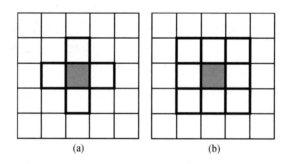

图 3.32　在矩形像素网格上关于连通的两种定义：（a）4 连通；（b）8 连通

While these definitions are easy to understand, they cause problematic behavior if the same definition is used on both the foreground and background. Figure 3.33 shows some of the problems that occur if 8-connectivity is used for the foreground and background. In Figure 3.33(a), there is clearly a single line in the foreground, which divides the background into two connected components. This is what we would intuitively

虽然上面的两个定义很容易被理解，但当对前景和背景都使用同一个定义时，上面的定义就会引出问题。图 3.33 显示的是当对前景和背景都使用 8 连通定义时出现的问题。在图 3.33（a）中，很明显在前景中有一条线，这条线将背景分成两个连通区域。这是我们希望得到的结果。但如图 3.33（b）所示，如果这条线稍微有

expect. However, as Figure 3.33(b) shows, if the line is slightly rotated we still obtain a single connected component in the foreground. However, now the background is also a single component. This is quite counterintuitive. Figure 3.33(c) shows another peculiarity. Again, the foreground region consists of a single connected component. Intuitively, we would say that the region contains a hole. However, the background is also a single connected component, indicating that the region contains no hole. The only remedy for this problem is to use opposite connectivities on the foreground and background. If, for example, 4-connectivity is used for the background in the examples in Figure 3.33, all of the above problems are solved. Likewise, if 4-connectivity is used for the foreground and 8-connectivity for the background, the inconsistencies are avoided.

些旋转，我们仍能在前景中得到一个连通区域，但此时背景中也只有一个连通区域，这个结果是与我们的直觉不符的。图 3.33（c）显示的是另一个怪异的现象。此图中前景仍然只有一个连通区域，直觉上，我们会说此区域包含一个洞，但背景也只是一个连通区域，这又表明区域中没有洞。解决此问题的唯一方法是对前景和背景使用不同的连通性定义。例如，如果在图 3.33 所有例子中使用 4 连通定义描述背景，所有问题都迎刃而解了。同理，如果 4 连通描述前景、8 连通描述背景，所有矛盾也被避免了。

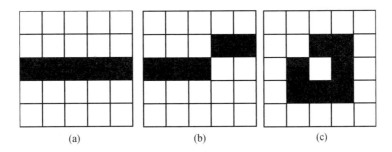

图 3.33　当对前景和背景使用同一个连通定义时的怪异现象，本例中使用的是 8 连通定义：（a）前景中的一条线很明显地将背景分为两个连通区域；（b）如果线的方向稍微有些旋转，前景仍然只有一条线，但背景也只有一个连通区域，这是与直觉不符的；（c）前景中一个区域中从直觉上讲包含一个洞，但背景也是只有一个连通区域，这又表明没有洞，这也是与直觉不符的

To compute the connected components on the run-length representation of a region, a classical depth-first search can be performed (Sedgewick, 1990). We can repeatedly search for the first unprocessed run, and then search for overlapping runs in the adjacent rows of the image. The used connectivity determines whether two runs overlap.

在用行程表示法描述的区域上计算连通区域，可以使用经典的深度优先搜索（Sedgewick, 1990）。我们反复搜索第一个未被处理的行程，然后在与此行程相邻的上下两行中搜索与此行程交叠的所有行程。判断两个行程是否交叠的依据就是连通性的定义。

For 4-connectivity, the runs must at least have one pixel in the same column, while for 8-connectivity, the runs must at least touch diagonally. An example of this procedure is shown in Figure 3.34. The run-length representation of the input region is shown in Figure 3.34(a), the search tree for the depth-first search using 8-connectivity is shown in Figure 3.34(b), and the resulting connected components are shown in Figure 3.34(c). For 8-connectivity, three connected components result. If 4-connectivity had been used, four connected components would have resulted.

对于 4 连通，两交叠的行程中必须至少有一个像素位于同一列。对于 8 连通，两行程必须至少对角接触。深度优先搜索过程的例子见图 3.34，用行程表示法描述的输入区域如图 3.34（a）所示，使用 8 连通进行深度优先搜索时的搜索树如图 3.34（b）所示。返回的连通区域结果如图 3.34（c）所示。对于 8 连通，搜索后返回 3 个连通区域。如果使用 4 连通，搜索后返回 4 个连通区域。

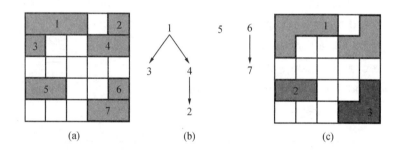

图 3.34 （a）行程表示法描述的一个区域，此区域中包含了 7 个行程；（b）用 8 连通对图（a）中的区域执行深度优先搜索时的搜索树。图中数字代表行程；（c）搜索后返回的连通区域结果

It should be noted that the connected components can also be computed from the representation of a region as a binary image. The output of this operation is a label image. Therefore, this operation is also called labeling or component labeling. For a description of algorithms that compute the connected components from a binary image, see (Haralick and Shapiro, 1992; Jain et al., 1995).

Figures 3.35(a) and (b) show the result of computing the connected components of the regions in Figures 3.26(e) and (f). Each character is a connected component. Furthermore, the noisy segmentation results are also returned as separate

应该注意的是对于以二值图像表示的区域也可以计算连通区域。处理后输出的是一幅标记图像。因此，此处理也被称为标记或区域标记。从二值图上计算连通区域的算法介绍参见（Haralick and Shapiro，1992；Jain et al.，1995）。

图 3.35（a）和图 3.35（b）显示的是在图 3.26（e）和图 3.36（f）所示区域中提取连通区域的结果。可以看到，每个字符都是一个连通区域，此外，分割得到的干扰区域也被作为独

components. Thus, it is easy to remove them from the segmentation, e.g., based on their area.

立的区域返回，可以非常容易地将这些区域从分割结果中剔除，比如，基于它们的面积。

(a) (b)

图 3.35 (a) 与 (b) 对图 3.24 (e) 和图 3.24 (f) 所示区域进行连通区域计算的结果。连通区域用 8 种不同灰度值循环显示出来

3.4.3 Subpixel-Precise Thresholding

All the thresholding operations we have discussed so far have been pixel-precise. In most cases, this precision is sufficient. However, some applications require a higher accuracy than the pixel grid. Therefore, an algorithm that returns a result with subpixel precision is sometimes required. Obviously, the result of this subpixel-precise thresholding operation cannot be a region, which is only pixel-precise. The appropriate data structure for this purpose therefore is a subpixel-precise contour (see Section 3.1.3). This contour will represent the boundary between regions in the image that have gray values above the gray value threshold g_{sub} and regions that have gray values below g_{sub}. To obtain this boundary, we must convert the discrete representation of the image into a continuous function. This can be done, for example, with bilinear interpolation (see Eq. (3.43) in Section 3.3.2.2). Once we have obtained a continuous representation of the image, the subpixel-precise thresholding operation conceptually consists of intersecting the image function $f(r,c)$ with the constant function $g(r,c) = g_{\mathrm{sub}}$. Figure 3.36 shows

3.4.3 亚像素精度阈值分割

到目前为止，我们已经讨论过的所有阈值分割处理都是像素精度的。在大多数情况下，这种精度是足够的。但某些应用需要的精度要高于像素级别。因此，有时需要能返回亚像素精度结果的算法。很显然，亚像素精度阈值分割处理的结果不能是一个区域，因为区域是像素精度的。为此，表示结果的适当数据结构应是亚像素精度轮廓（见 3.1.3 节）。此轮廓表示图像中两个区域之间的边界，这两个区域中一个区域的灰度值大于灰度值阈值 g_{sub} 而另一个区域的灰度值小于 g_{sub}。为获取这个边界，必须将图像的离散表示转换成一个连续函数。例如，可以通过双线性插值（见 3.3.2.2 节中的等式（3.43））完成这种转换。一旦获得了表示图像的一个连续函数，从概念上，亚像素精度阈值分割处理的结果就可以用常量函数 $g(r,c) = g_{\mathrm{sub}}$ 与图像函数 $f(r,c)$ 相交得到。图 3.36 给出了经过双线性内插处理后的图像 $f(r,c)$ 在一个由四个

238 | *3 Machine Vision Algorithms*

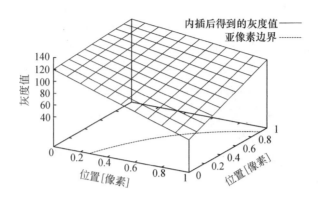

图 3.36 此图形显示的灰度值是四个像素间双线性插值得到的，四个像素的中心在图形的四个角上，以及在灰度值 $g_{\text{sub}} = 100$ 时的相交曲线，曲线在图形的下方。此曲线（双曲线的一部分）是灰度值 > 100 的区域和灰度值 < 100 的区域之间的边界

the bilinearly interpolated image $f(r,c)$ in a 2×2 block of the four closest pixel centers. The closest pixel centers lie at the corners of the graph. The bottom of the graph shows the intersection curve of the image $f(r,c)$ in this 2×2 block with the constant gray value $g_{\text{sub}} = 100$. Note that this curve is part of a hyperbola. Since this hyperbolic curve would be quite cumbersome to represent, we can simply substitute it with a straight line segment between the two points where the hyperbola leaves the 2×2 block. This line segment constitutes one segment of the subpixel contour we are interested in. Each 2×2 block in the image typically contains between zero and two of these line segments. If the 2×2 block contains an intersection of two contours, four line segments may occur. To obtain meaningful contours, these segments need to be linked. This can be done by repeatedly selecting the first unprocessed line segment in the image as the first segment of the contour and then tracing the adjacent line segments until the contour closes, reaches the image border, or reaches an intersection point. The result of this linking step is typically closed contours that enclose a region

邻近像素的中心构成的 2×2 局部。四个邻近像素的中心分别位于正方形的四个角上。图形底部显示的是在此 2×2 区域中用常量灰度值 $g_{\text{sub}} = 100$ 与图像 $f(r,c)$ 相交得到的曲线。注意此曲线是双曲线上的一部分。但是因为双曲线使用起来不方便，所以仅用一条直线来替代它，直线是由双曲线进出此 2×2 区域的两个点连接而成的。此线段构成了我们感兴趣的亚像素轮廓的一部分。一般情况下，图像中的每个 2×2 区域包含在零个、一个或两个这样的线段。如果一个 2×2 区域内两条轮廓相交于一点，那么将出现四条线段。为获取有意义的轮廓，这些线段要被连接起来。可以通过反复地选择图像中第一个未被处理的线段作为轮廓的第一段，然后跟踪邻近的线段直到轮廓闭合、到达图像边界或到达一个交点。此连线处理的结果通常是在图像内形成一个闭合的轮廓，此轮廓围绕的区域内部的灰度值大于或者小于阈值。注意如果这样的区域内包含孔洞，要为区域的外边界

in the image in which the gray values are either larger or smaller than the threshold. Note that, if such a region contains holes, one contour will be created for the outer boundary of the region and one for each hole.

Figure 3.37(a) shows an image of a PCB that contains a ball grid array (BGA) of solder pads. To ensure good electrical contact, it must be ensured that the pads have the correct shape and position. This requires high accuracy, and, since in this application typically the resolution of the image is small compared to the size of the balls and pads, the segmentation must be performed with subpixel accuracy. Figure 3.37(b) shows the result of performing a subpixel-precise thresholding operation on the image in Figure 3.37(a). To see enough details of the results, the part that corresponds to the white rectangle in Figure 3.37(a) is displayed. The boundary of the pads is extracted with very good accuracy. Figure 3.37(c) shows even more detail: the left pad in the center row of Figure 3.37(b), which contains an error that must be detected. As can be seen, the subpixel-precise contour correctly captures the erroneous region of the pad. We can also easily see the individual line segments in the subpixel-precise contour and how they are contained in the 2×2 pixel blocks. Note that each block lies between four pixel centers. Therefore, the contour's line segments end at the lines that connect the pixel centers. Note also that in this part of the image there is only one block in which two line segments are contained: at the position where the contour enters the error on the pad. All the other blocks contain one or no line segments.

和孔洞的外边界分别建立轮廓。

图 3.37（a）显示的图像中是一个包含 BGA 焊盘的 PCB。为保证最佳的电气连接，必须保证焊盘的形状和位置正确。这就需要高精度，并且由于在这个应用中，通常图像分辨率对于焊点和焊盘而言是不够高的，所以必须进行亚像素精度的分割处理。图 3.37（b）显示了对图 3.37（a）进行亚像素精度阈值分割处理后的结果。为看到结果中足够多的细节，在图 3.37（a）中白色矩形区域内的图像被放大显示出来。焊盘的边界被非常准确地提取出来。图 3.37（c）显示了更多的细节部分：图 3.37（b）中间一行最左侧的焊盘包含了必须要被检查出的缺陷。可以看到，亚像素精度轮廓正确地捕捉到了焊盘上的缺陷部分。我们也能容易地看到亚像素精度轮廓上的每条线段以及这些线段是如何被包含在这些 2×2 的区域中的。注意每个 2×2 的区域都位于四个相邻像素的中心组成的正方形内。因此，轮廓所包含的所有线段都截止于像素中心之间的连线上。也要注意在这部分的图像中，只有一个 2×2 的区域中包含了两条线段：这个区域位于焊盘轮廓进入缺陷的部分。其他的所有区域都只包含一条线段或根本不包含线段。

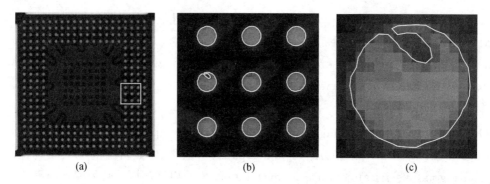

图 3.37 （a）一个带 BGA 焊盘的 PCB 图像；（b）对图（a）应用亚像素精度阈值分割的结果，图中显示的是与图（a）中白色矩形相对应的部分；（c）图（b）的中间一行最左侧焊盘的细节

3.5 Feature Extraction

In the previous sections, we have seen how to extract regions or subpixel-precise contours from an image. While the regions and contours are very useful, they may not be sufficient because they contain the raw description of the segmented data. Often, we must select certain regions or contours from the segmentation result, e.g., to remove unwanted parts of the segmentation. Furthermore, often we are interested in gauging the objects. In other applications, we might want to classify the objects, e.g., in OCR, to determine the type of the object. All these applications require that we determine one or more characteristic quantities from the regions or contours. The quantities we determine are called features. Typically they are real numbers. The process of determining the features is called feature extraction. There are different kinds of features. Region features are features that can be extracted from the regions themselves. In contrast, gray value features also use the gray values in the image within the region. Finally, contour features are based on the coordinates of the contour.

3.5 特征提取

在前面的各节中，已经看到了如何从图像中提取区域或亚像素精度轮廓。尽管区域和轮廓非常有用，但它们也许并不够用，因为它们只包含对分割结果的原始描述。通常，必须从分割结果中选出某些区域或轮廓，比如，为了去除分割结果中不想要的部分。而且，我们通常对物体测量感兴趣。在其他的应用中，我们或许想对物体进行分类以确定物体的类型，比如在 OCR 中就需要进行类似的处理。所有这些应用都需要能从区域或轮廓中确定一个或多个特征量。这些确定的特征量被称为特征，它们通常是实数，确定特征的过程被称为特征提取，存在着多种不同类型的特征。区域特征是能够从区域自身提取出来的特征。与之相比，灰度值特征还需要图像中区域内的灰度值。另外，轮廓特征是基于轮廓坐标的。

3.5.1 Region Features

3.5.1.1 Area

By far the simplest region feature is the area of the region:

$$a = |R| = \sum_{(r,c)^\top \in R} 1 = \sum_{i=1}^{n}(ce_i - cs_i + 1) \qquad (3.59)$$

Hence, the area a of the region is simply the number of points $|R|$ in the region. If the region is represented as a binary image, the first sum must be used to compute the area; whereas if a run-length representation is used, the second sum can be used. Recall from Eq. (3.2) that a region can be regarded as the union of its runs, and the area of a run is extremely simple to compute. Note that the second sum contains many fewer terms than the first sum, as discussed in Section 3.1.2. Hence, the run-length representation of a region will lead to a much faster computation of the area. This is true for almost all region features.

Figure 3.38 shows the result of selecting all regions with an area $\geqslant 20$ from the regions in Figures 3.35(a) and (b). Note that all the characters have been selected, while all noisy segmentation results have been removed. These regions could now be used as input for OCR.

3.5.1 区域特征

3.5.1.1 面积

到目前为止，最简单的区域特征是区域的面积：

由等式（3.59）可知，区域的面积 a 就是区域内的点数 $|R|$。如果区域是用一幅二值图像表示的，那么用等式（3.59）中的第一个求和公式计算区域的面积；如果区域是用行程编码表示的，那么用等式（3.59）中的第二个求和公式计算区域的面积。回顾等式（3.2），一个区域能够被视为其所有行程的一个并集，而每个行程的面积是极容易计算的。注意第二个累加式的项比第一个累加式的少很多，其原因在 3.1.2 节中已经讨论过了。所以，区域的行程表示法可使区域面积的计算速度快很多，这个特点对几乎所有的区域特征都适用。

图 3.38 显示了从图 3.35（a）和图 3.35（b）中提取出所有面积 $\geqslant 20$ 的区域后的结果。注意所有的字符都已被选出，同时所有分割出来的干扰区域已被剔除。这些选出的区域现在可以被用来作为 OCR 的输入。

(a)   (b)

图 3.38 （a）与（b）显示从图 3.35（a）和图 3.35（b）中提取所有面积 ≥20 的区域后的结果。循环使用 8 种不同灰度值显示连通区域

3.5.1.2 Moments

The area is a special case of a more general class of features called the moments of the region. The moment of order (p, q), with $p \geq 0$ and $q \geq 0$, is defined as

$$m_{p,q} = \sum_{(r,c)^\top \in R} r^p c^q \qquad (3.60)$$

Note that $m_{0,0}$ is the area of the region. As for the area, simple formulas to compute the moments solely based on the runs can be derived. Hence, the moments can be computed very efficiently in the run-length representation.

The moments in Eq. (3.60) depend on the size of the region. Often, it is desirable to have features that are invariant to the size of the objects. To obtain such features, we can simply divide the moments by the area of the region if $p + q \geq 1$ to get normalized moments:

$$n_{p,q} = \frac{1}{a} \sum_{(r,c)^\top \in R} r^p c^q \qquad (3.61)$$

The most interesting feature that can be derived from the normalized moments is the center of gravity of the region, which is given by $(n_{1,0}, n_{0,1})^\top$. It can be used to describe the position of the region. Note that the center of gravity is a subpixel-precise feature, even though it is computed from pixel-precise data.

The normalized moments depend on the position in the image. Often, it is useful to make the features invariant to the position of the region in the image. This can be done by calculating the moments relative to the center of gravity of the region. These central moments are given by $(p + q \geq 2)$:

3.5.1.2 矩

面积是被称为区域的矩的广义特征中一个特例。$p \geq 0$, $q \geq 0$ 时，(p, q) 阶矩被定义为：

注意 $m_{0,0}$ 就是区域的面积。与计算面积类似，因为能够推导出仅基于行程的简单公式来计算矩，所以使用行程表示法时就可以高效率地计算矩。

等式（3.60）中的矩依赖于区域的尺寸。通常我们期望有一些特征可不随物体尺寸变化而变化，为获取这样的特征，当 $p + q \geq 1$ 时，矩除以区域的面积就得到了归一化的矩：

从归一化的矩中推导得到的最令人感兴趣的特征是区域的重心，即 $(n_{1,0}, n_{0,1})^\top$。它能用来描述区域的位置。注意尽管重心是从像素精度的数据计算得到的，但它是一个亚像素精度特征。

归一化的矩是由图像中的位置决定的。通常，使特征不随图像中区域的位置变化而变化是很有用的。这可以通过计算相对于区域重心的矩来实现。这些中心矩是在 $(p + q \geq 2)$ 时由等式（3.62）计算得到的：

$$\mu_{p,q} = \frac{1}{a} \sum_{(r,c)^\top \in R} (r - n_{1,0})^p (c - n_{0,1})^q \qquad (3.62)$$

Note that they are also normalized.

注意这些中心矩也是归一化处理后的。

3.5.1.3 Ellipse Parameters

The second central moments ($p+q = 2$) are particularly interesting. They enable us to define an orientation and an extent for the region. This is done by assuming that the moments of order 1 and 2 of the region were obtained from an ellipse. Then, from these five moments, the five geometric parameters of the ellipse can be derived. Figure 3.39 displays the ellipse parameters graphically. The center of the ellipse is identical to the center of gravity of the region. The major and minor axes r_1 and r_2 and the angle of the ellipse with respect to the column axis are given by

3.5.1.3 椭圆参数

二阶中心矩 ($p+q = 2$) 尤其值得关注，它们可以用来定义区域的方位和区域的范围。这是通过假设从一个椭圆上获取区域的一阶矩和二阶矩而实现的。然后，从这五个矩推导出椭圆的五个几何参数。图 3.39 显示了椭圆的参数。椭圆的中心与区域的重心是一致的。椭圆的长轴 r_1 和短轴 r_2，以及相对于横轴的夹角 θ 可由等式 (3.63)～等式 (3.65) 计算得到：

$$r_1 = \sqrt{2\left(\mu_{2,0} + \mu_{0,2} + \sqrt{(\mu_{2,0} - \mu_{0,2})^2 + 4\mu_{1,1}^2}\right)} \qquad (3.63)$$

$$r_2 = \sqrt{2\left(\mu_{2,0} + \mu_{0,2} - \sqrt{(\mu_{2,0} - \mu_{0,2})^2 + 4\mu_{1,1}^2}\right)} \qquad (3.64)$$

$$\theta = -\frac{1}{2} \arctan \frac{2\mu_{1,1}}{\mu_{0,2} - \mu_{2,0}} \qquad (3.65)$$

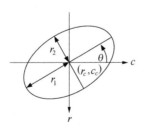

图 3.39 一个椭圆的几何参数

For a derivation of these results, see (Haralick and Shapiro, 1992) (note that, there, the diameters are used instead of the radii). From the ellipse para-

此结论的出处参见文献（Haralick and Shapiro, 1992）（注意文献中使用的是直径而不是半径）。通过椭圆的参

meters, we can derive another very useful feature: the anisometry r_1/r_2. This is scale-invariant and describes how elongated a region is.

The ellipse parameters are extremely useful in determining the orientations and sizes of regions. For example, the angle θ can be used to rectify rotated text. Figure 3.40(a) shows the result of thresholding the image in Figure 3.19(a). The segmentation result is treated as a single region, i.e., the connected components have not been computed. Figure 3.40(a) also displays the ellipse parameters by overlaying the major and minor axes of the equivalent ellipse. Note that the major axis is slightly longer than the region because the equivalent ellipse does not need to have the same area as the region. It only needs to have the same moments of orders 1 and 2. The angle of the major axis is a very good estimate for the rotation of the text. In fact, it has been used to rectify the images in Figures 3.19(b) and (c). Figure 3.40(b) shows the axes of the characters after the connected components have been computed. Note how well the orientation of the regions corresponds with our intuition.

数，能够推导出另一个非常有用的特征：各向异性 r_1/r_2。此特征量在区域缩放时是保持恒定不变的，它可以描述一个区域的细长程度。

椭圆的这些参数在确定区域的方位和尺寸时极其有用。例如角 θ 能被用来对经过旋转的文本进行校正。图 3.40（a）显示的是对图 3.19（a）进行阈值分割后的结果。分割结果被当作一个单一区域，也就是说还没有计算连通区域。图 3.40（a）中也叠加显示了等价椭圆的长轴和短轴。注意长轴比区域略长一些，这是因为等价椭圆不必与区域的面积相等，它仅需要与区域有同样的一阶矩和二阶矩即可。长轴与横轴的夹角很好地估计出了文本的旋转角度。事实上，在图 3.19（b）和图 3.19（c）中它已经被用来校正图像了。图 3.40（b）给出了计算连通区域后这些字符各自的长轴和短轴。注意这些区域的方位与人们的直觉是完全一致的。

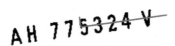

(a)

AH 775324 V

(b)

图 3.40　对图 3.19（a）进行阈值分割的结果，叠加显示了椭圆参数。浅灰色的线代表区域的长轴和短轴。这两个轴相交的点是区域的重心。(a) 分割结果被视为一个单独区域；(b) 计算后得到的区域内的连通区域。在图 3.19（b）和图 3.19（c）进行图像旋转校正时使用的是图（a）中得到的长轴与横轴夹角

While the ellipse parameters are extremely useful, they have two minor shortcomings. First of all, the orientation can be determined only if $r_1 \neq r_2$. Our first thought might be that this applies only to circles, which have no meaningful orientation anyway. Unfortunately, this is not true. There is a much larger class of objects for which $r_1 = r_2$. All objects that have a fourfold rotational symmetry, such as squares, have $r_1 = r_2$. Hence, their orientation cannot be determined with the ellipse parameters. The second slight problem is that, since the underlying model is an ellipse, the orientation θ can only be determined modulo π (180°). This problem can be solved by determining the point in the region that has the largest distance from the center of gravity and use it to select θ or $\theta + \pi$ as the correct orientation.

In the above discussion, we have used various transformations to make the moment-based features invariant to certain transformations, e.g., translation and scaling. Several approaches have been proposed to create moment-based features that are invariant to a larger class of transformations, e.g., translation, rotation, and scaling (Hu, 1962) or even general affine transformations (Flusser and Suk, 1993; Mamistvalov, 1998). They are primarily used to classify objects.

3.5.1.4 Enclosing Rectangles and Circles

Apart from the moment-based features, there are several other useful features that are based on the idea of finding an enclosing geometric primitive for the region. Figure 3.41(a) displays the smallest axis-parallel enclosing rectangle of a region. This rectangle is often also called the bounding box of the region. It can be calculated very easily

尽管椭圆的这些参数非常有用，但它们还是存在两个小问题。首先，只有在满足 $r_1 \neq r_2$ 时才可以确定区域的方位。第一感觉可能是只有对正圆不能使用这些参数，因为方位对正圆而言是没有意义的。不过，这个认识并不完整，还有很多物体也满足 $r_1 = r_2$。对于所有类似正方形那样的四折旋转对称物体，都是 $r_1 = r_2$。因此，这些物体的方位不能通过椭圆参数来确定。第二个问题是这样的，因为底层的模型是一个椭圆，所以仅能以 π (180°) 为模确定方位 θ。此问题可以这样解决，先确定区域内部距离重心最远的点，然后用此点来选择 θ 或 $\theta + \pi$ 作为正确方位。

在上述讨论中，已经使用不同的变换来让基于矩的特征对特定的变换结果保持不变。比如，平移和缩放后的结果。已经提出了一些新的方法，用这些方法建立的基于矩的特征可以在更多类型的变换结果上保持不变，比如平移、旋转、缩放（Hu，1962），甚至是仿射变换（Flusser and Suk，1993；Mamistvalov，1998）。这些方法优先被用来对物体进行分类。

3.5.1.4 外接矩形和外接圆

除了基于矩的特征外，还存在许多其他有用的特征，这些特征都基于为区域找到的一个外接几何基元。图 3.41（a）显示的是某个区域的最小平行轴外接矩形。此矩形也被称为区域的边框。它可基于区域横纵坐标的最大值和最小值计算得到。基于矩形

based on the minimum and maximum row and column coordinates of the region. Based on the parameters of the rectangle, other useful quantities like the width and height of the region and their ratio can be calculated. The parameters of the bounding box are particularly useful if we want to find out quickly whether two regions can intersect. Since the smallest axis-parallel enclosing rectangle sometimes is not very tight, we can also define a smallest enclosing rectangle of arbitrary orientation, as shown in Figure 3.41(b). Its computation is much more complicated than the computation of the bounding box, however, so we cannot give details here. An efficient implementation can be found in Toussaint (1983). Note that an arbitrarily oriented rectangle has the same parameters as an ellipse. Hence, it also enables us to define the position, size, and orientation of a region. Note that, in contrast to the ellipse parameters, a useful orientation for squares is returned. The final useful enclosing primitive is an enclosing circle, as shown in Figure 3.41(c). Its computation is also quite complex (Welzl, 1991). It also enables us to define the position and size of a region.

的参数，可计算出其他有用的特征量，如区域的宽度、高度、宽高比。当想要很快地判断两个区域是否存在相交的可能性时，边框的参数是非常有用的。因为最小平行轴外接矩形有时并不是非常紧地围绕在区域外，所以也需要定义任意方位的最小外接矩形，如图 3.41（b）所示。但它的计算较边框的计算复杂得多，所以不能在这里给出细节。一个高效率的实现算法可以参考文献（Toussaint，1983）。注意一个任意方位的矩形与椭圆有同样的参数。所以，它也能被用来定义区域的位置、尺寸和方向。注意与椭圆参数相比，它的好处是可以返回正方形的方向。最后一个有用的外接基元是外接圆，如图 3.41（c）。外接圆的计算也相当复杂（Welzl，1991）。外接圆也能定义区域的位置和尺寸。

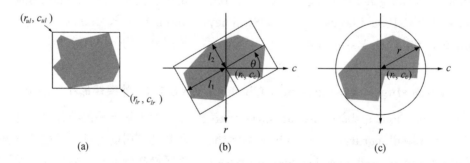

图 3.41 （a）某区域的最小平行轴外接矩形；（b）任意方位的最小外接矩形；（c）最小外接圆

The computations of the smallest enclosing rectangle of arbitrary orientation and the smallest

计算任意方位的最小外接矩形和最小外接圆的方法是基于首先计算区

enclosing circle are based on first computing the convex hull of the region. The convex hull of a set of points, and in particular a region, is the smallest convex set that contains all the points. A set is convex if, for any two points in the set, the straight line between them is completely contained in the set. The convex hull of a set of points can be computed efficiently (de Berg *et al.*, 2010; O'Rourke, 1998). The convex hull of a region is often useful to construct ROIs from regions that have been extracted from the image. Based on the convex hull of the region, another useful feature can be defined: the convexity, which is defined as the ratio of the area of the region to the area of its convex hull. It is a feature between 0 and 1 that measures how compact the region is. A convex region has a convexity of 1. The convexity can, for example, be used to remove unwanted segmentation results, which are often highly non-convex.

域的凸包。在一个特定区域里,一个点集的凸包就是包含了区域中所有的点的最小凸集。如果点集中任意两点连成的直线上的所有点都在此点集中,那么这个点集就是凸集。点集的凸包可以被高效率地计算出来(de Berg et al., 2010; O'Rourke, 1998)。在已经从图像中提取出的区域上构建 ROI 时,一个区域的凸包通常是很有用处的。基于此区域的凸包,能够定义另一个有用的特征:凸性。凸性被定义为某区域的面积和该区域凸包的面积之间的比值,它的值在 0 和 1 之间,可用来测量区域的紧凑程度。一个凸区域的凸性是 1。举个例子,凸性能被用来去除不想要的分割结果,这些不想要的结果通常是高度非凸的。

3.5.1.5 Contour Length

Another useful feature of a region is its contour length. To compute it, we need to trace the boundary of the region to get a linked contour of the boundary pixels (Haralick and Shapiro, 1992). Once the contour has been computed, we simply need to sum the Euclidean distances of the contour segments, which are 1 for horizontal and vertical segments and $\sqrt{2}$ for diagonal segments. Based on the contour length l and the area a of the region, we can define another measure for the compactness of a region: $c = l^2/(4\pi a)$. For circular regions, this feature is 1, while all other regions have larger values. The compactness has similar uses as the convexity.

3.5.1.5 轮廓长度

区域的另一个有用特征是区域的轮廓长度。为计算此特征量,必须跟踪区域的边界以获取一个轮廓,此轮廓将边界上的全部点连接在一起(Haralick and Shapiro, 1992)。一旦得到了区域的轮廓,仅需将全部轮廓线段的欧几里得距离求和即可。水平线段和垂直线段的欧几里得距离都是 1,而对角线段的距离是 $\sqrt{2}$。基于区域的轮廓长度 l 和区域的面积 a,可定义区域紧性的度量方法:$c = l^2/(4\pi a)$。所有圆形区域的紧性特征值都是 1,而其他区域的紧性特征值更大。紧性与凸性有着类似的用途。

3.5.2 Gray Value Features

3.5.2.1 Statistical Features

We have already seen some gray value features in Section 3.2.1.2, namely the minimum and maximum gray values within the region:

$$g_{\min} = \min_{(r,c)^\top \in R} g_{r,c}, \qquad g_{\max} = \max_{(r,c)^\top \in R} g_{r,c} \tag{3.66}$$

They are used for the contrast normalization in Section 3.2.1.2. Another obvious feature is the mean gray value within the region:

$$\bar{g} = \frac{1}{a} \sum_{(r,c)^\top \in R} g_{r,c} \tag{3.67}$$

Here, a is the area of the region, given by Eq. (3.59). The mean gray value is a measure of the brightness of the region. A single measurement within a reference region can be used to measure additive or multiplicative brightness changes with respect to the conditions when the system was set up. Two measurements within different reference regions can be used to measure linear brightness changes and thereby to compute a linear gray value transformation (see Section 3.2.1.1) that compensates for the brightness change, or to adapt segmentation thresholds.

The minimum, maximum, and mean gray values are statistical features. Another statistical feature is the variance of the gray values:

$$s^2 = \frac{1}{a-1} \sum_{(r,c)^\top \in R} (g_{r,c} - \bar{g})^2 \tag{3.68}$$

and the standard deviation $s = \sqrt{s^2}$. Measuring the mean and standard deviation within a reference region can also be used to construct a linear

3.5.2 灰度值特征

3.5.2.1 统计特征

在 3.2.1.2 节已经看到了一些灰度值特征：区域内的最大灰度值和最小灰度值：

在 3.2.1.2 节它们被用来进行对比度归一化处理。区域内灰度值的平均值是另一个明显的灰度值特征：

这里，a 是区域的面积，可由等式（3.59）计算得到。灰度值平均值是对区域内亮度的一个度量。对参考区域内灰度值的平均值进行测量可以用来确定加法或者乘法的亮度变化，此亮度变化是相对于系统最初被设置时的情况而言的。在两个不同参考区域内计算平均灰度值可测量出线性亮度变化，并且由此来计算一个线性灰度值变换（见 3.2.1.1 节），此变换可以用于补偿亮度的变化或调整分割阈值。

最小，最大和平均灰度值是一个统计特征。另一个统计特征是灰度值的方差：

和标准偏差 $s = \sqrt{s^2}$。在一个参考区域内测出的平均值和标准偏差也能被用来建立一个线性灰度值变换，此变

gray value transformation that compensates for brightness changes. The standard deviation can be used to adapt segmentation thresholds. Furthermore, the standard deviation is a measure of the amount of texture that is present within the region.

The gray value histogram (3.4) and the cumulative histogram (3.5), which we have already encountered in Section 3.2.1.3, are also gray value features. We have already used a feature that is based on the histogram for robust contrast normalization: the α-quantile

$$g_\alpha = \min\{g : c_g \geqslant \alpha\} \quad (3.69)$$

where c_g is defined in Eq. (3.5). It was used to obtain the robust minimum and maximum gray values in Section 3.2.1.3. The quantiles were called p_l and p_u there. Note that for $\alpha = 0.5$ we obtain the median gray value. It has similar uses to the mean gray value.

换可以补偿亮度的变化。标准偏差能够被用来调整分割阈值。而且，标准偏差可用来测量存在于区域内纹理的多少。

在 3.2.1.3 节中的灰度值直方图（3.4）和累积直方图（3.5）也属于灰度值特征。我们已经使用了直方图的一个特征来对图像进行可靠的对比度归一化处理，此特征就是 α-分位数：

其中 c_g 的定义见等式（3.5）。在 3.2.1.3 节中，此特征被用来获取可靠的最大灰度值和最小灰度值。分位数在 3.2.1.3 节中也被称为 p_l 和 p_u。注意当 $\alpha = 0.5$ 时，就得到了中值灰度值。它与平均值灰度值有着类似的用途。

3.5.2.2 Moments

In the previous section, we have seen that the region's moments are extremely useful features. They can be extended to gray value features in a natural manner. The gray value moment of order (p, q), with $p \geqslant 0$ and $q \geqslant 0$, is defined as

$$m_{p,q} = \sum_{(r,c)^\top \in R} g_{r,c}\, r^p c^q \quad (3.70)$$

This is the natural generalization of the region moments because we obtain the region moments from the gray value moments by using the characteristic function χ_R (3.1) of the region as the gray values. As for the region moments, the moment $a = m_{0,0}$ can be regarded as the gray value area

3.5.2.2 矩

在前面的小节中，已经看到了区域的矩是极其有用的特征量。它们也能很自然地被推广到灰度值特征中来。$p \geqslant 0$，$q \geqslant 0$ 时，(p, q) 阶灰度值矩定义为：

这是区域矩的自然而然的推广，因为在使用区域的特征函数 $\chi_R(3.1)$ 作为灰度值时，可以从灰度值矩中得到区域矩。与区域矩类似，矩 $a = m_{0,0}$ 能被视为区域的灰度值面积。它实质上是灰度值函数 $g_{r,c}$ 在区域内的"体

of the region. It is actually the "volume" of the gray value function $g_{r,c}$ within the region. As for the region moments, normalized moments can be defined by

$$n_{p,q} = \frac{1}{a} \sum_{(r,c)^\top \in R} g_{r,c} r^p c^q \qquad (3.71)$$

The moments $(n_{1,0}, n_{0,1})^\top$ define the gray value center of gravity of the region. With this, central gray value moments can be defined by

$$\mu_{p,q} = \frac{1}{a} \sum_{(r,c)^\top \in R} g_{r,c} (r - n_{1,0})^p (c - n_{0,1})^q \qquad (3.72)$$

3.5.2.3 Ellipse Parameters

As for the region moments, based on the second central moments, we can define the ellipse parameters, the major and minor axes, and the orientation. The formulas are identical to Eqs. (3.63)–(3.65). Furthermore, the anisometry can also be defined identically as for the regions.

3.5.2.4 Comparison of Region and Gray Value Moments

All the moment-based gray value features are very similar to their region-based counterparts. Therefore, it is interesting to look at their differences. As we saw, the gray value moments reduce to the region moments if the characteristic function of the region is used as the gray values. The characteristic function can be interpreted as the membership of a pixel to the region. A membership of 1 means that the pixel belongs to the region, while 0 means that the pixel does not belong to the region. This notion of belonging to the region is crisp, i.e., for every pixel a hard decision must

积"。与区域矩类似，归一化处理后的矩被定义为

矩 $(n_{1,0}, n_{0,1})^\top$ 定义的是区域的灰度值重心。通过此重心，中心灰度值矩被定义为

3.5.2.3 椭圆参数

与区域矩类似，在二阶中心矩的基础上我们能定义椭圆的长轴、短轴和方向等参数，公式与等式(3.63)～等式（3.65）一致。并且，各向异性的定义与区域矩中的也一样。

3.5.2.4 区域比对和灰度值矩

所有基于矩的灰度值特征与相应的基于矩的区域特性非常相似。因此，我们对它们之间的区别非常感兴趣。如上文所述，如果使用区域的特征函数作为灰度值，灰度值矩就简化为区域矩。特征函数可被解释为某像素对于此区域是否具备隶属关系。隶属关系为 1 时意味着此像素是区域内的，而 0 则表示此像素是区域外的。这个判断像素是否属于某区域的办法是脆弱的，因为针对每个像素点都必须做一次硬性判断。试想现在对每个像素

be made. Suppose now that, instead of making a hard decision for every pixel, we could make a "soft" or "fuzzy" decision about whether a pixel belongs to the region, and that we encode the degree of belonging to the region by a number $\in [0,1]$. We can interpret the degree of belonging as a fuzzy membership value, as opposed to the crisp binary membership value. With this, the gray value image can be regarded as a fuzzy set (Mendel, 1995).

The advantage of regarding the image as a fuzzy set is that we do not have to make a hard decision about whether a pixel belongs to the object or not. Instead, the fuzzy membership value determines what percentage of the pixel belongs to the object. This enables us to measure the position and size of the objects much more accurately, especially for small objects, because in the transition zone between the foreground and background there will be some mixed pixels that allow us to capture the geometry of the object more accurately. An example of this is shown in Figure 3.42. Here, a synthetically generated subpixel-precise ideal circle of radius 3 is shifted in subpixel increments. The gray values represent a fuzzy membership, scaled to values between 0 and 200 for display purposes. The figure displays a pixel-precise region, thresholded with a value of 100, which corresponds to a membership above 0.5, as well as two circles that have a center of gravity and area that were obtained from the region and gray value moments. The gray value moments were computed in the entire image. It can be seen that the area and center of gravity are computed much more accurately by the gray value moments because the decision about whether a pixel belongs

不使用硬性判断,而代以使用一个"软的"或"模糊的"判断来描述像素是否在区域内,这样对隶属于区域的程度进行编码,编码使用的数$\in [0,1]$。用一个模糊的隶属关系值来解释隶属的程度,而不使用脆弱的二值隶属关系。如此,灰度值图像就能被视为一个模糊集合(Mendel,1995)。

把图像看作一个模糊集合的好处是不必对一个像素点是否属于某物体做出硬性判断了,而是代以用模糊隶属关系值确定此像素属于此物体的百分比程度。这让在测定物体的位置和尺寸时更准确,特别是测定小的物体时。因为在前景与背景间的过渡区域中会存在一些混合像素,这些像素使在获取物体的几何信息时更准确。图3.42给出了一个这样的例子。此例中,以亚像素增量移动一个人为创建的、半径为3的亚像素精度理想圆。为方便显示,代表模糊隶属关系的灰度值按比例缩放到0~200。图中显示了以阈值100进行分割后得到的一个像素精度的区域,此区域对应的隶属关系高于0.5,图中同时显示了两个圆,其中一个圆的重心和面积是从区域矩获取的,另一个圆的重心和面积是从灰度值矩获取的。灰度值矩是在整图像上计算得到的。可以看到,因为避免了判断某个像素是否属于前景,所以由灰度值矩计算得到的面积和重心更准确。本例中,灰度值矩返回的面积误差总是小于0.25%,位置误差小于1/200像素。与之相比,区域矩求得

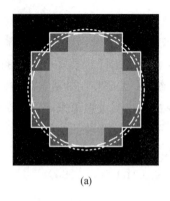 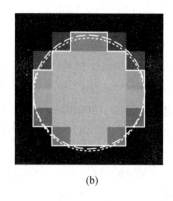 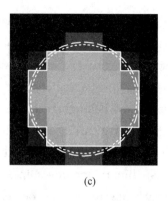

(a)   (b)   (c)

图 3.42　用灰度值矩和区域矩分别求得的亚像素精度圆的位置和面积。图像表示一个模糊隶属关系，为方便显示，代表模糊隶属关系的灰度值按比例缩放到 0~200。实线表示的区域是在隶属关系为 100 时分割得到的结果。点状虚线表示的圆与分割后的区域具有相同的重心和面积。段状虚线表示的圆是由灰度值矩计算得到的：（a）移动量，0；区域矩的面积误差，13.2%；灰度值矩的面积误差，−0.05%。（b）移动量，5/32 像素；区域矩的纵坐标误差，−0.129；灰度值矩的纵坐标误差，0.003。（c）移动量，1/2 像素；区域矩的面积误差，−8.0%；灰度值矩的面积误差，−0.015%。注意在处理小物体时，灰度值矩能得到准确度更好的处理结果

to the foreground or not has been avoided. In this example, the gray value moments result in an area error that is always smaller than 0.25% and a position error smaller than 1/200 of a pixel. In contrast, the area error for the region moments can be up to 13.2% and the position error can be up to 1/6 of a pixel. Note that both types of moments yield subpixel-accurate measurements. We can see that on ideal data it is possible to obtain an extremely high accuracy with the gray value moments, even for very small objects. On real data the accuracy will necessarily be somewhat lower. It should also be noted that the accuracy advantage of the gray value moments primarily occurs for small objects. Because the gray value moments must access every pixel within the region, whereas the region moments can be computed solely based on the run-length representation of the region, the region moments can be computed much more quickly. Hence, the gray moments are typically used only for relatively small regions.

的面积误差最高达到 13.2%，位置误差高达 1/6 像素。注意两种类型的矩得到的都是亚像素精度的测量结果。可以看到，尽管处理的是很小的物体，使用灰度值矩在理想的数据基础上还是能得到极高准确度的处理结果。基于实际数据计算时，准确度会稍微下降一些。同样应该注意的是灰度值矩的准确度优势主要体现在处理小的物体时。由于灰度值矩必须访问区域内的每个像素，而区域矩仅需要基于区域的行程编码就可以计算，所以计算区域矩的速度更快。因此，灰度值矩在一般情况下只用于处理相对小的区域。

The outstanding question we need to answer is how to define the fuzzy membership value of a pixel. If we assume that the camera has a fill factor of 100% and the gray value response of the image acquisition device and camera are linear, the gray value difference of a pixel from the background is proportional to the portion of the object that is covered by the pixel. Consequently, we can define a fuzzy membership relation as follows: every pixel that has a gray value below the background gray value $g_{\min}$ has a membership value of 0. Conversely, every pixel that has a gray value above the foreground gray value $g_{\max}$ has a membership value of 1. In between, the membership values are interpolated linearly. Since this procedure would require floating-point images, the membership is scaled to an integer image with b bits, typically 8 bits. Consequently, the fuzzy membership relation is a simple linear gray value scaling, as defined in Section 3.2.1.1. If we scale the fuzzy membership image in this manner, the gray value area needs to be divided by the maximum gray value, e.g., 255, to obtain the true area. The normalized and central gray value moments do not need to be modified in this manner since they are, by definition, invariant to a scaling of the gray values.

Figure 3.43 displays a real application where the above principles are used. In Figure 3.43(a) a BGA device with solder balls is displayed, along with two rectangles that indicate the image parts shown in Figures 3.43(b) and (c). The image in Figure 3.43(a) is first transformed into a fuzzy membership image using $g_{\min} = 40$ and $g_{\max} = 120$ with 8 bit resolution. The individual balls are segmented and then inspected for correct size and

必须要回答的未解决的问题是，如何定义某像素的模糊隶属关系值。若假定摄像机有 100% 的填充因子且摄像机和图像采集设备的灰度值响应是线性的，那么一个像素与背景的灰度值差异是与物体上被此像素所覆盖的面积成比例的。因此，能定义这样一个模糊隶属关系：灰度值低于背景灰度值 $g_{\min}$ 的每个像素，它们的隶属关系值都是 0。相反地，灰度值高于前景灰度值 $g_{\max}$ 的每个像素，它们的隶属关系值都是 1。灰度值落在此范围中的那些像素，它们的隶属关系值通过线性插值得到。由于这一计算过程需要使用浮点图像，所以通常将隶属关系值按比例放大到一个 b 位整数图像上，一般是 8 位整数图像。因此，隶属关系就成为了一个简单的线性灰度值比例缩放，如 3.2.1.1 节所述。当以此方式来按比例缩放隶属关系图像时，灰度值面积必须除以最大灰度值，比如除以 255，以得到真正的面积。归一化处理后的灰度值矩和中心灰度值矩无须此调整，因为由定义可知，它们在灰度值缩放时是恒定不变的。

图 3.43 给出应用上述理论的实例。图 3.43（a）显示的是 BGA 的焊点，图中用两个矩形指示的区域分别显示在图 3.43（b）和图 3.43（c）中。图 3.43（a）先在 $g_{\min} = 40$ 和 $g_{\max} = 120$ 时变换为一幅模糊隶属关系的 8 位图像。首先分割出每个独立的焊点。然后通过使用灰度值面积和灰度值各向异性对焊点的面积和形状

shape by using the gray value area and the gray value anisometry. The erroneous balls are displayed with dashed lines. To aid visual interpretation, the ellipses representing the segmented balls are scaled such that they have the same area as the gray value area. This is done because the gray value ellipse parameters typically return an ellipse with a different area than the gray value area, analogous to the region ellipse parameters (see the discussion in Section 3.5.1.3). As can be seen, all the balls that have an erroneous size or shape, indicating partially missing solder, have been correctly detected.

是否正确进行判断。不正确的焊点用虚线表示。为了更直观地看到效果，分割得到的焊点用椭圆表示，这些椭圆的面积已经按比例缩放到与灰度值面积相等。这样做是因为一般情况下由灰度值椭圆参数返回的椭圆，其面积与灰度值面积是不相等的，这与区域椭圆参数（见 3.5.1.3 节的讨论）类似。可以看到，所有的形状或尺寸错误的焊点都被正确地检测出来了，这些焊点表示部分缺少焊料。

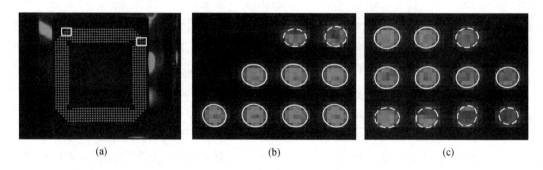

图 3.43 （a）BGA 的图像；两个矩形区域对应的图像见（b）和（c）。检测焊点是否具有正确尺寸（灰度值面积 ⩾ 20）和正确灰度值各向异性（⩽ 1.25），如图（b）和（c）所示。正确的焊点以实线椭圆表示，而有缺陷的焊点用虚线表示

3.5.3 Contour Features

3.5.3.1 Contour Length, Enclosing Rectangles and Circles

Many of the region features we have discussed in Section 3.5.1 can be transferred to subpixel-precise contour features in a straightforward manner. For example, the length of the subpixel-precise contour is even easier to compute because the contour is already represented explicitly by its control points $(r_i, c_i)^\top$, for $i = 1, \cdots, n$. It

3.5.3 轮廓特征

3.5.3.1 轮廓长度，外接矩形和外接圆

在 3.5.1 节讨论过的很多区域特征都能直接转换为亚像素精度轮廓特征。例如，亚像素精度轮廓长度的计算更容易些，因为轮廓已经用控制点 $(r_i, c_i)^\top$，$i = 1, \cdots, n$ 描述得很清楚了。计算轮廓的最小外接平行轴矩形（边框）也很简单。此外，计算轮

is also simple to compute the smallest enclosing axis-parallel rectangle (the bounding box) of the contour. Furthermore, the convex hull of the contour can be computed as for regions (de Berg et al., 2010; O'Rourke, 1998). From the convex hull, we can also derive the smallest enclosing circles (Welzl, 1991) and the smallest enclosing rectangles of arbitrary orientation (Toussaint, 1983).

3.5.3.2 Moments

In the previous two sections, we have seen that the moments are extremely useful features. An interesting question, therefore, is whether they can be defined for contours. In particular, it is interesting to see whether a contour has an area. Obviously, for this to be true, the contour must enclose a region, i.e., it must be closed and must not intersect itself. To simplify the formulas, let us assume that a closed contour is specified by $(r_1, c_1)^\top = (r_n, c_n)^\top$. Let the subpixel-precise region that the contour encloses be denoted by R. Then, the moment of order (p, q) is defined as

$$m_{p,q} = \iint\limits_{(r,c) \in R} r^p c^q \, dr \, dc \tag{3.73}$$

As for regions, we can define normalized and central moments. The formulas are identical to Eqs. (3.61) and (3.62) with the sums being replaced by integrals. It can be shown that these moments can be computed solely based on the control points of the contour (Steger, 1996). For example, the area and center of gravity of the contour are given by

廓的凸包与计算区域的类似（de Berg et al., 2010；O'Rouke, 1998）。根据轮廓的凸包，可推导出最小外接圆（Welzl, 1991）和任意方向最小外接矩形（Toussaint, 1983）。

3.5.3.2 矩

在前两节中，已经看到了矩特征是很有用。因此，是否也可以为轮廓定义矩特征就是我们感兴趣的问题。尤其令人感兴趣的是一个轮廓是否存在面积。很显然，必须是围绕一个区域的轮廓才存在这些矩特征，也就是说，轮廓必须是闭合的且不能自相交。为简化公式，假设一个闭合轮廓是通过 $(r_1, c_1)^\top = (r_n, c_n)^\top$ 来表示的。R 表示轮廓围绕的亚像素精度区域，则 (p, q) 阶矩被定义为

与区域类似，可定义归一化的矩和中心矩。该等式与等式（3.61）和等式（3.62）相同，只是将原式中的连加求和符号换成积分符号。可看到，以上这些矩都能仅基于轮廓上的控制点计算得到（Steger, 1996）。例如，轮廓的面积和重心能由等式（3.74）～等式（3.76）计算：

$$a = \frac{1}{2} \sum_{i=1}^{n} (r_{i-1} c_i - r_i c_{i-1}) \tag{3.74}$$

$$n_{1,0} = \frac{1}{6a} \sum_{i=1}^{n} (r_{i-1}c_i - r_i c_{i-1})(r_{i-1} + r_i) \qquad (3.75)$$

$$n_{0,1} = \frac{1}{6a} \sum_{i=1}^{n} (r_{i-1}c_i - r_i c_{i-1})(c_{i-1} + c_i) \qquad (3.76)$$

Analogous formulas can be derived for the second-order moments. Based on these, we can again compute the ellipse parameters, major axis, minor axis, and orientation. The formulas are identical to Eqs. (3.63)–(3.65). The moment-based contour features can be used for the same purposes as the corresponding region and gray value features. By performing an evaluation similar to that in Figure 3.42, it can be seen that the contour center of gravity and the ellipse parameters are equally as accurate as the gray value center of gravity. The accuracy of the contour area is slightly worse than that of the gray value area because we have approximated the hyperbolic segments with line segments. Since the true contour is a circle, the line segments always lie inside the true circle. Nevertheless, subpixel-thresholding and the contour moments could also have been used to detect the erroneous balls in Figure 3.43.

对于二阶矩能推导出类似的公式。根据这些公式，可计算出椭圆的长轴、短轴和方向等参数，该等式与等式（3.63）～等式（3.65）相同。基于矩的轮廓特征与基于矩的区域特征和基于矩的灰度值特征用途类似。通过执行一个与图 3.42 中类似的估计，可以看到轮廓的重心和椭圆参数的准确度与灰度值重心相同。轮廓面积的准确度比灰度值面积的准确度稍微差些，这是因为已经将双曲线线段近似处理为直线段。由于真正的轮廓是一圆周，而线段总是位于真正的轮廓圆周内部的。不过，亚像素阈值分割和轮廓矩也能用于检测图 3.43 中不正确的焊点。

3.6 Morphology

3.6 形态学

In Section 3.4, we discussed how to segment regions. We have already seen that segmentation results often contain unwanted noisy parts. Furthermore, sometimes the segmentation will contain parts in which the shape of the object we are interested in has been disturbed, e.g., because of reflections. Therefore, we often need to modify the shape of the segmented regions to obtain the desired results. This is the subject of the field

在 3.4 节已经讨论了如何分割区域。已经看到了分割结果中经常包含不想要的干扰。而且有时分割结果将包含这样一些部分，在这些部分内我们所感兴趣物体的形状已经被干扰了，例如，由于反射对分割结果造成的干扰。因此，通常必须调整分割后区域的形状以获取想要的结果。这是数学形态学领域的课题，数学形态学

of mathematical morphology, which can be defined as a theory for the analysis of spatial structures (Serra, 1982; Soille, 2003). For our purposes, mathematical morphology provides a set of extremely useful operations that enable us to modify or describe the shape of objects. Morphological operations can be defined on regions and gray value images. We will discuss both types of operations in this section.

3.6.1 Region Morphology

3.6.1.1 Set Operations

All region morphology operations can be defined in terms of six very simple operations: union, intersection, difference, complement, translation, and transposition. We will take a brief look at these operations first.

The union of two regions R and S is the set of points that lie in R or in S:

$$R \cup S = \{\boldsymbol{p} \mid \boldsymbol{p} \in R \vee \boldsymbol{p} \in S\} \tag{3.77}$$

One important property of the union is that it is commutative: $R \cup S = S \cup R$. Furthermore, it is associative: $(R \cup S) \cup T = R \cup (S \cup T)$. While this may seem like a trivial observation, it will enable us to derive very efficient implementations for the morphological operations below. The algorithm to compute the union of two binary images is obvious: we simply need to compute the logical OR of the two images. The run time complexity of this algorithm obviously is $O(wh)$, where w and h are the width and height of the binary image. In the run-length representation, the union can be computed with a lower complexity: $O(n + m)$, where n and m are the number of runs in R and S. The principle of the algorithm is to merge the

被定义为一种分析空间结构的理论（Serra，1982；Soille，2003）。为此，数学形态学提供了一组特别有用的方法，这些方法能调整或描述物体的形状。形态学的处理方法能够在区域和灰度值图像上被定义。本节将讨论这两种类型的形态学方法。

3.6.1 区域形态学

3.6.1.1 设置操作

所有的区域形态学处理能根据六个非常简单的操作来定义：并集，交集，差集，补集，平移和转置。首先简单浏览一下这些操作。

两个区域 R 和 S 的并集是所有位于 R 或者 S 内的点的集合：

并集的一个重要特性是其操作的可交换性：$R \cup S = S \cup R$。此外，并集操作是可结合的：$(R \cup S) \cup T = R \cup (S \cup T)$。尽管这看起来似乎是价值不大的发现，但稍后它能推导出非常高效率的形态学的处理实现算法。计算两幅二值图像并集的算法是显而易见的：仅需要对两图像进行逻辑或运算即可。计算的复杂度显然是 $O(wh)$，这里 w 和 h 分别是二值图像的宽度和高度。如果区域是用行程来表示的，并集计算时的复杂度更低：$O(n+m)$，此处 n 和 m 分别是区域 R 和 S 包含的行程数。计算的原理是观察行程的顺序（见 3.1.2 节）同时合并两个区域的

runs of the two regions while observing the order of the runs (see Section 3.1.2) and then to pack overlapping runs into single runs.

The intersection of two regions R and S is the set of points that lie in R and in S:

$$R \cap S = \{ \boldsymbol{p} \mid \boldsymbol{p} \in R \wedge \boldsymbol{p} \in S \} \tag{3.78}$$

Like the union, the intersection is commutative and associative. Again, the algorithm on binary images is obvious: we compute the logical AND of the two images. For the run-length representation, again an algorithm that has complexity $O(n+m)$ can be found.

The difference of two regions R and S is the set of points that lie in R but not in S:

$$R \setminus S = \{ \boldsymbol{p} \mid \boldsymbol{p} \in R \wedge \boldsymbol{p} \notin S \} = R \cap \overline{S} \tag{3.79}$$

The difference is neither commutative nor associative. Again, the algorithm on binary images is obvious: we compute the logical AND NOT of the two images. For the run-length representation, again an algorithm that has complexity $O(n+m)$ exists. Note that the difference can be defined in terms of the intersection and the complement of a region R, which is defined as all the points that do not lie in R:

$$\overline{R} = \{ \boldsymbol{p} \mid \boldsymbol{p} \notin R \} \tag{3.80}$$

Since the complement of a finite region is infinite, it is impossible to represent it as a binary image. Therefore, for the representation of regions as binary images, it is important to define the operations without the complement. It is, however, possible to represent it as a run-length-encoded region by adding a flag that indicates whether the

行程，然后将相互交叠的几个行程合并成一个行程。

两个区域 R 和 S 的交集是不但位于 R 且又位于 S 内的所有点的集合：

与并集类似，交集操作也是可交换的和可结合的。并且，在二值图像上求交集的算法也是明显的：仅对两图像进行逻辑与运算即可。对于行程法表示的区域，算法的复杂度也是 $O(n+m)$。

两个区域 R 和 S 的差集是位于 R 且不能位于 S 内的所有点的集合：

差集计算是不能交换的，也是不能结合的。此外，二值图像上的差集算法非常明显：仅对两幅图像进行逻辑与否运算。对于行程表示法，算法复杂度是 $O(n+m)$。注意差集可以根据交集和补集来定义。一个区域 R 的补集被定义为不位于 R 内的所有点的集合：

由于一个有限区域的补集是无限的，所以不可能用二值图像来表示。因此，对于以二值图像表示的区域，定义不含补集的操作是重要的。但以行程编码表示区域时是可以使用补集操作的，这是通过增加一个标记来指示保存的是区域还是区域的补集。这能被

region or its complement is being stored. This can be used to define a more general set of morphological operations.

There is an interesting relation between the number of connected components of the background $|C(\overline{R})|$ and the number of holes of the foreground $|H(R)|$: we have $|C(\overline{R})| = 1 + |H(R)|$. As discussed in Section 3.4.2, complementary connectivities must be used for the foreground and the background for this relation to hold.

Apart from the set operations, two basic geometric transformations are used in morphological operations. The translation of a region by a vector t is defined as

$$R_t = \{p \mid p - t \in R\} = \{q \mid q = p + t \text{ for } p \in R\} \qquad (3.81)$$

Finally, the transposition of a region is defined as a mirroring about the origin:

$$\check{R} = \{-p \mid p \in R\} \qquad (3.82)$$

Note that this is the only operation where a special point (the origin) is singled out. All the other operations do not depend on the origin of the coordinate system, i.e., they are translation-invariant.

3.6.1.2 Minkowski Addition and Dilation

With these building blocks, we can now take a look at the morphological operations. They typically involve two regions. One of these is the region we want to process, which will be denoted by R below. The other region has a special meaning. It is called the structuring element, and will be denoted by S. The structuring element is the means by which we can describe the shapes we are interested in.

用来定义一组更广义的形态学操作。

背景的连通区域数 $|C(\overline{R})|$ 与前景的孔洞数 $|H(R)|$ 之间存在着非常有趣的关系：$|C(\overline{R})| = 1 + |H(R)|$。如在 3.4.2 节中已经讨论的，对于前景和背景必须使用对立的连通性定义以保证此关系。

除了集合操作外，两种基础的几何变换也被用于形态学的处理中。根据一向量 t 来平移某个区域被定义为

最后，一个区域的转置被定义为关于原点的一个镜像：

注意以上操作中，转置是唯一需要挑选特殊点（原点）的操作。所有其他的操作都不依赖于坐标系的原点，也就是说，它们是平移不变的。

3.6.1.2 闵可夫斯基加法和膨胀

使用上面这些基本操作，现在就可以学习形态学的处理方法了。这些处理通常要涉及两个区域：其中一个是想要处理的那个区域，下文中用 R 表示；另一个区域有一个特殊的意义，它被称为结构元，用 S 表示。通过结构元来描述感兴趣的形状。

The first morphological operation we consider is Minkowski addition, which is defined by (see Serra, 1982; Soille, 2003)

我们考虑的第一个形态学的操作是闵可夫斯基加法，定义如下（参见 Serra，1982；Soille，2003）：

$$R \oplus S = \{r + s \mid r \in R, s \in S\} = \bigcup_{s \in S} R_s = \bigcup_{r \in R} S_r = \{t \mid R \cap (\check{S})_t \neq \emptyset\} \quad (3.83)$$

It is interesting to interpret the formulas. The first formula says that, to get the Minkowski addition of R with S, we take every point in R and every point in S and compute the vector sum of the points. The result of the Minkowski addition is the set of all points thus obtained. If we single out S, this can also be interpreted as taking all points in S, translating the region R by the vector corresponding to the point s from S, and computing the union of all the translated regions. Thus, we obtain the second formula. By symmetry, we can also translate S by all points in R to obtain the third formula. Another way to look at Minkowski addition is the fourth formula. It tells us that we move the transposed structuring element around in the plane. Whenever the translated transposed structuring element and the region have at least one point in common, we copy the translated reference point into the output. Figure 3.44 shows an example of Minkowski addition.

解释这个定义式是有趣的。第一个公式是说为得到闵可夫斯基加法 R 加 S 的结果，拿出 R 中的每个点以及 S 中的每个点，然后计算这些点的向量和。闵可夫斯基加法的结果是由此得到的所有点的集合。如果挑选 S，这也能被解释为取出 S 中所有的点，根据与从 S 中取出的点 s 所对应的向量来平移区域 R，然后对全部平移得到的区域取并集。这样，就获得了第二个公式。由于对称性，也能根据 R 中的全部点来平移 S 以得到第三个公式。另一个描述闵可夫斯基加法的方法是第四个公式：它告诉我们在平面内移动转置后的结构元，任何时刻当转置后的结构元平移到与区域存在至少一个公共点时，可拷贝此平移后的参考点到输出中。图 3.44 是闵可夫斯基加法的例子。

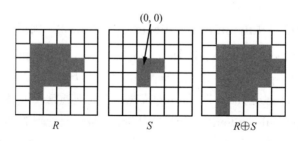

图 3.44　闵可夫斯基加法 $R \oplus S$ 的例子

If the structuring element S contains the origin $(0, 0)^\top$, the Minkowski addition is extensive,

如果结构元包含原点 $(0, 0)^\top$，则闵可夫斯基加法是外延的，即，$R \subseteq$

i.e., $R \subseteq R \oplus S$. If S contains more than one point, we have $R \subset R \oplus S$. If S does not contain the origin, there is a vector t for which $R \subseteq (R \oplus S)_t$. Therefore, Minkowski addition increases the size of the region R. Furthermore, Minkowski addition is increasing, i.e., if $R \subseteq S$ then $R \oplus T \subseteq S \oplus T$. Therefore, Minkowski addition preserves the inclusion relation of regions.

While Minkowski addition has a simple formula, it has one small drawback. Its geometric criterion is that the transposed structuring element has at least one point in common with the region. Ideally, we would like to have an operation that returns all translated reference points for which the structuring element itself has at least one point in common with the region. To achieve this, we only need to use the transposed structuring element in the Minkowski addition. This operation is called a dilation, and is defined by (Serra, 1982; Soille, 2003)

$R \oplus S$。如果 S 包含多于一个点，则 $R \subset R \oplus S$。如果 S 不包含原点，则对于每个 $R \subseteq (R \oplus S)_t$ 都存在一个向量 t。因此，闵可夫斯基加法增加了区域 R 的尺寸。此外，闵可夫斯基加法保持了区域的包含关系。

尽管闵可夫斯基加法的公式简单，但它还是有一个小的缺点。其几何学准则是转置后结构元至少与区域存在一个公共点。理想情况下，希望能有一种操作，此操作能返回所有平移后的参考点，在这些位置时结构元自身与区域存在至少一个公共点。要得到这样的操作，仅需在闵可夫斯基加法中使用转置后的结构元即可。此操作被称为膨胀，定义如下（Serra, 1982；Soille, 2003）

$$R \oplus \check{S} = \{t \mid R \cap S_t \neq \emptyset\} = \bigcup_{s \in S} R_{-s} \qquad (3.84)$$

Figure 3.45 shows an example of dilation. Note that the results of Minkowski addition and dilation are different. This is true whenever the structuring element is not symmetric with respect to the origin. If the structuring element is symmetric, the results of Minkowski addition and dilation are identical.

图 3.45 给出了膨胀的例子。注意只要结构元相对于原点是非对称的，那么闵可夫斯基加法的结果与膨胀的结果是不一样的；如果结构元是对称的，那么闵可夫斯基加法的结果和膨胀的结果是相同的。

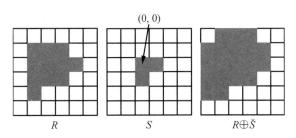

图 3.45　膨胀 $R \oplus \check{S}$ 的例子

Like Minkowski addition, dilation is extensive if the structuring element S contains the origin. The other extensive properties of Minkowski addition also hold for dilation. Furthermore, dilation is increasing.

The implementation of Minkowski addition for binary images is straightforward. As suggested by the second formula in Eq. (3.83), it can be implemented as a nonlinear filter with logical OR operations. The run time complexity is proportional to the size of the image times the number of pixels in the structuring element. The second factor can be reduced to roughly the number of boundary pixels in the structuring element (Van Droogenbroeck and Talbot, 1996). Also, for binary images represented with 1 bit per pixel, very efficient algorithms can be developed for special structuring elements (Bloomberg, 2002). Nevertheless, in both cases the run time complexity is proportional to the number of pixels in the image. To derive an implementation for the run-length representation of the regions, we first need to examine some algebraic properties of Minkowski addition. It is commutative: $R \oplus S = S \oplus R$. Furthermore, it is distributive with respect to the union: $(R \cup S) \oplus T = R \oplus T \cup S \oplus T$. Since a region can be regarded as the union of its runs, we can use the commutativity and distributivity to transform the Minkowski addition formula as follows:

与闵可夫斯基加法相同，如果结构元 S 包含了原点则膨胀算法也是外延的。闵可夫斯基加法的其他外延特性对膨胀算法也同样适用，此外，膨胀算法是增量运算。

闵可夫斯基加法在二值图像上的实现是简单易懂的，如等式（3.83）中第二个公式所示，可通过一个非线性滤波器的逻辑或运算来实现。运算复杂度与图像尺寸和结构元中的像素数的乘积成比例。结构元中的像素数可被结构元边界上的像素数粗略地取代以降低计算量（Van Droogenbroeck and Talbot, 1996）。对于每像素仅用一个二进制位表示的二值图像，可开发出针对特定结构元的高效率算法（Bloomberg, 2002）。不管怎样，在这两种情况下，算法运行时间的复杂度都与图像中的像素总数是成比例的。对于行程法描述的区域，为推导出适用的算法，首先需要研究闵可夫斯基加法的几个代数属性。它的可交换性 $R \oplus S = S \oplus R$，以及它对于并集操作的分配性 $(R \cup S) \oplus T = R \oplus T \cup S \oplus T$。由于一个区域能被视为其行程的一个并集，所以可用交换律和分配率来实现闵可夫斯基加法，如下所示：

$$R \oplus S = \left(\bigcup_{i=1}^{n} r_i\right) \oplus \left(\bigcup_{j=1}^{m} s_j\right) = \bigcup_{j=1}^{m}\left(\left(\bigcup_{i=1}^{n} r_i\right) \oplus s_j\right) = \bigcup_{i=1}^{n}\bigcup_{j=1}^{m} r_i \oplus s_j \quad (3.85)$$

Thus, Minkowski addition can be implemented as the union of nm dilations of single runs, which are trivial to compute. Since the union of the runs can be computed easily, the run time complexity

这样，闵可夫斯基加法能够通过对 nm 次单行程间的膨胀处理结果求并集实现，此实现法的计算量不大。由于行程的并集可被轻松地计算出来，

is $O(mn)$, which is better than for binary images.

As we have seen, dilation and Minkowski addition enlarge the input region. This can be used, for example, to merge separate parts of a region into a single part, and thus to obtain the correct connected components of objects. One example of this is shown in Figure 3.46. Here, we want to segment each character as a separate connected component. If we compute the connected components of the thresholded region in Figure 3.46(b), we can see that the characters and their dots are separate components (Figure 3.46(c)). To solve this problem, we first need to connect the dots with their characters. This can be achieved using a dilation with a circle of diameter 5 (Figure 3.46(d)). With this, the correct connected components are obtained (Figure 3.46(e)). Unfortunately, they have the wrong shape because of the dilation. This can be corrected by intersecting the components with the originally segmented

所以算法运行的复杂度只有 $O(mn)$，是优于基于二值图像的运算的。

正如上面已经看到的，膨胀和闵可夫斯基加法将输入区域扩大了。这可以用来将区域中彼此分开的几个部分合并成一个单一的部分，这样就得到了能正确表示物体的连通区域。图 3.46 给出了一个具体的例子。此例中，想将每个字符分割成一个连通区域。如果在阈值分割后得到的区域（如图 3.46（b））上计算连通区域的话，可看到字符与字符的点就变成了彼此分开的不同区域，见图 3.46（c）。为解决此问题，首先要把字符与它所包含的点连为一体，这可以通过用直径为 5 的圆对分割结果进行膨胀处理来实现，如图 3.46（d）。这样就得到了正确的连通区域，如图 3.46（e）所示。不幸的是这些区域的形状在经过膨胀处理后变得不正确了。这可以通过把这些区域与原分割区域进行相交

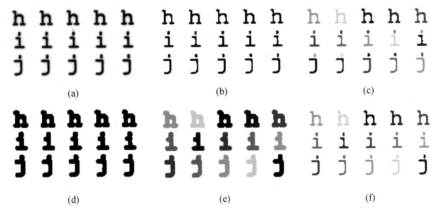

图 3.46 （a）印刷字符的图像；（b）对图（a）进行阈值分割的结果；（c）从图（b）中计算的连通区域，显示时用 6 种不同灰度值来表示不同的区域。注意字符和该字符包含的点已被处理成了相互独立的不同连通区域，这不是想要的结果；（d）对图（b）中区域进行膨胀，结构元是直径为 5 的圆；（e）从图（d）中计算的连通区域。注意字符与字符中的点变成了一个单一的连通区域；（f）将图（e）中的连通区域与图（b）中的原始分割结果进行相交处理后的结果。这将连通区域恢复成了正确的形状

region. Figure 3.46(f) shows that with thesesimple steps we have obtained one component with the correct shape for each character.

Dilation is also very useful for constructing ROIs based on regions that were extracted from the image. We will see an example of this in Section 3.7.3.5.

3.6.1.3 Minkowski Subtraction and Erosion

The second type of morphological operation is Minkowski subtraction. It is defined by (see Serra, 1982; Soille, 2003)

$$R \ominus S = \bigcap_{s \in S} R_s = \{r \mid \forall\, s \in S : r - s \in R\} = \{t \mid (\check{S})_t \subseteq R\} \qquad (3.86)$$

The first formula is similar to the second formula in Eq. (3.83) with the union having been replaced by an intersection. Hence, we can still think about moving the region R by all vectors s from S. However, now the points must be contained in all translated regions (instead of at least one translated region). This is what the second formula in Eq. (3.86) expresses. Finally, if we look at the third formula, we see that we can also move the transposed structuring element around in the plane. If it is completely contained in the region R, we add its reference point to the output. Again, note the similarity to Minkowski addition, where the structuring element had to have at least one point in common with the region. For Minkowski subtraction, it must lie completely within the region. Figure 3.47 shows an example of Minkowski subtraction.

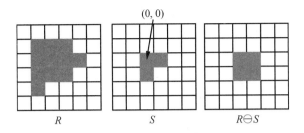

图 3.47 闵可夫斯基减法 $R \ominus S$ 的例子

If the structuring element S contains the origin, Minkowski subtraction is anti-extensive, i.e., $R \ominus S \subseteq R$. If S contains more than one point, we have $R \ominus S \subset R$. If S does not contain the origin, there is a vector t for which $(R \ominus S)_t \subseteq R$. Therefore, Minkowski subtraction decreases the size of the region R. Furthermore, Minkowski subtraction is increasing, i.e., if $R \subseteq S$ then $R \ominus T \subseteq S \ominus T$.

Minkowski subtraction has the same small drawback as Minkowski addition: its geometric criterion is that the transposed structuring element must lie completely within the region. As for dilation, we can use the transposed structuring element. This operation is called erosion and is defined by

如果结构元 S 包含原点,则闵可夫斯基减法是非外延的,即,$R \ominus S \subseteq R$。如果 S 包含多于一个点,则 $R \ominus S \subset R$。如果 S 不包含原点,则对于每个 $(R \ominus S)_t \subseteq R$ 都存在一个向量 t。因此,闵可夫斯基减法减小了区域 R 的尺寸。此外,闵可夫斯基减法增量运算,即如果 $R \subseteq S$,那么 $R \ominus T \subseteq S \ominus T$。

闵可夫斯基减法与闵可夫斯基加法一样都有一个小缺点:其几何准则是转置结构元必须完全落在区域内。在膨胀中,使用的是转置结构元。在闵可夫斯基减法中使用转置结构元被称为腐蚀,定义如下:

$$R \ominus \check{S} = \bigcap_{s \in S} R_{-s} = \{t \mid S_t \subseteq R\} \quad (3.87)$$

Figure 3.48 shows an example of erosion. Again, note that Minkowski subtraction and

图 3.48 是腐蚀的一个例子。再次注意闵可夫斯基减法和腐蚀只有在结

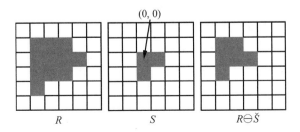

图 3.48 腐蚀 $R \ominus \check{S}$ 的例子

erosion produce identical results only if the structuring element is symmetric with respect to the origin.

Like Minkowski subtraction, erosion is antiextensive if the structuring element S contains the origin. The other anti-extensive properties of Minkowski subtraction also hold for erosion. Furthermore, erosion is increasing.

The fact that Minkowski subtraction and erosion shrink the input region can, for example, be used to separate objects that are attached to each other. Figure 3.49 shows an example. Here, the goal is to segment the individual globular objects. The result of thresholding the image is shown in Figure 3.49(b). If we compute the connected components of this region, an incorrect result is obtained because several objects touch each other (Figure 3.49(c)). The solution is to erode the region with a circle of diameter 15 (Figure 3.49(d)) before computing the connected components (Figure 3.49(e)). Unfortunately, the connected components have the wrong shape. Here, we cannot use the same strategy that we used for the dilation (intersecting the connected components with the original segmentation) because the erosion has shrunk the region. To approximately get the original shape back, we can dilate the connected components with the same structuring element that we used for the erosion (Figure 3.49(f)) (more precisely, the dilation is actually a Minkowski addition; however, since the structuring element is symmetric in this example, both operations return the same result).

We can see another use of erosion if we remember its definition: it returns the translated reference point of the structuring element S for

构元是基于原点对称时才会输出同样的处理结果。

与闵可夫斯基减法相同，如果结构元 S 包含了原点则腐蚀算法也是非外延的。闵可夫斯基减法的其他非外延特性对腐蚀算法也同样适用，此外，腐蚀算法是增量运算。

闵可夫斯基减法和腐蚀处理会将输入区域收缩。这可以用来将彼此相连的物体分开。图 3.49 给出了一个这样的例子。此例的目的是要分割出每个球状物体。对图像进行阈值分割后的结果如图 3.49（b）所示。如果我们在分割基础上计算连通区域，那么将会得到错误的结果。因为有一些物体是彼此相连的（见图 3.49（c））。处理方法是使用直径为 15 的圆对分割后的区域进行腐蚀处理（见图 3.49（d）），然后再进行连通区域的计算，结果如图 3.49（e）。不幸的是，这样得到的连通区域的形状是不正确的。这时，因为腐蚀已经使区域变小了，所以不能使用在膨胀处理中应用过的策略（即将连通区域与原分割结果进行相交）。为近似地恢复原来的形状，可采用腐蚀处理时使用的结构元对连通区域进行膨胀处理（见图 3.49（f））（更确切地说，膨胀实际上是一种闵可夫斯基加法；然而，由于本例中结构元是对称的，所以两种处理返回相同的结果）。

如果记得腐蚀的定义，就可看到腐蚀的另一个用途：腐蚀操作返回的是对结构元 S 平移后得到的参考点，

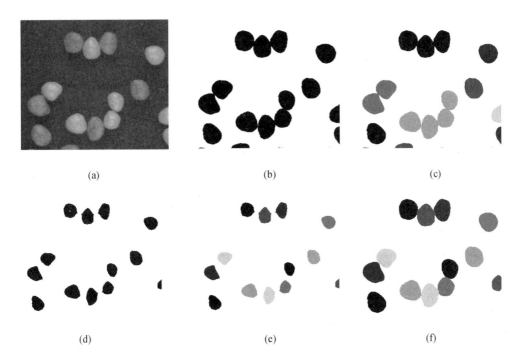

图 3.49 （a）一些圆形物体的图像；（b）对图（a）进行阈值分割的结果；（c）从图（b）计算的连通区域，用 6 种不同灰度值显示。注意彼此接触的物体求出连通区域也是相连的；（d）对图（b）中的区域进行腐蚀，采用的是直径 15 的圆；（e）从图（d）求出的连通区域；（f）对图（e）中的连通区域进行膨胀，仍用直径为 15 的圆。这是为了将连通区域修正为与正确形状接近的形状

every translation for which S_t completely fits into the region R. Hence, erosion acts like a template matching operation. An example of this use of erosion is shown in Figure 3.50. In Figure 3.50(a), we can see an image of a print of several letters, with the structuring element used for the erosion overlaid in white. The structuring element corresponds to the center line of the letter "e." The reference point of the structuring element is its center of gravity. The result of eroding the thresholded letters (Figure 3.50(b)) with the structuring element is shown in Figure 3.50(c). Note that all letters "e" have been correctly identified. In Figures 3.50(d)–(f), the experiment is repeated with another set of letters. The structuring element is the center line of the letter "o." Note that the

这些点是在每次平移后的结构元 S_t 彻底落在区域 R 内时得到的。因此，腐蚀处理的行为类似于模板匹配。腐蚀的这种应用案例见图 3.50。在图 3.50（a）中，可以看到许多印刷的字符，用于腐蚀操作的结构元用白色叠加显示在图上。此结构元对应的是字母 "e" 的中心线。结构元的参考点是此结构元的重心。对阈值分割后的字母（图 3.50（b））采用此结构元进行腐蚀后，结果如图 3.50（c）所示。注意所有的字符 "e" 都被正确地识别出来了。图 3.50（d）~（f）中，对另一组字符重复了同样的实验，使用的结构元是字母 "o"。注意腐蚀的结果能正确识别字母 "o"，但也包含了字母

erosion correctly finds the letters "o." However, in addition the circular parts of the letters "p" and "q" are found, since the structuring element completely fits into them.

"p"和"q"中的圆圈部分，这是因为结构元与这些部分同样匹配。

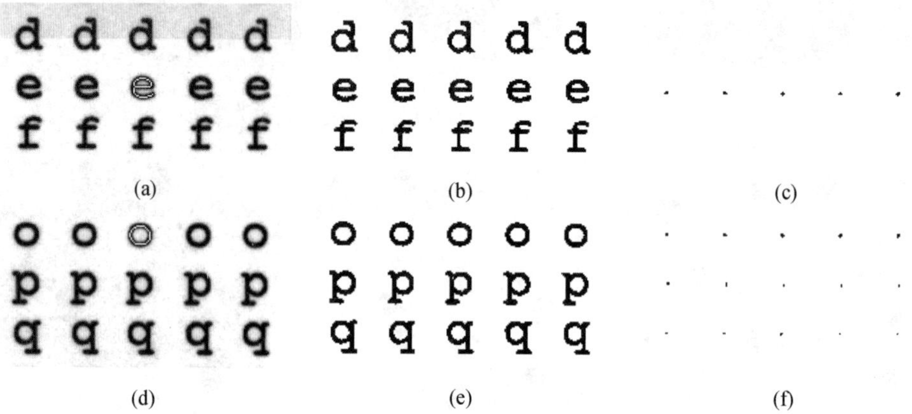

图 3.50　（a）若干印刷字符的图像，用于腐蚀操作的结构元用白色叠加显示在图上；（b）对图（a）进行阈值分割的结果；（c）使用图（a）中的结构元对图（b）进行腐蚀的结果。注意所有字母"e"的参考点都被找出来了；（d）另一组字符，用于腐蚀操作的结构元用白色叠加显示在图上；（e）对图（d）进行阈值分割的结果；（f）使用图（d）中结构元对图（e）进行腐蚀的结果。注意所有字母"o"的参考点被正确识别出来，此外，字母"p"和"q"的圆形部分也被提取出来了

An interesting property of Minkowski addition and subtraction as well as dilation and erosion is that they are dual to each other with respect to the complement operation. For Minkowski addition and subtraction, we have

闵可夫斯基加法和减法与膨胀和腐蚀都拥有一个有趣的属性，在进行补集操作时，它们彼此之间是互为对偶的。对于闵可夫斯基加法和减法，有：

$$R \oplus S = \overline{\overline{R} \ominus S} \quad \text{和} \quad R \ominus S = \overline{\overline{R} \oplus S} \tag{3.88}$$

The same identities hold for dilation and erosion. Hence, a dilation of the foreground is identical to an erosion of the background, and vice versa. We can make use of the duality whenever we want to avoid computing the complement explicitly, and hence to speed up some operations. Note that the duality holds only if the complement can be infinite. Hence, it does not hold for binary images, where the complemented region needs to be clipped to a certain image size.

同样的等式适用于膨胀操作和腐蚀操作。这样，对前景的一个膨胀处理等同于对背景的一个腐蚀处理。在任何想明确地避免大量的运算以提高处理速度时，都可以使用此对偶性。注意对偶性仅适用于当补集是无限的情况下。因此，它不适用于二值图，因为二值图描述的区域的补集必须被裁剪成为某种图像尺寸。

3.6.1.4 Region Boundaries

One extremely useful application of erosion and dilation is the calculation of the boundary of a region. The algorithm to compute the true boundary as a linked list of contour points is quite complicated (Haralick and Shapiro, 1992). However, an approximation to the boundary can be computed very easily. If we want to compute the inner boundary, we simply need to erode the region appropriately and to subtract the eroded region from the original region:

$$\partial R = R \setminus (R \ominus S) \tag{3.89}$$

By duality, the outer boundary (the inner boundary of the background) can be computed with a dilation:

$$\partial R = (R \oplus S) \setminus R \tag{3.90}$$

To get a suitable boundary, the structuring element S must be chosen appropriately. If we want to obtain an 8-connected boundary, we must use the structuring element S_8 in Figure 3.51. If we want a 4-connected boundary, we must use S_4.

3.6.1.4 区域边界

腐蚀和膨胀最有用处的应用是计算区域的边界。基于轮廓控制点构成的一个链表来求出真实的边界是相当复杂的算法（Haralick and Shapiro, 1992）。但是，计算出一个边界的近似值是非常容易的。如果想计算出内边界，仅需对区域进行适当的腐蚀，然后从原区域中减去腐蚀后得到的区域即可：

由对偶性可知，外边界（背景的内边界）可通过膨胀处理求出：

为得到合适的边界，必须适当地选取结构元 S。如果想得到一个 8 连通的边界，则必须使用图 3.51 中的结构元 S_8。如果想得到一个 4 连通的边界，则必须使用 S_4。

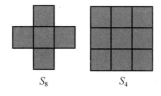

图 3.51　用于计算 8 连通区域边界的结构元 (S_8) 及用于计算 4 连通区域边界的结构元 (S_4)

Figure 3.52 shows an example of the computation of the inner boundary of a region. A small part of the input region is shown in Figure 3.52(a). The boundary of the region computed by Eq. (3.89) with S_8 is shown in Figure 3.52(b),

图 3.52 给出了区域内边界计算的一个例子。输入区域中的一个小部分如图 3.52（a）。使用 S_8 计算出的区域的边界如图 3.52（b），使用 S_4 计算的结果如图 3.52（c）。注意边界仅

while the result with S_4 is shown in Figure 3.52(c). Note that the boundary is only approximately 8- or 4-connected. For example, in the 8-connected boundary there are occasional 4-connected pixels. Finally, the boundary of the region as computed by an algorithm that traces around the boundary of the region and links the boundary points into contours is shown in Figure 3.52(d). Note that this is the true boundary of the region. Also note that, since only part of the region is displayed, there is no boundary at the bottom of the displayed part.

是近似为 8 连通或 4 连通。例如，在 8 连通的边界上有时出现 4 连通像素。最后，计算出来的区域边界显示在图 3.52（d）中，计算的方法是沿区域的边界跟踪所有的点然后将这些边界点连接成轮廓。注意因为图中仅显示了区域中的一个部分，所以在显示部分的最下方是没有边界的。

图 3.52 （a）一个大区域的局部；（b）图（a）的 8 连通边界，由等式（3.76）求出；（c）图（a）的 4 连通边界；（d）图（a）中边界的轮廓，由边界点连接而成

3.6.1.5 Hit-or-Miss Transform

As we have seen, erosion can be used as a template matching operation. However, sometimes it is not selective enough and returns too many matches. The reason for this is that erosion does not take the background into account. For this reason, an operation that explicitly models the background is needed. This operation is called the hit-or-miss transform. Since the foreground and background should be taken into account, it uses a structuring element that consists of two parts: $S = (S^f, S^b)$ with $S^f \cap S^b = \emptyset$. With this, the hit-or-miss transform is defined as (see Serra, 1982; Soille, 2003)

3.6.1.5 击中-击不中变换

正如已经看到的，腐蚀能被当成模板匹配来使用。但有时它的选择性不够，会返回太多的匹配项。原因是腐蚀操作并不会去考虑背景。这样就需要一个能将背景明确地模型化的方法。此处理方法被称为击中-击不中变换。由于前景和背景都要考虑，所以采用一个包含两个部分的结构元：$S = (S^f, S^b)$，其中 $S^f \cap S^b = \emptyset$。这样，击中-击不中变换被定义为（Serra, 1982；Soille, 2003）

$$R \circledast S = (R \ominus \check{S}^f) \cap (\overline{R} \ominus \check{S}^b) = (R \ominus \check{S}^f) \setminus (R \oplus \check{S}^b) \tag{3.91}$$

Hence, the hit-or-miss transform returns those translated reference points for which the foreground structuring element S^f completely lies within the foreground and the background structuring element S^b completely lies within the background. The second equation is especially useful from an implementation point of view since it avoids having to compute the complement. The hit-or-miss transform is dual to itself if the foreground and background structuring elements are exchanged: $R \circledast S = \overline{R} \circledast S'$, where $S' = (S^b, S^f)$.

Figure 3.53 shows the same image as Figure 3.50(d). The goal here is to match only the letters "o" in the image. To do so, we can define a structuring element that crosses the vertical strokes of the letters "p" and "q" (and also "b" and "d"). One possible structuring element for this purpose is shown in Figure 3.53(b). With the hit-or-miss transform, we are able to remove the found matches for the letters "p" and "q" from the result, as can be seen from Figure 3.53(c).

所以，击中-击不中变换返回的平移后参考点，在这些参考点位置上，前景结构元 S^f 彻底地落在前景内且背景结构元 S^b 彻底落在背景内。从实现角度看，第二个等式尤其有用，因为它避免了计算补集。当前景结构元和背景结构元互换时，击中-击不中变换与其自身是对偶的：$R \circledast S = \overline{R} \circledast S'$，其中 $S' = (S^b, S^f)$。

图 3.53 给出了与图 3.50（d）一样的图像。此处只想在图中匹配出字母"o"。为达到这个目标，可定义一个结构元，此结构元删除了字母"p"和"q"（当然也有"b"和"d"）中的垂直笔画。能达到要求的一个可能的结构元如图 3.53（b）所示。然后通过击中-击不中变换，能够从结果中去除字母"p"与"q"的匹配，见图 3.53（c）。

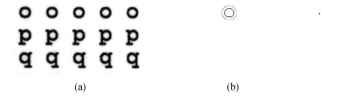

(a)   (b)   (c)

图 3.53 （a）一些印刷字符图像；（b）用于在击中-击不中变换中使用的结构元。黑色部分是前景结构元，浅灰部分是背景结构元；（c）使用图（b）的结构元对阈值分割后的图像（见图 3.50（e））应用击中-击不中变换后的结果。注意同腐蚀操作的结果相比（见图 3.50（f）），只有字母"o"的参考点被识别出来了

3.6.1.6 Opening and Closing

We now turn our attention to operations in which the basic operations we have discussed so far are executed in succession. The first such operation is opening (Serra, 1982; Soille, 2003):

3.6.1.6 开操作和闭操作

我们现在把注意力转移到已经讨论过的几种基本操作的综合应用。第一个这样的应用是开操作（Serra, 1982; Soille, 2003）：

$$R \circ S = (R \ominus \check{S}) \oplus S = \bigcup_{S_t \subseteq R} S_t \qquad (3.92)$$

Hence, opening is an erosion followed by a Minkowski addition with the same structuring element. The second equation tells us that we can visualize opening by moving the structuring element around the plane. Whenever the structuring element completely lies within the region, we add the entire translated structuring element to the output region (and not just the translated reference point as in erosion).

The opening's definition causes the location of the reference point to cancel out, which can be seen from the second equation. Therefore, opening is translation-invariant with respect to the structuring element, i.e., $R \circ S_t = R \circ S$. In contrast to erosion and dilation, opening is idempotent, i.e., applying it multiple times has the same effect as applying it once: $(R \circ S) \circ S = R \circ S$.

Like erosion, opening is anti-extensive. Since opening is translation-invariant, we do not have to require that the structuring element S contains the origin. Furthermore, opening is increasing.

Like erosion, opening can be used as a template matching operation. In contrast to erosion and the hit-or-miss transform, it returns all points of the input region into which the structuring element fits. Hence it preserves the shape of the object to find. An example of this is shown in Figure 3.54, where the same input images and structuring elements as in Figure 3.50 are used. Note that opening has found the same instances of the structuring elements as erosion but has preserved the shape of the matched structuring elements. Hence, in this example it also finds the

因此，开操作是先进行腐蚀操作后再紧接着进行一个使用同样结构元的闵可夫斯基加法。第二个等式告诉我们能通过在整个平面内移动结构元来直观地看出开操作的处理过程。任何时刻当结构元完全位于区域内时，将平移后的整个结构元加到输出区域内（而不是腐蚀操作中的平移后的参考点）。

从第二个定义式看，开操作无须考虑参考点的位置。因此，开操作相对于结构元是平移不变的，即，$R \circ S_t = R \circ S$ 与腐蚀操作和膨胀操作比较，开操作是幂等的，即多次使用与一次使用得到的结果一致：$(R \circ S) \circ S = R \circ S$。

与腐蚀操作类似，开操作是非外延的，由于开操作是平移不变的，不必要求结构元 S 包含原点，此外，开操作是增量操作。

与腐蚀操作类似，开操作能被用于模板匹配。与腐蚀操作和击中-击不中变换相比，它返回输入区域中能被结构元覆盖的全部的点。因此，它保持了要搜索的物体的形状。图 3.54 是此应用的一个例子，此例中使用的输入图像和结构元与图 3.50 中应用的一样。注意开操作同样找到了结构元的实例，此结果与使用腐蚀操作得到的一致，但开操作的结果保持了匹配到的结构元的形状。因此，在本例中也找出了字母 "p" 和 "q"。为了能只

letters "p" and "q." To find only the letters "o," we could combine the hit-or-miss transformation with a Minkowski addition to obtain a hit-or-miss opening: $R \odot S = (R \circledast S) \oplus S^{\text{f}}$.

找到字母"o",我们可以组合击中-击不中变换和闵可夫斯基加法以得到击中-击不中开操作:$R \odot S = (R \circledast S) \oplus S^{\text{f}}$。

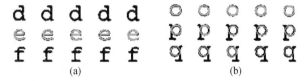

图 3.54　(a) 用图 3.50 (a) 中的结构元对图 3.50 (b) 中分割后的区域进行开操作处理的结果;(b) 用图 3.50 (d) 中的结构元对图 3.50 (e) 中分割后的区域进行开操作处理的结果。开操作的结果用浅灰叠加显示在输入区域上,输入区域用黑色显示。注意开操作找到了与腐蚀操作一致的结构元实例,但保持了匹配到结构元的形状

Another very useful property of opening results if structuring elements like circles or rectangles are used. If an opening with these structuring elements is performed, parts of the region that are smaller than the structuring element are removed from the region. This can be used to remove unwanted appendages from the region and to smooth the boundary of the region by removing small protrusions. Furthermore, small bridges between object parts can be removed, which can be used to separate objects. Finally, the opening can be used to suppress small objects. Figure 3.55 shows an example of using opening to remove unwanted appendages and small objects from the segmentation. In Figure 3.55(a), an image of a ball-bonded die is shown. The goal is to segment the balls on the pads. If the image is thresholded (Figure 3.55(b)), the wires that are attached to the balls are also extracted. Furthermore, there are extraneous small objects in the segmentation. By performing an opening with a circle of diameter 31, the wires and small objects are removed, and only smooth region parts that correspond to the balls are retained.

如果使用类似圆或矩形的结构元时,开操作就会呈现出另一个非常有用的特性。当一个开操作实施时采用了这样的结构元,那么区域中比结构元小的部分都会被从区域中去除掉。这个特性能用来从区域中去除不想要的附加物,也能用来去除边界上的突出部分以达到平滑区域边界的目的。此外,此特性还能用来去除物体间相连的部分以达到分开物体的目的。最后,开操作能用来抑制小物体。图 3.55 给出了使用开操作从分割结果中去除不想要的附加物和小物体的例子。图 3.55 (a) 中显示的是球焊晶片的图像。目的是分割出焊盘上的焊点。阈值分割后的图像 (图 3.55 (b)) 中线和焊点作为一个整体被提取出来,而且分割结果中还包含无关的小物体。通过使用直径为 31 的圆进行开操作处理,线和小物体都被消除了,且图中只保留下与焊点对应的平滑区域。

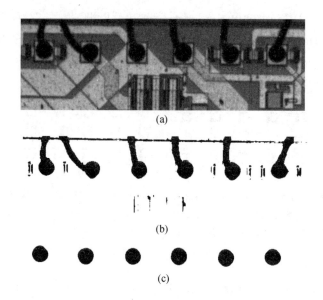

图 3.55　（a）一个球焊晶片的图像。目的是分割出焊点；（b）对图（a）进行阈值分割后的结果。分割结果包括焊盘上的线；（c）用直径为 31 的圆进行开操作处理的结果。通过开操作删除了线和其他无关的分割结果，只留下了焊点

The second interesting operation in which the basic morphological operations are executed in succession is closing (Serra, 1982; Soille, 2003):

由基本形态学操作组合而成的第二个有趣的形态学方法就是闭操作（Serra，1982；Soille，2003）：

$$R \bullet S = (R \oplus \check{S}) \ominus S = \overline{\bigcup_{S_t \subseteq \overline{R}} S_t} \tag{3.93}$$

Closing is a dilation followed by a Minkowski subtraction with the same structuring element. There is, unfortunately, no simple formula that tells us how closing can be visualized. The second formula is actually defined by the duality of opening and closing, namely a closing on the foreground is identical to an opening on the background, and vice versa:

闭操作是先执行一个膨胀操作后紧接着再用同一个结构元进行闵可夫斯基减法。不幸的是没有简单的公式能直观地表达闭操作的处理过程。第二个等式实际上是由开操作和闭操作的对偶性得到的，对前景的一个闭操作等同于对背景的一个开操作，反之亦然：

$$R \bullet S = \overline{\overline{R} \circ S} \quad \text{and} \quad R \circ S = \overline{\overline{R} \bullet S} \tag{3.94}$$

Like opening, closing is translation-invariant with respect to the structuring element, i.e., $R \bullet S_t = R \bullet S$, and idempotent, i.e., $(R \bullet S) \bullet S = R \bullet S$.

与开操作类似，闭操作相对于结构元是平移不变的，即，$R \bullet S_t = R \bullet S$，也是幂等的，即 $(R \bullet S) \bullet S = R \bullet S$

Like dilation, closing is extensive. Since closing is translation-invariant, we do not have to require that the structuring element S contains the origin. Furthermore, closing is increasing.

Since closing is dual to opening, it can be used to merge objects separated by gaps that are smaller than the structuring element. If structuring elements like circles or rectangles are used, closing can be used to close holes and to remove indentations that are smaller than the structuring element. The second property enables us to smooth the boundary of the region.

Figure 3.56 shows how closing can be used to remove indentations in a region. In Figure 3.56(a), a molded plastic part with a protrusion is shown. The goal is to detect the protrusion because it is a production error. Since the actual object is circular, if the entire part were visible, the protrusion could be detected by performing an opening with a circle that is almost as large as the object and then subtracting the opened region from the original segmentation. However, only a part of the object is visible, so the erosion in the opening would create artifacts or remove the object entirely. Therefore, by duality we can pursue the opposite approach: we can segment the background and perform a closing on it. Figure 3.56(b) shows the result of thresholding the background. The protrusion is now an indentation in the background. The result of performing a closing with a circle of diameter 801 is shown in Figure 3.56(c). The diameter of the circle was set to 801 because it is large enough to completely fill the indentation and to recover the circular shape of the object. If much smaller circles were used, e.g., with a diameter of 401, the indentation would not be filled

与膨胀操作类似，开操作是外延的，由于闭操作是平移不变的，不必要求结构元 S 包含原点，此外，闭操作是增量操作。

闭操作和开操作是对偶的，这可以用来合并彼此分开的物体，这些物体间的缝隙小于结构元。如果使用类似圆或矩形作为结构元，闭操作能用来填充孔洞及消除比结构元小的缺口。后一个特性能够平滑区域的边界。

图 3.56 显示了如何利用闭操作来消除区域内的缺口。图 3.56（a）中显示的是一个带有凸起的注塑零件。目的是要检查出这个凸起，因为这种凸起是一个生产缺陷。由于真实物体是圆的，所以如果图中能看到整个物体时，就能用与物体大小基本一致的圆来进行一个开操作，然后从原分割结果中减去开运算后得到的区域，这样就检测到了此凸起。但当图像中只显示了物体的一部分时，开操作中的腐蚀处理将会产生人为干扰或将整个物体删除。因此，通过对偶性可采用与之相对的方法：我们能分割出背景，然后在背景上进行闭操作。图 3.56（b）是阈值分割出的背景。零件上的凸起就变成了背景上的一个缺口。采用直径 801 的圆对背景进行闭操作的结果显示在图 3.56（c）中。圆的直径设定为 801 是因为这个尺寸对完全覆盖缺口以恢复圆形零件而言已经足够大了。如果使用太小的圆，比如直径为 401 的圆，缺口将不能被完全覆盖。要检测出此缺陷，在闭操作的结果

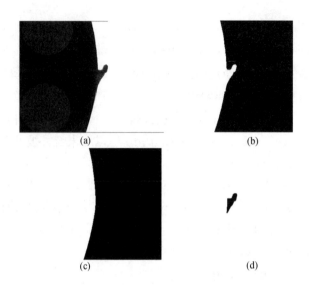

图 3.56　（a）尺寸是 768×576 的图像上有一个注塑零件，零件上带有一个凸起；（b）对图（a）中的背景进行阈值分割后的结果；（c）对图（b）用直径为 801 的圆进行闭操作的结果。注意凸起（即背景上的缺口）已被填充且注塑件的圆状外形已被恢复了；（d）在（c）和（b）之间求差，然后在计算出的差集上使用 5×5 矩形进行开操作消除较小的部分。最终结果中显示的就是注塑件上的凸起

completely. To detect the error itself, we can compute the difference between the closing and the original segmentation. To remove some noisy pixels that result because the boundary of the original segmentation is not as smooth as the closed region, the difference can be postprocessed with an opening, e.g., with a 5 × 5 rectangle, to remove the noisy pixels. The resulting error region is shown in Figure 3.56(d).

和背景分割结果间求差即可。由于阈值分割后得到的区域边界不像闭操作后得到的区域边界那样平滑，所以在求差后会产生干扰像素。可在后续的处理中使用开操作消除这些干扰像素，比如用 5×5 矩形进行的开操作来去除干扰像素。返回的缺陷区域见图 3.56（d）。

3.6.1.7　Skeleton

The operations we have discussed so far have been mostly concerned with the region as a 2D object. The only exception has been the calculation of the boundary of a region, which reduces a region to its 1D outline, and hence gives a more condensed description of the region. If the objects are mostly linear, i.e., are regions that have a much greater length than width, a more salient description of the object would be obtained if we could some-

3.6.1.7　骨架

到目前为止已经讨论的操作都更关注作为二维物体出现的区域。唯一的例外是计算一个区域的边界，此类计算把一个区域简化为一维轮廓，所以计算得到的结果是对区域的一个更精简的描述。如果物体几乎是线状的，即物体区域的长度比宽度大许多时，如果能设法得到物体的中心线，且线宽为一个像素时，那么就可以获得针

how capture its one-pixel-wide center line. This center line is called the skeleton or medial axis of the region. Several definitions of a skeleton can be given (Soille, 2003). One intuitive definition can be obtained if we imagine that we try to fit circles that are as large as possible into the region. More precisely, a circle C is maximal in the region R if there is no other circle in R that is a superset of C. The skeleton then is defined as the set of the centers of the maximal circles. Consequently, a point on the skeleton has at least two different points on the boundary of the region to which it has the same shortest distance. Algorithms to compute the skeleton are given in (Soille, 2003; Lam et al., 1992). They basically can be regarded as sequential hit-or-miss transforms that find points on the boundary of the region that cannot belong to the skeleton and delete them. The goal of skeletonization is to preserve the homotopy of the region, i.e., the number of connected components and holes. One set of structuring elements for computing an 8-connected skeleton is shown in Figure 3.57 (Soille, 2003). These structuring elements are used sequentially in all four possible orientations to find pixels with the hit-or-miss transform that can be deleted from the region. The iteration is continued until no changes occur. It should be noted that skeletonization is an example of an algorithm that can be implemented more efficiently on binary images than on the run-length representation.

对此物体的一个更直观的描述。这条中心线被称为骨架或区域的中轴。骨架存在着多种定义（Soille, 2003）。如果想象当试图将某区域拟合为一系列尽可能大的圆时，就能直观地得到骨架的定义。更精确些讲，某个圆 C 在某区域 R 内是极大圆就表明在 R 中不存在其他可以是 C 超集的圆。那么骨架被定义为所有极大圆圆心的集合。因此，骨架上的一个点至少对应着两个位于区域边界上的不同点，它到这两个点都有相等的最短距离。计算骨架的方法见文献（Soille, 2003; Lam et al., 1992）。这些算法基本被视为按顺序进行的击中-击不中变换，这些变换在区域的边界上找到不属于骨架的那些点然后删除它们。骨架化的目的是保持区域的同伦，即保持连通区域和孔洞的数量不变。用于计算 8 连通骨架的一组结构元见图 3.57（Soille, 2003）。使用这些结构元在全部四个可能的方向上反复进行击中-击不中变换，找到击中时的参考点并将这些点从此区域内删除。不断进行迭代直到结果不变为止。注意骨架化算法在二值图像上实现比在行程表示法上实现更有效率。

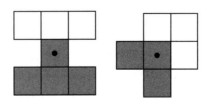

图 3.57 用于计算某区域 8 连通骨架的一组结构元。这些结构元被顺序地用来在全部四个可能方向上寻找能被删除的像素

Figure 3.58(a) shows a part of an image of a PCB with several tracks. The image is thresholded (Figure 3.58(b)), and the skeleton of the thresholded region is computed with the above algorithm (Figure 3.58(c)). Note that the skeleton contains several undesirable branches on the upper two tracks. For this reason, many different skeletonization algorithms have been proposed. An algorithm that produces relatively few unwanted branches is described by Eckhardt and Maderlechner (1993). The result of this algorithm is shown in Figure 3.58(d). Note that there are no undesirable branches in this case.

图 3.58（a）显示了一部分图像，图像中是带有多条金属线的印刷电路板。首先对图像进行阈值分割（图 3.58（b）），阈值分割后的区域对应的骨架是根据上面的算法计算得到的（图 3.58（c））。注意最上面两条金属线的骨架中包含一些多余的分叉。为了解决此问题，已经提出了许多不同的骨架化算法。产生相对较少的多余分叉的算法见文献（Eckhardt and Maderlechner，1993）。用此算法得到的处理结果见图 3.58（d）。注意在此例中多余的分叉已经被消除了。

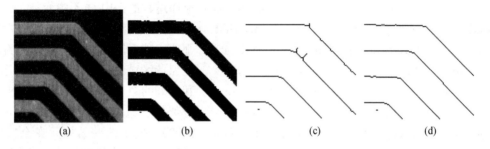

图 3.58　（a）带有若干金属线的印刷电路板的一个局部；(b) 对（a）进行阈值分割后的结果；(c) 通过基于图 3.57 中结构元的算法（Soille，2003）计算得到的 8 连通骨架；（d）使用能产生更少骨架分叉的算法求得的骨架（Eckhardt and Maderlechner，1993）

3.6.1.8 Distance Transform

The final region morphology operation that we will discuss is the distance transform, which returns an image instead of a region. This image contains, for each point in the region R, the shortest distance to a point outside the region (i.e., to $\overline{R}$). Consequently, all points on the inner boundary of the region have a distance of 1. Typically, the distance of the other points is obtained by considering paths that must be contained in the pixel grid. Thus, the chosen connectivity defines which paths are allowed. If 4-connectivity is used,

3.6.1.8　距离变换

接下来将要讨论的最后一个区域形态学操作是距离变换，距离变换返回的是一幅图像而不是一个区域。此图像的像素值是距离，这些距离代表的是区域 R 内每一个点到区域外（即到 $\overline{R}$）所有点的距离的最小值。因而，区域内边界上所有点的距离都是 1。对于区域内其他的点，通常计算它们的距离时要考虑在像素网格上对应的路径。因此，选定后的连通性将决定什么样的路径是被允许的。如果使用

the corresponding distance is called the city-block distance. Let $(r_1, c_1)^\top$ and $(r_2, c_2)^\top$ be two points. Then the city-block distance is given by

4 连通时，相应的距离被称为 4 连通距离。如果 $(r_1, c_1)^\top$ 和 $(r_2, c_2)^\top$ 是两个点，那么 4 连通距离可由下式计算得到：

$$d_4 = |r_2 - r_1| + |c_2 - c_1| \tag{3.95}$$

Figure 3.59(a) shows the city-block distance between two points. In the example, the city-block distance is 5. On the other hand, if 8-connectivity is used, the corresponding distance is called the chessboard distance. It is given by

图 3.59（a）显示的是两点间的 4 连通距离。在这个例子中，4 连通距离是 5。另一方面，如果使用 8 连通，那么对应的距离被称为 8 连通距离。8 连通距离由下式计算：

$$d_8 = \max(|r_2 - r_1|, |c_2 - c_1|) \tag{3.96}$$

In the example in Figure 3.59(b), the chessboard distance between the two points is 3. Both these distances are approximations to the Euclidean distance, given by

在图 3.59（b）的例子中，两点间的 8 连通距离是 3。这两个距离都是对欧几里得距离的近似，欧几里得距离计算公式是：

$$d_e = \sqrt{(r_2 - r_1)^2 + (c_2 - c_1)^2} \tag{3.97}$$

For the example in Figure 3.59(c), the Euclidean distance is $\sqrt{13}$.

图 3.59（c）中的欧几里得距离是 $\sqrt{13}$。

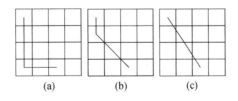

图 3.59　（a）两点的 4 连通距离；（b）8 连通距离；（c）欧几里得距离

Algorithms to compute the distance transform are described in Borgefors (1984). They work by initializing the distance image outside the region with 0 and within the region with a suitably chosen maximum distance, i.e., $2^b - 1$, where b is the number of bits in the distance image, e.g., $2^{16} - 1$. Then, two sequential line-by-line scans through the image are performed, one from the top left to the bottom right corner, and the second

距离变换的算法见文献（Borgefors，1984）。此算法先将距离图像初始化，即令区域外的所有距离值是 0，区域内的所有距离值是一个适当选定的最大距离值，即 $2^b - 1$，这里 b 是距离图像的像素位数，比如可用 $2^{16} - 1$ 作为初始时区域内全部点的距离值。然后，在全图上进行两种顺序的逐行扫描，一种扫描方向是从左上到右下，

in the opposite direction. In each case, a small mask is placed at the current pixel, and the minimum over the elements in the mask of the already computed distances plus the elements in the mask is computed. The two masks are shown in Figure 3.60. If $d_1 = 1$ and $d_2 = \infty$ are used (i.e., d_2 is ignored), the city-block distance is computed. For $d_1 = 1$ and $d_2 = 1$, the chessboard distance results. Interestingly, if $d_1 = 3$ and $d_2 = 4$ are used and the distance image is divided by 3, a very good approximation to the Euclidean distance results, which can be computed solely with integer operations. This distance is called the chamfer-3-4 distance (Borgefors, 1984). With slight modifications, the true Euclidean distance can be computed (Danielsson, 1980). The principle is to compute the number of horizontal and vertical steps to reach the boundary using masks similar to the ones in Figure 3.60, and then to compute the Euclidean distance from the number of steps.

另一种扫描的方向与第一种的相反。在每一种扫描情况中，都要在当前像素上先放置一个小的掩码，然后将掩码覆盖区域中已经有的距离值与掩码中此位置的元素值相加，然后再从掩码中所有计算结果中选择最小的一个值作为当前像素的距离值。参加计算的两个掩码见图 3.60。当 $d_1 = 1$ 且 $d_2 = \infty$（即忽略 d_2）时，计算出的是 4 连通距离。当 $d_1 = 1$ 且 $d_2 = 1$ 时，计算返回的是 8 连通距离。有趣的是，当 $d_1 = 3$ 且 $d_2 = 4$ 时，距离图像除以 3 得到的结果就是对欧几里得距离的一个很好近似，而此值是完全由纯整数运算得到的。这个距离被称为 chamfer-3-4 距离（Borgefors，1984）。对此方法稍加改动后就可以求出真正的欧几里得距离（Danielsson，1980）。原理如下：使用类似图 3.60 的掩码，分别在垂直方向和水平方向上计算到达边界的步长数，然后通过这些步长数来计算欧几里得距离。

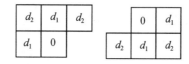

图 3.60　在两种顺序扫描中使用的掩码，这些掩码是用来计算距离变换的。左边的掩码被用在从左到右，从上到下的扫描中。右边的掩码被用在从右到左，从下到上的扫描中

The skeleton and the distance transform can be combined to compute the width of linear objects efficiently. In Figure 3.61(a), a PCB with tracks that have several errors is shown. The protrusions on the tracks are called spurs, while the indentations are called mouse bites (Moganti et al., 1996). They are deviations from the correct track width. Figure 3.61(b) shows the result

组合骨架和距离变换能够高效率地计算线状物体的宽度。在图 3.61(a) 中，PCB 上的若干金属线存在着一些缺陷。金属线上的凸起被称为"马刺"，而缺口被称为"鼠咬"（Moganti et al.，1996）。这些缺陷背离了正确的线宽。图 3.61（b）给出了在分割后的金属线上计算 chamfer-3-4 距离得到

of computing the distance transform with the chamfer-3-4 distance on the segmented tracks. The errors are clearly visible in the distance transform. To extract the width of the tracks, we need to calculate the skeleton of the segmented tracks (Figure 3.61(c)). If the skeleton is used as the ROI for the distance image, each point on the skeleton will have the corresponding distance to the border of the track. Since the skeleton is the center line of the track, this distance is the width of the track. Hence, to detect errors, we simply need to threshold the distance image within the skeleton. Note that in this example it is extremely useful that we have defined that images can have an arbitrary ROI. Figure 3.61(d) shows the result of drawing circles at the centers of gravity of the connected components of the error region. All major errors have been detected correctly.

的距离变换结果。在距离变换后得到的距离图像中可清晰地看到缺陷。为提取金属线的宽度，必须在分割后的金属线上计算骨架（图 3.61（c））。当骨架在距离图像中被当作 ROI 使用时，则骨架上的每个点都拥有到金属线边界的相应距离。由于骨架是金属线的中心线，故而相应的距离就是金属线的宽度。因此，要检测缺陷，我们仅需要在骨架内的距离图像上进行阈值分割即可。注意在本例中，首先说明在图像中可以任意设置感兴趣区域是非常有用的。图 3.61（d）给出了以缺陷区域所对应的连通区域重心为圆心画圆的结果。所有主要的缺陷都已经被正确检出。

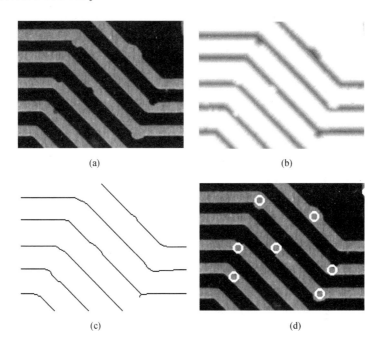

图 3.61 （a）图像显示了部分 PCB，PCB 上若干金属线上有"马刺"和"鼠咬"缺陷；（b）在对（a）进行阈值分割后的结果图上进行距离变换后的结果。距离图像是反向显示的（深灰色对应的是大的距离）；（c）分割后的区域的骨架；（d）通过在（b）图上将（c）作为 ROI 使用来对距离进行阈值分割，提取金属线上太宽和太窄的部分。以缺陷区域所对应的连通区域的重心为圆心画圆将全部缺陷显示出来

3.6.2 Gray Value Morphology

3.6.2.1 Minkowski Addition and Dilation

Because morphological operations are very versatile and useful, the question of whether they can be extended to gray value images arises quite naturally. This can indeed be done. In analogy to the region morphology, let $g(r,c)$ denote the image that should be processed, and let $s(r,c)$ be an image with ROI S. As in region morphology, the image s is called the structuring element. Gray value Minkowski addition is then defined as (see Soille, 2003)

$$g \oplus s = (g \oplus s)_{r,c} = \max_{(i,j)^\top \in S}\{g_{r-i,c-j} + s_{i,j}\} \tag{3.98}$$

This is a natural generalization because Minkowski addition for regions is obtained as a special case if the characteristic function of the region is used as the gray value image. If, additionally, an image with gray value 0 within the ROI S is used as the structuring element, Minkowski addition becomes

$$g \oplus s = \max_{(i,j)^\top \in S}\{g_{r-i,c-j}\} \tag{3.99}$$

For characteristic functions, the maximum operation corresponds to the union. Furthermore, $g_{r-i,c-j}$ corresponds to the translation of the image by the vector $(i,j)^\top$. Hence, Eq. (3.99) is equivalent to the second formula in Eq. (3.83).

As in region morphology, dilation can be obtained by transposing the structuring element. This results in the following definition (Soille, 2003):

$$g \oplus \check{s} = (g \oplus \check{s})_{r,c} = \max_{(i,j)^\top \in S}\{g_{r+i,c+j} + s_{i,j}\} \tag{3.100}$$

The typical choice for the structuring element in gray value morphology is the flat structuring element that was already used above: $s(r,c) = 0$ for $(r,c)^\top \in S$. With this, gray value Minkowski addition and dilation have the same properties as their corresponding region operations (see Section 3.6.1.2). If the structuring element S contains the origin, they are extensive, i.e., $g \leqslant g \oplus \check{s}$ (for simplicity, we will only list the formulas for dilation). Furthermore, they are increasing, i.e., if $g \leqslant h$ then $g \oplus \check{s} \leqslant h \oplus \check{s}$.

Therefore, gray value dilation has a similar effect to region dilation: it enlarges the foreground, i.e., parts in the image that are brighter than their surroundings, and shrinks the background, i.e., parts in the image that are darker than their surroundings. Hence, it can be used to connect disjoint parts of a bright object in the image. This is sometimes useful if the object cannot be segmented easily using region operations alone. Conversely, dilation can be used to split dark objects.

3.6.2.2 Minkowski Subtraction and Erosion

Minkowski subtraction for gray value images is given by (see Soille, 2003)

$$g \ominus s = (g \ominus s)_{r,c} = \min_{(i,j)^\top \in S} \{g_{r-i,c-j} - s_{i,j}\} \tag{3.101}$$

As above, by transposing the structuring element we obtain the gray value erosion (Soille, 2003):

$$g \ominus \check{s} = (g \ominus \check{s})_{r,c} = \min_{(i,j)^\top \in S} \{g_{r+i,c+j} - s_{i,j}\} \tag{3.102}$$

Again, the typical choice for the structuring element is a flat structuring element. With this, gray value Minkowski subtraction and erosion have the same properties as their corresponding region operations (see Section 3.6.1.3). If the structuring element S contains the origin, they are anti-extensive, i.e., $g \ominus \check{s} \leqslant g$. Furthermore, they are increasing, i.e., if $g \leqslant h$ then $g \ominus \check{s} \leqslant h \ominus \check{s}$.

Therefore, gray value erosion shrinks the foreground and enlarges the background. Hence, erosion can be used to split touching bright objects and to connect disjoint dark objects. In fact, dilation and erosion, as well as Minkowski addition and subtraction, are dual to each other, as for regions. For the duality, we need to define what the complement of an image should be. If the images are stored with b bits, the natural definition for the complement operation is (see Soille, 2003)

应的区域操作一样（见 3.6.1.3 节）。如果结构元 S 包含原点，它们是非外延的，即 $g \ominus \check{s} \leqslant g$。而且，它们是增量的，即如果 $g \leqslant h$，则 $g \ominus \check{s} \leqslant h \ominus \check{s}$。

因此，灰度值腐蚀收缩前景并扩大背景。所以，灰度值腐蚀能够被用来分开相互连接的亮物体和连接支离破碎的暗物体。事实上，灰度值膨胀和灰度值腐蚀之间，灰度值的闵可夫斯基加法和减法之间，都是彼此对偶的，这点与区域形态学是一样的。为了使用对偶性，必须定义一幅图像的补集应是什么样子的。如果图像按照 b 位存储的，那么补集操作的定义就自然是（Soille, 2003）：

$$\overline{g}_{r,c} = 2^b - 1 - g_{r,c} \tag{3.103}$$

With this, it can be easily shown that erosion and dilation are dual (Soille, 2003):

因此，根据此定义就能容易地将腐蚀和膨胀的对偶性显示出来（Soille, 2003）：

$$g \oplus \check{s} = \overline{\overline{g} \ominus \check{s}} \quad \text{and} \quad g \ominus \check{s} = \overline{\overline{g} \oplus \check{s}} \tag{3.104}$$

Therefore, all the properties that hold for one operation for bright objects hold for the other operation for dark objects, and vice versa.

Note that dilation and erosion can also be regarded as two special rank filters (see Section 3.2.3.9) if flat structuring elements are used. They select the minimum and maximum gray values within the domain of the structuring element, which can be regarded as the filter mask. Therefore, dilation and erosion are sometimes referred to as maximum and minimum filters (or max and min filters).

因此，对亮物体进行一个操作得到的结果与对暗物体进行一个对偶操作得到的结果完全一致，反之亦然。

注意当使用平坦结构元时，膨胀和腐蚀也能被视为两种特殊的顺序滤波器（见 3.2.3.9 节）。它们在结构元的范围内选择灰度值最小值和最大值，结构元可被视为是滤波器掩码。所以，膨胀和腐蚀有时也被称作最大值滤波和最小值滤波（或最大滤波和最小滤波）。

Efficient algorithms to compute dilation and erosion have been proposed by Van Droogenbroeck and Talbot (1996). Their run time complexity is $O(whn)$, where w and h are the dimensions of the image while n is roughly the number of points on the boundary of the domain of the structuring element for flat structuring elements. For rectangular structuring elements, algorithms with a run time complexity of $O(wh)$, i.e., with a constant number of operations per pixel, can be found (Gil and Kimmel, 2002). This is similar to the recursive implementation of a linear filter.

计算膨胀和腐蚀的高效率算法已经被 Van Droogenbroeck and Talbot（1996）所提出。这些算法的执行复杂度是 $O(whn)$，其中 w 和 h 是图像的宽度和高度，n 是平坦结构元区域边界上的点数的大概值。对于矩形结构元，文献（Gil and Kimmel，2002）提出算法的执行复杂度是 $O(wh)$，即每个像素有固定不变的操作次数。这与一个线性滤波器的递归实现类似。

3.6.2.3 Opening and Closing

With these building blocks, we can define a gray value opening, as for regions, as an erosion followed by a Minkowski addition (Soille, 2003)

3.6.2.3 开操作和闭操作

与区域形态学类似，通过这些基本模块，能定义灰度值开操作作为一个腐蚀操作后再执行一个闵可夫斯基加法（Soille，2003）

$$g \circ s = (g \ominus \check{s}) \oplus s \tag{3.105}$$

and a closing as a dilation followed by a Minkowski subtraction (Soille, 2003)

闭操作是一个膨胀操作后再执行一个闵可夫斯基减法

$$g \bullet s = (g \oplus \check{s}) \ominus s \tag{3.106}$$

If flat structuring elements are used, they have the same properties as the corresponding region operations (see Section 3.6.1.6). Opening is anti-extensive while closing is extensive. Furthermore, both operations are increasing and idempotent. In addition, with the definition of the complement in Eq. (3.103), they are dual to each other (Soille, 2003):

如果使用平坦结构元，则如它们对应的区域操作一样（见 3.6.1.6 节），开操作和闭操作具有相同的特性。开操作是非外延的而闭操作是外延的。而且，两个操作都是增量的和幂等的。尤其是根据等式（3.103）图像补集的定义，灰度值开操作和闭操作是互为对偶的（Soille，2003）：

$$g \circ s = \overline{\overline{g} \bullet s} \quad \text{and} \quad g \bullet s = \overline{\overline{g} \circ s} \tag{3.107}$$

Therefore, like the region operations, they can be used to fill in small holes or, by duality, to remove small objects. Furthermore, they can

因此，与区域操作类似，根据灰度值开闭操作间的对偶性，它们能够被用来填充小孔洞，或者删除小物体。

be used to join or separate objects and to smooth the inner and outer boundaries of objects in the image.

Figure 3.62 shows how gray value opening and closing can be used to detect errors in the tracks on a PCB. We have already seen in Figure 3.61 that some of these errors can be detected by looking at the width of the tracks with the distance transform and skeletonization. This technique is very useful because it enables us to detect relatively large areas with errors. However, small errors are harder to detect with this technique because the distance transform and skeleton are only pixel-precise, and consequently the width of the track only can be determined reliably with a precision of two pixels. Smaller errors can be detected more reliably with gray value morphology. Figure 3.62(a) shows part of a PCB with several tracks that have spurs, mouse bites, pinholes, spurious copper, and open and short circuits (Moganti et al., 1996). The results of performing a gray value opening and closing with an octagon of diameter 11 are shown in Figures 3.62(b) and (c). Because of the horizontal, vertical, and diagonal layout of the tracks, using an octagon as the structuring element is preferable. It can be seen that the opening smooths out the spurs, while the closing smooths out the mouse bites. Furthermore, the short circuit and spurious copper are removed by the opening, while the pinhole and open circuit are removed by the closing. To detect these errors, we can require that the opened and closed images should not differ too much. If there were no errors, the differences would solely be caused by the texture on the tracks. Since the gray values of the opened image are always smaller than those of the

此外，在灰度值图像中，它们还能够被用来连接或分开物体，平滑物体的内、外边界。

图 3.62 显示了如何运用灰度值开操作和闭操作来检测 PCB 金属线上的缺陷。在图 3.61 中，已经看到了一部分缺陷可以通过由距离变换和骨架求得的金属线宽度来检测。此技术很有用因为它可以检测面积相对较大的缺陷。但使用此技术很难检测到小的缺陷因为距离变换和骨架都是像素精度的，因此，只有在两像素以上时，金属线的宽度才能被可靠地检出。使用灰度值形态学可以可靠地检测出更小的缺陷。图 3.62（a）给出了 PCB 的一个部分，此 PCB 金属线上存在"马刺""鼠咬"、针孔、余铜、断路和短路（Moganti et al., 1996）。采用一个灰度值开操作和闭操作后的结果见图 3.62（b）和图 3.62（c）。由于金属线布局有水平、垂直、对角线三个方向，所以使用一个八边形作为结构元是更适合的。能看到开操作平滑掉了"马刺"，而闭操作平滑掉了"鼠咬"。而且，短路和余铜也通过开操作去除了，针孔和断路通过闭操作去除了。为检出这些缺陷，要求开操作和闭操作得到的两幅图像间的差异不能太大。当没有缺陷时，差异完全是由金属线上的纹理造成的。因为开操作得到的图像上的灰度值总是小于闭操作得到的图像上的灰度值，所以针对亮物体使用动态阈值分割处理（等式（3.51））以得到所需要的分割结果。灰度值差异大于 g_{diff} 的每个像素都被视为一个缺陷。图 3.62（d）显示了在动

closed image, we can use the dynamic threshold operation for bright objects (Eq. (3.51)) to perform the required segmentation. Every pixel that has a gray value difference greater than g_{diff} can be considered as an error. Figure 3.62(d) shows the result of segmenting the errors using a dynamic threshold $g_{\text{diff}} = 60$. This detects all the errors on the board.

态阈值 $g_{\text{diff}} = 60$ 时对缺陷进行分割处理后的结果。这就检测出了板上的所有缺陷。

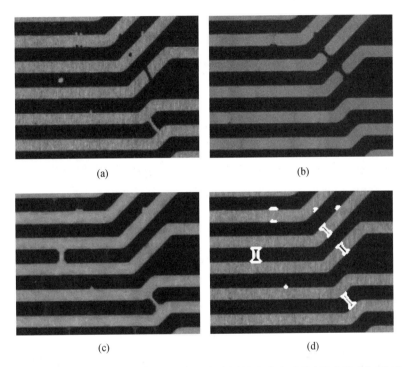

图 3.62 （a）图像中显示了 PCB 的一部分，PCB 的金属线上存在"马刺""鼠咬"、针孔、余铜、断路和短路等缺陷；（b）对图（a）用直径为 11 的八边形执行一个灰度值开操作后的结果；（c）对图（a）用直径为 11 的八边形执行一个灰度值闭操作后的结果；（d）通过对（b）和（c）使用一个动态阈值分割处理来分割出（a）中包括的各种缺陷

3.6.2.4 Morphological Gradient

We conclude this section with an operator that computes the range of gray values that occur within the structuring element. This can be obtained easily by calculating the difference between a dilation and an erosion (Soille, 2003):

3.6.2.4 形态学梯度算子

在本节最后部分介绍一个算子，此算子可计算得到出现在结构元内的灰度值范围。通过计算膨胀操作和腐蚀操作间的差就能轻松得到这个结果（Soille，2003）。

$$g \diamond s = (g \oplus \check{s}) - (g \ominus \check{s}) \tag{3.108}$$

Since this operator produces results similar to those of a gradient filter (see Section 3.7.3), it is sometimes called the morphological gradient.

Figure 3.63 shows how the gray range operator can be used to segment punched serial numbers. Because of the scratches, texture, and illumination, it is difficult to segment the characters in Figure 3.63(a) directly. In particular, the scratch next to the upper left part of the "2" cannot be separated from the "2" without splitting several of the other numbers. The result of computing the gray range within a 9×9 rectangle is shown in Figure 3.63(b). With this, it is easy to segment the numbers (Figure 3.63(c)) and to separate them from other segmentation results (Figure 3.63(d)).

由于与梯度滤波器给出的结果类似（见 3.7.3 节），此算子有时也被称为形态学梯度算子。

图 3.63 显示了如何使用灰度范围算子来分割序列号。由于受划痕、纹理和照明的影响，所以直接分割出图 3.63（a）中的字符是困难的。特别是字符"2"左上角的划痕很难在不打碎其他字符的前提下同"2"分开。在一个 9×9 矩形内计算灰度范围后的结果见图 3.63（b）。此时已经可以很容易地分割出序列号中的数字了（见图 3.63（c）），并能进一步将数字从其他分割结果中分离出来（见图 3.63（d））。

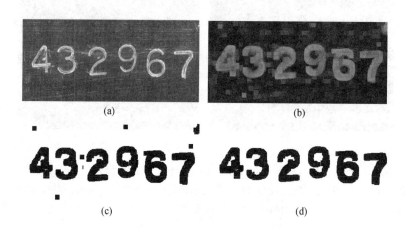

图 3.63 （a）一个冲压的序列号。由于划痕、纹理和照明的影响，直接分割出字符是很困难的；（b）在 9×9 的正方形内计算灰度范围的结果；（c）对（b）进行阈值分割的结果；（d）计算（c）中的连通区域并通过字符的尺寸将序列号选出

3.7 Edge Extraction

In Section 3.4, we discussed several segmentation algorithms. They have in common that they are

3.7 边缘提取

在 3.4 节中已经讨论了几种分割算法。它们的共同点是以对图像的像

based on thresholding the image, with either pixel or subpixel accuracy. It is possible to achieve very good accuracies with these approaches, as we saw in Section 3.5. However, in most cases the accuracy of the measurements that we can derive from the segmentation result critically depends on choosing the correct threshold for the segmentation. If the threshold is chosen incorrectly, the extracted objects typically become larger or smaller because of the smooth transition from the foreground to the background gray value. This problem is especially grave if the illumination can change, since in this case the adaptation of the thresholds to the changed illumination must be very accurate. Therefore, a segmentation algorithm that is robust with respect to illumination changes is extremely desirable. From the above discussion, we see that the boundary of the segmented region or subpixel-precise contour moves if the illumination changes or the thresholds are chosen inappropriately. Therefore, the goal of a robust segmentation algorithm must be to find the boundary of the objects as robustly and accurately as possible. The best way to describe the boundaries of the objects robustly is by regarding them as edges in the image. Therefore, in this section we will examine methods to extract edges.

3.7.1 Definition of Edges

3.7.1.1 Definition of Edges in 1D

To derive an edge extraction algorithm, we need to define what edges actually are. For the moment, let us make the simplifying assumption that the gray values in the object and in the background are constant. In particular, we assume that the

素精度或亚像素精度阈值分割为基础。如在 3.5 节中所见的那样，使用这些算法可以得到高准确度的结果。但基于阈值分割结果推导出的测量准确度在多数情况下是由分割时选定阈值的正确与否决定的。如果阈值选定错误，由于从前景灰度值到背景灰度值是平滑过渡的，那么提取出来的物体将通常会变得更大或者更小。当光照改变时，这个问题尤其严重，因为此时要求阈值必须非常准确地适应改变后的照明情况。因此，非常需要一个对于光照改变时仍然可靠的分割算法。从前面的讨论中我们看到当照明改变或者阈值选定不恰当时分割出的区域的边界或者亚像素精度轮廓会移动。所以，鲁棒分割算法的目的是尽可能不受影响地准确地找到物体的边界。描述物体边界的鲁棒性最好的方法是将边界视为图像中的边缘。因此，在本节中将考察提取边缘的方法。

3.7.1 边缘定义

3.7.1.1 一维边缘定义

要推导出一个边缘提取算法，必须定义边缘到底是什么。暂且做一个简单化的假设，即物体的灰度值和背景的灰度值都是常量。特别是要假设图像中没有噪声；并且图像不是离散

image contains no noise. Furthermore, let us assume that the image is not discretized, i.e., it is continuous. To illustrate this, Figure 3.64(b) shows an idealized gray value profile across the part of a workpiece that is indicated in Figure 3.64(a).

的,也就是说假设图像是连续的。举例说明,图 3.64(b)显示了一个理想化的灰度值剖面,此剖面穿过了图 3.64(a)中工件的一个部分。

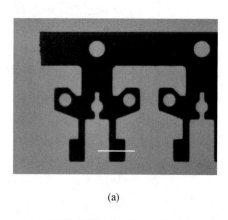
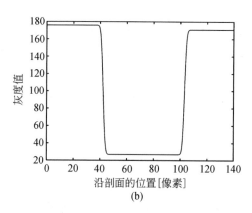

图 3.64　(a) 用背光照射的工件,工件上有一水平线。水平线所指位置的理想化的灰度值剖面如 (b) 所示

From the above example, we can see that edges are areas in the image in which the gray values change significantly. To formalize this, let us regard the image for the moment as a 1D function $f(x)$. From elementary calculus we know that the gray values change significantly if the first derivative of $f(x)$ differs significantly from 0, i.e., $|f'(x)| \gg 0$. Unfortunately, this alone is insufficient to define a unique edge location because there are typically many connected points for which this condition is true since the transition between the background and foreground gray values is smooth. This can be seen in Figure 3.65(a), where the first derivative $f'(x)$ of the ideal gray value profile in Figure 3.64(b) is displayed. Note, for example, that there is an extended range of points for which $|f'(x)| \geqslant 20$. Therefore, to obtain a unique edge position, we must additionally require that the absolute value of the first deriva-

从上面这个例子中,能够看到,边缘就是图像中的一些区域,在这些区域中灰度值变化得非常明显。为了得到更正式的描述,暂时将图像视为一个一维函数 $f(x)$。从初等积分学中得知,当 $f(x)$ 的一阶导数与 0 的差距非常大时灰度值就发生了显著的变化:$|f'(x)| \gg 0$。不幸的是仅此一点不足以定义一个唯一的边缘位置,因为通常有很多彼此相连的点都能满足此条件,因为背景灰度值到前景灰度值的过渡是平滑的。这点从图 3.65(a)中可以看出,此图中显示的是图 3.64(b)所示的理想灰度值剖面线的一阶导数 $f'(x)$。举例说明,注意在一个范围内的很多点都满足 $|f'(x)| \geqslant 20$,这就没法确定出一个唯一的位置作为边缘。因此,为获得一个唯一的边缘位置,必须加入额外的要求,即一阶导

tive $|f'(x)|$ is locally maximal. This is called non-maximum suppression.

From elementary calculus we know that, at the points where $|f'(x)|$ is locally maximal, the second derivative vanishes: $f''(x) = 0$. Hence, edges are given by the locations of inflection points of $f(x)$. To remove flat inflection points, we would additionally have to require that $f'(x)f'''(x) < 0$. However, this restriction is seldom observed. Therefore, in 1D, an alternative and equivalent definition to the maxima of the absolute value of the first derivative is to define edges as the locations of the zero-crossings of the second derivative. Figure 3.65(b) displays the second derivative $f''(x)$ of the ideal gray value profile in Figure 3.64(b). Clearly, the zero-crossings are in the same positions as the maxima of the absolute value of the first derivative in Figure 3.65(a).

数的绝对值 $|f'(x)|$ 是局部最大的。这也被称为非最大抑制。

从初等积分学中得知,当 $|f'(x)|$ 是局部最大时所对应的那些点的二阶导数等于零: $f''(x) = 0$。所以,边缘是由 $f(x)$ 的拐点位置给出的。为消除平坦拐点,必须额外地要求: $f'(x)f'''(x) < 0$。但此约束条件很少被注意到。因此,在一维中与一阶导数绝对值最大的对应的另一种等价定义是将边缘定义为二阶导数过零的那些位置。图 3.65(b) 显示的是图 3.64(b) 中理想灰度值剖面的二阶导数 $f''(x)$。很显然,二阶导数过零与图 3.65(a) 中一阶导数绝对值最大所对应的位置完全相同。

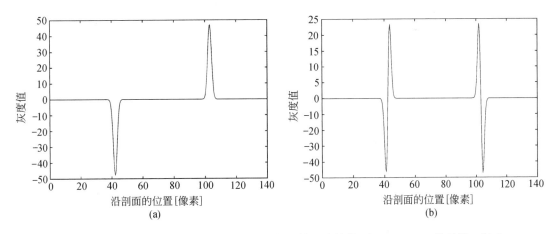

图 3.65 (a) 图 3.64(b) 中理想灰度值剖面的一阶导数 $f'(x)$;(b) 二阶导数 $f''(x)$

From Figure 3.65(a), we can also see that in 1D we can easily associate a polarity with an edge based on the sign of $f'(x)$. We speak of a positive edge if $f'(x) > 0$ and of a negative edge if $f'(x) < 0$.

从图 3.65(a) 中,也能发现在一维中能容易地根据 $f'(x)$ 的符号来将边缘同极性联系在一起,当 $f'(x) > 0$ 时是正边缘,$f'(x) < 0$ 时是负边缘。

3.7.1.2 Definition of Edges in 2D

We now turn to edges in continuous 2D images. Here, the edge itself is a curve $s(t) = (r(t), c(t))^\top$, which is parameterized by a parameter t, e.g., its arc length. At each point of the edge curve, the gray value profile perpendicular to the curve is a 1D edge profile. With this, we can adapt the first 1D edge definition above for the 2D case: we define an edge as the points in the image where the directional derivative in the direction perpendicular to the edge is locally maximal. From differential geometry we know that the direction $n(t)$ perpendicular to the edge curve $s(t)$ is given by $n(t) = s'(t)^\perp \parallel s''(t)$. Unfortunately, the edge definition seemingly requires us to know the edge position $s(t)$ already to obtain the direction perpendicular to the edge, and hence looks like a circular definition. Fortunately, the direction $n(t)$ perpendicular to the edge can be determined easily from the image itself. It is given by the gradient vector of the image, which points in the direction of steepest ascent of the image function $f(r, c)$. The gradient of the image is given by the vector of its first partial derivatives:

$$\nabla f = \nabla f(r, c) = \left(\frac{\partial f(r,c)}{\partial r}, \frac{\partial f(r,c)}{\partial c} \right) = (f_r, f_c) \qquad (3.109)$$

In the last equation, we have used a subscript to denote the partial derivative with respect to the subscripted variable. We will use this convention throughout this section. The Euclidean length

$$\|\nabla f\|_2 = \sqrt{f_r^2 + f_c^2} \qquad (3.110)$$

of the gradient vector is the equivalent of the absolute value of the first derivative $|f'(x)|$ in 1D. We will also call the length of the gradient vector its

3.7.1.2 二维边缘定义

现在讨论在连续二维图像中的边缘。此时，二维边缘是一个曲线 $s(t) = (r(t), c(t))^\top$，此曲线是用一个参数 t，如用曲线的弧长来描述的。在边缘曲线的每个点上与曲线垂直的灰度值剖面都是一个一维边缘剖面。这样，可将上面的一维边缘的第一个定义稍加改动后应用到二维边缘中：将一个边缘定义为图像中的若干个点，这些点的方向导数在垂直于边缘的方向上是局部最大的。从微分几何学中得知垂直于边缘曲线 $s(t)$ 的方向 $n(t)$ 可以这样计算 $n(t) = s'(t)^\perp \parallel s''(t)$。不幸的是如果使用此边缘定义来获得与边缘垂直的方向，看起来需要已知边缘位置 $s(t)$，这看起来似乎是一个循环定义。幸运的是，与边缘垂直的方向 $n(t)$ 能通过图像自身来确定。它可以由图像的梯度向量算出，图像的梯度向量指出图像函数 $f(r, c)$ 的最速上升方向。图像的梯度是由图像一阶偏导数的向量确定的：

在等式（3.109）中的最后一个等式中，采用脚标来表示偏导关系。将会在整节中都使用此约定。梯度向量的欧几里得长度

等价于一维中的一阶导数绝对值 $|f'(x)|$。也将梯度向量的长度称为它的量值，也常被称为幅度。当然梯度

magnitude. It is also often called the amplitude. The gradient direction is, of course, directly given by the gradient vector. We can also convert it to an angle by calculating $\phi = -\text{atan2}(f_r, f_c)$, where atan2 denotes the two-argument arctangent function that returns its result in the range $[-\pi, \pi]$. Note that ϕ increases in the mathematically positive direction (counterclockwise) starting at the column axis. This is the usual convention. With the above definitions, we can define edges in 2D as the points in the image where the gradient magnitude is locally maximal in the direction of the gradient. To illustrate this definition, Figure 3.66(a) shows a plot of the gray values of an idealized corner. The corresponding gradient magnitude is shown in Figure 3.66(b). The edges are the points at the top of the ridge in the gradient magnitude.

In 1D, we have seen that the second edge definition (the zero-crossings of the second derivative) is equivalent to the first definition. Therefore, it is natural to ask whether this definition can be adapted for the 2D case. Unfortunately, there is no direct equivalent for the second derivative in 2D, since there are three partial derivatives of order two. A suitable definition for the second derivative in 2D is the Laplacian operator (Laplacian for short), defined by

$$\Delta f = \Delta f(r,c) = \frac{\partial^2 f(r,c)}{\partial r^2} + \frac{\partial^2 f(r,c)}{\partial c^2} = f_{rr} + f_{cc} \tag{3.111}$$

2D equivalents of the additional 1D condition $f'(x)f'''(x) < 0$ have been proposed (Clark, 1989). However, they are very rarely used in practice. Therefore, edges also can be defined as the zero-crossings of the Laplacian: $\Delta f(r,c) = 0$. Figure 3.66(c) shows the Laplacian of the idealized corner in Figure 3.66(a). The results of the two

方向就是由梯度向量直接给出的。通过计算 $\phi = -\text{atan2}(f_r, f_c)$，也能将方向转换成一个角度，其中，atan2 表示含两参数的反正切函数，该函数返回的结果值为 $[-\pi, \pi]$。注意 ϕ 沿从横轴开始的算术正方向（逆时针方向）增大。这是通常的约定。有了上述的定义，可在二维中将边缘定义为图像中的若干点，在这些点上梯度量值在梯度的方向上局部最大。举例说明此定义，图3.66（a）给出了一个理想化拐角的灰度值图示。对应的梯度量值如图3.66（b）。此理想化拐角的边缘就是梯度量值岭上最高处的那些点。

可以看到，一维中边缘的第二个定义（二阶导数过零）是与第一个定义完全等效的。因此，很自然会问到是否在二维中也是如此呢？不幸的是，由于在二维中存在三种二阶偏导，所以二阶导数给出的结果并不是直接等效的。在二维中一个合适的二阶导数定义是拉普拉斯算子，定义如下：

附加的一维条件 $f'(x)f'''(x) < 0$ 的二维等价物已经提出（Clark，1989），然而在实际情况下很少使用。因此，边缘能通过拉普拉斯算子过零计算得到：$\Delta f(r,c) = 0$。图3.66（c）给出了使用拉普拉斯算子对图3.66（a）中理想化拐角处理后的结果。用两种边

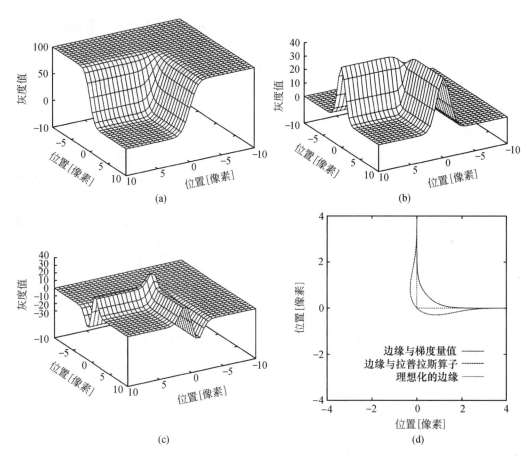

图 3.66 （a）理想化拐角，例如图 3.64（a）中所示工件的一个底角；（b）图（a）的梯度量值；
（c）对（a）使用拉普拉斯算子；（d）边缘在二维中的两种定义各自返回结果的对比

edge definitions are shown in Figure 3.66(d). It can be seen that, unlike for the 1D edges, the two definitions do not result in the same edge positions. The edge positions are identical only for straight edges. Whenever the edge is significantly curved, the two definitions return different results. It can be seen that the definition via the maxima of the gradient magnitude always lies inside the ideal corner, whereas the definition via the zero-crossings of the Laplacian always lies outside the corner and passes directly through the ideal corner (Berzins, 1984). The Laplacian edge is also in a different position from the true edge for a larger part of the edge. Therefore, in 2D the defi-

缘定义得到的边缘在图 3.66（d）中进行了对比。可以看出，与一维边缘不同，两个定义返回的边缘位置是不一样的。只有针对笔直的边缘时这两个定义各自返回的边缘位置结果才是一致的。对任何弯曲明显的边缘，两个定义都会返回不一样的结果。可以看到，通过梯度量值局部最大来定义的边缘总落在理想化拐角内，然而通过拉普拉斯算子过零来定义的边缘总落在理想化拐角外，且通过理想化拐角的顶点（Berzins，1984）。在边缘上的更大部分内，拉普拉斯算子得到的边缘与真实边缘的位置不同。因此，通

nition via the maxima of the gradient magnitude is usually preferred. However, in some applications the fact that the Laplacian edge passes through the corner can be used to measure objects with sharp corners more accurately (see Section 3.7.3.5).

3.7.2   1D Edge Extraction

We now turn our attention to edges in real images, which are discrete and contain noise. In this section, we will discuss how to extract edges from 1D gray value profiles. This is a very useful operation that is used frequently in machine vision applications because it is extremely fast. It is typically used to determine the position or diameter of an object.

3.7.2.1   Discrete Derivatives

The first problem we have to address is how to compute the derivatives of the discrete 1D gray value profile. Our first idea might be to use the differences of consecutive gray values on the profile: $f'_i = f_i - f_{i-1}$. Unfortunately, this definition is not symmetric. It would compute the derivative at the "half-pixel" positions $f_{i-\frac{1}{2}}$. A symmetric way to compute the first derivative is given by

$$f'_i = \tfrac{1}{2}(f_{i+1} - f_{i-1}) \qquad (3.112)$$

This formula is obtained by fitting a parabola through three consecutive points of the profile and computing the derivative of the parabola at the center point. The parabola is uniquely defined by the three points. With the same mechanism, we can also derive a formula for the second derivative:

$$f''_i = \tfrac{1}{2}(f_{i+1} - 2f_i + f_{i-1}) \qquad (3.113)$$

Note that the above methods to compute the first and second derivatives are linear filters, and hence can be regarded as the following two convolution masks:

注意上面用来计算一阶和二阶导数的方法是两个线性滤波器。因此能被视为如下两个卷积掩码：

$$\tfrac{1}{2} \cdot (1 \quad 0 \quad -1) \quad \text{and} \quad \tfrac{1}{2} \cdot (1 \quad -2 \quad 1) \tag{3.114}$$

Note that the -1 is the last element in the first derivative mask because the elements of the mask are mirrored in the convolution (see Eq. (3.18)).

注意一阶导数掩码中最后一个元素是 -1，这是因为掩码的元素在卷积处理中被镜像了（见等式（3.18））。

Figure 3.67(a) displays the true gray value profile taken from the horizontal line in the image in Figure 3.64(a). Its first derivative, computed with Eq. (3.112), is shown in Figure 3.67(b). The noise in the image causes a very large number of local maxima in the absolute value of the first derivative, and consequently also a large number of zero-crossings in the second derivative. The salient edges can easily be selected by thresholding the absolute value of the first derivative: $|f_i'| \geqslant t$. For the second derivative, the edges cannot be selected as easily. In fact, we must resort to calculating the first derivative as well to be able to select the relevant edges. Hence, the edge definition via the first derivative is preferable because it can be

图 3.67（a）给出了从图 3.64（a）所示的水平线上得到的真实灰度值剖面。此剖面的一阶导数由等式（3.112）计算得到，见图 3.67（b）。可看到图像中的噪声造成了图中非常多的位置上一阶导数绝对值都是局部最大，所以也相应地造成了二阶导数过零的次数非常多。通过对一阶导数绝对值进行阈值分割，最明显的边缘很容易地就被提取出来了：$|f_i'| \geqslant t$。但对于二阶导数，就不容易选出这些最明显的边缘了。事实上，为了在二阶导数上提取最明显的边缘，不得不借助于计算一阶导数。所以，由一阶导数给出的边缘定义是更优越的，因为使用

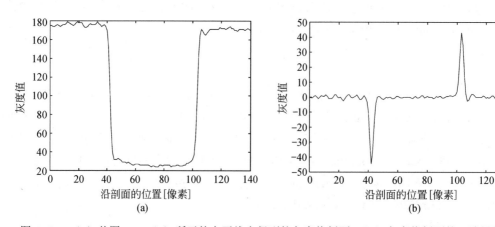

图 3.67 （a）从图 3.64（a）所示的水平线上得到的灰度值剖面；（b）灰度值剖面的一阶导数 f_i'

done with one filter operation instead of two, and consequently the edges can be extracted much faster.

3.7.2.2 Smoothing Perpendicular to a Profile

The gray value profile in Figure 3.67(a) already contains relatively little noise. Nevertheless, in most cases it is desirable to suppress the noise even further. If the object we are measuring has straight edges in the part in which we are performing the measurement, we can use the gray values perpendicular to the line along which we are extracting the gray value profile and average them in a suitable manner. The simplest way to do this is to compute the mean of the gray values perpendicular to the line. If, for example, the line along which we are extracting the gray value profile is horizontal, we can calculate the mean in the vertical direction as follows:

$$f_i = \frac{1}{2m+1} \sum_{j=-m}^{m} f_{r+j,\,c+i} \qquad (3.115)$$

This acts like a mean filter in one direction. Hence, the noise variance is reduced by a factor of $2m+1$. Of course, we could also use a 1D Gaussian filter to average the gray values. However, since this would require larger filter masks for the same noise reduction, and consequently would lead to longer execution times, in this case the mean filter is preferable.

If the line along which we want to extract the gray value profile is horizontal or vertical, the calculation of the profile is simple. If we want to extract the profile from inclined lines or from circles or ellipses, the computation is slightly more difficult. To enable meaningful measurements for distances, we must sample the line with a fixed distance, typically one pixel. Then, we need to generate lines perpendicular to the curve along which we want to extract the profile. This procedure is shown for an inclined line in Figure 3.68. Because of this, the points from which we must extract the gray values typically do not lie on pixel centers. Therefore, we will have to interpolate them. This can be done with the techniques discussed in Sections 3.3.2.1 and 3.3.2.2, i.e., with nearest-neighbor or bilinear interpolation.

要进行有意义的距离测量，必须用固定的距离在线上进行采样，通常是一个像素。然后，需要生成若干条直线，这些直线垂直于与获取灰度值剖面的那条曲线。这个过程在图 3.68 中以斜线显示。因为是斜线，所以线上的这些点通常都不落在像素的中心，然而又需要得到线上这些点的灰度值。所以，将必须进行插值处理以得到这些点对应的灰度值。完成插值的方法已经在 3.3.2.1 节和 3.3.2.2 节中讨论过了，也就是最近邻域插值和双线性插值。

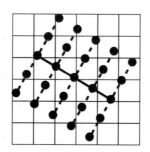

图 3.68　沿一条斜线建立灰度值剖面。此斜线在图中用粗实线表示。圆点表示那些被用来计算剖面的点。注意这些点不位于像素中心上。虚线表示的是计算一维平均值的方向

Figure 3.69 shows a gray value profile and its first derivative obtained by vertically averaging the gray values along the line shown in Figure 3.64(a). The size of the 1D mean filter was 21 pixels in this case. If we compare this with Figure 3.67, which shows the profile obtained from the same line without averaging, we can see that the noise in the profile has been reduced significantly. Because of this, the salient edges are even easier to select than without the averaging.

图 3.69 给出了一个灰度值剖面及其一阶导数，它们是通过沿着图 3.64（a）所示的直线并在直线的垂直方向上计算平均灰度值后得到的。本例中使用了 21 个像素的一维均值滤波器对垂直方向上的灰度值进行平均。与图 3.67 相比，虽然图中剖面也是沿同一直线获得的，但没有在垂直方向进行灰度值平均，可看到现在剖面中的噪声显著下降。正因为这样，与没有进行平均的情况相比，最明显的边缘能被更轻松地选择出来。

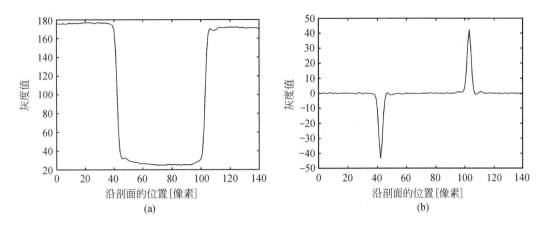

图 3.69　(a) 沿着图 3.64 (a) 中水平线方向并在垂直方向上用 21 个像素计算灰度值平均值后得到的灰度值剖面；(b) 此灰度值剖面的一阶导数 f'_i

3.7.2.3　Optimal Edge Filters

Unfortunately, averaging perpendicular to the curve along which the gray value profile is extracted is sometimes insufficient to smooth the profiles enough to enable us to extract the relevant edges easily. One example is shown in Figure 3.70. Here, the object to be measured has a significant amount of texture, which is not as random as noise and consequently does not average out completely. Note that, on the right side of the profile, there is a negative edge with an amplitude almost as large as the edges we want to

3.7.2.3　最优边缘滤波器

不幸的是在获取灰度值剖面的曲线的垂直方向上进行平均有时也不足以平滑剖面以保证可以容易地提取相关的边缘。见图 3.70 中的例子。此例中，被测物存在非常多的纹理，这些纹理与噪声不同，它们不是随机的，因此不能被完全平均掉。注意在剖面的右侧有一个负边缘，其幅度几乎与想要提取的边缘一样大。另一个没有完全消除噪声的原因可能是无法选择足够多的像素进行平均，因为物体的

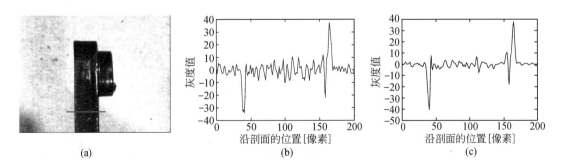

图 3.70　(a) 继电器的图像，水平线指示的是获取灰度值剖面的位置；(b) 未进行平均处理的灰度值剖面的一阶导数；(c) 在垂直方向上对 21 像素进行平均处理后的灰度值剖面的一阶导数

extract. Another reason for the noise not to cancel out completely may be that we cannot choose the size of the averaging large enough, e.g., because the object's boundary is curved.

To solve these problems, we must smooth the gray value profile itself to suppress the noise even further. This is done by convolving the profile with a smoothing filter: $f_s = f * h$. We can then extract the edges from the smoothed profile via its first derivative. This would involve two convolutions: one for the smoothing filter and the other for the derivative filter. Fortunately, the convolution has a very interesting property that we can use to save one convolution. The derivative of the smoothed function is identical to the convolution of the function with the derivative of the smoothing filter: $(f * h)' = f * h'$. We can regard h' as an edge filter.

As for the smoothing filters, the natural question to ask is which edge filter is optimal. This problem was addressed by Canny (1986). He proposed three criteria that an edge detector should fulfill. First of all, it should have a good detection quality, i.e., it should have a low probability of falsely detecting an edge point and also a low probability of erroneously missing an edge point. This criterion can be formalized as maximizing the SNR of the output of the edge filter. Second, the edge detector should have good localization quality, i.e., the extracted edges should be as close as possible to the true edges. This can be formalized by minimizing the variance of the extracted edge positions. Finally, the edge detector should return only a single edge for each true edge, i.e., it should avoid multiple responses. This criterion

边界是曲线。

为解决这个问题，必须对灰度值剖面自身进行平滑处理以进一步抑制噪声。这可以通过把剖面与一个平滑滤波器进行卷积来实现：$f_s = f * h$。这样，就能通过使用剖面的一阶导数从这个平滑后的剖面上提取边缘。这里将会涉及两种卷积：用于平滑处理的滤波器的卷积和用于求导的滤波器的卷积。幸运的是，卷积有一个非常有趣的特性，可以利用它来省去一个卷积。对函数平滑后再求导得到的结果与先对平滑滤波器求导后再与函数卷积得到的结果完全一样：$(f * h)' = f * h'$。可以将 h' 视为一个边缘滤波器。

与用于平滑处理的滤波器类似，我们自然会有这样一个问题，哪个边缘滤波器是最理想的？这个问题是由Canny（1986）提出的。他提出了一个边缘探测器应满足的 3 个准则。首先，它应该有一个好的检测质量，即它对一个边缘点的错检和漏检的可能性都要低。这条准则可被正式描述为边缘滤波器产生的输出的信噪比要最大化。第二，边缘探测器应该有好的局部化质量，即提取出来的边缘应该尽可能地靠近真正的边缘。这条准则可被正式描述为提取出来的边缘位置的方差要最小化。最后，边缘探测器应该对每个真正的边缘都只返回唯一的一个边缘，即它应该避免多重响应。这条准则可被正式描述为从纯噪声中

can be formalized by maximizing the distance between edge positions that are extracted from pure noise. Canny then combined these three criteria into one optimization criterion and solved it using the calculus of variations. To do so, he assumed that the edge filter has a finite extent (mask size). Since adapting the filter to a particular mask size involves solving a relatively complex optimization problem, Canny looked for a simple filter that could be written in closed form. He found that the optimal edge filter can be approximated very well with the first derivative of the Gaussian filter:

提取出来的边缘位置之间的距离要最大化。Canny 将这三条准则组合成一个最优化的准则并用变分积分学解决了这个问题。为解决这个问题，他假设边缘滤波器拥有一个有限的区域（掩码尺寸）。因为调整滤波器到一个特殊的掩码尺寸会陷入解决一个相对复杂的最优化问题的过程中，所以 Canny 寻找一个能以闭形式写出简单滤波器。他发现最理想的边缘滤波器能非常好地用高斯滤波器的一阶导数来近似。

$$g'_\sigma(x) = \frac{-x}{\sqrt{2\pi}\,\sigma^3} e^{-x^2/(2\sigma^2)} \tag{3.116}$$

One drawback of using the true derivative of the Gaussian filter is that the edge amplitudes become progressively smaller as σ is increased. Ideally, the edge filter should return the true edge amplitude independent of the smoothing. To achieve this for an idealized step edge, the output of the filter must be multiplied with $\sqrt{2\pi}\,\sigma$.

使用高斯滤波器真正的导数的一个缺点是当 σ 增加时边缘幅度逐步减小。理想情况下，边缘滤波器应该返回真正的边缘幅度，而不受平滑处理影响。为了做到这一点，对于一个理想化梯度边缘，滤波器的输出必须乘以 $\sqrt{2\pi}\,\sigma$。

Note that the optimal smoothing filter would be the integral of the optimal edge filter, i.e., the Gaussian smoothing filter. It is interesting to note that, like the criteria in Section 3.2.3.7, Canny's formulation indicates that the Gaussian filter is the optimal smoothing filter.

注意最理想的平滑滤波器是最理想的边缘滤波器的积分，即高斯平滑滤波器。请注意还有一点很有趣，类似 3.2.3.7 节中出现过的准则，Canny 的阐述表明高斯滤波器是最理想的平滑滤波器。

Since the Gaussian filter and its derivatives cannot be implemented recursively (see Section 3.2.3.7), Deriche (1987) used Canny's approach to find optimal edge filters that can be implemented recursively. He derived the following two filters:

由于高斯滤波和它的导数计算不能以递归的方式来实现（见 3.2.3.7 节），文献（Deriche, 1987）使用 Canny 的方法发现了能以递归方式实现的最理想的边缘滤波器。他推导出如下两个滤波器：

$$d'_\alpha(x) = -\alpha^2 x\, e^{-\alpha|x|} \tag{3.117}$$

$$e'_\alpha(x) = -2\alpha \sin(\alpha x)\, e^{-\alpha|x|} \tag{3.118}$$

The corresponding smoothing filters are:

对应的平滑滤波器是：

$$d_\alpha(x) = \tfrac{1}{4}\alpha(\alpha|x|+1)\,e^{-\alpha|x|} \tag{3.119}$$

$$e_\alpha(x) = \tfrac{1}{2}\alpha(\sin(\alpha|x|)+\cos(\alpha|x|))\,e^{-\alpha|x|} \tag{3.120}$$

In contrast to the Gaussian filter, where larger values for σ indicate more smoothing, in the Deriche filters smaller values for α indicate more smoothing. The Gaussian filter has the same SNR as the first Deriche filter for $\sigma = \sqrt{\pi}/\alpha$. For the second Deriche filter, the relation is $\sigma = \sqrt{\pi}/(2\alpha)$. Note that the Deriche filters are significantly different from the Canny filter. Figure 3.71 compares the Canny and Deriche smoothing and edge filters with equivalent filter parameters.

Deriche 滤波器同高斯滤波器的对比，对高斯滤波器而言 α 值越大则平滑程度越大，而对 Deriche 滤波器而言则是 α 值越小对应的平滑程度越大。当 $\sigma = \sqrt{\pi}/\alpha$ 时，高斯滤波器与第一个 Deriche 滤波器的效果类似。当 $\sigma = \sqrt{\pi}/(2\alpha)$ 时，高斯滤波器与第二个 Deriche 滤波器的效果类似。注意 Deriche 滤波器与 Canny 滤波器是明显不同的。这也能从图 3.71 中看出，此图是在设置等效滤波器参数的情况下，Canny 滤波器和 Deriche 滤波器之间的比较。

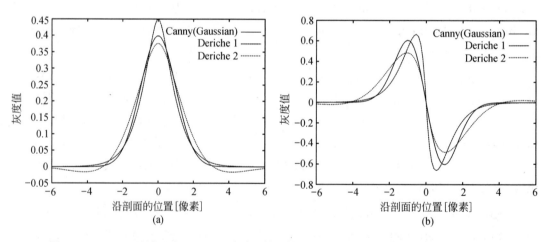

图 3.71 Canny 滤波器和 Deriche 滤波器的比较：（a）平滑滤波器；（b）边缘滤波器

Figure 3.72 shows the result of using the Canny edge detector with $\sigma = 1.5$ to compute the smoothed first derivative of the gray value profile in Figure 3.70(a). As in Figure 3.70(c), the profile was obtained by averaging over 21 pixels vertically. Note that the amplitude of the unwanted

图 3.72 给出了在 $\sigma = 1.5$ 时用 Canny 边缘检测器对从图 3.70（a）得到的灰度值剖面进行平滑一阶导数计算后的结果。与图 3.70（c）类似，计算时使用的灰度值剖面是通过对垂直方向上的 21 个像素进行平均处理

edge on the right side of the profile has been reduced significantly. This enables us to select the salient edges more easily.

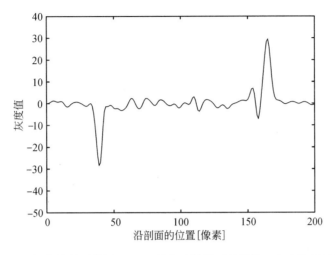

图 3.72 在 $s = 1.5$ 时对灰度值剖面使用 Canny 边缘滤波器后的结果。灰度值剖面是从图 3.70（a）中通过计算垂直方向上 21 个像素的平均值得到的

3.7.2.4　Pixel-Accurate Edge Extraction

To extract the edge position, we need to perform the non-maximum suppression. If we are only interested in the edge positions with pixel accuracy, we can proceed as follows. Let the output of the edge filter be denoted by $e_i = |f * h'|_i$, where h' denotes one of the above edge filters. Then, the local maxima of the edge amplitude are given by the points for which $e_i > e_{i-1} \wedge e_i > e_{i+1} \wedge e_i \geqslant t$, where t is the threshold to select the relevant edges.

3.7.2.5　Subpixel-Accurate Edge Extraction

Unfortunately, extracting the edges with pixel accuracy is often not accurate enough. To extract edges with subpixel accuracy, we can note that around the maximum, the edge amplitude can be approximated well with a parabola.

3.7.2.4　亚像素边缘提取

为提取边缘位置，需要进行非最大抑制。如果仅对像素精度的边缘位置感兴趣，可按如下方法进行。边缘滤波器的输出结果用 $e_i = |f * h'|_i$ 来表示，其中 h' 代表上述边缘滤波器中的一种。然后，边缘幅度的局部最大由若干点给出，对这些点满足 $e_i > e_{i-1} \wedge e_i > e_{i+1} \wedge e_i \geqslant t$，其中 t 是用来选择相关边缘的阈值。

3.7.2.5　亚像素精确边缘提取

不幸的是以像素精度来提取边缘所得到的结果准确度常常是不够的。为了以亚像素准确度来提取边缘，边缘幅度在最大值周围能被很好地拟合为一条抛物线。图 3.73 中显示了一个

Figure 3.73 illustrates this by showing a zoomed part of the edge amplitude around the right edge in Figure 3.72. If we fit a parabola through three points around the maximum edge amplitude and calculate the maximum of the parabola, we can obtain the edge position with subpixel accuracy. If an ideal camera system is assumed, this algorithm is as accurate as the precision with which the floating-point numbers are stored in the computer (Steger, 1998b).

抛物线拟合的实例，此图将图 3.72 中右侧边缘对应的边缘幅度的一部分放大显示。如果用边缘幅度最大值周围的三点来拟合抛物线，然后计算此抛物线的最大值，那么就可以得到亚像素准确度的边缘位置。在一个理想的摄像机系统中，此算法的精度与计算机储存浮点数所用精度相等（Steger, 1998b）。

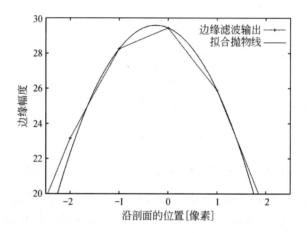

图 3.73 以亚像素精度提取边缘点的原理。首先，检测出边缘幅度的局部最大。然后，在局部最大附近找到三个点，通过此三点来拟合一条抛物线。此抛物线的最大值就是亚像素精度的边缘位置。边缘幅度是从图 3.72 中的右边缘上得到的

We conclude the discussion of 1D edge extraction by showing the results of edge extraction on the two examples we have used so far. Figure 3.74(a) shows the edges that have been extracted along the line shown in Figure 3.64(a) with the Canny filter with $\sigma = 1.0$. From the two zoomed parts around the extracted edge positions, we can see that, by coincidence, both edges lie very close to the pixel centers. Figure 3.74(b) displays the result of extracting edges along the line shown in Figure 3.70(a) with the Canny filter with $\sigma = 1.5$. In this case, the left edge is almost exactly in the middle of two pixel centers.

通过给出两个已用过的例子中边缘提取的结果来结束对一维边缘提取的讨论。图 3.74（a）显示的是沿图 3.64（a）中的水平线使用 $\sigma = 1.0$ 的 Canny 滤波器后提取到的边缘。在提取出的边缘位置附近进行局部放大显示，能看到两个边缘同样位于非常靠近像素中心的地方。图 3.74（b）显示的是沿图 3.70(a) 中水平线使用 $\sigma = 1.5$ 的 Canny 滤波器后提取到的边缘。此例中，左边缘几乎恰好在两个像素中心的正中间。因此，可看到上面的算法能成功提取亚像素精度

Hence, we can see that the algorithm is successful in extracting the edges with subpixel precision.

的边缘。

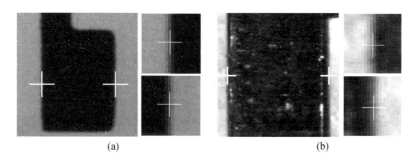

图 3.74　（a）沿图 3.64（a）中的直线提取一维边缘的结果。两个小图像中显示的是边缘位置附近的局部放大图。此例中，两个边缘都非常接近像素的中心。两边缘间的距离是 60.95 像素；（b）沿图 3.70（a）中的直线提取一维边缘的结果。左边缘的细节在右上图中显示，注意左边缘几乎正好位于两个像素中心的正中间。两边缘间的距离是 125.37 像素

3.7.3 2D Edge Extraction

As discussed in Section 3.7.1, there are two possible definitions for edges in 2D, which are not equivalent. As in the 1D case, the selection of salient edges will require us to perform a thresholding based on the gradient magnitude. Therefore, the definition via the zero-crossings of the Laplacian requires us to compute more partial derivatives than the definition via the maxima of the gradient magnitude. Consequently, we will concentrate on the maxima of the gradient magnitude for the 2D case. We will add some comments on the zero-crossings of the Laplacian at the end of this section.

3.7.3.1 Discrete Derivatives

As in the 1D case, the first question we need to answer is how to compute the partial derivatives of the image that are required to calculate the gradient. Similar to Eq. (3.112), we could use finite differences to calculate the partial derivatives. In 2D, they would be

3.7.3 二维边缘提取

如在 3.7.1 节中讨论过的那样，对二维边缘存在着两种可能的定义，这两种定义并不是等价的。与在一维中出现的情况类似，对那些明显边缘的提取将需要在梯度量值的基础上进行一个阈值分割处理。因此，与由梯度量值最大定义的边缘相比，由拉普拉斯算子过零定义的边缘需要计算更多的偏导数。所以，对于二维情况，需要将精力集中在梯度量值最大定义上。在本节结束部分，将加入一些对拉普拉斯算子过零定义的评论。

3.7.3.1 离散导数

与在一维中的情况类似，我们必须回答的第一个问题是如何计算图像的偏导数，因为计算梯度时必须使用它们。与等式（3.112）类似，可以用有限差分来计算偏导数。在二维中，有限差分是：

$$f_{r;i,j} = \tfrac{1}{2}(f_{i+1,j} - f_{i-1,j}) \quad \text{and} \quad f_{c;i,j} = \tfrac{1}{2}(f_{i,j+1} - f_{i,j-1}) \tag{3.121}$$

However, as we have seen above, typically the image must be smoothed to obtain good results. For time-critical applications, the filter masks should be as small as possible, i.e., 3×3. All 3×3 edge filters can be brought into the following form by scaling the coefficients appropriately (note that the filter masks are mirrored in the convolution):

但是，正如上面已经看到的，通常图像必须被平滑处理以便获得好的提取结果。对运行时间要求苛刻的那些应用，滤波器掩码应该尽可能的小，即 3×3。通过适当地对系数进行按比例缩放，所有 3×3 边缘滤波器都能用如下形式表示 (注意在卷积中滤波器掩码被镜像)：

$$\begin{pmatrix} 1 & 0 & -1 \\ a & 0 & -a \\ 1 & 0 & -1 \end{pmatrix} \qquad \begin{pmatrix} 1 & a & 1 \\ 0 & 0 & 0 \\ -1 & -a & -1 \end{pmatrix} \tag{3.122}$$

If we use $a = 1$, we obtain the Prewitt filter. Note that it performs a mean filter perpendicular to the derivative direction. For $a = \sqrt{2}$, the Frei filter is obtained, and for $a = 2$ we obtain the Sobel filter, which performs an approximation to a Gaussian smoothing perpendicular to the derivative direction. Of the above three filters, the Sobel filter returns the best results because it uses the best smoothing filter.

当 $a = 1$ 时，可得到 Prewitt 滤波器。注意它在垂直于导数的方向上进行一个均值滤波处理。当 $a = \sqrt{2}$ 时，就得到了 Frei 滤波器。当 $a = 2$ 时可得到 Sobel 滤波器，此滤波器在垂直于导数的方向上执行一个近似于高斯平滑的处理。在上面三个滤波器中，Sobel 滤波器返回的结果最好，因为它使用了最好的平滑滤波器。

Ando (2000) has proposed a 3×3 an edge filter that tries to minimize the artifacts that invariably are obtained with small filter masks. In our notation, his filter would correspond to $a = 2.435\,101$. Unfortunately, like the Frei filter, it requires floating-point calculations, which makes it unattractive for time-critical applications.

文献（Ando，2000）提出了一个 3×3 边缘滤波器，此滤波器设法将总是在使用小掩码滤波器时产生的人为干扰最小化。在等式（3.122）中，Ando 滤波器对应的 $a = 2.435\,101$。不幸的是与 Frei 滤波器类似，此滤波器要求浮点计算，这使得它在对运行时间要求苛刻的应用中没有吸引力。

The 3×3 edge filters are primarily used to quickly find edges with moderate accuracy in images of relatively good quality. Since speed is important and the calculation of the gradient magnitude via the Euclidean length (the 2-norm) of

在质量相对较好的图像中以中等准确度快速搜索边缘时主要使用 3×3 边缘滤波器。因为速度是重要的，而且由梯度向量的欧几里得长度 (2-范数)$\|\nabla f\|_2 = \sqrt{f_r^2 + f_c^2}$ 得到

the gradient vector ($\|\nabla f\|_2 = \sqrt{f_r^2 + f_c^2}$) requires an expensive square root calculation, the gradient magnitude is typically computed by one of the following norms: the 1-norm $\|\nabla f\|_1 = |f_r| + |f_c|$ or the maximum norm $\|\nabla f\|_\infty = \max(|f_r|, |f_c|)$. Note that the first norm corresponds to the city-block distance in the distance transform, while the second norm corresponds to the chessboard distance (see Section 3.6.1.8). Furthermore, the non-maximum suppression also is relatively expensive and is often omitted. Instead, the gradient magnitude is simply thresholded. Because this results in edges that are wider than one pixel, the thresholded edge regions are skeletonized. Note that this implicitly assumes that the edges are symmetric.

Figure 3.75 shows an example where this approach works well because the image is of good quality. Figure 3.75(a) displays the edge amplitude around the leftmost hole of the workpiece in Figure 3.64(a) computed with the Sobel filter and the 1-norm. The edge amplitude is thresholded (Figure 3.75(b)) and the skeleton of the resulting region is computed (Figure 3.75(c)). Since the assumption that the edges are symmetric is fulfilled in this example, the resulting edges are in the correct location.

的梯度量值计算需要进行耗时的求平方根运算，所以梯度量值通常是用如下向量范数中的一个来计算的：1-范数 $\|\nabla f\|_1 = |f_r| + |f_c|$ 或者最大值范数 $\|\nabla f\|_\infty = \max(|f_r|, |f_c|)$。注意第一个范数对应的是距离变换中的 4 连通距离，而第二个范数对应的是 8 连通距离（见 3.6.1.8 节）。此外，非最大抑制也是相对耗时较多的处理所以常被忽略掉，只对梯度量值进行阈值分割。由于这将返回宽度大于一个像素的边缘，所以阈值分割后的边缘区域要被骨架化。注意进行上述处理时默认边缘是对称的。

图 3.75 给出的例子中由于图像质量很好，所以应用上面给出的简单方法效果相当好。图 3.75（a）给出的是图 3.64（a）中工件上最左侧孔周围的边缘幅度，此边缘幅度是通过使用 Sobel 滤波器和计算 1-范数得到的。对边缘幅度进行阈值分割（图 3.75（b））然后提取阈值分割后区域的骨架（图 3.75（c））。由于这幅图像中的边缘完全满足边缘对称性的假设，所以得到的边缘都在正确的位置。

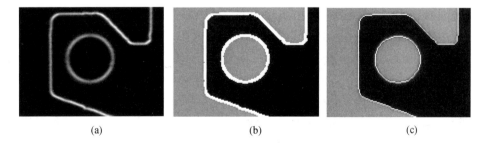

图 3.75　(a) 图 3.64（a）工件的最左侧孔周围的边缘幅度，此边缘幅度是使用 Sobel 滤波器和求 1-范数后得到的；(b) 阈值分割后的边缘区域；(c) 图 (b) 中区域的骨架

This approach fails to produce good results on the more difficult image of the relay in Figure 3.70(a). As can be seen from Figure 3.76(a), the texture on the relay causes many areas with high gradient magnitude, which are also present in the segmentation (Figure 3.76(b)) and the skeleton (Figure 3.76(c)). Another interesting thing to note is that the vertical edge at the right corner of the top edge of the relay is quite blurred and asymmetric. This produces holes in the segmented edge region, which are exacerbated by the skeletonization.

对于如图 3.70（a）继电器图像那样更复杂的图像，使用本方法就不能得到好的处理结果。如图 3.76（a）所示，继电器上的纹理导致出现了很多高梯度量值的区域，它们也会出现在分割结果中（图 3.76(b)）和骨架（图 3.76(c)）中。在图中另一个值得注意的是，继电器最顶端右角的垂直边缘是非常模糊的，而且是非对称的。这会在分割后的边缘区域内产生孔洞，并在骨架化处理后更明显。

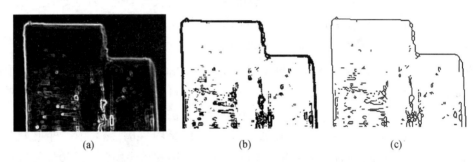

图 3.76　（a）由 Sobel 滤波器和 1-范数计算得到的图 3.70（a）继电器顶部的边缘幅度；（b）阈值分割后的边缘区域；（c）图（b）区域的骨架

3.7.3.2 Optimal Edge Filters

Because the 3 × 3 filters are not robust against noise and other disturbances, e.g., textures, we need to adapt the approach to optimal 1D edge extraction described in the previous section to the 2D case. In 2D, we can derive the optimal edge filters by calculating the partial derivatives of the optimal smoothing filters, since the properties of the convolution again allow us to move the derivative calculation into the filter. Consequently, Canny's optimal edge filters in 2D are given by the partial derivatives of the Gaussian filter. Because the Gaussian filter is separable, so are its derivatives:

3.7.3.2 最优边缘滤波器

因为 3×3 滤波器对噪声和其他干扰，如纹理，鲁棒性不好，必须将前一节中的最佳一维边缘提取的方法应用到二维情况中。在二维中，可通过计算最佳平滑滤波器的偏导数来推导出最佳的边缘滤波器，因为卷积操作的特性允许将求导计算移动到滤波器上。所以，Canny 的最理想二维边缘滤波器可由高斯滤波器的偏导数得到。因为高斯滤波器是可分的，所以其导数也是可分的：

$$g_r = \sqrt{2\pi}\,\sigma g'_\sigma(r)g_\sigma(c) \quad \text{and} \quad g_c = \sqrt{2\pi}\,\sigma g_\sigma(r)g'_\sigma(c) \qquad (3.123)$$

(see the discussion following Eq. (3.116) for the factors of $\sqrt{2\pi}\,\sigma$). To adapt the Deriche filters to the 2D case, the separability of the filters is postulated. Hence, the optimal 2D Deriche filters are given by $d'_\alpha(r)d_\alpha(c)$ and $d_\alpha(r)d'_\alpha(c)$ for the first Deriche filter, and by $e'_\alpha(r)e_\alpha(c)$ and $e_\alpha(r)e'_\alpha(c)$ for the second Deriche filter (see Eq. (3.118)).

The advantage of the Canny filter is that it is isotropic, i.e., rotation-invariant (see Section 3.2.3.7). Its disadvantage is that it cannot be implemented recursively. Therefore, the execution time depends on the amount of smoothing specified by σ. The Deriche filters, on the other hand, can be implemented recursively, and hence their run time is independent of the smoothing parameter α. However, they are anisotropic, i.e., the edge amplitude they calculate depends on the angle of the edge in the image. This is undesirable because it makes the selection of the relevant edges harder. Lanser has shown that the anisotropy of the Deriche filters can be corrected (Lanser and Eckstein, 1992). We will refer to the isotropic versions of the Deriche filters as the Lanser filters.

Figure 3.77 displays the result of computing the edge amplitude with the second Lanser filter with $\alpha = 0.5$. Compared to the the Sobel filter, the Lanser filter was able to suppress the noise and texture significantly better. This can be seen from the edge amplitude image (Figure 3.77(a)) as well as the thresholded edge region (Figure 3.77(b)). Note, however, that the edge region still contains

（见等式（3.116）下面对因数 $\sqrt{2\pi}\,\sigma$ 的讨论）。在二维中应用 Deriche 滤波器，也需要滤波器的可分性作为前提。因此，对第一种 Deriche 滤波器的最理想二维 Deriche 滤波器由 $d'_\alpha(r)d_\alpha(c)$ 和 $d_\alpha(r)d'_\alpha(c)$ 计算，而对第二种 Deriche 滤波器的最理想二维 Deriche 滤波器由 $e'_\alpha(r)e_\alpha(c)$ 和 $e_\alpha(r)e'_\alpha(c)$ 计算得到（见等式（3.118））。

Canny 滤波器的优点是具有各项同性和旋转不变（见 3.2.3.7 节）。它的缺点是不能以递归方式来计算。因此，其执行时间由 σ 决定的平滑程度来决定。另一方面，虽然 Deriche 滤波器能以递归方式计算，其运行时间不受平滑参数 α 影响。但是它们是各项异性的，即用其计算出的边缘幅度依赖于图像中边缘的角度。这不是期望达到的效果，因为它让选定相关的边缘变得更困难。Lanser 证明了 Deriche 滤波器的各项异性是可以修正的（Lanser and Eckstein, 1992）。我们称 Deriche 滤波器的各项同性版本为 Lanser 滤波器。

图 3.77 给出了用第二个 Lanser 滤波器在 $\alpha = 0.5$ 时计算出的边缘幅度。与使用 Sobel 滤波器计算的结果相比，此 Lanser 滤波器能更好地抑制噪声和纹理。这一点从边缘幅度的图像（见图 3.77（a））以及阈值分割后的边缘区域（见图 3.77（b））中都可以看到。但注意继电器最右上角

a hole for the vertical edge that starts at the right corner of the topmost edge of the relay. This happens because the edge amplitude only has been thresholded and the important step of non-maximum suppression has been omitted in order to compare the results of the Sobel and Lanser filters.

的垂直边缘所对应的边缘区域中仍然包含了一个孔。这是由于仅对边缘幅度进行了阈值分割处理,而将非最大抑制这个重要步骤省略了。

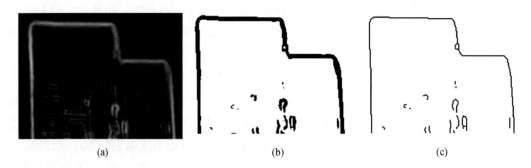

图 3.77 （a）图 3.70（a）中继电器顶部附近的边缘幅度,此图是用第二个 Lanser 滤波器在 $\alpha = 0.5$ 时计算得到的;（b）阈值分割后的边缘区域;（c）图（b）的骨架

3.7.3.3 Non-Maximum Suppression

As we saw in the above examples, thresholding the edge amplitude and then skeletonizing the region sometimes does not yield the desired results. To obtain the correct edge locations, we must perform a non-maximum suppression (see Section 3.7.1.2). In the 2D case, this can be done by examining the two neighboring pixels that lie closest to the gradient direction. Conceptually, we can think of transforming the gradient vector into an angle. Then, we divide the angle range into eight sectors. Figure 3.78 shows two examples of this. Unfortunately, with this approach, diagonal edges are often still two pixels wide. Consequently, the output of the non-maximum suppression must still be skeletonized.

3.7.3.3 非最大抑制处理

正如已经从上面例子中看到的那样,先对边缘幅度进行阈值分割,然后对分割出的区域进行骨架化处理,这个方法有时候不能给出预期的结果。为了获取正确的边缘位置,必须进行非最大抑制处理（见 3.7.1.2 节）。在二维情况下,通过比较梯度方向上最邻近的两个像素可实现非最大抑制。从概念上可想到将梯度向量变换为一个角度。然后,将角度范围分解到 8 个方格中。图 3.78 给出了此方法的两个例子,两个不同角度的向量被分解到 8 个格子中,然后找到与此方向最近的两个像素。不幸的是使用此方法,对角方向的边缘经常仍然是两个像素宽。因此,非最大抑制的输出仍需要被骨架化处理。

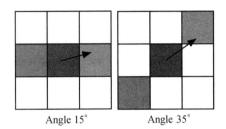

图 3.78 对两个不同的梯度方向，应用非最大抑制处理获得的像素

Figure 3.79 shows the result of applying non-maximum suppression to the edge amplitude image in Figure 3.77(a). From the thresholded edge region in Figure 3.79(b), it can be seen that the edges are now in the correct locations. In particular, the incorrect hole in Figure 3.77 is no longer present. We can also see that the few diagonal edges are sometimes two pixels wide. Therefore, their skeleton is computed and displayed in Figure 3.79(c).

图 3.79 显示了对图 3.77（a）中边缘幅度图像应用非最大抑制处理的结果。阈值分割后的边缘区域如图 3.79（b），现在能看到边缘出现在正确的位置。特别是图 3.77 中不正确的孔消失了。也能看到一些对角线方向的边缘是两个像素宽。因此，计算骨架后的结果如图 3.79（c）所示。

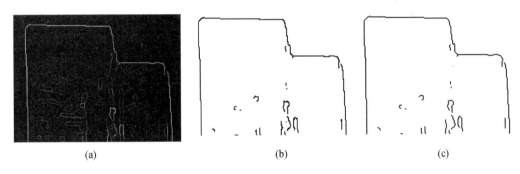

图 3.79　(a) 对图 3.77（a）边缘幅度图像使用非最大抑制处理后的结果；(b) 阈值分割后的边缘区域；(c) 图 (b) 的骨架

3.7.3.4 Hysteresis Thresholding

Up to now, we have been using simple thresholding to select the salient edges. This works well as long as the edges we are interested in have roughly the same contrast or have a contrast that is significantly different from the contrast of noise, texture, or other irrelevant objects in the image. In

3.7.3.4 滞后阈值分割

到目前为止，已经使用简单的阈值分割来选取显著边缘。只有在我们感兴趣的边缘上都有大致相等的对比度，或者边缘上的对比度与噪声、纹理或其他不相关物体的对比度有明显不同时，此方法才能取得好的效果。

many applications, however, we face the problem that, if we select the threshold so high that only the relevant edges are selected, they are often fragmented. If, on the other hand, we set the threshold so low that the edges we are interested in are not fragmented, we end up with many irrelevant edges. These two situations are illustrated in Figures 3.80(a) and (b). A solution to this problem was proposed by Canny (1986). He devised a special thresholding algorithm for segmenting edges: hysteresis thresholding. Instead of a single threshold, it uses two thresholds. Points with an edge amplitude greater than the higher threshold are immediately accepted as safe edge points. Points with an edge amplitude smaller than the lower threshold are immediately rejected. Points with an edge amplitude between the two thresholds are accepted only if they are connected to safe edge points via a path in which all points have an edge amplitude above the lower threshold. We can also think of this operation as first selecting the edge points with an amplitude above the upper threshold, and then extending the edges as far as possible while remaining above the lower threshold.

但在很多应用中会面临下面的问题，就是如果选择高的阈值以保证只将相关边缘选出时，边缘通常被割裂成若干段。另一方面，如果选择低的阈值以保证边缘不会断裂成一段一段时，最终的分隔结果中又会包含很多不相关边缘。这两种情况见图 3.80（a）和图 3.80（b）。文献（Canny，1986）提出了此问题的一种解决方案。他设计了一个特殊的阈值分割算法来分割边缘：滞后阈值分割。区别于使用单一阈值，滞后阈值分割使用两个阈值——高阈值和低阈值。边缘幅度比高阈值大的那些点立即作为安全边缘点被接受；边缘幅度比低阈值小的那些点被立即剔除。边缘幅度在高阈值和低阈值之间的那些点按如下原则处理，只有在这些点能按某一路径与安全边缘点相连时，它们才作为边缘点被接受。组成这一路径的所有点的边缘幅度都比低阈值要大。也可以把这个处理过程理解成：首先选定边缘幅度大于高阈值的所有边缘点，然后在边缘幅度大于低阈值的情况下尽可能延长

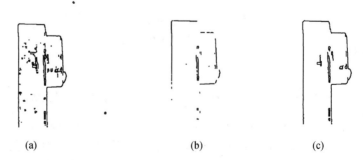

(a)　　　　　　　　(b)　　　　　　　　(c)

图 3.80　（a）对整个图 3.70（a）所示的继电器图像进行边缘幅度阈值分割处理的结果，使用的阈值是 60。此处理导致了许多不相关纹理边缘被选出了；（b）当阈值设为 140 时对边缘幅度进行阈值分割的结果。此处理虽然只选出了相关边缘，但也把边缘断裂成不完整的片段了；（c）当低阈值为 60 且高阈值为 140 时，滞后阈值分割的结果。只有相关边缘被选出，而且它们是完整的

Figure 3.80(c) shows that hysteresis thresholding enables us to select only the relevant edges without fragmenting them or missing edge points.

3.7.3.5 Subpixel-Accurate Edge Extraction

As in the 1D case, the pixel-accurate edges we have extracted so far are often not accurate enough. We can use a similar approach as for 1D edges to extract edges with subpixel accuracy: we can fit a 2D polynomial to the edge amplitude and extract its maximum in the direction of the gradient vector (Steger, 1998b, 2000). The fitting of the polynomial can be done with convolutions with special filter masks (the so-called facet model masks; Haralick and Shapiro, 1992; Haralick et al., 1983). To illustrate this, Figure 3.81(a) shows a 7 × 7 part of an edge amplitude image. The fitted 2D polynomial obtained from the central 3 × 3 amplitudes is shown in Figure 3.81(b), along with an arrow that indicates the gradient direction. Furthermore,

边缘。从图 3.80（c）可以看出滞后阈值分割处理能在不断裂边缘线也不丢失边缘点的情况下选择出相关边缘。

3.7.3.5　亚像素边缘提取

与在一维中的情况类似，到目前为止所提取的都是像素精度边缘，所以在很多时候其精度是不够的。可用类似于提取一维亚像素边缘时的办法来提取二维亚像素精度边缘：将边缘幅度拟合成一个二维多项式，然后在梯度向量方向上提取其最大值（Steger，1998b，2000）。多项式拟合能通过与一个专用滤波器掩码（facet 模型掩码，Haralick and Shapiro，1992；Haralick et al.，1983）进行卷积来实现。举例说明，图 3.81（a）显示一幅边缘幅度图像的一个 7×7 部分。使用中心 3×3 区域内的边缘幅度拟合得到的二维多项式见图 3.81（b），图中使用一个箭头来表示梯度方向。此

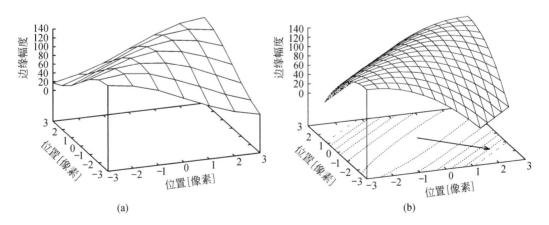

图 3.81　（a）某边缘幅度图像的 7×7 局部；（b）从图（a）的中心 3×3 幅度拟合得到的二维多项式。箭头指示的是梯度方向。图中的轮廓线指示出边缘点在箭头方向上大约偏移了 1/4 像素

contour lines of the polynomial are shown. They indicate that the edge point is offset by approximately a quarter of a pixel in the direction of the arrow.

The above procedure gives us one subpixel-accurate edge point per non-maximum suppressed pixel. These individual edge points must be linked into subpixel-precise contours. This can be done by repeatedly selecting the first unprocessed edge point to start the contour and then successively finding adjacent edge points until the contour closes, reaches the image border, or reaches an intersection point.

Figure 3.82 illustrates subpixel edge extraction along with a very useful strategy to increase the processing speed. The image in Figure 3.82(a) is the same workpiece as in Figure 3.64(a). Because subpixel edge extraction is relatively costly, we want to reduce the search space as much as possible. Since the workpiece is back-lit, we can threshold it easily (Figure 3.82(b)). If we calculate the inner boundary of the region with Eq. (3.89), the resulting points are close to the edge points we want to extract. We only need to dilate the boundary slightly, e.g., with a circle of diameter 5 (Figure 3.82(c)), to obtain an ROI for the edge extraction. Note that the ROI is only a small fraction of the entire image. Consequently, the edge extraction can be done an order of magnitude faster than on the entire image, without any loss of information. The resulting subpixel-accurate edges are shown in Figure 3.82(d) for the part of the image indicated by the rectangle in Figure 3.82(a). Note how well they capture the shape of the hole.

外，此多项式的轮廓也显示出来。这些轮廓指示出边缘点在箭头方向上大约偏移了 1/4 像素。

上面处理过程可以得到每个非最大抑制像素的一个亚像素准确度边缘点。这些单独的边缘点必须被连接成亚像素精度轮廓。实现方法如下：先选定第一个未处理的边缘点作为轮廓的起点，然后顺次找到相邻的边缘点直到轮廓闭合，或者到达图像边界，或者到达一个交叉点。

图 3.82 的例子在亚像素边缘提取过程中使用一个非常有用的处理策略来提高处理速度。图 3.82（a）显示的工件与图 3.64（a）中的相同。因为亚像素边缘提取相对耗时，所以希望能尽可能地缩小搜索区域。由于工件是用背光照射的，能很容易对其进行阈值分割（图 3.82（b））。如果用等式（3.89）计算区域的内边界，得到的这些点已经接近要提取的边缘点。仅需要对边界稍做膨胀，比如，用直径为 5 的圆来执行膨胀（图 3.82（c）），就得到一个区域，此区域可被用来作为边缘提取操作的感兴趣区域使用。注意此 ROI 仅是全部图像的一个小部分。所以，边缘提取处理的速度比在全图执行时有了数量级的提升，且处理结果中没有遗失任何信息。返回的亚像素准确度的边缘如图 3.82（d）所示，图中显示的是对 3.82（a）处理后的一个局部。可看到提取到的孔的形状非常好。

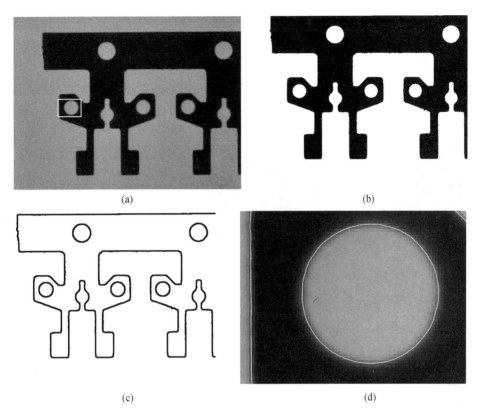

图 3.82 （a）图 3.64（a）的工件，矩形区域内的图像局部见（d）；（b）阈值分割后的工件；（c）用直径为 5 的圆对（b）进行膨胀处理，处理后的结果被当作执行亚像素边缘提取处理时用的 ROI；（d）用 $\sigma = 1$ 的 Canny 滤波器处理后得到的工件的亚像素精度边缘

We conclude this section with a look at the second edge definition via the zero-crossings of the Laplacian. Since the zero-crossings are just a special threshold, we can use the subpixel-precise thresholding operation, defined in Section 3.4.3, to extract edges with subpixel accuracy. To make this as efficient as possible, we must first compute the edge amplitude in the entire ROI of the image. Then, we threshold the edge amplitude and use the resulting region as the ROI for the computation of the Laplacian and for the subpixel-precise thresholding. The resulting edges for two parts of the workpiece image are compared to the gradient magnitude edge in Figure 3.83. Note that, since the Laplacian edges must follow the corners, they

最后讨论一下第二种边缘定义——拉普拉斯算子过零。由于过零也是一种特殊的阈值条件，所以我们能用亚像素精度阈值分割处理来以亚像素精度进行边缘提取，见 3.4.3 节。为尽可能使它更有效率，必须先在图像的整个 ROI 中计算边缘幅度，然后，对边缘幅度进行阈值分割，并把分割的结果作为计算拉普拉斯算子和计算亚像素精度阈值分割时的 ROI。在图 3.83 中对比了用此方法对工件图像的两个不同部分进行处理后得到的边缘和用梯度量值计算得到的边缘。从对比中可以注意到，由拉普拉斯算子求得的边缘比用梯度量值求得的边缘更

are much more curved than the gradient magnitude edges, and hence are more difficult to process further. This is another reason why the edge definition via the gradient magnitude is usually preferred.

弯曲些，这是因为拉普拉斯算子得到的边缘一定要经过真实边缘上所有角点，所以后续处理更困难些。这也是为什么更倾向于通过梯度量值来定义边缘的另一个原因。

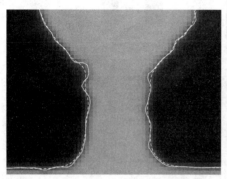
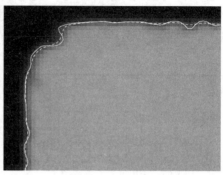

图 3.83　用梯度量值在梯度方向最大所提取到的亚像素精度边缘（虚线表示）和用拉普拉斯算子过零法提取的亚像素精度边缘间的比较图。在这两个处理中都是用了 $\sigma = 1$ 的高斯滤波器。注意拉普拉斯算子求得的边缘一定会通过各个角点，所以比梯度量值求得的边缘更曲折

Despite the above arguments, the property that the Laplacian edge exactly passes through corners in the image can be used advantageously in some applications. Figure 3.84(a) shows an image of a bolt for which the depth of the thread must be measured. Figures 3.84(b)–(d) display the results of extracting the border of the bolt with subpixel-precise thresholding, the gradient

除了上面的不足，拉普拉斯算子求得的边缘一定会恰好通过真实边缘上的角顶点的这一特性可以被用在一些应用中。图 3.84（a）显示的是一个螺丝钉，现在要测量螺纹的深度。图 3.84（b）～图 3.84（d）分别对应的是，用亚像素精度阈值分割提取螺钉边界后的结果，由 $\sigma = 0.7$ 的 Canny 滤波

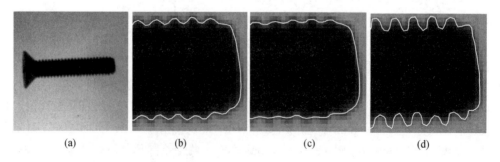

图 3.84　（a）螺钉图像，要从图像中测量螺钉的螺纹深度；（b）执行亚像素精度阈值分割处理的结果；（c）由 $\sigma = 0.7$ 的 Canny 滤波器提取的梯度量值边缘；（d）由 $\sigma = 0.7$ 的高斯滤波器提取的拉普拉斯算子边缘。注意对于这个应用，拉普拉斯边缘给出的结果更合适

magnitude edges with a Canny filter with $\sigma = 0.7$, and the Laplacian edges with a Gaussian filter with $\sigma = 0.7$. Note that in this case the most suitable results are obtained with the Laplacian edges.

3.7.4 Accuracy and Precision of Edges

In the previous two sections, we have seen that edges can be extracted with subpixel resolution. We have used the terms "subpixel-accurate" and "subpixel-precise" to describe these extraction mechanisms without actually justifying the use of the words "accurate" and "precise." Therefore, in this section we will examine whether the edges we can extract are actually subpixel-accurate and subpixel-precise.

3.7.4.1 Definition of Accuracy and Precision

Since the words "accuracy" and "precision" are often confused or used interchangeably, let us first define what we mean by them. By precision, we denote how close on average an extracted value is to its mean value (Haralick and Shapiro, 1993; JCGM 200:2012). Hence, precision measures how repeatably we can extract the value. By accuracy, on the other hand, we denote how close on average the extracted value is to its true value (Haralick and Shapiro, 1993; JCGM 200:2012). Note that the precision does not tell us anything about the accuracy of the extracted value. The measurements could, for example, be offset by a systematic bias, but still be very precise. Conversely, the accuracy does not necessarily tell us

器得到梯度量值边缘，由 $\sigma = 0.7$ 的高斯滤波器得到的拉普拉斯边缘。本例中最合适的结果是拉普拉斯边缘。

3.7.4 边缘的准确度和精确度

在前面的两节中，已经看到能够以亚像素的分辨率来提取边缘。我们已经用术语"亚像素-准确度"和"亚像素精度"来描述这种提取机制，但没有为"准确度"和"精确度"等词的使用提供真正的证明。因此，本节中将验证是否真的能以亚像素-准确度和亚像素精确度来提取边缘。

3.7.4.1 准确度和精确度定义

由于准确度和精确度两词常常被混淆，或者可交换使用，所以我们首先要定义使用它们时的含义。使用精确度时，说明的是平均起来看，某次提取值与若干次提取值的平均值接近的程度（Haralick and Shapiro, 1993; JCGM 200:2012）。所以，精确度估量的是能提取某值时可重复的程度。另一方面，使用准确度时，说明的是平均起来看，提取值与真实值接近的程度（Haralick and Shapiro, 1993; JCGM 200:2012）。注意精确度没有告诉我们任何有关提取值准确度的信息。比如，由于系统性偏置造成了测量结果偏离真值，但仍然可以是非常精确的。

how precise the extracted value is. The measurement could be quite accurate, but not very precise. Figure 3.85 shows the different situations that can occur. Also note that accuracy and precision are statements about the average distribution of the extracted values. From a single value, we cannot tell whether the measurements are accurate or precise.

相反地,准确度也没有说明提取值的精确程度。测量可以是相当准确的,但又是不可重复的。图 3.85 显示了可能发生的不同情况。也请注意准确度和精确度都是关于提取值的平均分布的表述。从一个单独的值,不能判断测量结果是否是准确的,或是否是精确的。

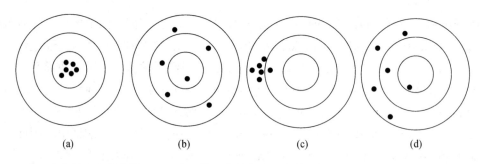

图 3.85　准确度和精确度的比较。圆心表示特征值的真值。点表示的是对特征值测量的结果;(a)准确且精确;(b)准确但不精确;(c)不准确但精确;(d)既不准确也不精确

If we adopt a statistical point of view, the extracted values can be regarded as random variables. With this, the precision of the values is given by the variance of the values: $V[x] = \sigma_x^2$. If the extracted values are precise, they have a small variance. On the other hand, the accuracy can be described by the difference of the expected value $E[x]$ from the true value T: $E[x]-T$. Since we typically do not know anything about the true probability distribution of the extracted values, and consequently cannot determine $E[x]$ and $V[x]$, we must estimate them with the empirical mean and variance of the extracted values.

如果采用统计学的观点,可将提取值视为随机变量。这样,值的精确度可由值的方差计算:$V[x] = \sigma_x^2$。如果提取值是精确的,那么其方差就小。另一方面,准确度能用期望值 $E[x]$ 与真值 T 之间的差来描述:$|E[x] - T|$。由于通常并不知道提取值真正的概率分布,因此也就不能确定 $E[x]$ 和 $V[x]$,必须用经验平均值和提取值的方差对这两个值进行估计。

3.7.4.2　Analytical Edge Accuracy and Precision

The accuracy and precision of edges is analyzed extensively in Steger (1998b,a). The precision of

3.7.4.2　边缘准确度和精确度分析

在文献(Steger,1998b,a)中对边缘的准确度和精确度进行了详尽地

ideal step edges extracted with the Canny filter can be derived analytically. If we denote the true edge amplitude by a and the noise variance in the image by σ_n^2, it can be shown that the variance of the edge positions σ_e^2 is given by

$$\sigma_e^2 = \frac{3}{8}\frac{\sigma_n^2}{a^2}$$

Even though this result was derived analytically for continuous images, it also holds in the discrete case. This result has also been verified empirically in Steger (1998b,a). Note that it is quite intuitive: the larger the noise in the image, the less precisely the edges can be located; furthermore, the larger the edge amplitude, the higher is the precision of the edges. Note also that increasing the smoothing does not increase the precision. This happens because the noise reduction achieved by the larger smoothing cancels out exactly with the weaker edge amplitude that results from the smoothing. From Eq. (3.124), we can see that the Canny filter is subpixel-precise ($\sigma_e \leqslant 1/2$) if the SNR $a^2/\sigma_n^2 \geqslant 3/2$. This can be achieved easily in practice. Consequently, we were justified in calling the Canny filter subpixel-precise.

The same derivation can also be performed for the Deriche and Lanser filters. For continuous images, the following variances result:

$$\sigma_e^2 = \frac{5}{64}\frac{\sigma_n^2}{a^2} \quad \text{和} \quad \sigma_e^2 = \frac{3}{16}\frac{\sigma_n^2}{a^2}$$

Note that the Deriche and Lanser filters are more precise than the Canny filter. As for the Canny filter, the smoothing parameter α has no influence on the precision. In the discrete case, this is, unfortunately, no longer true because of the discretization of the filter. Here, less smoothing (larger values of α) leads to slightly worse pre-

分析。用 Canny 滤波器提取出的理想化梯度边缘的精确度被解析分析出来。如果用 a 代表真实的边缘幅度，用 σ_n^2 代表图像中的噪声方差，则边缘位置的方差 σ_e^2 可由等式（3.124）计算：

(3.124)

尽管此结果是对连续图像解析推导得到的，但它也适用于离散图像。在文献（Steger，1998b,a）中也以经验为主地验证了这个结果。注意下面情况是相当直观的：图像中的噪声越大，边缘定位的精确度越差；而且，边缘幅度越大，边缘的精确度越好。也要注意可能与我们直觉相反的是，提高平滑处理的程度不能增加精确度。这是因为由提高平滑程度而实现的噪声降低会恰好由平滑处理得到的更弱的边缘幅度所抵消。从等式（3.124），能看到当信噪比 $a^2/\sigma_n^2 \geqslant 3/2$ 时，Canny 滤波器是亚像素精度（$\sigma_e \leqslant 1/2$）的。这个条件当然可以在实际应用中轻松满足。所以，可看到由此证明了调用 Canny 滤波器是亚像素精度的。

同样的推导也可用于 Deriche 滤波器和 Lanser 滤波器。对于连续图像，将返回如下方差：

(3.125)

注意 Deriche 滤波器和 Lanser 滤波器比 Canny 滤波器更精确。与 Canny 滤波器类似，平滑参数 α 对精确度没有影响。不幸的是，在离散情况下，此规律就不适用了，这是由滤波器的离散化造成的。这样，较少的平滑（α 值更大）得到结果的精确度比由

cision than predicted by Eq. (3.125). However, for practical purposes, we can assume that the smoothing for all the edge filters that we have discussed has no influence on the precision of the edges. Consequently, if we want to control the precision of the edges, we must maximize the SNR by using suitable lighting and cameras. In particular, digital cameras should be used. If, for some reason, analog cameras must be used, the frame grabber should have a line jitter that is as small as possible.

For ideal step edges, it is also easy to convince oneself that the expected position of the edge under noise corresponds to its true position. This happens because both the ideal step edge and the above filters are symmetric with respect to the true edge positions. Therefore, the edges that are extracted from noisy ideal step edges must be distributed symmetrically around the true edge position. Consequently, their mean value is the true edge position. This is also verified empirically for the Canny filter in Steger (1998b,a). It can also be verified for the Deriche and Lanser filters.

3.7.4.3 Edge Accuracy and Precision on Real Images

While it is easy to show that edges are very accurate for ideal step edges, we must also perform experiments on real images to test the accuracy on real data. This is important because some of the assumptions that are used in the edge extraction algorithms may not hold in practice. Because these assumptions are seldom stated explicitly, we should examine them carefully here. Let us focus on straight edges because, as we have seen from the discussion in Section 3.7.1.2, especially

等式（3.125）计算得稍微差些。但为了实用，假设讨论的所有对边缘滤波器的平滑对边缘的精确度都没有影响。因此，如果想控制边缘的精确度，必须使用正确的照明、摄像机、图像采集卡来使信噪比最大化。而且，如果使用的是模拟摄像机，图像采集卡的行抖动应该尽可能小。

对于理想化梯度边缘，人们可以相信噪声下边缘的期望位置对应的是其真实位置。这是因为理想化梯度边缘和滤波器都是关于真实边缘位置对称的。所以，从带有噪声的理想化梯度边缘提取出的边缘一定对称分布在真实边缘位置周围。因此，它们的平均值就是真实边缘位置。对于 Canny 滤波器，这是被以经验为主验证过的（Steger，1998b,a），对于 Deriche 滤波器和 Lanser 滤波器这点也是可以被验证的。

3.7.4.3 实际图像中的边缘准确度和精确度

尽管很容易证明对于理想化梯度边缘，边缘提取是准确的。我们还必须在实际的图像上进行实验来测试对于实际数据的准确度。这一点很重要，因为在边缘提取算法中使用的一些假设条件可能不适用于实际情况。因为这些假设很少被明确地表述出来，所以此时应该仔细地进行检查。关注直线边缘，因为如在 3.7.1.2 节的讨论中，特别是图 3.66 中看到的，锐利的

Figure 3.66, sharply curved edges will necessarily lie in incorrect positions. See also Berzins (1984) for a thorough discussion on the positional errors of the Laplacian edge detector for ideal corners of two straight edges with varying angles. Because we are concentrating on straight edges, we can reduce the edge detection to the 1D case, which is simpler to analyze. From Section 3.7.1.1 we know that 1D edges are given by the inflection points of the gray value profiles. This implicitly assumes that the gray value profile, and consequently its derivatives, are symmetric with respect to the true edge. Furthermore, to obtain subpixel positions, the edge detection implicitly assumes that the gray values at the edge change smoothly and continuously as the edge moves in subpixel increments through a pixel. For example, if an edge covers 25% of a pixel, we would assume that the gray value in the pixel is a mixture of 25% of the foreground gray value and 75% of the background gray value. We will see whether these assumptions hold in real images below.

To test the accuracy of the edge extraction on real images, it is instructive to repeat the experiments in Steger (1998b,a) with a different camera. In Steger (1998b,a), a print of an edge is mounted on an xy-stage and shifted in 50 µm increments, which corresponds to approximately 1/10 of a pixel, for a total of 1 mm. The goals are to determine whether the shifts of 1/10 of a pixel can be detected reliably and to obtain information about the absolute accuracy of the edges. Figure 3.86(a) shows an image used in this experiment. We are not going to repeat the test to see whether the subpixel shifts can be detected reliably here. The 1/10 pixel shifts can be detected with a very high

曲线边缘会不可避免地位于错误的位置。也可参见文献（Berzins，1984），其中详尽讨论了位置误差，此位置误差是用拉普拉斯算子边缘检测器对两直线组成的不同角度的成角进行提取后得到的。因为我们把注意力集中在直线边缘上，所以能将边缘检测简化到一维的情况，这样分析起来更容易。从 3.7.1.1 节中可知，一维边缘是由灰度值剖面的拐点确定的，这实际上是建立在一个默认假设基础上的，即灰度值剖面及其导数都相对于真实边缘是对称的。而且，为得到亚像素位置，边缘检测的默认假设是，当边缘在一个像素内以亚像素增量移动时，边缘上的灰度值变化是平滑的和连续的。例如，如果某边缘覆盖了一个像素的 25%，我们将假设在此像素上的灰度值是 25% 的前景灰度值和 75% 的背景灰度值混合成的。下面将讨论此假设前提是否适用于真实图像。

为了测试在真实图像上边缘提取的准确度，使用一个不同的摄像机重复（Steger，1998b,a）中的实验是有帮助的。在文献（Steger，1998b,a）中，一个打印出来的边缘被固定在一个 xy 双向移动平台上，它在平台上能以 50 µm 的步长移动，最多移动 1 mm，50 µm 大致对应为 1/10 像素。实验的目的是确定能否可靠地检测出这些 1/10 像素的位移，得到边缘绝对准确度的信息。图 3.86（a）显示的是此实验中的一幅图像。这里不打算重复此测试以观察亚像素的位移能否被可靠地检测出来。这些 1/10 像素

confidence (more than 99.999 99%). What is more interesting is to look at the absolute accuracy. Since we do not know the true edge position, we must get an estimate for it. Because the edge was shifted in linear increments in the test images, such an estimate can be obtained by fitting a straight line through the extracted edge positions and subtracting the line from the measured edge positions.

的位移能被以非常高的置信度（超过99.999 99%）检出。观察绝对准确度就更有趣了。由于不知道真实边缘位置在哪，所以必须对其进行估计。因为在测试图像中是线性增量的移动，所以通过将提取出的边缘位置拟合成直线，然后从测到的边缘中扣除这条线就完成了对真实边缘位置的估计。

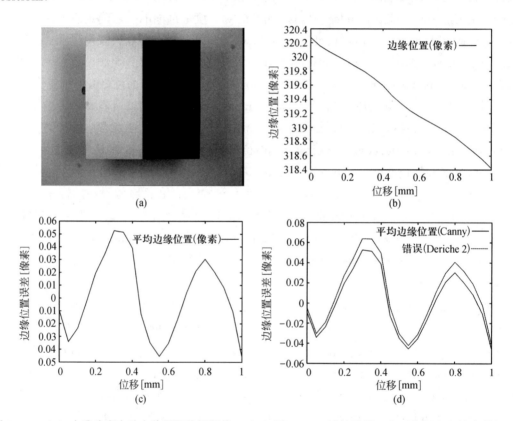

图 3.86　（a）在准确度实验中使用的边缘图像；（b）用 Canny 滤波器沿一水平线提取出的边缘位置。此边缘位置是用函数表示的，纵轴用像素表示的边缘位置，横轴是以毫米为单位的真实位移值；（c）边缘位置的误差，误差是通过将（b）中边缘位置拟合成一直线，然后用（b）中值减去拟合线上的值得到的；（d）用 Canny 滤波器和第二种 Deriche 滤波器处理后的误差比较图

Figure 3.86(b) displays the result of extracting the edge in Figure 3.86(a) along a horizontal line with a Canny filter with $\sigma = 1$. The edge position error is shown in Figure 3.86(c). We can see that there are errors of up to $\approx 1/22$ pixel.

图 3.86（b）显示的是在图 3.86（a）中沿一水平线使用 $\sigma = 1$ 的 Canny 滤波器进行边缘提取的结果。边缘位置的误差见图 3.86（c）。我们能看到的最大误差大约是 1/22 像素。

What causes these errors? As we discussed above, for ideal cameras, no error occurs, so one of the assumptions must be violated. In this case, the assumption that is violated is that the gray value is a mixture of the foreground and background gray values that is proportional to the area of the pixel covered by the object. This happens because the camera did not have a fill factor of 100%, i.e., the light-sensitive area of a pixel on the sensor was much smaller than the total area of the pixel. Consider what happens when the edge moves across the pixel and the image is perfectly focused. In the light-sensitive area of the pixel, the gray value changes as expected when the edge moves across the pixel because the sensor integrates the incoming light. However, when the edge enters the light-insensitive area, the gray value no longer changes (Lenz and Fritsch, 1990). Consequently, the edge does not move in the image. In the real image, the focus is not perfect. Hence, the light is spread slightly over adjacent sensor elements. Therefore, the edges do not jump as they would in a perfectly focused image, but shift continuously. Nevertheless, the poor fill factor causes errors in the edge positions. This can be seen very clearly from Figure 3.86(c). Recall that a shift of 50 μm corresponds to 1/10 of a pixel. Consequently, the entire shift of 1 mm corresponds to two pixels. This is why we see a sine wave with two periods in Figure 3.86(c). Each period corresponds exactly to one pixel. That these effects are caused by the fill factor can also be seen if the lens is defocused. In this case, the light is spread over more sensor elements. This helps to create an artificially increased fill factor, which causes smaller errors.

是什么造成这些误差呢？如上面讨论的那样，对理想摄像机是没有误差发生的，所以上面假设的前提中至少有一个被违背了。本例中，灰度值是由前景灰度值和背景灰度值按物体所覆盖的面积成比例混合而成的这一假设是不满足的。这是因为摄像机并不具备 100% 的填充因子，也就是说，摄像机传感器上的一个像素的感光区面积比此像素的全部区域面积更小。思考一下当图像聚焦极佳且边缘穿过某个像素时会发生什么。当边缘在像素的感光区域内时，边缘穿过此区域时灰度值的变化与期望的一样，这是因为传感器能对输入的光进行积分。但是，当此边缘进入到像素的不感光区域时，灰度值就不再变化了（Lenz and Fritsch, 1990）。因此，边缘就不在图像中移动了。对真实图像，聚焦不是极佳的。因此，光会稍微覆盖到传感器多个相邻像素上。所以，边缘不会像它在聚焦极佳的图像上那样跳跃，而是连续地平移。不过，差的填充因子引起边缘位置的误差。这从图 3.86（c）中可清楚看到。回想一下 50 μm 位移对应的是 1/10 像素。所以，移动全部 1 mm 距离就对应的是 2 个像素。这就是为什么图 3.86（c）的正弦波包含两个周期。每个周期恰好对应一个像素。由于填充因子引发的此效应也可以在镜头聚焦不清楚时看到。在此情况下，光在传感器上会分散覆盖多个像素单元。这可以帮助建立一种人工增加后的填充因子，这会导致较小的误差。

From the above discussion, it would appear that the edge position can be extracted with an accuracy of 1/22 of a pixel. To check whether this is true, let us repeat the experiment with the second Deriche filter. Figure 3.86(d) shows the result of extracting the edges with $\alpha = 1$ and computing the errors with the line fitted through the Canny edge positions. The last part is done to make the errors comparable. We can see, surprisingly, that the Deriche edge positions are systematically shifted in one direction. Does this mean that the Deriche filter is less accurate than the Canny filter? Of course, it does not, since on ideal data both filters return the same result. It shows that another assumption must be violated. In this case, it is the assumption that the edge profile is symmetric with respect to the true edge position. This is the only reason why the two filters, which are symmetric themselves, can return different results.

There are many reasons why edge profiles may become asymmetric. One reason is that the gray value responses of the camera and image acquisition device are nonlinear. Figure 3.87 illustrates that an originally symmetric edge profile becomes asymmetric by a nonlinear gray value response function. It can be seen that the edge position accuracy is severely degraded by the nonlinear response. To correct the nonlinear response of the camera, it must be calibrated radiometrically using the methods described in Section 3.2.2.

Unfortunately, even if the camera has a linear response or is calibrated radiometrically, other factors may cause the edge profiles to become asymmetric. In particular, lens aberrations like coma, astigmatism, and chromatic aberrations

从上述讨论中可看到，边缘被提取出来的准确度是 1/22 像素。为检测此结论是否正确，让我们用第二种 Deriche 滤波器来重复实验。图 3.86 (d) 显示的是先在 $\alpha = 1$ 时进行提取边缘，然后用 Canny 边缘位置来拟合一直线以用于计算误差。后面一步的目的是使误差可以比较。可看到令人惊奇的结果，Deriche 边缘位置在一个方向上系统地偏移。这是否意味着 Deriche 滤波器比 Canny 滤波器的准确度低呢？当然不是，因为对于理想数据时两种滤波器返回的结果一模一样。这表明另一个假设前提也一定被违背了。这个假设是：边缘剖面是相对于真实边缘位置而对称的。这是两个自身都是对称的滤波器返回不同结果的唯一原因。

导致边缘剖面变为非对称的原因有很多。一个原因是摄像机和图像采集设备的灰度值响应是非线性的。图 3.87 显示的是由于一个非线性灰度值响应函数使原本关于原点对称的边缘剖面变成非对称的。能看到由于非线性的响应，边缘位置的准确度严重地下降。为修正摄像机的非线性响应，必须使用在 3.2.2 节中介绍的方法进行辐射标定。

不幸的是，就算是摄像机具备线性响应或者对其进行了辐射标定，其他因素也可能造成边缘剖面变为非对称的。特别是如彗差，像散以及色差等镜头的像差可能会导致非对称的剖

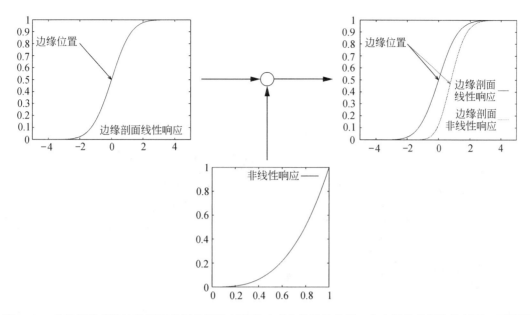

图 3.87 对理想的对称边缘剖面应用非线性灰度值响应曲线后的结果。左上图是理想边缘剖面，下图是非线性响应。右上图显示的是调整前和调整后的灰度值剖面。注意非线性响应严重影响了边缘位置

may cause asymmetric profiles (see Section 2.2.6). Since lens aberrations cannot be corrected easily with image processing algorithms, they should be as small as possible.

While all the error sources discussed above influence the edge accuracy, we have so far neglected the largest source of errors. If the camera is not calibrated geometrically, extracting edges with subpixel accuracy is pointless because the lens distortions alone are sufficient to render any subpixel position meaningless. Let us, for example, assume that the lens has a distortion that is smaller than 1% in the entire field of view. At the corners of the image, this means that the edges are offset by 4 pixels for a 640×480 image. We can see that extracting edges with subpixel accuracy is an exercise in futility if the lens distortions are not corrected, even for this relatively small distortion. This is illustrated in Figure 3.88, where the result of correcting the lens distortions after calibrating

面（见 2.2.6 节）。由于镜头像差不容易通过图像处理算法进行修正，所以应该让它们尽可能小。

到目前为止，在上面已经讨论过的影响边缘准确度的所有误差来源中，忽略了一个最大的误差来源。如果摄像机没有进行几何标定，以亚像素准确度提取边缘是无意义的，因为仅镜头畸变就可以宣布任何亚像素位置是无意义的。举例说明，假设在整个视野内某镜头的畸变小于 1%，这意味着对于一幅 640×480 图像，边缘偏移 4 像素。所以能够看到，就算是相对而言如此小的镜头畸变，如果不被修正，那么以亚像素准确度来提取边缘就是无效劳动，见图 3.88 中的例子，图中显示了校正镜头畸变后进行边缘提取的结果，修正是通过标定摄像机完成的，方法描述见 3.9 节。

the camera as described in Section 3.9 is shown. Note that, despite the fact that the application used a very high-quality lens, the lens distortions cause an error of approximately 2 pixels.

注意除了在实际应用中使用极高质量镜头的情况外，由镜头畸变引发的误差大约是 2 像素。

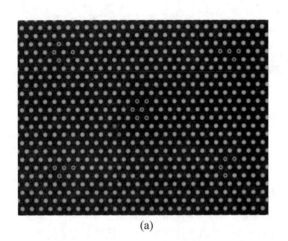
(a)

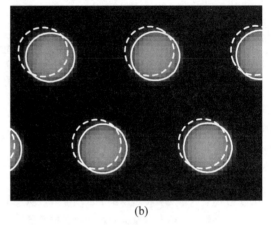
(b)

图 3.88　(a) 标定板的图像；(b) 提取出的亚像素准确度边缘 (实线) 和修正镜头畸变后提取的边缘 (虚线)。注意镜头畸变引发的误差约是 2 像素

Another detrimental influence on the accuracy of the extracted edges is caused by the perspective distortions in the image. They happen whenever we cannot mount the camera perpendicular to the objects we want to measure. Figure 3.89(a) shows the result of extracting the 1D edges along the ruler markings on a caliper. Because of the severe perspective distortions, the distances between the ruler markings vary greatly throughout the image. If the camera is calibrated, i.e., its interior orientation and the exterior orientation of the plane in which the objects to be measured lie have been determined with the approach described in Section 3.9, the measurements in the image can be converted into measurements in world coordinates in the plane determined by the calibration. This is done by intersecting the optical ray that corresponds to each edge point in

对边缘提取准确度的另一个有害的影响是由图像的透视失真造成的。任何时候如果在安装摄像机时不能保证其垂直于被测物体，那么都会引发透视失真。图 3.89 (a) 显示的是提取卡尺刻线上的一维边缘的结果。由于严重的透视失真，图中刻线间的距离变化的非常严重。如果对摄像机进行了标定，即按照 3.9 节中介绍的方法确定了摄像机内方位参数以及被测物所在平面的外方位参数，那么通过此标定就能将在图像中得到的测量结果转换成在世界坐标上的测量结果。这是通过将图像中每个边缘点所对应的光学射线与世界平面相交结得到的。图 3.89 (b) 显示的是用此方法将图 3.89 (a) 中的图像测量转换成以毫米表示的测量值后的结果。注意尽管

the image with the plane in the world. Figure 3.89(b) displays the results of converting the measurements in Figure 3.89(a) into millimeters with this approach—note that the measurements are extremely accurate even in the presence of severe perspective distortions.

存在着很严重的透视失真，但测量结果仍然很准确。

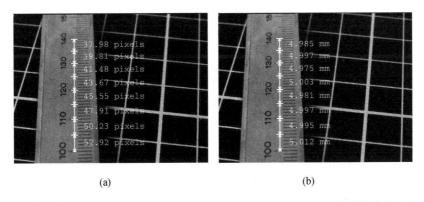

(a)   (b)

图 3.89 提取卡尺上刻线的一维边缘：（a）刻度间的像素距离；（b）应用摄像机标定后将刻线间距离转换为用毫米表示

From the above discussion, we can see that extracting edges with subpixel accuracy relies on careful selection of the hardware components. First of all, the gray value responses of the camera and image acquisition device should be linear. To ensure this, the camera should be calibrated radiometrically. Furthermore, lenses with very small aberrations (such as coma and astigmatism) should be chosen. Also, monochromatic light should be used to avoid the effects of chromatic aberrations. In addition, the fill factor of the camera should be as large as possible to avoid the effects of "blind spots." Finally, the camera should be calibrated geometrically to obtain meaningful results. All these requirements for the hardware components are, of course, also valid for other subpixel algorithms, e.g., subpixel-precise thresholding (see Section 3.4.3), gray value moments (see Sections 3.5.2.2–3.5.2.4), and contour features (see Section 3.5.3).

从上面的讨论中可看到，以亚像素准确度提取边缘依赖于对图像采集设备的精心挑选。首先，摄像机和图像采集设备的灰度值响应应该是线性的。为保证这个前提，应该对摄像机进行辐射标定。此外，被选用的镜头其像差（如彗差和像散）应该非常小。还应该使用单色光以避免色差的影响。另外，摄像机的填充因子应该尽可能的大以避免"盲点"的影响。最后，应该对摄像机进行几何标定以获取有意义的测量结果。所有这些针对图像采集硬件的要求当然也适用于其他亚像素算法，如亚像素精度阈值分割（见 3.4.3 节），灰度值矩（见 3.5.2.2~3.5.2.4 节），以及轮廓特征（见 3.5.3 节）。

3.8 Segmentation and Fitting of Geometric Primitives

In Sections 3.4 and 3.7.3, we have seen how to segment images by thresholding and edge extraction. In both cases, the boundary of objects either is returned explicitly or can be derived by some postprocessing (see Section 3.6.1.4). Therefore, for the purposes of this section, we can assume that the result of the segmentation is a contour containing the points of the boundary, which may be subpixel-accurate. This approach often creates an enormous amount of data. For example, the subpixel-accurate edge of the hole in the workpiece in Figure 3.82(d) contains 172 contour points. However, we are typically not interested in such a large amount of information. For example, in the application in Figure 3.82(d), we would probably be content with knowing the position and radius of the hole, which can be described with just three parameters. Therefore, in this section we will discuss methods to fit geometric primitives to contour data. We will only examine the most relevant geometric primitives: lines, circles, and ellipses. Furthermore, we will examine how contours can be segmented automatically into parts that correspond to the geometric primitives. This will enable us to substantially reduce the amount of data that needs to be processed, while also providing us with a symbolic description of the data. Furthermore, the fitting of the geometric primitives will enable us to reduce the influence of incorrectly or inaccurately extracted points (so-called outliers). We will start by examining the fitting of the geometric primitives in Sections 3.8.1–3.8.3. In each case, we will assume

3.8 几何基元的分割和拟合

在 3.4 节和 3.7.3 节中，已经看到了如何通过阈值分割和边缘提取来分割图像。使用此两种方法时，物体的边界都是明确返回得到或通过一些后处理推导得到（见 3.6.1.4 节）。所以，为了本节的目标，我们假设分割的结果就是由边界点组成的轮廓，此轮廓可以是亚像素-准确度的。这种方法常会生成大量的数据。例如，图 3.82（d）工件上的孔的亚像素 -准确度边缘包含了 172 个轮廓点。但是，我们通常并不感兴趣如此大数据量的信息。例如，在图 3.82（d）的应用中，可能仅满足于知道孔的半径和位置，而这些信息用三个参数就足以描述了。因此，本节将讨论将轮廓数据拟合成几何基元的方法。我们仅讨论最相关的几种几何基元：直线、圆和椭圆。此外，我们还将分析如何将轮廓自动分割成多个部分，每部分都有相对应的几何基元。这能充分地减小需要被处理的数据量，并提供一种数据的符号化描述方式。而且，对几何基元的拟合将使我们受不正确或不准确的提取点（也称为离群值）影响更少。我们将从 3.8.1 节~3.8.3 节中对几何基元拟合的分析开始。在每种情况的讨论中，将假设正在分析的轮廓或轮廓上的一部分与正要拟合成的基元相互对应，即，假定将轮廓分割成为不同基元的处理已经完成。而分割处理本身将在 3.8.4 节中讨论。

that the contour or part of the contour we are examining corresponds to the primitive we are trying to fit, i.e., we are assuming that the segmentation into the different primitives has already been performed. The segmentation itself will be discussed in Section 3.8.4.

3.8.1 Fitting Lines

If we want to fit lines, we first need to think about the representation of lines. In images, lines can occur in any orientation. Therefore, we must use a representation that enables us to represent all lines. For example, the common representation $y = mx + b$ does not allow us to do this since vertical lines cannot be represented. One representation that can be used is the Hessian normal form of the line, given by

$$\alpha r + \beta c + \gamma = 0 \qquad (3.126)$$

This is actually an over-parameterization, since the parameters (α, β, γ) are homogeneous (Hartley and Zisserman, 2003; Faugeras and Luong, 2001). Therefore, they are defined only up to a scale factor. The scale factor in the Hessian normal form is fixed by requiring that $\alpha^2 + \beta^2 = 1$. This has the advantage that the distance of a point to the line can simply be obtained by substituting its coordinates into Eq. (3.126).

3.8.1.1 Least-Squares Line Fitting

To fit a line through a set of points $(r_i, c_i)^\top$, $i = 1, \cdots, n$, we can minimize the sum of the squared distances of the points to the line:

3.8.1 直线拟合

在进行直线拟合前必须首先考虑直线的表示方法。在图像中，直线可以出现在任何方位。因此，不得不使用一种可以描述所有直线的表示方法。例如，由于无法表示纵向垂线，通常的表示法 $y = mx + b$ 不能满足要求。可用的一种表示法是直线的黑塞范式，表示为：

这实际上是一种过度参数化的表达，因为参数 (α, β, γ) 是齐次的（Hartley and Zisserman，2003；Faugeras and Luong，2001）。因此，这些参数仅需被定义到一个比例因子。此比例因子在黑塞范式中是通过令 $\alpha^2 + \beta^2 = 1$ 来固定下来的。这样做的好处是某点到直线的距离能通过将该点的坐标直接代入等式（3.126）得到。

3.8.1.1 最小二乘法直线拟合

为了从一系列点 $(r_i, c_i)^\top$, $i = 1, \cdots, n$ 中来拟合出一条直线，我们能对这些点到这条直线的距离的平方和进行最小化处理：

$$\varepsilon^2 = \sum_{i=1}^{n}(\alpha r_i + \beta c_i + \gamma)^2 \qquad (3.127)$$

While this is correct in principle, it does not work in practice, because we can achieve a zero error if we select $\alpha = \beta = \gamma = 0$. This is caused by the over-parameterization of the line. Therefore, we must add the constraint $\alpha^2 + \beta^2 = 1$ as a Lagrange multiplier, and hence must minimize the following error:

尽管等式（3.127）理论上是正确的，但实际中它是不能工作的，因为当 $\alpha = \beta = \gamma = 0$ 时，将得到一个零误差。这是由直线的过度参数化造成的。所以，必须加入约束条件 $\alpha^2 + \beta^2 = 1$ 作为拉格朗日乘子，从而必须将下式的误差最小化：

$$\varepsilon^2 = \sum_{i=1}^{n}(\alpha r_i + \beta c_i + \gamma)^2 - \lambda(\alpha^2 + \beta^2 - 1)n \tag{3.128}$$

The solution to this optimization problem is derived in Haralick and Shapiro (1992). It can be shown that $(\alpha, \beta)^\top$ is the eigenvector corresponding to the smaller eigenvalue of the following matrix:

解决此最优化问题的方法推导参见文献（Haralick and Shapiro, 1992）。可以看到 $(\alpha, \beta)^\top$ 是与如下矩阵的较小本征值相对应的本征向量：

$$\begin{pmatrix} \mu_{2,0} & \mu_{1,1} \\ \mu_{1,1} & \mu_{0,2} \end{pmatrix} \tag{3.129}$$

With this, γ is given by $\gamma = -(\alpha n_{1,0} + \beta n_{0,1})$. Here, $\mu_{2,0}$, $\mu_{1,1}$, and $\mu_{0,2}$ are the second-order central moments of the point set $(r_i, c_i)^\top$, while $n_{1,0}$ and $n_{0,1}$ are the normalized first-order moments (the center of gravity) of the point set. If we replace the area a of a region with the number n of points and sum over the points in the point set instead of the points in the region, the formulas to compute these moments are identical to the region moments of Eqs. (3.61) and (3.62) in Section 3.5.1.2. It is interesting to note that the vector $(\alpha, \beta)^\top$ thus obtained, which is the normal vector of the line, is the minor axis that would be obtained from the ellipse parameters of the point set. Consequently, the major axis of the ellipse is the direction of the line. This is a very interesting connection between the ellipse parameters and the line fitting, because the results were derived using different approaches and models.

此时，由 $\gamma = -(\alpha n_{1,0} + \beta n_{0,1})$ 得到。式中，$\mu_{2,0}$, $\mu_{1,1}$ 和 $\mu_{0,2}$ 都是点集 $(r_i, c_i)^\top$ 的二阶中心矩，而 $n_{1,0}$ 和 $n_{0,1}$ 是此点集归一化后的一阶矩（重心）。如果将 3.5.1.2 节中计算区域矩的等式（3.61）和等式（3.62）里的区域面积 a 换成点数 n，把对区域中点的连加求和换成对点集中点的连加求和，那么得到的公式就可以计算上述的这些矩。请注意令人感兴趣的一点是，用此方法得到的向量 $(\alpha, \beta)^\top$，它既是直线的法向量，也是从点集的椭圆参数中得到的短轴。因此，椭圆的长轴就是直线的方向。这是椭圆参数与拟合直线之间非常有趣的联系，因为这些结果是用不同的方法和模型推导得到的。

Figure 3.90(b) illustrates the line fitting procedure for an oblique edge of the workpiece shown in Figure 3.90(a). Note that, by fitting the line, we were able to reduce the effects of the small protrusion on the workpiece. As mentioned above, by inserting the coordinates of the edge points into the line equation (3.126), we can easily calculate the distances from the edge points to the line. Therefore, by thresholding the distances, the protrusion easily can be detected.

图 3.90（b）给出的是对图 3.90（a）中工件的斜边进行直线拟合的过程。注意通过直线拟合，我们能降低工件上小凸起的影响。如上面讲述过的，通过将边缘上的点的坐标代入到等式（3.126）中，我们可以很容易计算出边缘上的点到直线的距离。因此，通过对距离进行阈值分割，可以很容易检测到凸起。

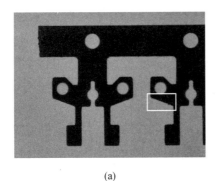

(a)   (b)

图 3.90　（a）一幅工件的图像，其中白色矩形指示部分的放大图像示于（b）；（b）对工件斜边的周围区域的边缘提取（虚线）及边缘拟合后的直线（实线）

As can be seen from the above example, the line fit is robust to small deviations from the assumed model (small outliers). However, Figure 3.91 shows that large outliers severely affect the quality of the fitted line. In this example, the line is fitted through the straight edge as well as the large arc caused by the relay contact. Since the line fit must minimize the sum of the squared distances of the contour points, the fitted line has a direction that deviates from that of the straight edge.

从上个例子我们可以看出，直线拟合对于假定模型上小的背离（小的离群值）而言是可靠的。但是，图 3.91 显示了大的离群值严重影响拟合后直线的质量。在这个例子里，拟合后的直线既穿过了直线边缘也穿过了由继电器触点造成的大圆弧。由于拟合直线必须满足轮廓上所有的点到此直线的平方距离的加和最小，所以拟合后直线的方向从直边缘上偏离了。

3.8.1.2　Robust Line Fitting

The least-squares line fit is not robust to large outliers since points that lie far from the line have

3.8.1.2　可靠的直线拟合

最小二乘法直线拟合对于大的离群值是不足够可靠的，由于采用的是

 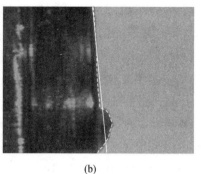

图 3.91　（a）一个继电器的图像，其中浅灰色矩形指示部分的放大图像示于（b）；（b）对继电器的垂直边缘的周围区域的边缘提取（虚线）及边缘拟合后的直线（实线）。为更清楚地显示出边缘和直线，图像（b）的对比度被降低了

a very large weight in the optimization because of the squared distances. To reduce the influence of distant points, we can introduce a weight w_i for each point. The weight should be $\ll 1$ for distant points.

Let us assume for the moment that we have a way to compute these weights. Then, the minimization becomes

平方距离，所以与直线距离远的那些点在最优化过程中将会拥有非常大的权重。为减轻这些远离点的影响，我们可以为每个点引入权重 w_i。对于远离线的那些点，其权重应该 $\ll 1$。

让我们先暂时假设已经知道了计算这些权重的方法。那么，最小化过程将变为：

$$\varepsilon^2 = \sum_{i=1}^{n} w_i(\alpha r_i + \beta c_i + \gamma)^2 - \lambda(\alpha^2 + \beta^2 - 1)n \tag{3.130}$$

The solution of this optimization problem is again given by the eigenvector corresponding to the smaller eigenvalue of a moment matrix like in Eq. (3.129) (Lanser, 1997). The only difference is that the moments are computed by taking the weights w_i into account. If we interpret the weights as gray values, the moments are identical to the gray value center of gravity and the second-order central gray value moments (see Eqs. (3.71) and (3.72) in Section 3.5.2.2). As above, the fitted line corresponds to the major axis of the ellipse obtained from the weighted moments of the point set. Hence, there is an interesting connection to the gray value moments.

此最优化问题的解仍然是由与等式（3.129）类似的矩量矩阵的更小本征值相对应的本征向量决定的（Lanser, 1997）。唯一的区别是在矩的计算时考虑了权重 w_i。如果我们将权重解释为灰度值，那么这些计算出的矩将等同于灰度值重心和灰度值二阶中心矩（见 3.5.2.2 节的等式（3.71）和等式（3.72））。如上，拟合后的直线与从点集的权重矩中计算得到的椭圆的长轴相对应。因此，这里存在着与灰度值矩的一个有趣联系。

The only remaining problem is how to define the weights w_i. Since we want to give smaller weights to points with large distances, the weights must be based on the distances $\delta_i = |\alpha r_i + \beta c_i + \gamma|$ of the points to the line. Unfortunately, we do not know the distances without fitting the line, so this seems an impossible requirement. The solution is to fit the line in several iterations. In the first iteration, $w_i = 1$ is used, i.e., a normal line fit is performed to calculate the distances δ_i. They are used to define weights for the following iterations by using a weight function $w(\delta)$. This method is called iteratively reweighted least-squares (IRLS) (Holland and Welsch, 1977; Stewart, 1999). In practice, often one of the following two weight functions is used. They both work very well. The first weight function was proposed by Huber (Huber, 1981; Lanser, 1997). It is given by

$$w(\delta) = \begin{cases} 1, & |\delta| \leqslant \tau \\ \tau/|\delta|, & |\delta| > \tau \end{cases} \tag{3.131}$$

The parameter τ is the clipping factor. It defines which points should be regarded as outliers. We will see how it is computed below. For now, note that all points with a distance $\leqslant \tau$ receive a weight of 1. This means that, for small distances, the squared distance is used in the minimization. Points with a distance $> \tau$, on the other hand, receive a progressively smaller weight. In fact, the weight function is chosen such that points with large distances use the distance itself and not the squared distance in the optimization. Sometimes, these weights are not small enough to suppress outliers completely. In this case, the Tukey weight function can be used (Lanser, 1997; Mosteller and Tukey, 1977). It is given by

仅剩的一个问题就是如何定义权重 w_i。因为我们想给距离远的那些点较小的权重，所以权重必须是基于点到线的距离 $\delta_i = |\alpha r_i + \beta c_i + \gamma|$。不幸的是，没有拟合出直线时我们并不能得到这些距离，所以这看起来是个不可能的要求。解决的方法是用多次的迭代来拟合直线。在第一次迭代中使用 $w_i = 1$，即执行一个标准的直线拟合来计算出距离 δ_i。通过一个权重函数 $w(\delta)$ 可用已经计算出的距离来定义权重，这些权重用于后续的迭代处理中，这个方法称为迭代重加权最小二乘法（IRLS）（Holland and Welsch, 1977; Stewart, 1999）。在实践中，可以使用以下两个权重函数中的任何一个。这两个权重函数的应用效果都很好。第一个权重函数是由 Huber 提出的（Huber, 1981; Lanser, 1997），其定义为：

参数 τ 是削波因数，它定义哪些点应被视为离群值。我们将在后面看到它是如何被计算出来的。现在，请记住所有距离 $\leqslant \tau$ 的点对应的权重都是 1。这意味着，对于小的距离的点，在极小化处理中就直接使用其平方距离。另一方面，对于距离 $> \tau$ 的点，将获得一个更小些的权重。事实上，在最优化计算中，此权重函数为那些距离远的点选定了其距离值而不是平方距离值参加运算。有时，选定的权重值不足够的小所以不能完全抑制离群值。这种情况下，可以使用 Tukey 权重函数（Lanser, 1997; Mosteller and Tukey, 1977），其定义为：

$$w(\delta) = \begin{cases} (1-(\delta/\tau)^2)^2, & |\delta| \leqslant \tau \\ 0, & |\delta| > \tau \end{cases} \quad (3.132)$$

Again, τ is the clipping factor. Note that this weight function completely disregards points that have a distance $> \tau$. For distances $\leqslant \tau$, the weight changes smoothly from 1 to 0.

In the above two weight functions, the clipping factor specifies which points should be regarded as outliers. Since the clipping factor is a distance, it could simply be set manually. However, this would ignore the distribution of the noise and the outliers in the data, and consequently would have to be adapted for each application. It is more convenient to derive the clipping factor from the data itself. This is typically done based on the standard deviation of the distances to the line. Since we expect outliers in the data, we cannot use the normal standard deviation, but must use a standard deviation that is robust to outliers. Typically, the following formula is used to compute the robust standard deviation:

$$\sigma_\delta = \frac{\text{median} |\delta_i|}{0.6745} \quad (3.133)$$

The constant in the denominator is chosen such that, for normally distributed distances, the standard deviation of the normal distribution is computed. The clipping factor is then set to a small multiple of σ_δ, e.g., $\tau = 2\sigma_\delta$.

In addition to the Huber and Tukey weight functions, other weight functions can be defined. Several other possibilities are discussed in Hartley and Zisserman (2003).

Figure 3.92 displays the result of fitting a line robustly to the edge of the relay using the Tukey

τ 在这里也是削波因数。注意此权重函数完全忽略那些距离 $> \tau$ 的点。而对于距离 $\leqslant \tau$ 的点，其权重值在 1 到 0 之间平滑变化。

在上面的两个权重函数中，削波因数指明哪些点应被视为离群值。因为削波因数代表的是一个距离，所以它可以被手动设置。但是，这将忽视噪声的分布和数据中的离群值，所以不得不针对每个应用来进行调整。从数据自身来推出削波因数会更方便。这通常可以基于到直线的这些距离值的标准偏差来实现。因为我们想得到的是数据中的离群值，所以我们不能使用正规标准偏差，而必须使用对于离群值是可靠的标准偏差。一般情况下，下式用来计算这个可靠的标准偏差：

对于正态分布的距离值，分母上的常量被选定为适用于基于正态分布计算的标准偏差。削波因数被设置为 σ_δ 的一个小倍数，如 $\tau = 2\sigma_\delta$。

除了 Huber 权重函数和 Tukey 权重函数外，还可以定义其他的权重函数。对这些其他可能性的讨论参见文献（Hartley and Zisserman，2003）。

图 3.92 显示的是使用 Tukey 权重函数在削波因数时，经过五次迭代

weightfunction with a clipping factor of $\tau = 2\sigma_\delta$ and five iterations. If we compare this with the standard least-squares line fit in Figure 3.91(b), we see that with the robust fit, the line is now fitted to the straight-line part of the edge, and the outliers caused by the relay contact have been suppressed.

对继电器边缘进行鲁棒地直线拟合后的结果。若我们将此结果与图 3.91（b）中最小二乘法直线拟合结果进行对比，我们能看到采用了可靠的拟合方法后，得到的直线与直线边缘部分重合得很好，而且由继电器触点引起的离群值也已经被抑制住了。

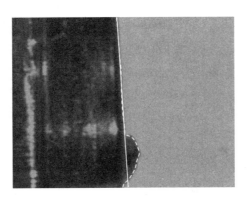

图 3.92　对垂直边缘（虚线）进行鲁棒地拟合后得到的直线（实线）。这里，使用 Tukey 权重函数在削波因数 $\tau = 2\sigma_\delta$ 时进行了五次迭代。与图 3.91（b）比较，这里拟合后得到的直线更好地反映了继电器的直线边缘部分

It should also be noted that the above approach to outlier suppression by weighting down the influence of points with large distances can sometimes fail because the initial fit, which is a standard least-squares fit, can produce a solution that is dominated by the outliers. Consequently, the weight function will drop inliers. In this case, other robust methods must be used. The most important approach is the random sample consensus (RANSAC) algorithm, proposed by Fischler and Bolles (1981). Instead of dropping outliers successively, it constructs a solution (e.g., a line fit) from the minimum number of points (e.g., two for lines), which are selected randomly, and then checks how many points are consistent with the solution. The process of randomly selecting points, constructing the solution, and checking the num-

还应该指出的是上述通过降低那些距离远的点的权重来对离群值进行抑制的方法有时会失败，这是因为迭代中的初始拟合仍然是标准的最小平方拟合，它给出的结果是能被离群值干扰的。因此，权重函数可能会丢掉正常值。此时，必须使用其他可靠的拟合方法，其中最重要的方法就是随机采样一致性（RANSAC）算法，此算法是由文献（Fischler and Bolles，1981）提出的。这个算法并不是采用相继丢弃离群值的处理，此方法是通过随机选择的最少数量的点（如两个点）来构造一个解（如拟合一条直线），然后检查存在多少与此解相一致的点。随机选择点、构造一个解、检测与此解一致的点数，由此三个步骤

ber of consistent points is continued until a certain probability of having found the correct solution, e.g., 99%, is achieved. At the end, the solution with the largest number of consistent points is selected.

3.8.2 Fitting Circles

3.8.2.1 Least-Squares Circle Fitting

Fitting circles or circular arcs to a contour uses the same idea as fitting lines: we want to minimize the sum of the squared distances of the contour points to the circle:

3.8.2 圆拟合

3.8.2.1 最小二乘法圆拟合

将轮廓拟合成圆或圆弧采用的是与直线拟合一样的思路：我们想先将轮廓上的所有点到拟合圆的平方距离进行连加求和，然后使求得的总和最小化。

$$\varepsilon^2 = \sum_{i=1}^{n} \left(\sqrt{(r_i - \alpha)^2 + (c_i - \beta)^2} - \rho \right)^2 \tag{3.134}$$

Here, $(\alpha, \beta)^\top$ is the center of the circle and ρ is its radius. Unlike line fitting, this leads to a nonlinear optimization problem, which can only be solved iteratively using nonlinear optimization techniques. Details can be found in (Haralick and Shapiro, 1992; Joseph, 1994; Ahn et al., 2001).

Figure 3.93(a) shows the result of fitting circles to the edges of the holes of a workpiece,

式中，$(\alpha, \beta)^\top$ 是圆心，ρ 是圆的半径。与直线拟合不同的是，这是一个非线性最优化问题，只能采用非线性最优化技术的迭代来解决。算法细节部分请参见文献（Haralick and Shapiro, 1992; Joseph, 1994; Ahn et al., 2001）。

图 3.93（a）给出的是将工件上孔的边缘拟合成圆的结果，图中还显示

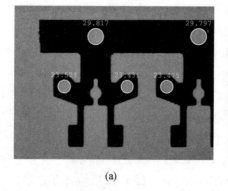
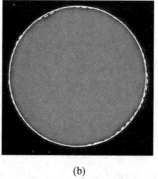

(a)　　　　　　　　(b)

图 3.93 （a）将工件上孔的边缘拟合成圆；（b）右上角孔的细节，提取出的边缘（虚线）和拟合后的圆（实线）

along with the extracted radii in pixels. In Figure 3.93(b), details of the upper right hole are shown. Note how well the circle fits the extracted edges.

3.8.2.2 Robust Circle Fitting

Like the least-squares line fit, the least-squares circle fit is not robust to outliers. To make the circle fit robust, we can use the same approach that we used for line fitting: we can introduce a weight that is used to reduce the influence of the outliers. Again, this requires that we perform a normal least-squares fit first and then use the distances that result from it to calculate the weights in later iterations. Since it is possible that large outliers can prevent this algorithm from converging to the correct solution, a RANSAC approach might be necessary in extreme cases.

Figure 3.94 compares standard circle fitting with robust circle fitting using the BGA example of Figure 3.37. With standard fitting (Figure 3.94(b)), the circle is affected by the error in the pad, which acts like an outlier. This is corrected with the robust fitting (Figure 3.94(c)).

To conclude this section, we should give some thought to what happens when a circle is fitted to a contour that only represents a part of a circle (a circular arc). In this case, the accuracy of the parameters becomes progressively worse as the angle of the circular arc becomes smaller. An excellent analysis of this effect is given by Joseph (1994). This effect is obvious from the geometry of the problem. Simply think about a contour that only represents a 5° arc. If the contour points are

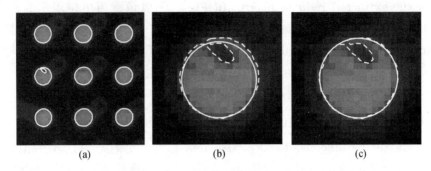

图 3.94 （a）某 BGA 的图像，用亚像素精度阈值分割提取出了焊盘（见图 3.37）；（b）对（a）中中间一行最左边的焊盘进行拟合后得到的圆。拟合后的圆用实线显示，提取出的轮廓用虚线显示。焊盘中的缺陷类似于离群值，对拟合圆产生了影响；（c）使用可靠的拟合算法得到的圆。拟合后的圆对应着焊盘的真实边界

disturbed by noise, we have a very large range of radii and centers that lead to almost the same fitting error. On the other hand, if we fit to a complete circle, the geometry of the circle is much more constrained.

都可得到同样的拟合误差。另一方面，如果我们对完整的圆进行拟合，那么此圆的几何形状将为拟合处理提供更多的限制。上述的问题是由于拟合问题的几何学特性引起的，并不是由某特定的拟合算法造成的，也就是说，这个问题会出现在所有拟合算法上。

3.8.3 Fitting Ellipses

3.8.3.1 Least-Squares Ellipse Fitting

To fit an ellipse to a contour, we would like to use the same principles as for lines and circles: i.e., minimize the distance of the contour points to the ellipse. This requires us to determine the closest point to each contour point on the ellipse. While this can be determined easily for lines and circles, for ellipses it requires finding the roots of a fourth-degree polynomial. Since this is quite complicated and expensive, ellipses are often fitted by minimizing a different kind of distance. The principle is similar to the line fitting approach: we write down an implicit equation for ellipses (for lines, the implicit equation is given by Eq. (3.126)), and then

3.8.3 椭圆拟合

3.8.3.1 最小二乘法椭圆拟合

为了将椭圆拟合到轮廓上，我们将使用与直线拟合和圆拟合同样的原理：即将轮廓点到此椭圆的距离最小化。这要求我们在椭圆上确定与每个轮廓点最接近的点。对于直线和圆，这些点是很容易被确定的，对于椭圆，这需要找到四次多项式的根。因为此法相当复杂且耗时，通常采用对一不同类型的距离进行最小化来完成椭圆的拟合。这个原理与直线拟合的方法类似：我们为椭圆写下隐式方程（针对直线的隐式方程由等式（3.126）给出），然后将点的坐标代入此隐式方

substitute the point coordinates into the implicit equation to get a distance measure for the points to the ellipse. For the line fitting problem, this procedure returns the true distance to the line. For ellipse fitting, it only returns a value that has the same properties as a distance, but is not the true distance. Therefore, this distance is called the algebraic distance. Ellipses are described by the following implicit equation:

$$\alpha r^2 + \beta rc + \gamma c^2 + \delta r + \zeta c + \eta = 0 \qquad (3.135)$$

As for lines, the set of parameters is a homogeneous quantity, i.e., only defined up to scale. Furthermore, Eq. (3.135) also describes hyperbolas and parabolas. Ellipses require $\beta^2 - 4\alpha\gamma < 0$. We can solve both problems by requiring $\beta^2 - 4\alpha\gamma = -1$. An elegant solution to fitting ellipses by minimizing the algebraic error with a linear method was proposed by Fitzgibbon. The interested reader is referred to Fitzgibbon *et al.* (1999) for details. Unfortunately, minimizing the algebraic error can result in biased ellipse parameters. Therefore, if the ellipse parameters are to be determined with maximum accuracy, the geometric error should be used. A nonlinear approach for fitting ellipses based on the geometric error is proposed by Ahn *et al.* (2001). It is significantly more complicated than the linear approach by Fitzgibbon *et al.* (1999).

3.8.3.2 Robust Ellipse Fitting

Like the least-squares line and circle fits, fitting ellipses via the algebraic or geometric distance is not robust to outliers. We can again introduce weights to create a robust fitting procedure. If the ellipses are fitted with the algebraic distance, this again results in a linear algorithm in each iteration of the robust fit (Lanser, 1997). In applications with a very large number of outliers or with very large outliers, a RANSAC approach might be necessary.

Ellipse fitting is very useful in camera calibration, where circular marks often are used on the calibration targets (see Section 3.9; Lenz and Fritsch, 1990; Lanser *et al.*, 1995; Heikkila, 2000; Steger, 2017). Since circles project to ellipses, fitting ellipses to the edges in the image is the natural first step in the calibration process. Figure 3.95(a) displays a part of an image of a calibration target. The ellipses fitted to the extracted edges of the calibration marks are shown in Figure 3.95(b). In Figure 3.95(c), a detailed view of the center mark with the fitted ellipse is shown. Since the subpixel edge extraction is very accurate, there is hardly any visible difference between the edge and the ellipse, and therefore the edge is not shown in the figure.

在进行拟合时的每次迭代处理中都可以采用线性算法（Lanser，1997）。对于离群值非常多或者离群值非常大的应用，可能需要 RANSAC 算法。

椭圆拟合在摄像机标定中非常有用，因为标定板上经常使用圆形标记点（见 3.9 节和 Lenz and Fritsch, 1990；Lanser *et al.*, 1995；Heikkila, 2000；Steger, 2017）。由于圆形会投射成椭圆，所以拟合椭圆到图像中的标记点边缘上自然成为了标定过程的第一步。图 3.95（a）显示了一幅标定板的图像。将椭圆拟合到提取出的标定点的边缘上之后的图像见图 3.95（b）。图 3.95（c）中显示的是中心标记点的细节及由其拟合出的椭圆。由于亚像素边缘提取的结果非常准确，因此很难看出提取的边缘与拟合得到的椭圆之间的差别，所以边缘也就没有在图中显示出来。

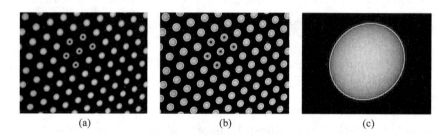

图 3.95　（a）一幅定标板的图像；（b）拟合椭圆到标定板上提取出的圆形标志点的边缘上；（c）标定板上中心标志点的细节及拟合后的椭圆

To conclude this section, we should note that, if ellipses are fitted to contours that only represent a part of an ellipse, the same comments that were made for circular arcs at the end of the last section apply: the accuracy of the parameters will become worse as the angle that the arc subtends becomes smaller.

在本节结束部分我们应注意，只将代表一部分椭圆的轮廓拟合成椭圆时，在上节中讨论拟合圆弧时做出的结论仍然适用：弧所包含角度越小，拟合得到参数的准确度越差。产生此现象的原因来自于拟合问题的几何学方面，而非源于我们使用的拟合算法。

3.8.4 Segmentation of Contours

So far, we have assumed that the contours to which we are fitting the geometric primitives correspond to a single primitive of the correct type, e.g., a line segment. Of course, a single contour may correspond to multiple primitives of different types. Therefore, in this section we will discuss how contours can be segmented into the different primitives.

3.8.4.1 Segmentation of Contours into Lines

We will start by examining how a contour can be segmented into lines. To do so, we would like to find a polygon that approximates the contour sufficiently well. Let us call the contour points $p_i = (r_i, c_i)^\top$, for $i = 1, \ldots, n$. Approximating the contour by a polygon means that we want to find a subset p_{i_j}, for $j = 1, \cdots, m$ with $m \leqslant n$, of the control points of the contour that describes the contour reasonably well. Once we have found the approximating polygon, each line segment $(p_{i_j}, p_{i_{j+1}})$ of the polygon is a part of the contour that can be approximated well with a line. Hence, we can fit lines to each line segment afterward to obtain a very accurate geometric representation of the line segments.

The question we need to answer is this: How do we define whether a polygon approximates the contour sufficiently well? A large number of different definitions have been proposed over the years. A very good evaluation of many polygonal approximation methods has been carried out by Rosin (1997, 2003). In both cases, it was established that

3.8.4 轮廓分割

到目前为止，我们都假设进行几何基元拟合的轮廓各自对应的几何基元类型都是正确的，比如将直线拟合到一个线段上。当然，某一轮廓可能对应多种不同类型的基元。所以，本节中我们将讨论如何将轮廓分割成不同的基元。

3.8.4.1 轮廓分割成直线

我们从分析怎样将轮廓分割成直线开始。为将轮廓分割成直线，我们希望找到一个多边形，此多边形可以充分地近似该轮廓。将轮廓点称为 $p_i = (r_i, c_i)^\top$ $i = 1, \ldots, n$。用一多边形对轮廓进行近似意味着我们想找到轮廓上控制点的一个子集 p_{i_j}, $j = 1, \cdots, m, m \leqslant n$ 此子集可以非常好地描述该轮廓。一旦我们发现了近似的多边形，则此多边形的每条线段 $(p_{i_j}, p_{i_{j+1}})$ 就是轮廓中可以用直线很好地近似的一部分。所以，我们就能拟合直线到每个线段上，然后就能得到这些线段的非常准确的几何表示。

我们必须回答的问题是，怎样定义一个多边形对轮廓近似得是否非常好。在这些年里，已经有非常多的不同定义被提出来了。文献（Rosin，1997；2003）对许多多边形逼近方法进行了非常好的评估。在所有逼近算法中，文献（Ramer，1972）提出的方

the algorithm proposed by Ramer (1972), which curiously enough is one of the oldest algorithms, is the best overall method.

The Ramer algorithm performs a recursive subdivision of the contour until the resulting line segments have a maximum distance to the respective contour segments that is lower than a user-specified threshold $d_{\max}$. Figure 3.96 illustrates how the Ramer algorithm works. We start out by constructing a single line segment between the first and last contour points. If the contour is closed, we construct two segments: one from the first point to the point with index $n/2$, and the second one from $n/2$ to n. We then compute the distances of all the contour points to the line segment and find the point with the maximum distance to the line segment. If its distance is larger than the threshold we have specified, we subdivide the line segment into two segments at the point with the maximum distance. Then, this procedure is applied recursively to the new segments until no more subdivisions occur, i.e., until all segments fulfill the maximum distance criterion.

法在所有方法中最好，最奇怪的是此算法也是最早提出的几个逼近算法中的一个。

Ramer 算法对轮廓进行递归细分，直到得到的全部线段到各自对应的轮廓段的最大距离小于某一用户指定的阈值 $d_{\max}$ 为止。图 3.96 阐明了 Ramer 算法是如何工作的。开始时我们在轮廓的起点和终点间建立一条线段。如果轮廓是闭合的，我们建立两条线段：一条是从第一点到索引为 $n/2$ 的点，另一条是从 $n/2$ 到 n。然后我们计算所有轮廓点到线段的距离并找到与线段距离最大的那个轮廓点。如果此距离比我们指定的阈值要大，那么在具有最大距离的轮廓点处，我们再将当前线段分成两条线段。然后在新得到的线段上重复进行此处理直到不能再细分为止，即直到所有线段都满足最大距离约束条件。

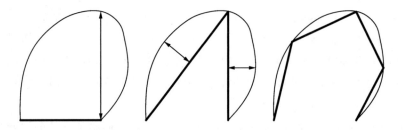

图 3.96　Ramer 算法中使用递归细分的例子。细线给出的是轮廓，逼近多边形用粗线表示

Figure 3.97 illustrates the use of the polygonal approximation in a real application. In Figure 3.97(a), a back-lit cutting tool is shown. In the application, the dimensions and angles of the cutting tool must be inspected. Since the

图 3.97 给出的是在实际应用中使用多边形逼近法的例子。在图 3.97（a）中显示的是一个用背光照射的切削工具。此例中，必须检测此切削工具的尺寸和角度。因为此工具中包含

3.8 Segmentation and Fitting of Geometric Primitives 343

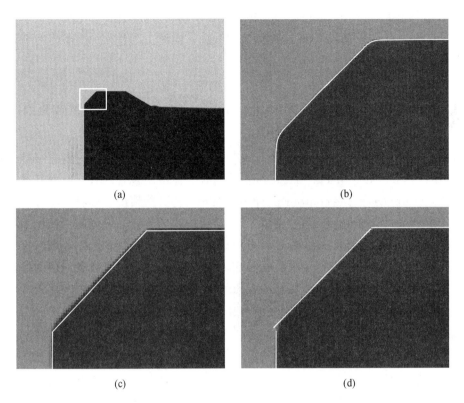

图 3.97 （a）用背光照射的切削工具图像；（b）～（d）中显示的部分用白色矩形标出。为了更好地显示结果，（b）～（d）的图像对比度被降低了；（b）用 $\alpha = 0.7$ 的 Lanser 滤波器进行亚像素-准确度的边缘提取；（c）用 $d_{\max} = 2$ 的 Ramer 算法提取到的多边形；（d）用 Tukey 权重函数以高鲁棒性的算法拟合直线到多边形线段上

tool consists of straight edges, the obvious approach is to extract edges with subpixel accuracy (Figure 3.97(b)) and to approximate them with a polygon using the Ramer algorithm. From Figure 3.97(c) we can see that the Ramer algorithm splits the edges correctly. We can also see the only slight drawback of the Ramer algorithm: it sometimes places the polygon control points into positions that are slightly offset from the true corners. In this application, this poses no problem, since to achieve maximum accuracy we must fit lines to the contour segments robustly anyway (Figure 3.97(d)). This enables us to obtain a concise and very accurate geometric description of the

直线边缘，所以最明显的处理方法是先以亚像素准确度提取边缘（图 3.97（b）），然后用 Ramer 算法以多边形对提取后的边缘进行逼近。从图 3.97（c）中我们能看到 Ramer 算法将边缘正确地分割开了。我们也能看到 Ramer 算法的一个小缺点：算法有时会将多边形控制点放置到稍微偏离真实拐角的位置上。在此应用中，这些偏离的位置不会带来问题，因为要得到最大的准确度，我们还会用鲁棒性好的算法拟合直线到这些轮廓线段上（图 3.97（d））。这使我们得到了此切削工具的一个简洁准确的几何描述。

cutting tool. With the resulting geometric parameters, it can easily be checked whether the tool has the required dimensions.

3.8.4.2 Segmentation of Contours into Lines, Circles, and Ellipses

While lines are often the only geometric primitive that occurs for the objects that should be inspected, in several cases the contour must be split into several types of primitives. For example, machined tools often consist of lines and circular arcs or lines and elliptical arcs. Therefore, we will now discuss how such a contour segmentation can be performed.

The approaches to segmenting contours into lines and circles can be classified into two broad categories. The first type of algorithm tries to identify breakpoints on the contour that correspond to semantically meaningful entities. For example, if two straight lines with different angles are next to each other, the tangent direction of the curve will contain a discontinuity. On the other hand, if two circular arcs with different radii meet smoothly, there will be a discontinuity in the curvature of the contour. Therefore, the breakpoints typically are defined as discontinuities in the contour angle, which are equivalent to maxima of curvature, and as discontinuities in the curvature itself. The first definition covers straight lines or circular arcs that meet at a sharp angle. The second definition covers smoothly joining circles or lines and circles (Wuescher and Boyer, 1991; Sheu and Hu, 1999). Since the curvature depends on the second derivative of the contour, it is an unstable feature that is very prone to even small errors in the contour coordinates. Therefore, to

有了这些几何参数的结果就可以轻松检测工具是否拥有需要的尺寸。

3.8.4.2 轮廓分割成直线，圆形和椭圆

尽管在很多时候直线都是被测物体中所包含的唯一几何基元，但还是有一些情况必须将轮廓分割成几种类型的基元。例如，机械加工的工具通常是由直线和圆弧，或者直线和椭圆弧组成的。所以我们现在将讨论如何对轮廓进行这样的分割。

将轮廓分割成直线和圆的算法可以分为两大类。第一类算法试图将那些与明确实体相对应的轮廓上的断点识别出来。例如，如果彼此相邻的两直线拥有不同的角度，那么此曲线的切线方向上将包含一个不连续点。另一方面，如果拥有不同半径的两个圆弧平滑地相交，那么在轮廓的弯曲部分存在一个不连续点。因此，断点通常被定义为轮廓角上的不连续点，也等价于曲率最大处，以及曲线部分自身的不连续点。前一个定义覆盖了直线或圆弧以锐角相交的情况。后一个定义覆盖了圆和圆之间，或者直线和圆直线平滑相接的情况（Wuescher and Boyer, 1991; Sheu and Hu, 1999）。由于曲率是由轮廓的二阶导数决定的，所以是一个不稳定的特征，很容易受轮廓坐标中的小误差的影响。因此，为使这些算法能够正常地工作，轮廓必须被充分地平滑过。这样可能导致断点从想得到的位置上发生偏移。

enable these algorithms to function properly, the contour must be smoothed substantially. This, in turn, can cause the breakpoints to shift from their desired positions. Furthermore, some breakpoints may be missed. Therefore, these approaches are often followed by an additional splitting and merging stage and a refinement of the breakpoint positions (Chen et al., 1996; Sheu and Hu, 1999).

While the above algorithms work well for splitting contours into lines and circles, they are quite difficult to extend to lines and ellipses because ellipses do not have constant curvature like circles. In fact, the two points on an ellipse on the major axis have locally maximal curvature and consequently would be classified as breakpoints by the above algorithms. Therefore, if we want to have a unified approach to segmenting contours into lines and circles or ellipses, the second type of algorithm is more appropriate. This type of algorithm is characterized by initially performing a segmentation of the contour into lines only. This produces an over-segmentation in the areas of the contour that correspond to circles and ellipses since here many line segments are required to approximate the contour. Therefore, the line segments are examined in a second phase as to whether they can be merged into circles or ellipses (Lanser, 1997; Rosin and West, 1995). For example, the algorithm by Lanser (1997) initially performs a polygonal approximation with the Ramer algorithm. Then, it checks each pair of adjacent line segments to see whether it can be better approximated by an ellipse (or, alternatively, a circle). This is done by fitting an ellipse to the part of the contour that corresponds to the two line segments. If the fitting error of the ellipse is smaller

此外，一些断点可能被丢失了。所以，通常在这些算法后进行额外的分裂和合并处理，以及对断点位置的改进操作（Chen et al., 1996; Sheu and Hu, 1999）。

尽管上述算法在分割轮廓为直线和圆时工作得很好，但它们很难适用于分割成直线和椭圆的情况，因为椭圆不像圆那样拥有常数曲率。事实上，椭圆在长轴上的两点拥有局部的最大曲率，相应地可被上述算法分类为断点。所以，如果我们想用统一的方法来把轮廓分割成直线和圆或者椭圆，第二类算法是更适当的。其特点是一开始时仅将轮廓分割成直线。这就会在与圆或椭圆相对应的轮廓所在区域内产生出一个过度分割的结果，因为这里是用很多的线段来逼近轮廓。所以，在第二步中就检查这些线段能否被合并成圆或者椭圆（Lanser, 1997; Rosin and West, 1995）。例如，文献（Lanser, 1997）中的算法先用 Ramer 算法执行一个多边形逼近。然后，它检查每一对彼此相邻的线段，看它们能否用一个椭圆（或一个圆）来更好地近似。这是通过将椭圆拟合到这两条线段所对应的那部分轮廓上实现的。如果椭圆的拟合误差比这两条直线的最大误差还要小，那么这两条线段就被标记成合并处理的候选对象。在检查完所有线段对后，具有最小的拟合误差的那对直线被合并。在后续的迭代处理中，算法也会考察直线对

than the maximum error of the two lines, the two line segments are marked as candidates for merging. After examining all pairs of line segments, the pair with the smallest fitting error is merged. In the following iterations, the algorithm also considers pairs of line and ellipse segments. The iterative merging is continued until there are no more segments that can be merged.

Figure 3.98 illustrates the segmentation into lines and circles. As in the previous example, the application is the inspection of cutting tools. Figure 3.98(a) displays a cutting tool that consists of two linear parts and a circular part. The result of

以及椭圆线段。迭代合并处理直到再也没有能被合并的线段时结束。

图 3.98 给出了分割轮廓为直线和圆的例子。与前一个例子类似，此例中的应用也是检测切削工具。图 3.98（a）显示的切削工具里包括两个直线部分和一个圆形部分。初始分割

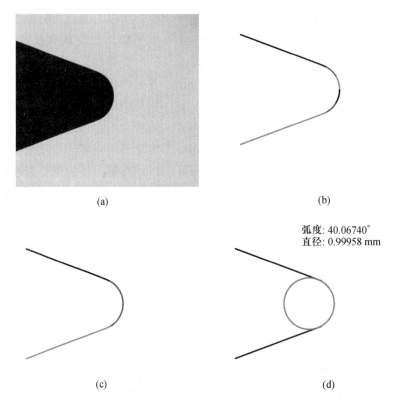

图 3.98 （a）背光照射的切削工具图像；（b）用 Ramer 算法初始分割得到的直线在轮廓上的相应部分。轮廓上的这些部分被用三种不同灰度值显示出来；（c）线和圆分割算法的合并阶段处理结果。本例中，合并处理后返回的是两条直线和一个圆弧；（d）通过拟合直线到轮廓的直线部分和拟合圆到轮廓的圆弧部分得到的几何测量结果。因为摄像机是标定过的，所以计算出的半径用毫米表示

the initial segmentation into lines with the Ramer algorithm is shown in Figure 3.98(b). Note that the circular arc is represented by four contour parts. The iterative merging stage of the algorithm successfully merges these four contour parts into a circular arc, as shown in Figure 3.98(c). Finally, the angle between the linear parts of the tool is measured by fitting lines to the corresponding contour parts, while the radius of the circular arc is determined by fitting a circle to the circular arc part. Since the camera is calibrated in this application, the fitting is actually performed in world coordinates. Hence, the radius of the arc is calculated in millimeters.

3.9 Camera Calibration

At the end of Section 3.7.4.3, we already discussed briefly that camera calibration is essential to obtain accurate measurements of objects. When the camera is calibrated, it is possible to correct lens distortions, which occur with different magnitudes for every lens, and to obtain the object's coordinates in metric units, e.g., in meters or millimeters. Since the previous sections have described the methods required for geometric camera calibration, we can now discuss how the calibration is performed.

To calibrate a camera, a model for the mapping of the three-dimensional (3D) points of the world to the 2D image generated by the camera, lens, and frame grabber (if used) is necessary.

Several different types of lenses are relevant for machine vision tasks. If the image plane is perpendicular to the optical axis, we will call this configuration a camera with a regular lens. As described in Section 2.2.4.4, we can ignore the image-side projection characteristics of the lens in this case. Consequently, we must differentiate between perspective and telecentric lenses, with the understanding that in this case the terms refer to the object-side projection characteristics. Perspective lenses perform a perspective projection of the world into the image. We will call a camera with a perspective lens a perspective camera. Telecentric lenses perform a parallel projection of the world into the image. We will call a camera with a telecentric lens a telecentric camera.

If the lens is tilted, the image-side projection characteristics of the lens become essential. We must distinguish all four lens types in this case: entocentric, image-side telecentric, object-side telecentric, and bilateral telecentric lenses. We will call cameras with a tilt lens of type t a "t tilt camera." For example, a camera with an entocentric tilt lens will be called an entocentric tilt camera.

In addition to the different lens types, two types of sensors need to be considered: area sensors and line sensors (see Section 2.3.1). For area sensors, all six of the lens types discussed above are in common use. For line sensors, only perspective lenses are commonly used. Therefore, we will not introduce additional labels for the perspective, telecentric, and tilt cameras. Instead, it will be silently assumed that these types of cameras use area sensors. For line sensors with perspective lenses, we will simply use the term line scan camera.

For all camera models, the mapping from 3D to 2D performed by the camera can be described

镜头摄像机。如 2.2.4.4 节所述，这种情况下，我们可以忽略镜头在像方的投影特征。因此，我们必须区分透视镜头和远心镜头，在这种情况下，远心镜头指的是物方远心。透视镜头通过透视投影把世界坐标系投影到图像坐标系。我们把将摄像机与透视镜头组合称作透视摄像机模型。远心镜头实现世界坐标系到图像坐标系的平行投影，我们称摄像机与远心镜头的组合为远心摄像机模型。

如果镜头是倾斜的，镜头的像方投影特征就变得重要了。在这种情况下，我们必须区分所有四种镜头类型：普通镜头、像方远心镜头、物方远心镜头和双远心镜头。我们把使用类型 t 倾斜镜头的摄像机称为"t 倾斜摄像机"。例如，使用近心（entocentric）倾斜镜头的摄像机被称作近心（entocentric）倾斜摄像机。

除了不同的镜头类型之外，还需要考虑两种不同的图像传感器：面阵图像传感器和线阵图像传感器（见 2.3.1 节）。对于面阵传感器来说，上面讨论的所有六种镜头类型都是常用的。对于线阵传感器来说，只有透视镜头是常用的。因此，当我们提到透视、远心和倾斜摄像机时，不再额外介绍是使用哪种传感器，而是默认假设这些模型都使用面阵传感器。对于使用透视镜头的线阵摄像机，我们简单的称之为线扫摄像机。

对于所有摄像机模型，从三维空间坐标到二维图像坐标的映射关系

by a certain number of parameters:

都可以使用一个固定数量的参数来表示：

$$p = \pi(P_w, \theta_1, \cdots, \theta_n) \qquad (3.136)$$

Here, p is the 2D image coordinate of the 3D point P_w produced by the projection π. Camera calibration is the process of determining the camera parameters $\theta_1, \cdots, \theta_n$.

式中，p 是三维空间点 P_w 通过投影 π 生成的二维图像坐标。摄像机标定就是确定摄像机参数 θ_1 到 θ_n 的过程。

3.9.1 Camera Models for Area Scan Cameras with Regular Lenses

Figure 3.99 displays the perspective projection performed by a perspective camera. The world point P_w is projected through the projection center of the lens to the point p in the image plane. As discussed in Section 2.2.2.4, the projection center corresponds to the center of the entrance pupil of the lens. Furthermore, for the purposes of the perspective camera model, we have applied the construction that is described in Section 2.2.2.4: the exit pupil has been shifted virtually to the projection center, and the image plane has been moved to the principal distance to cause the ray angles in object and image space to be identical.

3.9.1 普通镜头与面阵摄像机组成的摄像机模型

图 3.99 所示为一个透视摄像机的透视投影关系。世界坐标点 P_w 通过镜头的投影中心投影到像平面的点 p。如 2.2.2.4 节所述，投影中心相当于镜头入射光瞳的中心。而且，对于透视摄像机模型的用途，我们应用了在 2.2.2.4 节中所描述的结构：出射光瞳已经几乎移动到投影中心，而且像平面已经移动到主距从而使物方与像方空间的光线角度是相同的。从而，我们得到了针孔摄像机的几何结构。因此，如图中虚线表示，如果镜头没有畸变，则点 p 位于 P_w 点与投影中

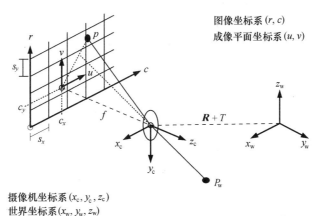

图 3.99　透视摄像机的摄像机模型

Consequently, we obtain the geometry of the pinhole camera. Therefore, if there were no lens distortions, p would lie on a straight line from P_w through the projection center, indicated by the dotted line. Lens distortions cause the point p to lie at a different position.

The image plane is located at a distance of f behind the projection center. As explained in Section 2.2.2.4, f is the camera constant or principal distance (which is called c in Section 2.2.2.4) and not the focal length of the lens. Nevertheless, we use f to denote the principal distance since we use c to denote column coordinates in the image.

Although the image plane in reality lies behind the projection center of the lens, it is easier to pretend that it lies at a distance of f in front of the projection center, as shown in Figure 3.100. This causes the image coordinate system to be aligned with the pixel coordinate system (row coordinates increase downward and column coordinates to the right) and simplifies many calculations.

心连线的延长线上。镜头的畸变会使点 p 的位置发生偏移。

像平面位于投影中心后距离为 f 的位置。如 2.2.2.4 节所解释的那样，f 是摄像机常量或主距（在 2.2.2.4 节中称之为 c）并不是镜头的焦距。不过，用 f 表示主距是因为我们用 c 表示图像的列坐标。

尽管像平面实际上是位于镜头投影中心后端，但我们可以假设它在投影中心前端 f 处（如图 3.100）。这可以使图像坐标系与像素坐标系对齐（行坐标向下递增，列坐标向右递增），这种模型可以简化很多计算。

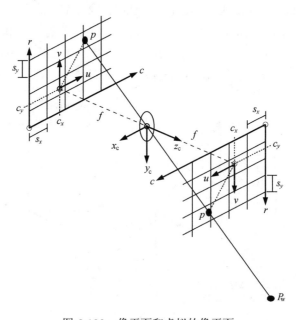

图 3.100　像平面和虚拟的像平面

3.9.1.1 Exterior Orientation

We are now ready to describe the mapping of objects in 3D world coordinates to the 2D image plane and the corresponding camera parameters. First, we should note that the world points $\boldsymbol{P}_\mathrm{w}$ are given in a world coordinate system (WCS). To make the projection into the image plane possible, they need to be transformed into the camera coordinate system (CCS). The CCS is defined such that its origin lies at the projection center. Its x and y axes are parallel to the column and row axes of the image, respectively, and the z axis is perpendicular to the image plane and is oriented such that points in front of the camera have positive z coordinates. The transformation from the WCS to the CCS is a rigid transformation, i.e., a rotation followed by a translation. Therefore, the point $\boldsymbol{P}_\mathrm{w} = (x_\mathrm{w}, y_\mathrm{w}, z_\mathrm{w})^\top$ in the WCS is given by the point $\boldsymbol{P}_\mathrm{c} = (x_\mathrm{c}, y_\mathrm{c}, z_\mathrm{c})^\top$ in the CCS, where

$$\boldsymbol{P}_\mathrm{c} = \mathrm{R}\boldsymbol{P}_\mathrm{w} + \boldsymbol{T} \tag{3.137}$$

Here, $\boldsymbol{T} = (t_x, t_y, t_z)^\top$ is a translation vector and $\mathrm{R} = \mathrm{R}(\alpha, \beta, \gamma)$ is a rotation matrix, which is determined by the three rotation angles γ (around the z axis of the CCS), β (around the y axis), and α (around the x axis):

$$\mathrm{R}(\alpha, \beta, \gamma) = \begin{pmatrix} 1 & 0 & 0 \\ 0 & \cos\alpha & -\sin\alpha \\ 0 & \sin\alpha & \cos\alpha \end{pmatrix} \begin{pmatrix} \cos\beta & 0 & \sin\beta \\ 0 & 1 & 0 \\ -\sin\beta & 0 & \cos\beta \end{pmatrix} \begin{pmatrix} \cos\gamma & -\sin\gamma & 0 \\ \sin\gamma & \cos\gamma & 0 \\ 0 & 0 & 1 \end{pmatrix} \tag{3.138}$$

The six parameters $(\alpha, \beta, \gamma, t_x, t_y, t_z)$ of R and $\boldsymbol{T}$ are called the exterior camera parameters, exterior orientation, or camera pose, because they determine the position of the camera with respect to

3.9.1.1 位姿

我们现在开始描述三维世界坐标系中的物体如何投影到二维成像平面以及其相关的摄像机参数。首先,我们应注意世界坐标点 $\boldsymbol{P}_\mathrm{w}$ 是在世界坐标系(WCS)中。为了能够投影到像平面上,需要将其转换到摄像机坐标系(CCS)中。CCS定义原点位于投影中心,其 x 轴和 y 轴分别平行于图像的列轴和行轴,其 z 轴垂直于像平面且确定其方向为:在摄像机前面的所有点的 z 坐标为正数。从WCS到CCS的变换为刚性变换,即平移加旋转。因此,WCS中的点 $\boldsymbol{P}_\mathrm{w} = (x_\mathrm{w}, y_\mathrm{w}, z_\mathrm{w})^\top$ 可由CCS中的点 $\boldsymbol{P}_\mathrm{c} = (x_\mathrm{c}, y_\mathrm{c}, z_\mathrm{c})^\top$ 来确定,它们之间的关系是

式中,$\boldsymbol{T} = (t_x, t_y, t_z)^\top$ 是一个平移向量,$\mathrm{R} = \mathrm{R}(\alpha, \beta, \gamma)$ 是一个旋转矩阵,由三个旋转角度确定,γ(绕CCS的 z 轴旋转),β(绕 y 轴),α(绕 x 轴):

R 和 $\boldsymbol{T}$ 的六个参数(α、β、γ、t_x、t_y、t_z)称为摄像机外部参数、外方位参数,或者摄像机姿态,因为这些参数决定了摄像机坐标系与世界坐标系之

the world. As for 2D transformations, it is possible to represent rigid 3D transformations, and hence poses, as homogenous transformation matrices. In homogeneous form, Eq. (3.137) becomes

$$\boldsymbol{P}'_c = \mathrm{H} \boldsymbol{P}'_w \tag{3.139}$$

with the 4 × 4 matrix

$$\mathrm{H} = \begin{pmatrix} \mathrm{R} & \boldsymbol{t} \\ \boldsymbol{0}^\top & 1 \end{pmatrix} \tag{3.140}$$

and the homogenous 4-vectors $\boldsymbol{P}'_c = (\boldsymbol{P}_c^\top, 1)^\top$ and $\boldsymbol{P}'_w = (\boldsymbol{P}_w^\top, 1)^\top$.

3.9.1.2 Projection From 3D to 2D

The next step of the mapping is the projection of the 3D point $\boldsymbol{P}_c$ into the image plane coordinate system (IPCS). For the perspective camera model, the projection is a perspective projection, which is given by

$$\begin{pmatrix} u \\ v \end{pmatrix} = \frac{f}{z_c} \begin{pmatrix} x_c \\ y_c \end{pmatrix} \tag{3.141}$$

For the telecentric camera model, the projection is a parallel projection, which is given by

$$\begin{pmatrix} u \\ v \end{pmatrix} = m \begin{pmatrix} x_c \\ y_c \end{pmatrix} \tag{3.142}$$

where m is the magnification of the lens (which was called β in Section 2.2.4). Because of the parallel projection, the distance z_c of the object to the camera has no influence on the image coordinates.

3.9.1.3 Lens Distortions

After the projection to the image plane, lens distortions cause the coordinates $(u, v)^\top$ to be modified. Lens distortions are a transformation that can be modeled in the image plane alone, i.e., 3D information is unnecessary. For many lenses, the distortion can be approximated sufficiently well by a radial distortion using the division model (Lenz, 1988; Lenz and Fritsch, 1990; Lanser, 1997; Lanser *et al.*, 1995; Steger, 2017), which is given by

是一种可以单独在像平面上建模的变换，也就是说，不需要三维信息。对于大多数镜头而言，畸变可以充分近似为基于除法模型的径向畸变（Lenz，1988；Lenz and Fritsch，1990；Lanser，1997；Lanser *et al.*，1995；Steger，2017）。如下：

$$\begin{pmatrix}\tilde{u}\\\tilde{v}\end{pmatrix}=\frac{2}{1+\sqrt{1-4\kappa(u^2+v^2)}}\begin{pmatrix}u\\v\end{pmatrix}=\frac{2}{1+\sqrt{1-4\kappa r^2}}\begin{pmatrix}u\\v\end{pmatrix} \quad (3.143)$$

where $r^2 = u^2 + v^2$. The parameter κ models the magnitude of the radial distortions. If κ is negative, the distortion is barrel-shaped; while for positive κ it is pincushion-shaped.

这里 $r^2 = u^2 + v^2$。参数 κ 表示了径向畸变量级。如果 κ 为负值，畸变为桶形畸变。如果 κ 为正值，畸变则为枕形畸变。

Figure 3.101 shows the effect of κ for an image of a calibration target that was used in previous versions of HALCON. The calibration target that is currently used by HALCON will be described in Section 3.9.4.1. The calibration target shown in Figure 3.101 has a straight border, which has the advantage that it clearly shows the effects of the distortions. Compare Figure 3.101 to Figure 2.39, which shows the effect of lens distortions for circular and rectangular grids.

图 3.101 表示了 κ 值对标定板图像的影响，这些标定板是以前 HALCON 版本所使用的。现在 HALCON 使用的标定板会在 3.9.4.1 节中进行描述。图 3.101 所示的标定板有直的外边框，它的优点是能够很清楚地看到畸变的影响。图 3.101 和图 2.39 显示了镜头畸变对于圆形和矩形栅格的影响。

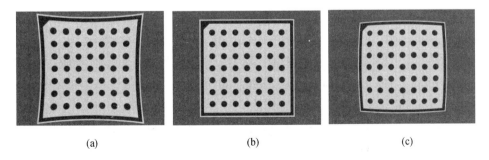

图 3.101 在除法模型中，畸变系数 κ 的影响；（a）枕形畸变：$\kappa > 0$；（b）无畸变：$\kappa = 0$；（c）枕形畸变：$\kappa < 0$

The division model has the great advantage that the rectification of the distortion can be calculated analytically by

$$\begin{pmatrix} u \\ v \end{pmatrix} = \frac{1}{1 + \kappa(\tilde{u}^2 + \tilde{v}^2)} \begin{pmatrix} \tilde{u} \\ \tilde{v} \end{pmatrix} = \frac{1}{1 + \kappa \tilde{r}^2} \begin{pmatrix} \tilde{u} \\ \tilde{v} \end{pmatrix} \quad (3.144)$$

where $\tilde{r}^2 = \tilde{u}^2 + \tilde{v}^2$. This will be important when we compute world coordinates from image coordinates.

If the division model is not sufficiently accurate for a particular lens, a polynomial distortion model that is able to model radial as well as decentering distortions can be used (Brown, 1966, 1971; Gruen and Huang, 2001; Heikkilä, 2000; Steger, 2017). Here, the rectification of the distortion is modeled by

$$\begin{pmatrix} u \\ v \end{pmatrix} = \begin{pmatrix} \tilde{u}(1 + K_1 \tilde{r}^2 + K_2 \tilde{r}^4 + K_3 \tilde{r}^6 + \cdots) \\ \quad + (P_1(\tilde{r}^2 + 2\tilde{u}^2) + 2P_2 \tilde{u}\tilde{v})(1 + P_3 \tilde{r}^2 + \cdots) \\ \tilde{v}(1 + K_1 \tilde{r}^2 + K_2 \tilde{r}^4 + K_3 \tilde{r}^6 + \cdots) \\ \quad + (2P_1 \tilde{u}\tilde{v} + P_2(\tilde{r}^2 + 2\tilde{v}^2))(1 + P_3 \tilde{r}^2 + \cdots) \end{pmatrix} \quad (3.145)$$

The terms K_i describe a radial distortion, while the terms P_i describe a decentering distortion, which may occur if the optical axes of the individual lenses are not aligned perfectly with each other. In practice, the terms K_1, K_2, K_3, P_1, and P_2 are typically used, while higher order terms are neglected.

Note that (3.145) models the rectification of the distortion, i.e., the analog of (3.144). In the polynomial model, the distortion (the analog of (3.143)) cannot be computed analytically. Instead, it must be computed numerically by a root-finding algorithm. This is not a drawback since in applications we are typically interested in transforming image coordinates to measurements in the

除法模型在矫正畸变上有很大的优点，可以通过如下计算分析：

这里 $\tilde{r}^2 = \tilde{u}^2 + \tilde{v}^2$。在我们通过图像坐标计算世界坐标时，这个非常重要。

对于某些特殊镜头，如果除法模型不足够精确，可以使用多项式畸变，此模型可以模拟径向畸变和偏心畸变（Brown；1966；1971；Gruen and Huang，2001；Heikkilä；2000；Steger，2017）。此时，畸变校正可通过下式建模：

K_i 描述了径向畸变，而 P_i 描述了偏心畸变。如果单个镜头的光轴没有完全对齐，就会出现偏心畸变。事实上，参数 K_1，K_2，K_3，P_1 和 P_2 常用，而更高阶项常被忽略。

注意，等式（3.145）畸变校正模型与等式（3.144）类似。在多项式模型中，畸变（与等式（3.143）类似）无法计算分析。相反，它必须通过求根算法来进行数值计算。这并不是缺点，因为在实际应用中我们通常感兴趣的是把图像坐标系变换到世界坐标系中测量（见 3.9.5 节和 3.10.1 节）。因

world (see Sections 3.9.5 and 3.10.1). Therefore, it is advantageous if the rectification can be computed analytically.

Figure 3.102 shows the effect of the parameters of the polynomial model on the distortion. In contrast to the division model, where $\kappa > 0$ leads to pincushion distortion and $\kappa < 0$ to barrel distortion, in the polynomial model, $K_i > 0$ leads to barrel distortion and $K_i < 0$ to pincushion distortion. Furthermore, higher order terms lead to very strong distortions at the edges of the image, while they have a progressively smaller effect in the center of the image (if the distortions at the corners of the image are approximately the same). Decentering distortions cause an effect that is somewhat similar to perspective distortion. However, they additionally bend the image in the horizontal or vertical direction.

此，如果矫正可以计算分析，则是有优势的。

图 3.102 显示了多项式模型参数的影响。除法模型当 $\kappa > 0$ 时是枕形畸变而 $\kappa < 0$ 时是桶形畸变，与除法模型相比，在多项式模型当中，$K_i > 0$ 导致桶形畸变而 $K_i < 0$ 导致枕形畸变。而且，更高阶的项在图像边缘会产生更严重的畸变，而在图像中心影响逐渐变小（如果畸变在图像的角上，规律基本一样）。偏心畸变产生的影响有点像透视畸变。然而，除此之外它们会使图像在水平或者竖直方向上变得弯曲。

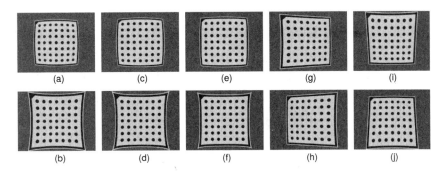

图 3.102 多项式模型中影像畸变的系数，没有明确说明的系数为 0：（a）$K_1 > 0$；（b）$K_1 < 0$；（c）$K_2 > 0$；（d）$K_2 < 0$；（e）$K_3 > 0$；（f）$K_3 < 0$；（g）$P_1 > 0$；（h）$P_1 < 0$；（i）$P_2 > 0$；（j）$P_2 < 0$

There is a deeper connection between the radial distortion coefficients in the division and polynomial models, which can be seen by expanding the rectification factor $1/(1+\kappa\tilde{r}^2)$ in Eq. (3.144) into a geometric series:

通过扩展等式（3.144）中校正因子 $1/(1+\kappa\tilde{r}^2)$ 为几何级数，可以看出在除法模型和多项式模型的径向畸变系数之间有很深的联系

$$\frac{1}{1+\kappa\tilde{r}^2} = \sum_{i=0}^{\infty}\left(-\kappa\tilde{r}^2\right)^i = 1 - \kappa\tilde{r}^2 + \kappa^2\tilde{r}^4 - \kappa^3\tilde{r}^6 + \cdots \qquad (3.146)$$

Therefore, the division model corresponds to the polynomial model without decentering distortions and with infinitely many radial distortion terms K_i that all depend functionally on the single distortion coefficient κ: $K_i = (-\kappa)^i$.

Because of the complexity of the polynomial model, we will only use the division model in the discussion below. However, everything that will be discussed also holds if the division model is replaced by the polynomial model.

因此，除法模型相当于没有偏心畸变且有无穷多的径向畸变项 K_i 的多项式模型，其中所有 K_i 从功能上都依赖于单一畸变系数 κ：$K_i = (-\kappa)^i$。

因为多项式模型的复杂性，在下面讨论中我们只用除法模型。但是，如果把除法模型替换为多项式模型，所有的讨论也会成立。

3.9.1.4 Image Coordinates

The final step in the mapping is to transform the point $(u,v)^\top$ is transformed from the IPCS into the image coordinate system (ICS):

$$\begin{pmatrix} r \\ c \end{pmatrix} = \begin{pmatrix} \dfrac{\tilde{v}}{s_y} + c_y \\ \dfrac{\tilde{u}}{s_x} + c_x \end{pmatrix} \tag{3.147}$$

Here, s_x and s_y are scaling factors. They represent the horizontal and vertical pixel pitch on the sensor (see Section 2.3.4). The point $(c_x, c_y)^\top$ is the principal point of the image. For perspective cameras, this is the perpendicular projection of the projection center onto the image plane, i.e., the point in the image from which a ray through the projection center is perpendicular to the image plane. It also defines the center of the distortions. For telecentric cameras, the projection center is at infinity. This causes all rays to be perpendicular to the image plane. Consequently, the principal point is solely defined by the distortions.

The six parameters $(f, \kappa, s_x, s_y, c_x, c_y)$ of the perspective camera and the six parameters $(m, \kappa, s_x, s_y, c_x, c_y)$ of the telecentric camera are

3.9.1.4 图像坐标

映射的最后一步是把从像平面坐标系 IPCS 中转换过来的点 $(u,v)^\top$ 转换到图像坐标系（ICS）中：

这里，s_x 和 s_y 是缩放比例因子。它们表示图像传感器上水平和垂直方向上相邻像素之间的距离（见 2.3.4 节）。点 $(c_x, c_y)^\top$ 是图像的主点。对于透视摄像机而言，这个点是投影中心在成像平面上的垂直投影，也就是说图像中的这个点与投影中心的连线与成像平面垂直。同时这个点也是径向畸变的中心。对于远心摄像机来说，投影中心在无限远处。使得所有光线都垂直于成像平面。因此，图像中的主点只表示径向畸变的中心。

透视摄像机的六个参数 $(f, \kappa, s_x, s_y, c_x, c_y)$ 和远心摄像机的六个参数 $(m, \kappa, s_x, s_y, c_x, c_y)$ 称为摄像机内参

called the interior camera parameters or interior orientation, because they determine the projection from 3D to 2D performed by the camera.

3.9.2 Camera Models for Area Scan Cameras with Tilt Lenses

We will now extend the camera models of Section 3.9.1 to handle tilt lenses correctly. To do so, we must model the transformation that occurs when the image plane is tilted for lenses that are telecentric in image space (image-side telecentric and bilateral telecentric lenses) and for lenses that are perspective in image space (entocentric and object-side telecentric lenses).

3.9.2.1 Lens Distortions

The camera models in Section 3.9.1 have proven their ability to model standard cameras correctly for many years. Therefore, the tilt camera models should reduce to the standard models if the image plane is not tilted. From the discussion in Section 2.2.2.4, we can see that the principal distance f models the ray angles in object space correctly for perspective lenses. This is also obviously the case for telecentric lenses, where the magnification m determines the spacing of the rays in object space instead of their angles. Furthermore, as discussed by Sturm et al. (2010), the distortion models discussed above essentially model distortions of ray angles with respect to the optical axis. The rays in image space are represented by their intersections with a plane that is perpendicular to the optical axis. This is convenient for untilted image planes, since this plane is already available: it is the image plane. Since the optical axis of the lens is unaffected by a tilt of the image plane, we can still use the above mechanism to represent the distortions, which models the distortions of ray angles with respect to the optical axis by way of their intersections with a plane that is perpendicular to the optical axis. Since the actual image plane is now tilted, the untilted image plane becomes a virtual image plane that is used solely for the purpose of representing ray angles with respect to the optical axis and thus to compute the distortions. Consequently, these two parts of the model, i.e., the modeling of the ray angles or spacing in object space by f or m and the modeling of the distortions in a plane that is perpendicular to the optical axis, can remain unchanged.

3.9.2.2 Modeling the Pose of the Tilted Image Plane

Both tilt models therefore work by projecting a point from a virtual image plane that is perpendicular to the optical axis to the tilted image plane. Therefore, we first describe how we can model the pose of the tilted image plane in a manner that is easy to understand. Almost all tilt lenses work by first selecting the direction in which to tilt the lens and then tilting the lens (Steger, 2017). The selection of the direction in which to tilt essentially determines a rotation axis n in a plane orthogonal to the optical axis, i.e., in the untilted image plane, and then rotating the image plane around that axis. Let the image coordinate system of the untilted image plane be given by the axes $\tilde{u}$ and $\tilde{v}$. We can extend this image coordinate system to a 3D coordinate system by the axis $\tilde{w}$, which points back to the scene along the optical axis. Figures 3.103 and 3.105 display the

通过光线和光轴与垂直于光轴的平面相交，来建立相对于光轴的光线角度的畸变模型。现在，因为实际的像平面是倾斜的，非倾斜的像平面变成了一个虚拟的像平面，只用来表示光线相对于光轴的角度，从而计算畸变。因此，模型中的这两部分，即物方空间中 f 或者 m 表示的光线角度和间距的模型，以及垂直于光轴的平面上的畸变的模型，可以保持不改变。

3.9.2.2 对倾斜像平面的位姿建模

两种倾斜模型都是通过把垂直于光轴的虚拟像平面上的点投影到倾斜像平面上。因此，我们首先描述如何以一种简单易懂的方式来对倾斜像平面的位姿进行建模。几乎所有的倾斜镜头都是首先选择倾斜镜头的方向，然后再倾斜镜头（Steger，2017）。倾斜镜头方向的选择，本质上是在正交于光轴的平面上，也就是，在未倾斜的像平面上，选择一个旋转轴 n，然后绕着轴 n 旋转像平面。通过轴 $\tilde{u}$ 和 $\tilde{v}$ 来确定未倾斜像平面的图像坐标系。我们可以通过轴 $\tilde{w}$ 把图像坐标系扩展为三维坐标系，轴 n 上的点是沿着光轴相反的方向。图 3.103 和图 3.105 所示为非倾斜像平面和它们的坐标系用灰色表示。倾斜成像平面的旋转轴 n 可以用角度 ρ $(0 \leqslant \rho < 2\pi)$

untilted image plane and this coordinate system in medium gray. The rotation axis $\boldsymbol{n}$ around which the image plane is tilted can be parameterized by the angle ρ ($0 \leqslant \rho < 2\pi$) as follows:

来参数化：

$$\boldsymbol{n} = \begin{pmatrix} \cos\rho \\ \sin\rho \\ 0 \end{pmatrix} \tag{3.148}$$

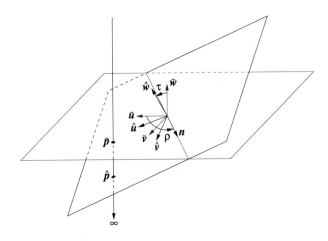

图 3.103　像方远心摄像机，非倾斜像平面上的点到倾斜平面上的点的投影

If we rotate the coordinate system $(\tilde{\boldsymbol{u}}, \tilde{\boldsymbol{v}}, \tilde{\boldsymbol{w}})$ by the tilt angle τ ($0 \leqslant \tau < \pi/2$) around $\boldsymbol{n}$, the coordinate axes $(\hat{\boldsymbol{u}}, \hat{\boldsymbol{v}}, \hat{\boldsymbol{w}})$ of the tilted image plane are given in the coordinate system of the untilted image plane by (see Steger, 2017):

如果我们绕着 $\boldsymbol{n}$ 把坐标系 $(\tilde{\boldsymbol{u}}, \tilde{\boldsymbol{v}}, \tilde{\boldsymbol{w}})$ 旋转一个角度 τ ($0 \leqslant \tau < \pi/2$)，那么倾斜成像平面坐标轴 $(\hat{\boldsymbol{u}}, \hat{\boldsymbol{v}}, \hat{\boldsymbol{w}})$ 可以在非倾斜平面坐标系中确定（参见 Steger，2017）：

$$R_t = \begin{pmatrix} c_\rho^2(1-c_\tau)+c_\tau & c_\rho s_\rho(1-c_\tau) & s_\rho s_\tau \\ c_\rho s_\rho(1-c_\tau) & s_\rho^2(1-c_\tau)+c_\tau & -c_\rho s_\tau \\ -s_\rho s_\tau & c_\rho s_\tau & c_\tau \end{pmatrix} \tag{3.149}$$

where c_θ and s_θ are abbreviations for $\cos\theta$ and $\sin\theta$, respectively, and θ is either ρ or τ.

这里，c_θ 和 s_θ 分别是 $\cos\theta$ 和 $\sin\theta$ 的缩写，θ 是 ρ 或者 τ。

Note that the semantics of the tilt parameters are quite easy to understand. A rotation angle $\rho = 0°$ means that the lens (i.e., the optical axis) is tilted downwards by τ with respect to the camera

注意，倾斜参数的语义非常容易理解。旋转角度 $\rho = 0°$ 意味着镜头（光轴）相对于摄像机外壳向下倾斜 τ。$\rho = 90°$，向左倾斜；$\rho = 180°$，向

housing; for $\rho = 90°$, it is tilted leftwards; for $\rho = 180°$ upwards; and for $\rho = 270°$ rightwards.

3.9.2.3 Image-Space Telecentric Lenses

The transformation from untilted image coordinates $\tilde{\boldsymbol{p}} = (\tilde{u}, \tilde{v})^\top$ to tilted image coordinates $\hat{\boldsymbol{p}} = (\hat{u}, \hat{v})^\top$ for cameras that are telecentric in image space (image-side telecentric and bilateral telecentric lenses) is given by (see Steger, 2017):

$$\begin{pmatrix} \hat{u} \\ \hat{v} \end{pmatrix} = \frac{1}{c_\tau} \begin{pmatrix} c_\rho^2 c_\tau + s_\rho^2 & c_\rho s_\rho (c_\tau - 1) \\ c_\rho s_\rho (c_\tau - 1) & s_\rho^2 c_\tau + c_\rho^2 \end{pmatrix} \begin{pmatrix} \tilde{u} \\ \tilde{v} \end{pmatrix} \qquad (3.150)$$

We insert this tilt transformation into the camera model between the distortion in Eqs. (3.143)–(3.145) and the transformation to the image coordinate system in Eq. (3.147), i.e., we use $(\hat{u}, \hat{v})^\top$ in Eq. (3.147) instead of $(\tilde{u}, \tilde{v})^\top$. Hence, the world points are projected to the untilted image plane, distorted within that plane, transformed to the tilted image plane, and transformed to the image coordinate system.

3.9.2.4 Image-Space Perspective Lenses

We now turn to cameras with lenses that are perspective in image space (entocentric and object-side telecentric lenses). Here, we must be able to model the different ray angles in object and image space correctly. From the discussion in Section 2.2.2.4, it is evident what must be done to model the different ray angles correctly: we must locate the untilted image plane at the true distance from the center of the exit pupil. This distance was called a' in Figure 2.23. For simplicity, we will call it d from now on and will refer to d as the image plane distance. We require that $0 < d < \infty$. Figure 3.104 displays this geometry.

Points in object space are first projected into a virtual image plane that is orthogonal to the optical axis and lies at a distance f from the projection center O. This causes the object and image space ray angles to be identical ($\omega'' = \omega$) and therefore causes the object space ray angles ω to be modeled correctly. To model the image space ray angles correctly, the virtual image plane is shifted to a distance d (which corresponds to a' in Figure 2.23), resulting in the correct image space ray angles $\omega' \neq \omega$. This shift does not change the virtual image in any way. Next, the points are distorted in the virtual image plane. With the virtual image at its correct distance d, the plane can now be tilted by the correct tilt angle τ.

先被投影到一个虚拟像平面，该平面与光轴正交，位于距离投影中心 O 为 f 的位置。这使得物方空间和像方空间的光线角度是一致的 ($\omega'' = \omega$)，也因此使物方空间光线角度 ω 可以正确建模。为了正确地对像方空间的光线角度进行建模，虚拟像平面移动到距离为 d 的位置（相对于图 2.23 中的 a'），导致正确成像空间光线角度 $\omega' \neq \omega$。这个偏移不会改变虚拟像平面。接下来，虚拟像平面中的点发生畸变。基于在正确距离为 d 的虚拟图像，虚拟平面现在可以倾斜一个正确的角度 τ。

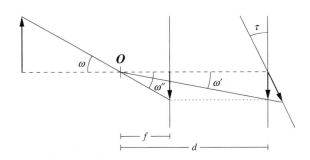

图 3.104　透视倾斜摄像机模型的光线几何形状

Consequently, as shown in Figure 3.105, the major difference to the telecentric model is that the image-side projection center (the center of the exit pupil) lies at a finite distance d in front of the image plane.

因此，如图 3.105 所示，与远心模型的主要区别是像方投影中心（出瞳中心）位于成像平面前面有限的距离为 d 的位置。

The transformation from untilted image coordinates $\tilde{\boldsymbol{p}} = (\tilde{u}, \tilde{v})^\top$ to tilted image coordinates $\hat{\boldsymbol{p}} = (\hat{u}, \hat{v})^\top$ is given by the following perspective transformation (see Steger, 2017):

非倾斜图像坐标 $\tilde{\boldsymbol{p}} = (\tilde{u}, \tilde{v})^\top$ 到倾斜图像坐标 $\hat{\boldsymbol{p}} = (\hat{u}, \hat{v})^\top$ 的变换由下面的投影变换（Steger, 2017）给出：

$$\begin{pmatrix} \hat{u} \\ \hat{v} \\ \hat{t} \end{pmatrix} = \begin{pmatrix} c_\rho^2 c_\tau + s_\rho^2 & c_\rho s_\rho (c_\tau - 1) & 0 \\ c_\rho s_\rho (c_\tau - 1) & s_\rho^2 c_\tau + c_\rho^2 & 0 \\ s_\rho s_\tau / d & -c_\rho s_\tau / d & c_\tau \end{pmatrix} \begin{pmatrix} \tilde{u} \\ \tilde{v} \\ 1 \end{pmatrix} \quad (3.151)$$

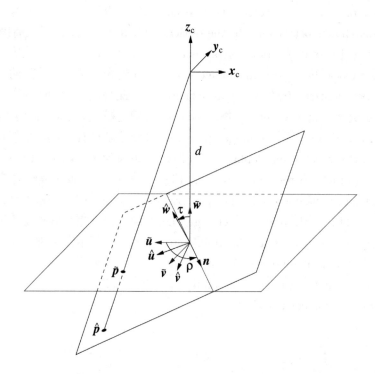

图 3.105　像方空间透视摄像机模型，从非倾斜像平面的点到倾斜平面的点的投影

Note that in this equation, $\hat{t}$ refers to the homogeneous coordinate of $\hat{\boldsymbol{p}}$. As described in Section 3.3.1.1, we must divide $\hat{u}$ and $\hat{v}$ by $\hat{t}$ to obtain the coordinates of the perspectively transformed point.

As above, we insert this tilt transformation into the camera model between the distortion in Eqs. (3.143)–(3.145) and the transformation to the image coordinate system in Eq. (3.147), i.e., we use $(\hat{u}, \hat{v})^\top$ in Eq. (3.147) instead of $(\tilde{u}, \tilde{v})^\top$.

We conclude this section by noting that f and d are related to the pupil magnification factor in Eqs. (2.7) and (2.8) by (see Steger, 2017):

注意，在这个等式中，$\hat{t}$ 指的是 $\hat{\boldsymbol{p}}$ 的齐次坐标。如 3.3.1.1 节所述，我们必须通过把 $\hat{u}$ 和 $\hat{v}$ 除以 $\hat{t}$ 来得到经过透视变换的点的坐标。

如上描述，在等式（3.143）～（3.145）的畸变和等式（3.147）图像坐标系的变换之间，我们把倾斜变换插入到摄像机模型中。即我们在等式（3.147）中使用 $(\hat{u}, \hat{v})^\top$ 代替 $(\tilde{u}, \tilde{v})^\top$。

通过标注 f 和 d 与等式（2.7）和等式（2.8）中的光瞳放大系数 f 有关（Steger，2017）来结束此节内容，如下：

$$\beta_\mathrm{p} = \frac{d}{f} \tag{3.152}$$

3.9.3 Camera Model for Line Scan Cameras

3.9.3.1 Camera Motion

As described in Section 2.3.1.1, line scan cameras must move with respect to the object to acquire a useful image. The relative motion between the camera and the object is part of the interior orientation. By far the most frequent motion is a linear motion of the camera, i.e., the camera moves with constant velocity along a straight line relative to the object, the orientation of the camera is constant with respect to the object, and the motion is equal for all images (Gupta and Hartley, 1997). In this case, the motion can be described by the motion vector $\boldsymbol{V} = (v_x, v_y, v_z)^\top$, as shown in Figure 3.106. The vector $\boldsymbol{V}$ is best described in units of meters per scan line in the camera coordinate system. As shown in Figure 3.106, the definition

3.9.3 线阵摄像机的摄像机模型

3.9.3.1 摄像机运动

如 2.3.1.1 节中所述，线阵摄像机必须与被拍摄物体之间有相对移动才能够拍摄得到一幅有用的图像。摄像机与目标物之间的相对运动是摄像机内参的一部分。到目前为止，最常见的运动是摄像机线性运动，也就是说摄像机相对于目标物体沿一条直线匀速运动，摄像机相对于目标物的方向固定，并且采集所有图像时摄像机运动状态相同（Gupta and Hartley, 1997）。这种情况下，这种运动可以表示为运动向量 $\boldsymbol{V} = (v_x, v_y, v_z)^\top$，如图 3.106。向量 $\boldsymbol{V}$ 极佳地描述了每扫描一行在摄像机坐标系的各个方向上移动的距离（米）。如图 3.106 所示，

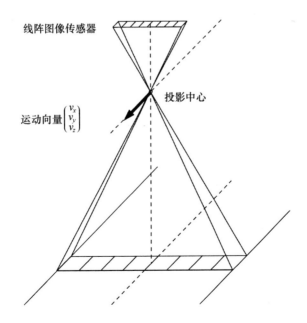

图 3.106　线阵摄像机拍摄图像的原理

of V assumes a moving camera and a fixed object. If the camera is stationary and the object is moving, e.g., on a conveyor belt, we can simply use $-V$ as the motion vector.

The camera model for line scan cameras is shown in Figure 3.107. The origin of the CCS is the projection center. The z axis is identical to the optical axis and is oriented such that points in front of the camera have positive z coordinates. The y axis is perpendicular to the sensor line and to the z axis. It is oriented such that the motion vector has a positive y component, i.e., if a fixed object is assumed, the y axis points in the direc-

V 的定义假设摄像机移动并且目标物体静止。如果摄像机静止而目标物移动，例如目标物在传送带上移动，我们可以简单地使用 $-V$ 作为运动向量。

线阵摄像机模型如图 3.107 所示。摄像机坐标系的原点在投影中心。z 轴与光轴一致，并且设置 z 轴的方向使摄像机前端所有点的 z 坐标为正。y 轴垂直于 z 轴和线阵传感器形成的平面，设置 y 轴的方向已使运动向量的 y 元素值为正，也就是说如果假设目标物体静止，y 轴指向摄像机移动的方向。x 轴垂直于 y 轴和 z 轴形成

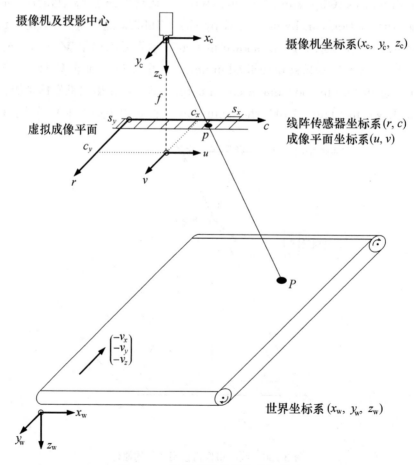

图 3.107　线阵摄像机的摄像机模型

tion in which the camera is moving. The x axis is perpendicular to the y and z axes such that the x, y, and z axes form a right-handed coordinate system.

3.9.3.2 Exterior Orientation

Similar to area scan cameras, the projection of a point given in world coordinates into the image is modeled in two steps: first, the point is transformed from the WCS into the CCS; then, it is projected into the image.

As the camera moves over the object during the image acquisition, the CCS moves with respect to the object, i.e., each image is imaged from a different position. This means that each line has a different pose. To make things easier, we can use the fact that the motion of the camera is linear. Hence, it suffices to know the transformation from the WCS to the CCS for the first line of the image. The poses of the remaining lines can be computed from the motion vector, i.e., the motion vector V is taken into account during the projection of P_c into the image. With this, the transformation from the WCS to the CCS is identical to Eq. (3.137). As for area scan cameras, the six parameters $(\alpha, \beta, \gamma, t_x, t_y, t_z)$ are called the exterior camera parameters or exterior orientation, because they determine the position of the camera with respect to the world.

3.9.3.3 Interior Orientation

To obtain a model for the interior geometry of a line scan camera, we can regard a line sensor as one particular line of an area sensor. Therefore, as in area scan cameras, there is an IPCS that lies at a distance of f (the principal distance) behind the projection center. Again, the computations can be simplified if we pretend that the image plane lies in front of the projection center, as shown in Figure 3.107. We will defer the description of the projection performed by the line scan camera until later, since it is more complicated than for area scan cameras.

Let us assume that the point P_c has been projected to the point $(u,v)^\top$ in the IPCS. As for area scan cameras, the point is now distorted by the radial distortion (3.143), which results in a distorted point $(\tilde{u},\tilde{v})^\top$.

Finally, as for area scan cameras, $(\tilde{u},\tilde{v})^\top$ is transformed into the ICS, resulting in the coordinates $(r,c)^\top$. Since we want to model the fact that the line sensor may not be mounted exactly behind the projection center, which often occurs in practice, we again have to introduce a principal point $(c_x,c_y)^\top$ that models how the line sensor is shifted with respect to the projection center, i.e., it describes the relative position of the principal point with respect to the line sensor. Since $(\tilde{u},\tilde{v})^\top$ is given in metric units, e.g., meters, we need to introduce two scale factors s_x and s_y that determine how the IPCS units are converted to ICS units (i.e., pixels). As for area scan cameras, s_x represents the horizontal pixel pitch on the sensor. As we will see below, s_y only serves as a scaling factor that enables us to specify the principal point in pixel coordinates. The values of s_x and s_y cannot be calibrated and must be set to the pixel size of the line sensor in the horizontal and vertical directions, respectively.

是面阵图像传感器的某一行。这样就可以如面阵摄像机一样，建立一个像平面坐标系，这个坐标系在投影中心后端 f（主距）处。同样，为了简化此后的计算，可以假设这个像平面在投影中心前端如图 3.107 所示。由于线阵摄像机的投影关系比面阵摄像机的投影关系更复杂，因此我们将在后面再详细论述。

首先假设点 P_c 已经被投影到像平面坐标系中的点 $(u,v)^\top$。与面阵摄像机相似，镜头产生的径向畸变导致该点坐标变为 $(\tilde{u},\tilde{v})^\top$（等式 3.143）。

最后，与面阵摄像机相似，点 $(\tilde{u},\tilde{v})^\top$ 被转换到图像坐标系，得到坐标为 $(r,c)^\top$。在现实中通常线阵图像传感器并不是正好在投影中心后端，为了在模型中考虑这个因素，我们同样使用主点 $(c_x,c_y)^\top$ 表示线阵图像传感器相对于投影中心的平移量，也就是表示了主点与线阵图像传感器之间的相对位置关系。由于 $(\tilde{u},\tilde{v})^\top$ 坐标的单位是米制单位，我们需要引入两个缩放比例因子 s_x 和 s_y，它们确定了如何从像平面坐标系中单位转换到像平面坐标系中的单位（例如：像素）。与面阵摄像机相似，s_x 表示图像传感器水平方向上相邻两像素之间的距离。如同我们随后将提到的，缩放比例因子 s_y 只是为了使我们将主点坐标转换为像素坐标。s_x 和 s_y 的值不能够被标定并且只能被相应的设置为线阵图像传感器单个像素在水平方向和垂直方向上的尺寸。

To determine the projection of the point $P_c = (x_c, y_c, z_c)^\top$ (specified in the CCS), we first consider the case where there are no radial distortions ($\kappa = 0$), the line sensor is mounted precisely behind the projection center ($c_y = 0$), and the motion is purely in the y direction of the CCS ($V = (0, v_y, 0)^\top$). In this case, the row coordinate of the projected point p is proportional to the time it takes for the point P_c to appear directly under the sensor, i.e., to appear in the xz plane of the CCS. To determine this, we must solve $x_c - tv_y = 0$ for the "time" t (since V is specified in meters per scan line, the units of t are actually scan lines, i.e., pixels). Hence, $r = t = x_c/v_y$. Since $v_x = 0$, we also have $u = fx_c/z_c$ and $c = u/s_x + c_x$. Therefore, the projection is a perspective projection in the direction of the line sensor and a parallel projection perpendicular to the line sensor (i.e., in the special motion direction $(0, v_y, 0)^\top$).

For general motion vectors (i.e., $v_x \neq 0$ or $v_z \neq 0$), non-perfectly aligned line sensors ($c_y \neq 0$), and radial distortions ($\kappa \neq 0$), the equations become significantly more complicated. As above, we need to determine the "time" t when the point P_c appears in the "plane" spanned by the projection center and the line sensor. We put "plane" in quotes since the radial distortion will cause the back-projection of the line sensor to be a curved surface in space whenever $c_y \neq 0$ and $\kappa \neq 0$. To solve this problem, we construct the optical ray through the projection center and the projected point $p = (r, c)^\top$. Let us assume that we

为了确定摄像机坐标系中的点 $P_c = (x_c, y_c, z_c)^\top$ 的投影，我们首先考虑如下这种较理想的情况，不存在径向畸变 ($\kappa = 0$)，线阵图像传感器安装在投影中心正后方 ($c_y = 0$)，并且摄像机运动完全与摄像机坐标系 y 轴方向一致 ($V = (0, v_y, 0)^\top$)。这种情况下，投影得到点 p 的行坐标与点 P_c 在传感器正下方的时间成正比，也就是点 P_c 出现在摄像机坐标系的 xz 平面上的时间。为了得到这个时间我们可以使用公式 $x_c - tv_y = 0$ 计算"时间" t，（由于运动向量 V 的单位为每扫描一行多少米，因此时间 t 的单位实际上是第几行，即像素）。因此，$r = t = x_c/v_y$。由于 $v_x = 0$，我们同样可以得出 $u = fx_c/z_c$ 和 $c = u/s_x + c_x$。因此，这种理想情况时的投影就相当于在线阵图像传感器的方向上为透视投影，而在垂直于线阵图像传感器的方向（也就是与运动向量 $(0, v_y, 0)^\top$ 方向相同）上为平行投影。

一般情况下，运动向量并不是完全沿 y 轴方向 ($v_x \neq 0$ 或 $v_z \neq 0$)，线阵图像传感器并不是正好对齐 ($c_y \neq 0$)，镜头也存在径向畸变 ($\kappa \neq 0$)，此时投影的等式就明显复杂得多。从上述论述中可以看出，我们在摄像机坐标系中点 P_c 出现在由投影中心和线阵传感器组成的"平面"上时确定"时间" t。我们将"平面"两个字放在引号里面，这是因为由于径向畸变将导致线阵传感器的反投影在空间中为曲面。为了解决这个问题，我们构造一条穿过投影中心的光线，

have transformed $(r, c)^\top$ into the distorted IPCS, where we have coordinates $(\tilde{u}, \tilde{v})^\top$. Here, $\tilde{v} = s_y c_y$ is the coordinate of the principal point in metric units. Then, we rectify $(\tilde{u}, \tilde{v})^\top$ by Eq. (3.144), i.e., $(u, v)^\top = d(\tilde{u}, \tilde{v})^\top$, where $d = 1/(1 + \kappa(\tilde{u}^2 + \tilde{v}^2))$ is the rectification factor from Eq. (3.144). The optical ray is now given by the line equation $\lambda(u, v, f)^\top = \lambda(d\tilde{u}, d\tilde{v}, f)^\top$. The point $\boldsymbol{P}_c$ moves along the line given by $(x_c, y_c, z_c)^\top - t(v_x, v_y, v_z)^\top$ during the acquisition of the image. If $\boldsymbol{p}$ is the projection of $\boldsymbol{P}_c$, both lines must intersect. Therefore, to determine the projection of $\boldsymbol{P}_c$, we must solve the following nonlinear set of equations:

通过这条光线投影得到的点 $\boldsymbol{p} = (r, c)^\top$。假设我们已经将 $(r, c)^\top$ 转换到受畸变影响后的像平面坐标系中，得到坐标值为 $(\tilde{u}, \tilde{v})^\top$。这里 $\tilde{v} = s_y c_y$ 是主点的米制坐标值。然后我们使用等式（3.144）求取无畸变时的坐标 $(u, v)^\top = d(\tilde{u}, \tilde{v})^\top$，公式中 $d = 1/(1 + \kappa(\tilde{u}^2 + \tilde{v}^2))$ 是等式（3.144）中的校正系数。这时就可以求出构造的光线的方程为 $\lambda(u, v, f)^\top = \lambda(d\tilde{u}, d\tilde{v}, f)^\top$。点 $\boldsymbol{P}_c$ 在图像拍摄过程中沿方程为 $(x_c, y_c, z_c)^\top - t(v_x, v_y, v_z)^\top$ 的直线移动。如果 $\boldsymbol{p}$ 是 $\boldsymbol{P}_c$ 点的投影，两条线必须相交。因此，为了确定 $\boldsymbol{P}_c$ 点的投影，我们必须求解下述的非线性方程组：

$$\begin{aligned} \lambda d\tilde{u} &= x_c - tv_x \\ \lambda d\tilde{v} &= y_c - tv_y \\ \lambda f &= z_c - tv_z \end{aligned} \quad (3.153)$$

for λ, $\tilde{u}$, and t, where d and $\tilde{v}$ are defined above. From $\tilde{u}$ and t, the pixel coordinates can be computed by

来得到 λ，$\tilde{u}$ 和 t，其中 d 和 $\tilde{v}$ 在上面已经定义。通过 $\tilde{u}$ 和 t 我们可以利用下面等式计算像素坐标：

$$\begin{pmatrix} r \\ c \end{pmatrix} = \begin{pmatrix} t \\ \dfrac{\tilde{u}}{s_x} + c_x \end{pmatrix} \quad (3.154)$$

The nine parameters $(f, \kappa, s_x, s_y, c_x, c_y, v_x, v_y, v_z)$ of the line scan camera are called the interior orientation because they determine the projection from 3D to 2D performed by the camera.

对于线阵摄像机来讲，九个参数 $(f, \kappa, s_x, s_y, c_x, c_y, v_x, v_y, v_z)$ 被称为摄像机内参，因为它们确定了摄像机从三维空间到二维图像的投影关系。

3.9.3.4 Nonlinearities of the Line Scan Camera Model

Although the line scan camera geometry is conceptually simply a mixture of a perspective and a

3.9.3.4 线阵摄像机模型的非线性

虽然线阵摄像机几何模型在原理上只是一个透视镜头和一个远心镜

telecentric lens, precisely this mixture makes the camera geometry much more complex than the area scan geometries. Figure 3.108 displays some of the effects that can occur. In Figure 3.108(a), the pixels are non-square because the motion is not tuned to the line frequency of the camera. In this example, either the line frequency would need to be increased or the motion speed would need to be decreased in order to obtain square pixels. Figure 3.108(b) shows the effect of a motion vector of the form $\boldsymbol{V} = (v_x, v_y, 0)^\top$ with $v_x = v_y/10$. We obtain skew pixels. In this case, the camera would need to be better aligned with the motion vector. Figures 3.108(c) and (d) show that straight lines can be projected to hyperbolic arcs, even if the lens has no distortions ($\kappa = 0$) (Gupta and Hartley, 1997). This effect occurs even if $\boldsymbol{V} = (0, v_y, 0)^\top$, i.e., if the line sensor is perfectly aligned perpendicular to the motion vector. The hyperbolic arcs are more pronounced if the motion vector has nonzero v_z. Figures 3.108(e)–(h) show

头的组合，但正好这种组合导致线阵摄像机几何模型比面阵摄像机几何模型要复杂得多。图 3.108 中所示图像为一些使用线阵摄像机拍摄图像时可能发生的情况。在图 3.108（a）中图像中的像素都不是正方的，因为摄像机的运动与摄像机采集频率不符，在这个例子中可以通过提高采集频率或者通过降低运动速度来使得到的每个像素为正方。图 3.108（b）表示了 $\boldsymbol{V} = (v_x, v_y, 0)^\top$ 的运动向量 $v_x = v_y/10$ 对图像的影响，如图所示拍摄得到的图像发生倾斜。这种情况下，需要更好地安装摄像机以保证摄像机 $\boldsymbol{y}$ 轴与运动向量对齐。如图 3.108（c）和图 3.108（d）中所示，在镜头不存在畸变的情况下，直线也有可能在图像中显示为双曲线 ($\kappa = 0$)（Gupta and Hartley，1997）。另外甚至在运动向量为 $\boldsymbol{V} = (0, v_y, 0)^\top$ 也就是说线阵图像传感器完全垂直于运动

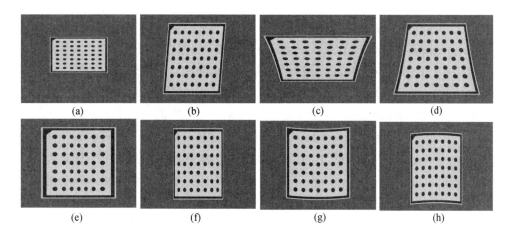

图 3.108　线阵摄像机几何模型对成像的影响：（a）由于摄像机运动与图像采集频率不符导致像素呈非正方形；（b）由于摄像机运动方向与摄像机坐标系 $\boldsymbol{y}$ 轴不平行导致图像发生倾斜；（c）甚至在线阵图像传感器垂直于运动向量时，直线也可能被投影为双曲线；（d）如果运动向量的 z 分量不等于 0，就算镜头不存在任何畸变，图（c）中的情况也会更加明显；（e）枕形畸变 ($\kappa > 0$，$c_y = 0$)；（f）桶形畸变 ($\kappa < 0$，$c_y = 0$)；（g）枕形畸变 ($\kappa > 0$，$c_y > 0$)；（h）桶形畸变 ($\kappa < 0$，$c_y > 0$)

the effect of a lens with distortions ($\kappa \neq 0$) for the case where the sensor is located perfectly behind the projection center ($c_y = 0$) and for the case where the sensor is not perfectly aligned (here $c_y > 0$). For $c_y = 0$, the pincushion and barrel distortions only cause distortions within each row of the image. For $c_y \neq 0$, the rows are also bent.

3.9.4 Calibration Process

3.9.4.1 Calibration Target

From the above discussion, we can see that camera calibration is the process of determining the interior and exterior camera parameters. To perform the calibration, it is necessary to know the location of a sufficiently large number of 3D points in world coordinates, and to be able to determine the correspondence between the world points and their projections in the image. To meet the first requirement, usually objects or marks that are easy to extract, e.g., circles or linear grids, must be placed at known locations. If the location of a camera is to be known with respect to a given coordinate system, e.g., with respect to the building plan of, say, a factory building, then each mark location must be measured very carefully within this coordinate system. Fortunately, it is often sufficient to know the position of a reference object with respect to the camera to be able to measure the object precisely, since the absolute position of the object in world coordinates is unimportant. Therefore, a movable calibration target that has been measured accurately can be used to

向量时，也可能出现这种情况。这种双曲线效果在运动向量的 z 分量不为 0 时更为明显。图 3.108（e）～（h）中表示了镜头畸变对线阵摄像机成像的影响（$\kappa \neq 0$），第一种情况是当线阵图像传感器在投影中心的正后方（$c_y = 0$），这时只导致图像的每一行中发生枕形和桶形畸变；第二种情况是线阵图像传感器并不在投影中心的正后方（$c_y > 0$），对于 $c_y = 0$，图像中的所有行都会发生弯曲。

3.9.4 标定过程

3.9.4.1 标定板

通过上面的讨论，我们可以看出摄像机标定其实就是确定摄像机内参和外参的过程。为了进行摄像机标定，必须已知世界坐标系中足够多三维空间点的坐标，找到这些空间点在图像中的投影点的二维图像坐标，并建立对应关系。为了满足第一个需求，通常将容易提取特征的目标物体或标志（如圆点或线性网格）放置在一个已知位置上。如果必须知道摄像机相对于某个给定坐标系的位姿，例如必须知道摄像机相对于工厂建筑的一个建筑规划的位姿，此时就必须非常精细地测量该坐标系中各个标志点在这个给定坐标系中的坐标。幸运的是，一般情况下知道一个参考物相对于摄像机的位置已经足够用来精确测量物体，因为物体在世界坐标系中的绝对位置一般不重要。因此，可以使用已经事先精确测量过的已知尺寸的可移动的标定板来标定摄像机。这种方式的优势在于可以在摄像机固定的情况下对

calibrate the camera. This has the advantage that the calibration can be performed with the camera in place, e.g., already mounted in the machine. Furthermore, the position of the camera with respect to the objects can be recalibrated if required, e.g., if the object type to be inspected changes.

To make it easier for the user to handle the calibration target, to provide more possibilities for 3D movement of the calibration target in tight spaces, and to make it as simple as possible to determine a plane in the world in which measurements are performed (see Section 3.9.5), it is advantageous to use planar calibration targets. Furthermore, they can be manufactured very accurately and easily can be used for back light applications if a transparent medium is used as the carrier for the marks.

The second requirement, i.e., the necessity to determine the correspondence of the known world points and their projections in the image, is in general a hard problem. Therefore, calibration targets are usually constructed in such a way that this correspondence can be determined easily. Often, a planar calibration target with a rectangular layout of circular marks is used (Lenz, 1988; Lenz and Fritsch, 1990; Lanser et al., 1995; Lanser, 1997). Circular marks are used because their center point can be determined with high accuracy. This design was also used in previous HALCON versions. We have already seen examples of this kind of calibration target in Figures 3.101, 3.102, and 3.108. The old design has a rectangular border around the calibration marks that includes a triangular orientation mark in one of its corners to make the orientation of the calibration target unique. Due to the regular matrix layout, the correspondence between the marks and their image points can be determined easily.

摄像机进行标定，例如摄像机已经安装在机器上。另外，如果需要，摄像机相对于目标物的位置也可以进行重新标定，例如被测物体类型发生变化时。

为了让用户能够更容易地使用标定板，在紧凑的空间中移动标定板提供更多的可能性，同时也为了在执行测量的空间中能够尽可能简单地确定一个平面（见 3.9.5 节），使用平面标定板是非常有利的。而且加工精度高，如果标定板使用透明材质作为标志点底盘，可以非常方便地应用在背光照明应用中。

第二个需求，也就是确定世界坐标系中已知点与它们在图像中投影的对应关系。这是个比较困难的问题。因此，标定板的结构一般都尽量使确定对应关系的过程更简单。经常会使用矩形布局带有圆形标志点的平面标定板（Lenz，1988；Lenz and Fritsch，1990；Lanser et al.，1995；Lanser，1997）。使用圆形标志点是因为可以非常精确地提取出来圆的中心点坐标，这种设计也用在之前的 HALCON 版本中。我们已经在图 3.101、图 3.102、图 3.108 中看到了这类标定板的例子。旧的标定板在标志点周围有矩形边框，另外在边框的某个角上有一个三角形方向标志用来唯一确定标定板的方向。由于使用规则的矩阵布局，很容易确定标志点和图像坐标之间的对应关系。

The old calibration target design has worked well for many years. However, experience has shown that it has some drawbacks. The border and the orientation mark on it imply that the entire calibration target must be visible in the image. Furthermore, the calibration target only has 7×7 calibration marks, which is adequate for sensors with a smaller resolution, e.g., 640×480, but is less suitable for high-resolution sensors. Both properties led to the fact that several images were required to cover the entire field of view and especially the corners of the images, where the distortions typically have the largest effect. As a consequence, a large number of calibration images had to be captured to achieve a certain accuracy. We will look at how the number of calibration images influences the accuracy in more detail in Section 3.9.6.1.

The improved calibration target design that is currently used by HALCON is shown in Figure 3.109. It uses a hexagonal layout of circular marks, which provides the highest possible density of marks. There are 27 rows with 31 marks each for a total of 837 control points. Five finder patterns, indicated by the white circles that contain smaller black circles, provide the means to compute the correspondence of the marks and their projections uniquely as long as at least one finder pattern is completely visible in the image. Consequently, the calibration target does not need to be contained completely in the image. Therefore, the entire field of view can be covered with a single image of the calibration target, and the number of calibration images can be reduced with respect to the old design. The center fiducial defines the calibration target. The x axis points WCS of the

旧的标定板设计已经使用好多年，然而经验表明，旧的标定板存在一些缺陷。标定板上的边框以及方向标志表明整个标定板在图像中必须是可见的。而且，标定板仅有 7×7 个标志点，对于低分辨率的传感器来说是足够的，例如 640×480，但是对于高分辨率传感器来说不太适合。旧标定板的这两个特点导致需要拍摄许多图像来覆盖整个视野尤其是图像的边角区域，因为这些区域畸变影响最大。因此，为了实现某个精度，则需要拍摄大量的标定板图像。3.9.6.1 节详细描述了标定板图像数量如何影响精度，我们可以参考。

图 3.109 为目前 HALCON 所使用的改进后的标定板设计。新的标定板采用圆形标志点六边形布局。标定板有 27 行，每行 31 个标志点，一共 837 个标志点。某些白色标志点内部含有更小的黑色标志点，这决定了标定板有 5 种探测模式。通过这些模式，只要至少一个探测模式在图像中完整可见，就可以计算标志点及其投影之间的对应关系。因此，标定板不需要完整出现在图像中，视野中可以只用一幅标定板图像所覆盖，和旧的设计相比，标定板图像数量明显减少。中心基准定义了标定板的世界坐标系，x 轴从中心标志点，即世界坐标系原点，指向没有黑色圆点的标志点（图 3.109 中心标志点右侧的标志点），y 轴则相对于 x 轴顺时针转 90 度（即

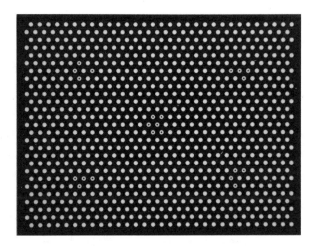

图 3.109　HALCON 标定板图像

from the center mark, which is the origin of the WCS, to the mark without the black circle (the mark to the right of the center mark in Figure 3.109). The y axis is turned by 90° clockwise with respect to the x axis (i.e., straight down in Figure 3.109). The z axis forms a right-handed 3D coordinate system with the x and y axes, i.e., it points away from the camera (through the calibration target).

The borders of the calibration marks can be extracted with subpixel-accurate edge extraction (see Section 3.7.3.5). Since the projections of the circular marks are ellipses, ellipses can then be fitted to the extracted edges with the algorithms described in Section 3.8.3.2 to obtain robustness against outliers in the edge points and to increase the accuracy. An example of the extraction of the calibration marks was already shown in Figure 3.95.

3.9.4.2　Single-Image Camera Calibration

Based on the correspondence between the marks and their projections, the camera can be calibrated. Let us denote the 3D positions of the

图 3.109 中垂直向下方向），z 轴与 x 轴和 y 轴形成一个右手三维坐标系，即指向远离摄像机的方向（穿过标定板）。

标定板各个圆点的边缘可以利用亚像素边缘提取方法来提取（见 3.7.3.5 节）。因为圆形标注点的投影为椭圆，我们使用 3.8.3.2 节介绍的算法将所有提取出的边缘拟合为椭圆，该拟合算法非常健壮，它可以抑制边缘中孤立点对边缘拟合的影响从而提高准确度。图 3.95 中所示为一个提取标志点的示例。

3.9.4.2　单一图像摄像机标定

基于标志点与它们在图像中投影之间的对应关系，就可以进行摄像机标定。我们将标志点在世界坐标系中

centers of the marks by $\boldsymbol{M}_j$. Since the calibration target is planar, we can place the calibration target in the plane $z = 0$. However, what we describe in the following is completely general and can be used for arbitrary calibration targets. Furthermore, let us denote the projections of the centers of the marks in the image by $\boldsymbol{m}_j$. Here, we must take into account that the projection of the center of the circle is not the center of the ellipse (Steger, 2017; Heikkila, 2000; Ahn et al., 1999). Finally, let us denote the camera parameters by a vector $\boldsymbol{\theta}$ that consists of the interior and exterior orientation parameters of the respective camera model (cf. Sections 3.9.1–3.9.3). For example $\boldsymbol{\theta} = (f, \kappa, s_x, s_y, c_x, c_y, \alpha, \beta, \gamma, t_x, t_y, t_z)$ for perspective cameras. As described above, the exterior orientation is determined by attaching the WCS to the calibration target. Then, the camera parameters can be determined by minimizing the distance of the extracted mark centers $\boldsymbol{m}_j$ and their projections $\pi(\boldsymbol{M}_j, \boldsymbol{\theta})$:

坐标表示为 $\boldsymbol{M}_j$。因为标定板是平面的，我们可以认为标定板放置在世界坐标系的平面 $z = 0$ 中。然而下面将要讨论的对使用任意标定对象都可用。另外将标志点中心点投影到图像中的坐标表示为 $\boldsymbol{m}_j$。这里必须考虑到圆的中心点投影到图像中并不是椭圆的中心点（Steger, 2017; Heikkila, 2000; Ahn et al., 1999）。最后，我们将摄像机参数表示为向量，其包含了相应摄像机模型的内参和外参（见 3.9.1~3.9.3 节）。例如对针孔摄像机模型来讲 $\boldsymbol{\theta} = (f, \kappa, s_x, s_y, c_x, c_y, \alpha, \beta, \gamma, t_x, t_y, t_z)$。如上所述，外参通过把世界坐标系和标定板相结合来确定，然后，通过使提取出的标志点中心点坐标 $\boldsymbol{m}_j$ 与通过投影计算得到的坐标 $\pi(\boldsymbol{M}_j, \boldsymbol{\theta})$ 之间的距离最小化来确定摄像机参数。

$$d(\boldsymbol{\theta}) = \sum_{j=1}^{n_\mathrm{m}} v_j \|\boldsymbol{m}_j - \pi(\boldsymbol{M}_j, \boldsymbol{\theta})\|^2 \to \min \quad (3.155)$$

Here, n_m is the number of calibration marks and v_j is a variable that is 1 if the calibration mark $\boldsymbol{M}_j$ is visible in the image and 0 otherwise.

While Eq. (3.155) is conceptually simple, it has the problem that it only works for the division model since we are only able to compute the distortions analytically for this model. Since we only support the division model for line scan cameras, Eq. (3.155) is what is used for this camera type. For area scan cameras, we must use a different approach since, as described in Section 3.9.1.3, for the polynomial distortion model, we are only

式中，n_m 是标定板上标志点的数量。v_j 是一个变量，如果标志点 $\boldsymbol{M}_j$ 可见，则是 1，否则为 0。

尽管等式（3.155）从概念上讲很简单，但存在的问题是仅适用于除法模型，因为我们只能解析计算除法模型的畸变。由于线阵摄像机仅支持除法模型，等式（3.155）是用于线阵摄像机模型的。对于面阵摄像机模型，必须使用不同的方法，如 3.9.1.3 节所描述，对于多项式畸变模型，只能解析计算畸变校正。因此，实际上最小

able to compute the rectification of the distortion analytically. Therefore, the error that is actually minimized is the error in the IPCS. Conceptually, we compute the projection $\pi_i(M_j, \theta_i)$ from world coordinates to image plane coordinates using (3.137) and (3.141) or (3.142). Here, θ_i denotes the subset of the camera parameters that influence this projection. On the other hand, we compute the rectification $\pi_r(m_j, \theta_r)$ of the image points to the IPCS using the inverse of (3.147), the inverse of (3.150) or (3.151) if tilt lenses are used, and (3.144) or (3.145). Here, θ_r denotes the subset of the camera parameters that influence this rectification. Then, the following error is minimized:

化的误差是像平面坐标系误差。概念上，我们使用等式（3.137）和等式（3.141）或者等式（3.142）来计算从世界坐标系到像平面坐标系的投影 $\pi_i(M_j, \theta_i)$，这里，θ_i 表示摄像机内参的子集，内参会影响投影。另一方面，使用等式（3.147）、等式（3.150）或者等式（3.151）的逆矩阵来计算图像点到 IPCS 的校正 $\pi_r(m_j, \theta_r)$，如果使用倾斜镜头，则用到等式（3.144）或等式（3.145）的逆矩阵。这里，θ_r 表示摄像机内参的子集，内参会影响校正。那么，下面所示误差是最小化的：

$$d(\theta) = \sum_{j=1}^{n_m} v_j \|\pi_r(m_j, \theta_r) - \pi_i(M_j, \theta_i)\|^2 \to \min \quad (3.156)$$

This is a difficult nonlinear optimization problem. Therefore, good starting values are required for the parameters. The interior orientation parameters can be determined from the specifications of the image sensor and the lens. The starting values for the exterior orientation are in general harder to obtain. For the planar calibration target described above, good starting values can be obtained based on the geometry and size of the projected circles (Lanser, 1997; Lanser et al., 1995).

这个问题是一个非常复杂的非线性最优化问题。因此，需要为这些参数提供更好的初始值。摄像机内参的初始值可以在图像传感器以及镜头说明书中得到。摄像机外参的初始值一般很难得到。对于上面描述的平面标定板，可以通过几何学以及标志点投影得到椭圆的尺寸来得到一个好的初始值（Lanser, 1997; Lanser et al., 1995）。

3.9.4.3 Degeneracies When Calibrating With a Single Image

The optimization in Eq. (3.156) cannot determine all camera parameters because the physically motivated camera models we have chosen are over-parameterized (Steger, 2017). For the perspective camera, f, s_x, and s_y cannot be determined

3.9.4.3 单幅图像标定时的退化问题

等式（3.156）的最优化过程并不能得到所有摄像机参数，因为选择的摄像机模型的待解参数过多（Stegere, 2017）。例如对于针孔摄像机模型，如图 3.110 所示参数 f、s_x、s_y 不能得

uniquely since they contain a common scale factor, as shown in Figure 3.110. For example, making the pixels twice as large and increasing the principal distance by a factor of 2 results in the same image. The same problem occurs for telecentric cameras: m, s_x, and s_y cannot be determined uniquely because they contain a common scale factor. The solution to this problem is to keep s_y fixed in the optimization since the image is transmitted row-by-row in the video signal (see Section 2.4). This fixes the common scale factor. On the other hand, s_x cannot be kept fixed in general since for analog frame grabbers the video signal may not be sampled pixel-synchronously (see Section 2.4.1). A similar effect happens for line scan cameras. Here, s_x and f cannot be determined uniquely. Consequently, s_x must be kept fixed in the calibration. Furthermore, as described in Section 3.9.3.3, s_y cannot be determined for line scan cameras and therefore also must be kept fixed.

到唯一解，这是因为它们有一个共同的缩放因子。例如将像素尺寸变大两倍同时将摄像机主距也变大两倍可以得到相同的图像。对于远心摄像机模型也有同样的问题：参数 m、s_x 和 s_y 同样不能得到唯一解，因为它们包含共同的缩放因子。解决这个问题的方法就是在最优化的过程中保持 s_y 不变，因为图像在视频信号中是一行一行传输的（见 2.4 节）。这样将使这三个参数的共有缩放因子固定。另一方面，对于模拟图像采集卡来说一般情况下 s_x 不可以保持不变因为视频信号可能不是像素同步采样（见 2.4.1 节）。对于线阵摄像机有相似的效果，这里 s_x 和 f 不能得到唯一解，因此，在标定过程中 s_x 必须保持固定，而且，如 3.9.3.3 节所描述，对于线阵摄像机，s_y 也无法确定，因此也必须保持不变。

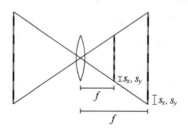

图 3.110　针孔摄像机模型参数中，f、s_x、s_y 不能计算得到唯一解

Even if the above problem is solved, some degeneracies remain because we are using a planar calibration target. For example, as shown in Figure 3.111, if the calibration target is parallel to the image plane, f and t_z cannot be determined uniquely since they contain a common scale factor. This problem also occurs if the calibration target is rotated around the x or y axes of the camera

由于我们使用的是一个平面标定板，因此就算上面的问题被解决了，仍然还会存在一些其他退化问题。例如，图 3.111 所示，如果标定板平行于像平面，f 和 t_z 不能得到唯一解因为它们包含一个共有的缩放因子。如果标定板沿摄像机坐标系的 x 轴和 y 轴旋转，这个问题仍然存在（Steger，

coordinate system (Steger, 2017; Sturm and Maybank, 1999). Here, f and a combination of parameters from the exterior orientation can be determined only up to a one-parameter ambiguity. For example, if the calibration target is rotated around the x axis, f, t_z, and α cannot be determined at the same time. For telecentric cameras, it is generally impossible to determine s_x, s_y, and the rotation angles from a single image. For example, a rotation of the calibration target around the x axis can be compensated by a corresponding change in s_x. For line scan cameras, there are similar degeneracies as the ones described above plus degeneracies that include the motion vector.

2017; Sturm and Maybank, 1999)。这种情况下 f 和外参中一组参数不能得到唯一解。例如，标定板绕 x 轴有一定的旋转，那 f、t_z 和 α 就不能同时计算得到。对于远心摄像机来说，一般情况下都不可能从单幅图像中计算得到 s_x、s_y 以及旋转角度。例如，标定板绕 x 轴旋转造成图像的变化可以通过相应更改 s_x 的值来抵消。对于线阵摄像机来说，除了存在与上面提到的相似退化问题外，还存在一些由于运动向量造成的退化问题。

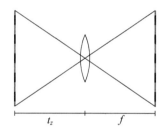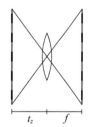

图 3.111　透视摄像机模型参数中，f 和 t_z 不能计算得到唯一解

3.9.4.4 Multi-Image Camera Calibration

To prevent the above degeneracies, the camera must be calibrated from multiple images in which the calibration target is positioned to avoid the degeneracies. For example, for perspective cameras, the calibration targets must not be parallel to each other in all images; while for telecentric cameras, the calibration target must be rotated around all of its axes to avoid the above degeneracies. Suppose we use n_o images for the calibration. Then, we also have n_o sets of exterior orientation parameters $(\alpha_k, \beta_k, \gamma_k, t_{x,k}, t_{y,k}, t_{z,k})$, $k = 1, \cdots, n_o$, that must be determined. As above, we collect

3.9.4.4 多幅图像摄像机标定

为了解决上述退化问题，摄像机必须使用多幅图像进行标定，在这些图像中标定板的位置要不同，这样才能避免上述问题发生。例如，对透视摄像机而言标定板均不能在所有拍摄的标定图像中相互平行。而对于远心摄像机来说，为了解决上述的问题，标定板必须绕所有轴旋转。假设我们使用 n_o 幅图像进行标定，那么就必须求得 n_o 组外参 $(\alpha_k, \beta_k, \gamma_k, t_{x,k}, t_{y,k}, t_{z,k})$，$k = 1, \cdots, n_o$。将内参和 n_o 组外参放在摄像机参数向量 θ 中。然后我们

the interior orientation parameters and the n_o sets of exterior orientation parameters into the camera parameter vector $\boldsymbol{\theta}$. Then, to calibrate the camera, we must solve the following optimization problem:

$$d(\boldsymbol{\theta}) = \sum_{k=1}^{n_\mathrm{o}} \sum_{j=1}^{n_\mathrm{m}} v_{j,k} \|\pi_\mathrm{r}(\boldsymbol{m}_{j,k}, \boldsymbol{\theta}_\mathrm{r}) - \pi_\mathrm{i}(\boldsymbol{M}_j, \boldsymbol{\theta}_\mathrm{i})\|^2 \to \min \qquad (3.157)$$

Here, $\boldsymbol{m}_{j,k}$ denotes the projection of the jth calibration mark in the kth image and $v_{j,k}$ is a variable that is 1 if the calibration mark $\boldsymbol{M}_j$ is visible in image k and 0 otherwise. If the calibration targets are placed and oriented suitably in the images, this will determine all the camera parameters uniquely. To ensure high accuracy of the camera parameters so determined, the calibration target should be placed such that the entire field of view is covered (by the union of all calibration images). Since the distortion is largest in the image corners, this will facilitate the determination of the distortion coefficient(s) with the highest possible accuracy.

3.9.4.5 Degeneracies Occurring With Tilt Lenses

Even when calibrating with multiple images, the transformations that model tilted image planes that were described in Section 3.9.2 have a few degeneracies that we need to be aware of (Steger, 2017). For all tilt lens types, the tilt parameters can be determined robustly only if there are enough distortions. If there are no or only small distortions, the tilt can be modeled by a shift of the principal point. For lenses that are perspective in image space, if $\tau = 0°$, the value of d cannot be determined. This, combined with the fact that

必须解决下面的最优化问题来标定摄像机：

式中，$\boldsymbol{m}_{j,k}$ 表示在第 k 幅标定图像中的第 j 个标志点的坐标，$v_{j,k}$ 是一个变量，当标志点 $\boldsymbol{M}_j$ 在图像 k 中可见时，$v_{j,k}$ 为 1，否则为 0。如果在图像中标定板放置的位置和方向合适，就可以求出所有摄像机参数的唯一解。为了使求得的摄像机参数准确度更高，标定板需要覆盖整个视野（组合所有标定图像），这主要是因为角落处的镜头畸变最大，这样就可以得到畸变系数 k 最准确的值。

3.9.4.5 倾斜镜头出现的退化现象

即便使用多幅图像进行标定，对于 3.9.2 节中提到的倾斜像平面变换，我们需要意识到仍然存在退化现象 (Steger, 2017)。对于所有倾斜镜头类型，只有当它们有足够畸变时，倾斜参数才能比较可靠的确定。如果没有或者只有很小的畸变，可以通过偏移主点来进行建模。对于透视镜头，如果 $\tau = 0°$，则 d 无法求解。对于 $d \to \infty$，成像空间中透视镜头的倾斜镜头模型会收敛到远心镜头的倾斜镜

the model for tilt lenses that are perspective in image space converges to the model for tilt lenses that are telecentric in image space for $d \to \infty$, is also the reason why we don't need to distinguish between the image-side projection characteristics for regular lenses. The fact that d cannot be determined if $\tau = 0°$ also implies that the smaller τ is, the less precisely d can be determined. Furthermore, if a lens that is perspective in image space is tilted around the horizontal or vertical axis, i.e., if $\rho \in \{0°, 90°, 180°, 270°\}$, the values of τ, d, s_x, and s_y cannot be determined uniquely, even if s_y is fixed. In this case, s_x also must be kept fixed. For lenses that are telecentric in image space, the values of f or m, ρ, τ, s_x, and s_y cannot be determined uniquely, even if s_y is fixed. In this case, s_x must also be kept fixed.

3.9.4.6 Excluding Parameters From the Optimization

A flexible calibration algorithm will enable one to specify a subset of the parameters that should be determined. For example, from the mechanical setup, some of the parameters of the exterior orientation may be known. One frequently encountered example is that the mechanical setup ensures with high accuracy that the calibration target is parallel to the image plane. In this case, we should set $\alpha = \beta = 0$, and the camera calibration should leave these parameters fixed. Another example is a camera with square pixels that transmits the video signal digitally. Here, $s_x = s_y$, and cameras both parameters should be kept fixed.

We will see in the next section that the calibration target determines the world plane in which measurements from a single image can be per-

头模型，这也是为什么我们不需要区分普通镜头像方投影特征的原因。$\tau = 0°$ 则 d 无法求解也意味着 τ 越小，d 求解的精确性也越低。而且，如果成像空间中的透视镜头沿着横轴或者纵轴倾斜，即 $\rho \in \{0°, 90°, 180°, 270°\}$，则 τ, d, s_x 和 s_y 无法唯一求解，即使 s_y 固定，这种情况，s_x 也必须固定。对于远心镜头来讲，则 f 或者 m，ρ, τ, s_x 和 s_y 不能唯一求解，即使 s_y 固定下，这种情况，s_x 也必须固定。

3.9.4.6 从优化中排除参数

一个灵活的图像标定算法可以允许我们只求摄像机参数中需要计算的某一个子集。例如，在机械安装的过程中摄像机参数中某些外参已知。一种常见的例子就是通过机械安装来保证标定板平面与像平面绝对平行。这种情况下，应该设置 $\alpha = \beta = 0$，并且摄像机标定过程中应该保持这些参数不变。另一个例子是当使用的摄像机中将视觉信号数字化的图像传感器上的像素均为正方形，此时 $s_x = s_y$，对于摄像机模型而言这两个参数都应该在摄像机标定的过程中保持不变。

我们将在下节中介绍如何使用标定板确定世界坐标系上一个平面，在这个平面上使用单幅图像即可进行测

formed. In some applications, there is not enough space for the calibration target to be turned in 3D if the camera is mounted in its final position. Here, a two-step approach can be used. First, the interior orientation of the camera is determined with the camera not mounted in its final position. This ensures that the calibration target can be moved freely. (Note that this also determines the exterior orientation of the calibration target; however, this information is discarded.) Then, the camera is mounted in its final position. Here, it must be ensured that the focus and diaphragm settings of the lens are not changed, since this will change the interior orientation. In the final position, a single image of the calibration target is taken, and only the exterior orientation is optimized to determine the pose of the camera with respect to the measurement plane.

3.9.5 World Coordinates from Single Images

As mentioned above and at the end of Section 3.7.4.3, if the camera is calibrated, it is possible in principle to obtain undistorted measurements in world coordinates. In general, this can be done only if two or more images of the same object are taken at the same time with cameras at different spatial positions. This is called stereo reconstruction. With this approach, discussed in Section 3.10.1, the reconstruction of 3D positions for corresponding points in the two images is possible because the two optical rays defined by the two optical centers of the cameras and the points in the image plane defined by the two image points can be intersected in 3D space to give the 3D position of that point. In some applications, however, it is impossible to use two cameras, e.g., because there is not enough space to mount two cameras. Nevertheless, it is possible to obtain measurements in world coordinates for objects acquired through telecentric lenses and for objects that lie in a known plane, e.g., on a conveyor belt, for perspective and line scan cameras. Both of these problems can be solved by intersecting an optical ray (also called the line of sight) with a plane. With this, it is possible to measure objects that lie in a plane, even if the plane is tilted with respect to the optical axis.

3.9.5.1 Telecentric Cameras

Let us first look at the problem of determining world coordinates for telecentric cameras. In this case, the parallel projection in Eq. (3.142) discards any depth information completely. Therefore, we cannot hope to discover the distance of the object from the camera, i.e., its z coordinate in the CCS. What we can recover, however, are the x and y coordinates of the object in the CCS (x_c and y_c in Eq. (3.142)), i.e., the dimensions of the object in world units. Since the z coordinate of $\boldsymbol{P}_c$ cannot be recovered, in most cases it is unnecessary to transform $\boldsymbol{P}_c$ into world coordinates by inverting Eq. (3.137). Instead, the point $\boldsymbol{P}_c$ is regarded as a point in world coordinates. To recover $\boldsymbol{P}_c$, we can start by inverting Eq. (3.147) to transform the coordinates from the ICS into the IPCS:

$$\begin{pmatrix} \tilde{u} \\ \tilde{v} \end{pmatrix} = \begin{pmatrix} s_x(c - c_x) \\ s_y(r - c_y) \end{pmatrix} \tag{3.158}$$

Then, we can rectify the lens distortions by applying Eqs. (3.144) or (3.145) to obtain the rectified

机。例如没有足够的空间安装两个摄像机。然而下面两种情况也可以得到被测物体在世界坐标系中的尺寸，第一就是使用远心镜头拍摄被测物体；第二就是如果使用透视摄像机或线阵摄像机，此时必须将被测物体放在一个已知平面上（如传送带）上。这两种情况都可以通过光线（视线）与已知被测平面相交来解决问题。就算测量平面与光轴之间有一定的倾斜角度，也可以进行测量。

3.9.5.1 远心摄像机

让我们首先看一看如何使用远心摄像机来测量被测物体在世界坐标系中的尺寸。这种情况下，平行投影（等式 3.142）完全不受深度信息的影响。因此，我们不可能使用这种方式得到物体与摄像机之间的距离，也就是说不能得到物体在摄像机坐标系中的 z 坐标。然而我们可以得到物体在摄像机坐标系中 x 和 y 坐标（等式 3.142 中的 x_c 和 y_c），也就得到了物体在世界坐标系中的尺寸。由于点 $\boldsymbol{P}_c$ 的 z 坐标不可求，大多数情况下没有必要通过反转换等式（3.137）将 $\boldsymbol{P}_c$ 转换到世界坐标系中，只需要将 $\boldsymbol{P}_c$ 看作是世界坐标系中的一个点即可。为了求得 $\boldsymbol{P}_c$ 的坐标，我们可以首先将该点坐标从图像坐标系转换到像平面坐标系：

然后，可以应用等式（3.144）或等式（3.145）校正镜头畸变，得到像平面

coordinates $(u, v)^\top$ in the image plane. Finally, the coordinates of $\boldsymbol{P}_c$ are given by

$$\boldsymbol{P}_c = (x_c, y_c, z_c)^\top = (u, v, 0)^\top \tag{3.159}$$

Note that the above procedure is equivalent to intersecting the optical ray given by the point $(u, v, 0)^\top$ and the direction perpendicular to the image plane, i.e., $(0, 0, 1)^\top$, with the plane $z = 0$.

3.9.5.2 Perspective Cameras

The determination of world coordinates for perspective cameras is slightly more complicated, but uses the same principle of intersecting an optical ray with a known plane. Let us look at this problem by using the application where this procedure is most useful. In many applications, the objects to be measured lie in a plane in front of the camera, e.g., a conveyor belt. Let us assume for the moment that the location and orientation of this plane (its pose) are known. We will describe below how to obtain this pose. The plane can be described by its origin and a local coordinate system, i.e., three orthogonal vectors, one of which is perpendicular to the plane. To transform the coordinates in the coordinate system of the plane (the WCS) into the CCS, a rigid transformation given by Eq. (3.137) must be used. Its six parameters $(\alpha, \beta, \gamma, t_x, t_y, t_z)$ describe the pose of the plane. If we want to measure objects in this plane, we need to determine the object coordinates in the WCS defined by the plane. Conceptually, we need to intersect the optical ray corresponding to an image point with the plane. To do so, we need to know two points that define the optical ray.

Recalling Figures 3.99 and 3.100, obviously the first point is given by the projection center,

上校正后的坐标 $(u, v)^\top$。最终计算得到点 $\boldsymbol{P}_c$ 的坐标为：

注意上面的方法其实相当于通过点 $(u, v, 0)^\top$，并且与像平面 $(0, 0, 1)^\top$ 垂直的光线和平面 $z = 0$ 相交得到交点。

3.9.5.2 透视摄像机

使用透视摄像机确定目标物体的世界坐标稍微复杂一些，不过原理相同，也是将光线与已知平面相交。让我们通过应用来看看这个问题，在这些应用中这个方法最有用。在很多应用中，被测物体都放在摄像机前面的一个平面上，如传送带上。我们假设此时这个平面的位置与方向（位姿）已知。我们稍后将介绍如何得到这个位姿。这个测量平面由它的原点以及一个局部坐标系来表示，也就是三个直角向量，其中一个向量垂直于测量平面。必须使用等式（3.137）给出的刚性变换将测量平面坐标系中的坐标转换为摄像机坐标系中的坐标。其中的六个参数 $(\alpha, \beta, \gamma, t_x, t_y, t_z)$ 表示了测量平面的位姿。如果我们想在这个平面上测量物体，那么就需要得到这个平面所定义的世界坐标系中的坐标。从原理上讲，我们需要使用图像上某点所对应的光线与该测量平面相交。所以我们需要知道定义该光线的两个点。

回顾图 3.99 和图 3.100 可以看出，很显然第一个点就是摄像机的投

which has the coordinates $(0,0,0)^\top$ in the CCS. To obtain the second point, we need to transform the point $(r,c)^\top$ from the ICS into the IPCS. This transformation is given by Eqs. (3.158) and (3.144) or (3.145). To obtain the 3D point that corresponds to this point in the image plane, we need to take into account that the image plane lies at a distance of f in front of the optical center. Hence, the coordinates of the second point on the optical ray are given by $(u,v,f)^\top$. Therefore, we can describe the optical ray in the CCS by

$$\boldsymbol{L}_\mathrm{c} = (0,0,0)^\top + \lambda(u,v,f)^\top \qquad (3.160)$$

To intersect this line with the plane, it is best to express the line $\boldsymbol{L}_\mathrm{c}$ in the WCS of the plane, since in the WCS the plane is given by the equation $z=0$. Therefore, we need to transform the two points $(0,0,0)^\top$ and $(u,v,f)^\top$ into the WCS. This can be done by inverting Eq. (3.137) to obtain

$$\boldsymbol{P}_\mathrm{w} = \boldsymbol{R}^{-1}(\boldsymbol{P}_\mathrm{c} - \boldsymbol{T}) = \boldsymbol{R}^\top(\boldsymbol{P}_\mathrm{c} - \boldsymbol{T}) \qquad (3.161)$$

Here, $\boldsymbol{R}^{-1} = \boldsymbol{R}^\top$ is the inverse of the rotation matrix R in Eq. (3.137). Let us call the transformed optical center $\boldsymbol{O}_\mathrm{w}$, i.e., $\boldsymbol{O}_\mathrm{w} = \boldsymbol{R}^\top((0,0,0)^\top - T) = -\boldsymbol{R}^\top T$, and the transformed point in the image plane $\boldsymbol{I}_\mathrm{w}$, i.e., $\boldsymbol{I}_\mathrm{w} = \boldsymbol{R}^\top((u,v,f)^\top - T)$. With this, the optical ray is given by

$$\boldsymbol{L}_\mathrm{w} = \boldsymbol{O}_\mathrm{w} + \lambda(\boldsymbol{I}_\mathrm{w} - \boldsymbol{O}_\mathrm{w}) = \boldsymbol{O}_\mathrm{w} + \lambda \boldsymbol{D}_\mathrm{w} \qquad (3.162)$$

in the WCS. Here, $\boldsymbol{D}_\mathrm{w}$ denotes the direction vector of the optical ray. With this, it is a simple matter to determine the intersection of the optical ray in Eq. (3.162) with the plane $z=0$. The intersection point is given by

$$P_\mathrm{w} = \begin{pmatrix} o_x - o_z\, d_x/d_z \\ o_y - o_z\, d_y/d_z \\ 0 \end{pmatrix} \quad (3.163)$$

where $\boldsymbol{O}_\mathrm{w} = (o_x, o_y, o_z)^\top$ and $\boldsymbol{D}_\mathrm{w} = (d_x, d_y, d_z)^\top$.

Up to now, we have assumed that the pose of the plane in which we want to measure objects is known. Fortunately, the camera calibration gives us this pose almost immediately since a planar calibration target is used. If the calibration target is placed on the plane, e.g., the conveyor belt, in one of the images used for calibration, the exterior orientation of the calibration target in that image almost defines the pose of the plane we need in the above derivation. The pose would be the true pose of the plane if the calibration target were infinitely thin. To take the thickness of the calibration target into account, the WCS defined by the exterior orientation must be moved by the thickness of the calibration target in the positive z direction. This modifies the transformation from the WCS to the CCS in Eq. (3.137) as follows: the translation $\boldsymbol{T}$ simply becomes $\boldsymbol{RD} + \boldsymbol{T}$, where $\boldsymbol{D} = d(0,0,1)^\top$, and d is the thickness of the calibration target. This is the pose that must be used in Eq. (3.161) to transform the optical ray into the WCS. An example of computing edge positions in world coordinates for perspective cameras is given at the end of Section 3.7.4.3 (see Figure 3.89).

3.9.5.3 Line-Scan Cameras

For line scan cameras, the procedure to obtain world coordinates in a given plane is conceptually similar to the approaches described above. First, the optical ray is constructed from Eqs. (3.154) and (3.153). Then, it is transformed into the WCS and intersected with the plane $z = 0$.

3.9.5.4 Image Rectification

In addition to transforming image points, e.g., 1D edge positions or subpixel-precise contours, into world coordinates, sometimes it is also useful to transform the image itself into world coordinates. This creates an image that would have resulted if the camera had looked perfectly perpendicularly without distortions onto the world plane. This image rectification is useful for applications that must work on the image data itself, e.g., region processing, template matching, or OCR. It can be used whenever the camera cannot be mounted perpendicular to the measurement plane. To rectify the image, we conceptually cut out a rectangular region of the world plane $z = 0$ and sample it with a specified distance, e.g., 200 μm. We then project each sample point into the image with the equations of the relevant camera model, and obtain the gray value through interpolation, e.g., bilinear interpolation. Figure 3.112 shows an example of this process. In Figure 3.112(a), the image of a caliper together with the calibration target that defines the world plane is shown. An old calibration target was used to show that the calibration target is placed onto the object to define the world plane (the new calibration target typically fills the entire field of view). The unrectified and rectified images of the caliper are shown in Figures 3.112(b) and (c), respectively. Note that the rectification has removed the perspective and radial distortions from the image.

式（3.153）构造光线，然后将该光线转换到世界坐标系中并求该光线与平面 $z = 0$ 的交点。

3.9.5.4 图像校正

上面讨论了如何将图像中某些点转换到世界坐标系中，例如将一维边缘位置或亚像素精度轮廓线转换到世界坐标系中。除此之外，有些情况下将图像本身转换到世界坐标系中也非常有用。这个转换得到的图像就相当于摄像机在与世界平面绝对垂直并且镜头不存在任何畸变的情况下拍摄得到的图像。这种图像校正在一些需要使用图像本身进行处理的应用中非常有用，例如需要在校正后的图像中进行区域处理、模板匹配或光学字符识别（OCR）等。在任何情况下，只要摄像机不能安装为垂直于测量平面都可以进行图像校正。为了进行图像校正，原理上我们在世界坐标系中平面 $z = 0$ 中截取一个矩形区域，然后在矩形区域中每隔一个指定距离（如 200 μm）抽取一个点。将这些抽样点使用相应的摄像机模型投影到图像中，然后通过插值算法（如双线性内插算法）得到该点的灰度值。图 3.112 中是一个图像校正的实例。在图 3.112（a）中，图像中有一把尺子和一个标定板，其中这个标定板定义了世界平面。例子中使用了旧的标定板，从图中可以看出，标定板放到了物体上用来定义世界平面（新的标定板则需要覆盖整个视野）。图 3.112（b）和图 3.112（c）中分别显示的是校正前与校正后的图像。注意图像校正已经从图像中消除了透视畸变和径向畸变。

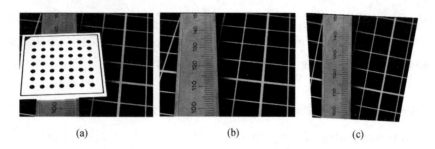

图 3.112 （a）尺子与标定板的图像；（b）校正前尺子图像；（c）校正后尺子图像

3.9.6 Accuracy of the Camera Parameters

We conclude the discussion of camera calibration by discussing two different aspects: the accuracy of the camera parameters, and the changes in the camera parameters that result from adjusting the focus and diaphragm settings on the lens.

3.9.6.1 Influence of the Number of Calibration Images on the Accuracy

As was already noted in Section 3.9.4.2, there are some cases where an inappropriate placement of the calibration target can result in degenerate configurations where one of the camera parameters or a combination of some of the parameters cannot be determined. These configurations must obviously be avoided if the camera parameters are to be determined with high accuracy. Apart from this, the main influencing factor for the accuracy of the camera parameters is the number of images that are used to calibrate the camera. This is illustrated in Figure 3.113, where the standard deviations of the principal distance f, the radial distortion coefficient κ, and the principal point $(c_x, c_y)^\top$ are plotted as functions of the number of images that are used for calibration. To obtain this data, 20 images of a calibration target

3.9.6 摄像机参数的准确度

作为摄像机标定的最后部分，我们讨论下面两个方面的问题：一是摄像机参数的准确度，二是调整摄像机焦距和光圈对摄像机参数的影响。

3.9.6.1 标定图像数量对准确度的影响

我们已经在 3.9.4.2 节中强调过，摄像机标定过程中如果标定板放置不适当将导致退化配置（degeneracies configurations），此时摄像机参数中的某个参数或某些参数将不能得到唯一值。为了得到高准确度的摄像机参数必须避免这种情况发生。除了这个影响外，影响摄像机参数准确度的最主要的因素就是用于进行标定摄像机的图像数量。在图 3.113 中显示了用于摄像机标定的图像数量对参数的影响，图中分别显示了主距 f 的标准差、径向畸变系数 κ 的标准差和主点 $(c_x, c_y)^\top$ 的标准差与用于标定图像数量的关系。为了总结这些数据，我们拍摄了 20 幅标定板的图像。然后使用 20 幅图像中所有子集的 l 幅图像

were taken. Then, every possible subset of l images ($l = 2, \cdots, 19$) from the 20 images was used to calibrate the camera. The standard deviations were calculated from the resulting camera parameters when l of the 20 images were used for the calibration. From Figure 3.113 it is obvious that the accuracy of the camera parameters increases significantly as the number of images l increases. This is not surprising when we consider that each image serves to constrain the parameters. If the images were independent measurements, we could expect the standard deviation to decrease proportionally to $l^{-0.5}$. In this particular example, the standard deviation of f decreases roughly proportionally to l^{-2} initially and to $l^{-1.3}$ for larger l, that of κ decreases roughly proportionally to $l^{-0.7}$ initially and faster for larger l, and that of c_x and c_y decreases roughly proportionally to $l^{-1.2}$. Thus, all standard deviations decrease much faster than $l^{-0.5}$.

($l = 2, \cdots, 19$) 来标定摄像机。通过使用 20 幅图像中所有含有 l 幅图像的子集标定得到的摄像机参数来计算标准差。从图 3.113 中可以看出摄像机参数的准确度随使用图像数量的增加而明显增加。如果我们考虑到其实每幅图像都可以用来约束摄像机参数，就会觉得这个结果理所应当了。如果这些图像之间都是无关度量，我们可以期望摄像机参数的标准差按 $l^{-0.5}$ 比例下降。在当前这个例子中，f 的标准差大约以最初是 l^{-2} 到 $l^{-1.3}$（对于较大 l）的比例减少，κ 的标准差大约以最初 $l^{-0.7}$ 到更快的（对于较大 l）比例减少，c_x 和 c_y 的标准差大约以 $l^{-1.2}$ 比例减少。因此，所有标准差减少比例比 $l^{-0.5}$ 要快很多。

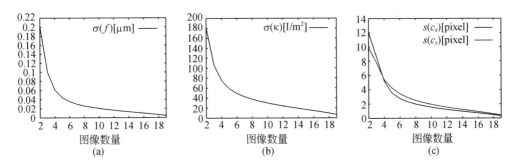

图 3.113 （a）主距 f 的标准差；（b）径向畸变系数 k 的标准差；（c）主点 (c_x, c_y) 的标准差与标定摄像机过程中使用的图像数量之间的函数关系

From Figure 3.113, it can also be seen that a comparatively large number of calibration images is required to determine the camera parameters accurately. This happens because there are non-negligible correlations between the camera parameters, which can only be resolved through multiple

从图 3.113 中还可以看出，为了使得到的摄像机参数更加准确，就需要有相当多的标定图像。这主要是因为摄像机参数之间存在不容忽视的相关性，这些参数只能通过多次无关测量进行求解。标定板在标定图像中

independent measurements. To obtain accurate camera parameters, it is important that the calibration target covers the entire field of view and that it covers the range of exterior orientations as well as possible (over all calibration images). In particular, all parameters can be determined more accurately if the calibration target covers a large depth range. This can be achieved by turning the calibration target around its x and y axes, and by placing it at different depths relative to the camera. Furthermore, to determine the principal point accurately, it is necessary that the calibration target is rotated around its z axis. Finally, as was already noted, to determine κ as accurately as possible, it is necessary that the calibration target covers as much of the field of view of each single image as possible (ideally, the entire field of view).

3.9.6.2 Influence of the Focus Setting on the Camera Parameters

We now turn to a discussion of whether changes in the lens settings change the camera parameters. Figure 3.114 displays the effect of changing the focus setting of the lens. Here, a 12.5 mm lens with a 1 mm extension tube was used. The lens was set to the nearest and farthest focal settings. In the near focus setting, the camera was calibrated with a calibration target of size 4 cm × 3 cm, while for the far focus setting a calibration target of size 8 cm × 6 cm was used. This resulted in the same size of the calibration target in the focusing plane for both settings. Care was taken to use images of calibration targets with approximately the same range of depths and positions in the images. For each setting, 20 images of the calibration target were taken. To be able to evaluate statistically

最好能够覆盖整个视野并尽可能覆盖摄像机外参的范围（对于所有标定图像），为了得到准确的摄像机参数这点非常重要。尤其是如果标定板可以覆盖较大的深度范围，那么所有的摄像机参数都会更准确。可以通过将标定板绕它的 x 轴和 y 轴旋转或者将标定板放置在与摄像机不同距离的位置上使标定板覆盖较大的深度范围。而且，为了准确地确定主点，有必要将标定板沿 z 轴旋转。最后，正如已经提到过的，为了尽可能准确的确定 κ，标定板必须尽可能在每幅图像中覆盖尽可能多的视野（比较理想的是整个视野）。

3.9.6.2 焦距对摄像机参数的影响

现在我们开始讨论调整镜头会不会影响摄像机参数。图 3.114 显示了焦距变化对摄像机参数的影响。这个实验中使用 12.5mm 镜头与一个 1mm 近拍接圈。将镜头焦点调至最远和最近。在近焦时摄像机使用 4cm×3cm 大小的标定板进行标定，在远焦时使用 8cm×6cm 大小的标定板进行标定。这样在两种设置下焦平面上标定板大小一致。注意我们在拍摄标定图像时，两种情况下标定板大约覆盖同样范围的景深，并且放置的位置也大约相同。两种情况下分别拍摄标定板的 20 幅图像。为了从统计学上评估摄像机参数是否不同，我们取 20 幅图像中包含任意 19 幅图像

whether the camera parameters are different, all 20 subsets of 19 of the 20 images were used to calibrate the camera. As can be expected from the discussion in Section 2.2.2, changing the focus will change the principal distance. From Figure 3.114(a), we can see that this clearly is the case. Furthermore, the radial distortion coefficient κ also changes significantly. From Figure 3.114(b), it is also obvious that the principal point changes.

的 20 个子集来标定摄像机。通过 2.2.2 节的讨论我们可以预料到调节焦点将会改变主距。从图 3.114（a）中可以明显看到这个现象。另外，径向畸变系数 κ 的变化也非常明显。同样从图 3.114（b）中可以看到主点也发生了变化。

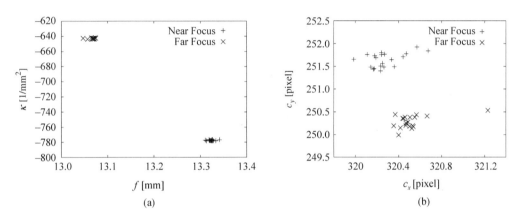

图 3.114　镜头在焦点不同两种情况下的（a）主距与径向畸变系数；（b）主点位置

3.9.6.3 Influence of the Diaphragm Setting on the Camera Parameters

3.9.6.3 光圈对摄像机参数的影响

Finally, we examine what can happen when the diaphragm on the lens is changed. To test this, a similar setup as above was used. The camera was set to f-numbers $f/4$ and $f/11$. For each setting, 20 images of a 4 cm × 3 cm calibration target were taken. Care was taken to position the calibration targets in similar positions for the two settings. The lens is an 8.5 mm lens with a 1 mm extension tube. Again, all 20 subsets of 19 of the 20 images were used to calibrate the camera. Figure 3.115(a) displays the principal distance and

最后，我们将分析镜头上光圈调节时会发生什么情况。测试的环境与上面的实验类似。摄像机分别设置为 $f/4$ 和 $f/11$。每个设置拍摄 20 幅 4 cm×3 cm 标定板的图像。注意两种设置时拍摄标定板的图像时尽量使标定板在图像中的位置相似。摄像机使用 8.5 mm 镜头和一个 1 mm 的近摄接圈。同样，我们取 20 幅图像中包含任意 19 幅图像的 20 个子集来标定摄像机。图 3.115（a）中显示了主距和

radial distortion coefficient. Clearly, the two parameters change in a statistically significant way. Figure 3.115(b) also shows that the principal point changes significantly. The changes in the parameters mean that there is an overall difference in the point coordinates across the image diagonal of approximately 1.5 pixels. Therefore, we can see that changing the f-number on the lens requires a recalibration, at least for some lenses.

径向畸变系数。很明显，从统计学上看这两个参数都发生了明显的变化。图 3.115（b）中显示了主点的明显变化。这种变化就意味着主点在坐标系中沿图像对角线的总体差异大约为 1.5 个像素。因此，我们可以看出调节镜头的 f 值后需要重新对摄像机进行标定，至少对某些镜头而言是必须重新标定的。

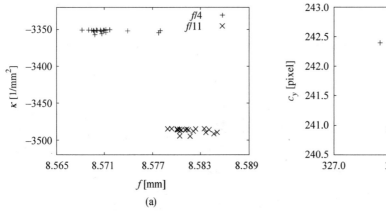
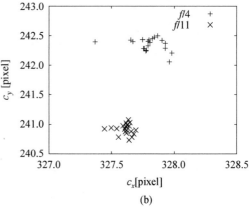

图 3.115　镜头不同光圈设置（f 值：$f/4$ 和 $f/11$）情况下的（a）主距与径向畸变系数系数和；（b）主点位置

3.10　3D Reconstruction

As discussed in Section 2.5, there are several technologies that can be used to acquire a 3D reconstruction of a scene. In this section, we will discuss the algorithms that are used for stereo, sheet of light, and structured light reconstruction.

3.10.1　Stereo Reconstruction

In Sections 3.9 and 3.7.4.3, we have seen that we can perform very accurate measurements from a single image by calibrating the camera and by determining its exterior orientation with respect to a plane in the world. We could then convert the

3.10　三维重构

如 2.5 节所介绍，有许多技术可以用来获取场景的三维重构结果。本节，我们要讨论立体视觉、激光三角测量以及结构光重构等算法。

3.10.1　立体重构

通过 3.9 节与 3.7.4.3 节的介绍，我们已经了解到，如果将摄像机进行标定并且得到摄像机与测量平面之间的位姿关系，就可以使用该平面上的单幅图像进行精确测量。然后就可以

image measurements to world coordinates within the plane by intersecting optical rays with the plane. Note, however, that these measurements are still 2D measurements within the world plane. In fact, from a single image we cannot reconstruct the 3D geometry of the scene because we can only determine the optical ray for each point in the image. We do not know at which distance on the optical ray the point lies in the world. In the approach in Section 3.9.5, we had to assume a special geometry in the world to be able to determine the distance of a point along the optical ray. Note that this is not a true 3D reconstruction. To perform a 3D reconstruction, we must use at least two images of the same scene taken from different positions. Typically, this is done by simultaneously taking the images with two cameras (see Section 2.5.1). This process is called stereo reconstruction. In this section, we will examine the case of binocular stereo, i.e., the two-camera case. Throughout this section, we will assume that the cameras have been calibrated, i.e, their interior orientations and relative orientation are known. Note that uncalibrated stereo (Hartley and Zisserman, 2003; Faugeras and Luong, 2001) and multi-view stereo (Seitz *et al.*, 2006; Szeliski, 2011; Seitz *et al.*, 2017) are also possible. However, these methods are beyond the scope of this book.

3.10.1.1 Stereo Geometry

Before we can discuss stereo reconstruction, we must examine the geometry of the two cameras, as shown in Figure 3.116. Since the cameras are assumed to be calibrated, we know their interior

通过光线与该平面相交来将图像中测量结果转换到世界坐标系中。然而需要注意的是这些测量结果仍然是世界坐标系中某平面上二维测量。实际上，我们不可能通过单幅图像来重构场景的三维几何信息，因为我们只能得到图像中每单个点的视线，但并不知道这条视线上世界坐标系中相应的三维空间点距离摄像机多远。我们曾经在 3.9.5 节介绍过一种方法，通过世界场景中一个特殊的几何形状就可以计算得到某一点的视线上相应空间点与摄像机之间的距离。注意这种方法并不是真正的三维重构。进行真正的三维重构我们必须至少使用从不同角度拍摄同样场景的两幅图像。典型情况下，这两幅图像通过两个摄像机在不同位置上同时拍摄得到（见 2.5.1 节）。这个过程就是立体重构。在这节中，我们将主要分析使用两个摄像机的情况，也就是双目立体视觉。在本节中，我们假设摄像机都已经经过标定，也就是说摄像机的内参以及两个摄像机之间的相对位姿关系都已知。注意，未标定的立体视觉（Hartley and Zisserman，2003；Faugeras and Luong，2001）和多目立体视觉（Seitz *et al.*，2006；Szeliski，2011；Seitz *et al.*，2017）也是可以实现的。但是，这些方法超出了本书的范围。

3.10.1.1 立体几何结构

在开始讨论立体重构之前，我们必须分析两个摄像机的立体几何结构，如图 3.116 所示。由于假设摄像机已经经过标定，也就是说我们知道摄

orientations, i.e., their principal points, principal distances, pixel size, distortion coefficient(s), and, if applicable, tilt parameters. In Figure 3.116, the principal points are shown by the points C_1 and C_2 in the first and second image, respectively. Furthermore, the projection centers (i.e, the centers of the entrance pupils) are shown by the points O_1 and O_2. The dashed lines between the projection centers and principal points show the principal distances. Note that, since the image planes physically lie behind the projection centers, the image is turned upside down. Consequently, the origin of the ICS lies in the lower right corner, with the row axis pointing upward and the column axis pointing leftward. The CCS axes are defined such that the x axis points to the right, the y axis points downwards, and the z axis points forward from the image plane, i.e., along the viewing direction.

像机的内参，其中包括两个摄像机的主点、主距、像素尺寸和畸变系数，对于某些情况，还有倾斜参数。两幅图像中的主点在图 3.116 中分别表示为 C_1 和 C_2。另外投影中心（即，入射光瞳的中心）表示为 O_1 和 O_2。投影中心与主点之间的虚线为两个摄像机的主距。注意由于像平面实际上位于成像中心后端，图像上下翻转。从而图像坐标系的原点就在像平面的右下角，纵轴向上横轴向左。摄像机坐标系的 x 轴向右，y 轴向下，z 轴背离像平面向外，也就是沿视角方向。

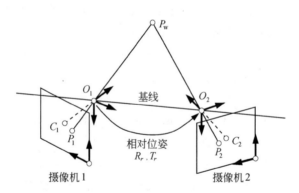

图 3.116　两个摄像机的立体几何结构

The position and orientation of the two cameras with respect to each other are given by the relative orientation, which is a rigid 3D transformation specified by the rotation matrix R_r and the translation vector T_r. The relative orientation transforms point coordinates in the CCS of the first camera into point coordinates of the CCS

两个摄像机之间的相对位置以及相对方位表示为摄像机之间的相对位姿，这个位姿是一个由旋转矩阵 R_r 和平移向量 T_r 构成的刚性三维转换。相对位姿把第一个摄像机坐标系的点坐标转换到第二个摄像机坐标系中：$P_{c2} = R_r P_{c1} + T_r$，也可以认为是第二

of the second camera: $P_{c2} = R_r P_{c1} + T_r$. This is visualized by the arrow in Figure 3.116. The relative orientation can also be interpreted as the transformation of the camera coordinate system of the second camera into the camera coordinate system of the first camera. The translation vector T_r, which specifies the translation between the two projection centers, is also called the base.

With this, we can see that a point P_w in the world is mapped to a point P_1 in the first image and to a point P_2 in the second image. If there are no lens distortions (which we will assume for the moment), the points P_w, O_1, O_2, P_1, and P_2 all lie in a single plane.

3.10.1.2 Stereo Calibration

To calibrate the stereo system, we can extend the method of Section 3.9.4.4 as follows. Let M_j denote the positions of the calibration marks. We extract their projections in both images with the methods described in Section 3.9.4.1. Let us denote the projection of the centers of the marks in the first set of calibration images by $m_{j,k,1}$ and in the second set by $m_{j,k,2}$. Furthermore, let us denote the camera parameters by a vector θ. The camera parameters θ include the interior orientations of the first and second cameras, the exterior orientations of the n_o calibration targets in the second image, and the relative orientation of the two cameras. From the above discussion of the relative orientation, it follows that these parameters determine the mappings into the first and second images completely. Hence, to calibrate the stereo system, the following optimization problem must be solved:

个摄像机坐标系到第一个摄像机坐标系的转换。我们定义两个投影中心之间的平移为平移向量 T_r，也被称为基线。

基于此，可以看出世界坐标系中的一点 P_w 投影为第一个图像中的 P_1 点与第二个图像中 P_2 点。如果我们暂时假设镜头没有畸变，点 P_w、O_1、O_2、P_1、P_2 在同一平面上。

3.10.1.2 立体视觉标定

为了标定立体视觉系统，可以按下面的方式扩展 3.9.4.4 节中介绍的标定方法。使用 M_j 表示标定板上标记点的位置。然后使用 3.9.4.1 节中介绍的方法提取标定板中标记点在两幅图像中的投影点坐标。将第一个摄像机拍摄得到的标定图像看作第一组，其中标志点的中心点表示为 $m_{j,k,1}$；第二个摄像机拍摄得到的标定图像看作第二组，其中标志点的中心点表示为 $m_{j,k,2}$。另外将摄像机参数表示为向量 θ。摄像机参数向量 θ 中包含第一个和第二个摄像机的内参，第二幅图像中 n_o 个标定板的外方位参数，以及两个摄像机之间的相对位姿。从上述关于相对位姿的讨论中可以看出，这些参数可以确定到第一和第二幅图像中的映射。因此，标定立体视觉系统就必须解决下面的最优化问题：

$$d(\boldsymbol{\theta}) = \sum_{l=1}^{2} \sum_{k=1}^{n_o} \sum_{j=1}^{n_m} v_{j,k,l} \|\pi_r(\boldsymbol{m}_{j,k,l}, \boldsymbol{\theta}_{r,l}) - \pi_i(\boldsymbol{M}_j, \boldsymbol{\theta}_{i,l})\|^2 \to \min \qquad (3.164)$$

As in Section 3.9.4.4, π_i denotes the projection of a calibration mark into the image plane coordinate system, $\boldsymbol{\theta}_{i,l}$ are the subset of camera parameters that influence this projection for camera l, π_r denotes the rectification of an image point into the image plane coordinate system, $\boldsymbol{\theta}_{r,l}$ are the subset of camera parameters that influence this rectification for camera l, and $v_{j,k,l}$ is a variable that is 1 if the calibration mark $\boldsymbol{M}_j$ is visible in calibration image k of camera l.

To illustrate the relative orientation and the stereo calibration, Figure 3.117 shows an image pair taken from a sequence of 15 image pairs that were used to calibrate a binocular stereo system. Since we wanted to display the relative orientation in an easy-to-understand manner, and since perspective distortions are visible more clearly based on the rectangular border of the old calibration target, we used this kind of target to calibrate the cameras for the purposes of this example. In a real application, we would use the new calibration target, of course. The calibration returns a translation vector of $(0.1534\,\text{m}, -0.0037\,\text{m}, 0.0449\,\text{m})^\top$ between the cameras, i.e., the second camera is 15.34 cm to the right, 0.37 cm above, and 4.49 cm in front of the first camera, expressed in the camera coordinates of the first camera. Furthermore, the calibration returns a rotation angle of 40.1139° around the axis $(-0.0035, 1.0000, 0.0008)^\top$, i.e., almost around the vertical y axis of the CCS. Hence, the cameras are rotated inward, as in Figure 3.116.

如 3.9.4.4 节所描述，π_i 表示标定板某个标志点到像平面坐标系的投影，$\boldsymbol{\theta}_{i,l}$ 是摄像机 l 中影响此投影的摄像机参数的子集，π_r 表示图像中某个点到像平面坐标系的校正，$\boldsymbol{\theta}_{r,l}$ 表示摄像机 l 中影响此校正的摄像机参数的子集，$v_{j,k,l}$ 是一个变量，当标志点 $\boldsymbol{M}_j$ 在图像 k 中可见时，$v_{j,k}$ 为 1。

为了解释相对位姿与立体视觉系统标定，图 3.117 中显示了用来标定双目立体视觉系统的 15 个立体图像对中的其中一个图像对。因为我们想用一种较容易理解的方式展示相对位姿，同时由于旧款标定板矩形边框可以很清楚地显示透视畸变，基于这些原因在此例中我们使用了旧款标定板，在实际应用中，我们当然会使用新的标定板。标定返回了两个摄像机之间的平移向量 $(0.1534\,\text{m}, -0.0037\,\text{m}, 0.0449\,\text{m})^\top$，也就是说第二个摄像机的位置在第一个摄像机右方 15.34 cm、上方 0.37 cm、前方 4.49 cm，这个坐标是在第一个摄像机坐标系中的坐标。此外，立体视觉系统标定返回一个绕轴 $(-0.0035, 1.0000, 0.0008)^\top$ 旋转角度 40.1139，也就是说几乎绕摄像机坐标系竖直方向的 y 轴旋转。因此，两个摄像机向内倾斜，与图 3.116 所示情况相似。

图 3.117 用来标定双目立体视觉系统的 15 个图像对中的一个图像对。标定返回了两个摄像机之间的平移向量 $(0.1534\,\text{m}, -0.0037\,\text{m}, 0.0449\,\text{m})^\top$ 以及一个绕轴 $(-0.0035, 1.0000, 0.0008)^\top$ 旋转角度 40.1139，也就是说几乎绕摄像机坐标系竖直方向的 y 轴旋转。因此，两个摄像机向内倾斜

3.10.1.3 Epipolar Geometry

To reconstruct 3D points, we must find corresponding points in the two images. "Corresponding" means that the two points P_1 and P_2 in the images belong to the same point P_w in the world. At first, it might seem that, given a point P_1 in the first image, we would have to search in the entire second image for the corresponding point P_2. Fortunately, this is not the case. In Figure 3.116 we already noted that the points P_w, O_1, O_2, P_1, and P_2 all lie in a single plane. The situation of trying to find a corresponding point for P_1 is shown in Figure 3.118. We can note that we know P_1, O_1, and O_2. We do not know at what distance the point P_w lies on the optical ray defined by P_1 and O_1. However, we know that P_w is coplanar with the plane spanned by P_1, O_1, and O_2 (the epipolar plane). Hence, we can see that the point P_2 can only lie on the projection of the epipolar plane onto the second image. Since O_2 lies on the epipolar plane, the projection of the epipolar plane is a line called the epipolar line.

It is obvious that the above construction is symmetric for both images, as shown in Figure 3.119. Hence, given a point P_2 in the second

3.10.1.3 外极线几何结构

为了重构三维空间点，我们必须在两幅图像中找到对应点。这里"对应"意思是指两幅图像中点 P_1 和 P_2 都是世界坐标系中同一点 P_w 的投影。我们可能会认为对第一幅图像中的一点 P_1，需要在第二幅图像的全图中搜索对应点 P_2，幸运的是实际中我们并不需要这样做。在图 3.116 中我们已经说明点 P_w、O_1、O_2、P_1、P_2 都在同一个平面上。图 3.118 显示了如何为点 $P1$ 寻找对应点。我们已知 P_1、O_1 和 O_2 点，并且已知 P_w 点位于 P_1 和 O_1 点定义的视线上，但不知道 P_w 与摄像机之间的距离。不过我们已知 P_w 在由 P_1、O_1、O_2 三点定义的平面（外极平面）上。因此点 P_2 只可能位于极平面在第二幅图像上的投影上。由于 O_2 点在外极平面上，因此外极平面在第二幅图像上的投影为一条直线，我们称这条直线为外极线。

非常明显的是上面的结构对两幅图像是对称的，如图 3.119。因此如果给定第二幅图像中点 P_2，第一幅图

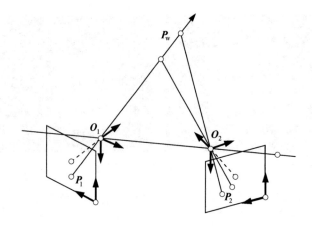

图 3.118　两个摄像机外极线几何结构。给定第一幅图像中一点 P_1，第二幅图像中对应点 P_2 只可能在 P_1 点外极线上，该外极线是由 P_1、O_1、O_2 定义的外极平面在第二幅图像上的投影

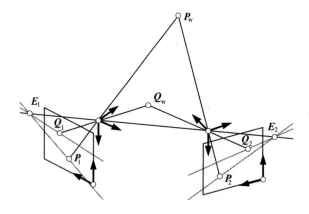

图 3.119　外极线几何结构对两幅图像是对称的。另外一般情况下不同的点对应不同的外极线。每幅图像中的所有外极线都分别相交在外极点 E_1 和 E_2，外极点位置就是某个摄像机投影中心在另一幅图像上的投影

image, the corresponding point can only lie on the epipolar line in the first image. Furthermore, from Figure 3.119, we can see that different points typically define different epipolar lines. We can also see that all epipolar lines of one image intersect at a single point called the epipole. The epipoles are the projections of the opposite projective centers onto the respective image. Note that, since all epipolar planes contain O_1 and O_2, the epipoles lie on the line defined by the two projection centers (the base line).

像中对应点 P_1 也一定位于 P_2 在第一幅图像中的外极线上。从图 3.119 中我们可以看出一般情况下图中不同的点对应不同的外极线。我们还可以看出一幅图像中所有外极线相交在同一点，这个点被称为外极点。外极点是另一个摄像机的投影中心在各自图像中的投影。注意由于所有外极平面都包含 O_1 和 O_2 点，因此外极点都位于由两个投影中心定义的直线（基线）上。

Figure 3.120 shows an example of the epipolar lines. The stereo geometry is identical to Figure 3.117. The images show a PCB. In Figure 3.120(a), four points are marked—they have been selected manually to lie at the tips of the triangles on the four small ICs, as shown in the detailed view in Figure 3.120(c). The corresponding epipolar lines in the second image are shown in Figures 3.120(b) and (d). Note that the epipolar lines pass through the tips of the triangles in the second image.

图 3.120 为外极线的一个示例。立体几何结构与图 3.117 相同。图中为一个 PCB 板。图 3.120（a）中标记出了四个点。从图 3.120（c）中显示的细节图可以看出，它们是通过手动选取四个小芯片上的三角形顶点得到的。图 3.120（b）和图 3.120（d）中显示的是这四个点在第二幅图像中对应的极线位置。注意第二幅图像中的极线通过了相应的四个小芯片上三角形的顶点。

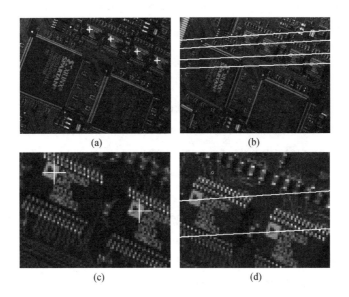

图 3.120 一个 PCB 板的立体图像对：（a）第一幅图像中标记的四个点；（b）第二幅图中显示四个点相应的外极线；（c）图（a）的细节；（d）图（b）的细节；（a）图中四个点是手动选取的四个小芯片上的三角形顶点。注意第二幅图像中的极线通过了相应的四个小芯片上三角形的顶点

As noted above, we have so far assumed that the lenses have no distortions. In reality, this is very rarely true. In fact, by looking closely at Figure 3.120(b), we can already perceive a curvature in the epipolar lines because the camera calibration has determined the radial distortion coefficient for us. If we set the displayed image part as in Figure 3.121, we can clearly see the curvature of

上面我们提到过，到目前为止我们都假设镜头没有畸变。事实上基本不可能存在这种情况。实际上如果仔细观察图 3.120（b），我们可以感觉到外极线是弯曲的，这是由于摄像机标定时我们已经得到了一个径向畸变系数。如果我们像图 3.121 中一样设置图像显示部分，就可以在实际图像中

the epipolar lines in real images. Furthermore, we can see the epipole of the image clearly.

明显看出外极线是弯曲的。另外我们也可以清楚看到该图像的外极点。

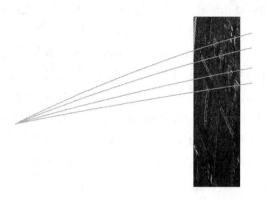

图 3.121　由于镜头畸变影响，外极线并不是直线。这幅图像显示的是图 3.120（b）中同一幅图像。通过缩放使显示图像的同时显示出外极点。纵横比的设置使更容易看出外极线的弯曲

From the above discussion, we can see that the epipolar lines are different for different points. Furthermore, because of lens distortions, typically they are not even straight. This means that, when we try to find corresponding points, we must compute a new, complicated epipolar line for each point that we are trying to match, typically for all points in the first image. The construction of the curved epipolar lines would be much too time consuming for real-time applications. Hence, we can ask ourselves whether the construction of the epipolar lines can be simplified for particular stereo geometries. This is indeed the case for the stereo geometry shown in Figure 3.122. Here, both image planes lie in the same plane. The common plane of both images must be parallel to the base. Additionally, the two images must be vertically aligned. Furthermore, it is assumed that there are no lens distortions. Note that this implies that the two principal distances are identical, that the principal points have the same row coordinate, that the images are rotated such that the column axis is parallel to the base, and that the

从上面的论述中，我们可以看出不同的点对应不同的外极线。并且镜头的畸变导致这些外极线并不是直线。这就意味着如果我们试图找到某点的对应点，我们就必须计算出该点对应的复杂的外极线，一般情况下要为第一幅图像中所有点找到它们对应的外极线。对于实时应用来说求取弯曲的外极线是非常耗时的。因此我们问自己是否能够针对某种特殊的立体几何结构来简化外极线的求取过程。对图 3.122 所示的立体几何结构就可以简化外极线的求解。在这个几何结构中，两个像平面在同一平面，两幅图像的公共平面必须平行于基线，另外，两幅图像必须垂直对齐。此外，假设这里不存在镜头畸变影响。这就意味着两个摄像机的主距相等，两幅图像上主点的行坐标相等，两幅图像旋转到使列轴与基线平行，两个摄像机之间的相对位姿只包含一个 x 方向的平移而不存在任何旋转。由于像平面相互平行，因此外极点与基线之间的

relative orientation contains only a translation in the x direction and no rotation. Since the image planes are parallel to each other, the epipoles lie infinitely far away on the base line. It is easy to see that this stereo geometry implies that the epipolar line for a point is simply the line that has the same row coordinate as the point, i.e., the epipolar lines are horizontal and vertically aligned. Hence, they can be computed without any overhead at all. Since almost all stereo matching algorithms assume this particular geometry, we can call it the epipolar standard geometry.

距离为无穷远。可以看出这种立体几何结构中某点的极线就是与该点行坐标相同的直线，也就是说所有的外极线水平对齐并且垂直对齐。因此，此时计算外极线位置根本不需要耗费任何时间。由于几乎所有的立体匹配算法都假定符合这种特殊几何结构，因此我们称这个几何结构为标准的外极线几何结构。

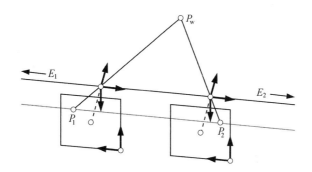

图 3.122　标准的外极线几何结构是指在这种几何结构中，两个像平面在同一平面上，两幅图像的公共平面必须平行于基线，此外，两幅图像必须垂直对齐。另外假设这里不存在镜头畸变的影响。这种几何结构中，某点的极线就是与该点行坐标相同的直线，也就是说所有的极线水平对齐并且垂直对齐

3.10.1.4　Image Rectification

While the epipolar standard geometry results in very simple epipolar lines, it is extremely difficult to align real cameras into this configuration. Furthermore, it is quite difficult and expensive to obtain distortion-free lenses. Fortunately, almost any stereo configuration can be transformed into the epipolar standard geometry, as indicated in Figure 3.123 (Faugeras, 1993). The only exceptions to lie within one of the images. This typically does not occur in are if an epipole happens

3.10.1.4　图像校正

虽然外极线标准几何结构可以使求取极线非常简单，但是实际中将摄像机按该几何结构对齐非常困难。另外得到没有畸变的镜头也非常困难并且也非常昂贵。幸运的是，几乎任何立体视觉几何结构都可以转换为图 3.123 所示的外极线标准几何结构（Faugeras，1993）。唯一一种不能转换的情况就是当其中一个外极点在图像上，这种情况一般不会出现在实际

practical stereo configurations. The process of transforming the images to the epipolar standard geometry is called image rectification. To rectify the images, we need to construct two new image planes that lie in the same plane. To keep the 3D geometry identical, the projective centers must remain at the same positions in space, i.e., $O_{r1} = O_1$ and $O_{r2} = O_2$.

应用中。将图像转换为外极线标准几何结构的过程称为图像校正。为了校正图像需要将两个新的像平面放置在同一平面上。为了保持三维几何结构一致性,两个投影中心的空间位置不能变,也就是说 $O_{r1} = O_1$、$O_{r2} = O_2$。

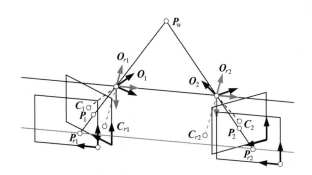

图 3.123 将一个立体视觉几何结构转换为标准的外极线几何结构

Note, however, that we need to rotate the CCSs such that their x axes become identical to the base line. Furthermore, we need to construct two new principal points C_{r1} and C_{r2}. Their connecting vector must be parallel to the base. Furthermore, the vectors from the principal points to the projection centers must be perpendicular to the base. This leaves us two degrees of freedom. First of all, we must choose a common principal distance. Second, we can rotate the common plane in which the image planes lie around the base. These parameters can be chosen by requiring that the image distortion should be minimized (Faugeras, 1993). The image dimensions are then typically chosen such that the original images are completely contained within the rectified images. Of course, we must also remove the lens distortions in the rectification.

然而需要注意的是我们需要旋转摄像机坐标系使它们的 x 轴与基线方向相同。另外,需要创建两个新的主点 C_{r1} 和 C_{r2}。它们之间的连接向量必须与基线平行。另外从主点到各自投影中心的向量必须垂直于基线。这样将留给我们两个自由度。首先必须选择一个共同的主距,然后可以绕基线旋转包含两个像平面在内的公共平面。这些参数可以通过最小化图像畸变来选择(Faugeras,1993)。一般情况下选取的图像尺寸都需要使原始图像完全包含在校正后的图像中。当然,我们同样必须在校正过程中移除镜头畸变。

To obtain the gray value for a pixel in the rectified image, we construct the optical ray for this pixel and intersect it with the original image plane. This is shown, for example, for the points P_{r1} and P_1 in Figure 3.123. Since this typically results in subpixel coordinates, the gray values must be interpolated with the techniques described in Section 3.3.2.2.

While it may seem that image rectification is a very time-consuming process, the entire transformation can be computed once offline and stored in a table. Hence, images can be rectified very efficiently online.

Figure 3.124 shows an example of image rectification. The input image pair is shown in Figures 3.124(a) and (b). The images have the same relative orientation as the images in Figure 3.117.

为了得到校正后图像中某个像素的灰度值，我们构造该像素的视线并将其与原始像平面相交，这种方法可以见图 3.123 中点 P_{r1} 和 P_1。由于一般情况下结果为亚像素坐标，因此灰度值必须使用 3.3.2.2 节中介绍的插值技术得到。

虽然看起来图像校正似乎是一个非常耗时的过程，但是整个转换可以在离线的情况下计算一次并且保存在一个表中。因此图像在线校正的速度非常快。

图 3.124 中显示了一个图像校正的例子。图 3.124（a）和图 3.124（b）中显示的是输入的立体图像对。两幅图像之间的相对位姿与图 3.117 的图像对相同。

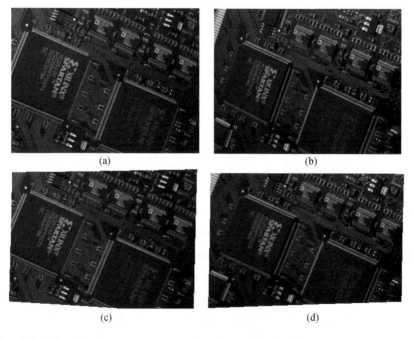

图 3.124 校正立体图像对的示例：（a）和（b）中的两幅图像之间的相对位姿与图 3.117 的图像对相同；（c）和（d）显示的是校正后的图像。注意校正后图像的边缘为梯形。同时校正后图像比原始图像稍宽

The principal distances of the cameras are 13.05 mm and 13.16 mm, respectively. Both images have dimensions 320 × 240. Their principal points are $(155.91, 126.72)^\top$ and $(163.67, 119.20)^\top$, i.e, they are very close to the image center. Finally, the images have a slight barrel-shaped distortion. The rectified images are shown in Figures 3.124(c) and (d). Their relative orientation is given by the translation vector $(0.1599\,\text{m}, 0\,\text{m}, 0\,\text{m})^\top$. As expected, the translation is solely along the x axis. Of course, the length of the translation vector is identical to that in Figure 3.117, since the position of the projective centers has not changed. The new principal distance of both images is 12.27 mm. The new principal points are given by $(-88.26, 121.36)^\top$ and $(567.38, 121.36)^\top$. As can be expected from Figure 3.123, they lie well outside the rectified images. Also, as expected, the row coordinates of the principal points are identical. The rectified images have dimensions 336 × 242 and 367 × 242, respectively. Note that they exhibit a trapezoidal shape that is characteristic of the verging camera configuration. The barrel-shaped distortion has been removed from the images. Clearly, the epipolar lines are horizontal in both images.

3.10.1.5 Disparity

Apart from the fact that rectifying the images results in a particularly simple structure for the epipolar lines, it also results in a very simple reconstruction of the depth, as shown in Figure 3.125. In this figure, the stereo configuration is displayed as viewed along the direction of the row axis of the images, i.e., the y axis of the camera coordinate system. Hence, the image planes are shown as the lines at the bottom of the figure. The depth of a point is quite naturally defined as its z coordinate in the camera coordinate system. By examining the similar triangles $O_1O_2P_w$ and $P_1P_2P_w$, we can see that the depth of P_w depends only on the difference of the column coordinates of the points P_1 and P_2 as follows. From the similarity of the triangles, we have $z/b = (z+f)/(d_w+b)$. Hence, the depth is given by $z = bf/d_w$. Here, b is the length of the base, f is the principal distance, and d_w is the sum of the signed distances of the points P_1 and P_2 to the principal points C_1 and C_2. Since the coordinates of the principal points are given in pixels, but d_w is given in world units, e.g., meters, we have to convert d_w to pixel coordinates by scaling it with the size of the pixels in the x direction: $d_p = d_w/s_x$.

在摄像机坐标系中的 z 坐标。通过分析相似三角形 $O_1O_2P_w$ 和 $P_1P_2P_w$，可以看出点 P_w 的深度只取决于点 P_1 和点 P_2 的列坐标的差值。通过三角形的相似性可以得到 $z/b = (z+f)/(d_w+b)$。因此，深度可以通过下式得到：$z = bf/d_w$。公式中 b 是基线长度，f 是主距，d_w 是点 P_1 到 C_1 与点 P_2 到 C_2 的有符号的距离的总和。由于主点坐标是以像素为单位，而 d_w 的单位是世界单位 (如：米)，因此我们必须通过使用 x 方向上的像素尺寸缩放 d_w 将其转换到像素坐标系中：$d_p = d_w/s_x$。

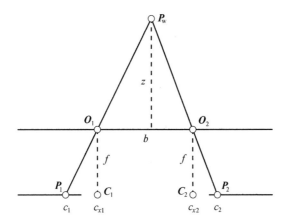

图 3.125　重构某空间点深度 z 只需要该点的视差 $d = c_2-c_1$，也就是在两个校正后图像中相应两点列坐标之差

Now, we can easily see that $d_p = (c_{x1} - c_1) + (c_2 - c_{x2})$, where c_1 and c_2 denote the column coordinates of the points P_1 and P_2, while c_{x1} and c_{x2} denote the column coordinates of the principal points. Rearranging the terms, we find

此时我们可以容易地看出 $d_p = (c_{x1} - c_1) + (c_2 - c_{x2})$，式中 c_1 和 c_2 表示点 P_1 和 P_2 的列坐标，c_{x1} 和 c_{x2} 表示主点的列坐标。将等式重新排列可以发现

$$d_p = (c_{x1} - c_{x2}) + (c_2 - c_1) \qquad (3.165)$$

Since $c_{x1} - c_{x2}$ is constant for all points and known from the calibration and rectification, we can see that the depth z depends only on the difference of the column coordinates $d = c_2 - c_1$. This difference is called the disparity. Hence, we can see that, to reconstruct the depth of a point, we must determine its disparity.

由于 $c_{x1} - c_{x2}$ 对所有点都是常量,并且通过摄像机标定及校正就可得到,我们可以看到深度 z 只与两点列坐标的差 $d = c_2 - c_1$ 相关。这个差值被称为视差。因此,可以看出为了重构某点的深度信息就必须求出它的视差。

3.10.1.6 Stereo Matching

As we have seen in the previous section, the main step in stereo reconstruction is the determination of the disparity of each point in one of the images, typically the first image. Since one calculates, or at least attempts to calculate, a disparity for each point, these algorithms are called dense reconstruction algorithms. It should be noted that there is another class of algorithms that only tries to reconstruct the depth for selected features, e.g., straight lines or points. Since these algorithms require a typically expensive feature extraction, they are seldom used in industrial applications. Therefore, we will concentrate on dense reconstruction algorithms.

Reviews of dense stereo reconstruction algorithms are given in (Scharstein and Szeliski, 2002; Brown et al., 2003; Szeliski, 2011). Many new stereo algorithms are regularly published. Evaluations of newly proposed algorithms are constantly updated on various stereo vision benchmark web sites (see, e.g., Geiger et al., 2017; Scharstein et al., 2017). While many of these algorithms offer stereo reconstructions of somewhat better quality than the algorithms we will discuss below, they are also often much too slow or have too demanding memory requirements for industrial applications.

3.10.1.6 立体匹配

从上节中可以看出,进行立体重构主要的一步就是得到图像中每个点的视差,一般是指第一幅图的所有点。由于需要计算每个点的视差或者至少需要尝试计算每个点的视差,因此这种算法被称为密集型重构算法。应该注意还存在其他类型的算法,这些算法只试图重构某些选中特征的深度信息,例如某些直线或点。由于这些算法一般需要花费较多的时间进行特征提取,它们很少应用在工业应用中。因此我们将集中介绍密集型重构算法。

文献(Scharstein and Szeliski, 2002;Brown et al., 2003;Szeliski, 2011)中介绍了密集型立体重构算法。许多新的立体重构算法也在定期出版,新提出算法的评估也在许多立体视觉基准评测网站上定期更新(Scharstein et al., 2017;Geiger et al., 2017)。尽管这些立体重构算法某种程度上比我们下面要讨论的算法效果更好,但对于工业应用来说,这些算法经常太慢或者对内存需求太高。

Since the goal of dense reconstruction is to find the disparity for each point in the image, the determination of the disparity can be regarded as a template matching problem. Given a rectangular window of size $(2n+1) \times (2n+1)$ around the current point in the first image, we must find the most similar window along the epipolar line in the second image. Hence, we can use the techniques that will be described in greater detail in Section 3.11 to match a point. The gray value matching methods described in Section 3.11.1 are of particular interest because they do not require a costly model generation step, which would have to be performed for each point in the first image. Therefore, gray value matching methods are typically the fastest methods for stereo reconstruction. The simplest similarity measures are the sum of absolute gray value differences (SAD) and sum of squared gray value differences (SSD) measures described later (see Eqs. (3.174) and (3.175)). For the stereo matching problem, they are given by

由于密集型重构的目的是找到图像中每个点的视差值，求视差的过程可以看做模板匹配问题。在第一幅图像中当前点周围给定一个尺寸为 $(2n+1) \times (2n+1)$ 的矩形窗口，我们必须沿第二幅图像中外极线找到一个最相似的窗口。因此，我们可以使用 3.11 节详细讨论的技术来进行匹配。在 3.11.1 节中介绍的灰度值匹配方法非常重要，因为它不需要一个非常耗时的创建模板的步骤。因为如果需要创建模板，就必须为第一幅图像中的所有点创建模板。因此一般情况下基于灰度值匹配方法是立体重构中最快的方法。最简单的计算相似度量的方法是 SAD（aum of absolute gray value differences）和 SSD（sum of squared gray value difference），我们将在下节中详细介绍这两种方法（见等式（3.174）和等式（3.175））。对于立体匹配问题来讲，可使用如下等式表示：

$$\mathrm{SAD}(r,c,d) = \frac{1}{(2n+1)^2} \sum_{i=-n}^{n} \sum_{j=-n}^{n} |g_1(r+i, c+j) - g_2(r+i, c+j+d)| \qquad (3.166)$$

and 和

$$\mathrm{SSD}(r,c,d) = \frac{1}{(2n+1)^2} \sum_{i=-n}^{n} \sum_{j=-n}^{n} (g_1(r+i, c+j) - g_2(r+i, c+j+d))^2 \qquad (3.167)$$

As will be discussed in Section 3.11.1, these two similarity measures can be computed very quickly. Fast implementations for stereo matching using the SAD are given in Hirschmüller et al. (2002); Mühlmann et al. (2002). Unfortunately, these similarity measures have the disadvantage that they are not robust against illumination changes,

与将在 3.11.1 节讨论的一样，这两种相似测量方法的运算速度非常快。使用 SAD 实现快速立体匹配可参见文献（Hirschmuller et al., 2002; Muhlmann et al., 2002）。不幸的是，这两个方法有一个缺点，就是光照变化对算法影响比较大，而由于两个摄

which frequently happen in stereo reconstruction because of the different viewing angles along the optical rays. One way to deal with this problem is to perform a suitable preprocessing of the stereo images to remove illumination variations (Hirschmüller and Scharstein, 2009). The preprocessing, however, is rarely invariant to arbitrary illumination changes. Consequently, in some applications it may be necessary to use the normalized cross-correlation (NCC) described later (see Eq. (3.176)) as the similarity measure, which has been shown to be robust to a very large range of illumination changes that can occur in stereo reconstruction (Hirschmüller and Scharstein, 2009). For the stereo matching problem, it is given by

像机视角不同，光照变化在立体重构的应用中经常出现。解决这个问题的一个方法是对立体图像采用一些适当的预处理算法来消除光照变化带来的影响（Hirschmuller and Scharstein, 2009）。然而，对于任意的光照变化预处理几乎是不变的。因此，在一些应用中可能必须使用后面将要介绍的归一化互相关方法（NCC）（见等式 (3.176)）作为相似度量，此方法已经证明对于立体重构中出现的大范围光照变化具有鲁棒性（Hirschmuller and Scharstein, 2009）。在解决立体匹配问题时，这个方法可以使用下面等式表示：

$$\mathrm{NCC}(r,c,d) = \frac{1}{(2n+1)^2} \sum_{i=-n}^{n} \sum_{j=-n}^{n} \frac{g_1(r+i,c+j) - m_1(r+i,c+j)}{\sqrt{s_1^2(r+i,c+j)}} \cdot \frac{g_2(r+i,c+j+d) - m_2(r+i,c+j+d)}{\sqrt{s_2^2(r+i,c+j+d)}} \quad (3.168)$$

Here, m_i and s_i ($i = 1, 2$) denote the mean and standard deviation of the window in the first and second images. They are calculated analogously to their template matching counterparts in Eqs. (3.177)–(3.180). The advantage of the NCC is that it is invariant to linear illumination changes. However, it is more expensive to compute.

From the above discussion, it might appear that, to match a point, we would have to compute the similarity measure along the entire epipolar line in the second image. Fortunately, this is not the case. Since the disparity is inversely related to the depth of a point, and we typically know in which range of distances the objects we are inter-

式中，m_i 和 s_i ($i = 1, 2$) 分别表示第一幅和第二幅图像中窗口部分的均差和标准差。它们与模板匹配中使用的相关技术的计算方法（等式 (3.177) ~ (3.180)）相似。使用归一化互相关系数的优点就是线性的光照变化不会导致该系数发生变化，然而计算过程更耗时。

从上面的讨论中可以看出，为了找到第一幅图像中某一个点的匹配点，我们应该在第二幅图像中沿整条外极线计算相似度量。幸运的是，实际应用中并不需要这样做。由于一个点的视差与其深度有关，并且随深度的增大而减小。一般情况下我们事先

ested in occur, we can restrict the disparity search space to a much smaller interval than the entire epipolar line. Hence, we have $d \in [d_{\min}, d_{\max}]$, where $d_{\min}$ and $d_{\max}$ can be computed from the minimum and maximum expected distance in the images. Therefore, the length of the disparity search space is given by $l = d_{\max} - d_{\min} + 1$.

After we have computed the similarity measure for the disparity search space for a point to be matched, we might be tempted to simply use the disparity with the minimum (SAD and SSD) or maximum (NCC) similarity measure as the match for the current point. However, typically this will lead to many false matches, since some windows may not have a good match in the second image. In particular, this happens if the current point is occluded because of perspective effects in the second image. Therefore, it is necessary to threshold the similarity measure, i.e., to accept matches only if their similarity measure is below (SAD and SSD) or above (NCC) a threshold. Obviously, if we perform this thresholding, some points will not have a reconstruction, and consequently the reconstruction will not be completely dense.

With the above search strategy, the matching process has a run-time complexity of $O(whln^2)$. This is much too expensive for real-time performance. Fortunately, it can be shown that with a clever implementation, the above similarity measures can be computed recursively. With this, the complexity can be made independent of the window size n, and becomes $O(whl)$. Thus, real-time performance becomes possible. The interested reader is referred to Faugeras et al. (1993) for details.

为一个点在整个视差搜索区域内计算相似度量后，我们可能倾向于简单的使用最小（SAD 和 SSD）或最大（NCC）度量值时的视差作为该点的匹配点。然而，一般情况下这将导致非常多的错误匹配，由于某些窗口可能不能在第二幅图像中找到匹配窗口。实际中这种情况可能是由于透视投影导致当前点在第二幅图像中被遮挡造成。因此，有必要为相似度量设置阈值，也就是只在它们的相似度量低于（SAD 和 SSD）或高于（NCC）某个阈值时才接受该匹配结果。显然如果我们设置这个阈值，图中某些点将不能得到重构结果，因此这个立体重构也不再是完全密集。

使用上述的搜索策略，匹配过程的运行时间复杂度为 $O(whln^2)$。这个复杂度对于需要实时性能的场合来讲过于耗时。幸运的是使用一种巧妙的实现方式，就可以将上述的相似度量进行递归计算。这种算法与匹配窗口尺寸 n 无关，复杂度为 $O(whl)$。此时匹配算法就基本满足实时处理的需要。关于这个算法的详细内容感兴趣的读者可以参见文献（Faugeras et al., 1993）。

Once we have computed the match with an accuracy of one disparity step from the extremum (minimum or maximum) of the similarity measure, the accuracy can be refined with an approach similar to the subpixel extraction of matches that will be described in Section 3.11.3. Since the search space is 1D in the stereo matching, a parabola can be fitted through the three points around the extremum, and the extremum of the parabola can be extracted analytically. Obviously, this will also result in a more accurate reconstruction of the depth of the points.

3.10.1.7 Effect of Window Size

To perform stereo matching, we need to set one parameter: the size of the gray value windows n. This has a major influence on the result of the matching, as shown by the reconstructed depths in Figure 3.126. Here, window sizes of 3×3, 17×17, and 31×31 have been used with the NCC as the similarity measure. We can see that, if the window size is too small, many erroneous results will be found, despite the fact that a threshold of 0.4 has been used to select good matches. This happens because the matching requires a sufficiently distinctive texture within the window. If the window is too small, the texture is not distinctive enough, leading to erroneous matches. From Figure 3.126(b), we see that the erroneous matches are mostly removed by the 17×17 window. However, because there is no texture in some parts of the image, especially in the lower left corners of the two large ICs, some parts of the image cannot be reconstructed. Note also that the areas of the leads around the large ICs are broader than in Figure 3.126(a). This happens because

一旦我们通过相似度量的极值（最大或最小值）得到准确度为一个视差等级的匹配结果，就可以使用将在 3.11.3 节讨论的提取亚像素精度的匹配结果的方法，将准确度进一步提高。由于在立体匹配过程中的搜索范围是一维空间，因此可以使用极值附近的三个点拟合一个抛物线，然后使用抛物线的顶点作为匹配结果。显然这可以使我们重构得到的深度信息更准确。

3.10.1.7 窗口尺寸的影响

为了实现立体匹配，我们需要设置一个参数：灰度值窗口尺寸 n。从图 3.126 中显示的图像得到的深度信息可以看出，这个参数对匹配结果影响很大。在这个示例中使用相似度量的方法为 NCC，窗口尺寸分别为 3×3、17×17 和 31×31。我们可以看出如果窗口尺寸过小，就算我们使用 0.4 作为相似度量的阈值，结果中仍出现很多错误匹配结果。这主要是因为匹配时需要匹配窗口内有足以辨别的特征。如果窗口过小那么特征不容易辨别，就一定会导致匹配结果出现错误。从图 3.126（b）中可以看出，使用 17×17 的窗口可以去除大部分的匹配错误。然而由于图像中某些部分（尤其在两个大芯片表面左下角）没有任何特征，因此这些部分仍不能被重构。同时注意大芯片的管脚部分比图 3.126（a）中的更宽，这是因为此时匹配窗口在图像中横跨高度不连续的区域更大。由于管脚的特征比芯片

the windows now straddle height discontinuities in a larger part of the image. Since the texture of the leads is more significant than the texture on the ICs, the matching finds the best matches at the depth of the leads. To fill the gaps in the reconstruction, we could try to increase the window size further, since this leads to more positions in which the windows have a significant texture. The result of setting the window size to 31 × 31 is shown in Figure 3.126(c). Note that now most of the image can be reconstructed. Unfortunately, the lead area has broadened even more, which is undesirable.

表面的特征更为明显，因此匹配过程中得到最佳匹配部分是在管脚部分。为了填充重构结果中的缺口部分，我们可以进一步增加匹配窗口尺寸，这可以使窗口在图像更多位置都有明显特征。图 3.126(c) 中显示了使用 31×31 匹配窗口的结果。注意此时大部分图像可以被重构。不幸的是，管脚部分变得更宽，而这个结果是不希望得到的。

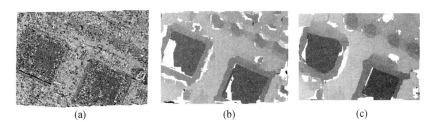

图 3.126　使用 NCC 方法对图 3.124（c）和（d）显示的校正后的立体图像对进行深度重构：（a）匹配窗口尺寸 3×3；（b）窗口尺寸 17×17；（c）窗口尺寸 31×31。图中白色区域部分的点是指由于相似度太低而不能找到匹配对象的

From the above example, we can see that too small window sizes lead to many erroneous matches. In contrast, larger window sizes generally lead to fewer erroneous matches and a more complete reconstruction in areas with little texture. Furthermore, larger window sizes lead to a smoothing of the result, which may sometimes be desirable. However, larger window sizes lead to worse results at height discontinuities, which effectively limits the window sizes that can be used in practice.

从上面的示例中，我们可以看出太小的匹配窗口会导致出现很多错误匹配。相对的，更大的匹配窗口尺寸一般情况下导致更少的错误匹配结果，并且能够将含较少特征的区域进行重构。此外，更大的窗口尺寸将导致结果的平滑，这种平滑效果有些情况是可以接受的。然而窗口尺寸过大也会导致在高度不连续位置结果更加不准确，这使得在实际应用中使用匹配窗口的大小受到限制。

3.10.1.8　Robust Stereo Matching

Despite the fact that larger window sizes generally lead to fewer erroneous matches, they typically

3.10.1.8　鲁棒的立体匹配

尽管将匹配窗口尺寸调大可以减少错误匹配的数量，但是一般情况

cannot be excluded completely based on the window size alone. Therefore, additional techniques are sometimes desirable to reduce the number of erroneous matches even further. An overview of methods to detect unreliable and erroneous matches is given by Hu and Mordohai (2012).

Erroneous matches occur mainly for two reasons: weak texture and occlusions. Erroneous matches caused by weak texture can sometimes be eliminated based on the matching score. However, in general it is best to exclude windows with weak texture a priori from the matching. Whether a window contains a weak texture can be decided on the basis of the output of a texture filter. Typically, the standard deviation of the gray values within the window is used as the texture filter. This has the advantage that it is computed in the NCC anyway, while it can be computed with just a few extra operations in the SAD and SSD. Therefore, to exclude windows with weak textures, we require that the standard deviation of the gray values within the window should be large.

The second reason why erroneous matches can occur are perspective occlusions, which, for example, occur at height discontinuities. To remove these errors, we can perform a consistency check that works as follows. First, we find the match from the first to the second image as usual. We then check whether matching the window around the match in the second image results in the same disparity, i.e., finds the original point in the first image. If this is implemented naively, the run time increases by a factor of 2. Fortunately, with a little extra bookkeeping the disparity consistency check can be performed with very few extra operations, since most of the required data

下不能只基于调节窗口尺寸来减少错误。因此，有时就需要其他额外的技术来进一步减少错误数量。文献（Hu and Mordohai，2012）中概要描述了检测不可靠匹配和错误匹配的方法。

错误匹配的出现主要有两个原因：缺乏特征和遮挡。由缺乏特征导致的匹配错误有时可以通过调整匹配相似度阈值进行消除。然而一般情况下最好在匹配之前就将缺少特征的窗口剔除。一个匹配窗口中是否包含足够的特征可以基于一个纹理滤波器的输出决定。一般情况匹配窗口中所有灰度值的标准差可用作纹理滤波器。这个方法的优势在于 NCC 中无论如何都需要计算该值，在使用 SAD 和 SSD 时也只需要几个额外的操作就可以计算得到该值。因此，为了排除缺乏特征的窗口，我们需要窗口中灰度值标准差足够大。

第二种可能导致出现错误匹配的情况就是透视投影造成的遮挡，比如这种遮挡很可能出现在高度不连续情况。为了消除这种错误我们可以按下面方式进行视差一致性校验：首先，按普通方法在第二幅图像中搜索第一幅图像中某点的匹配点，然后测试第二幅图像中匹配结果附近窗口在第一幅图像中找到的匹配点是否视差值相同，也就是说反向匹配。如果这个算法使用最平常的方法实现，那么运算时间会是以前的 2 倍。幸运的是，使用一个额外的记录就可以使视差一致性校验的工作通过非常少的额外的

have already been computed during the matching from the first to the second image.

Figure 3.127 shows the results of the different methods to increase robustness. For comparison, Figure 3.127(a) displays the result of the standard matching from the first to the second image with a window size of 17×17 using the NCC. The result of applying a texture threshold of 5 is shown in Figure 3.127(b). It mainly removes untextured areas on the two large ICs. Figure 3.127(c) shows the result of applying the disparity consistency check. Note that it mainly removes matches in the areas where occlusions occur.

操作实现，这是由于许多需要的数据已经在第一次匹配过程中计算得到了。

图 3.127 显示使用不同方法增加算法鲁棒性的结果。为了进行比较，图 3.127（a）中显示了标准的使用 NCC 相似度量、匹配窗口为 17×17 时在第二幅图像中直接搜索第一幅图像中各点匹配点的结果。图 3.127（b）中显示的是使用 5 作为纹理阈值的结果。这种方法基本上可以去除两个大芯片上的无特征区域。图 3.127（c）中显示应用了视差一致性校验的结果。注意这种方法主要去除存在遮挡的区域。

 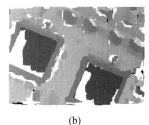 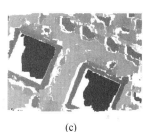

(a)　　　　　　(b)　　　　　　(c)

图 3.127　提高立体匹配算法的健壮性：（a）使用 NCC 相似度量、匹配窗口为 17×17 时在第二幅图像中直接搜索第一幅图像中各点匹配点的结果；（b）使用 5 作为纹理阈值后的结果；（c）应用视差一致性校验后的结果

3.10.1.9　Spacetime Stereo Matching

As can be seen from the above discussion, stereo matching using rectangular windows has to deal with two main problems: weak texture and height discontinuities. As mentioned in Section 2.5.1, the first problem can be solved by projecting a random texture onto the scene. This provides matchable windows in all parts of the scene that are visible from the projector and both cameras. To solve the second problem, the window size must be

3.10.1.9　时空立体匹配

通过上面的讨论我们可以看出，立体匹配使用矩形窗口必须处理两个主要问题：弱纹理和高度不连续性。如 2.5.1 节提到的，第一个问题可以通过在场景上投影一个随机纹理来解决。这为场景的各个部分提供了可匹配的窗口，这些窗口对投影仪和两只摄像机都是可见的。为解决第二个问题，窗口的尺寸必须缩小。如 2.5.1 节

decreased. As described in Section 2.5.1, this can be achieved by projecting a larger number of random patterns onto the scene. This approach is called spacetime stereo (Davis *et al.*, 2005). The greater the number of patterns, the smaller the windows can be made. If a sufficiently large number of patterns is projected, a window size of 1×1 can be used (Davis *et al.*, 2005). This provides a more accurate 3D reconstruction at depth discontinuities.

All of the similarity measures that were described above can be extended to handle multiple images by adding a further sum over the m images. Let the stereo images resulting from projecting m random textures be denoted by $g_1(r,c;t)$ and $g_2(r,c;t)$, $t = 1, \cdots, m$. Then, the spacetime SAD is given by

所述，这可以通过投影更多数量的随机图案到场景上来得到。这个方法被称为时空立体视觉（Davis *et al.*, 2005）。图案数量越多，窗口就能做得越小。如果能够投影足够多的图案，窗口的最小尺寸可以做到 1×1（Davis *et al.*, 2005）。这在深度不连续的情况下可以提供更高精度的三维重建。

上面描述的所有相似度量，可以通过增加 m 张图像求和来扩展到处理多幅图像。用 $g_1(r,c;t)$, $g_2(r,c;t)$, $t = 1, \cdots, m$ 表示基于投影 m 个随机图案所得到的立体图像，则时空 SAD 可由下式给出：

$$\text{SSAD}(r,c,d) = \frac{1}{m(2n+1)^2} \sum_{i=-n}^{n} \sum_{j=-n}^{n} \sum_{t=1}^{m} |g_1(r+i, c+j;t) - g_2(r+i, c+j+d;t)| \quad (3.169)$$

The SSD and NCC can be extended in an identical manner.

SSD 和 NCC 也可以通过相同的方式扩展。

3.10.2 Sheet of Light Reconstruction

3.10.2 激光三角测量法（片光）重建

As described in Section 2.5.2, sheet of light sensors project a laser plane onto the objects in the scene. This creates a bright laser line in the camera image. The 3D shape of the object is related to the vertical displacement (parallax) of the laser line in the image.

Figures 3.128(a)–(j) display 10 images out of a sequence of 290 images of a connecting rod acquired with a sheet of light sensor. The displacement of the laser line can be seen clearly in these images.

如 2.5.2 节所述，激光三角测量（片光）传感器在场景中的物体上投射一个激光平面。这在摄像机图像上会产生一条亮的激光线。物体的三维形状和图像中激光线的垂直位移（视差）是相关的。

图 3.128（a）～（j）所示为通过激光三角测量（片光）传感器采集的连杆的 290 张图像序列中的 10 张图像，在这些图像中可以清晰看到激光线的位移。

图 3.128 （a）～（j）通过激光三角测量（片光）传感器采集的连杆的 290 张图像序列中的 10 张图像。图像已经被裁减的只显示激光线；（k）连杆的三维重建结果

3.10.2.1 Extraction of the Laser Line

To perform the 3D reconstruction, the laser line must be extracted from the images. Since the only relevant information is the vertical position of the laser line, a simple method to extract the laser line is to threshold the image column-by-column using a global threshold with a minimum gray value g_{min} (see Section 3.4.1.1). Next, the connected components of the segmentation result are computed for each column (see Section 3.4.2). In most cases, there only will be a single connected component that corresponds to the laser line or no connected component if the laser line is occluded. However,

3.10.2.1 激光线的提取

为了实现三维重建，激光线必须从图像中提取出来。因为唯一有关的信息就是激光线的垂直位置，一个简单提取激光线的方法是用最小灰度值 g_{min} 作为全局阈值（见 3.4.1.1 节）一行一行地对图像进行阈值分割。然后，对于每一列计算分割结果的连通区域（见 3.4.2 节）。多数情况下，只有一个连通区域和激光线对应，如果激光线被遮挡，则没有连通区域。然而，在激光线被部分反射和部分散射的情况下，可能会出现多个连通区

it might happen that there are multiple connected components if the laser line is partly reflected by the object and partly scattered. In this case, the correct connected component must be selected to resolve the ambiguity. One simple strategy is to select the topmost or bottommost connected component. A more sophisticated strategy is to select the brightest connected component. To determine the brightness of each component, its mean gray value is used (see Section 3.5.2.1). Once the correct connected component has been determined, the position of the laser line can be computed with subpixel accuracy as the gray value center of gravity of the connected component (see Section 3.5.2.2).

3.10.2.2 Sensor Calibration and 3D Reconstruction

The task of reconstructing the 3D position of a point on the laser line within the laser plane is identical to the problem of computing the position of an image point within a world plane from a single image and can be solved with the approach described in Section 3.9.5: the optical ray of a point on the laser line is computed and intersected with the laser plane in 3D. To be able to do this, the interior orientation of the camera must be calibrated as described in Section 3.9.4.4. The calibration also is used to determine the exterior orientation of the camera with respect to the WCS.

The only difference to the approach in Section 3.9.5 is that the exterior orientation of the laser plane must be determined by a different method. This is done by measuring the 3D position of points on the laser plane at different

域。在这种情况下，必须选择正确的连通区域来解决歧义。一个简单的策略就是选择顶端的或者底端的连通区域。更复杂点的策略是选择最亮的连通区域。为确定每个区域的亮度，可以使用平均灰度值（见 3.5.2.1 节）。一旦正确的连通区域确定下来，激光线的位置就可以作为连通区域的重心以亚像素精度计算出来（见 3.5.2.2 节）。

3.10.2.2 传感器标定和三维重建

重建激光平面内激光线上某点的三维位置，和通过单张图像计算世界平面中某个图像点的位置相同，都可以通过 3.9.5 节描述的方法解决：在激光线上某个点的光线可以计算出且与三维空间中激光平面相交。要做到这点，摄像机的内参必须如 3.9.4.4 节所述的方法标定。标定也可以用来确定摄像机相对于世界坐标系的位姿。

与 3.9.5 节方法唯一不同的是激光平面的外参必须通过不同的方法来确定。可以通过测量不同位置激光平面上点的三维位置以及通过测量拟合一个平面来实现（使用 3.8.1 节所述

places within the laser plane and fitting a plane through the measurements (using a trivial extension of the approach described in Section 3.8.1). The individual measurements of points within the laser plane can be obtained as follows with the approach of Section 3.9.5. The calibration target is placed into the scene such that the laser plane is projected onto the calibration target. An image with the laser projector switched off is acquired. Then, the laser projector is switched on. Since the laser line is typically much brighter than the rest of the scene, when acquiring the second image it might be necessary to reduce the brightness of this image to avoid overexposure. As discussed in Section 3.9.6.3, this must be done by changing the exposure time and not the diaphragm. The first image is used to extract the exterior orientation of the calibration target. The second image is used to extract the laser line. Then, the points on the laser line are extracted and the algorithm of Section 3.9.5 is used to determine the 3D coordinates of the points as the intersection of the optical rays corresponding to the laser line points and the plane of the calibration object. Effectively, this calculates the intersection of the laser plane and the plane of the calibration target. Using multiple image pairs of this type in which the calibration target is moved or rotated to different poses results in a set of 3D measurements from which the exterior orientation of the laser plane can be determined by fitting a plane through the 3D points.

It only remains to calibrate the relative motion of the sensor and the scene. This can be done by acquiring two images of the calibration target with the laser projector switched off. For increased accuracy, the two images should not correspond

方法的一个简单扩展）。激光平面内每个点的测量可以通过 3.9.5 节的方法实现。标定板放置到场景中以便激光平面可以投射到标定板上。先获取一幅关闭激光器情况下的图像。然后，开启激光器。因为激光线的亮度要比场景中其他区域亮得多，当采集第二幅图像的时候，可能需要降低图像亮度来避免过曝光。如 3.9.6.3 节讨论的，必须通过降低曝光时间而不是减小光圈来实现。第一幅图像用来提取标定板的位姿。第二幅图像用来提取激光线。然后，提取激光线上的点，并用 3.9.5 节的算法确定点的三维坐标，这些点是激光线上点对应的光线与标定板平面相交的点。实际上，是计算标定板平面和激光平面的交线。移动或者旋转标定板到不同的位置，获取多个这种类型的图像对，使用这些图像对会产生一组三维测量，测量中激光平面的外参可以通过拟合一个通过这些三维点的平面来确定。

剩下的操作就是标定传感器和场景的相对运动。这可以在激光器关闭状态下，通过采集两幅标定板图像来完成。为增加精度，这两幅图像不能是相连的两帧，而是间隔一个较大运

to subsequent frames, but to frames with a larger number n of motion steps. The exterior orientation of the calibration target is computed from both images. This gives the relative motion for n motion steps, which can be converted easily to the relative motion for a single motion step. The relative motion can then be used to transform the 3D points that were computed from a single sheet of light image into the WCS. Figure 3.128(k) displays the result of reconstructing a connecting rod in 3D with the algorithm described above.

3.10.3 Structured Light Reconstruction

As discussed in Section 2.5.3, a structured light sensor consists of a camera and a projector in a geometric configuration that is equivalent to a stereo configuration (see Section 3.10.1.1). The projector projects striped patterns onto the scene, which are used to determine the projector column that corresponds to a point in the scene. This defines a plane in space, which can be intersected with the optical ray of the corresponding point in the camera image to perform the 3D reconstruction. In this section, we will describe how the stripes can be decoded, i.e., how the projector column can be determined, and how the structured light system can be calibrated geometrically as well as radiometrically.

3.10.3.1 Decoding the Stripes

As mentioned in Section 2.5.3.4, we will assume a hybrid system that projects Gray codes as well as phase-shifted fringes. Furthermore, we assume that a completely dark and a completely bright pattern are projected.

动步长为 n 的两帧。标定板的外参通过两幅图像计算。给出步长为 n 的相对运动，可以很容易地转化为单一运动步长的相对运动。相对运动可以用来把从激光三角测量传感器单幅图像计算得出的三维点转换到世界坐标系。图 3.128(k) 显示了基于上面描述的算法三维重建连杆的结果。

3.10.3 结构光重建

如 2.5.3 节中所讨论的，结构光传感器由一只摄像机和一个投影仪以一定的几何结构组成，相当于一个双目立体结构（见 3.10.1.1 节）。投影仪投影条纹图案到场景上，这些条纹用来确定对应场景中某一个点的投影仪列。投影仪列在空间中定义了一个平面，此平面与摄像机图像上对应点的光线相交来实现三维重建。本节，我们将描述如何解码条纹，也就是，如何确定投影仪列，如何在几何以及辐射上标定结构光系统。

3.10.3.1 解码条纹

如 2.5.3.4 节所述，我们假设一个投射格雷码和相移条纹的混合系统。此外，我们假设投射一个全黑和全白的图案。

The first task that must be solved is the decoding of the Gray code images. For this purpose, we must decide whether a pixel in the camera image is illuminated by the projector. This decision reduces to a thresholding operation. Since the decoding algorithm must be able to handle objects with varying reflectance, the threshold must obviously vary locally, depending on the reflectance of the objects in the scene. Since we assume that images of a completely dark and a completely bright pattern have been acquired, the threshold can simply be determined as the mean of the dark and bright images (Sansoni *et al.*, 1997):

$$t_{r,c} = (d_{r,c} + b_{r,c})/2 \qquad (3.170)$$

where $d_{r,c}$ and $b_{r,c}$ denote the dark and bright images, respectively, and $t_{r,c}$ denotes the threshold image. All pixels whose gray value is $\geq t_{r,c}$ are classified as bright (a Gray code bit of 1), all others as dark (a Gray code bit of 0). Note that this is equivalent to the dynamic threshold operation in Eq. (3.51) with $g_{\text{diff}} = 0$. Thresholding the n Gray code images in the above manner allows the Gray code for each pixel to be determined. The Gray code can then be easily decoded into the associated code word number, i.e., the integer column of the code word.

We mention that another strategy to project Gray codes that makes the decoding invariant to the scene reflectance is to project each Gray code pattern as well as its inverse (Sansoni *et al.*, 1997). A Gray code bit of 1 is decoded whenever the regular image is brighter than the inverse image. However, this requires more patterns to be projected than the above approach.

In structured light systems, it may happen that some points in the scene are not illuminated

首先必须解决的任务是格雷码图像的解码。为此，我们必须决定摄像机图像中某个像素是否被投影仪照亮。这个决策简化为一个阈值分割操作。因为解码算法必须能够处理具有不同反射率的物体，根据场景中物体的反射率，阈值必须有明显的局部变化。因为我们假设全黑和全亮图案的图像已经获取，阈值可以简单地确定为黑图和亮图的平均值（Sansoni *et al.*, 1997）：

式中 $d_{r,c}$ 和 $b_{r,c}$ 分别表示黑图和亮图，$t_{r,c}$ 表示阈值图。所有灰度值 $\geq t_{r,c}$ 的像素被归为亮（格雷码的1），其他的像素则为黑（格雷码的0）。注意这和等式（3.51）中动态阈值分割算子在 $g_{\text{diff}} = 0$ 时等价。使用以上方式分割 n 张格雷码图片可确定每个像素的格雷码。然后，格雷码可以轻易被解码成相关的码字编号，也就是码字的整数列。

我们已经提到，对场景反射解码不变的投射格雷码的策略，是投射格雷码图案和它的相反的图案（Sansoni *et al.*, 1997）。无论何时，只要正常图像比相反图像亮，格雷码即被解码为1。因此，比起上面的方法，这要求投射更多的图案。

在结构光系统中，可能会因为遮挡，存在场景中一些像素没有被投影

by the projector because of occlusions. Obviously, these points cannot be reconstructed. Furthermore, parts of the scene that have a very low reflectance also cannot be reconstructed reliably. Therefore, the reliability of the reconstruction can be increased by excluding all pixels for which the difference between the bright and dark images is too small, i.e., only those pixels for which $b_{r,c} - d_{r,c} \geq g_{\min}$ are actually decoded and reconstructed.

仪照亮的情况。显然，这些像素不能重建。此外，场景中低反射率的部分也不能可靠重建。因此，重建的可靠性可以通过排除亮暗图像相差太小的像素来提高，也就是说，只有满足 $b_{r,c} - d_{r,c} \geq g_{\min}$ 的那些像素实际上才会解码和重建。

The only step that remains to be done is to decode the phase-shifted fringes. As already described in Section 2.5.3.3, the phase ϕ of the fringes can be detemined by (see, e.g., Sansoni et al., 1999)

唯一剩下的步骤就是解码相移条纹。如 2.5.3.3 节所述，条纹的相 ϕ 可以通过下式确定（Sansoni et al., 1999）

$$\phi(r,c) = \operatorname{atan2}(I_1(r,c) - I_3(r,c), I_0(r,c) - I_2(r,c)) \tag{3.171}$$

where atan2 denotes the two-argument arctangent function that returns its result in the full 2π range, and we assume that the result has been normalized to $[0, 2\pi)$.

式中，atan2 表示双参数反正切函数，它的返回值范围为 2π，我们假定结果已经归一化到 $[0, 2\pi)$ 区间。

Suppose the pixel-precise integer column of the code word decoded from the Gray code is given by $p(r, c)$. Since the fringe frequency is one cycle per two Gray code words (see Section 2.5.3.4), the subpixel-precise column $s(r, c)$ can be computed as

假设从格雷码解码的码字的像素精度整数列是 $p(r, c)$。因为条纹频率是每两个格雷码字一个周期（见 2.5.3.4 节），亚像素精度列 $s(r, c)$ 可通过如下计算：

$$s(r,c) = p(r,c) + \frac{1}{\pi}(\phi(r,c) \bmod \pi) \tag{3.172}$$

We conclude this section with an example for stripe decoding. Figure 3.129 shows four Gray code images of an M1 screw that has been sprinkled with titanium dioxide powder to reduce specular reflections. The image in Figure 3.129(a) was acquired with a completely bright pattern.

我们通过一个条纹解码的例子来结束本节内容。图 3.129 显示了一个 M1 螺丝的四幅格雷码图像，螺丝已经撒上二氧化钛粉末来减少镜面反射。图 3.129（a）中的图像是全亮图案。图 3.129（b）~（d）显示了最佳

Figures 3.129(b)–(d) show the finest three resolutions of the Gray code patterns. Figure 3.130 displays four images of the screw with fringes that are phase-shifted by 90° each. The system uses a DMD projector with a diamond pixel array layout. As discussed in Section 2.5.3.1, it is advantageous to align the stripes with the pixel grid. Therefore, the stripes are rotated by 45°.

的三种分辨率的格雷码图案。图 3.130 显示了四幅带有条纹的螺丝图像，每个条纹分别相移 90°。该系统使用了一个菱形像素阵列布局的 DMD（数字微镜头设备）投影仪。如 2.5.3.1 节所述，将条纹和像素网格对齐是有利的。因此，条纹被旋转 45°。

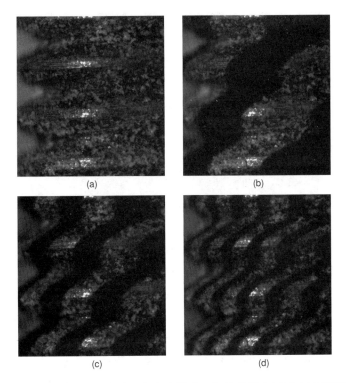

图 3.129　一个 M1 螺丝的四幅格雷码图像，螺丝上撒有二氧化钛粉末以减少镜面反射：（a）全亮图案的图像；（b）～(d) 格雷码图案的三个最好分辨率的图像

The images in Figures 3.129 and 3.130 show a field of view of approximately 1 mm². The camera is equipped with a telecentric lens with a magnification of 3. Because of the large magnification of the system, the projector uses a Scheimpflug configuration in which the DMD is tilted by 39.5°. The projector is mounted at an angle of 30° with respect to the camera.

图 3.129 和图 3.130 中的图像显示的视野大约为 1 mm²。摄像机上安装了一个放大率为 3 的远心镜头。因为系统的放大倍率较大，所以投影仪使用了沙姆结构，其中 DMD 倾斜了 39.5°。投影仪相对于摄像机的安装角度是 30°。

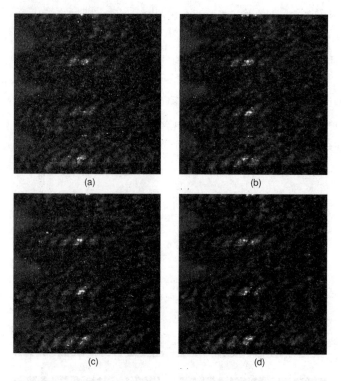

图 3.130 图 3.129 中的四幅螺丝图像，带有相移 90° 的条纹

Figure 3.131(a) displays the integer projector column decoded from the Gray code images in Figure 3.129. The phase decoded from the phase-shifted fringes in Figure 3.130 using Eq. (3.171) is shown in Figure 3.131(b). The result of computing the subpixel-precise projector column by Eq. (3.172) from Figure 3.131(a) and (b) can be seen in Figure 3.131(c). Finally, Figure 3.131(d) displays the 3D reconstruction of the screw. Note that the reconstruction is so accurate that even the titanium dioxide powder is visible in the reconstruction.

图 3.131（a）显示的是从图 3.129 所示的格雷码图像解码的整数投影仪列。使用等式（3.171）从图 3.130 所示相移条纹解码的相如图 3.131（b）所示。图 3.131（a）和图 3.131（b）中用等式（3.172）计算的亚像素投影仪列的结果如图 3.131（c）所示。最后，3.131（d）显示了螺丝的三维重构结果。注意，重建精度是如此准确，甚至二氧化钛粉末在重构结果中都可以看到。

3.10.3.2 Sensor Calibration and 3D Reconstruction

As already mentioned, the geometry of a structured light system is equivalent to that of a stereo

3.10.3.2 传感器标定和三维重建

正如已经提到的那样，结构光系统的几何结构类似于一个双目立体系

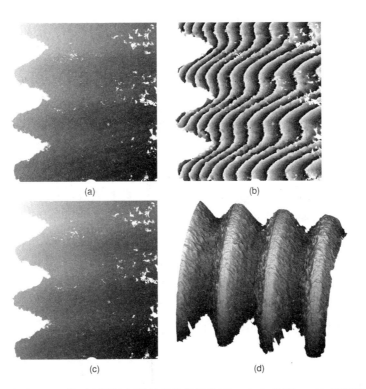

图 3.131 （a）从图 3.129 格雷码图片解码的整数投影仪列；（b）从图 3.130 相移图像解码的相位；（c）从（a）和（b）派生出来的亚像素精度的投影仪列；（d）M1 螺丝的三维重建。注意重建的精度非常高以至于上面的二氧化钛粉末都可以看到

system. The only slight difference is that the light direction of the projector is inverse to that of the second camera in a stereo system. Therefore, technically, the projection center of the projector is the center of the exit pupil of the projector's lens and not its entrance pupil, as in a stereo system. This is, however, only a notational difference. Geometrically, the exit pupil of the projector is the same as the entrance pupil of the second camera of a stereo system.

Ideally, we would like to use the stereo calibration algorithm of Section 3.10.1.1 to calibrate the structured light system. In the following, we will present an algorithm based on the approach by Chen *et al.* (2009) that facilitates this.

统。唯一细微的差别就是投影仪的光线方向与双目立体系统中的第二只摄像机的光线方向相反。因此，技术上，与双目立体系统一样，投影仪的投影中心是投影仪镜头的出射光瞳中心而非入射光瞳中心。然而，这只是标记上的差异。几何机构上，投影仪的出射光瞳和双目立体系统中第二只摄像机的入射光瞳一样。

理想情况下，我们愿意使用 3.10.1.1 节的双目立体标定算法来标定结构光系统。下面，我们会提出一个基于文献（Chen *et al.*, 2009）中方法的算法。

The basic problem that we must solve is that there is no projection from a calibration target into the projector. Therefore, we must find a way to determine the projector coordinates of the center of a calibration mark. To obtain these coordinates, we can proceed as follows. We acquire multiple images of the calibration target in different poses, as in the stereo calibration approach. For each pose of the calibration target, we project Gray codes and phase-shifted fringes in orthogonal directions, e.g., horizontally and vertically (or at 45° and 135° if a DMD with a diamond pixel array layout is used). It is essential that the calibration marks are bright since we must be able to measure their coordinates accurately in the stripe images.

As discussed above, the projected codes will contain an image with a completely bright pattern. These images can be used to extract the calibration mark coordinates and to calibrate the interior orientation of the camera and the exterior orientation of the calibration targets, as described in Section 3.9.4.4. The orientation parameters can then be used to project the mark center points into the camera images to obtain unbiased subpixel-accurate image coordinates of the mark centers.

The only thing that remains to be done is to determine the coordinates of the mark centers in the projector. This can be done using the orthogonal stripes. For simplicity, we assume that the stripes are horizontal and vertical. The projector column coordinate of each mark center in the camera image can be decoded based on the vertical stripes. The same algorithm can be used to decode the row coordinate based on the horizontal

我们必须解决的基本问题是从标定板到投影仪没有投影。因此，必须找到一种方法来确定标定板标志中心的投影仪坐标。为了得到这些坐标，可以如下进行操作。和双目立体系统中的方法一样，采集不同姿态的标定板的多幅图像。对标定板的每个姿态来说，我们在正交方向投影格雷码和相移条纹，也就是，水平方向和垂直方向（如果使用菱形像素阵列布局方式的DMD，则是45°和135°方向）。标定板的标记点必须是亮的因为我们必须在条纹图像上准确测量它们的坐标。

如我们上面所讨论的，投影编码包含一幅带有完全明亮图案的图像。这些图像用来提取标定板的标记点坐标、标定摄像机内参和标定板位姿，如3.9.4.4节所述。位姿参数用来投影标记中心点到摄像机图像来获取标记点中心的无偏差的亚像素精度图像坐标。

唯一还要做的就是确定投影仪中标记点中心的坐标。这可以用正交条纹来完成。为简单起见，假设条纹是水平和垂直的。摄像机图像中，每个标记点中心的投影仪列坐标可以基于垂直条纹解码。同样的算法可以基于水平条纹来解码行坐标。因为摄像机图像上的标记点中心是亚像素精度的，所以可以通过对解码出来的摄

stripes. Since the mark centers in the camera image are subpixel-accurate, we can increase the accuracy of the projector coordinates by bilinear interpolation (cf. Section 3.3.2.2) of the decoded row and column coordinates from 2×2 pixels around each mark coordinate in the camera image. This gives us all the data that we need to use the stereo calibration algorithm of Section 3.10.1.1 to calibrate the structured light system. The relevant data that is obtained from the calibration is the interior orientation of the camera and projector and their relative orientation.

Once the camera and projector have been calibrated, the optical rays of the camera image can in principle be intersected with the corresponding planes of the projector. However, there is one slight difficulty that must be solved: if the projector exhibits lens distortions, the back-projection of a column in the projector is actually no longer a plane, but a curved surface. In this case, the 3D reconstruction is slightly more complicated. An algorithm that solves the reconstruction problem for the polynomial distortion model by a numeric root finding algorithm is described by Chen et al. (2009). For the division model, the same approach leads to a quadratic equation that can be solved analytically.

It is well known that a structured light sensor that uses fringe projection must have a linear gray value response in the projector and the camera (Huang et al., 2003; Gorthi and Rastog, 2010; Wang et al., 2010; Zhang, 2010). If any of these components exhibits a nonlinear behavior, the pure sinusoidal waves will be transformed into waves that contain higher-order harmonic frequencies. These lead to errors in the phase decod-

像机图像上每个标记点坐标周围 2×2 像素范围内的行和列进行双线性插值（见 3.3.2.2 节）来增加投影仪坐标的精度。这样做提供给我们所有需要的数据，借助这些数据使用 3.10.1.1 节中双目标定算法来标定结构光系统。标定中获取的相关数据是摄像机和投影仪的内参以及它们的相对关系。

摄像机和投影仪一旦标定好，摄像机图像的光线原则上可以与投影仪对应的平面相交。但是，还有一个小问题要解决：如果投影仪出现镜头畸变，投影仪中列的后投影实际上不再是一个平面，而是个曲面。这种情况下，三维重建就稍微有点复杂。文献（Chen et al., 2009）中描述了一个通过数字求根算法解决多项式畸变模型重建问题的算法。对于除法模型，同样的方法会引出一个可以分析解决的二次方程。

众所周知，使用条纹投影的结构光传感器在投影仪和摄像机中必须有线性的灰度响应值（Huang et al., 2003; Gorthi and Rastog, 2010; Wang et al., 2010; Zhang, 2010）。如果这些组件中任何一个表现出非线性的情况，纯的正弦波将会转换为含有高阶谐波频率的波。这会导致相位解码中的错误，表现为高频重构误差。因此，

ing, which manifest themselves as high-frequency reconstruction errors. Therefore, any component that exhibits a nonlinear behavior must be calibrated radiometrically. The camera can be calibrated radiometrically as described in Section 3.2.2. The same algorithm can also be used to calibrate the projector. Alternatively, the projector can be calibrated as described by Huang *et al.* (2003). The projector projects each possible gray value in turn onto a uniform surface of high reflectance. The gray values are measured and averaged in a small window in the center of the camera image. The inverse projector response can be constructed from these measurements. It can then be used to correct the fringe images before sending them to the projector (Wang *et al.*, 2010).

3.11 Template Matching

In the previous sections, we have discussed various techniques that can be combined to write algorithms to find objects in an image. While these techniques can in principle be used to find any kind of object, writing a robust recognition algorithm for a particular type of object can be quite cumbersome. Furthermore, if the objects to be recognized change frequently, a new algorithm must be developed for each type of object. Therefore, a method to find any kind of object that can be configured simply by showing the system a prototype of the class of objects to be found would be extremely useful.

The above goal can be achieved by template matching. Here, we describe the object to be found by a template image. Conceptually, the

任何表现出非线性行为的组件必须进行辐射标定。摄像机可按 3.2.2 节所述方法进行辐射标定。同样的算法也可用来标定投影仪。作为选择，投影仪也可以用文献（Huang *et al.*, 2003）所述方法标定。投影仪依次投影每个可能的灰度值到一个高反射率的均匀表面。在摄像机图像中心一个很小的窗口里对灰度值进行测量和平均。这些测量可以构建逆投影仪响应。然后，在它们发送到投影仪之前（Wang *et al.*, 2010），可以用来矫正条纹图像。

3.11 模板匹配

前面的章节中我们讨论了各种各样的技术，将这些技术合理组合就可以实现在一幅图像中搜索目标物体的算法。虽然从理论上讲使用这些技术可以搜索所有类型的物体，但对某种特殊类型的物体来说，实现一个可靠的识别算法是非常复杂的。另外如果被识别物体经常发生变化，就必须为每种物体开发一个新的识别算法。因此，通过为系统提供此类被测物体原型即可对系统进行简单配置，从而可在图像中寻找所有类型的目标物的方法就非常有用。

通过模板匹配技术可以实现上述目的。我们使用一个模板图像描述被搜索物体。从概念上为了在图像中找

template is found in the image by computing the similarity between the template and the image for all relevant poses of the template. If the similarity is high, an instance of the template has been found. Note that the term "similarity" is used here in a very general sense. We will see below that it can be defined in various ways, e.g., based on the gray values of the template and the image, or based on the closeness of template edges to image edges.

Template matching can be used for several purposes. First of all, it can be used to perform completeness checks. Here, the goal is to detect the presence or absence of the object. Furthermore, template matching can be used for object discrimination, i.e., to distinguish between different types of objects. In most cases, however, we already know which type of object is present in the image. In these cases, template matching is used to determine the pose of the object in the image. If the orientation of the objects can be fixed mechanically, the pose is described by a translation. In most applications, however, the orientation cannot be fixed completely, if at all. Therefore, often the orientation of the object, described by rotation, must also be determined. Hence, the complete pose of the object is described by a translation and a rotation. This type of transformation is called a rigid transformation. In some applications, additionally, the size of the objects in the image can change. This can happen if the distance of the objects to the camera cannot be kept fixed, or if the real size of the objects can change. Hence, a uniform scaling must be added to the pose in these applications. This type of pose (translation, rotation, and uniform scaling) is called a similar-

到模板位置，必须计算模板的所有相关位姿与图像各个位置之间的相似度。如果相似度很高，那么就意味着找到了该模板的一个示例。注意这里提到的相似度是一个广义的概念。我们下面将看到相似度可以使用不同方式定义，例如基于模板和图像的灰度值或基于模板边缘与图像边缘的接近程度。

使用模板匹配可以应用在下面几种场合。首先，它可以用来实现完整性检测。完整性检测的目的是为了检测某个物体存在与否。另外模板匹配还可以用来做物体识别，也就是区分不同类型的物体。然而，大多数情况下我们已经知道图像中是哪类物体。在这种情况下，模板匹配可以用来得到目标物体在图像中的位姿。如果通过机械手段可以保证目标物体的方向不变，此时得到的位姿就可表示为一个平移。然而在大多数情况下，目标物的方向不能保证完全不变，因此必须通过匹配同时确定目标物体的方向，该方向可以表示一个旋转矩阵。因此目标物体的完整位姿可以表示为一个平移和一个旋转。这种类型的变换被称为刚性变换。在一些应用中，图像中目标物体的尺寸也可能发生变化。这种情况可能是因为目标物与摄像机的距离不能保持不变或者目标物本身的尺寸发生变化。因此，这些应用中目标物的位姿中必须加入一个缩放系数。这种类型的位姿（平移、旋转和统一缩放）称为相似变换。如果连摄像机与目标物之间的相对三维方位都

ity transformation. If even the 3D orientation of the camera with respect to the objects can change and the objects to be recognized are planar, the pose is described by a projective transformation (see Section 3.3.1.1). Consequently, for the purposes of this chapter, we can regard the pose of the objects as a specialization of an affine or projective transformation.

In most applications, a single object is present in the search image and the goal of template matching is to find this single instance. In some applications, more than one object is present in the image. If we know a priori how many objects are present, we want to find exactly this number of objects. If we do not have this knowledge, we typically must find all instances of the template in the image. In this mode, one of the goals is also to determine how many objects are present in the image.

3.11.1 Gray-Value-Based Template Matching

In this section, we will examine the simplest kind of template matching algorithms, which are based on the raw gray values in the template and the image. As mentioned above, template matching is based on computing a similarity between the template and the image. Let us formalize this notion. For the moment, we will assume that the object's pose is described by a translation. The template is specified by an image $t(r,c)$ and its corresponding ROI T. To perform the template matching, the template is moved over all positions in the image and a similarity measure s is computed at each position. Hence, the similarity measure s is a function that takes the gray values in the template $t(r, c)$ and the gray values in the shifted ROI of the template at the current position in the image $f(r + u, c + v)$ and calculates a scalar value that measures the similarity based on the gray values within the respective ROI. With this approach, a similarity measure is returned for each point in the transformation space, which for translations can be regarded as an image. Hence, formally, we have

移到图像当前位置时感兴趣区域中的灰度值 $f(r + u, c + v)$，然后基于这些灰度值计算一个标量值作为相似度量。使用这个方法，在变换空间中每个点都会得到一个相似度量，我们可以把这个结果看作一幅图像。公式为：

$$s(r,c) = \mathbf{s}\{t(u,v), f(r+u, c+v); (u,v) \in T\} \quad (3.173)$$

To make this abstract notation concrete, we will discuss several possible gray-value-based similarity measures (Brown, 1992).

为了使这个抽象的公式具体化，我们下面将讨论几种可用的基于灰度值的相似度量方法（Brown，1992）。

3.11.1.1 Similarity Measures Based on Gray Value Differences

The simplest similarity measures are to sum the absolute or squared gray value differences between the template and the image (SAD and SSD). They are given by

3.11.1.1 基于灰度值差值的相似度量方法

最简单的相似度量方法是计算模板与图像之间差值的绝对值的总和或所有差值的平方和（SAD 和 SSD）。它们的等式分别为：

$$\text{SAD}(r,c) = \frac{1}{n} \sum_{(u,v) \in T} |t(u,v) - f(r+u, c+v)| \quad (3.174)$$

and

和

$$\text{SSD}(r,c) = \frac{1}{n} \sum_{(u,v) \in T} (t(u,v) - f(r+u, c+v))^2 \quad (3.175)$$

In both cases, n is the number of points in the template ROI, i.e., $n = |T|$. Note that both similarity measures can be computed very efficiently with just two operations per pixel. These similarity measures have similar properties: if the template and the image are identical, they return a similarity measure of 0. If the image and template are not identical, a value greater than 0 is returned.

在这两个等式中，n 是模板感兴趣区域中点的数量，即 $n = |T|$。注意这两种相似度量的计算效率非常高，因为对每个像素只需要两个操作。这两个相似度量的属性很相似：如果模板和图像是相同的话，它们得到的相似度量为 0。如果图像与模板是不同的，那么相似度量将大于 0。模板与图像

As the dissimilarity increases, the value of the similarity measure increases. Hence, in this case the similarity measure should probably be better called a dissimilarity measure. To find instances of the template in the search image, we can threshold the similarity image $SAD(r,c)$ with a certain upper threshold. This typically gives us a region that contains several adjacent pixels. To obtain a unique location for the template, we must select the local minima of the similarity image within each connected component of the thresholded region.

Figure 3.132 shows a typical application for template matching. Here, the goal is to locate the position of a fiducial mark on a PCB. The ROI used for the template is displayed in Figure 3.132(a). The similarity computed with the SAD, given by Eq. (3.174), is shown in Figure 3.132(b). For this example, the SAD was computed with the same image from which the template was generated. If the similarity is thresholded with a threshold of 20, only a region around the position of the fiducial mark is returned (Figure 3.132(c)). Within this region, the local minimum of the SAD must be computed (not shown) to obtain the position of the fiducial mark.

The SAD and SSD similarity measures work very well as long as the illumination can be kept constant. However, if the illumination can change, they both return larger values, even if the same object is contained in the image, because the gray values are no longer identical. This effect is illustrated in Figure 3.133. Here, a darker and brighter image of the fiducial mark are shown. They were obtained by adjusting the illumination intensity. The SAD computed with the template of Figure 3.132(a) is displayed in Figures 3.133(b) and (e).

之间的区别越大，相似度量的值也越大。因此，这种相似度量更适合被称为不相似度量。为了在图像中找到模板的实例，我们可以使用一个给定的上限对相似性图像 $SAD(r,c)$ 进行阈值分割。一般情况这个阈值分割得到的是包含一些邻接像素的区域。为了得到模板的唯一位置，我们必须在阈值分割得到的每个连通区域中选择相似性图像的最小值。

图 3.132 显示了一个模板匹配的典型应用。目的是在一个 PCB 板上找到基准标志的位置。图 3.132（a）中显示了模板图像使用的感兴趣区域。图 3.132（b）中显示的图像为使用等式（3.174）中 SAD 方法计算得到相似度量。这个例子中，使用与创建模板的图像相同的图像来计算 SAD 值。如果使用 20 作为阈值来对相似性图像进行阈值分割，将只得到在基准标志周围的一个区域（见图 3.132（c））。然后必须在这个区域中计算 SAD 的局部最小值来得到基准标志的位置。

在光照情况保持不变的情况下，SAD 和 SSD 相似度量的结果非常好。但是如果光照发生变化，甚至在图像中存在相同物体的情况下，它们都将返回非常大的结果。这主要是因为图像中的灰度值已经不再相等。图 3.133 显示了这种影响。其中显示了包含基准标志的一个较暗的图像和较亮的图像。它们是通过调节光源亮度拍摄得到的。使用图 3.132（a）中的模板计算得到的 SAD 值显示在图 3.132（b）

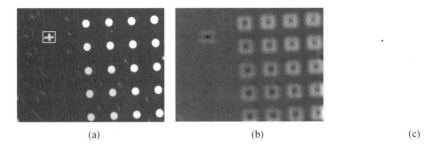

图 3.132 （a）包含一个基准标志的 PCB 板的图像，该图像的白色矩形框中的部分用作模板；（b）图 (a) 中模板与图像 (a) 之间的 SAD 值；（c）使用阈值 20 对图 (b) 进行阈值分割；如图所示结果中只包括基准标志周围的一个区域

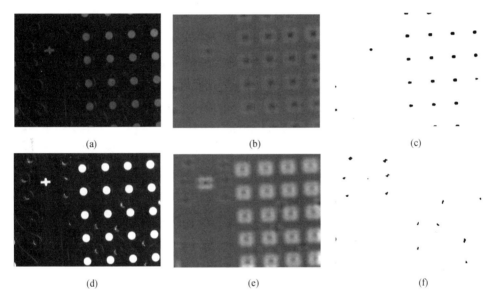

图 3.133 （a）包含一个基准标志的 PCB 板的对比度较低的图像；（b）使用 SAD 方法计算图 3.132 (a) 中模板与图 (a) 的相似度量；（c）使用阈值 35 对图 (b) 进行阈值分割的结果；（d）包含一个基准标志的 PCB 板的对比度较高的图像；（e）使用 SAD 方法计算图 3.132 (a) 中模板与图 (d) 的相似度量；（f）使用阈值 35 对图 (e) 进行阈值分割的结果。这两种情况下，都不能选择一个合适的阈值，使阈值分割只返回包含正确基准标志位置的区域

The result of thresholding them with a threshold of 35 is shown in Figures 3.133(c) and (f). The threshold was chosen such that the true fiducial mark is extracted in both cases. Note that, because of the contrast change, many extraneous instances of the template have been found.

和图 3.132（e）中。使用 35 作为阈值得到的阈值分割结果分别显示在图 3.133（c）和图 3.133（f）中。选择 35 作为阈值是为了保证两种情况下真正的基准标志位置都能够提取出。注意由于对比度发生变化，因此在图像中找到了非常多的错误位置。

3.11.1.2 Normalized Cross-Correlation

As we can see from the above examples, the SAD and SSD similarity measures work well as long as the illumination can be kept constant. In applications where this cannot be ensured, a different kind of similarity measure is required. Ideally, this similarity measure should be invariant to all linear illumination changes (see Section 3.2.1.1). A similarity measure that achieves this is the NCC, given by

$$\mathrm{NCC}(r,c) = \frac{1}{n} \sum_{(u,v) \in T} \frac{t(u,v) - m_t}{\sqrt{s_t^2}} \cdot \frac{f(r+u, c+v) - m_f(r,c)}{\sqrt{s_f^2(r,c)}} \qquad (3.176)$$

Here, m_t is the mean gray value of the template and s_t^2 is the variance of the gray values, i.e.,

$$m_t = \frac{1}{n} \sum_{(u,v) \in T} t(u,v) \qquad (3.177)$$

$$s_t^2 = \frac{1}{n} \sum_{(u,v) \in T} (t(u,v) - m_t)^2 \qquad (3.178)$$

Analogously, $m_f(r,c)$ and $s_f^2(r,c)$ are the mean value and variance in the image at a shifted position of the template ROI:

$$m_f(r,c) = \frac{1}{n} \sum_{(u,v) \in T} f(r+u, c+v) \qquad (3.179)$$

$$s_f^2(r,c) = \frac{1}{n} \sum_{(u,v) \in T} (f(r+u, c+v) - m_f(r,c))^2 \qquad (3.180)$$

The NCC has a very intuitive interpretation. It holds that $-1 \leqslant \mathrm{NCC}(r,c) \leqslant 1$. If $\mathrm{NCC}(r,c) = \pm 1$, the image is a linearly scaled version of the template:

$$\mathrm{NCC}(r,c) = \pm 1 \quad \Leftrightarrow \quad f(r+u, c+v) = at(u,v) + b \qquad (3.181)$$

3.11.1.2 归一化互相关系数

从上面的例子中我们可以看出，SAD 和 SSD 相似度量只有在光照情况不发生变化的情况下可以使用。在不能保证光照稳定的情况下，必须使用另外一种相似度量方法。理想的情况下这个相似度量应该不随任何线性的光照变化而变化（见 3.2.1.1 节）。能够达到这种要求的一种相似度量是归一化互相关系数（NCC），等式为：

式中，m_t 是模板的平均灰度值，s_t^2 是模板所有像素灰度值的方差。也就是说：

与之相似，$m_f(r,c)$ 和 $s_f^2(r,c)$ 是平移到图像当前位置的模板 ROI 中图像所有点的平均灰度值与方差：

归一化互相关系数有一个非常直观的解释。NCC 认为 $-1 \leqslant \mathrm{NCC}(r,c) \leqslant 1$。如果 $\mathrm{NCC}(r,c) = \pm 1$，图像就是模板的一个线性比例版本：

For $NCC(r, c) = 1$ we have $a > 0$, i.e., the template and the image have the same polarity; while $NCC(r, c) = -1$ implies that $a < 0$, i.e., the polarity of the template and image are reversed. Note that this property of the NCC implies the desired invariance against linear illumination changes. The invariance is achieved by explicitly subtracting the mean gray values, which cancels additive changes, and by dividing by the standard deviation of the gray values, which cancels multiplicative changes.

While the template matches the image perfectly only if $NCC(r, c) = \pm 1$, large absolute values of the NCC generally indicate that the template closely corresponds to the image part under examination, while values close to zero indicate that the template and image do not correspond well.

Figure 3.134 displays the results of computing the NCC for the template in Figure 3.132(a) (reproduced in Figure 3.134(a)). The NCC is shown in Figure 3.134(b), while the result of thresholding the NCC with a threshold of 0.75 is shown in Figure 3.134(c). This selects only a region around the fiducial mark. In this region, the local maxi-

如果 $NCC(r, c) = 1$，那么 $a > 0$，也就是说模板与图像的极性相同；如果 $NCC(r, c) = -1$，那么 $a < 0$，也就是说模板与图像的极性相反。注意归一化互相关系数的这个属性就意味着线性光照变化不会影响它的结果。这个不变性通过下面方法实现，通过直接减去平均灰度值可以消除加法对图像的影响，通过用灰度值的标准方差除来消除乘法对图像的影响。

只有在 $NCC(r, c) = \pm 1$ 的情况下，模板与图像之间才完全匹配。一般情况下，归一化互相关系数的绝对值越大就表示模板与正在检测的部分图像之间越接近，归一化互相关系数的绝对值越接近零就表示模板与图像越不一致。

图 3.134 显示了使用图 3.132（a）的模板（重现在图 3.134（a）中）计算 NCC 的结果。图 3.134（b）显示了 NCC 结果，同时图 3.134（c）中显示了使用 0.75 作为阈值对图 3.134（c）进行阈值分割的结果。在结果图中可以看出只选出了基准标志周围的一

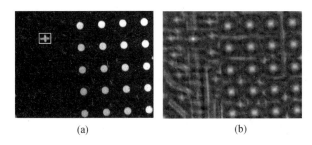

(a) (b) (c)

图 3.134 （a）包含一个基准标志的 PCB 板的图像，该图像的白色矩形框中的部分用作模板；这个图像与图 3.132（a）中的图像相同；（b）图（a）与（a）图中模板的归一化互相关系数；（c）使用阈值 0.75 对图（b）进行阈值分割的结果。图中没有显示对图 3.133 中较暗和较亮的图像求归一化互相关系数的结果，因为它们实际上与图 3.133（b）和（c）中的结果基本一致

mum of the NCC must be computed to derive the location of the fiducial mark (not shown). The results for the darker and brighter images in Figure 3.133 are not shown because they are virtually indistinguishable from the results in Figures 3.134(b) and (c).

3.11.1.3 Efficient Evaluation of the Similarity Measures

In the above discussion, we have assumed that the similarity measures must be evaluated completely for every translation. This is, in fact, unnecessary, since the result of calculating the similarity measure will be thresholded with a threshold t_s later on. For example, thresholding the SAD in Eq. (3.174) means that we require

$$\text{SAD}(r,c) = \frac{1}{n}\sum_{i=1}^{n}|t(u_i,v_i) - f(r+u_i, c+v_i)| \leqslant t_s \qquad (3.182)$$

Here, we have explicitly numbered the points $(u,v)^\top \in T$ by $(u_i,v_i)^\top$. We can multiply both sides by n to obtain

$$\text{SAD}'(r,c) = \sum_{i=1}^{n}|t(u_i,v_i) - f(r+u_i, c+v_i)| \leqslant nt_s \qquad (3.183)$$

Suppose we have already evaluated the first j terms in the sum in Eq. (3.183). Let us call this partial result $\text{SAD}'_j(r,c)$. Then, we have

$$\text{SAD}'(r,c) = \text{SAD}'_j(r,c) + \underbrace{\sum_{i=j+1}^{n}|t(u_i,v_i) - f(r+u_i, c+v_i)|}_{\geqslant 0} \leqslant nt_s \qquad (3.184)$$

Hence, we can stop the evaluation as soon as $\text{SAD}'_j(r,c) > nt_s$ because we are certain that we can no longer achieve the threshold. If we are

3.11.1.3 相似度量的有效评估

在上述讨论中，我们假设必须对所有平移位置都计算相似度量。事实上这样做是不必要的，因为计算相似度量的结果将在随后使用阈值 t_s 进行阈值分割。例如将式 3.174 求出的 SAD 结果进行阈值分割意味着我们需要：

式中使用 $(u,v)^\top \in T$ 对 $(u_i,v_i)^\top$ 进行清楚地编号。然后我们可以在公式两端同时乘以 n 得到下式：

假设我们已经计算得到式 3.183 中前 j 项的总和。我们将这个部分结果称为 $\text{SAD}'_j(r,c)$。然后，我们可以得到公式：

因此我们可以在 $\text{SAD}'_j(r,c) > nt_s$ 的情况下停止计算，这是由于此时该点的相似度量已经不可能再达到阈值。

looking for a maximum number of m instances of the template, we can even adapt the threshold t_s based on the instance with the mth best similarity found so far. For example, if we are looking for a single instance with $t_s = 20$ and we have already found a candidate with $\text{SAD}(r, c) = 10$, we can set $t_s = 10$ for the remaining poses that need to be checked. Of course, we need to calculate the local minima of $\text{SAD}(r, c)$ and use the corresponding similarity values to ensure that this approach works correctly if more than one instance should be found.

For the NCC, there is no simple criterion to stop the evaluation of the terms. Of course, we can use the fact that the mean m_t and standard deviation $\sqrt{s_t^2}$ of the template can be computed once offline because they are identical for every translation of the template. The only other optimization we can make, analogous to the SAD, is that we can adapt the threshold t_s based on the matches we have found so far (Di Stefano et al., 2003).

The above stopping criteria enable us to stop the evaluation of the similarity measure as soon as we are certain that the threshold can no longer be reached. Hence, they prune unwanted parts of the space of allowed poses. Further improvements for pruning the search space have been proposed. For example, Di Stefano et al. (2003) describe additional optimizations that can be used with NCC. Gharavi-Alkhansari (2001) and Hel-Or and Hel-Or (2003) discuss different strategies for pruning the search space when using SAD or SSD. They rely on transforming the image into a representation in which a large portion of the SAD and SSD can be computed with very few evaluations

如果我们正在图像中寻找模板的实例最多 m 个，那么可以基于目前为止在图中找到最相似的 m 个实例来调整阈值 t_s。例如，如果我们正在寻找单个 $t_s = 20$ 实例，并且我们已经找到一个 $\text{SAD}(r, c) = 10$ 的候选者，此时就可以对剩下需要被检查的位姿设置 $t_s = 10$。当然如果在图像中需要找到多于一个的实例，我们就需要计算 $\text{SAD}(r, c)$ 的局部最小值并使用相应的相似值来确保这种方法的正确运行。

对于归一化互相关系数来说，没有简单的标准来判断是否可以停止当前计算。当然，由于模板每次平移后它的平均灰度值 m_t 和标准方差 $\sqrt{s_t^2}$ 都不变，因此这两个值只需要离线计算一次即可。其他我们可以进行的唯一优化与 SAD 相似，就是可以基于当前已经找到的实例来调整阈值 t_s（Di Stefano et al., 2003）。

上面提到的停止标准可以使我们在确定不可能达到阈值的情况下停止相似度量的计算。因此，他们删除了位姿区间中不必要的部分。进一步改进裁减搜索区间方法的研究仍在积极地进行中。例如，文献（Di Stefano et al., 2003）中介绍了可以和归一化互相关系数一起使用的额外优化方法，文献（Gharavi-Alkhansari, 2001）和 (Hel-Or, 2003) 中讨论了使用 SAD 或 SSD 时裁减搜索区间的不同策略。他们首先将图像转换为另一种表示方法，使用这种表示方法时 SAD 和 SSD 的一大部分可以使用非常少的计

so that the above stopping criteria can be reached as soon as possible.

3.11.2 Matching Using Image Pyramids

The evaluation of the similarity measures on the entire image is very time consuming, even if the stopping criteria discussed above are used. If they are not used, the run time complexity is $O(whn)$, where w and h are the width and height of the image and n is the number of points in the template. The stopping criteria typically result in a constant factor for the speed-up, but do not change the complexity. Therefore, a method to further speed up the search is necessary to be able to find the template in real time.

3.11.2.1 Image Pyramids

To derive a faster search strategy, we note that the run time complexity of the template matching depends on the number of translations, i.e., poses, that need to be checked. This is the $O(wh)$ part of the complexity. Furthermore, it depends on the number of points in the template. This is the $O(n)$ part. Therefore, to gain a speed-up, we can try to reduce the number of poses that need to be checked as well as the number of template points. Since the templates typically are large, one way to do this would be to take into account only every ith point of the image and template in order to obtain an approximate pose of the template, which could later be refined by a search with a finer step size around the approximate pose. This strategy is identical to subsampling the image and template. Since subsampling can cause aliasing

3.11.2 使用图形金字塔进行匹配

使用上述讨论的停止标准的情况下，在整个图像中计算相似度量也是一个非常耗时的工作。如果不使用停止标准，那么算法的复杂度为 $O(whn)$，其中 w 和 h 是图像的宽和高，n 是模板中点的数量。使用停止标准一般情况下可以提速的比例为一个常数，但是并不能改变算法复杂度。因此，为了能够在图像中实时找到模板，有必要提出能够进一步提高搜索速度的方法。

3.11.2.1 图像金字塔

为了得到一个更快的搜索策略，我们注意到模板匹配的运行时间的复杂度取决于需要检查的平移数量，也就是位姿的数量。这是算法复杂度中的 $O(wh)$ 部分。另外，算法复杂度还取决于模板中点的数量，这是复杂度中的 $O(n)$ 部分。因此，为了提高算法速度，我们需要试图减少需要检查的位姿数量以及模板中点的数量。由于一般情况下模板非常大，因此其中一种提高速度的方法就是首先只考虑图像和模板中间隔为 i 的点集，此时可以得到模板的大概位姿，随后使用间隔更小的点集在这个大概的位姿周围进行进一步搜索得到更准确的结果。这种策略就相当于对模板和图像进行二次抽样。由于二次抽样可能导

effects (see Sections 3.2.4.2 and 3.3.2.4), this is not a very good strategy because we might miss instances of the template because of the aliasing effects. We have seen in Section 3.3.2.4 that we must smooth the image to avoid aliasing effects. Furthermore, typically it is better to scale the image down multiple times by a factor of 2 than only once by a factor of $i > 2$. Scaling down the image (and template) multiple times by a factor of 2 creates a data structure that is called an image pyramid. Figure 3.135 displays why the name was chosen: we can visualize the smaller versions of the image stacked on top of each other. Since their width and height are halved in each step, they form a pyramid.

致锯齿效应（见 3.2.4.2 节与 3.3.2.4 节），因此这并不是一个好的策略，它可能由于锯齿效应造成某些模板的实例的遗漏。我们在 3.3.2.4 节中已经提到过必须利用平滑图像来消除锯齿影响。另外，一般情况下图像多次缩小 2 倍比直接缩小大于 2 倍的效果更好。将图像与模板多次缩小 2 倍建立起来的数据结构被称为图像金字塔。从图 3.135 中可以看出为什么将这个数据结构用图像金字塔来命名：我们将图像从大到小依次向上堆放。由于每层图像的宽高都比上层减半，因此形成了金字塔形状。

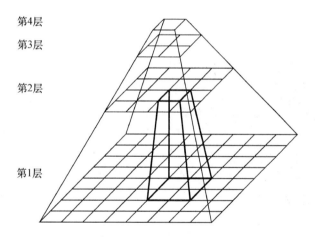

图 3.135 一个图像金字塔是由分辨率连续减半的图像序列组成，并且在高分辨率图像中 2×2 区域中的像素结合为下一层低分辨率图像中的一个像素

When constructing the pyramid, speed is essential. Therefore, the smoothing is performed by applying a 2×2 mean filter, i.e., by averaging the gray value of each 2×2 block of pixels (Tanimoto, 1981). The smoothing could also be performed by a Gaussian filter (Glazer et al., 1983). Note, however, that, in order to avoid the introduction of unwanted shifts into the image pyramid, the Gaus-

在构建金字塔的过程中，速度是非常重要的。因此使用 2×2 的均值滤波器来平滑图像，也就是说求 2×2 区域中的所有像素的平均灰度值（Tanimoto，1981）。也可以使用一个高斯滤波器实现图像平滑（Glazer et al.，1983）。然而，需要注意为了避免在图像金字塔中造成图像平移，高斯

sian filter must have an even mask size. Therefore, the smallest mask size would be 4 × 4. Hence, using the Gaussian filter would incur a severe speed penalty in the construction of the image pyramid. Furthermore, the 2 × 2 mean filter does not have the frequency response problems that the larger versions of the filter have (see Section 3.2.3.6): it drops off smoothly toward a zero response for the highest frequencies, like the Gaussian filter. Finally, it simulates the effects of a perfect camera with a fill factor of 100%. Therefore, the mean filter is the preferred filter for constructing image pyramids.

Figure 3.136 displays the image pyramid levels 2–5 of the image in Figure 3.132(a). We can see that on levels 1–4 the fiducial mark can still be discerned from the BGA pads. This is no longer

滤波器的尺寸必须是偶数。因此高斯滤波器最小为 4×4。这样在创建图像金字塔过程中如果使用高斯滤波器将导致耗时严重增加。另外，2×2 的均值滤波器没有频率响应问题，而较大的滤波器存在这个问题（见 3.2.3.6 节）。事实上与高斯滤波器相似，它平稳减小到对最高频率接近零响应。最后，它模拟了一个填充因数为 100% 的完美摄像机的效果。因此，均值滤波器是创建图像金字塔首选滤波器。

图 3.136 显示图 3.132（a）中图像的图像金字塔的 2~5 层。我们可以看出在图像金字塔的 1~4 层中，图中可以明显看出基准标志与 BGA pads

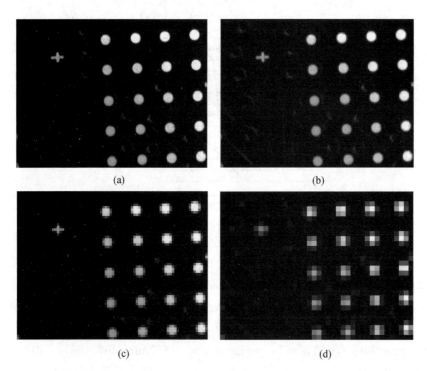

图 3.136　（a）~（d）是图 3.132（a）的图像金字塔的第 2~5 层；注意在第 5 层中基准标志与 BGA pads 之间已经难以辨别

the case on level 5. Therefore, if we want to find an approximate location of the template, we can start the search on level 4.

The above example produces the expected result: the image is progressively smoothed and subsampled. The fiducial mark we are interested in can no longer be recognized as soon as the resolution becomes too low. Sometimes, however, creating an image pyramid can produce results that are unexpected at first glance. One example of this behavior is shown in Figure 3.137. Here, the image pyramid levels 1–4 of an image of a PCB are shown. We can see that on pyramid level 4, all the tracks are suddenly merged into large components with identical gray values. This happens

的不同。但在图像金字塔的第 5 层中已经难以辨别基准标志与 BGA pads。因此，如果希望找到模板的大概位置，我们可以从图像金字塔的第 4 层开始搜索。

上面例子的结果是在意料中的：图像持续平滑和二次采样。在图像分辨率变得太低的时候我们感兴趣的基准标志就不能被识别出了。然而有时创建一个图像金字塔的结果第一眼看上去与期望结果不同。图 3.137 中显示的就是这样的一个例子。显示了一个 PCB 图像的图像金字塔的 1~4 层。我们可以看出在金字塔第 4 层上，PCB 板上所有的导线突然合并为一个灰度值相同的大区域。这种情况出现主要是因为在构建图像金字塔的

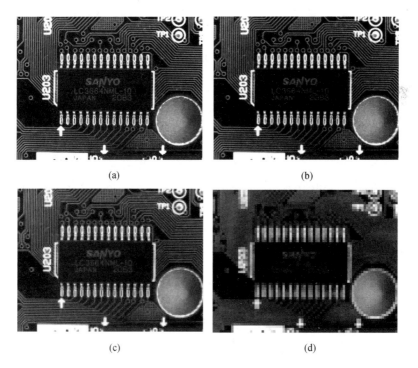

图 3.137　（a）～（d）是 PCB 图像的图像金字塔的 1~4 层；注意在金字塔第 4 层上，PCB 板上所有的导线突然合并为一个灰度值相同的大区域，这是由构建图像金字塔的过程中的图像平滑造成的

because of the smoothing that is performed when the pyramid is constructed. Here, the neighboring thin lines start to interact with each other once the smoothing is large enough, i.e., once we reach a pyramid level that is large enough. Hence, we can see that sometimes valuable information is destroyed by the construction of an image pyramid. If we were interested in matching, say, the corners of the tracks, we could only go as high as level 3 in the image pyramid.

3.11.2.2 Hierarchical Search

Based on image pyramids, we can define a hierarchical search strategy as follows. First, we calculate an image pyramid on the template and search image with an appropriate number of levels. How many levels can be used is mainly defined by the objects we are trying to find. On the highest pyramid level, the relevant structures of the object must still be discernible. Then, a complete matching is performed on the highest pyramid level. Here, of course, we take the appropriate stopping criterion into account. What gain does this give us? In each pyramid level, we reduce the number of image points and template points by a factor of 4. Hence, each pyramid level results in a speed-up of a factor of 16. Therefore, if we perform the complete matching, for example, on level 4, we reduce the amount of computations by a factor of 4096.

All instances of the template that have been found on the highest pyramid level are then tracked to the lowest pyramid level. This is done by projecting the match to the next lower pyramid level, i.e., by multiplying the coordinates of the found match by 2. Since there is an uncertainty in the location of the match, a search area is constructed around the match in the lower pyramid level (e.g., a 5×5 rectangle). Then, the matching is performed within this small ROI, i.e., the similarity measure is computed and thresholded, and the local extrema are extracted. This procedure is continued until the match is lost or tracked to the lowest level. Since the search spaces for the larger templates are very small, tracking the match down to the lowest level is very efficient.

While matching the template on the higher pyramid levels, we need to take the following effect into account: the gray values at the border of the object can change substantially on the highest pyramid level depending on where the object lies on the lowest pyramid level. This happens because a single pixel shift of the object translates to a subpixel shift on higher pyramid levels, which manifests itself as a change in the gray values on the higher pyramid levels. Therefore, on the higher pyramid levels we need to be more lenient with the matching threshold to ensure that all potential matches are being found. Hence, for the SAD and SSD similarity measures we need to use slightly higher thresholds, and for the NCC similarity measure we need to use slightly lower thresholds on the higher pyramid levels.

The hierarchical search is shown in Figure 3.138. The template is the fiducial mark shown in Figure 3.132(a). The template is searched in the same image from which the template was created. As discussed above, four pyramid levels are used in this case. The search starts on level 4. Here, the ROI is the entire image. The NCC and found matches on level 4 are displayed in Figure 3.138(a). As we can see, 12 potential matches

定性，在下一层中的搜索区域定为匹配结果周围的一个区域（如一个 5×5 的矩阵）。然后在这个小的感兴趣区域中进行匹配，也就是说在这个区域内计算相似度量、进行阈值分割以及提取局部极值。这个过程直到找不到匹配对象或到金字塔的最底层结束。由于对于较大模板的搜索区域非常小，跟踪匹配位置到图像金字塔最底层的过程非常高效。

当在金字塔较高层上进行匹配时，我们需要注意以下几点：最高层上目标物体边界上的灰度值可能发生实质性的变化，这种变化主要取决于在最底层图像中目标物体的位置。这种情况主要是因为目标物体上一个像素通过变换后在高层金字塔的图像中发生亚像素级的偏移，这样很明显会造成高层图像金字塔中灰度值发生变化。因此，在图像金字塔中层数越高，就需要使匹配阈值越宽松，这样才能够保证找到所有可能的匹配位置。因此，在图像金字塔的较高层中，对于 SAD 和 SSD 相似度量我们需要使用稍微高一些的阈值，对于 NCC 相似度量我们需要使用稍微低一些的阈值。

在图 3.138 中显示了如何进行分层搜索。模板是图 3.132（a）中显示的基准标志。这里我们在创建模板的图像中进行搜索。与上面讨论的一样，这个情况中我们使用的金字塔层数为 4。搜索从第 4 层金字塔开始。这层中感兴趣区域是整幅图像。对该层图像求得的归一化互相关系数以及找到的匹配点位置如图 3.138（a）所示。我

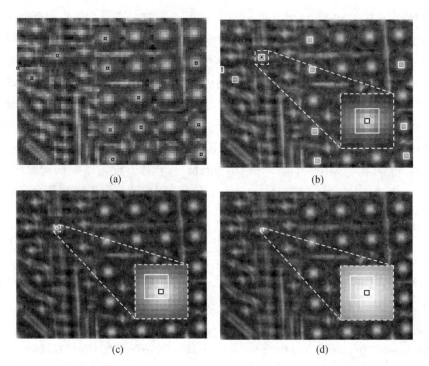

图 3.138　使用图像金字塔进行分级模板匹配；模板是图 3.132（a）中显示的基准标志；为了能够看得更加清楚，在图像金字塔的每层中都显示整幅图像的归一化互相关系数；然而实际上每层中只在适当的感兴趣区域（图中白色矩形框）中计算归一化互相关系数；找到的匹配位置显示为黑色框；（a）在金字塔第 4 层上，在整幅图像中进行匹配并得到 12 个可能的匹配位置；（b）紧接着在第 3 层的白色矩形表示的感兴趣区域中进行匹配，在 12 个感兴趣区域中只找到了一个可能的匹配位置；匹配结果周围的感兴趣区域中的相似度量放大显示在图像的右下角；（c），（d）匹配跟踪到金字塔第 2 层和第 1 层

are initially found. They are tracked down to level 3 (Figure 3.138(b)). The ROIs created from the matches on level 4 are shown in white. For visualization purposes, the NCC is displayed for the entire image. In reality, it is, of course, only computed within the ROIs, i.e., for a total of $12 \times 25 = 300$ translations. Note that at this level the true match turns out to be the only viable match. It is tracked down through levels 2 and 1 (Figures 3.138(c) and (d)). In both cases, only 25 translations need to be checked. Therefore, the match is found extremely efficiently. The zoomed part of the NCC in Figures 3.138(b)–(d) also shows that the pose of the match is progres-

们可以看到找到了 12 个可能的匹配位置。然后跟踪这些位置到第 3 层（图 3.138（b））。通过在第 4 层中找到的大概位置创建的感兴趣区域显示为白色矩形框。为了能够看得更加清楚，图（b）显示的是整幅图像的归一化互相关系数。当然在实际计算过程中，只计算感兴趣区域中的归一化互相关系数，也就是一共计算 $12 \times 25 = 300$ 个平移位置。注意在这层中的匹配只返回正确的匹配位置。然后跟踪这些正确的位置到第 2 层和第 1 层（图 3.138（c）和（d））。在这两层中，只需要检测 25 个平移位置。因此可

sively refined as the match is tracked down the pyramid.

3.11.3 Subpixel-Accurate Gray-Value-Based Matching

So far, we have located the pose of the template with pixel precision. This has been done by extracting the local minima (SAD, SSD) or maxima (NCC) of the similarity measure. To obtain the pose of the template with higher accuracy, the local minima or maxima can be extracted with subpixel precision. This can be done in a manner that is analogous to the method we have used in edge extraction (see Section 3.7.3.5): we simply fit a polynomial to the similarity measure in a 3×3 neighborhood around the local minimum or maximum. Then, we extract the local minimum or maximum of the polynomial analytically. Another approach is to perform a least-squares matching of the gray values of the template and the image (Tian and Huhns, 1986). Since least-squares matching of gray values is not invariant to illumination changes, the illumination changes must be modeled explicitly, and their parameters must be determined in the least-squares fitting in order to achieve robustness to illumination changes (Lai and Fang, 1999).

3.11.4 Template Matching with Rotations and Scalings

Up to now, we have implicitly restricted the template matching to the case where the object must

以非常高效地找到正确的匹配位置。从图3.138（b）～（d）中归一化互相关系数的放大部分同样可以看出，随着图像金字塔的层数越来越低，匹配的结果也越来越精确。

3.11.3 基于灰度值的亚像素精度匹配

目前为止，我们通过匹配得到的模板位置为像素精度。这个位置是通过提取相似度量的局部最大值（SAD，SSD）或局部最小值（NCC）得到。为了使得到的模板位姿的准确度更高，可以提取亚像素精度的局部最小值或最大值。这可以通过与边缘提取（见3.7.3.5节）中所使用的相似的方法实现：我们只需要将在局部最小值或最大值附近 3×3 的邻域内的相似度量拟合为一个多项式。然后求解得到该多项式的局部最大值或最小值。另外一种方法是在模板和图像灰度值之间使用最小二乘匹配（Tian and Huhns，1986）。由于灰度值之间的最小二乘匹配会受光照变化的影响，因此必须为光照变化建立明确的模型，并且它们的参数必须在最小二乘法拟合中确定，从而使算法不受光照变化影响（Lai and Fang，1999）。

3.11.4 带旋转与缩放的模板匹配

目前为止，我们将模板匹配限定为模板与图像中的目标物体方向与尺

have the same orientation and scale in the template and the image, i.e., the space of possible poses was assumed to be the space of translations. The similarity measures we have discussed above only can tolerate small rotations and scalings of the object in the image. Therefore, if the object does not have the same orientation and size as the template, the object will not be found. If we want to be able to handle a larger class of transformations, e.g., rigid or similarity transformations, we must modify the matching approach. For simplicity, we will only discuss rotations, but the method can be extended to scalings and even more general classes of transformations in an analogous manner.

To find a rotated object, we can create the template in multiple orientations, i.e., we discretize the search space of rotations in a manner that is analogous to the discretization of the translations that is imposed by the pixel grid (Anisimov and Gorsky, 1993). Unlike for translations, the discretization of the orientations of the template depends on the size of the template, since the similarity measures are less tolerant to small angle changes for large templates. For example, a typical value is to use an angle step size of 1° for templates with a radius of 100 pixels. Larger templates must use smaller angle steps, while smaller templates can use larger angle steps. To find the template, we simply match all rotations of the template with the image. Of course, this is done only on the highest pyramid level. To make the matching in the pyramid more efficient, we can also use the fact that the templates become smaller by a factor of 2 on each pyramid level. Consequently, the angle step size can be increased

寸一致的情况，也就是说可能位姿的空间被限定为平移的空间。上面我们讨论到的相似度量方法只能够容忍图像中的目标物有很小的旋转和缩放。因此，如果图像中目标物体的方向或大小与模板中不同，那么该目标物体将不能被找到。如果我们希望处理更多类型的变换，例如刚性变换或相似变换，就必须更新匹配方法。为了简单起见，我们将只讨论存在旋转的情况，但这个方法也可以扩展到存在缩放的情况，甚至类似的广义变换。

为了在图像中找到发生旋转的目标物体，我们可以创建多个方向的模板。也就是说，我们将搜索空间进行离散化，这种离散方式就类似于平移的情况下利用像素格进行离散（Anisimov and Gorsky, 1993）。与平移情况不同的是，模板方向的离散取决于模板的大小，这是因为对模板越大就越能够区别更小角度的变化。例如，一般情况下，对于半径为100像素大小的模板设置角度步幅为1°。更大的模板就必须使用更小的角度步幅，更小的模板必须使用更大的角度步幅。为了在图像中找到模板，我们在图像中匹配所有可能旋转角度的模板。当然，只是在图像金字塔的最高层这样做。为了使在图像金字塔中匹配更高效，我们可以利用模板在每层金字塔中都会缩小2倍这个事实。从而，每层金字塔上模板的角度步幅也会增大2倍。因此如果在金字塔最底

by a factor of 2 for each pyramid level. Hence, if an angle step size of 1° is used on the lowest pyramid level, a step size of 8° can be used on the fourth pyramid level.

While tracking potential matches through the pyramid, we also need to construct a small search space for the angles in the next lower pyramid level, analogous to the small search space that we already use for the translations. Once we have tracked the match to the lowest pyramid level, we typically want to refine the pose to an accuracy that is higher than the resolution of the search space we have used. In particular, if rotations are used, the pose should consist of a subpixel translation and an angle that is more accurate than the angle step size we have chosen. The techniques for subpixel-precise localization of the template described above can easily be extended for this purpose.

3.11.5 Robust Template Matching

The above template matching algorithms have served for many years as the methods of choice to find objects in machine vision applications. Over time, however, there has been an increasing demand to find objects in images even if they are occluded or disturbed in other ways so that parts of the object are missing. Furthermore, the objects should be found even if there are a large number of disturbances on the object itself. These disturbances are often referred to as clutter. Finally, objects should be found even if there are severe nonlinear illumination changes. The gray-value-based template matching algorithms we have discussed so far cannot handle these kinds of disturbances. Therefore, in the remainder of this section, we

层上使用的角度步幅为 1°，那么在第 4 层金字塔上可以用 8° 作为角度步幅。

在图像金字塔各层跟踪可能的匹配位置时，我们同样需要在金字塔下一层上为角度创建一个小的搜索区域，这与我们在讨论平移的情况相似。一旦我们跟踪匹配到金字塔的最底层，一般希望使得到的位姿准确度高于所使用的搜索空间的分辨率。特别情况下，如果存在旋转的情况下，位姿应该包含一个亚像素精度的平移位置和一个角度，这个角度值应该比我们选择的角度步幅更精确。实现亚像素精度模板定位的技术在上面已经讨论过，这个技术可以非常容易地扩展到角度应用中。

3.11.5 可靠的模板匹配算法

很多年来，机器视觉应用中都选用上面讨论的模板匹配算法。然而，随着时间流逝越来越多的应用中要求即使目标物体由于遮挡或其他方式的干扰导致只有部分出现在图像中时也能找到模板的位置，另外甚至目标物体本身存在大的干扰的情况下也能够找到目标物体的位置。这些干扰通常称为混乱。另外在图像中存在严重的非线性光照变化时，也应该能够找到目标物体。目前为止我们讨论的基于灰度值的模板匹配算法不能够处理这些类型的干扰。因此，这章节中剩余的部分我们将讨论几种其他的方法，使用这些方法可以在存在遮挡、混乱

will discuss several approaches that have been designed to find objects in the presence of occlusion, clutter, and nonlinear illumination changes.

We have already discussed a feature that is robust to nonlinear illumination changes in Section 3.7: edges are not (or at least very little) affected by illumination changes. Therefore, they are frequently used in robust matching algorithms. The only problem when using edges is the selection of a suitable threshold to segment the edges. If the threshold is chosen too low, there will be many clutter edges in the image. If it is chosen too high, important edges of the object will be missing. This has the same effect as if parts of the object are occluded. Since the threshold can never be chosen perfectly, this is another reason why the matching must be able to handle occlusions and clutter robustly.

To match objects using edges, several strategies exist. First of all, we can use the raw edge points, possibly augmented with some features per edge point, for the matching (see Figure 3.139(b)). Another strategy is to derive geometric primitives by segmenting the edges with the algorithms discussed in Section 3.8.4, and to match these to segmented geometric primitives in the image (see Figure 3.139(c)). Finally, based on a segmentation of the edges, we can derive salient points and match them to salient points in the image (see Figure 3.139(d)). It should be noted that the salient points can also be extracted directly from the image without extracting edges first (Förstner, 1994; Schmid et al., 2000).

3.11.5.1 Mean Squared Edge Distance

A large class of algorithms for edge matching is based on the distance of the edges in the tem-

和非线性光照变化的情况下找到目标物体。

我们已经在 3.7 节中提到了一个不会受非线性光照变化影响的特性：图像中的边缘不会（至少很少）受光线变化的影响。因此，它们经常被应用在可靠的模板匹配算法中。使用边缘时的唯一问题就是如何选择合适阈值来分割边缘。如果这个阈值太小，那么图像中将会有非常多乱七八糟的边缘。如果这个阈值太大，目标物体重要的边缘可能丢失，这种情况与物体发生遮挡时效果相同。由于这个阈值不可能选择得非常完美，这就是为什么需要匹配算法足够可靠能够处理遮挡和混乱情况。

使用边缘匹配物体存在几种策略。首先，我们可以在匹配中使用原始的边缘点，或者增加每个点的一些特性（图 3.139（b））。另外一种策略是使用 3.8.4 节中介绍的算法将边缘分割为多个几何基元，然后在图像中匹配这些分割得到的几何基元（图 3.139（c））。最后一种策略是基于边缘的分割，我们可以得到边缘上的突变点然后在图像中匹配这些突变点（图 3.139（d））。应该注意的是这些突变点可以直接从图像中提取，而不需要首先提取边缘（Forstner, 1994; Schmid et al., 2000）。

3.11.5.1 均方边缘距离

边缘匹配算法中的一大类是基于模板边缘与图像边缘之间的距离。这

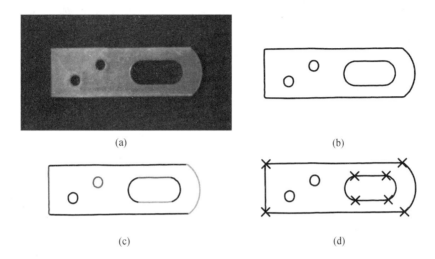

图 3.139 （a）模板对象的图像；（b）图（a）的边缘；（c）将图（b）中的边缘分割为线段与圆弧；（d）通过图（c）中对边缘的分割得到的突变点

plate to the edges in the image. These algorithms typically use the raw edge points for the matching. One natural similarity measure based on this idea is to minimize the mean squared distance between the template edge points and the closest image edge points (Borgefors, 1988). Hence, it appears that we must determine the closest image edge point for every template edge point, which would be extremely costly. Fortunately, since we are only interested in the distance to the closest edge point and not in which point is the closest point, this can be done in an efficient manner by calculating the distance transform of the complement of the segmented edges in the search image (Borgefors, 1988). See Figures 3.140(b) and (d) for examples of the distance transform. A model is considered as being found if the mean distance of the template edge points to the image edge points is below a threshold. Of course, to obtain a unique location of the template, we must calculate the local minimum of this similarity measure. If we want to formalize this similarity measure, we can

些算法一般使用原始边缘点进行匹配。一个基于这种想法的非常普通的相似度量，是使模板边缘点与离它最近的图像边缘点之间的均方距离最小（Borgefors，1988）。因此，显然我们必须为模板边缘上每个点确定离它最近的图像边缘点，这个过程非常耗时。幸运的是，由于我们感兴趣的是与最近图像边缘点之间的距离，而不需要知道哪个点是最近的点，因此可以使用一种效率很高的方式，就是计算搜索图像分割后边缘补集的距离变换（Borgefors，1988）。距离变换的例子请参考图 3.140（b）和图 3.140（d）。如果模板边缘点与图像边缘点之间的平均距离小于一个阈值，我们就认为找到了一个模板的实例。当然，为了得到模板的唯一位置，我们必须计算相似度量的局部最小值。如果我们希望用公式表示这个相似度量，将模板中边缘点表示为 T，分割后搜索图像背景的距离变换（也就是在搜索图像

denote the edge points in the model by T and the distance transform of the complement of the segmented edge region in the search image by $d(r, c)$. Hence, the mean squared edge distance (SED) for the case of translations is given by

中边缘区域以外的部分）表示为 $d(r,c)$。因此，平移情况的均方边缘距离（SED）可表示为：

$$\mathrm{SED}(r,c) = \frac{1}{n} \sum_{(u,v)^\top \in T} d(r+u, c+v)^2 \qquad (3.185)$$

Note that this is very similar to the SSD similarity measure in Eq. (3.175) if we set $t(u,v) = 0$ there and use the distance transform image for $f(u,v)$. Consequently, the SED matching algorithm can be implemented very easily if we already have an implementation of the SSD matching algorithm. Of course, if we use the mean distance instead of the mean squared distance, we could use an existing implementation of the SAD matching, given by Eq. (3.174), for the edge matching.

注意，如果我们令 $t(u,v) = 0$ 并使用距离变换图像替代 $f(u,v)$，这样就与 SED 等式（3.175）相同。因此，如果我们已经实现了 SSD 匹配算法，那么 SED 匹配算法可以非常容易实现。当然，如果我们不使用均方距离，而使用平均距离，可以使用已有的表示 SAD 匹配算法的等式（3.174）来进行边缘匹配。

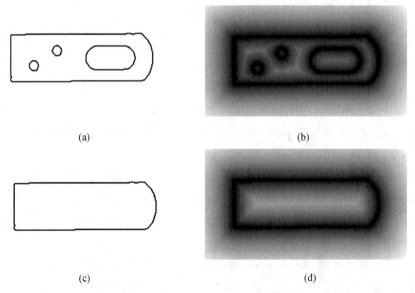

图 3.140 （a）模板边缘；（b）图（a）背景的距离变换；为了看得更加清楚，在图像中应用了一个平方根查找表；（c）有部分边缘没有出现的搜索图像；（d）图（c）背景的距离变换；如果在搜索图像中边缘是完整的，或者包含比模板边缘更多的边缘，此时可以在该搜索图像中找到模板；但是如果搜索图像中模板边缘有部分丢失，那么在该搜索图像中可能找不到模板，这是由于丢失的边缘与最近存在边缘的距离过大

We can now ask ourselves whether the SED fulfills the above criteria for robust matching. Since it is based on edges, it is robust to arbitrary illumination changes. Furthermore, since clutter, i.e., extra edges in the search image, can only decrease the distance to the closest edge in the search image, it is robust to clutter. However, if edges are missing in the search image, the distance of the missing template edges to the closest image edges may become very large, and consequently the model may not be found. This is illustrated in Figure 3.140. Imagine what happens when the model in Figure 3.140(a) is searched in a search image in which some of the edges are missing (Figures 3.140(c) and (d)). Here, the missing edges will have a very large squared distance, which will increase the SED significantly. This will make it quite difficult to find the correct match.

3.11.5.2 Hausdorff Distance

Because of the above problems of the SED, edge matching algorithms using a different distance have been proposed. They are based on the Hausdorff distance of two point sets. Let us call the edge points in the template T and the edge points in the image E. Then, the Hausdorff distance of the two point sets is given by

$$H(T, E) = \max(h(T, E), h(E, T)) \quad (3.186)$$

where

$$h(T, E) = \max_{t \in T} \min_{e \in E} \|t - e\| \quad (3.187)$$

and $h(E, T)$ is defined symmetrically. Hence, the Hausdorff distance consists of determining the maximum of two distances: the maximum distance of the template edges to the closest image edges, and the maximum distance of the image edges to the closest template edges (Rucklidge, 1997). It is immediately clear that, to achieve a low overall distance, every template edge point must be close to an image edge point and vice versa. Therefore, the Hausdorff distance is robust to neither occlusion nor clutter. With a slight modification, however, we can achieve the desired robustness. The reason for the bad performance for occlusion and clutter is that in Eq. (3.187) the maximum distance of the template edges to the image edges is calculated. If we want to achieve robustness to occlusion, instead of computing the largest distance, we can compute a distance with a different rank, e.g., the fth largest distance, where $f = 0$ denotes the largest distance. With this, the Hausdorff distance will be robust to $100f/n\%$ occlusion, where n is the number of edge points in the template. To make the Hausdorff distance robust to clutter, we can similarly modify $h(E,T)$ to use the rth largest distance. However, normally the model covers only a small part of the search image. Consequently, typically there are many more image edge points than template edge points, and hence r would have to be chosen very large to achieve the desired robustness against clutter. Therefore, $h(E,T)$ must be modified to be calculated only within a small ROI around the template. With this, the Hausdorff distance can be made robust to $100r/m\%$ clutter, where m is the number of edge points in the ROI around the template (Rucklidge, 1997). Like the SED, the Hausdorff distance can be computed based on distance transforms: one for the edge region in the image and one for each pose (excluding transla-

最大值组成：一是模板边缘与最近图像边缘之间的最大距离，二是图像边缘与最近模板边缘之间的最大距离（Rucklidge, 1997）。很明显，为了得到一个低的总距离，必须保证每个模板边缘点与一个图像边缘点非常接近，同时也要保证每个图像边缘点都与一个模板边缘点非常接近。因此，使用Hausdorff距离的话，图像中的遮挡和混乱情况都会影响匹配。然而我们可以通过一个小的改进来实现期望的可靠性。在图像中存在遮挡和混乱情况下算法效果比较差主要是由于等式（3.187）中是计算模板边缘到图像边缘的最大距离。如果希望在遮挡的情况下实现足够可靠的算法，我们可以求另一级的距离而不求最大距离，例如我们可以求第 f 大的距离，在 $f = 0$ 时表示最大距离。此时，Hausdorff距离将对 $100f/n\%$ 的遮挡情况是可靠的，其中 n 表示模板中边缘点的数量。为了使 Hausdorff 距离在存在混乱的情况下是可靠的，我们可以类似地使用第 r 大距离来修改 $h(E,T)$。然而，通常模板只能覆盖搜索图像的一小部分，从而图像中边缘点的数量比模板中边缘点数量要大得多，因此为了使Hausdorff距离达到期望的不受混乱影响的效果，选择的 r 值必须非常大。所以，必须改进 $h(E,T)$ 只在模板周围的一个小的感兴趣区域内计算。此时 Hausdorff 距离将对 $100r/m\%$ 的混乱情况是可靠的，其中 m 是指模板周围的感兴趣区域中图像边缘点数量（Rucklidge, 1997）。与 SED 方法相似，也可以基于距离变换来计算Hausdorff 距离：一是为图像中边缘区

tions) of the template edge region. Therefore, we must compute either a very large number of distance transforms offline, which requires an enormous amount of memory, or the distance transforms of the model during the search, which requires a large amount of computation.

As we can see, one of the drawbacks of the Hausdorff distance is the enormous computational load that is required for the matching. In Rucklidge (1997), several possibilities are discussed to reduce the computational load, including pruning regions of the search space that cannot contain the template. Furthermore, a hierarchical subdivision of the search space is proposed. This is similar to the effect that is achieved with image pyramids. However, the method in Rucklidge (1997) only subdivides the search space, but does not scale the template or image. Therefore, it is still very slow. A Hausdorff distance matching method using image pyramids is proposed by Kwon *et al.* (2001).

The major drawback of the Hausdorff distance, however, is that, even with very moderate amounts of occlusion, many false instances of the template will be detected in the image (Olson and Huttenlocher, 1997). To reduce the false detection rate, Olson and Huttenlocher (1997) propose a modification of the Hausdorff distance that takes the orientation of the edge pixels into account. Conceptually, the edge points are augmented with a third coordinate that represents the edge orientation. Then, the distance of these augmented 3D points and the corresponding augmented 3D image points is calculated as the modified Hausdorff distance. Unfortunately, this requires the cal-

域计算距离变换，二是为模板边缘区域的每个位姿（不包括平移）计算距离变换。因此，我们可以在离线情况下计算一个非常大数量的距离变换，这就需要巨大的存储空间，否则就只能在搜索的过程中计算模板的距离变换，这需要非常大的计算量。

我们可以看出，使用 Hausdorff 距离的一个缺点是匹配过程中巨大的计算量。在文献（Rucklidge, 1997）中讨论了几种减少计算量的可能性，其中包括裁减搜索空间中不包含模板在内的区域。另外，提出了一种搜索空间分级细分的方法。这种方法与使用图像金字塔实现的效果类似。然而，文献（Rucklidge, 1997）中的方法只将搜索空间细分，但并没有对模板或图像进行缩放。因此，这个方法还是非常慢。文献（Kwon *et al.*, 2001）中提出了一个使用图像金字塔的 Hausdorff 距离匹配方法。

然而，Hausdorff 距离的主要缺点是就算图像中待搜索的模板实例只存在稍许的遮挡，也会导致在图像中找到很多错误的实例（Olson and Huttenlocher, 1997）。为了减少误判率，文献（Olson and Huttenlocher, 1997）中提出了对 Hausdorff 距离的一种改进方法，这种方法将边缘像素的方向考虑在内。从原理上讲，边缘点增加了表示方向的第 3 个坐标。于是，改进后 Hausdorff 距离就是计算这些扩展的三维点与图像中相应的扩展的三维点之间距离。不幸的是这就需要计算三维距离变换，这将使这种

culation of a 3D distance transform, which makes the algorithm too expensive for machine vision applications. A further drawback of all approaches based on the Hausdorff distance is that it is quite difficult to obtain the pose with subpixel accuracy based on the interpolation of the similarity measure.

3.11.5.3 Generalized Hough Transform

Another algorithm to find objects that is based on the edge pixels themselves is the generalized Hough transform (GHT) proposed by Ballard (1981). The original Hough transform is a method that was designed to find straight lines in segmented edges (Hough, 1962; Duda and Hart, 1972). It was later extended to detect other shapes that can be described analytically, e.g., circles or ellipses. The principle of the GHT can be best explained by looking at a simple case. Let us try to find circles with a known radius in an edge image. Since circles are rotationally symmetric, we only need to consider translations in this case. If we want to find the circles as efficiently as possible, we can observe that, for circles that are brighter than the background, the gradient vector of the edge of the circle is perpendicular to the circle. This means that it points in the direction of the center of the circle. If the circle is darker than its background, the negative gradient vector points toward the center of the circle. Therefore, since we know the radius of the circle, we can theoretically determine the center of the circle from a single point on the circle. Unfortunately, we do not know which points lie on the circle (this is actually the task we would like to solve). However, we can detect the circle by observing that all points

算法对机器视觉应用来讲过于耗时。基于 Hausdorff 距离的所有方法的另外一个缺点就是非常难以基于相似度量的内插算法得到亚像素精度的位姿。

3.11.5.3 广义霍夫变换

另外一种基于边缘像素点本身搜索目标物体的算法是由 Ballard（1981）提出的广义霍夫变换。原始霍夫变换是定义为在分割后的边缘中寻找直线的一种方法（Hough，1962；Duda and Hart，1972）。后来霍夫变换经过扩展可以用来检测其他可解析描述的形状，如圆或椭圆。广义霍夫变换的原理可以通过一个简单例子来解释。如果我们试着在一个边缘图像中寻找一个已知半径的圆。因为圆是旋转对称的，这种情况只需要考虑平移。如果希望能够在边缘图像中尽可能迅速地找到圆的位置，我们可以注意到比背景亮的圆，它的边缘的梯度向量垂直于圆的边界，并且梯度向量的方向指向圆的中心。如果圆比背景更暗，梯度向量方向则背离圆的中心向外。因此，由于我们已知圆的半径，理论上我们就可以从圆上单个点确定圆的中心。但是不幸的是我们不知道哪个点在圆上（这也正是我们真正需要解决的问题）。然而，我们注意到圆上所有像素点拥有同样的特性，这个特性就是基于梯度向量可以构造圆的中心，这样我们就可以找到圆的位置。我们可以累计图像中所有边缘点提供

on the circle will have the property that, based on the gradient vector, we can construct the circle center. Therefore, we can accumulate evidence provided by all edge points in the image to determine the circle. This can be done as follows. Since we want to determine the circle center (i.e., the translation of the circle), we can set up an array that accumulates the evidence that a circle is present as a particular translation. We initialize this array with zeros. Then, we loop through all the edge points in the image and construct the potential circle center based on the edge position, the gradient direction, and the known circle radius. With this information, we increment the accumulator array at the potential circle center by one. After we have processed all the edge points, the accumulator array should contain a large amount of evidence, i.e., a large number of votes, at the locations of the circle centers. We can then threshold the accumulator array and compute the local maxima to determine the circle centers in the image.

An example of this algorithm is shown in Figure 3.141. Suppose we want to locate the circle on top of the capacitor in Figure 3.141(a) and that we know that it has a radius of 39 pixels. The edges extracted with a Canny filter with $\sigma = 2$ and hysteresis thresholds of 80 and 20 are shown in Figure 3.141(b). Furthermore, for every eighth edge point, the gradient vector is shown. Note that for the circle they all point toward the circle center. The accumulator array that is obtained with the algorithm described above is displayed in Figure 3.141(c). Note that there is only one significant peak. In fact, most of the cells in the accumulator array have received so few votes that

的证据来得到圆的位置。可以按照下面步骤实现这个方法：由于我们希望得到圆的中心点坐标（也就是圆的平移），因此可以创建一个数组用来累计圆出现在某个特殊平移位置的次数。首先将该数组中的元素都初始化为 0。然后我们遍历图像中所有边缘点并且基于边缘点位置、该点处的梯度向量以及已知的半径求出一个潜在的圆心位置。使用这些信息，我们将累计数组中对应于这个潜在的圆心位置的元素加 1。在我们遍历图像中所有边缘点后，累计数组中将包含所寻找圆心坐标的非常多的证据，这也相当于非常多数量的投票。然后我们可以将累计数组进行阈值分割，计算局部最大值来确定图像中圆心的位置。

图 3.141 中显示了使用这种算法的一个例子。假设我们希望定位图 3.141（a）中所示电容顶端的圆圈，并且我们已知它的半径为 39 个像素。使用 $\sigma = 2$ 的 Canny 滤波器以及一个高低阈值分别为 80 和 20 的 Hysteresisthresholds 算法对图像进行边缘提取，提取出的边缘显示在 3.141（b）中。另外，每 8 个边缘像素点显示一个梯度向量。注意，对于这个圆来说所有的梯度向量都指向圆的中心。使用上面介绍的算法求出的累计数组显示在图 3.141（c）中。注意累计数组中只有一个主要的波峰。实际上，累计数组中

a square root LUT had to be used to visualize whether there are any votes at all in the rest of the accumulator array. If the accumulator array is thresholded and the local maxima are calculated, the circle in Figure 3.141(d) is obtained.

大部分元素的投票都太少，因此我们显示时使用了一个平方根查找表，这样就可以看出累计数组中其他所有元素都有一些投票。通过在累计数组中进行阈值分割并求出局部最大值，就可以得到图 3.141（d）中显示的圆。

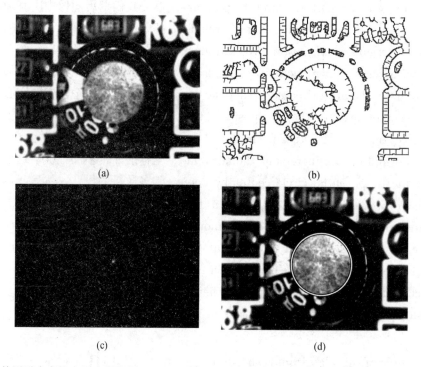

图 3.141　使用霍夫变换在图中提取圆；（a）一个 PCB 板上电容的图像；（b）边缘提取。对边缘上每八个点显示一个梯度向量表示相应的方向；（c）使用边缘点和方向通过霍夫变换得到累计数组；显示时使用平方根查找表可以使累计空间中值较小的区域更清楚。如果使用线性查找表我们只能看到最高峰；（d）对图（c）进行阈值分割并计算局部最大值得到圆的位置

From the above example, we can see that we can find circles in the image extremely efficiently. If we know the polarity of the circle, i.e., whether it is brighter or darker than the background, we only need to perform a single increment of the accumulator array per edge point in the image. If we do not know the polarity of the edge, we need to perform two increments per edge point. Hence, the run time is proportional to the number

从上面例子中可以看出，我们可以在图像中非常高效地找到圆的位置。如果已知圆的极性，也就是知道它比背景暗还是亮，此时我们只需要在图像中每个边缘点处对累计数组执行一个简单加法即可。如果事先不知道边缘的极性，我们只需要在每个边缘点处执行两个加法即可。因此，算法的运算时间与图像中边缘点数量成

of edge points in the image and not to the size of the template, i.e., the size of the circle. Ideally, we would like to find an algorithm that is equally efficient for arbitrary objects.

What can we learn from the above example? First of all, it is clear that, for arbitrary objects, the gradient direction does not necessarily point to a reference point of the object as it did for circles. Nevertheless, the gradient direction of the edge point provides a constraint on where the reference point of the object can be, even for arbitrarily shaped objects. This is shown in Figure 3.142. Suppose we have singled out the reference point o of the object. For the circle, the natural choice would be its center. For an arbitrary object, we can, for example, use the center of gravity of the edge points. Now consider an edge point e_i. We can see that the gradient vector ∇f_i and the vector r_i from e_i to o always enclose the same angle, no matter how the object is translated, rotated, and scaled. For simplicity, let us consider only translations for the moment. Then, if we find an edge point in the image with a certain gradient direction or gradient angle ϕ_i, we could calculate the possible location of the template with the vector r_i

比例，而与模板大小无关，也就是与圆大小无关。理想情况下，我们当然希望能够找到一种搜索任意目标物体的算法可以达到同样的效率。

从上面的例子中我们可以学到什么呢？首先，非常明显，任意物体的边界上的梯度向量不是必然指向一个参考点，这点与圆形不同。不过，边缘点的梯度方向约束了物体的参考点可能存在的位置，甚至对任意形状的物体也是一样。关于这点可以参看图3.142。假设我们已经选定对象参考点为 o。对于圆形来将，自然应该选择它的中心点。对于任意形状的物体，我们可以选择例如边缘点的重心。现在考虑其中一个边缘点 e_i。我们可以看出无论物体如何平移、旋转或缩放，梯度向量 ∇f_i 与从 e_i 到 o 的向量 r_i 之间的夹角始终保持不变。简单起见，我们现在只考虑平移的问题。于是，如果在图像中找到某个边缘点的梯度方向或梯度角度为 ϕ_i，我们可以使用向量 r_i 计算模板可能的位置并相应增加累计数组中的值。注意对于圆形来将，梯度向量 ∇f_i 与向量 r_i 的方

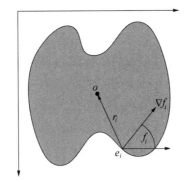

j	ϕ_j	r_i
0	0	$\{r_i \mid \phi_i = 0\}$
1	$\Delta\phi$	$\{r_i \mid \phi_i = \Delta\phi\}$
2	$2\Delta\phi$	$\{r_i \mid \phi_i = 2\Delta\phi\}$
$\vdots$	$\vdots$	$\vdots$

图 3.142　在广义霍夫变换 (GHT) 中构建 R-table 的原理；右边的 R-table 基于模板物体的每个边缘点的梯度角度 ϕ_i 以及每个边缘点到参考点 o 之间的向量 r_i 创建得到

and increment the accumulator array accordingly. Note that for circles, the gradient vector ∇f_i has the same direction as the vector r_i. For arbitrary shapes this no longer holds. From Figure 3.142, we can also see that the edge direction does not necessarily uniquely constrain the reference point, since there may be multiple points on the edges of the template that have the same orientation. For circles, this is not the case. For example, in the lower left part of the object in Figure 3.142, there is a second point that has the same gradient direction as the point labeled e_i, which has a different offset vector to the reference point. Therefore, in the search we have to increment all accumulator array elements that correspond to the edge points in the template with the same edge direction.

Hence, during the search we must be able to quickly determine all the offset vectors that correspond to a given edge direction in the image. This can be achieved in a preprocessing step in the template generation that is performed offline. Basically, we construct a table, called the R-table, that is indexed by the gradient angle ϕ. Each table entry contains all the offset vectors r_i of the template edges that have the gradient angle ϕ. Since the table must be discrete to enable efficient indexing, the gradient angles are discretized with a certain step size $\Delta\phi$. The concept of the R-table is also shown in Figure 3.142. With the R-table, it is very simple to find the offset vectors for incrementing the accumulator array in the search: we simply calculate the gradient angle in the image and use it as an index into the R-table. After the construction of the accumulator array, we threshold the array and calculate the local maxima to find the possible locations of the object.

向相同。但对于任意形状来说并没有这个特性。从图 3.142 中可以看出边缘的方向不是一定可以唯一确定参考点位置，因为在模板的边缘中可能有多个边缘点的方向相同。对于圆形来说没有这个问题。例如，在图 3.142 中目标物体的左下部分有第二个点与点 e_i 的梯度方向相同，但这个点与参考点之间的偏移向量不同。因此，在搜索过程中，我们必须将累计数组中与模板中边缘方向相同的点对应的所有元素加 1。

因此必须能够快速确定图像中一个给定边缘方向对应的所有偏移向量。这个可以在离线创建模板过程中通过一个预处理步骤实现。基本上我们是创建了一个表格，这个表格被称为 R-table，它使用梯度角度 ϕ 作为索引。每个表格条目中包含模板边缘中梯度角度 f 对应的所有偏移向量 r_i。由于这个表格必须离散化以使其能够高效索引，梯度角度使用一个固定间隔大小 $\Delta\phi$ 进行离散化。图 3.142 中图示了 R-table 的概念。使用 R-table 就可以在搜索过程中非常简单地找到偏移向量，然后增加累计数组中相应的元素：我们简单计算图像中边缘的梯度方向并且用这个方向作为 R-table 的索引。在创建了累计数组后，对数组进行阈值分割并提取局部最大值作为目标物的可能位置。这种方法非常容易扩展，可用来搜索

This approach can also be extended easily to deal with rotated and scaled objects (Ballard, 1981). In real images, we also need to consider that there are uncertainties in the location of the edges in the image and in the edge orientations. We have already seen in Eq. (3.124) that the precision of the Canny edges depends on the SNR. Using similar techniques, it can be shown that the precision of the edge angle ϕ for the Canny filter is given by $\sigma_\phi^2 = \sigma_n^2/(4\sigma^2 a^2)$. These values must be used in the online phase to determine a range of cells in the accumulator array that must be incremented to ensure that the cell corresponding to the true reference point is incremented.

The GHT described above is already quite efficient. On average, it increments a constant number of accumulator cells. Therefore, its run time depends only on the number of edge points in the image. However, it is still not fast enough for machine vision applications because the accumulator space that must be searched to find the objects can quickly become very large, especially if rotations and scalings of the object are allowed. Furthermore, the accumulator uses an enormous amount of memory. Consider, for example, an object that should be found in a 640×480 image with an angle range of $360°$, discretized in $1°$ steps. Let us suppose that two bytes are sufficient to store the accumulator array entries without overflow. Then, the accumulator array requires $640 \times 480 \times 360 \times 2 = 221\,184\,000$ bytes of memory, i.e., 211 MB. While this amount of memory is no problem for PCs, it is unacceptably large for many embedded computers. Furthermore, this means that initializing this array alone will require a significant amount of processing time. For this

存在旋转和缩放的目标物体(Ballard, 1981)。在实际图像中,我们同样需要认识到图像中边缘位置以及边缘方向的不确定性。从等式(3.142)中可以看出 Canny 边缘的精度取决于图像的信噪比。使用相似的技术可以得到,使用 Canny 滤波器求出的边缘角度 ϕ 的精度为 $\sigma_\phi^2 = \sigma_n^2/(4\sigma^2 a^2)$。这些值必须在在线匹配的过程中使用,用来确定累计数组中需要增加的元素的范围,这样才可以保证累计数组中对应真正参考点的元素被增加。

上述讨论的广义霍夫变换效率已经非常高。平均起来,它增加固定数量的累计数组元素。因此,它的运行时间只取决于图像中边缘点的数量。然而,这种算法对于机器视觉应用还是不够快,因为用来搜索对象的累计空间会在很短时间内变得非常大,尤其在允许目标物体旋转或缩放的应用中。另外,这个累计器还需要一个巨大的存储空间。想象一下例如我们需要在 640×480 的图像中寻找一个允许角度范围为 360° 的目标物体,并使用 1° 作为步幅。我们假设累计数组中每个元素使用 2 个字节就可以保证不发生数据越界。此时累计数组需要 640×480×360×2 = 221 184 000 字节的存储空间,也就是 211 MB。尽管对于 PC 来说这种内存消耗问题不大,但对于许多嵌入式计算机来说无法接受这种消耗。另外这也意味着单是累计数组的初始化就需要很长的时间。针对这种情况,在文献(Ulrich

reason, a hierarchical GHT is proposed by Ulrich et al. (2003). It uses image pyramids to speed up the search and to reduce the size of the accumulator array by using matches found on higher pyramid levels to constrain the search on lower pyramid levels. The interested reader is referred to Ulrich et al. (2003) for details of the implementation. With this hierarchical GHT, objects can be found in real time even under severe occlusions, clutter, and almost arbitrary illumination changes.

3.11.5.4 Geometric Hashing

The algorithms we have discussed so far were based on matching edge points directly. Another class of algorithms is based on matching geometric primitives, e.g., points, lines, and circles. These algorithms typically follow the hypothesize-and-test paradigm, i.e., they hypothesize a match, typically from a small number of primitives, and then test whether the hypothetical match has enough evidence in the image.

The biggest challenge that this type of algorithm must solve is the exponential complexity of the correspondence problem. Let us, for the moment, suppose that we are only using one type of geometric primitive, e.g., lines. Furthermore, let us suppose that all template primitives are visible in the image, so that potentially there is a subset of the primitives in the image that corresponds exactly to the primitives in the template. If the template consists of m primitives and there are n primitives in the image, there are $\binom{n}{m}$, i.e., $O(n^m)$, potential correspondences between the template and image primitives. If the objects in the search image can be occluded, the number of potential

3.11.5.4 几何哈希法

目前为止上面讨论的算法都是基于直接匹配边缘上的点。另一类边缘匹配算法基于匹配几何基元，例如点、线、圆。这些算法一般是猜测和测试的组合，也就是一般为少数几何基元假设一个匹配位置，然后测试这些假定的匹配位置在图像中有没有足够的证据。

这种类型算法的最大挑战就是它必须解决指数级的复杂度问题。我们现在假设图像中只有一种几何基元，例如线段。另外假设模板中所有基元在图像中可见，也就是说图像中有一个几何基元的子集与模板的几何基元完全一致。如果模板中包含 m 个几何基元，而图像中基元数量为 n，此时在模板与图像基元之间存在 $\binom{n}{m}$（也就是 $O(n^m)$）个潜在的对应关系。如果目标物在图像中可以被部分遮挡，那么潜在的对应关系就会更多，因此此时我们必须允许图像中多个基元匹配模板中的单个基元，这是因为模板

matches is even larger, since we must allow that multiple primitives in the image can match a single primitive in the template, because a single primitive in the template may break up into several pieces, and that some primitives in the template are not present in the search image. It is clear that, even for moderately large values of m and n, the cost of exhaustively checking all possible correspondences is prohibitive. Therefore, geometric constraints and strong heuristics must be used to perform the matching in an acceptable time.

One approach to perform the matching efficiently is called geometric hashing (Lamdan et al., 1990). It was originally described for points as primitives, but can equally well be used with lines. Furthermore, the original description uses affine transformations as the set of allowable transformations. We will follow the original presentation and will note where modifications are necessary for other classes of transformations and lines as primitives. Geometric hashing is based on the observation that three points define an affine basis of the 2D plane. Thus, once we select three points e_{00}, e_{10}, and e_{01} in general position, i.e., not collinear, we can represent every other point as a linear combination of these three points:

$$q = e_{00} + \alpha(e_{10} - e_{00}) + \beta(e_{01} - e_{00}) \qquad (3.188)$$

The interesting property of this representation is that it is invariant to affine transformations, i.e., (α, β) depend only on the three basis points (the basis triplet), but not on the affine transformation, i.e., they are affine invariants. With this, the values (α, β) can be regarded as the affine coordinates of the point $\boldsymbol{q}$. This property holds equally well

中的某单个基元可能被分割成几份，并且有可能模板中的某些几何基元没有出现在图像中。可以看出即使 m 和 n 的值不是特别大，彻底检测所有潜在对应关系的耗时也是不容许的。因此，必须使用几何约束和强大的试探法来保证匹配的耗时可以接受。

一种实现高效匹配的方法称为几何哈希法（Lamdan et al., 1990）。这个方法最初提出时使用点作为基元，但与使用线作为基元时方法相同。另外，最初介绍时使用仿射变换作为允许的变换空间。我们将介绍最原始的算法，同时会强调为了使算法适用于其他类型变换及线段基元，在什么地方需要对算法进行修改。几何哈希法是基于三个点可以定义二维平面的仿射基准的事实。因此一旦我们选择三个普通位置的点 e_{00}、e_{10} 和 e_{01}，也就是不共线的三个点，此时我们可以使用这三个点的线性组合表示其他每个点：

这个表达式的一个有趣的特性就是仿射变换时不会使公式发生变化，也就是说 (α, β) 的值只取决于三个基点，与仿射变换无关。这样 (α, β) 值可以看作点 $\boldsymbol{q}$ 的仿射坐标。这个特性对线段同样适用：三个不平行的非相交于一点的线段可以用来定义一个仿射基

for lines: three non-parallel lines that do not intersect in a single point can be used to define an affine basis. If we use a more restricted class of transformations, fewer points are sufficient to define a basis. For example, if we restrict the transformations to similarity transformations, two points are sufficient to define a basis. Note, however, that two lines are sufficient to determine only a rigid transformation.

The aim of geometric hashing is to reduce the amount of work that has to be performed to establish the correspondences between the template and image points. To do this, it constructs a hash table that enables the algorithm to determine quickly the potential matches for the template. This hash table is constructed as follows. For every combination of three non-collinear points in the template, the affine coordinates (α, β) of the remaining $m - 3$ points of the template are calculated. The affine coordinates (α, β) serve as the index into the hash table. For every point, the index of the current basis triplet is stored in the hash table. Should more than one template be found, the template index also is stored; however, we will not consider this case further, so for our purposes only the index of the basis triplet is stored.

To find the template in the image, we randomly select three points in the image and construct the affine coordinates (α, β) of the remaining $n - 3$ points. We then use (α, β) as an index into the hash table. This returns us the index of the basis triplet. With this, we obtain a vote for the presence of a particular basis triplet in the image. If the randomly selected points do not correspond to a basis triplet of the template, the votes points will not agree. However, if they

准。如果我们使用进一步限制变换类型，那么更少的点就足够定义基准。例如，如果我们限制变换为相似变换，那么两个点足以定义一个基准。然而需要注意两条线段只能定义一个刚性变换。

几何哈希法的目的是减少确定模板与图像点之间对应关系的工作量。因此，它创建了一个哈希表，这个哈希表可以使算法能够快速确定模板的潜在匹配位置。这个哈希表通过下面的步骤创建：对于模板边缘上每 3 个不共线的点，计算模板中其他 $m - 3$ 个点的仿射坐标 (α, β)。这个仿射坐标 (α, β) 将作为哈希表的索引。对于每个点，哈希表中都会保存了当前的三个基准点。如果需要在图像中搜索多个模板，必须在哈希表中保存额外的模板序号；然而，我们不考虑这种情况，因此只需要在哈希表中保存仿射基准的序号即可。

在图像中寻找模板，我们在图像中随机选择 3 个点然后构造图像中其他 $n - 3$ 个点的仿射坐标 (α, β)。我们将使用 (α, β) 为哈希表的索引，这样可以得到 3 个基准点的序号。这样我们可以得到在图像中这 3 个基准点出现的一个投票。如果随机选择的点计算结果没有 3 个基准点与之对应，那么这个投票不接受。然而，如果随机选择的点能够找到 3 个基准点与之

of all the correspond to a basis triplet of the template, many of the votes will agree and will indicate the index of the basis triplet. Therefore, if enough votes agree, we have a strong indication for the presence of the model. The presence of the model is then verified as described below. Since there is a certain probability that we have selected an inappropriate basis triplet in the image, the algorithm iterates until it has reached a certain probability of having found the correct match. Here, we can make use of the fact that we only need to find one correct basis triplet to find the model. Therefore, if k of the m template points are present in the image, the probability of having selected at least one correct basis triplet in t trials is approximately

$$p = 1 - \left(1 - \left(\frac{k}{n}\right)^3\right)^t \tag{3.189}$$

If similarity transforms are used, only two points are necessary to determine the affine basis. Therefore, the inner exponent will change from 3 to 2 in this case. For example, if the ratio of visible template points to image points k/n is 0.2 and we want to find the template with a probability of 99% (i.e., $p = 0.99$), 574 trials are sufficient if affine transformations are used. For similarity transformations, 113 trials would suffice. Hence, geometric hashing can be quite efficient in finding the correct correspondences, depending on how many extra features are present in the image.

After a potential match has been obtained with the algorithm described above, it must be verified in the image. In Lamdan *et al.* (1990), this is done by establishing point correspondences for the remaining template points based on the affine transformation given by the selected

对应，投票将会被接受并且将指出基准点的序号。因此如果有足够多的投票被接受，那么我们将有足够的证据表示图像中存在模板。然后将按照下面的方法检查模板是否真正存在。因为我们非常可能在图像中选中了 3 个不合适的基准点，因此算法不断重复直到确定已经找到正确匹配点。在这里，事实上为了找到模板，我们只需要找到一个正确的。因此，如果 m 个模板边缘点中有 k 个在图像中出现，在 t 次试验中至少找到一个正确基准点组的可能性大约为：

如果使用相似变换，两个点就足够定义仿射基准。因此，这种情况下括号里面的指数将从 3 变成 2。例如，如果图中出现的模板上边缘点的数量与图像上所有边缘点的数量比值 k/n 为 0.2，我们希望找到模板的可能性是 99%（也就是说 $p = 0.99$），那么如果使用仿射变换的话，574 次试验即可。如果使用相似变换，113 次试验就足够了。因此，几何哈希法可以非常高效的找到正确的对应关系，它还取决与图像中存在多少额外的特征。

通过上述描述的算法得到一个潜在匹配位置后，必须在图像中进行验证。在文献（Lammdan *et al.*, 1990）中，利用选中的三个基准点建立的仿射变换，建立模板上其他点的对应关系。基于这些对应关系，使用最小二

basis triplet. Based on these correspondences, an improved affine transformation is computed by a least-squares minimization over all corresponding points. This, in turn, is used to map all the edge points of the template, i.e., not only the characteristic points that were used for the geometric hashing, to the pose of the template in the image. The transformed edges are compared to the image edges. If there is sufficient overlap between the template and image edges, the match is accepted and the corresponding points and edges are removed from the segmentation. If more than one instance of the template is to be found, the entire process is repeated.

The algorithm described so far works well as long as the geometric primitives can be extracted with sufficient accuracy. If there are errors in the point coordinates, an erroneous affine transformation will result from the basis triplet. Therefore, all the affine coordinates (α, β) will contain errors, and hence the hashing in the online phase will access the wrong entry in the hash table. This is probably the largest drawback of the geometric hashing algorithm in practice. To circumvent this problem, the template points must be stored in multiple adjacent entries of the hash table. Which hash table entries should be used can in theory be derived through error propagation (Lamdan et al., 1990). However, in practice, the accuracy of the geometric primitives is seldom known. Therefore, estimates have to be used, which must be very conservative in order for the algorithm not to miss any matches in the online phase. This, in turn, makes the algorithm slightly less efficient because more votes will have to be evaluated during the search.

乘法通过所有对应点计算得到改进后的仿射变换。这时轮流投影模板上所有边缘点到图像中模板的位置上，而不是只映射那些用在几何哈希法中的几个特征点。将变换后的边缘与图像中的边缘进行对比。如果模板边缘与图像边缘之间有足够多的交迭，那么这个匹配被接受并且图像上相应的点和边缘会从分割的图像中移除。如果在图像中需要找到多个模板，上面的整个过程就可以重复执行。

目前为止介绍的算法当图像基元能够足够准确提取出时可以正常工作。如果点的坐标存在一定误差，那么三个基准点的误差将导致仿射变换出现误差。因此，所有的仿射坐标 (α, β) 中都将包含误差，因此在线状态时哈希法将找到哈希表中错误的条目。这大概是实际应用中几何哈希算法最大的缺点。为了绕开这个问题，模板上的点必须保存在哈希表中相邻的多个条目中。理论上可以通过误差传递来得到哪些哈希表条目会被使用（Lamdan et al., 1990）。然而，实际中几何基元的准确度很少能够预料。因此必须进行估计，这个估计必须能够保证在线状态算法不会丢失任何匹配点。这样会导致算法会稍微低效一些，因为搜索过程中不得不评估更多的投票。

3.11.5.5 Matching Geometric Primitives

The final class of algorithms we will discuss tries to match geometric primitives themselves to the image. Most of these algorithms use only line segments as the primitives (Ayache and Faugeras, 1986; Grimson and Lozano-Pérez, 1987; Koch and Kashyap, 1987). One of the few exceptions to this rule is the approach by Ventura and Wan (1997), which uses line segments and circular arcs. Furthermore, in 3D object recognition, sometimes line segments and elliptical arcs are used (Costa and Shapiro, 2000). As already discussed, exhaustively enumerating all potential correspondences between the template and image primitives is prohibitively slow. Therefore, it is interesting to look at examples of different strategies that are employed to make the correspondence search tractable.

The approach by Ayache and Faugeras (1986) segments the contours of the model object and the search image into line segments. Depending on the lighting conditions, the contours are obtained by thresholding or by edge detection. The 10 longest line segments in the template are singled out as privileged. Furthermore, the line segments in the model are ordered by adjacency as they trace the boundary of the model object. To generate a hypothesis, a privileged template line segment is matched to a line segment in the image. Since the approach is designed to handle similarity transforms, the angle to the preceding line segment in the image, which is invariant under these transforms, is compared with the angle to the preceding line segment in the template. If they are not close enough, the potential match is rejected. Furthermore, the length ratio of these two segments, which also is invariant, is used to check the validity of the hypothesis. The algorithm generates a certain number of hypotheses in this manner. These hypotheses are then verified by trying to match additional segments. The quality of the hypotheses, including the additionally matched segments, is then evaluated based on the ratio of the lengths of the matched segments to the length of the segments in the template. The matching is stopped once a high-quality match has been found or if enough hypotheses have been evaluated. Hence, we can see that the complexity is kept manageable by using privileged segments in conjunction with their neighboring segments.

In Koch and Kashyap (1987), a similar method is proposed. In contrast to Ayache and Faugeras (1986), corners (combinations of two adjacent line segments of the boundary of the template that enclose a significant angle) are matched first. To generate a matching hypothesis, two corners must be matched to the image. Geometric constraints between the corners are used to reject false matches. The algorithm then attempts to extend the hypotheses with other segments in the image. The hypotheses are evaluated based on a dissimilarity criterion; if the dissimilarity is below a threshold, the match is accepted. Hence, the complexity of this approach is reduced by matching features that have distinctive geometric characteristics first.

The approach by Grimson and Lozano-Pérez (1987) also generates matching hypotheses and tries to verify them in the image. Here, a tree of possible correspondences is generated and evaluated in a depth-first search. This search tree

也不会发生变化，这可以用来确定假设的有效性。这个算法通过这种方式确定一些潜在的匹配位置。然后通过匹配其他线段来验证这些假设。这些假设（包括其他匹配的线段）的质量可以随后基于图像中匹配得到的线段长度与模板中相应线段之间的比值来进行评估。在找到一个非常高质量的匹配对象或评估了足够多的假设后停止匹配。因此，我们可以看出通过在使用特殊线段与它们相邻的线段，算法的复杂性是可控的。

在文献（Koch and Kashyap, 1987）中提出了一个相似的方法。与文献（Ayache and Faugeras, 1986）中提出的方法相比，这个方法中首先匹配拐角（在模板边界中相邻的两条有一定夹角的线段的组合）。为了得到一个匹配假设，必须在图像中首先匹配两个拐角。两个拐角之间的几何约束用来排除错误匹配位置。算法然后尝试使用图像中其他线段来扩展这个假设。这些假设的评估是基于一个不同的标准。如果两者之间的差异不超过一定的阈值，那么这个匹配位置被接受。因此，这个算法通过首先匹配拥有独特几何特征的部分来降低复杂度。

文献（Grimson and Lozano-Perez，1987）中的方法同样是首先生成匹配假设并尝试在图像中验证它们。这个方法中创建了一个可能对应关系的搜索树并且使用一个深度优先

is called the interpretation tree. A node in the interpretation tree encodes a correspondence between a model line segment and an image line segment. Hence, the interpretation tree would exhaustively enumerate all correspondences, which would be prohibitively expensive. Therefore, the interpretation tree must be pruned as much as possible. To do this, the algorithm uses geometric constraints between the template line segments and the image line segments. Specifically, the distances and angles between pairs of line segments in the image and in the template must be consistent. This angle is checked by using normal vectors of the line segments that take the polarity of the edges into account. This consistency check prunes a large number of branches of the interpretation tree. However, since a large number of possible matchings still remain, a heuristic is used to explore the most promising hypotheses first. This is useful because the search is terminated once an acceptable match has been found. This early search termination is criticized by Joseph (1999), and various strategies to speed up the search for all instances of the template in the image are discussed. The interested reader is referred to Joseph (1999) for details.

To make the principles of the geometric matching algorithms clearer, let us examine a prototypical matching procedure on an example. The template to be found is shown in Figure 3.143(a). It consists of five line segments and five circular arcs. They were segmented automatically from the image in Figure 3.139(a) using a subpixel-precise Canny filter with $\sigma = 1$ and by splitting the edge contours into line segments and circular arcs using the method described in Section 3.8.4.2.

搜索进行评估。这个搜索树被称为鉴定树。鉴定树中的一个节点表示了一个模板线段与图像线段之间的对应关系。因此，鉴定树将完全列举所有可能的对应关系，这种方式应该是不容许的。因此，解释树必须尽可能多地被裁减。为了实现这个裁减，算法使用模板线段与图像线段之间的几何约束。特定情况下，图像中两个线段之间的距离和角度与模板中相应两个线段之间的距离和角度必须是一致的。这个角度通过使用线段的法向量并将该边缘的极性考虑在内进行检查。这种一致性检查可以裁减掉解释树中很多的分支。然而，由于仍然剩余很大数量的匹配可能性，因此使用试探法保证首先检测最有可能的假设。由于在找一个可接受的匹配后匹配过程将停止，因此这种方法非常有用。在文献（Joseph，1999）中批判了这种过早停止搜索的方法，并且论述了很多可以加快在图像中搜索模板的所有实例的策略。关于这些策略的详细内容有兴趣的读者可以参考文献（Joseph，1999）。

为了使几何匹配算法的原理更加清晰，我们首先通过一个例子了解匹配过程的原型。需要在图像中搜索的模板显示在图 3.143(a) 中，其中包含 5 条线段和 5 个圆弧。首先使用一个 $\sigma = 1$ 的亚像素精度 Canny 滤波器从图 3.139(a) 中提取边缘，并且使用 3.8.4.2 节中介绍的方法将边缘分割为线段和圆弧，即得到了图中的结果。模板中包含这些图像基元的几何参数

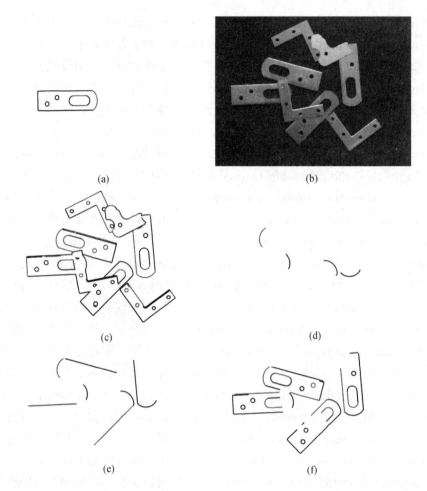

图 3.143 使用几何基元在图像中匹配对象的一个示例；(a) 使用图 3.139 (a) 创建得到的模板，该模板中包含 5 条线段和 5 个圆弧；(b) 搜索的图像中包含 4 个部分被遮挡的模板实例，同时还包含 4 个其他对象；(c) 使用 $\sigma = 1$ 的亚像素精度的 Canny 滤波器对图 (b) 进行边缘提取并将提取得到的边缘分割成线段和圆弧；(d) 这个例子中的匹配首先尝试匹配模板中最长的圆弧并且在图像中找到 4 个可能的匹配位置；(e) 使用图 (a) 中下方的长线段来扩展这些假设，使用这两个几何基元足以估计模板到图像中实例的刚性变换；(f) 将模板中剩余的基元与图像进行匹配；图 (f) 中显示了匹配得到的结果

The template consists of the geometric parameters of these primitives as well as the segmented contours themselves. The image in which the template should be found is shown in Figure 3.143(b). It contains four partially occluded instances of the model along with four clutter objects. The matching starts by extracting edges in the search image and by segmenting them into line segments and circular arcs (Figure 3.143(c)). As for the

同时也包括这些图像基元本身。我们将在图 3.143（b）中搜索该模板。这个图像中包含 4 个部分被遮挡的实例，也包含 4 个其他对象。首先在图像中进行边缘提取并且将边缘分割为线段和圆弧（图 3.143（c））。与创建模板相似，这些图像基元的几何参数需要计算得到。此时匹配过程开始对模板中所有的基元确定可能的匹配位

template, the geometric parameters of the image primitives are calculated. The matching now determines possible matches for all of the primitives in the template. Of these, the largest circular arc is examined first because of a heuristic that rates moderately long circular arcs as more distinctive even than long line segments. The resulting matching hypotheses are shown in Figure 3.143(d). Of course, the line segments could also have been examined first. Because in this case only rigid transformations are allowed, the matching of the circular arcs uses the radii of the circles as a matching constraint.

Since the matching should be robust to occlusions, the opening angle of the circular arcs is not used as a constraint. Because of this, the matched circles are not sufficient to determine a rigid transformation between the template and the image. Therefore, the algorithm tries to match an adjacent line segment (the long lower line segment in Figure 3.143(a)) to the image primitives while using the angle of intersection between the circle and the line as a geometric constraint. The resulting matches are shown in Figure 3.143(e). With these hypotheses, it is possible to compute a rigid transformation that transforms the template to the image. Based on this, the remaining primitives can be matched to the image based on the distances of the image primitives and the transformed template primitives. The resulting matches are shown in Figure 3.143(f). Note that, because of specular reflections, sometimes multiple parallel line segments are matched to a single line segment in the template. This could be fixed by taking the polarity of the edges into account. To obtain the rigid transformation between the

置。这些几何基元中，首先检查最长的圆弧，这是由于在探索方法中认为长的圆弧比同长度的直线更有特点。得到的匹配假设显示在图 3.143（d）中。当然，也可以首先检查直线。因为这个例子中，只可能存在刚性变换，因此我们还可以使用圆弧的半径作为匹配约束。

由于匹配应该能够不受遮挡的影响，圆弧的张角不能作为匹配约束。因此，匹配得到的圆弧不足以确定模板与图像之间的刚性变换。所以，算法中尝试在图像基元中匹配一个与圆弧相邻的线段（图 3.143（a）中下面的长线段），此时使用线段与圆弧相交角度作为几何约束。图 3.143（e）中显示匹配结果。此时可以计算这些假设时模板到图像的刚性变换。基于这些刚性变换，通过计算图像中几何基元与变换后模板中的几何基元之间的距离可以在图像中匹配模板中其他的几何基元。图 3.143（f）中显示了最终的匹配结果。需要注意的是由于镜面反射的影响，有时会在图像中找到多条平行线段与模板中一条线段匹配。通过考虑边缘的极性可以解决这个问题。为了能够尽可能准确地得到模板与图像中匹配对象之间的刚性变换，可以使用最小二乘法优化模板中边缘与图像中相应边缘之间的距离。另一种方法是文献（Ventura and Wan,

template and the matches in the image as accurately as possible, a least-squares optimization of the distances between the edges in the template and the edges in the image can be used. An alternative is the minimal tolerance error zone optimization described by Ventura and Wan (1997). Note that the matching has already found the four correct instances of the template. For the algorithm, the search is not finished, however, since there might be more instances of the template in the image, especially instances for which the large circular arc is occluded more than in the leftmost instance in the image. Hence, the search is continued with other primitives as the first primitives to try. In this case, however, the search does not discover new viable matches.

3.11.5.6 Shape-Based Matching

After having discussed some of the approaches for robustly finding templates in an image, the question as to which of these algorithms should be used in practice naturally arises. We will say more on this topic below. From the above discussion, however, we can see that the effectiveness of a particular approach greatly depends on the shape of the template itself. Generally, geometric matching algorithms have an advantage if the template and image contain only a few salient geometric primitives, like in the example in Figure 3.143. Here, the combinatorics of the geometric matching algorithms can work to their advantage. On the other hand, they work to their disadvantage if the template or search image contains a large number of geometric primitives.

A difficult model image is shown in Figure 3.144. The template contains fine structures that

1997）中介绍的最优化最小公差误差带（minimal tolerance error zone optimization）。注意，通过匹配已经找到了4个模板的正确实例。然而对于这个算法来讲，搜索并没有完成，因为图像中可能有更多的模板实例，尤其是那些大圆弧比最左边实例遮挡还严重的实例。因此，需要使用其他的几何基元作为第一个搜索的基元再进行搜索。然而这个例子中，后续的搜索找不到新的匹配位置。

3.11.5.6 基于形状的模板匹配

上面我们论述了一些可靠的模板匹配方法，自然就存在这样一个问题，在实际应用中应该使用哪个算法呢？我们下面将更多地陈述这个主题。然而从上面的论述中，我们可以看出一个特定的模板匹配方法的效率主要取决于模板的形状。一般情况下，如果与图 3.143 中的示例相似，模板和图像中只包含很少几个明显的几何基元，这时使用几何匹配算法有一定的优势。这时的条件非常利于几何匹配算法的组合工作。然而，如果模板或搜索图像中包含非常多的几何基元，这时不适合使用这种几何匹配算法。

图 3.144 中显示了一个复杂的模板图像。这个包含很多细节结构的模

3.11 Template Matching 467

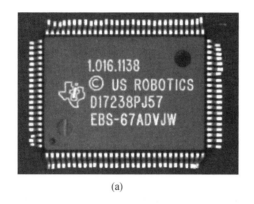

(a)　　　　　　　　　　　(b)

图 3.144　（a）不适合使用几何匹配算法的模板图像；（b）将模板边缘分割得到的线段和圆弧中包含的像素点大约是图 3.143 中模板中像素点数量的 3 倍，但是它包含的几何基元的数量是图 3.143 中模板的 35 倍；也就是 350 个

result in 350 geometric primitives, which are not particularly salient. Consequently, the search would have to examine an extremely large number of hypotheses that could be dismissed only after examining a large number of additional primitives. Note that the model contains 35 times as many primitives as the model in Figure 3.143, but only approximately 3 times as many edge points. Consequently, it could easily be found with pixel-based approaches like the GHT.

A difficult search image is shown in Figure 3.145. Here, the goal is to find the circular fiducial mark. Since the contrast of the fiducial mark is very low, a small segmentation threshold must be used in the edge detection to find the relevant edges of the circle. This causes a very large number of edges and broken fragments that must be examined. Again, pixel-based algorithms will have little trouble with this image.

From the above examples, we can see that pixel-based algorithms have the advantage that

板中包含 350 个几何基元，这些几何基元在图像中还不是特别的显眼。因此，使用几何匹配算法进行匹配将不得不检查非常多的假设，并且只能通过检查过非常多额外的几何基元后才能确定是否应该抛弃该假设。注意这个模板中包含几何基元的数量是图 3.143 中模板中几何基元数量的 35 倍，但是这个模板中的边缘点的数量才大约是图 3.143 中模板中边缘点数量的 3 倍。因此，这个应用中使用与广义霍夫变换类似的基于像素的方法更容易些。

图 3.145 中显示了一个非常复杂的搜索图像。这个应用的目的是在图像中找到圆形基准标志。由于基准标志的对比度非常低，因此在图像中提取圆的相关边缘时必须使用一个非常小的分割阈值。这将会导致匹配过程中需要匹配非常多的边缘以及断掉的边缘片段。同样这种情况下在图像中应用基于像素点的匹配算法问题会比较少。

从上面这些例子中我们可以看出，基于像素的算法的优势在于它们

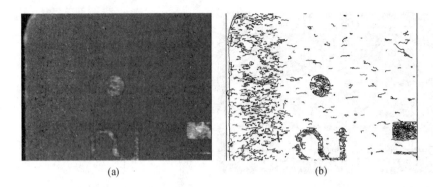

图 3.145　（a）对于几何匹配算法非常困难的一幅搜索图像；由于圆形基准标志的对比度非常差，因此在图像中提取圆的相关边缘必须使用一个非常小的分割阈值，这样图（b）中包含几何基元的数量就会非常多，而这些基元在搜索过程中都必须被检查

they can represent arbitrarily shaped templates without problems. Geometric matching algorithms, on the other hand, are restricted to relatively simple shapes that can be represented with a very small number of primitives. Therefore, in the remainder of this section, we will discuss a pixel-based robust template matching algorithm called shape-based matching (Steger, 2001, 2002, 2005, 2006a,b) that works very well in practice (Ulrich and Steger, 2001, 2002).

One of the drawbacks of all the algorithms that we have discussed so far is that they segment the edge image. This makes the object recognition algorithm invariant only to a narrow range of illumination changes. If the image contrast is lowered, progressively fewer edge points will be segmented, which has the same effect as progressively larger occlusion. Consequently, the object may not be found for low-contrast images. To overcome this problem, a similarity measure that is robust against occlusion, clutter, and nonlinear illumination changes must be used. This similarity measure can then be used in the pyramid-based recognition strategy described in Sections 3.11.2 and 3.11.4.

可以表示任意形状的模板。而另一方面，几何匹配算法受限于一些可以使用少量几何基元组成的形状相对简单的模板。因此，在这章节下面的部分我们将讨论一种基于像素的可靠的模板匹配算法，称为基于形状的模板匹配（Steger，2001，2002，2005，2006a,b），这种方法能够在实际应用中很好地工作（Ulrich and Steger，2001，2002）。

上面我们论述过的所有算法都存在一个缺点，就是它们都需要在图像中提取出边缘。这就导致目标识别算法只能够适用于非常小范围的光照变化。如果降低图像对比度，提取得到的边缘点就越来越少，更多的遮蔽也会造成相似的影响。因此，在低对比度的图像中可能找不到目标物体。为了解决这个问题，必须使用一种能够不受遮蔽、混乱和非线性光照变化的相似度量。这个相似度量然后可以用在 3.11.2 节和 3.11.4 节中介绍的基于图像金字塔的分层识别策略中。

To define the similarity measure, we first define the model of an object as a set of points $p_i = (r_i, c_i)^\top$ and associated direction vectors $d_i = (t_i, u_i)^\top$, with $i = 1, \cdots, n$. The direction vectors can be generated by a number of different image processing operations. However, typically edge extraction (see Section 3.7.3) is used. The model is generated from an image of the object, where an arbitrary ROI specifies the part of the image in which the object is located. It is advantageous to specify the coordinates p_i relative to the center of gravity of the ROI of the model or to the center of gravity of the points of the model.

The image in which the model should be found can be transformed into a representation in which a direction vector $e_{r,c} = (v_{r,c}, w_{r,c})^\top$ is obtained for each image point $(r, c)^\top$. In the matching process, a transformed model must be compared with the image at a particular location. In the most general case considered here, the transformation is an arbitrary affine transformation (see Section 3.3.1). It is useful to separate the translation part of the affine transformation from the linear part. Therefore, a linearly transformed model is given by the points $p'_i = Ap_i$ and the accordingly transformed direction vectors $d'_i = (A^{-1})^\top d_i$, where

$$A = \begin{pmatrix} a_{11} & a_{12} \\ a_{21} & a_{22} \end{pmatrix} \tag{3.190}$$

As discussed above, the similarity measure by which the transformed model is compared with the image must be robust to occlusions, clutter, and illumination changes. One such measure is to sum the (unnormalized) dot product of the

为了定义这个相似度量，我们首先将一个目标对象的模板定义为点集 $p_i = (r_i, c_i)^\top$ 和每个点关联的方向向量 $d_i = (t_i, u_i)^\top$，其中 $i = 1, \cdots, n$。方向向量可以通过许多不同的图像处理操作得到。然而一般情况下都使用边缘提取方法计算方向向量（见 3.7.3 节）。使用目标物体的一幅图像生成模板，在这幅图像中可以指定包含目标物体在内的任意的一个感兴趣区域。将坐标 p_i 定义为相对于感兴趣区域重心的相对坐标或相对于模板中所有点重心的相对坐标对后面的操作非常有利。

首先我们将需要在其中搜索模板的图像转换为另外一种表示法，就是为图像中每个点 $(r, c)^\top$ 计算出一个方向向量 $e_{r,c} = (v_{r,c}, w_{r,c})^\top$。在匹配过程中，变换后的模板必须与图像中某个特定的位置进行比较。在大多数一般的应用中，这种变换是一个任意的仿射变换（见 3.3.1 节）。我们从线性部分将仿射变换中的平移分离出来非常有用。因此，一个线性变换模型可以通过点 $p'_i = Ap_i$ 以及相应的变换后的方向向量 $d'_i = (A^{-1})^\top d_i$ 来给出。

与上面论述的一样，将转换后的模板与图像进行对比的相似度量必须不受遮挡、混乱以及光照变化的影响。能够达到要求的一种测量方式是在图像中某个特定点 $q = (r, c)^\top$ 处，计算变

direction vectors of the transformed model and the image over all points of the model to compute a matching score at a particular point $q = (r,c)^\top$ of the image. That is, the similarity measure of the transformed model at the point q, which corresponds to the translation part of the affine transformation, is computed as follows:

换后模板中所有点的方向向量与图像中对应点处的方向向量的（未归一化）点积的总和，并以此作为匹配分值。这也就是变换后模板在点 q 处的相似度量，这对应于仿射变换中的平移部分，计算这个相似度量的公式如下：

$$s = \frac{1}{n}\sum_{i=1}^{n}{d'_i}^\top e_{q+p'} = \frac{1}{n}\sum_{i=1}^{n}(t'_i v_{r+r'_i,\,c+c'_i} + u'_i w_{r+r'_i,\,c+c'_i}) \qquad (3.191)$$

If the model is generated by edge filtering and the image is preprocessed in the same manner, this similarity measure fulfills the requirements of robustness to occlusion and clutter. If parts of the object are missing in the image, there will be no edges at the corresponding positions of the model in the image, i.e., the direction vectors will have a small length and hence contribute little to the sum. Likewise, if there are clutter edges in the image, there will either be no point in the model at the clutter position or it will have a small length, which means it will contribute little to the sum.

如果使用边缘滤波器生成模板并且在搜索图像中使用同样的方式进行预处理，相似度量能够完全不受遮挡和混乱的影响。如果部分目标物体在图像中消失，模板上相应位置的边缘就没有出现在图像中，也就是说这些点方向向量非常短，因此它基本不影响总和。同样的，如果图像中有非常多其他的边缘，此时要不在其他边缘上没有对应于模板上的点，要不有对应点但方向向量非常短，这也就表示它基本不影响总和。

The similarity measure in Eq. (3.191) is not truly invariant to illumination changes, however, because the length of the direction vectors depends on the brightness of the image if edge detection is used to extract the direction vectors. However, if a user specifies a threshold on the similarity measure to determine whether the model is present in the image, a similarity measure with a well-defined range of values is desirable. The following similarity measure achieves this goal:

然而等式（3.191）中提供的相似度量算法并不真正能够不受光照变化的影响，这是因为如果使用边缘提取算法来计算方向向量的话，那么方向向量的长短取决于图像的亮度。可是，如果在相似度量中指定一个阈值来确定模板是否出现在图像中，则需要一个已经定义了合适范围的相似度量。下式中的相似度量可以实现这个要求：

$$s = \frac{1}{n}\sum_{i=1}^{n}\frac{{d'_i}^\top e_{q+p'}}{\|d'_i\|\|e_{q+p'}\|} = \frac{1}{n}\sum_{i=1}^{n}\frac{t'_i v_{r+r'_i,\,c+c'_i} + u'_i w_{r+r'_i,\,c+c'_i}}{\sqrt{t'^2_i + u'^2_i}\sqrt{v^2_{r+r'_i,\,c+c'_i} + w^2_{r+r'_i,\,c+c'_i}}} \qquad (3.192)$$

Because of the normalization of the direction vectors, this similarity measure is additionally invari-

因为将方向向量进行归一化，相似度量可以不受任意光照变化的影响，这

ant to arbitrary illumination changes, since all vectors are scaled to a length of 1. What makes this measure robust against occlusion and clutter is the fact that, if a feature is missing, either in the model or in the image, noise will lead to random direction vectors, which, on average, will contribute nothing to the sum.

The similarity measure in Eq. (3.192) will return a high score if all the direction vectors of the model and the image align, i.e., point in the same direction. If edges are used to generate the model and image vectors, this means that the model and image must have the same contrast direction for each edge. Sometimes it is desirable to be able to detect the object even if its contrast is reversed. This is achieved by:

$$s = \left| \frac{1}{n} \sum_{i=1}^{n} \frac{\boldsymbol{d}_i'^{\top} \boldsymbol{e}_{\boldsymbol{q}+\boldsymbol{p}'}}{\|\boldsymbol{d}_i'\| \|\boldsymbol{e}_{\boldsymbol{q}+\boldsymbol{p}'}\|} \right| \qquad (3.193)$$

In rare circumstances, it might be necessary to ignore even local contrast changes. In this case, the similarity measure can be modified as follows:

$$s = \frac{1}{n} \sum_{i=1}^{n} \frac{|\boldsymbol{d}_i'^{\top} \boldsymbol{e}_{\boldsymbol{q}+\boldsymbol{p}'}|}{\|\boldsymbol{d}_i'\| \|\boldsymbol{e}_{\boldsymbol{q}+\boldsymbol{p}'}\|} \qquad (3.194)$$

The normalized similarity measures in Eqs. (3.192)–(3.194) have the property that they return a number smaller than 1 as the score of a potential match. In all cases, a score of 1 indicates a perfect match between the model and the image. Furthermore, the score roughly corresponds to the portion of the model that is visible in the image. For example, if the object is 50% occluded, the score (on average) cannot exceed 0.5. This is a highly desirable property, because it gives the

是因为所有的向量的长度都变成 1。这个相似度量也不受遮挡和混乱的影响，这是由于实际上如果一个模板或图像中某个特征丢失，噪声将导致一个随机的方向向量，这些方向向量平均起来不会对总和造成影响。

如果模板与图像中所有方向向量都对齐（所有点方向一致），等式（3.192）中的相似度量将返回非常高的分值。如果使用边缘提取方法生成模板和图像向量，就需要模板和图像中在每个边缘处的明暗变化方向一致。但有的应用中需要在目标物体明暗对比颠倒的情况下也能够搜索到目标物体。这可以通过：

在非常特殊的应用场合，可能甚至需要忽略局部的明暗对比方向变化。这种情况下，需要按下式修改相似度量：

等式（3.192）～（3.194）中的归一化相似度量都将返回一个小于 1 的数作为潜在匹配对象的分值。在所有情况下，如果分值为 1 则表示模板与图像之间完美一致。另外，这个分值大约与模板中多少部分在图像中出现相关。例如，如果物体有 50% 被遮挡，（平均）分值不会超过 0.5。匹配分值的这个属性非常需要，因为它提供给使用者一个有意义的数据，使用

user the means to select an intuitive threshold for when an object should be considered as recognized.

A desirable feature of the above similarity measures in Eqs. (3.192)–(3.194) is that they do not need to be evaluated completely when object recognition is based on a user-defined threshold $s_{\min}$ for the similarity measure that a potential match must achieve. Let s_j denote the partial sum of the dot products up to the jth element of the model. For the match metric that uses the sum of the normalized dot products, this is

$$s_j = \frac{1}{n}\sum_{i=1}^{j}\frac{{d'_i}^{\top} e_{q+p'}}{\|d'_i\|\|e_{q+p'}\|} \tag{3.195}$$

Obviously, all the remaining terms of the sum are all ≤ 1. Therefore, the partial score can never achieve the required score $s_{\min}$ if $s_j < s_{\min} - 1 + j/n$, and hence the evaluation of the sum can be discontinued after the jth element whenever this condition is fulfilled. This criterion speeds up the recognition process considerably.

As mentioned above, to recognize the model, an image pyramid is constructed for the image in which the model is to be found (see Section 3.11.2.1). For each level of the pyramid, the same filtering operation that was used to generate the model, e.g., edge filtering, is applied to the image. This returns a direction vector for each image point. Note that the image is not segmented, i.e., thresholding or other operations are not performed. This results in true robustness to illumination changes.

As discussed in Sections 3.11.2.2 and 3.11.4, to identify potential matches, an exhaustive

者可以选择一个直觉上的阈值来决定什么时候应该认为找到了匹配对象。

等式（3.192）～（3.194）中的相似度量有一个非常好的特性，就是当要求在潜在的匹配位置计算得到的相似度量必须达到一个用户定义的阈值 $s_{\min}$ 时，这个相似度量不需要完全地求出。我们使用 s_j 表示累计到模板的第 j 个元素时点积的总和。对于使用归一化点积总和作为匹配度量时：

显然，总和中剩下的所有剩余项都小于或等于 1。因此，如果 $s_{\min}$ if $s_j < s_{\min} - 1 + j/n$，那么匹配分值就不可能达到需要达到的阈值 $s_{\min}$，因此当这个条件满足时就可以在第 j 个元素后停止计算。这个标准大大提高了识别过程的速度。

上面我们已经提到过，为了识别模板可以为图像创建一个图像金字塔，然后在金字塔中搜索模板（见 3.11.2.1 节）。在图像金字塔的每层都在图中应用与创建模板时相同的滤波运算，例如边缘滤波。这个滤波运算将得到每个图像点的方向向量。注意图像并没有被分割，也就是说没有进行阈值分割或其他操作。这就可以使算法真正不受光照变化的影响。

在 3.11.2.2 节和 3.11.4 节中我们曾经讨论过，为了确定潜在的匹配位

search is performed for the top level of the pyramid, i.e., all possible poses of the model are used on the top level of the image pyramid to compute the similarity measure via Eqs. (3.192), (3.193), or (3.194). A potential match must have a score larger than $s_{\min}$, and the corresponding score must be a local maximum with respect to neighboring scores. The threshold $s_{\min}$ is used to speed up the search by terminating the evaluation of the similarity measure as early as possible. Therefore, this seemingly brute-force strategy actually becomes extremely efficient.

After the potential matches have been identified, they are tracked through the resolution hierarchy until they are found at the lowest level of the image pyramid. Once the object has been recognized on the lowest level of the image pyramid, its pose is extracted with a resolution better than the discretization of the search space with the approach described in Section 3.11.3.

While the pose obtained by the extrapolation algorithm is accurate enough for most applications, in some applications an even higher accuracy is desirable. This can be achieved through a least-squares adjustment of the pose parameters. To achieve a better accuracy than the extrapolation, it is necessary to extract the model points as well as the feature points in the image with subpixel accuracy. Then, the algorithm finds the closest image point for each model point, and then minimizes the sum of the squared distances of the image points to a line defined by their corresponding model point and the corresponding tangent to the model point, i.e., the directions of the model points are taken to be correct and are assumed to

置，我们需要在金字塔的顶层进行彻底的搜索，也就是对图像金字塔顶层的图像需要使用等式（3.192）、（3.193）或（3.194）计算模板在所有可能位姿上的相似度量。一个潜在的匹配位置的匹配分值必须大于 $s_{\min}$，并且这个位置相应的匹配分值还是局部的最大值，也就是说它比临近位置的匹配分值都大。设置阈值 $s_{\min}$ 通过尽早结束无用的相似度量的计算来加快搜索速度。因此，这个从表面上看起来需要完全遍历增幅图像的强力搜索策略实际上非常高效。

在确定了潜在匹配位置后，我们会跟踪这些匹配位置到金字塔的更低层，直到在图像金字塔最底层找到它们。一旦在图像金字塔最底层找到目标物体，将使用 3.11.3 节中介绍的方法得到比离散化的搜索空间分辨率更精确的最终位姿。

虽然通过外推算法得到的位置对大多数应用已经足够准确，但是一些应用中希望得到更准确的结果。这可以通过最小二乘法调整位姿参数来实现。为了达到比外推法更高的精度，在提取模板点与图像中特征点时必须达到亚像素精度。然后，通过算法找到每个模板点在图像中的最近点，使这些图像点与相应的线段之间距离平方和最小化，其中图像点相应的线段是由相应的模板点与该模板点的切线所定义的。也就是说我们认为模板点的方向是正确的并且假设它们表示了物体边界的方向。如果使用一个边缘提取算法，模板点的方向向量与物体

describe the direction of the object's border. If an edge detector is used, the direction vectors of the model are perpendicular to the object boundary, and hence the equation of a line through a model point tangent to the object boundary is given by

$$t_i(r - r_i) + u_i(c - c_i) = 0 \tag{3.196}$$

Let $q_i = (r'_i, c'_i)^\top$ denote the matched image points corresponding to the model points p_i. Then, the following function is minimized to refine the pose a:

$$d(\boldsymbol{a}) = \sum_{i=1}^{n} \bigl(t_i\bigl(r'_i(\boldsymbol{a}) - r_i\bigr) + u_i\bigl(c'_i(\boldsymbol{a}) - c_i\bigr)\bigr)^2 \to \min \tag{3.197}$$

The potential corresponding image points in the search image are obtained without thresholding by a non-maximum suppression and are extrapolated to subpixel accuracy. By this, a segmentation of the search image is avoided, which is important to preserve the invariance against arbitrary illumination changes. For each model point, the corresponding image point in the search image is chosen as the potential image point with the smallest Euclidian distance using the pose obtained by the extrapolation to transform the model to the search image. Since the point correspondences may change through the pose refinement, an even higher accuracy can be gained by iterating the correspondence search and pose refinement. Typically, after three iterations the accuracy of the pose no longer improves.

Figure 3.146 shows six examples in which the shape-based matching algorithm finds the print on the IC shown in Figure 3.144. Note that the object is found despite severe occlusions and clutter.

边界垂直，因此通过模板点并与物体边界相切的直线的等式为

如果使用 $q_i = (r'_i, c'_i)^\top$ 表示与模板点 p_i 相应的图像点。那么就可以将下面函数最小化来使位姿 a 的参数更准确：

搜索图像中潜在对应的图像点可以不通过非最大值抑止的阈值分割得到，而通过外推法得到亚像素精度。这样就避免在搜索图像中应用阈值分割，这点非常重要，可以使算法保证不受任意光照变化影响。对于每个模板点，使用通过外推法得到的位姿将模板变换到图像中，求出搜索图像中与变换后模板之间欧几里得距离最小的点作为潜在图像点。由于将位姿参数求取更加准确后可能影响点的对应关系，因此通过迭代进行优化位姿参数和对应关系搜索就可以得到更高的准确度。一般情况下，在 3 次迭代后位姿的准确度基本就不会有更高的改善。

图 3.146 中显示了使用基于形状匹配算法在图 3.144 中所示 IC 上搜索印刷体的 6 个例子。注意尽管图像中存在严重的遮挡与混乱现象，仍然可以在图中找到目标物体。

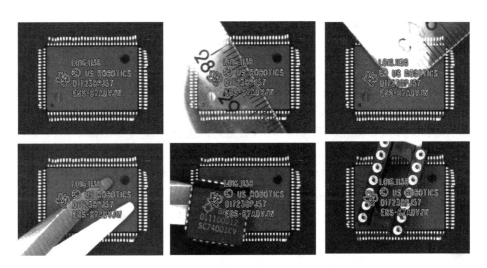

图 3.146　在图像中虽然存在严重的遮挡和混乱现象，使用基于形状模板匹配算法还是可以准确找到目标物体（图 3.144 中所示 IC 上的印刷体）

Extensive tests with shape-based matching have been carried out by Ulrich and Steger (2001, 2002). The results show that shape-based matching provides extremely high recognition rates in the presence of severe occlusion and clutter as well as in the presence of nonlinear illumination changes. Furthermore, accuracies better than 1/30 of a pixel and better than 1/50 degree can be achieved.

The basic principle of the shape-based matching algorithm can be extended in various ways to handle larger classes of deformations and objects. A method to recognize objects that consist of multiple parts that can move with respect to each other by rigid 2D transformations is described in (Ulrich *et al.*, 2002; Ulrich, 2003; Ulrich *et al.*, 2004; Ulrich and Steger, 2007, 2009, 2011, 2013a,b). Further extensions of shape-based matching for 3D object recognition will be introduced in Section 3.12.

From the above discussion, we can see that the basic algorithms for implementing robust template matching are already fairly complex. In

在文献（Ulrich and Steger, 2001, 2002）中对基于形状的模板匹配算法进行了更多的测试。测试的结果显示基于形状的模板匹配算法在存在严重遮挡、混乱或非线性光照变化的情况下实现了极高的识别率。另外，能够达到的准确度超过 1/30 像素和 1/50 度。

基于形状匹配算法的基本原理可以扩展为不同的形式来处理更大的形变和物体类别。在文献（Ulrich *et al.*, 2002；Ulrich, 2003；Ulrich *et al.*, 2004 和 Ulrich and Steger, 2007, 2009, 2011, 2013a,b）中介绍了识别物体的一种方法，这里，物体由多个部件组成，而且部件之间能够通过二维刚性变换相互移动。基于形状匹配的更进一步扩展是针对三维物体的识别，会在 3.12 节中介绍。

通过上面的论述，我们可以看出实现一个稳定可靠的模板匹配的基本算法已经相当复杂。然而实际上，花

reality, however, the complexity additionally resides in the time and effort that needs to be spent in making the algorithms very robust and fast. Additional complexity comes from the fact that, on the one hand, templates with arbitrary ROIs should be possible to exclude undesired parts from the template, while on the other hand, for speed reasons, it should also be possible to specify arbitrarily shaped ROIs for the search space in the search images. Consequently, these algorithms cannot be implemented easily. Therefore, wise machine vision users rely on standard software packages to provide this functionality rather than attempting to implement it themselves.

费时间和精力使这些匹配算法更加稳定快速的过程更加复杂。这些复杂性主要是由于：一方面应该可以在模板图像中设置任意的感兴趣区域，从而将其他不需要的部分排除在外；另一方面，在搜索图像中应该能够设置任意形状的感兴趣区域作为搜索区域来提高搜索速度。因此，这些算法实现起来并不容易。所以，聪明的机器视觉用户都依赖标准软件包来提供这些功能而不会试着自己实现这些算法。

3.12 3D Object Recognition

3.12 三维物体识别

In the previous section, we discussed various template matching approaches that can recognize planar (two-dimensional) objects in images. Most of the proposed methods assume that the objects to be recognized appear (approximately) parallel to a known plane in object space. Hence, the 2D object pose in the image often can be described by a similarity transformation. While this assumption is valid for many applications, there are some applications where the object might occur with an unknown tilt. This results in perspective distortions of the object in the image. Typical applications are the grasping of objects by a robot, where the 3D pose of the object must be determined, or the rectification of a perspectively distorted object before performing subsequent inspections. Unfortunately, a straightforward extension of one of the methods described in Section 3.11 to handle perspective distortions would be far too slow for practical applications.

在上一节中，我们讨论了各种各样的，可识别图像中平面（二维）物体的模板匹配算法。我们之前提到的大多数算法都假设被测物在空间中平行于某个已知的平面。因此，在这个平面上，物体的姿态通常可以描述为一个相似变换。虽然这个假设可以应用于很多场景，但有时我们无法确定被测物的倾斜程度。在此情况下，我们所拍摄的图像中，被测物会产生透视畸变。典型的应用包括，通过视觉获取物体三维姿态并进行抓取以及在进行图像检测之前根据其姿态进行畸变矫正。如果我们只是简单地外延之前在 3.11 节中提到的算法（例如，对各种角度情况创建相应的模板）来处理当前的问题，算法执行时间将会大大增加，不能满足实际应用。

Real objects are always three-dimensional. However, many objects or object parts exhibit only a small depth. For example, the print on a planar side of an object, shallow metal parts, or printed circuit boards can often be assumed to be two-dimensional or planar. One important characteristic of planar objects is that their mapping into the image plane of the camera can be modeled as a 2D transformation. Perspective effects like self-occlusions do not occur. Therefore, from an algorithmic point of view, recognizing planar objects in images is much easier than recognizing general 3D objects. This justifies the use of the highly efficient template matching approaches that were described in Section 3.11. For objects with more distinct 3D structures, however, template matching approaches cannot be used.

The methods described in Section 3.11 expect an image as input data. For 3D object recognition, the benefit of using depth or 3D data is evident. 3D data can be obtained by various well-known 3D reconstruction methods (see Section 3.10) or from 3D image acquisition devices (see Section 2.5).

In this section, we describe methods for 3D object recognition. First, in Section 3.12.1, we will look at methods that recognize planar objects that might be tilted in 3D space, and hence show perspective deformations. Because the basic principle of the proposed methods is general with respect to the deformation type, they can also be extended to handle smooth local object deformations. Therefore, we will summarize these methods under the term deformable matching.

In Section 3.12.2, the basic principle of shape-based matching is used as the foundation for

在实际应用中，物体一般是立体的，但很多物体或者其某个局部只有小范围的深度变化。例如，印刷在被测物某个平面上的文字、冲压工件以及印刷电路板，它们都可以被视为二维平面。平面物体最大的特征在于它们可以被二维仿射变换模型映射到图像上。此类物体并不会出现自遮挡的现象，因此，从算法角度上来看，在图像中识别平面物体要比三维物体更容易。在 3.11 节中所提及的高效的模板匹配是可行的，但是对于三维物体是不适用的。

3.11 节阐述的算法采用图像作为数据输入，而在三维物体识别过程中，我们则使用深度数据和三维数据进行运算，显然在识别方面更具优势。这些三维数据可以通过一些成熟的三维重构算法（见 3.10 节）或者三维采集设备（见 2.5 节）来获得。

在本节我们将讨论三维物体的识别。首先在 3.12.1 节中我们将讨论倾斜的平面物体识别。该算法适用于存在透视畸变得被测物识别，而此类畸变一般是由于物体在三维空间中的倾斜导致。因为上述畸变类型处理算法的理论基础有一定的通用性。所以，我们可以把该原理推广到存在平滑局部变形的被测物识别上。因此，在下文中我们将这些算法统称为变形匹配。

在 3.12.2 节中，我们将介绍基于单幅图像的三维物体识别算法。该算

shape-based 3D matching, a sophisticated algorithm that is able to recognize the pose of 3D objects from single images. Compared to the recognition of planar objects, we must deal with additional challenges like view-dependent object appearance, self-occlusions, degenerate views, and a 6D pose space along with high computational costs.

Finally, in Section 3.12.3, we will introduce surface-based 3D matching, which exploits the benefits of using 3D data to recognized the 3D pose of an object.

3.12.1 Deformable Matching

We will first discuss an approach for recognizing planar objects under perspective distortions. As mentioned above, a straightforward extension of one of the methods described in the previous sections to handle perspective distortions by simply increasing the dimensionality of the search space would result in impractically long run times. Therefore, we will discuss an efficient extension of shape-based matching (see Section 3.11.5.6). The approach is described in more detail in (Hofhauser et al., 2008; Hofhauser and Steger, 2010, 2011, 2012, 2013).

Figure 3.147 shows two examples where perspectively distorted planar objects must be recognized. In the following, we will distinguish an uncalibrated and a calibrated case. In the uncalibrated case, the result of the matching is the parameters of a projective 2D transformation (see Section 3.3.1.1). They can be used to rectify the image and subsequently perform actions such as a completeness check or print inspection of the

法以二维形状匹配为基础，经过一系列运算求解被测物的三维姿态。相比于平面物体识别，我们需要更多复杂的计算过程来应对诸如视点位置判断、物体自遮挡、退化视图以及六自由度位姿确定等问题。

最后，在 3.12.3 节中，我们将介绍基于物体表面三维数据的匹配算法。该算法借助被测物在空间中的三维数据对物体进行识别和位姿确定。

3.12.1 变形匹配

首先我们来讨论如何从存在透视畸变的图像中识别三维平面物体。正如我们之前所讨论的，直接将前一节中提到的某个算法进行推广，通过简单的增加搜索平面的维度来处理透视畸变问题，是非常耗时的，也是不实际的。因此我们将讨论一个更高效的，基于形状模板匹配原理推广而来的算法（见 3.11.5.6）。该算法相关细节详见文献（Hofhauser et al., 2008; Hofhauser and Steger, 2010, 2011, 2012, 2013）。

如图 3.147 所示，我们需要识别图中存在透视畸变的平面被测物。在下文中，我们将分别在标定和未标定两种情况下讨论这个问题。在未标定的情况下，匹配结果为二维投影变换参数组（见 3.3.1.1 节）。该参数可以用来矫正图像或图像处理数据，比如，对倾斜的表面进行检查或印刷检测。在标定的情况下，我们可以对被

tilted surfaces. In the calibrated case, the 3D pose of the object can be calculated, which can be used in applications such as bin picking.

测物的三维姿态进行估计，此类算法可以应用在机械手抓取等场景下。

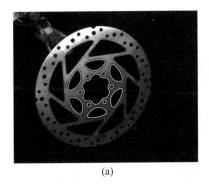 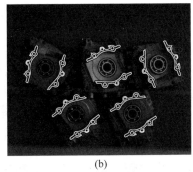

(a) (b)

图 3.147 识别存在透视畸变的被测物，图中白色为已匹配模板边缘。（a）刹车盘零件；（b）引擎部件

3.12.1.1 Principle of Deformable Matching

The idea of deformable matching is based on the fact that large deformations become smaller on higher pyramid levels with respect to the pixel size. This effect is illustrated in Figure 3.148 by comparing an image (top row) with its perspectively distorted version (bottom row). The perspective distortions were created by applying a projective transformation to the image. While the differences, measured in pixels, are large in the original image resolution (left column of Figure 3.148), they become smaller on pyramid level 4 (center column), and are hardly observable on pyramid level 6 (right column). This observation allows us to perform an exhaustive search on the top pyramid level with the shape-based matching approach as described in Section 3.11.5.6. Let us assume, for example, that a model of the IC was generated for shape-based matching from the image of Figure 3.148(a) and should be searched in

3.12.1.1 变形匹配原理

变形匹配的处理思路基于这样一个事实：在考虑像素大小的情况下，在高层金字塔中，透视畸变对图像的影响会变小。如图 3.148 所示，通过对比原始图像（图 3.148 第一行）和带有透视畸变的图像（图 3.148 第二行），我们可以明显地看到这个效果。其中透视畸变图像是由原图经过透视变换处理所得。虽然两幅图像（图 3.148 左侧第一列）在原始分辨率下差异较大，但是它们的差异在第 4 层金字塔图像（图 3.148 中间列）中有所减小，而在第 6 层金字塔图像（图 3.148 右侧第一列）中，通过观察我们已经很难找到它们的差异。这个观察结果使得我们可以在高层的金字塔图像上应用 3.11.5.6 节中所提到的基于形状的匹配算法进行全面的搜索。假如我们希望使用基于形状匹配算法，通过

the image of Figure 3.148(d). If the search is started on pyramid level 6, the matching on this level should yield a high score for the correct match despite the perspective distortions.

图 3.148（a）创建模板，并在图 3.148（d）中找到该模板。这时，如果我们能够从金字塔第 6 层开始搜索，尽管存在透视畸变，但是匹配算法在该层的得分依然很高。

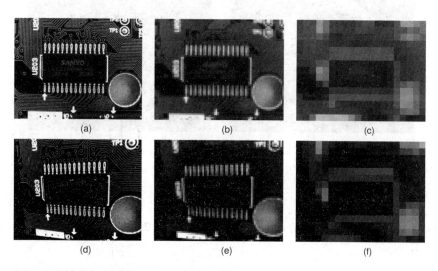

图 3.148　金字塔图像中的透视畸变效果。(a) 为原始图像；(b) 和 (c) 为第 4 和第 6 层金字塔图像；(d) 为应用透视投影后的图像；(e) 和 (f) 为 (d) 图的金字塔第 4 和第 6 层；值得注意的是，在考虑像素尺寸的情况下，在较高层金字塔中，透视变换的影响也较小

After the potential matches have been identified on the top pyramid level, they are tracked through the resolution hierarchy. During the tracking, the deformations of the model become larger on lower pyramid levels, and hence must be determined and taken into account. The idea is to compensate for the deformations by deforming the model accordingly. Stronger deformations are compensated on higher pyramid levels, whilesmaller deformations are compensated on lower pyramid levels.

To robustly determine the deformations on a pyramid level, the model on this level is split into smaller parts. Each part is searched independently in a local neighborhood of the initial pose

在顶层金字塔确定候选匹配对象之后，它们将通过分辨率层级结构进行跟踪，在跟踪过程中，随着金字塔层级的降低，模板的变形也会不断增大，因此我们需要对模板的变形程度进行计算。这一处理过程的思路是通过模板的变形来补偿形变。相应地，较大的畸变会在金字塔较高层中进行补偿，而较小的畸变则在金字塔低层进行补偿。

为了更准确地计算某一层金字塔中的形变，将模板拆分为若干部分。每个部分都在初始区域的邻域内进行局部搜索。各部分的初始区域则通过

of the part. The initial pose is obtained from the object pose and the model deformations, both of which were determined on the next higher pyramid level. The reference points of each found part in the image and in the model yield one correspondence. From the set of correspondences, the deformation of the model on the current pyramid level is refined and propagated to the next lower level until the lowest level is reached.

3.12.1.2 Model Generation

The splitting of the model into small parts is performed during model generation. It is motivated by the observation that a small model part, on the one hand, is more descriptive than a single model point, and, on the other hand, its shape is hardly affected by model deformations. This enables a robust establishment of corresponding features between the model and the search image, especially in the presence of deformations (see Figure 3.149).

上一层金字塔中对象的位置以及模板的形变确定。模板每一部分的参考点都会在图像上搜索其对应点。这组点将精确地表示当前金字塔层级中模板的形变状态。与此同时，它们将向下传递，直至金字塔最底层。

3.12.1.2 创建模板

在创建过程中，模板将被分割为若干部分。模板的分割遵循以下两点，一方面该部分模板要比单个匹配点包含更多的描述信息，另一方面它所受畸变的影响较小。通过局部模板的匹配，即使图像中存在畸变也可以在模板和被测图像之间找到稳定的匹配特征（见图 3.149）。

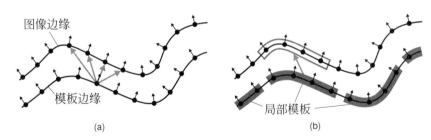

图 3.149 将模板拆分为若干部分：（a）在模板中使用单个点进行一致性搜索很难有准确的结果；如图，从梯度方向上判断，一个模板的采样点和 4 个图像上的采样点都有很好的匹配；因此，它们都可以作为匹配候选点；图中，黑色箭头表示梯度方向，灰色箭头表示匹配点；（b）对于局部模板（灰色粗线部分），我们更容易在图像中找到唯一的对应位置；图中灰色框表示被测图像中与局部模板相对应的部分

The model parts are determined by spatially clustering the model points on each pyramid level with the k-means clustering algorithm (Shapiro and Stockman, 2001). The resulting assignment of each model point to a cluster is represented by

通过对每层金字塔中模板边缘点集进行 k 均值聚类，我们可以得到各层金字塔中的局部模板（Shapiro and Stockman, 2001）。经过聚类算法，每一簇模板点的分配状况可以用标志

the indicator matrix C. If the model point p_i ($i = 1, \cdots, n$) is contained in cluster j ($j = 1, \cdots, k$), the corresponding matrix element c_{ij} is set to 1, and to 0 otherwise. Typically, k is chosen such that a cluster contains 5 to 10 model points on average. In addition, each cluster, or model part, is assigned a label λ that indicates whether the contained model points form a circular or a linear structure. While circular structures yield position information in two dimensions, linear structures yield information only perpendicular to the line direction. This information is later used to increase the robustness of the matching. The measure

矩阵 C 来表示。如果模板点 p_i ($i = 1, \cdots, n$) 属于簇 j ($j = 1, \cdots, k$)，则与之对应的矩阵元素 c_{ij} 为 1 反之为 0。一般来说，聚类的每一簇包含模板点集中的 5 到 10 个点，并以此来选择 k 的大小。此外，每一簇，或者说是每个局部模板还会被分配一个标签，用来表示该模板取自一个圆形或线形的边缘结构。对于圆形结构我们将求取二维位置信息，而线形结构则只会计算该线的垂直方向。在后面的匹配运算中我们会通过这些数据提高算法的稳定性。

$$\lambda_j = \left\| \frac{\sum_{i=1}^{n} \frac{d_i}{\|d_i\|} c_{ij}}{\sum_{i=1}^{n} c_{ij}} \right\| \tag{3.198}$$

indicates whether the model points with associated direction vectors d_i in cluster j have a dominant gradient direction. For a cluster with a perfect linear structure, $\lambda_j = 1$; for a cluster with a perfect circular structure, $\lambda_j = 0$.

上式可用来判断簇 j 的模板点集是否存在明显的梯度方向。当簇中点集趋近于直线，λ_j 趋近于 1；而当簇中点集趋近于圆形，λ_j 趋近于 0。

3.12.1.3 Similarity Measure

The similarity measure that is calculated in shape-based matching (see Section 3.11.5.6) is also the basis for deformable matching. The key idea is to assume a global model transformation with few degrees of freedom and to additionally allow model parts to locally deviate from the global transformation. The global transformation might be, for example, a similarity transformation. This ensures that the recognition is still fast enough for practical applications. The local deviations of the model parts from the global similarity transformation cover the remaining degrees

3.12.1.3 相似度量

在基于形状匹配（见 3.11.5.6 节）中用到的相似计算同样也是变形匹配的基础。与之不同的是，我们在假设模板存在小角度全局变换的同时，还允许局部模板相对于全局模板存在一定偏差。其中全局变换类似于相似变换。全局变换可以大大降低算法的处理时间，使该算法可以被一般的应用场景所接受。在相对于全局变换的局部模板偏差数据中包含了剩余的角度旋转，例如，存在透视畸变情况下的投影变换。由于使用了聚类算

of freedom of, for example, a projective 2D transformation in the case of perspective distortions of planar objects. Because of the clustering, the individual model parts only have a small spatial extent. Therefore, it is sufficient to model the deviations by a local translation. Following this idea, the similarity measure s is the sum of contributions s_j of all model parts

$$s = \frac{1}{n} \sum_{j=1}^{k} s_j \qquad (3.199)$$

with

$$s_j = \max_{\boldsymbol{t} \in T} \sum_{i=1}^{n} c_{ij} \frac{\boldsymbol{d}_i'^\top \boldsymbol{e}_{q+p'+t}}{\|\boldsymbol{d}_i'^\top\| \|\boldsymbol{e}_{q+p'+t}\|} \qquad (3.200)$$

where e are the direction vectors in the image (see Eq. (3.192)) and T is the set of local translations that are examined. For model parts that, according to Eq. (3.198), have a circular structure, T covers translations within a local neighborhood of 5×5 pixels. For model parts with a linear structure, T covers translations within ± 2 pixels in the direction of λ_j. For each model part, the translation that yields the highest score is selected. Tests where T additionally covers small local rotations showed no significant improvements. Therefore, for speed reasons we only examine translations.

The similarity measure in Eq. (3.200) assumes that the edges in the model and in the image have the same contrast direction. As for shape-based matching, the similarity measure easily can be extended to cases where the contrast might be reversed globally or locally. This is achieved by adapting the similarity measure in a

法，每个独立的局部模板只会在一定范围内进行匹配运算，这足以应付局部的模板偏差。由此，我们定义相似度 s 由各局部模板相似度 s_j 共同确定。

上式中 e 为被测图像方向向量（见等式 (3.192)），T 为某组局部变换计算结果。对于每个局部模板，根据等式（3.198）进行判断，如果模板为圆形结构，T 中包含当前位置邻域内的 5×5 像素的平移变换。如果模板形状为线形结构，T 中包含沿该线垂直方向 ±2 个像素的平移变换。对于每一个局部模板我们选择平移变换后得分最高者作为 λ_j。除了平移变换，我们还尝试在 T 中增加旋转变换。但是经过测试，我们发现旋转变换对得分影响较弱。因此，考虑到算法的处理速度，我们只计算平移变换。

在等式（3.200）中，我们假设模板和被测图像中对比度变化的方向是一致的。在基于形状的模板匹配中，我们可以很容易地通过类似等式（3.193）和等式（3.194）的方式，将相似度算法进行推广，使其可以支持全局或局部明暗变化方向相反的情况。因此，

manner analogous to Eqs. (3.193) and (3.194). Therefore, deformable matching inherits the properties of shape-based matching, i.e., robustness to occlusions, clutter, and illumination changes. In contrast to shape-based matching, however, the existence of model parts allows us to introduce another similarity measure: it permits a global contrast reversal of each model part by replacing Eq. (3.200) with

$$s_j = \max_{t \in T} \left| \sum_{i=1}^{n} c_{ij} \frac{{d'_i}^\top e_{q+p'+t}}{\|{d'_i}^\top\| \|e_{q+p'+t}\|} \right| \quad (3.201)$$

On the one hand, this measure is more distinctive, and hence less sensitive to noise, than the measure that is invariant to local contrast reversals because it requires that nearby model points have a consistent contrast direction. On the other hand, it is more tolerant than the measure that is invariant to a global contrast reversal because it allows contrast reversals between different model parts. Therefore, it should be preferred when metallic or plastic parts with specular reflections must be recognized, for example.

变形匹配继承了基于形状的模板匹配的特性，例如在稳定性方面，可以抗遮挡，抗混乱背景以及抗光照变化。相比于形状模板匹配，当算法推广到局部匹配后，我们可以通过将等式（3.200）替换为等式（3.201）来描述一种新的相似度：在每个局部模板中可以存在全局的反向对比度。

这种独特的处理方式，一方面，相比于只使用允许局部对比度方向颠倒的检测，该处理可以降低算法对噪声的敏感度，因为它要求相邻的模板点在匹配时拥有相同的对比度方向。另一方面，相比于只使用允许全局对比度方向颠倒的检测，该处理容忍度更高，因为，在匹配过程中不同的局部模板之间的全局对比度方向是相互独立的。因此，不得不说它是检测诸如金属、塑料等镜面反光材质被测物的首选。

3.12.1.4 Hierarchical Search

A coarse-to-fine search within an image pyramid is applied, similar to that of shape-based matching described in Section 3.11.2.2. First, at the top pyramid level, the score is evaluated for all global model transformations of the chosen transformation class, e.g., the class of similarity transformations. Because of potential model deformations, the threshold that is applied to the score on the top pyramid level to identify match candidates should be slightly lowered in comparison

3.12.1.4 分层搜索

在一个图像金字塔中，我们使用一种由粗到细的搜索方式，该算法类似 3.11.2.2 节中所描述的形状模板匹配。首先在金字塔顶层，我们将为所选变换类型的所有全局变换的模板计算得分，例如，相似变换的某个类型。由于候选模板存在变形，相比于识别未变形的被测物，应用于顶层金字塔中候选模板得分筛选的阈值要有所降低。虽然这一结果会通过金字塔在更

to the recognition of non-deformable objects. Although this results in more candidates to be tracked through the pyramid, it also increases the robustness.

On lower pyramid levels, the deformations that were determined on the next higher level are used to initialize a deformed model on the current pyramid level. For this, the model points on the current level are transformed according to the deformation before the refinement is performed: From the deformation of the candidate on the next higher level we know the relative local translation of each model part that yielded the maximum score (e.g., in Eq. (3.200)). The original and the translated cluster centers are used as point correspondences to estimate the eight parameters of a projective 2D transformation (see Section 3.3.1.1). Then, the projective transformation is applied to the model points on the current level. It is assumed that the transformed model represents the true appearance of the object better. Hence, it is used for refining the candidate on the current pyramid level. This process assumes that the magnitudes of any still existing inaccuracies of the projective deformation lie in the range of the local neighborhood in which the local search is performed, i.e., in the range of 5 pixels in our case. Therefore, strong deformations of the object are already compensated on higher pyramid levels while on lower pyramid levels, only small adaptations of the model geometry are possible.

3.12.1.5 Least-Squares Pose Refinement

After the candidates have been tracked to the lowest level, a least-squares pose refinement is applied to obtain an even higher accuracy, similar to

多的候选模板中进行跟踪传递，但他依然可以在一定程度上增加算法的稳定性。

在较低的金字塔中，我们将通过上一层金字塔的变形为当前金字塔初始化一个变形模板。为此，在细化处理之前，在当前层模板中的点将根据此变形进行变换：从上一层金字塔候选模板的变形中，我们可以获得的局部相关变换。该变换可通过对每个局部模板最大得分进行计算获得（例如，等式（3.200））, 簇最初的中心将进行平移变换并基于该点对二维投影的8个参数进行估计（见3.3.1.1节）。这个投影变换将应用于金字塔当前层的模板点上。通过这个处理我们可以更好地描述模板的状态。因此，它将被用于细化金字塔当前层中的候选模板。在此过程中我们假设，在进行局部搜索时，依然存在任意大小的投影变换误差，例如在5个像素的范围内。因此，由于较大的变形已经在更高层中得到补偿，当我们进行金字塔低层的处理时，模板只可能存在较小的几何变化。

3.12.1.5 最小二乘法位置精确计算

当候选模板跟踪传递到金字塔最底层时，我们将使用类似3.11.5.6节中所描述的最小二乘法来提高匹配

the approach described in Section 3.11.5.6. However, instead of an affine transformation, a projective transformation must be estimated in the case of perspective distortions. There are two important differences to the estimation of the projective transformation during tracking that is described in the previous section. First, to obtain the highest possible accuracy, correspondences are not simply determined for the cluster centers but for all model points. Second, to be independent of the point sampling, the distances between the points in the image and the tangents of the model points are minimized (see Eq. (3.197)) instead of minimizing point-to-point distances that are used during the tracking.

3.12.1.6 3D Pose Estimation

Up to now, we have discussed the uncalibrated case of deformable matching. In many machine vision applications, it is sufficient to know the projective transformation that maps the model to the search image. However, in some applications, e.g., bin picking, the 3D pose of the object with respect to the camera is required. To compute the 3D pose, the interior orientation of the camera must be known, i.e., the camera must be calibrated as described in Section 3.9.4. Furthermore, the model must be known in world coordinates. A detailed description of the approach can be found in (Hofhauser et al., 2009b; Hofhauser and Steger, 2010, 2011, 2012, 2013).

In some cases, the model is created based on a computer-aided design (CAD) file that directly provides the metric information of the object. Otherwise, the metric properties of the model can be determined by acquiring an image of the object

运算的精度。然而，由于存在透视变形，我们将使用投影变换代替原有放射变换运算。在金字塔跟踪传递过程中，本章中所介绍的投影变换与前面章节主要有两点不同之处。第一，为了获得最高的精度，相关性不是简单的由簇的中心来确定，而是取决于所有的模板点。第二，采样点是相对独立的，相比于原有的点对点的最小距离计算，我们将使用图像中的点和模板点的切向进行最小距离运算（见等式（3.197））。

3.12.1.6 三维位姿估计

至此我们已经讨论了未标定状态下的变形模板匹配。这种在图像搜索中通过映射模板获取投影变换参数的算法，足以应付大多数视觉应用。然而在一些应用中，如机器人抓取，我们需要获取被测物相对于摄像机的三维位姿。为了获取三维位姿，我们先要已知摄像机的内参，例如，摄像机需要经过 3.9.4 节中的方法进行标定。此外，还要已知模板在世界坐标系中的三维数据。该算法的详细描述可以参考文献（Hofhauser et al., 2009b; Hofhauser and Steger, 2010, 2011, 2012, 2013）。

在某些情况下，模板是基于 CAD 文件创建的，此类文件可以直接提供被测物的公制尺寸信息。否则，尺寸参数只能从所采集图像中与被测物相邻的刻度或标定板获取。如图 3.150

next to a scale or a calibration target. Figure 3.150 illustrates the transformation of a model of a planar object to world coordinates. The process is based on the computation of world coordinates from a single image, which was described in Section 3.9.5. The final model coordinates are given in the model coordinate system (MCS). The 3D object pose that is returned by the matching describes the rigid 3D transformation from the MCS to the CCS.

中所示，一个平面被测物中的模板到世界坐标系的转换。这一处理过程是，基于 3.9.5 节中描述的，单幅图像计算世界坐标的方法。三维模板的位置将在模板坐标系（MCS）下给出。而被测物的三维位姿则在匹配运算后获得，并使用 MCS 到 CCS 的三维刚性转换关系表示。

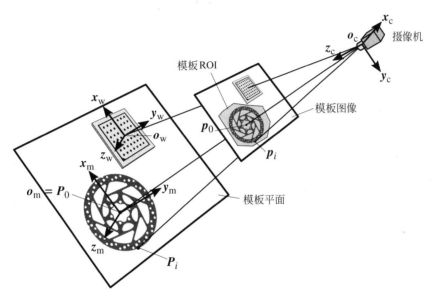

图 3.150 通过模板 ROI 定义一个模板，并将其从模板图像中转换到世界坐标系中；我们在模板图像上提取点 P_i，为了计算该点在世界坐标系中的对应点 p_i，我们需要已知摄像机坐标系下模板平面的位姿；这个平面可以通过在模板旁边摆放的标定板来确定；同时，我们还可以通过该标定板来确定世界坐标系 WCS (o_c, x_c, y_c, z_c)；在世界坐标系中，点的位置可以通过，模板平面上经过 p_i 点的光线获得；这些坐标通常是相对于模板参考点 P_0 的，且它们都是模板平面上的投影点；其中 P_0 对应模板图像中所定义的模板中心 p_0；该点被定义为模板坐标系 MCS $(o_m, x_m, y_m, z_m)^\top$ 的原点；MCS 的 xy 平面和 WCS 的 xy 平面是共面的；我们可以随意选择 WCS 平面的 Z 方向。例如，所有模板点的 z 坐标均为 0，如 $P_i = (x_i, y_i, 0)^\top$；最后，我们将三维对象的位姿定义为从 MCS 到 CCS 的刚性三维变换

There are two analytic methods to determine the 3D object pose. The first is to decompose the projective transformation that was obtained from the least-squares pose refinement of Section 3.12.1.5 into its 3D translation and 3D rotation

我们可以通过以下两种方法确定物体的三维位姿。第一种方法，我们可以将投影变换分解为三维平移和三维旋转两部分（Faugeras, 1993; Zhang, 2000）。其中投影变换参数可通过

components (Faugeras, 1993; Zhang, 2000). The second is to directly estimate the 3D pose from point correspondences with a non-iterative algorithm (Moreno-Noguer et al., 2007). However, practical evaluations have shown that the highest accuracy is obtained by minimizing the geometric distances between the points in the image and the tangents of the projected model points within an iterative nonlinear optimization. Essentially, the optimization is comparable to camera calibration, which was described in Section 3.9.4.2. However, instead of the centers of the calibration marks, the model points are used as correspondences. The optimization can be initialized with the result of one of the two analytic methods, for example.

Figure 3.151 shows an example application where the 3D pose of a car door is determined by using the described approach. Note that the model is created only from a small planar part of the car door because the entire door has a distinct 3D structure, and hence would violate the planarity assumption.

3.12.1.7 Recognition of Locally Deformed Objects

In some applications, arbitrarily deformed objects must be recognized, like the print on a T-shirt or on a foil package. Therefore, in contrast to the previous sections, where the distortions were modeled by a projective 2D transformation or a 3D pose, for deformed objects we must cope with arbitrary local object deformations. The approach is described in detail in (Hofhauser et al., 2009a; Hofhauser and Steger, 2010, 2011, 2012, 2013).

3.12.1.5 节中使用二乘法所求的精确位姿计算获得。第二种方法，我们可以使用非迭代算法（Moreno-Noguer et al.，2007）通过特征点直接计算物体的三维位姿。然而，在实际运算中我们发现，存在另一精度极高的算法。该算法通过非线性迭代优化获取图像点集与模板经过投影变换后的目标点集的最小几何距离。从本质上讲，这个优化相当于 3.9.4.2 节中所提到的摄像机标定。然而，我们在算法中使用的对应点，从标定板上的靶标中心换成了模板点集。例如，这一优化的初始状态可以通过我们上述两种方法中的任意一个进行计算。

图 3.151 展示了一个通过上述算法确定车门三维姿态的应用案例。值得注意的是在创建模板时，只需选取车门上一小部分的平面区域。因为整个车门是立体的，这有悖于平面性假设。

3.12.1.7 局部变形物体识别

在一些应用中，我们需要识别任意变形的被测物，如印刷在 T 恤或者铝箔包装上的图案。因此，相比于前面章节中所提及的使用二维投影变换或者三维位姿对物体的畸变进行建模的方法，对于变形物体，我们需要对被测物进行裁剪，而在裁剪后的每个局部对象中都包含任意的变形。算法相关细节详见文献（Hofhauser et al.，2009a；Hofhauser and Steger，2010，2011，2012，2013）。

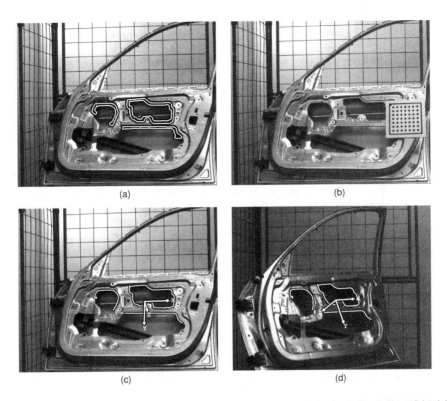

图 3.151 确定车门的三维位姿：（a）模板图像。我们只使用车门上包含平面部分的区域创建模板（白色轮廓）；（b）通过标定板确定模板平面；（c）被转换到世界坐标系下的模板点集（白色）和 MCS（白色向量）通过投影显示在模板图像中；（d）通过投影模板点和 MCS 标识搜索到的实例

When recognizing deformable objects, one essential question is how to model the object deformations. The chosen deformation model must be general with regard to the object and deformation type because in most applications, the physical deformation properties of the objects are unknown. Furthermore, the determination of the deformations must be robust to ambiguous matches of model parts. In particular, the position of linear model parts only can be determined perpendicular to their linear extent. Also, the effect of false correspondences, i.e., outliers, should be dampened by the deformation model. Finally, the deformation model should be able to allow the interpolation and extrapolation of deformation

当进行变形物体识别时，一个最重要的问题是如何对物体的变形进行建模。我们所选择的变形模型需要兼顾被测物和变形类型两个方面。因为，在大多数的应用中被测物的物理变形是未知的。此外，对于变形的确定，在我们进行局部模板匹配时，需要足够稳定的模糊匹配算法。尤其是在线性局部模板的位置确定方面，因为该位置只能在模板垂直方向范围内进行计算。对于误匹配的数据，如离群值，也需要通过变形模板进行抑制。最后，变形模板还必须支持变形信息的插值和外推。这对于候选模板在金字塔各层中的跟踪传递非常重要。为了使得

information. This is important during the tracking of the candidates through the pyramid. To refine a candidate on the current pyramid level, the model of the current level must be deformed in accordance with the deformation determined on the next higher level (see Section 3.12.1.4). Because of the properties of image pyramids (see Section 3.11.2.1), the model on the current pyramid level might contain points in areas where its lower-resolution version on the next higher level did not contain any points. Therefore, in these areas, no deformation information is explicitly available. Instead, it must be calculated from neighboring deformation information by interpolation or extrapolation. Furthermore, as in the case of perspective distortions, it is often desirable to rectify the image after the matching in order to eliminate the deformations. For the rectification, deformation information for each pixel is necessary, which requires interpolation as well as extrapolation.

The method of recognizing locally deformed objects is summarized in Figure 3.152. The local translation of each cluster that yielded the maximum score is obtained from Eq. (3.199). Each model point within the cluster adopts the local translation of the cluster. Thus, the local translation represents the local deformation in the row and column directions at each model point. The deformation values are visualized in Figures 3.152(c) and (d), respectively. In the next step, a dense deformation field is computed, i.e., two images that contain the deformations in the row and column directions at each pixel. For this, the row and column deformations at the model points are propagated to all image pixels by applying a harmonic interpolation (Aubert and

当前金字塔中候选模板精度更高，我们需要使用在上一层金字塔中所计算的变形参数对该层金字塔中的模板进行畸变矫正（见 3.12.1.4 节）。由于金字塔的特性（见 3.11.2.1 节），在金字塔当前层上模板所包含的点中，可能存在这样一些区域，因为上一层金字塔图像的分辨率较低，在这些区域中并不包含任何的点。因此在这些区域中并没有明确有效的变形信息可用。取而代之的是通过插值和外推算法结合相邻区域数据对该区域的变形信息进行计算所获得的填充数据。因此，存在透视畸变的情况下，该算法通常适用于在匹配后对图像进行矫正，以此来去除图像变形。而为了对图像进行矫正，我们需要获取每个像素的变形信息，这时就需要通过插值和外推算法对它们进行求解。

在图 3.152 中我们对这种识别局部变形物体的方法进行了总结。每一簇局部平移变换的最大得分可通过等式（3.199）计算。每一个簇中的模板点都应用该簇的局部平移变换进行运算。由此，局部的平移变换表示了每个模板点在行和列两个方向上的局部变形。在图 3.152（c）和图 3.152（d）中我们使用图像亮度分别表示这两个方向上的变形程度。接下来，我们将计算一个密集的变形场，即两张图中每个像素在行和列方向上的变形。为了进行这一运算，模板点在行和列上的变形将通过赫米特插值运算填充推导每一个像素（Aubert and Kornprobst, 2006）。赫米特插值假设插入值符合

Kornprobst, 2006). The harmonic interpolation assumes that the interpolated values satisfy the Laplacian equation $\Delta f = 0$ (see Eq. (3.111)), with f being the deformations in row and column directions, respectively. This ensures a smooth interpolation result.

拉普拉斯等式 $\Delta f = 0$（见等式 (3.111)），其中 f 为行和列的变形分量。由此，我们就可以获得一个平滑的插值结果。

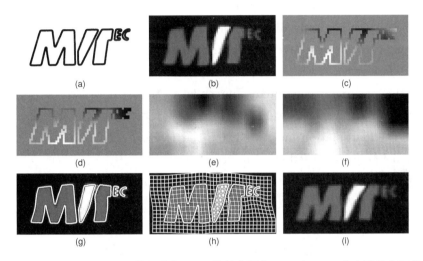

图 3.152　识别局部变形物体：(a) 模板点集 (b) 待搜索图像 (c) 和 (d) 为在被搜索图像中行和列方向上模板点的局部变形（中间灰度级表示没有变形，黑色灰阶表示负方向变形，白色灰阶表示正方向变形）；(e) 和 (f) 为对 (c)、(d) 两图进行赫米特插值填充后所得的行和列方向的图像；(g)~(i) 为识别结果；(g) 变形模板点集和搜索图像的叠加显示；(h) 通过规则网格变形实现的可视化的变形场；(i) 矫正后的搜索图像

Because of local mismatches, the deformations at the model points might contain outliers. Outliers might result in locally contradicting deformations, which in turn could introduce undesirable discontinuities or even crossings in the interpolated deformations. Fortunately, the use of the harmonic interpolation model prevents such discontinuities and crossings. The effect of outliers is further reduced by applying a median filter to the deformation images. Finally, by applying a smoothing filter to the deformation images, noise and high-frequency deformations are eliminated. The final deformation images are shown in Figure 3.152(e) and (f).

由于局部的不匹配，在模板点的变形中有可能存在离群值。离群值可能导致在候选变形中引入不连续间断点，甚至使得变形插值出现交叉。幸运的是赫米特插值模型可以预防不连续和交叉问题。我们可以通过在图像上使用中值滤波进一步降低离群值的影响。最后再使用平滑滤波去除变形中的噪声和高频分量。图 3.152（e）和图 3.152（f）为经过上述处理后的变形信息图像。

The harmonic deformation model fulfills the requirements stated above and can be computed efficiently. More complex deformation models like the thin-plate splines model of Bookstein (1989), for example, are often too slow or do not fulfill all the requirements.

Hierarchical search is performed as described in Section 3.12.1.4. The only difference is that instead of computing a projective transformation for each candidate on the lower pyramid levels, the two deformation images are computed. The model points on the current pyramid level are deformed according to the deformation images. Then, the deformed model points are used to refine the pose and the deformation of the candidate on the current level.

Figure 3.152(g) shows the model points of the deformed model found. For visualization purposes, Figure 3.152(h) additionally shows the model deformations as a deformed regular grid. Finally, the search image can be rectified by applying the inverse deformation with the techniques described in Section 3.3.2.2.

In Figure 3.153, an example application is shown where a deformed gasket must be inspected

赫米特变形模型可以满足上述应用需求且运算效率高。当然，还有很多更复杂的变形模型，但它们通常运算过于缓慢或者不能满足应用要求，例如，文献（Bookstein，1989）的薄板样条函数。

在运算过程中，我们会进行与3.12.1.4 节中描述相同的分层搜索。不同的是，这次我们不是要为金字塔低层的候选模板计算投影变换，而是为两幅图像计算。当前金字塔中的模板点将根据变形图像进行变换。随后，变形后的模板点将被用于矫正金字塔当前层中候选模板的位置和变形。

图 3.152（g）展示了被搜索到的变形模板点集。为了看起来更加直观，图 3.152（h）在图 3.152（g）图的基础上附加了变形网格，以此显示模板的变形。最后我们使用 3.3.2.2 节中所描述的技术进行逆向变形，以此达到图像矫正的目的。

图 3.153 中为我们展示了对存在局部变形的垫圈进行缺陷检测的实

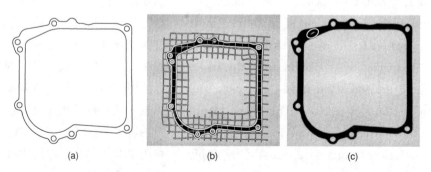

图 3.153 局部变形垫圈的识别：（a）模板点集；（b）搜索图像，识别结果和变形场；（c）缺陷检测（用白色椭圆标记）

for defects. The defect detection is performed by comparing the gray values of the gasket with the gray values of a perfect reference image (see Section 3.4.1.3). Because of the local deformations of the gasket, the search image must be rectified onto the reference image to eliminate the deformations before performing the comparison.

In the example shown in Figure 3.154, locally deformable foil packages must be recognized. Possible applications are print inspection or OCR in the rectified image or order picking in a warehouse.

例。该缺陷检测需要通过对比垫圈图像和垫圈良品参考图像的灰度值来完成（见 3.4.1.3 节）。由于垫圈存在局部形变，在进行对比之前需要先将被测图像矫正到参考图像上，以此消除变形对检测结果的影响。

在图 3.154 所示的例子中，我们需要对存在局部变形的铝箔包装进行识别。通过匹配后的矫正图像，我们可以进行印刷检测或 OCR 识别。同时，匹配结果还可以为仓库拣选系统提供待抓取物品的位置信息。

  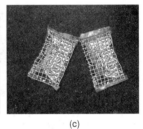

(a) (b) (c)

图 3.154　局部变形铝箔包装的识别：（a）模板图像和模板点集；（b）待搜索图像；（c）识别结果和变形场

3.12.2　Shape-Based 3D Matching

After having discussed several approaches to finding planar objects robustly in Sections 3.11.5 and 3.12.1, in this section we will discuss how to recognize 3D objects in a single image. The approach we will describe is an extension of shape-based matching, which was introduced in Section 3.11.5.6 and is described in more detail in (Wiedemann et al., 2008; Ulrich et al., 2009, 2012; Wiedemann et al., 2009, 2011, 2013a,b).

Determining the pose of 3D objects is important in many industrial applications like bin pick-

3.12.2　基于形状的三维匹配

在 3.11.5 节和 3.12.1 节中我们已经讨论了关于平面物体的稳定的搜索算法。在本节中我们将讨论如何通过单幅图像进行三维物体的识别。我们将为大家介绍一个由 3.11.5.6 节中所介绍的形状模板匹配算法推广而来的新的算法。该算法相关细节详见文献（Wiedemann et al.，2008；Ulrich et al.，2009，2012；Wiedemann et al.，2009，2011，2013a,b）。

三维物体位姿的确定在很多工业应用场合都有及其重要的作用，如机

ing or 3D inspection tasks. In particular, the increasing demand for autonomous robots in general and for vision-guided robots in particular calls for versatile and robust 3D object recognition methods. In many applications, only a single (monocular) image is available for the recognition because a setup of two or more cameras is either too expensive, too cumbersome to calibrate, or simply too bulky. And even though 3D image acquisition devices are becoming increasingly important (see Section 2.5), conventional cameras are still much more widespread in industrial applications. Furthermore, the objects that must be recognized in industrial applications often have an untextured metallic or plastic surface like the two metallic clamps shown in Figure 3.155(a). Consequently, the only significant information that is available for recognition are the geometry edges of the object in the image. For a geometry edge, its two adjacent faces have a different normal vector. For many objects in industrial applications, a CAD model that contains the object's geometry edges is available. Figure 3.155(b) shows a CAD model of the metallic clamps of Figure 3.155(a). Note that in addition to the geometry edges, the model also contains edges that result from the triangulation of the surface. Therefore, it is natural to match the geometry edges of the CAD model with the image edges for object recognition. With shape-based matching, there is already a robust edge-based template matching approach available that is able to determine the 2D pose of a planar object in an image. In the following sections, we will discuss how to extend shape-based matching to the recognition of 3D objects in monocular images and to determining the 3D object pose.

械手抓取或三维检测。特别是为了应付日益增长的普通的自动机器人和特种的视觉引导机器人的需求，我们需要更多稳定的三维物体识别算法。在许多应用场景中，我们只有单张图像（单目）可用，因为采用双目或多目的采集设备会存在价格昂贵、标定复杂、结构笨重等问题。虽然三维图像采集设备在项目应用中正在变得越来越重要（见 2.5 节），但在工业应用中普遍使用的，更多的还是传统摄像机。此外，在工业应用中，需要识别的物体通常会拥有不带纹理的金属或塑料的表面，如图 3.155（a）中的金属夹具。因此在图像中我们所能利用的信息只有被测物的几何边缘。几何边缘由两个相邻的，具有不同法向量的平面组成。在工业应用中，很多被测物都有自己的 CAD 模型，其中包含了有效的被测物几何边缘。图 3.155（b）为在图 3.155（a）中所看到夹具的 CAD 模型。值得注意的是，在模型中，除了包含几何边缘外，还包括通过三角化算法所获得的平面边缘。因此，很自然的，我们可以使用 CAD 模型中的几何边缘和被测图像中的边缘进行匹配运算，以此达到被识别测物的目的。基于形状的模板匹配，是现有的一种稳定的基于边缘的模板匹配算法，它可以在单幅图像中确定一个平面物体的二维姿态。在接下来的章节中，我们将讨论如何对基于形状的匹配进行推广，使之可以通过单幅图像进行三维物体的识别和位姿确定。

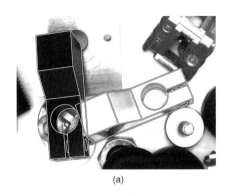 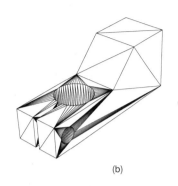

图 3.155 （a）两金属夹具图像，通过基于形状的三维匹配在图像中进行搜索，将找到的被测物用白色边缘标记；（b）在基于形状的三维匹配算法中，用于创建模板的 CAD 模型。在图（a）中，为了显示更直观，我们去除了共面的模板边缘

3.12.2.1　View-Based Approach

Shape-based 3D matching uses a view-based recognition approach: 2D views of the object are precomputed and searched in the image using template matching. To compute the 3D object pose, the interior orientation of the camera must be known, i.e., the camera must be calibrated as described in Section 3.9.4. For the highest accuracy it is assumed that lens distortions have been eliminated from the images in advance (see Sections 3.9.1.3).

One major problem of these approaches is the six degrees of freedom of the object pose, which lead to an enormous number of 2D views that would have to be compared to the image. Therefore, most view-based approaches reduce the run time complexity by using a virtual view sphere (see Figure 3.156). The object is assumed to lie at the center of the sphere, and virtual cameras looking at the center are placed on the sphere's surface. To obtain the views, the object is projected into the image plane of each virtual camera. Thus, for each virtual camera, a template image

3.12.2.1　基于视图的方法

基于形状的三维匹配使用一种基于视图的识别方法：使用模板匹配算法对事先计算好的被测物二维视图进行搜索。为了计算物体三维位姿，我们需要已知摄像机的内参。例如，摄像机需要通过 3.9.4 节中所描述的算法进行标定。为了获得更高的处理精度，我们假设镜头的畸变在处理之前已经从图像中去除（见 3.9.1.3 节）。

一个关键的问题是，算法需要对物体六自由度的位姿进行处理，这将导致我们需要对庞大的二维视图图像进行比对处理。因此，大多数基于视图的方法，都会通过使用虚拟球面视图的方式，降低算法执行时间复杂度（见图 3.156）。我们假设将物体置于球心，而虚拟摄像机被放置于球面上，并对球心进行观测。我们可以通过将物体投影到每一个虚拟摄像机的像平面上获取当前摄像机的视图。如此，对于每个虚拟摄像机，我们都将获得一个模板图

is obtained, which is used to search the view in the search image using a template matching approach.

For the perspective projection of the object into the image plane of a virtual camera, the object parts that are visible from the current virtual camera must be determined. All parts that are invisible due to self occlusions must be excluded from the projection. This visibility calculation is computationally expensive. One advantage of the view-based approach is that the projection of the CAD model into the image plane, and hence the visibility calculation, can be performed offline during model generation, which speeds up object recognition.

If the application requires that the distance between camera and object is allowed to vary, typically several view spheres with different radii are generated. Hence, virtual cameras are created within a spherical shell in this case.

像。它们将用于在搜索图像中通过模板匹配算法对视图进行查找。

为了将物体投影到虚拟摄像机的像平面上，我们需要确定在当前虚拟摄像机中，物体的哪个部分是可见的。我们将从投影数据中剔除所有由于物体自遮挡而不可见的部分。这一物体可见部分的评估算法是非常复杂的。基于视图算法的优势在于，使用 CAD 模型进行像平面的投影运算。由此，可以在创建模板时进行离线处理，这将大大提高物体的识别速度。

如果在应用中，摄像机和被测物的距离是不定的，我们可以生成多个典型的半径不同的球面视图。此后，我们将会根据当前的球面状况创建虚拟摄像机。

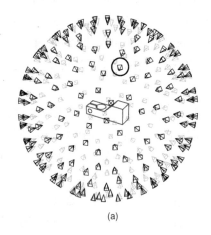
(a)

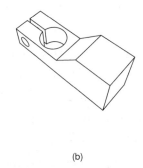
(b)

图 3.156　(a) 球面视图；图中黑色锥形金字塔为假想的虚拟摄像机，它们从各个角度对球心处的物体进行观察；(b) 物体在其中一个视图的观测结果；该图像是我们通过将物体投影到图 (a) 所圈选的虚拟摄像机的像平面所得

As a first result of the view-based approach, the 3D pose $(\alpha, \beta, \gamma, t_x, t_y, t_z)$ (see Section 3.9.1.1) of the virtual camera whose view yielded the high-

首先，通过搜索图像，我们将获得与之相似度最高的虚拟摄像机的三维位姿 $(\alpha, \beta, \gamma, t_x, t_y, t_z)$（见 3.9.1.1

est similarity with the search image is obtained. As a second result, the 2D matching pose of this view in the search image is obtained, which is returned by shape-based matching. The matching pose is represented by the parameters of a similarity transformation (see Section 3.11), i.e., (r, c, ϕ, s), where r and c are the row and column position of the object in the image, ϕ is its rotation in the image plane, and s is its uniform scaling. Instead of returning only the result with the highest similarity, the 3D and 2D poses of all matches that yield a similarity measure above a user-given threshold are obtained. As we will see below, the 3D object pose can be reconstructed from the 3D pose of the virtual camera and the 2D matching pose.

When applying the described view-based approach, the 3D object pose is decomposed into a 3D pose of the virtual camera and the 2D matching pose. What is the advantage of this decomposition? The six degrees of freedom of the 3D object pose are split into two degrees of freedom of the position of the virtual camera on the surface of the sphere, three degrees of freedom of the 2D matching parameters r, c, and ϕ, and one degree of freedom for the distance between the camera and the object. The latter is jointly modeled by the radius of the view sphere and the 2D matching parameter s. Splitting the object–camera distance is justified by the observation that a larger distance change results in a changed perspective, which must be represented by an additional virtual camera. Note that the changed perspective also requires performing the expensive visibility calculation. In contrast, a smaller distance change can be approximated by a uniform scaling of the

节）。紧接着，通过二维匹配，我们将从搜索图像中获得当前视图中的模板位姿。匹配位姿被表示为一个相似变换参数组（见 3.11 节），例如 (r, c, ϕ, s)，其中 r 和 c 分别为物体在图像中行和列的坐标，ϕ 为物体在像平面中的旋转角度，s 为物体均匀缩放系数。匹配算法将为我们返回所有高于用户设置阈值的三维和二维位姿，而不是只返回最佳的匹配结果。正如我们接下来将要看到的，物体的三维位姿可以通过虚拟摄像机的三维位姿和二维匹配位姿来重建。

当应用我们所描述的基于视图的算法时，物体三维位姿将被分解为虚拟摄像机的三维位姿和二维匹配位姿。这样分解有什么好处呢？三维物体的六自由度位姿被分割成三个部分，在球面上虚拟摄像机的位置存在两个自由度，二维匹配参数组 r, c 和 ϕ 存在三个自由度，最后摄像机到物体的距离也是一个自由度。这最后一个自由度，是由球面视图的半径和二维匹配参数 s 共同建模获取的。对于摄像机和物体之间距离的分割调整，我们遵循如下规则。当我们发现增加一段距离将导致视角变化时，就在此处添加一个额外的虚拟摄像机。值得注意的是，为应对视角的变化，我们需要增加大量的运算。与之相反的，当物体和像平面之间距离变化较小时，我们可以通过均匀缩放来近似表示一个物投影在像平面上的变化。因此，

projected object in the image plane. Consequently, a computationally expensive perspective projection only needs to be performed for each of the resulting virtual cameras, i.e., only for two and a half degrees of freedom, in order to obtain the views. Furthermore, the features for the shape-based matching only need to be extracted for each of these views. The remaining three and a half degrees of freedom are efficiently covered by the shape-based matching, where the extracted features of each view are translated, rotated, and scaled in 2D only.

After the matching, the 3D object pose must be reconstructed from the 3D pose of the virtual camera and the 2D matching pose. Let H_v be the homogenous transformation matrix that represents the 3D pose of the virtual camera. Hence, H_v transforms points from the MCS (see Figure 3.157) into the CCS. Now, H_v must be adapted in accordance with the 2D matching parameters r, c, φ, and s. First, the 2D scaling s is taken into account by assuming that the scaling approximately corresponds to an inverse scaling of the object–camera distance. Therefore, an isotropic scaling matrix S with a scaling factor of $1/s$ is applied. Then, the 2D rotation is applied, which corresponds to a 3D rotation R_z of the camera about its z axis. Finally, the position $(r, c)^\top$ is interpreted as a 3D rotation of the camera about its x and y axis by the angle α and β, respectively (see Ulrich et al., 2012) for the calculation of α and β). Then, the homogeneous matrix H_o that represents the 3D object pose finally is given by

$$H_o = R_y(\beta) R_x(\alpha) R_z(-\varphi) S(1/s) H_v \qquad (3.202)$$

There are some drawbacks of view-based approaches that limit their application in practice.

我们只需要使用之前提到的复杂运算求解虚拟摄像机。例如，只计算 2.5 维的自由度，并以此来获取视图。此外，我们只需在这些视图中提取基于形状匹配所需的特征。通过基于形状的匹配，我们将获得剩余的 3.5 维自由度。它们可以通过算法从二维数据中所提取出的视图平移、旋转和缩放参数计算获得。

在匹配之后，我们可以通过三维虚拟摄像机的位姿和二维匹配所得位姿重构物体三维位姿。我们使用齐次变换矩阵 H_v 表示虚拟摄像机的三维位姿。因此，H_v 可将 MCS（见图 3.157）中的点，变换到 CCS 中。现在，我们需要根据二维匹配所得参数 r, c, φ 和 s 对 H_v 进行修改。首先我们应用二维缩放 s。我们假设缩放可以表示为一个沿物体到摄像机距离的反方向缩放。则我们可将 $1/s$ 作为各向同性缩放矩阵 S 的缩放系数。随后，我们应用二维旋转，它所对应的矩阵为关于 z 轴的，摄像机的三维旋转矩阵 R_z。最后，位置 $(r,c)^\top$ 可以分别表示摄像机关于 x 轴和 y 轴的三维旋转角度 α 和 β（α 和 β 求解过程见（Ulrich et al., 2012）。综上，用于表示三维物体位姿的齐次矩阵 H_o 最终以下式给出

在实际应用中，基于视图的算法，存在一定局限性。尽管使用了球面视

Despite the use of the view sphere, typically several thousands of views are still necessary to ensure robust recognition. Searching several thousands of templates is much too slow for practical applications. Furthermore, the achievable accuracy depends on the density of the views. Unfortunately, the views cannot be chosen to be arbitrarily dense for run time reasons. Finally, the use of a view sphere assumes that the object lies at the center of the sphere. Hence, it requires that the object lies in the center of the image, too. Objects that occur in the image corners are perspectively distorted in comparison to the centered view, which might result in reduced robustness of the recognition. In the following sections, we will describe how shape-based 3D matching overcomes the above drawbacks.

3.12.2.2 Restricting the Pose Range

In most industrial applications, the objects to be recognized appear only within a limited range in front of the camera. In some applications, for example, the camera is mounted above a conveyer belt and the objects are always upright on the belt. Obviously, the run time of the recognition, which depends linearly on the number of views, could be reduced by restricting the pose range for which the object is searched. The view sphere allows an intuitive determination of this pose range. In Figure 3.157, an example pose range is visualized. The object in the center of the sphere defines the Cartesian MCS. It is also the origin of a spherical coordinate system with the xz-plane being the equatorial plane and the y axis pointing to the south pole. In the spherical coordinate system, the position of a virtual camera, which is repre-

图，为了识别的稳定性，我们依然要对数以千计的有代表性的视图进行计算。对数以千计的模板进行搜索是十分耗时的，并不能满足实际应用需求。此外，算法所能达到的精度取决于视图的密度。不幸的是，由于处理时间的原因，我们不能随意地选择视图密度。最后，在算法中我们假设物体被置于球面视图的球心上。因此，我们也需要物体被置于图像的中心。在图像角落出现的物体相比于图像中心的物体会存在透视畸变，这有可能导致算法稳定性下降。在随后的章节中，我们将讨论基于形状的三维匹配是如何克服这些缺陷的。

3.12.2.2 限制位姿区域

在大多数工业应用中，需要识别的物体只会出现在摄像机前面有限的范围内。在一些应用中，例如，摄像机安装在传送带上方，且一般情况下，物体直立于传送带上。显然，识别算法的运行时间取决于视图的数目，并与之成线性关系。我们可以通过限制物体可能被搜索到的位姿范围缩短处理时间。我们可以通过观察，直观地确定球面视图中的位姿范围。我们以如图 3.157 中所示的位姿范围为例。物体被置于定义在笛卡尔坐标系 MCS 下的球心上。同时这个球心也是球坐标系的原点。且赤道平面为 xz 平面，y 轴指向南极。虚拟摄像机可通过 CCS 坐标系表示。在球面坐标系下，摄像机的位置可以通过球坐标系下的经度

sented by the CCS, is described by the spherical coordinates longitude λ, latitude φ, and distance d. Then, the pose range, in which virtual cameras are generated, can be easily defined by specifying intervals for the three parameters. Figure 3.157 shows the resulting pose range as a section of a spherical shell.

λ, 纬度 φ 和距离 d 来表示。而后, 生成虚拟摄像机所需的位姿范围, 可以很容易地通过为这三个参数选取指定间隔的方式来确定。如图 3.157 所示, 位姿范围的选择结果为一个球体外壳的截面。

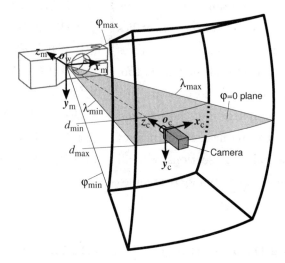

图 3.157　通过球面坐标系限制位姿范围；在模板坐标系 $(o_m, x_m, y_m, z_m)^\top$ 中定义的物体，被置于球面坐标系中，球面视图的中心；摄像机的位置可以通过球面坐标系的经度 λ 和纬度 φ 以及距离 d 表示；图中粗线所示为，我们通过最小和最大的 λ, φ 以及 d 确定的位姿区域

3.12.2.3 Hierarchical Model

In addition to restricting the pose range, the number of virtual cameras can be reduced further by using a hierarchical model. After the pose range is specified, it must be decided how many virtual cameras should be generated within this range to obtain the best tradeoff between speed and robustness. In general, the speed of the recognition can be increased by creating the model on multiple levels of an image pyramid (see Section 3.11.2.1).

In the first step, an over-sampling is applied, i.e., virtual cameras are generated within the pose range very densely. The virtual cameras are

3.12.2.3 分层模板

除了限制位姿范围，我们还可以通过分层模板减少虚拟摄像机的数目。当位姿范围确定好之后，我们将权衡处理速度和稳定性，以此来确定在位姿范围内需要使用多少个虚拟摄像机。通常我们可以通过在一个多层图像金字塔上创建模板的方式提高识别速度（见 3.11.2.1 节）。

第一步，我们将使用一个过采样算法，例如，在位姿范围中创建密集的虚拟摄像机。通过虚拟摄像机的放

placed such that after projecting the model edges into the image planes of two neighboring cameras, the projected edges differ by at most one pixel. This ensures that shape-based matching will be able to find the object even if it is viewed from a camera that lies in the center between two sampled virtual cameras.

In the second step, the similarity between the views of all pairs of neighboring virtual cameras is computed. For this, the same similarity measure is applied to the pair of views that is used during object recognition, i.e., one of the similarity measures of shape-based matching of Eqs. (3.192), (3.193), (3.194), or (3.199). When applying the similarity measure of Eq. (3.199), the model is split into model parts of small connected groups of model points. Because of the 3D geometry of the object and the often unknown illumination, it cannot be predicted whether the direction vectors in the model and in the search image point in the same or in opposite directions. Therefore, the similarity measure that is invariant to local contrast reversals (Eq. (3.194)) or to global contrast reversals of model parts (Eq. (3.199)) should be preferred for the shape-based 3D matching. Neighboring virtual cameras with a similarity that exceeds a certain threshold, e.g., 0.9, are successively merged into a single virtual camera. Only the virtual cameras that remain after the merging process are included in the matching model and used to create views for template matching.

Like shape-based matching (see Section 3.11.2.2), shape-based 3D matching uses image pyramids to speed up the search. Consequently, the model must be generated on multiple pyramid levels. On higher pyramid levels, a change in

置，使得物体在相邻两只摄像机的像平面上，模板边缘投影只相差一个像素。这样可以确保基于形状的匹配能够正确地找到被测物，即使它所匹配的视图是从两个虚拟采样摄像机中间进行观测的。

第二步，对所有相邻的虚拟摄像机视图进行成对的相似性比较。为此，在被测物识别过程中，我们将对每一对视图使用相同的相似度算法。例如，使用基于形状匹配等式（3.192）、等式（3.193）、等式（3.194）或等式（3.199）中任意一种相似度计算方法。当使用等式（3.199）进行相似度计算时，模板将被分割成若干局部模板。每个局部模板都由一小组连续的模板点构成。由于被测物体是立体的以及未知的光照条件，我们无法确定在模板中是否存在畸变向量，也无法确定，在搜索的图像点中，它们的光线是同向或反向。因此，对于局部对比度反向（等式（3.194））或局部模板全局对比度反向（等式（3.199））存在不变性的相似度算法就成为了基于形状的三维匹配的首选。在相邻的虚拟摄像机中如果存在超过某个阈值相似度，例如 0.9，则这些虚拟摄像机将陆续被合并为一个新的虚拟摄像机。只有这些通过合并过程生成的虚拟摄像机才会被包含在匹配模板中，并且通过它们来创建模板匹配视图。

如同基于形状的模板匹配（见3.11.2.2节）一样，基于形状的三维模板匹配也可以使用图像金字塔进行搜索提速。因此，我们需要在多层金字塔中创建模板，在金字塔最顶层，投

perspective has a smaller effect in the image (measured in pixels). This is similar to the effect illustrated in Figure 3.148 of Section 3.12.1.1. Therefore, the virtual cameras can be thinned out further on higher pyramid levels. Starting with the virtual cameras from the next lower pyramid level, the similarity computation between the views of neighboring cameras is continued on the current level. Note that the views in which the similarity is computed now have a lower resolution in accordance with the resolution of the current pyramid level. Again, the virtual cameras are merged if their similarity exceeds the threshold. Figure 3.158 shows the resulting virtual cameras of the final hierarchical model, which contains four pyramid levels in this example.

影对图像的（像素测量）影响很小。这一影响类似于 3.12.1.1 节中的图 3.148。因此，在上一层金字塔中虚拟摄像机的数目将进一步减少。从低层金字塔中的虚拟摄像机开始，我们不停地对金字塔当前层中的摄像机视图进行相似度计算。值得注意的是，在进行相似度计算时，我们会根据当前金字塔层级为所用到的视图设置一个较低的分辨率。当有相似度超过阈值时，虚拟摄像机将被合并，不断地重复这一步骤直至没有可合并的摄像机。图 3.158 中是我们最终生成的多层模板的虚拟摄像机，在这个示例中包含了 4 层金字塔。

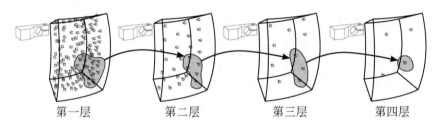

图 3.158　分层模板。图中我们使用粗线表示位姿范围。在位姿范围内，我们用灰色锥形表示以上 4 层金字塔中的虚拟摄像机。在灰色区域内的虚拟摄像机通过合并，在上层金字塔中生成单个虚拟摄像机。在本例中最顶层金字塔上，通过合并，最终只剩下 4 个虚拟摄像机

The resulting hierarchy of virtual cameras can be represented as a set of trees, i.e., a forest (see Figure 3.159). For each virtual camera, a view is available that is used for matching. Therefore, the terms "virtual camera" and "view" are used interchangeably in the following. Root views are searched exhaustively using shape-based matching, i.e., they are searched in the full unrestricted 2D matching pose range: The parameters r and c vary over the pixels of the domain of the search image, ϕ varies over the full range of orientations,

虚拟摄像机的分层结果可以被表示为一组树状结构，就像一片森林（见图 3.159）。每一个虚拟摄像机都对应一个可用于匹配的视图。因此，在下文中，"虚拟摄像机"和"视图"是可以相互替换的。我将使用基于形状的匹配，在根视图中最大限度地进行搜索，例如，它们将在二维匹配范围内进行全面的搜索：参数 r 和 c 取决于图像中搜索区域内的像素，ϕ 取决于方向范围，例如 $[0, 2\pi)$，s 取决

i.e., $[0, 2\pi)$, and s varies over the scale range of the respective view. Matches with a similarity that exceeds a user-specified threshold are refined on the next lower pyramid level. For the refinement, the child views are selected from the tree and searched within a small restricted pose range, where the restriction applies to all four matching parameters r, c, ϕ, and s. The use of the hierarchical model combines a high accuracy with a fast computation because not all views on the lowest pyramid level need to be searched. Typical run times of shape-based 3D matching are in the range of a few 100 milliseconds.

于各个视图的缩放范围。在我们所进行的匹配中，相似度达到用户设置阈值的匹配将会在下一层金字塔中进行更精确的计算。为了进行精确计算，我们将从树中选择对应的子视图，并在一个较小的位姿范围内进行搜索，我们可以通过 r、c、ϕ 和 s 这些匹配参数对位姿范围进行限制。运用分层模板算法，可以将高精度与高速运算有效的结合。因为在金字塔底层不是所有的视图都需要进行搜索。基于形状的三维匹配通常处理时间在 100 ms 以内。

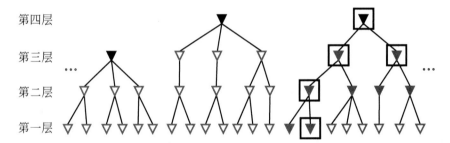

图 3.159 用树状结构表示一个使用分层模板进行物体识别的过程；图中，结构树上每一个三角形节点表示在这 4 层金字塔上的视图；树中的连线表示，在模板生成过程中，该层金字塔上的单个视图是由下一层金字塔中哪些视图合并而来的；黑色实心三角表示在搜索时使用二维匹配算法，对位姿范围内全部位姿进行计算的视图；灰色实心三角表示只需要在较小的二维匹配位姿范围内进行精确搜索的视图。方框表示经过计算相似度达到用户指定阈值的视图；最底层金字塔上的方框表示找到的匹配结果

3.12.2.4　2D Model Generation

Up to now, we have not answered the question of how the shape-based matching features are derived from the views of the 3D CAD model. A possible way would be to project the edges of the CAD model into the image plane of the virtual camera and sample them according to the pixel grid.

One drawback of this method is that features that are not visible in the image would be

3.12.2.4　生成二维模板

到目前为止，我们还没有讨论过如何从三维 CAD 模型中获取基于形状的匹配特征。有一个可行的方法，将 CAD 模型的边缘投影到虚拟摄像机的像平面上，并通过像素网格进行采样。

但这种方法有一个缺陷，模型中包含了图像中不可见的特征。例如，

included in the model. For example, the edges between coplanar surfaces in the CAD model of Figure 3.155(b) are not visible in the image of Figure 3.155(a). Furthermore, very fine structures in the CAD model might not be visible in the image because their size is too small with respect to the image resolution. The invisible structures would decrease the similarity measure because they would have the same effect as partial object occlusions. Consequently, the robustness of the matching would decrease.

Another drawback of this method is that it ignores the smoothing effects that are introduced when using an image pyramid. Some of the effects are illustrated in Section 3.11.2.1. For example, objects might change their shape, or neighboring objects might merge into a single object on higher pyramid levels. Therefore, features that were derived directly from the CAD model might not match the image features on higher pyramid levels well. This would further decrease the robustness of the matching.

Both problems can be solved by projecting the CAD model for each view into a three-channel model image. The three channels of the model image contain the three elements of the unit normal vector of the faces of the projected CAD model. Then, an image pyramid can be computed from the model image and features can be extracted from the respective pyramid level, similar to the approach described in Section 3.11.5.6. Figures 3.160(a)–(c) show the three channels of the model image of a selected view.

Ulrich et al. (2012) showed that the multi-channel edge amplitude in the three-channel image is related to the angle in 3D space between

在图 3.155（b）的 CAD 模型上，一些存在于共面平面之间的边缘在图 3.155（a）中是不可见的。此外，在 CAD 模型中非常精细的结构在图像中也可能是不可见的，因为对于图像分辨率来说它们的尺寸实在是太小了。不可见的结构会降低匹配度，因为它们和物体的局部遮挡一样，都会计算多余的边缘。因此匹配算法的鲁棒性将会降低。

这种方法的另一个缺陷在于，算法忽略了运用金字塔时带来的平滑效果。有关金字塔的部分特性已经在 3.11.2.1 节中说明。例如，在高层金字塔上，物体的形状有可能发生改变，或者两个相邻的物体将被合并为一个物体。因此，从 CAD 模型中直接提取出的特征，在高层金字塔上可能无法正常地匹配。这将进一步降低匹配算法的鲁棒性。

为了解决两个问题，我们将 CAD 模型对每个视图进行 3 个通道的模板图像投影。模板图像的 3 个通道包含了所投影的 CAD 模型表面单位法向量 3 个方向的分量。而后，我们可以从模板图像中计算金字塔图像，并在各层金字塔中提取特征，类似于 3.11.5.6 节中所描述的方法。图 3.160(a)~(c) 表示某个被选中的视图所对应的三通道模板图像。

文献（Ulrich et al., 2012）表明，在三维空间中，CAD 模型两个相邻表面法向量之间的角度与我们之前提

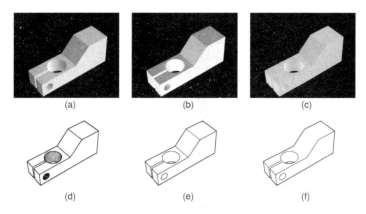

图 3.160 （a）～（c）这 3 个通道的模板图像分别表示模板表面点单位法向量在 x(a), y(b), z(c) 3 个方向的分量；（b）～(d) 通过对多通道图像进行边缘幅度的阈值分割，可以对图像中的边缘进行提取。我们可以通过两平面最小角度差异对边缘进行筛选，图（d）、（e）、（f）分别对应最小角度差异 5°, 30°, 50°

the normal vectors of the two neighboring faces of the CAD model. The edge amplitude in a multi-channel image is obtained by computing the eigenvalues of the multi-channel edge tensor C (Di Zenzo, 1986; Steger, 2000):

及的三通道图像中的多通道边缘幅度有关。在一幅多通道图像中，边缘幅度可以通过多通道边缘张量 C 的特征值计算获得（Di Zenzo, 1986; Steger, 2000）：

$$C = \begin{pmatrix} C_{11} & C_{12} \\ C_{12} & C_{22} \end{pmatrix} \tag{3.203}$$

For the three-channel image, the matrix elements contain the derivatives of the x, y, and z components of the normal vector in the row and column directions:

对于三通道图像，矩阵中的每个元素都由单位向量在 x、y 和 z 各个方向分量图像沿着行和列的偏导数组成。

$$C_{11} = \left(\frac{\partial x}{\partial r}\right)^2 + \left(\frac{\partial y}{\partial r}\right)^2 + \left(\frac{\partial z}{\partial r}\right)^2 \tag{3.204}$$

$$C_{12} = \frac{\partial x}{\partial r}\frac{\partial x}{\partial c} + \frac{\partial y}{\partial r}\frac{\partial y}{\partial c} + \frac{\partial z}{\partial r}\frac{\partial z}{\partial c} \tag{3.205}$$

$$C_{22} = \left(\frac{\partial x}{\partial c}\right)^2 + \left(\frac{\partial y}{\partial c}\right)^2 + \left(\frac{\partial z}{\partial c}\right)^2 \tag{3.206}$$

The edge amplitude A is the square root of the largest eigenvalue of C. It is related to the angle δ between the two neighboring faces by the following simple equation (see Ulrich et al. (2012)):

边缘幅度 A 为 C 中最大特征值的平方根。它和两相邻平面的角度 δ 有如下关系（详见（Ulrich et al., 2012））：

$$\delta = 2\arcsin(A/2) \qquad (3.207)$$

By extracting the features for the shape-based matching from the image pyramid of the three-channel model image, similar smoothing effects are introduced in the model as in the pyramid of the search image. Consequently, the similarity measure also will return high values for correct matches on higher pyramid levels. Furthermore, by restricting the feature extraction to points with an edge amplitude that exceeds a certain threshold, invisible edges with a small face angle easily can be suppressed in the model. Figures 3.160(d)–(f) show the resulting model points when requiring a minimum face angle of 5° (edges between coplanar faces are suppressed), 30° (edges that approximate the cylinders are suppressed), and 50° (the two 45° edges are suppressed). The corresponding three threshold values for the edge amplitude were computed via Eq. (3.207). For most applications, a minimum face angle of 30° works well. Excluding invisible features from the model also increases the similarity measure, and hence the robustness of the recognition.

3.12.2.5 Perspective Correction

View-based approaches that support cameras with a perspective lens assume that the objects appear in the image center. This is because the appearance of the object changes depending on the object position in the image. The effect is illustrated in Figure 3.161. As described above, the 2D models are created by using virtual cameras that look at the center of the object (black projection in the image center of Figure 3.161). In the search image, the object may appear at arbi-

对于基于形状的模板匹配，通过对三通道模板图像的金字塔图像进行特征提取，诸如在搜索图像金字塔中存在的平滑效果将在模板创建过程中进行抑制。因此，在高层金字塔中正确的匹配依然会返回较高的匹配度。此外，通过为指定边缘幅度的阈值，我们可以限制特征点的提取。如此，在模板中，由于较小表面角度变化所产生的边缘将受到抑制。图3.160（d）～（f）分别为需要最小平面角度达到5°（抑制两共面平面之间的边缘），30°（抑制近似圆柱体表面的边缘）以及50°（抑制两个存在45°夹角平面的边缘）所提取的模板点集。我们可以通过等式（3.207）计算边缘幅度在3个通道图像对应的阈值。大多数应用中，我们可以设置最小平面角度为30°。通过剔除不可见的特征，也可以增加正确匹配时的相似度，增强识别算法的鲁棒性。

3.12.2.5 投影校正

基于视图的处理算法，适用于用摄像机和镜头组成的投影模型，与此同时我们还需要确保物体出现在图像的中心。这是因为物体的外形会随着它在图像中的位置变化而变化。我们可以在图3.161中看到上述效果。正如上文中提及的，二维模板是通过一个对物体中心进行观测的虚拟摄像机创建的（图3.161中，在图像中心的黑色投影）。但在搜索图像中，被测物

trary positions (gray projections in the image corners of Figure 3.161, for example). Note that all projections represent the same view of the object, i.e., the position of the camera with respect to the object is the same. Nevertheless, the appearance of the object changes considerably. This effect is larger for lenses with a smaller focal length.

有可能出现在视场任意位置（例如，图 3.161 中，在图像角落的灰色投影）。值得注意的是所有的投影都属于物体的同一个视图，例如摄像机相对于物体的位置是一致的，只不过物体的外形发生了变化。对于这种现象镜头的焦距越短变形就越明显。

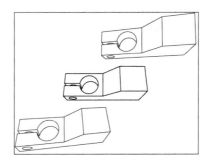

图 3.161　三维物体在图像中的形状取决于该物体在图像中的位置；物体在图像中心的投影用黑色表示，该物体在图像两角落的二维投影变换用灰色表示；值得注意的是，所有投影都对应相同的视图

Changing the object position in the image can be accomplished by rotating the camera about its optical center, or, to be more specific, by rotating the camera about its x and y axes. Consequently, all projections of the same view are related by a 2D projective transformation (Hartley and Zisserman, 2003). Therefore, a projective transformation is applied to the features of the 2D model of the respective view before matching them with the image feature. The transformation parameters of the projective transformation are computed from the approximate position of the match candidate on the next higher pyramid level (Ulrich et al., 2012). This ensures a higher value of the similarity measure for correct matches, and hence also increases the robustness of the recognition.

On the top pyramid level, knowledge from a previous level for computing the projective transformation is unavailable. Furthermore, computing

我们可以通过摄像机绕光学中心的旋转来改变物体在图像中的位置，具体来说，就是将摄像机沿 x 轴和 y 轴旋转。因此，同一视图中所有投影的旋转都与一个二维投影变换相对应（Hartley and Zisserman，2003）。在进行特征匹配之前，我们将对每个视图中二维模板的特征进行一次投影变换。其中的变换参数，可以由上一层金字塔中的候选匹配的粗略位置计算获得（Ulrich et al.，2012）。这样可以确保正确的匹配结果拥有较高的相似度，同时，也可以提高识别算法的鲁棒性。

在顶层金字塔上，我们无法从更高层金字塔中获取数据，对投影变换进行计算。而为图像中每个位置都计

an individual projective transformation at each image position would be too expensive. Fortunately, on the top pyramid level the projective distortion of the imaged object with respect to the model is small. This is also similar to the effect illustrated in Figure 3.148 of Section 3.12.1.1. Nevertheless, the distortion can be further reduced by applying a spherical projection of the top pyramid level of both the model and the search image and performing the matching in the spherical projection. The idea is that when rotating the camera about its optical center, the projection of a 3D object onto the surface of a sphere does not result in a projective transformation but in a translation only. For details, the interested reader is referred to Ulrich et al. (2012).

3.12.2.6 Least-Squares Pose Refinement

The accuracy of the resulting 3D object pose H_o, which is obtained from Eq. (3.202), is limited by the density of the sampled views and the approximation of the object–camera distance by the 2D matching parameter s (see Section 3.12.2.1). In practical applications, a higher accuracy is often desirable. Therefore, H_o is used to provide initial values for a nonlinear least-squares optimization. For each projected model edge point l_i, a corresponding edge point p_i in the image is searched. The optimized object pose H_{opt} is obtained by minimizing the squared distances d of the n image edge points to their corresponding projected model edges:

$$\sum_{i=1}^{n} d(p_i, \pi(l_i, H_{opt}))^2 \to \min \quad (3.208)$$

The function $\pi(l, H)$ represents the transformation of the model edge l from the MCS to the

算一个投影变换会大大增加算法的复杂度。幸运的是，在金字塔顶层图像中，物体的透视畸变相对于模板而言很小。这个效果就像3.12.1.1节中的图3.148一样。与之不同的是，我们还可以通过对顶层金字塔的模板和搜索图像进行球面投影的方法进一步降低畸变的影响，并且将匹配过程也引入到球面投影图像中。其中算法的工作原理是，当摄像机关于光心进行旋转时，三维物体在球面上的投影只有平移变换，而没有投影变换。对算法细节感兴趣的读者可以参考（Ulrich et al.，2012）。

3.12.2.6 最小二乘法位姿优化

我们可以通过等式（3.202）求得物体三维位姿 H_o。该位姿的精度与采样视图的密度和物体到摄像机的近似距离有关，其中近似距离可通过二维匹配参数 s 计算（见3.12.2.1节）。但在实际应用中，我们可能需要更高的精度。因此，我们以 H_o 为初始值，对位姿数据进行非线性最小二乘法优化，以此获得更高精确的数据。对于每个投影模板边缘点 l_i，我们在图像中都会进行搜索，并找到其对应点 p_i。优化后的位姿 H_{opt} 满足图像边缘点到其对应模板边缘点的距离平方和最小：

函数 $\pi(l, H)$ 表示边缘 l 从 MCS 到 CCS 的转换。这一转换通过位姿 H

CCS, which is represented by the pose H, and its subsequent projection into the image. New correspondences may arise from the optimized pose. Therefore, the search for correspondences and the optimization are iterated until the optimized pose no longer changes. Figure 3.162 visualizes the object poses that are obtained after different numbers of iterations.

来表示，并应用于后续的图像投影中。由于在优化位姿的过程中，有可能出现新的边缘对应点。因此对应点的搜索和位姿的优化是一个迭代过程。我们会不停地进行匹配运算，直至位姿不再变化为止。图 3.162 为迭代运算过程中不同次数的物体位姿。

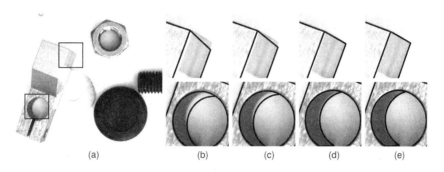

图 3.162 最小二乘法位姿迭代优化：（a）搜索图像；（b）~(e) 通过模板边缘投影的方式在图中显示物体位姿（黑色边框）；为了更好地观察处理结果，我们将（a）图中的两个局部矩形区域进行放大；（b）初始位姿；（c）进行一次迭代运算后的位姿；（d）三次迭代运算后的位姿；（e）迭代收敛后的位姿

The accuracy of the object pose depends on several factors, e.g., the sharpness of the object edges (round edges yield worse accuracy), the complexity of the object (the risk of falsely recognizing very simple objects in cluttered scenes is higher), and the accuracy of the CAD model. Practical evaluations have shown that the accuracy of the object position is in the order of 0.3% of the object–camera distance, and the accuracy of the object rotation is in the order of 0.5°(Ulrich et al., 2012).

物体位姿的精度取决于以下几个因素。如，物体边缘的形状（圆形的边缘会导致位姿精度下降），物体的复杂度（在复杂的场景中识别形状简单的物体，误识率较高）以及 CAD 模型的精度。根据实际测试，一般物体定位精度约为物体到摄像机距离的 0.3%，物体的旋转精度约为 0.5°（Ulrich et al., 2012）。

3.12.2.7 Examples

Figure 3.163 shows some example applications in which shape-based 3D matching is used to recognize objects in an image. In Figure 3.163(a), two

3.12.2.7 实例

图 3.163 中为我们展示了一些通过基于形状的三维匹配算法，在单幅图像中进行物体识别的应用案例。在

instances of the metallic clamp of Figure 3.155(b) are recognized. Note the robustness of the recognition to different surface finishes of the object (polished and brushed). Figure 3.163(b) shows another metallic object with a highly reflective surface. Because shape-based 3D matching is based on shape-based matching, it is also robust to challenging lighting conditions (see Figure 3.163(c)) as well as clutter and partial object occlusions (see Figure 3.163(d)). Furthermore, it is suited to bin picking applications as shown in Figure 3.163(e) and (f).

图 3.163（a）中，我们对前面章节中图 3.155（b）所示的金属夹具进行识别，并成功地搜索到本图中的两个被测物。值得注意的是对于不同的物体表面材质，算法的鲁棒性也是不同的（如，抛光的表面或刷漆的表面）。在图 3.163（b）中，作为被测物的金属工件表面反光极好，在金属表面甚至可以映照出周边环境的明暗变化，但是匹配算法依然可以找到物体。这是因为，基于形状的三维模板匹配有一定的抗光照变化的能力（见图 3.163（c））。与此同时，匹配算法还可以应付较为复杂的背景以及被遮挡的物体（见图 3.163（d））。此外，该算法还可以应用于机械手引导抓取中（见图 3.163（e）和图 3.163（f））。

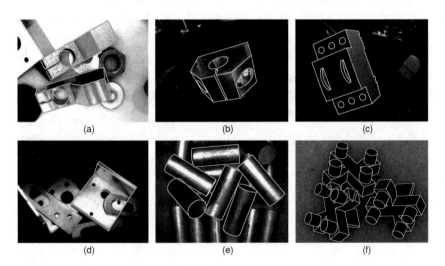

图 3.163 实例，在单幅图像中使用基于形状的三维匹配算法进行三维物体的识别；在上图中，我们用模板边缘（白色）表示搜索到的物体；(a) 金属夹具，(b) 镜面金属夹具，(c) 空气开关，(d) 金属支架，(e) 金属圆柱，(f) 瓷砖十字架

3.12.3 Surface-Based 3D Matching

In this section, we introduce methods for recognizing 3D objects in 3D data and for determining

3.12.3 基于表面的三维匹配

在本节中，将为大家介绍通过三维数据进行三维物体的识别和位姿确

their 3D pose. The methods exploit the benefits of the additional dimension of 3D data compared to image data. The 3D data either can be obtained from 3D reconstruction methods (see Section 3.10) or from 3D image acquisition devices (see Section 2.5). Because of the multitude of possible sources for 3D data, there is a broad range of data characteristics of 3D data. In general, the differences in the characteristics of 3D data are much more pronounced than those of image data, and hence pose a greater challenge for robust generic algorithms. Some examples are different noise characteristics, different numbers of measurement outliers, different data representations (e.g., depth images, xyz-images, point clouds), different behaviors at depth discontinuities, different coverages of the scene (depending, for example, on the texture, the illumination, or the reflection properties of the objects), and different resolutions.

In the following, we will describe surface-based 3D matching, an algorithm that is able to recognize rigid 3D objects in 3D data and determine their 3D pose (Drost et al., 2010; Drost and Ulrich, 2012, 2014, 2015a,b). It is robust to object occlusions and to clutter and noise in the 3D data. Subsequently, we will briefly describe two extensions of surface-based 3D matching: a method to find deformable 3D objects in 3D data (Drost and Ilic, 2015) and an extension to find rigid 3D objects in multimodal data, i.e., combining 3D and image data for recognition (Drost and Ilic, 2012; Drost and Ulrich, 2015c,d). For more details about surface-based 3D matching and its extensions, the interested reader is referred to Drost (2017).

定。相比于二维图像的处理，这些算法很好地利用了三维图像在维度上的优势。计算中所用到的三维数据可以通过三维重构算法（见 3.10 节）或三维图像采集设备（见 2.5 节）获得。由于三维数据来源的多样性，不同的采集设备或者算法所获得的数据特性也不尽相同。一般来说，三维数据的特征差异比二维图像数据更大，这对于算法的稳定性是一个极大的挑战。在常见的一些案例中，通常三维数据会存在诸多不同的噪声，不同的离群值，不同的表现形式（例如，深度图像、xyz 图像、点云），在深度方向上存在不同程度的断差，对采集场景覆盖程度的差异（点云对场景的覆盖程度通常取决于物体的纹理，环境光以及物体表面材质的反射率）以及不同的分辨率。

在下文中，我们将为大家介绍基于表面的三维匹配，这是一种可以在三维数据中进行刚性物体识别以及位姿确定的算法（Drost et al., 2010; Drost and Ulrich, 2012, 2014, 2015a,b）。该算法对于存在遮挡、混淆以及噪声的三维数据都有不错的鲁棒性。稍后我们还会为大家介绍两个基于表面的三维匹配算法的推广：用于在三维数据中进行变形物体匹配的算法（Drost and Ilic, 2015）以及刚性物体搜索在多模态数据中的推广（Drost and Ilic, 2012; Drost and Ulrich, 2015c,d）。对于基于表面的三维匹配及其推广算法细节感兴趣的读者请参考文献（Drost, 2017）。

Figure 3.164 shows an example application in which pipe joints must be recognized in 3D data. The 3D data is obtained by using stereo reconstruction (see Section 3.10.1). Figure 3.164(a) shows an image from the first camera of the stereo setup. For the matching, a model of the object must be provided. In this case, a CAD model of the pipe joint is available (see Figure 3.164(b)). Alternatively, the reconstructed 3D data of the object can be provided. The result of the stereo reconstruction of the search scene in which the objects must be recognized is shown in Figure 3.164(c). The found object instances are visualized in Figure 3.164(d).

图 3.164 为我们展示了一个从三维数据中进行管件接头识别的应用案例，其中，三维数据由三维重构算法获得（见 3.10.1 节）。图 3.164（a）为三维重构系统中摄像机 1 采集的图像。在进行匹配时，我们需要为算法提供物体的模型。在此例中，我们使用管件接头的 CAD 模型作为匹配的模板（见图 3.164（b））。此外，我们也可以使用工件的 3D 点云数据作为模板。经过一系列的处理，待搜索场景的重构结果如图 3.164（c）。搜索结果如图 3.164（d）。

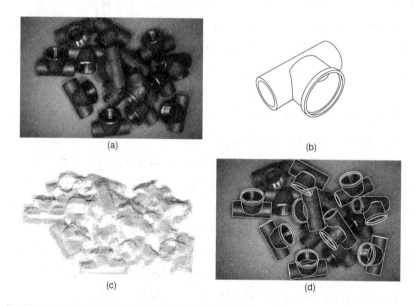

图 3.164　使用基于表面的三维匹配识别三维场景数据中的管件接头；其中，三维数据通过三维重构获得；（a）三维重构系统中摄像机 1 的图像；（b）用于为基于表面的三维匹配创建模板的 CAD 模型；（c）在场景中通过重建获得的三维数据；（d）基于表面三维匹配算法的搜索结果，图中白色边框为搜索到的物体，我们将 CAD 模型根据匹配后的位姿投影到摄像机 1 上以方便结果的观测

3.12.3.1　Global Model Description

For surface-based 3D matching, the object is modeled as a set of point pairs. A point pair consists of two points on the surface of the object

3.12.3.1　全局模型描述

为了进行基于表面的三维匹配，我们使用一组点对，对物体进行建模。每一个点对都包含物体表面上的两个

and their two corresponding surface normal vectors. Figure 3.165 shows one example point pair m_1 and m_2 and their normal vectors n_1 and n_2. Let d be the vector $m_2 - m_1$. The geometry of the point pair is described by the point pair feature $F(m_1, m_2) = (F_1, F_2, F_3, F_4)$ with

- the distance $F_1 = \|d\|_2$ between the points m_1 and m_2,
- the two angles F_2 and F_3 of each normal vector n_1 and n_2 with the vector d, and
- the angle F_4 between both normal vectors.

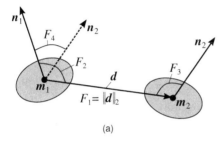 

图 3.165 （a）物体表面上，点 m_1 和 m_2 的点对特征 $F = (F_1, F_2, F_3, F_4)$；我们通过 n_1 和 n_2 定义其对应点上的切平面（图中灰色圆）；F_1 为两点间距离。角 F_2, F_3 为向量 n_1 和 n_2 与向量 d 所成夹角；F_4 为两向量夹角；（b）样本模型；对于模型中的每一个点对，我们都会进行图（a）所示的特征值计算

To create a global model description, the point pair feature is computed for each point pair on the object surface. A high robustness to noise and errors in the 3D data is obtained by grouping similar point pair features. This is achieved by discretizing the four elements of F, resulting in $F_d = (\lfloor F_1/\delta_1 \rfloor, \lfloor F_2/\delta_a \rfloor, \lfloor F_3/\delta_a \rfloor, \lfloor F_4/\delta_a \rfloor)$. Practical evaluations have shown that suitable values are $\delta_1 = D/20$ and $\delta_a = 360°/30 = 12°$, where D is the diameter of the model.

The global model description can either be created from a CAD model of the object, from

3D data obtained from a 3D image acquisition device (see Section 2.5), or from a 3D reconstruction of the object (see Section 3.10). To be able to create a global model description from a CAD model and to be independent of the resolution of the 3D data, the surface of the object is uniformly sampled to obtain a fixed number of points. The number of sampled points is an important parameter of the approach because it balances, on the one hand, memory consumption of the model and computation time of the recognition and, on the other hand, the robustness of the recognition. Values that are typically chosen in practice lie in the range between 500 and 3000 points. Figure 3.165(b) shows the sampled model points of the CAD model of Figure 3.164(b).

For triangulated input data, the normal vectors can be directly obtained from the triangulation. If no triangulation is available, the normal vectors must be derived from the 3D points. One way to compute the normal vector of a 3D point is to fit a plane to the points in its local neighborhood. The normal vector of the fitted plane is used as the normal vector of the point. Additionally, the orientation of the normal (inward or outward) must be determined, for example, by orienting the normal vectors towards the sensor. Finally, for all pairs of sampled points, a point pair feature is computed as described above.

During the recognition phase, it is necessary to determine, for a pair of points in the search scene, all similar point pairs on the model, i.e., point pairs on the model that have the same point pair feature as the point pair in the search scene. To allow an efficient search, the point pairs of the model are stored in a hash table (see Figure 3.166).

获得的三维数据或者通过三维重构算法获得的物体三维数据，创建全局模型描述。为了在使用 CAD 模型创建全局模型描述时，不会受到三维数据分辨率的影响，我们会对物体的表面进行固定点数的均匀采样。采样点数是一个重要的算法参数，一方面它会影响模板的内存使用量以及识别算法的执行时间，另一方面它会影响识别的稳定性。因此在设置时，我们需要权衡以上两个方面。在实际应用中，我们通常将其设置为 300 到 5000 点之间。图 3.165（b）为图 3.164（b）中的 CAD 模型采样后的模板点。

对于三角化后的输入数据，法向量可以从三角化网格中直接获取。如果没有三角化网格可用，则需要从三维数据点中进行计算。计算法向量的其中一种方法是利用该点周围的数据点拟合一个平面，并以这个平面的法向量作为该点的法向量。此外，我们还需要定义法向量的方向（向内或向外），例如，定义法向量指向传感器为正方向。最后，根据上述步骤，求取所有采样点对的特征值。

在识别阶段，我们需要确认，在搜索场景中哪些点对和我们模型上的点对是相似的，即，它们拥有相同的特征值。为了进行高效的搜索，我们事先将模板的点对存入一张哈希表中（见图 3.166）。我们将离散的点对特征作为哈希表的关键码值。因此，拥

The discretized feature of the point pair serves as the key to the hash table. Consequently, model point pairs with similar point pair features are stored in the same slot in the hash table. The hash table represents the global model description.

有相似特征值的点对将被保存在哈希表的同一位置。我们可以通过局模型描述创建哈希表。

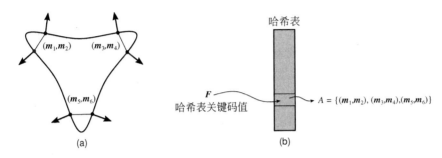

图 3.166　点对特征值将在哈希表中进行存储：（a）三个特征值相似的点对，例如，$F_d(m_1, m_2) = F_d(m_3, m_4) = F_d(m_5, m_6) = F$；（b）哈希表；由于我们以 F 作为关键码值，因此，这三个点对都将保存在哈希表中 A 位置

This model description has some important advantages: During the recognition, all model features F_m that are similar to a given scene feature F_s can be searched in constant time by using F_s as the key to access the hash table. Furthermore, by using the proposed point pair features, the global model description is invariant to rigid 3D transformations. Finally, two corresponding point pairs of the search scene and the model are sufficient to estimate the 3D pose of the object. Note that the size of the global model description increases quadratically with the number of sampled model points. However, this is not an issue if the number of sampled model points is kept at a constant value of reasonable size as proposed above.

上述的模板描述拥有如下优点：在识别过程中，由于我们使用了哈希表，所有模板特征值 F_m 在搜索其场景中对应特征值 F_s 时，可直接使用 F_s 进行查表，如此，既可以提高搜索效率也可以保证算法处理时间相对稳定。此外，使用上文中的点对特征值，使得全局模板描述对于刚性三维变换具有不变性。最后，两两对应的点对，为我们进行物体的三维位姿估计提供了充足的数据。值得注意的是，全局模型描述特征数据量的大小与模型采样点数的平方成正比。然而，正如上文中提到的，为了权衡处理时间和算法的稳定性，我们无法使用一个定值作为模型采样点数。

3.12.3.2 Local Parameters

As already pointed out in Section 3.12.2.1, one major challenge of 3D object recognition approaches are the six degrees of freedom of the

3.12.3.2　局部参数

在 3.12.2.1 节中我们指出，在三维物体识别中，对于处理算法最大的挑战就是物体六自由度的计算，例如

object pose, i.e., three translation and three rotation parameters. Instead of determining the six pose parameters of the object, in surface-based 3D matching, so-called local parameters are used. A point in the search scene is selected that we will denote as the reference point. It is assumed that the reference point lies on the surface of the object. If this assumption holds, the pose of the object can be determined by

- identifying the model point that corresponds to the reference point (see Figure 3.167(a)) and
- fixing the rotation about the normal vector of the reference point point (see Figure 3.167(b)).

3 个平移和 3 个旋转参数。与物体六自由度参数不同的是，在基于表面的三维匹配中，我们将使用一种被称为局部参数的数据表示被识别的物体。我们将选取场景中特定的点，并将其指定为参考点。我们假设该参考点位于物体表面上。如果这个假设成立，我们就可以通过如下方法确定物体的位姿：

- 识别与参考点对应的模型点（见图 3.167（a））。
- 修正该点关于参考点法向量的旋转（见图 3.167（b））。

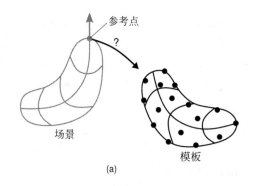 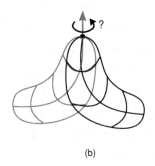

图 3.167　局部参数；假设场景中的参考点（灰色）位于物体表面（黑色）；（a）局部参数中，第一个变量为该点所对应的模板点的序号；（b）当我们识别到对应的模板点，该点和其对应的模板点将进行对齐（包括它们的法向量）；局部参数中，第二个变量为关于法向量的旋转角度

The index of the corresponding model point and the rotation angle are the local parameters with respect to the selected reference point. They can be determined by applying a voting scheme that will be described in Section 3.12.3.3.

If the scene contains clutter objects or if the background is included in the scene, randomly selecting a reference point does not ensure that

对于所选的参考点，我们将其对应模板点的序号以及旋转角度作为该点的局部参数。我们可以通过投票的方式确定参考点的取舍。在 3.12.3.3 节中我们将对投票算法进行详细的介绍。

我们也可以通过随机采样的方式选择参考点。但场景中存在其他易混淆的物体或者背景时，我们无法确保

the reference point lies on the surface of the object. Therefore, multiple scene points are successively selected as the reference point. To robustly recognize an object, it is sufficient that at least one point on the object is selected as the reference point. Figure 3.168 shows the selected reference points in the scene of Figure 3.164(c).

通过单次采样得到的参考点是否位于物体的表面。因此,我们在选择参考点时需要进行多次随机采样。为了保证算法的鲁棒性,在每个物体上,我们至少要选取一个点作为参考点。图 3.168 为对图 3.164(c) 中的场景进行参考点选择后的处理结果。

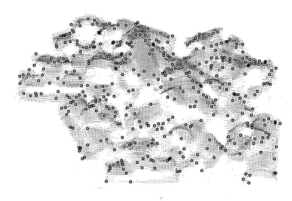

图 3.168 在图 3.164(c) 场景中我们选择了多个参考点(用圆形标记)。以此确保在场景中的每个物体上至少有一个参考点被选中。在这个例子中,我们随机选择场景中 10% 的点作为参考点

3.12.3.3 Voting

For a selected reference point, the optimum local parameters must be determined. The optimum local parameters are the parameters that maximize the number of points in the scene that lie on the model. To determine the parameters, a voting scheme that is similar to the generalized Hough transform (see Section 3.11.5.3) is applied. The parameters of the Hough transform are the local parameters, i.e., the index of the model point and the rotation angle. As for the generalized Hough transform, the two-dimensional parameter space must be sampled. Therefore, the rotation angle is uniformly sampled within the interval $[0°, 360°)$, for example in steps of $12°$. Then, an accumulator

3.12.3.3 投票

为了进行参考点的选择,我们需要确定最佳的局部参数。我们将能够使得场景中的点最大程度映射到模板上的参数定义为最佳局部参数。我们使用投票的方式进行参数的确定。这种方式类似于广义的霍夫变换(见 3.11.5.3 节)。我们以局部参数作为霍夫变换的参数,例如,模板点的序号以及旋转角度。对于广义霍夫变换,我们需要在二维参数空间中进行采样。因此,我们需要对旋转角度在 $[0°, 360°)$ 上进行均匀采样,例如,我们可以进行采样间隔为 $12°$ 角度采样。而后,我们生成一个全零的累加

array of size $n \times m$ is initialized with zeros, where n is the number of sampled model points and m is the number of sampled rotations. The accumulator array represents the discrete space of local paramters for a given reference point.

The voting is performed by pairing the reference point s_r with every other point s_i in the scene. The process is illustrated in Figure 3.169. In the first step, for each such point pair, the point pair feature $F(s_r, s_i)$ is computed as described in Section 3.12.3.1. Similar point pairs in the model are queried by using the hash table. For each entry in the hash table, the resulting local parameters can be computed, i.e., the index of the model point and the rotation angle that aligns the model point pair with the scene point pair. Note that the rotation angle can be computed very efficiently from the two corresponding point pairs as described by Drost *et al.* (2010). The local parameters obtained are discretized according to the sampling of the accumulator array, and the corresponding accumulator cell is incremented. After all scene points have been paired with the reference point, the number of votes in a cell corresponds to the number of scene points that coincide with the model when aligning the model in accordance with the local parameters that the cell represents. The highest peak in the accumulator array corresponds to the optimum local parameters for the current reference point. From the optimum local parameters, the 3D pose of the object can be easily calculated. To obtain a higher robustness, all peaks that have a certain amount of votes relative to the highest peak can be selected and further processed as described below.

器数组，数组大小为 $n \times m$，其中 m 为模板点的数目，n 为采样旋转角度的数目。通过这个数组，我们可以表示一个给出的局部参数的二维离散空间。

在投票过程中，我们需要以场景中的参考点 s_r 以及周边场景点 s_i 组成的点对对作为算法的输入，如图 3.169 所示。首先，通过 3.12.3.1 节中的算法，对每一对点进行计算并求得特征值 $F(s_r, s_i)$。而后，通过哈希表对在模板中与该特征值相似的点对进行查找。在哈希表中，对于每一个输入我们都可以找到与其对应的局部参数组，例如，我们可以为场景中的点对确定一组模板点的序号以及旋转角度。正如文献（Drost *et al.*, 2010）中描述的，我们很难通过两个对应的点对计算出模板旋转角度。因此，我们根据累加器数组对查表所得的局部参数组进行离散化采样，并对累加器中相应的单元的投票进行自加。在场景中所有的点都与其参考点进行配对后，累加单元中的票数即为使用相应的局部参数时模板与场景中一致的点的数目。累加器数组中峰值所对应的数据为当前参考点的最佳局部参数。通过最佳局部参数，我们可以很容易地计算出物体的位姿。为了使识别更加稳定，在所有的峰值中只要其票数相对于最高峰值达到一定的数目，我们就将其选出，并进行如下处理。

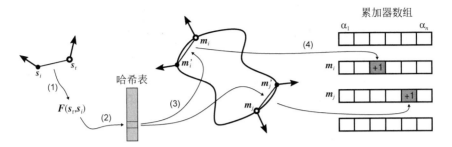

图 3.169 投票过程：（1）场景中被选中的参考点 s_r 以及与之配对的其他场景点 s_i，图中 $F(s_r, s_i)$ 为每对点的特征值；（2）使用 $F(s_r, s_i)$ 作为关键码值进行查表；（3）通过查表，我们可以获得所有与之相似的模板点对；（4）对每个模板点对的局部参数进行采样计算并对相应的累加器数组单元进行自加

After the voting process, for each selected reference point in the scene, at least one pose candidate along with a score value, i.e., the number of votes, is available. To remove duplicate pose candidates, a non-maximum suppression is performed: for each selected candidate, all similar pose candidates are removed. Two pose candidates are similar if their rotation and translation parts differ by at most two predefined thresholds.

3.12.3.4 Least-Squares Pose Refinement

The accuracy of the object poses that are obtained from the voting process is limited by the sampling of the parameter space and by the sampling of the feature space. The sampling of the parameter space is determined by the sampling of the accumulator array, i.e., the sampling of the rotation and of the model points. The sampling of the feature space is determined by the sampling of the point pair features in the hash table. When choosing the sampling as suggested in Sections 3.12.3.1 and 3.12.3.3, the accuracy is about 5% of the object diameter in translation and 12° in rotation. In most applications, higher accuracies are required. For example, if the surface of an object is to be inspected by comparing it to a reference model,

在投票后，我们需要确保每一个选中的参考点至少存在一个候选分值，例如，我们可以用累加器数组中的票数作为该点分值。而后，我们会使用非极大值抑制算法，剔除重复位姿：在所有被选中的候选者中，移除位姿相似的候选者。我们可以通过事先设定好的差值阈值判断两个位姿在旋转和平移上是否相似。

3.12.3.4 最小二乘法位姿优化

我们通过投票方式所获得物体位姿的精度是有限的。因为在处理之前，我们对输入的数据在参数空间以及特征空间中进行了采样。其中，参数空间的采样取决于累加器数组，例如，我们会根据模板点以及旋转角度进行采样。特征空间的采样取决于在哈希表中点对特征值的采样。如果我们使用 3.12.3.1 节和 3.12.3.3 节中推荐的采样间隔，则平移精度为物体直径的 5%，旋转精度为 12°。但在很多应用场景中，我们需要更高的精度。例如，如果一个物体需要和模板对比并进行缺陷检测时，物体位姿的误差有可能导致其与参考模板对位的偏差，而这

small inaccuracies in the object pose might result in deviations from the reference model that might wrongly be interpreted as a defect. Therefore, the object poses are refined by using a least-squares optimization. For the optimization, the poses that are obtained from the non-maximum suppression are used as initial values.

To perform a least-squares optimization, the correspondences between the model points and the scene points must be determined. The determination of the correspondences and the least-squares optimization are both performed within the framework of the iterative closest point (ICP) algorithm (Besl and McKay, 1992). The ICP algorithm is able to align two point clouds (e.g., model and scene) if a sufficiently accurate initial pose is provided. As the name suggests, ICP is an iterative algorithm. It iteratively repeats the following two steps:

- Correspondence search: for each point in the first point cloud, the closest point in the second point cloud is determined.
- Pose estimation: the pose that minimizes the sum of the squared distances between the correspondences is computed.

Let s_i be the points in the scene and m_j the points in the model. Furthermore, let the object pose at iteration k be represented by the rigid 3D transformation matrix T_k. The matrix T_0 is computed from the initial pose values that result from the non-maximum suppression. Note that T_k transforms points from the scene to the model in our notation. Then, the correspondence search in step k determines for each scene point s_i the closest model point $m_{c(i)}$:

些偏差部分很可能被识别为被测物的表面缺陷。因此，我们需要使用最小二乘法对物体位姿进行优化。我们以非极大值抑制算法的处理结果作为该算法位姿的初值。

为了进行最小二乘法优化，我们需要确定模板点集和场景点集的对应关系。对应点的确定以及最小二乘法优化都会在迭代最近点（ICP）这一算法（Besl and McKay, 1992）框架下执行。ICP 算法在已知足够准确初始位姿的情况下，可以将两点云（例如，模板和场景中的点云）进行对齐。正如它名称中提到的，ICP 是一个迭代算法。该算法会进行如下两步迭代：

- 对应搜索：对于第一片点云中的每个点，在第二片点云中确定其最近的点。
- 位姿估计：计算使得所有对应点之间距离平方和最小的位姿。

我们令 s_i 为场景点集，m_j 为模板点集。此外，我们定义 T_k 为物体位姿在第 k 次迭代时的刚性变换矩阵。其中 T_0 可以通过非最大抑制算法获得的初始位姿进行计算。值得注意的是，T_k 为从场景到模板的变换。综上，在第 k 次迭代中，对于每一个场景点 s_i 的最近模板点 $m_{c(i)}$ 对应点搜索可应用如下等式：

$$c(i) = \arg\min_{j} \|\boldsymbol{T}_k \boldsymbol{s}_i - \boldsymbol{m}_j\| \qquad (3.209)$$

In the subsequent pose estimation step, the squared distances between the corresponding points are minimized:

在随后的位姿估计步骤中，我们取对应点之间距离平方值最小的变换矩阵

$$\boldsymbol{T}_{k+1} = \arg\min_{\mathrm{T}} \sum_i \|\mathrm{T}\boldsymbol{s}_i - \boldsymbol{m}_{c(i)}\|^2 \qquad (3.210)$$

The iteration is stopped as soon as the desired accuracy or a maximum number of iterations is reached. Figure 3.170 illustrates the improvement of the pose accuracy after different numbers of iterations.

迭代过程将在位姿达到预设的精度或最大迭代次数后停止。如图 3.170 所示，在不同次数的迭代处理后物体的位姿将得到相应的改善。

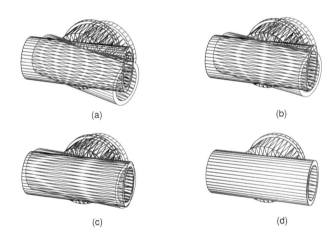

图 3.170　经过若干次 ICP 算法迭代处理后，物体的位姿变化；图中黑色为模板位姿，灰色为逆变换后的场景位姿。（a）初始位姿；（b）～（d）经过 1、2、5 次 ICP 迭代后的位姿

The run time of the ICP algorithm is dominated by the run time of the correspondence search. Therefore, an efficient data structure for the determination of the nearest point neighbor is crucial for practical applications. Well-known efficient methods are based on k-d-trees (Bentley, 1975) or voxel grids (Yan and Bowyer, 2007), for example. Drost and Ilic (2013) combine the advantages of tree-based and voxel-based

ICP 算法的处理时间取决于对应点搜索算法的执行时间。因此，在实际应用中，对于最近邻域点的确定，一个能够高效运算的数据结构是尤为重要的。有很多知名的高效的算法，例如基于 k-d 树的搜索算法（Bentley，1975）以及基于体素网格的搜索算法（Yan and Bowyer，2007）。此外，在文献（Drost and Ilic，2013）中提出

approaches to propose a method that is faster than traditional *k*-d-tree approaches.

There are several extensions of the ICP algorithm that speed up the computation or improve its accuracy. For example, instead of minimizing point distances as in Eq. (3.210), distances of points to the tangent planes at the corresponding points can be minimized (Chen and Medioni, 1992). This allows a lower sampling density of the two point clouds. Furthermore, the robustness of the ICP algorithm to outliers can be improved by applying the IRLS (Stewart, 1999) algorithm, as described in Section 3.8.1.2.

Figure 3.164(d) shows the final poses that are obtained after the pose refinement step for the illustrated example.

3.12.3.5 Extension for Recognizing Deformed Objects

Drost and Ilic (2015) introduce an extension of surface-based 3D matching that is able to recognize deformable objects. Deformable in this context either means that one object instance itself is deformable, like objects made of rubber or silicone, or that objects vary their shape over different instances, like vegetables or fruits.

The range of possible deformations is learned from training examples. Each example instance, which shows one particular deformation of the object, is registered to a reference pose. This registration can be performed by using surface-based 3D matching, as described in the previous sections. Then, for each point pair in the model, the range of possible deformations over the example

的结合基于树和基于体素算法优点的新的算法拥有比传统 *k*-d 树更快的处理速度。

我们对 ICP 算法进行了一些扩展，以此提高处理速度和精度。例如，将等式（3.210）中的最小距离替换为到对应点三角平面的最小距离（Chen and Medioni，1992）。这样，我们就可以使用降采样以后两点云进行距离运算。此外，如在 3.8.1.2 节中所描述的，在离群值方面，我们可以通过应用 IRLS（Stewart，1999）算法提高 ICP 算法的鲁棒性。

图 3.164（d）为经过位姿优化后获得的最终位姿。

3.12.3.5 算法扩展——变形物体识别

文献（Drost and Ilic，2015）为我们描述了一个基于形状的三维匹配的扩展算法。该算法可用于识别存在变形的物体。本文中我们所说的变形是指物体本身会产生形变。如，一些橡胶或硅胶制品，又比如水果或蔬菜等，存在个体差异的物体。

我们可以从训练样本中对物体可能的变形范围进行训练。对于每个存在特定变形的物体样本，我们都会为其注册一个参考位姿。我们可以使用前面章节中提及的基于表面的三维匹配算法对这些参考位姿进行处理。经过处理，对于模板上的任意点对，我们都可以确定出可能在样本实例中出

instances are determined. For all point pairs and for all deformations, the point pair features are computed and stored in the hash table, as described in Section 3.12.3.1. Thus, in contrast to the recognition of rigid objects, each point pair is contained in the hash table multiple times.

In contrast to the voting for rigid objects as described in Section 3.12.3.3, the voting for deformable objects is performed iteratively using a graph structure. Each vertex represents a correspondence between a scene point and a model point. An edge between two vertices in the graph indicates that there exists a consistent non-rigid transformation between scene and model that aligns the two scene points with their corresponding model points. Which edges must be created can be efficiently queried from the hash table.

If a perfect instance of the model was present in the scene, all vertices representing correspondences between model points and their corresponding scene points would be connected with edges, thus forming a dense subgraph. The extraction of this subgraph is performed in two steps: In the first step, weights are assigned to the vertices and initialized with 1. Iterative voting is performed to amplify the dense subgraph: over several voting rounds, each vertex votes for its connected vertices, using its weighting vote obtained in the previous round. The iterative voting leads to high voting counts for the subgraph that represents the correspondences between the object instance in the scene and the model. In the second step, the most dominant consistent subgraph is extracted, yielding a consistent set of point correspondences.

现形变的范围。如 3.12.3.1 节中所述，对于所有的点对以及所有的形变，它们的点对特征将被计算，并存储在哈希表中。因此，相比于刚性物体的识别，在变形物体识别中每个点对将在哈希表中多次出现。

与 3.12.3.3 节中描述的刚性物体投票算法相比，变形物体的投票则采用图状结构迭代执行。结构中的每一个定点代表一组对应的场景点和模板点。图状结构中两顶点之间的边表示在场景和模板之间存在一个可以使它们点集对齐的非刚性变换。我们可以通过高效的哈希查表算法创建图状结构中的弧。

如果一个完整的被测物在场景中出现，则代表模板点与其对应场景点之间对应关系的图状结构的所有顶点将被连接，从而形成一个密集的子图。我们通过以下两步进行子图的提取：第一步，我们为每个顶点设置初始权重 1。通过迭代投票对密集子图进行增强：在若干次的投票过程中，每一个顶点会使用该顶点在上一轮投票中获得的权值为与之相连的顶点投票。这种迭代式的投票会使得表示场景与模板间存在相关性的特征子图获得较高的票数。第二步，我们将从图状结构中提取存在显著一致性的子图，并生成一致性对应点集合。

From the set of correspondences, an approximate rigid 3D transformation can be computed. Furthermore, the rigid 3D transformation can be used to initialize a deformable ICP (Myronenko and Song, 2010) to obtain a model that is deformed in accordance with the scene points. Figure 3.171 shows an example application where deformable silicone baking molds are recognized and grasped by a robot.

根据这个点集，我们可以计算出一个大致的三维刚性变换。此外，这个三维刚性变换将用于初始化一个可变形的 ICP（Myronenko and Song, 2010）算法，通过该算法，我们可以从存在变形的场景点集中，搜索到已经训练的模板。图 3.171 为存在变形的硅胶烘焙模具的识别和机械手抓取。

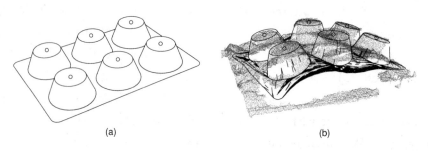

图 3.171　通过三维数据识别变形的硅胶烘焙模具：（a）为模具 CAD 模型，图中机器人的 6 个抓取位置用圆形标出；（b）图中，灰色为场景三维点云，黑色为搜索到的变形模板以及根据模板变形重新确定的抓取位置

3.12.3.6　Extension for Multimodal Data

Some 3D image acquisition devices return both 3D data and 2D image data. Furthermore, some 3D reconstruction methods like stereo reconstruction are based on 2D image data. In these cases, multimodal data, i.e., 3D data and gray value or color information, of the reconstructed scene are available. Drost and Ilic (2012) and Drost and Ulrich (2015c,d) introduce an extension of surface-based 3D matching that exploits the benefits of using such multimodal data, which increases the robustness of the recognition in many applications.

The use of edge information for object recognition in image data has proven to be a powerful feature (see Sections 3.11.5.6 and 3.12.2). Howe-

3.12.3.6　算法扩展——多模态数据

一些三维图像采集设备在采集三维数据的同时还可以获取到二维图像数据。此外，一些三维重构算法，例如立体重构算法，它们本身就是基于二维图像数据进行处理的。在这种情况下，我们将获得多模态的场景重构数据，例如，三维数据、灰度数据以及彩色信息。文献（Drost and Ilic，2012）和（Drost and Ulrich，2015c,d）为我们描述了一种基于形状的三维匹配的扩展算法，并阐述了多模态数据在多种应用场合下对算法鲁棒性的增强。

对于图像中物体的识别，其边缘信息是尤为重要的（见 3.11.5.6 节以及 3.12.2 节）。然而，在图像中我们

ver, image edges cannot distinguish between texture and geometric edges. In contrast, edges or depth discontinuities in 3D data only occur at geometric edges, and hence provide additional information. Unfortunately, edges in 3D data are often error-prone. 3D image acquisition devices often have problems accurately reconstructing points that are close to 3D edges. Stereo reconstruction or structured light methods suffer from occlusion problems around edges. Furthermore, sharp edges are often smoothed in the reconstruction because of the spatial filter that is applied in stereo reconstruction methods based on gray value matching (see Section 3.10.1.7). TOF cameras also tend to smooth over edges and introduce interpolation points that do not correspond to real points in the scene. Therefore, Drost and Ulrich (2015c,d) present a method that combines the accuracy of image edges with the higher information content of 3D data. In the following, we will briefly summarize the basic idea.

To create the global model description (see Section 3.12.3.1), a multimodal point pair feature is computed. The feature uses geometric edges in the intensity image and depth information from the 3D data, and hence combines the stable information of both modalities. First, for one viewpoint of the object, geometric edges are extracted. For this, edges are extracted in the intensity image of the viewpoint. An edge point is accepted as a geometric edge only if the 3D data contains a depth discontinuity at the corresponding position. Then, features are computed from the point pairs similar to the method described in Section 3.12.3.1. Here, however, a 3D reference point on the object surface is paired with a

很难区分出纹理边缘和几何边缘。相比之下，三维数据中的边缘和断差通常只出现在物体的几何边缘上，因此，它们可以为识别算法提供更多的信息。不幸的是，三维数据通常很容易出现误差。三维图像采集设备通常无法准确地重建三维边缘附近的点。立体重构或结构光算法在物体边缘附近会存在遮挡问题。此外，由于使用了空间滤波器（如，在基于灰度匹配的立体重构算法中用到的平滑算法，见3.10.1.7节），锋利的边缘在重构过程中通常会被平滑。TOF摄像机采集的三维数据在物体边缘附近会较为平滑，并会使用差值算法引入一些与实际场景不符的点。因此，文献（Drost and Ulrich，2015c,d）提出了一种新的算法，该算法结合图像边缘的精度以及包含更大信息量的三维数据，进行物体的识别。在下文中，我们将简略地为大家概述算法的基本思路。

为了创建全局模板描述（见3.12.3.1节），我们将计算一个多模态的点对特征。这些特征来自于灰度图像中的几何边缘和三维图像的深度信息，因此它是结合了两种数据形式的稳定的信息。首先，对于物体的某一个视角提取聚合边缘。这一过程是通过物体在该视角的灰度图像完成的。在三维数据中，如果某点对应位置的深度数据存在断差，则我们将该点作为几何边缘并进行存储。而后，我们从点对中计算特征，计算的过程类似于3.12.3.1节中描述的算法。本算法中，与前面章节不同的是，每个三维的参考点都存在一个与之对应的二维

2D geometric edge point. For each point pair, a four-dimensional feature vector is computed. It contains the distance between the two points in the image transformed to metric units, the angle in the image plane between the difference vector of the two points and the edge gradient direction, the angle in the image plane between the difference vector of the two points and the normal vector of the reference point, and the angle in 3D between the normal vector of the reference point and the direction towards the camera.

The model is created by rendering or acquiring the object from various viewpoints yielding a set of template images (about 300 are sufficient in most applications). For each viewpoint, the geometric edges are extracted in the template image. A set of 3D model reference points is paired with each 2D edge point in the template image and the above described multimodal feature is computed for each pair and stored in a hash table as described in Section 3.12.3.1. The model created is used to recognize the instances of the object in a search scene by applying the voting scheme of Section 3.12.3.3 and the least-squares pose refinement of Section 3.12.3.4.

One major advantage of the method described over conventional surface-based 3D matching is that by the use of the geometric edges, even planar objects in front of a planar background can be robustly detected.

3.13 Hand–Eye Calibration

One important application area of the 3D object recognition approaches discussed in Section 3.12 is grasping of the recognized objects by a robot, e.g., to assemble objects or to place the objects at a predefined pose. Object recognition allows us to determine the object pose in the camera coordinate system. To be able to grasp the object, the object pose must first be transformed into the coordinate system of the robot. For this, the pose of the camera relative to the robot must be known. The process of determining this pose is called hand–eye calibration.

3.13.1 Introduction

A robot consists of multiple links that are connected by joints (ISO 8373:2012; ISO 9787:2013). A joint allows the two adjacent links to perform a motion relative to each other: for example, a prismatic joint allows a linear motion, a rotary joint allows a rotary motion about a fixed axis, and a spherical joint allows a rotary motion about a fixed point in three degrees of freedom. Figure 3.172 shows a model of an articulated six-axis robot with three rotary joints and one spherical joint. The spherical joint is constructed as a combination of three rotary joints with their rotation axes intersecting in a single point. The base of an industrial robot is the structure to which the origin of the first link is attached. The robot base is mounted on the base mounting surface. The last

物体。例如，进行组装或将物体放置在预定位置。通过对物体的识别，可以确定摄像机坐标系下物体的位姿。但为了完成物体的抓取，需要将物体的位姿转换到机器人坐标系下。为此，需要求取已知摄像机到机器人的转换关系。而这个位姿的确定过程，一般称之为手眼标定。

3.13.1 前言

一个机器人一般由多个连杆以及连接它们的关节组成（ISO 8373:2012；ISO9787:2013）。一个关节可以允许两个相邻的连杆进行相对运动：例如，柱状关节可以允许线性运动，旋转关节允许沿固定轴的旋转运动，球面关节允许沿固定点的三自由度旋转。图 3.172 展示了一个拥有三个旋转关节和一个球形关节的六自由度多关节机器人。其中，球形关节由三个旋转轴交于一点的旋转关节组成。机器人固定在安装平面上，它的末端连杆提供一个机械接口，用于安装执行机构（工具）。工具可以帮助机器人完成实际任务，例如，工具可以是机械式的或真空吸附式的夹钳。机械接口的位置

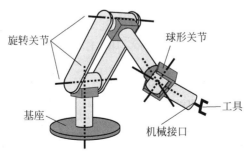

图 3.172　多关节六轴机械手模型。机械手的旋转轴用粗线标出

link of the robot provides the mechanical interface, to which the end effector (tool) is attached. The tool enables the robot to perform its actual task and might be a mechanical or a vacuum gripper, for example. The pose of the mechanical interface can be calculated from the joint positions and the geometry of the links and joints. This mathematical relationship is called forward kinematics. Furthermore, if the pose of the tool relative to the mechanical interface is provided to the robot controller, the pose of the tool also can be calculated by using the forward kinematics.

In general, there are two different configurations of a vision-guided robot: the camera can be mounted either at the robot's tool and is moved to different positions by the robot (moving camera scenario, see Figure 3.173(a)) or outside the robot without moving with respect to the robot base while observing its workspace (stationary camera scenario, see Figure 3.173(b)).

可以通过关节的位置以及连杆和关节的几何结构计算获得。对于这一数学关系的研究，被称为正向运动学。此外，如果工具和机械接口的位置关系已经输入到机器人控制器中，依然可以使用正向运动学计算工具的位姿。

一般情况下，视觉引导机器人有两种不同的配置：将摄像机安装在机器人的工具上，并随着机器人运动到不同的位置进行图像采集（运动摄像机方案，见图 3.173（a）），或者将摄像机安置在机器人外部，并相对于机器人的基座静止，从而观测机器人的工作空间（固定摄像机方案，见图 3.173（b））。

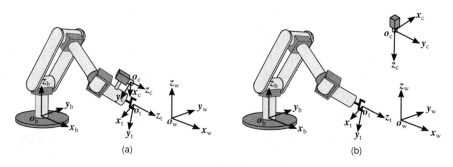

图 3.173　两种可能出现的视觉引导机器人配置：（a）运动摄像机；（b）固定摄像机。相关坐标系使用原点 o 和坐标轴 x, y, z 表示：WCS（下标 w），BCS（下标 b），TCS（下标 t），CCS（下标 c）

In the hand–eye calibration process, the following four coordinate systems are involved (see Figures 3.173(a) and (b)): The world coordinate system (WCS) (o_w, x_w, y_w, z_w) is independent of the robot movement. The base coordinate system (BCS) (o_b, x_b, y_b, z_b) is typically located at

在手眼标定的过程中，将涉及下列 4 个坐标系（见图 3.173（a）和（b））：世界坐标系（WCS）(o_w, x_w, y_w, z_w) 独立于机械手的运动。基坐标系（BCS）(o_b, x_b, y_b, z_b)，通常位于机器人基座位置，其中 $x - y$ 平面平行

the robot base with its xy-plane parallel to the base mounting surface and its z-axis pointing upwards (ISO 9787:2013). The tool coordinate system (TCS) (o_t, x_t, y_t, z_t) is defined by the tool that is attached to the mechanical interface. The origin of the TCS is often denoted as the tool center point. Finally, the camera coordinate system (CCS) (o_c, x_c, y_c, z_c) is defined as described in Section 3.9.1.

For the case of a moving camera, hand–eye calibration basically determines the pose of the CCS relative to the TCS. For the case of a stationary camera, the pose of the CCS relative to the BCS is determined. This allows us to grasp objects with the robot by transforming the object pose, which is determined in the CCS, into the robot BCS. Note that there exist further camera configurations and systems that can be calibrated by using a hand–eye calibration method, e.g., pan-tilt cameras, endoscopes (Schmidt *et al.*, 2003; Schmidt and Niemann, 2008), X-ray systems (Mitschke and Navab, 2000), and augmented reality systems (Baillot *et al.*, 2003).

3.13.2 Problem Definition

Like conventional camera calibration (see Section 3.9.4), hand–eye calibration is typically performed by using a calibration object. For this, the tool of the robot is moved to n different robot poses. At each robot pose, the camera acquires an image of the calibration object. For moving cameras, the calibration object is placed at a fixed position within the workspace of the robot (see Figure 3.174(a)). For stationary cameras, the calibration object is rigidly attached to the tool, and hence moves with the robot (see Figure 3.174(b)).

于基座安装平面且 z 轴向上（ISO 9787:2013）。工具坐标系（TCS）(o_t, x_t, y_t, z_t)，由机器人工具确定并与机械接口绑定。TCS 的原点通常表示工具的中心点。最后，摄像机坐标系（CCS）(o_c, x_c, y_c, z_c) 定义详见 3.9.1 节。

对于运动摄像机的情况，手眼标定的主要任务是确定 CCS 和 TCS 之间的关系。对于固定摄像机的标定，需要确定 CCS 和 BCS 之间的关系。通过上述标定，可以将在 CCS 中确定的位姿转换到机器人 BCS 下，并引导机器人进行抓取。值得注意的是，除了上述的应用，还有很多摄像机配置和系统的标定也可以使用手眼标定算法，例如，云台摄像机、内窥镜（Schmidt *et al.*, 2003; Schmidt and Niemann, 2008）、X 光系统（Mitschke and Navab, 2000）以及 AR 系统（Baillot *et al.*, 2003）。

3.13.2 问题定义

就像传统摄像机标定（见 3.9.4 节）一样，手眼标定通常会通过标定板进行。为此，机器人的工具将会被移动到 n 个不同的机器人位姿。在每个位姿下，摄像机都会对标定板进行一次图像采集。对于运动摄像机，会将标定板放置于机器人工作空间中的一个固定位置（见图 3.174（a））。而对于固定摄像机，标定板需要与工具建立刚性的物理连接，并随机器人一起运动（见图 3.174（b））。如果

If the interior orientation parameters of the camera are unknown, the calibration images are used to fully calibrate the camera, i.e., to determine the interior camera orientation as well as the exterior orientation of each calibration image. If the interior orientation is already known, the calibration images are used to determine the exterior orientation of each calibration image. The camera calibration is performed as described in Section 3.9.4.4.

摄像机的内参是未知的，可以使用标定图像做完整的标定，即，确定摄像机的内参以及每张标定图像的外参。如果内参是已知的，那么只需要确定每张标定图像的外参即可。其中摄像机的标定已在 3.9.4.4 节中阐述，这里就不再赘述了。

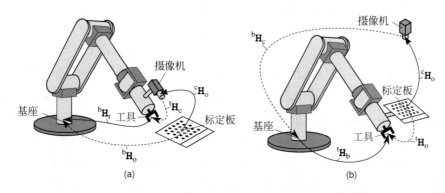

图 3.174　为在运动摄像机（a）和固定摄像机（b）两种情况下的四坐标系（摄像机、基座、工具以及标定板）的转换。图中实线表示在手眼标定过程中作为输入的已知的转换关系。虚线表示需要通过手眼标定求解的未知的转换关系。值得注意的是，在上述的两种情况下，这 4 个转换关系都是以闭环形式存在的

Note that it is also possible to perform hand–eye calibration by using an arbitrary object instead of a calibration object. In this case, the 3D pose of the object relative to the sensor is determined by using a 3D object recognition approach. For example, if a camera is used as the sensor, one of the approaches described in Sections 3.12.1 or 3.12.2 can be applied. If a 3D image acquisition device is used instead of a camera, one of the approaches described in Section 3.12.3 can be applied.

After this step, the pose $^c\mathbf{H}_o$ of the calibration object relative to the camera is known for all n calibration images. Here, $^{c2}\mathbf{H}_{c1}$ denotes a

值得注意的是，一般可以使用任意物体代替标定板进行标定。这种情况下，可以使用三维物体识别算法来确定物体和传感器的三维位姿关系。例如，如果传感器为单目摄像机，可以使用 3.12.1 或 3.12.2 节中的变形匹配算法。如果使用三维图像采集设备代替单目摄像机，可以使用 3.12.3 节中描述的基于表面的三维匹配算法。

经过这一步，可以获取每张图像中标定板相对于摄像机的位姿 $^c\mathbf{H}_o$。在这里使用符号 $^{c2}\mathbf{H}_{c1}$ 表示一个刚性

rigid 3D transformation or pose, represented by a 4×4 homogeneous transformation matrix, that transforms 3D points from the coordinate system c1 into c2 (see Section 3.9.1.1). Furthermore, the pose $^{b}H_{t}$ of the tool relative to the robot base (or its inverse $^{t}H_{b}$) can be queried from the robot controller, and hence is known for all n images as well. If the robot controller only returns the pose of the mechanical interface, the pose of the tool with respect to the mechanical interface must be added manually. For moving cameras, the fixed and unknown transformations are $^{c}H_{t}$, i.e., the pose of the camera relative to the tool, and $^{o}H_{b}$, i.e., the pose of the calibration object relative to the base (see Figure 3.174(a)). For stationary cameras, the fixed and unknown transformations are $^{c}H_{b}$, i.e., the pose of the camera relative to the base, and $^{t}H_{o}$, i.e., the pose of the tool relative to the calibration object (see Figure 3.174(b)). The transformations can be concatenated to form a closed chain:

的三维变换或位姿，它为一个可以将三维点从坐标系 c1 转换到 c2 的 4×4 齐次转换矩阵（见 3.9.1.1 节）。此外，工具相对于基座的位姿 $^{b}H_{t}$（或它的逆 $^{t}H_{b}$）可以从机器人控制器获得，因此对于全部的 n 张图像，它们都是已知的。如果机器人控制器只返回机械接口的位姿，则需要人为添加工具相对于机械接口的位姿。对于运动摄像机，固定的未知转换关系为 $^{c}H_{t}$。即摄像机到工具的位置关系，除此之外还有 $^{o}H_{b}$，即标定板相对于基座的位姿（图 3.174（a））。对于固定摄像机，固定的未知转换关系为 $^{c}H_{b}$，即摄像机相对于基座的位姿，以及 $^{t}H_{o}$，即，工具相对于标定板的位姿（见图 3.174（b））。这些转换可以相互连接并形成一个闭环：

$$^{b}H_{o} = {^{b}H_{t}}\,{^{t}H_{c}}\,{^{c}H_{o}} \tag{3.211}$$

for moving cameras and

对于运动摄像机有

$$^{t}H_{o} = {^{t}H_{b}}\,{^{b}H_{c}}\,{^{c}H_{o}} \tag{3.212}$$

for stationary cameras. Note that both equations have the same structure

对于固定摄像机。值得注意的是这两个等式都拥有相同的结构

$$Y = A_{i}XB_{i} \tag{3.213}$$

where A_i is the pose of the tool relative to the base (or vice versa) and B_i is the pose of the camera relative to the calibration object for robot pose i, with $i = 1, \cdots, n$. In the equations, the essential unknown is X, which represents the pose of the camera relative to the tool for moving cameras and the pose of the camera relative to the base for stationary cameras.

式中 A_i 为工具相对于基座的位姿，B_i 为摄像机相对于标定板的第 i 个位姿。$i = 1, \cdots, n$。在等式中，未知项 X，对于运动摄像机表示摄像机相对于工具的位姿，对于固定摄像机表示摄像机相对于基座的位姿。

For a pair of different robot poses i and j, we obtain the two equations $Y = A_i X B_i$ and $Y = A_j X B_j$. This allows us to eliminate the unknown pose Y:

$$A_i X B_i = A_j X B_j \tag{3.214}$$

By rearranging, we obtain

$$A_j^{-1} A_i X = X B_j B_i^{-1} \tag{3.215}$$

After substituting $A_j^{-1} A_i$ by A and $B_j B_i^{-1}$ by B, the equation to be solved finally becomes

$$AX = XB \tag{3.216}$$

Here, A represents the movement of the tool and B represents the movement of the camera or the calibration object, respectively, when moving the robot from pose i to j.

There are linear algorithms that solve Eq. (3.216) by handling the rotation and translation parts of the three matrices separately (Tsai and Lenz, 1989; Chou and Kamel, 1991). As a consequence, rotation errors propagate and increase translation errors. More advanced linear methods avoid this drawback. They are typically based on screw theory (see Section 3.13.3 for an introduction to screw theory) and solve for rotation and translation simultaneously (Chen, 1991; Daniilidis, 1999; Horaud and Dornaika, 1995). In general, linear approaches have the advantage that they provide a direct solution in a single step, i.e., no initial values for the unknowns need to be provided and there is no risk of getting stuck in local minima of the error function. Furthermore, because no iterations are involved, linear approaches are typically faster than nonlinear methods. However, linear approaches often minimize an alge-

对于一对不同的机器人位姿 i 和 j，可以得到两个等式 $Y = A_i X B_i$ 和 $Y = A_j X B_j$。通过这两个等式可以消除位姿 Y：

通过代换得到

令 A 为 $A_j^{-1} A_i$，B 为 $B_j B_i^{-1}$，最终，等式化简为

当机器人从位姿 i 移动至位姿 j 时，式中 A 表示工具的移动，而 B 表示摄像机的移动或者标定板的移动。

这里可以通过线性算法求解等式。由于，以上三个矩阵中的旋转和平移是分开的（Tsai and Lenz, 1989; Chou and Kamel, 1991），因此，旋转矩阵的误差会传递累加到平移矩阵上。有一些更先进的线性算法可以避免这个缺陷。它们通常基于螺旋理论（见 3.13.3 节）求解旋转矩阵和平移矩阵（Chen, 1991; Horaud and Dornaika, 1995; Daniilidis, 1999）。通常情况下，线性求解方法拥有如下优势，只需一步求解便能直接获得处理结果，例如，不提供未知数的初值也不会存在误差函数局部震荡的问题。此外，由于没有迭代的过程，线性解法比非线性解法处理速度更快。然而，线性解通常是令一个代数误差最小，因此解法本身并没有几何意义，所以，这有可能导致处理结果精度的降低。

braic error that lacks any geometric meaning, which sometimes leads to reduced accuracy. Therefore, to obtain a higher accuracy, the result of the linear methods is often used to initialize a subsequent nonlinear optimization (Horaud and Dornaika, 1995; Daniilidis, 1999; Kaiser et al., 2008).

Because of the advantages described above, we will focus for the linear approach on methods that simultaneously solve for rotation and translation, and especially on the approach of Daniilidis (1999). Because this approach is based on screw theory and uses dual quaternions, we will give a brief introduction to dual quaternions and screw theory in the following section.

3.13.3 Dual Quaternions and Screw Theory

3.13.3.1 Quaternions

Quaternions are 4D vectors $\boldsymbol{q} = (q_0, q_1, q_2, q_3) = (q_0, \boldsymbol{q})$ with scalar part q_0 and vector part $\boldsymbol{q}$. The multiplication of two quaternions $\boldsymbol{p}$ and $\boldsymbol{q}$ is defined as

$$\boldsymbol{pq} = \begin{pmatrix} p_0 q_0 - p_1 q_1 - p_2 q_2 - p_3 q_3 \\ p_1 q_0 + p_0 q_1 + p_2 q_3 - p_3 q_2 \\ p_2 q_0 + p_0 q_2 + p_3 q_1 - p_1 q_3 \\ p_3 q_0 + p_0 q_3 + p_1 q_2 - p_2 q_1 \end{pmatrix} = (p_0 q_0 - \boldsymbol{p} \cdot \boldsymbol{q}, p_0 \boldsymbol{q} + q_0 \boldsymbol{p} + \boldsymbol{p} \times \boldsymbol{q}) \quad (3.217)$$

It should be noted that the multiplication is not commutative, i.e., $\boldsymbol{pq} \neq \boldsymbol{qp}$ in general. The conjugation of a quaternion is obtained by reversing the sign of its vector part:

$$\bar{\boldsymbol{q}} = (q_0, -q_1, -q_2, -q_3) = (q_0, -\boldsymbol{q}) \quad (3.218)$$

Unit quaternions are quaternions with norm one, i.e, they satisfy $\boldsymbol{q}\bar{\boldsymbol{q}} = 1$. Quaternions can be con-

为了获取更高的精度，线性算法的处理结果通常用于初始化一个非线性优化算法（Horaud and Dornaika, 1995; Daniilidis, 1999; Kaiser et al., 2008）。

鉴于上述优势，在接下来的文章中将着手于如何通过线性算法同时计算旋转和平移矩阵，尤其是文献（Daniilidis, 1999）的算法。因为该算法是基于之前提到的螺旋理论以及对偶四元数的求解方法。在下一章中将为大家简述对偶四元数以及螺旋理论。

3.13.3 对偶四元数和螺旋理论

3.13.3.1 四元数

四元数是一个四维矩阵，$\boldsymbol{q} = (q_0, q_1, q_2, q_3) = (q_0, \boldsymbol{q})$，它由一个标量 q_0 以及一个向量 $\boldsymbol{q}$ 组成。两个四元数的 $\boldsymbol{p}$ 和 $\boldsymbol{q}$ 的乘积定义为

值得注意的是，这个乘法不满足交换律，即，通常 $\boldsymbol{pq} \neq \boldsymbol{qp}$。通过变换向量部分的符号，可以获得该四元数的共轭四元数：

单位四元数是绝对值为 1 的四元数，即，满足 $\boldsymbol{q}\bar{\boldsymbol{q}} = 1$。四元数可以用超复

sidered as a generalization of complex numbers. The basis elements of the quaternion vector space are

数表示。四元数向量空间中的每一个基础元素为：

$$\mathbf{1} = (1,0,0,0), \quad \mathbf{i} = (0,1,0,0), \quad \mathbf{j} = (0,0,1,0), \quad \mathbf{k} = (0,0,0,1) \tag{3.219}$$

When applying the definition of quaternion multiplication (3.217) to the basis elements, we obtain

当应用四元数乘法定义（3.217）到这些项时，可得

$$\mathbf{1}^2 = 1, \quad \mathbf{i}^2 = \mathbf{j}^2 = \mathbf{k}^2 = \mathbf{ijk} = -1 \tag{3.220}$$

A quaternion with a scalar part $q_0 = 0$ is called a pure quaternion. The quaternion inversion is given by

如果一个四元数的 $q_0 = 0$，则被称为纯四元数。四元数的逆定义为

$$q^{-1} = \frac{\bar{q}}{q\bar{q}} \tag{3.221}$$

It is clear from the above definitions that for a unit quaternion $q^{-1} = \bar{q}$.

根据上式，单位四元数的逆可表示为 $q^{-1} = \bar{q}$。

Every unit quaternion can be written as $q = (\cos(\theta/2), \mathbf{n}\sin(\theta/2))$ with $0 \leqslant \theta < 4\pi$ and a unit 3D vector $\mathbf{n}$. Let $\mathbf{p} = (0, \mathbf{p})$ be a pure quaternion. Then, it can be shown that the quaternion product

每个单位四元数都可以被表示为 $q = (\cos(\theta/2), \mathbf{n}\sin(\theta/2))$，其中 $0 \leqslant \theta < 4\pi$，$\mathbf{n}$ 为三维单位向量。令一个纯四元数 $\mathbf{p} = (0, \mathbf{p})$。则可产生如下公式

$$qpq^{-1} = qp\bar{q} \tag{3.222}$$

can be interpreted geometrically as a rotation of the point $\mathbf{p}$ about the axis $\mathbf{n}$ by the angle θ and

其几何意义为，点 $\mathbf{p}$ 关于坐标轴 $\mathbf{n}$ 进行角度为 θ 的旋转。同理可得公式

$$q^{-1}pq = \bar{q}pq \tag{3.223}$$

as a rotation of the coordinate system about the same axis and by the same angle (Rooney, 1977; Kuipers, 1999). Hence, the product $qp\bar{q}$ is a linear operator with respect to $\mathbf{p}$, which can be converted into a rotation matrix:

该公式表示对系统坐标系中的坐标轴 $\mathbf{n}$ 进行角度为 θ 的旋转（Rooney, 1977; Kuipers, 1999）。因此乘积 $qp\bar{q}$ 是一个关于 $\mathbf{p}$ 的线性运算，同时，可以转换为旋转矩阵：

$$\mathbf{R}_q = \begin{pmatrix} q_0^2 + q_1^2 - q_2^2 - q_3^2 & 2q_1q_2 - 2q_0q_3 & 2q_1q_3 + 2q_0q_2 \\ 2q_1q_2 + 2q_0q_3 & q_0^2 - q_1^2 + q_2^2 - q_3^2 & 2q_2q_3 - 2q_0q_1 \\ 2q_1q_3 - 2q_0q_2 & 2q_2q_3 + 2q_0q_1 & q_0^2 - q_1^2 - q_2^2 + q_3^2 \end{pmatrix} \tag{3.224}$$

The consecutive execution of two rotations can be represented by multiplying the two corresponding unit quaternions by analogy to multiplying the two corresponding rotation matrices.

Like complex numbers, quaternions can also be represented in exponential form:

两个连续的旋转可以表示为两个对应的单位四元数的相乘，就像两个对应的旋转矩阵相乘一样。

如复数一样，四元数也可表示为指数形式：

$$q = e^{\boldsymbol{I} \cdot \boldsymbol{n} \theta/2} = \cos(\theta/2) + \boldsymbol{I} \cdot \boldsymbol{n} \sin(\theta/2) \tag{3.225}$$

with $\boldsymbol{I} = (\boldsymbol{i}, \boldsymbol{j}, \boldsymbol{k})$. A rotation matrix $\boldsymbol{R}$ can be converted into a quaternion by

上式中 $\boldsymbol{I} = (\boldsymbol{i}, \boldsymbol{j}, \boldsymbol{k})$，这里可以通过下式将矩阵 $\boldsymbol{R}$ 转换为四元数

$$q_0 = \frac{1}{2}\sqrt{\mathrm{Tr}(\mathrm{R}) + 1}$$

$$\begin{pmatrix} q_1 \\ q_2 \\ q_3 \end{pmatrix} = \frac{1}{4q_0} \begin{pmatrix} r_{32} - r_{23} \\ r_{13} - r_{31} \\ r_{21} - r_{12} \end{pmatrix} \tag{3.226}$$

for a rotation matrix with a positive trace ($\mathrm{Tr}(\boldsymbol{R}) > 0$). The case $\mathrm{Tr}(\boldsymbol{R}) \leqslant 0$ must be handled more carefully and is described in Hanson (2006). Note that the representation of rotations as quaternions is ambiguous, i.e., $\boldsymbol{q}$ and $-\boldsymbol{q}$ represent the same rotation $\boldsymbol{R_q} = \boldsymbol{R_{-q}}$.

What is the advantage of representing rotations as quaternions? Compared to rotation matrices (nine elements), unit quaternions (four elements) allow a more compact representation of rotations. Nevertheless, the four elements of a quaternion still overparameterize 3D rotations, which only have three degrees of freedom. The fourth degree of freedom is eliminated by taking the unity constraint $q\bar{q} = 1$ into account. Furthermore, rotations that are represented as quaternions directly encode the rotation axis and angle (see Eq. (3.225)), and hence can be interpreted more easily than rotation matrices or Euler angles.

上式为一个存在正迹 ($\mathrm{Tr}(\boldsymbol{R}) > 0$) 的矩阵转换过程。对于 $\mathrm{Tr}(\boldsymbol{R}) \leqslant 0$ 的情况则需要特殊处理，详见（Hanson, 2006）。值得注意的是，四元数对旋转矩阵的表示不是唯一的，例如，$\boldsymbol{q}$ 和 $-\boldsymbol{q}$ 表示的是同一个旋转矩阵 $\boldsymbol{R_q} = \boldsymbol{R_{-q}}$。

使用四元数表示旋转有什么好处呢？相比于旋转矩阵（9 个元素），单位四元数（4 个元素）对于旋转的表示更为紧凑。然而，对于只存在三个自由度的三维旋转问题，四元数的四个元素在表示三维旋转时依然是过参数化的。第四个自由度由于使用了单位四元数，因此受到 $q\bar{q} = 1$ 的限制而被抹掉了。因此，四元数相当于直接对坐标轴和角度进行编码（见等式（3.225）），这也使得它比旋转矩阵或欧拉角更容易理解。四元数的另一个优点是它可以在旋转之间进行任意

Another advantage of representing rotations as quaternions is that quaternions allow a meaningful interpolation between rotations (Hanson, 2006).

3.13.3.2 Screws

Unfortunately, rigid 3D transformations (or poses) such as in Eq. (3.216), which in addition to a 3D rotation also include a 3D translation, cannot be represented as quaternions in a comparably compact and elegant way. The question is whether there exists an alternative equivalent representation. The first step towards answering this question is the insight that a rigid 3D transformation can be represented as a screw, which is known as Chasles' theorem (Rooney, 1978). A screw is a rotation about a screw axis by an angle θ followed by a translation by d along this axis (see Figure 3.175(a)). The direction of the screw axis is defined by its direction vector l. The position of the screw axis is defined by its moment vector m with respect to the origin. The moment m is perpendicular to l and to a vector from the origin to an arbitrary point p on the screw axis (see Figure 3.175(b)).

的差值 (Hanson, 2006)。

3.13.3.2 螺旋

不幸的是，如等式（3.216）所示，刚性的三维变换（或位姿）除了三维旋转以外还包括三维的平移。问题是，是否存在一种等价的表示呢？为了回答这个问题，首先需要认识到，刚性的三维转换可以表示为一个螺旋变换，这也正是著名的沙勒定理（Rooney，1978）。一个螺旋是由一个围绕螺旋轴的旋转 θ 和一个沿着这个轴的平移 d 组成的（见图 3.175（a））。定义方向向量 l 为旋转轴的方向。旋转轴的位置，使用与原点相关的矩向量 m 表示。其中 m 垂直于 l，以及为原点指向螺旋轴上任意一点 p 的向量（见图 3.175（b））。

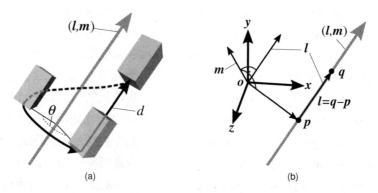

图 3.175 (a) 一个刚性的三维变换可以被表示为一个螺旋，螺旋轴线用浅灰色表示；(b) 螺旋轴的位置由其关于原点的矩确定

The vectors l and m are called the Plücker coordinates of the straight line that represents the screw axis. A line in 3D that passes through the two different points p and q can be transformed to Plücker coordinates by $l = q - p$ and $m = p \times l = p \times q$ with $l^\top m = 0$ and $\|l\| = 1$.

Furthermore, a rigid 3D transformation that is represented as a rotation matrix R and a translation vector t can be converted to a screw and vice versa. From the rotation matrix, the vector l and the screw angle θ can be computed as described in (Hanson, 2006). The screw translation d is the projection of t onto l, i.e., $d = t^\top l$. Finally, the screw moment is calculated as $m = \frac{1}{2}(t \times l + (t - dl)\cot\frac{\theta}{2})$ (Daniilidis, 1999). Conversely, the screw parameters l and θ can be converted to a rotation matrix as described in (Hanson, 2006). The translation vector is obtained as $t = (I - R)(l \times m) + dl$ (Daniilidis, 1999).

3.13.3.3 Dual Numbers

For the second step towards an efficient representation of rigid 3D transformations, we must introduce the concept of dual numbers. By analogy to complex numbers, a dual number $\hat{z}$ is defined as $\hat{z} = a + \varepsilon b$ with the real part a, the dual part b, and the dual unit $\varepsilon^2 = 0$. In the same way, it is possible to define dual vectors, where a and b represent vectors instead of scalar values. With this, we can express a straight line that is represented in Plücker coordinates as a dual vector $\hat{l} = l + \varepsilon m$ with l and m being the direction and the moment of the line, respectively.

3.13.3.4 Dual Quaternions

To obtain the desired compact representation of

向量 l 和 m 被称为螺旋轴线的普吕克坐标。在三维空间中，穿过两个不同点 p 和 q 的直线可以通过如下公式转换到普吕克坐标系中等式为 $l = q - p$，$m = p \times l = p \times q$ 令 $l^\top m = 0$，$\|l\| = 1$。

此外一个由旋转矩阵 R 和平移向量 t 表示的刚性三维变换可以转换为一个螺旋，反之亦然。这里可以从旋转矩阵中求得向量 l 和螺旋角度 θ，详见（Hanson，2006）。螺旋平移距离 d 为 t 在 l 上的投影，即 $d = t^\top l$。最后，螺旋的矩 m=$m = \frac{1}{2}(t \times l + (t - dl)\cot\frac{\theta}{2})$（Daniilidis，1999）。相反，螺旋的参数 l 和 θ 可以转换为旋转矩阵，详见（Hanson，2006）。平移向量 t 可以通过公式 $t = (I - R)(l \times m) + dl$ 获得（Daniilidis，1999）。

3.13.3.3 对偶数

为了在第二步中能够有效地表示刚性三维变换，引入了对偶数概念。类似于复数，对偶数 $\hat{z}$ 被定义为 $\hat{z} = a + \varepsilon b$，其中 a 为实部，b 为对偶部，且对偶单位 $\varepsilon^2 = 0$。采用同样方式，只要将 a 和 b 两个标量替换为向量，即可得到对偶向量的定义。借助这个定义，就可以在普吕克坐标系下表示一条直线 $\hat{l} = l + \varepsilon m$，其中 l 和 m 分别表示直线的方向和直线的矩。

3.13.3.4 对偶四元数

为了获得紧凑的三维刚性变换

rigid 3D transformations, we finally combine the concept of dual numbers with the concept of quaternions to obtain dual quaternions. For a general introduction to dual quaternions, the interested reader is referred to Rooney (1978). A brief outline can be found in Daniilidis (1999).

A dual quaternion $\hat{\mathbf{q}} = \mathbf{q} + \varepsilon \mathbf{q}'$ consists of the quaternions $\mathbf{q}$ (real part) and $\mathbf{q}'$ (dual part). The rules for multiplication, conjugation, and inversion of quaternions also apply to dual quaternions, i.e., Eqs. (3.217), (3.218), and (3.221) also hold after replacing all quantities by their dual variants. Analogously, unit dual quaternions are dual quaternions that satisfy $\hat{\mathbf{q}}\bar{\hat{\mathbf{q}}} = 1$. Consequently, a dual quaternion $\hat{\mathbf{q}} = \mathbf{q} + \varepsilon \mathbf{q}'$ is a unit dual quaternion if the following two conditions hold:

的表示方法，我们最后结合对偶数以及四元数的概念提出新的数学概念，对偶四元数。对于对偶四元数的有关介绍感兴趣的读者请参见（Rooney, 1978）。可以在文章 Daniilidis（1999）找到此概念的概述。

一个对偶四元数 $\hat{\mathbf{q}} = \mathbf{q} + \varepsilon \mathbf{q}'$ 由四元数 $\mathbf{q}$（实部）和 $\mathbf{q}'$（对偶部）构成。四元数中的乘法、共轭、求逆运算同样适用于对偶四元数，例如，将等式（3.217）、（3.218）和（3.221）中的四元数都替换为与之对应的对偶形式，这些公式依然有效。同样，单位对偶四元数同样需要满足 $\hat{\mathbf{q}}\bar{\hat{\mathbf{q}}} = 1$。因此，如果 $\hat{\mathbf{q}} = \mathbf{q} + \varepsilon \mathbf{q}'$ 为一个单位对偶四元数，则需满足如下条件：

$$\begin{aligned} \mathbf{q}\bar{\mathbf{q}} &= 1 \\ \bar{\mathbf{q}}\mathbf{q}' + \mathbf{q}\bar{\mathbf{q}}' &= 0 \end{aligned} \qquad (3.227)$$

This means that for a unit dual quaternion, its real part is a unit quaternion that is orthogonal to its dual part. Also, the inverse of a unit dual quaternion is its conjugate

这意味着，一个单位对偶四元数，它的实部为一个单位四元数且正交于它的对偶部。同样，单位对偶四元数的逆为它的共轭

$$\hat{\mathbf{q}}^{-1} = \bar{\hat{\mathbf{q}}}$$

A dual quaternion with its real and dual part being pure quaternions, i.e., $q_0 = 0$ and $q'_0 = 0$, is called a pure dual quaternion. Like dual vectors, pure dual quaternions can be used to represent straight lines in Plücker coordinates:

这里称实部和对偶部均为纯四元数，即（$q_0 = 0$）且（$q'_0 = 0$）的对偶四元数为纯对偶四元数。在普吕克坐标系中，纯对偶数可用于表示直线：

$$\hat{\boldsymbol{l}} = (0, \hat{\boldsymbol{l}}) = (0, \boldsymbol{l}) + \varepsilon(0, \boldsymbol{m}) \qquad (3.228)$$

A screw, and hence a rigid 3D transformation, can be represented by a unit dual quaternion

一个螺旋，即一个刚性三维变换，可以表示为一个单位对偶四元数。

$$\hat{\mathbf{q}} = \begin{pmatrix} \cos\frac{\hat{\theta}}{2} \\ \hat{\boldsymbol{n}}\sin\frac{\hat{\theta}}{2} \end{pmatrix} = \begin{pmatrix} \cos\frac{\theta}{2} \\ \boldsymbol{n}\sin\frac{\theta}{2} \end{pmatrix} + \varepsilon \begin{pmatrix} -\frac{d}{2}\sin\frac{\theta}{2} \\ \boldsymbol{o}\sin\frac{\theta}{2} + \boldsymbol{n}\frac{d}{2}\cos\frac{\theta}{2} \end{pmatrix} \qquad (3.229)$$

with the dual number $\hat{\theta} = \theta + \varepsilon d$ representing the angle and the translation of the screw and the dual vector $\hat{\boldsymbol{n}} = \boldsymbol{n} + \varepsilon \boldsymbol{o}$ representing the direction vector $\boldsymbol{n}$ and the moment vector $\boldsymbol{o}$ of the screw axis (with $\|\boldsymbol{n}\| = 1$ and $\boldsymbol{n} \cdot \boldsymbol{o} = 0$). Then, a straight line $\hat{\mathbf{l}}$ can be transformed by a rigid 3D transformation $\hat{\mathbf{q}}$ as (Rooney, 1978; Daniilidis, 1999):

$$\hat{\mathbf{k}} = \hat{\mathbf{q}} \hat{\mathbf{l}} \bar{\hat{\mathbf{q}}}$$

A unit dual quaternion can be transformed into a homogeneous transformation matrix that represents a rigid 3D transformation, and vice versa (Daniilidis, 1999). For the transformation of lines, the dual quaternion representation is more efficient, and hence should be preferred over the representation with homogenous transformation matrices (Rooney, 1978). In the same way as the quaternions $\mathbf{q}$ and $-\mathbf{q}$ represent the same rotation, the dual quaternions $\hat{\mathbf{q}}$ and $-\hat{\mathbf{q}}$ represent the same rigid 3D transformation. Compared to homogenous transformation matrices (twelve elements), unit dual quaternions (eight elements) allow a more compact representation of rigid 3D transformations. Note that the eight parameters of a unit dual quaternion still overparameterize rigid 3D transformations, which have six degrees of freedom. The two remaining degrees of freedom are fixed by the unity constraint (see Eq.(3.227)).

The consecutive execution of two rigid 3D transformations can be represented by multiplying the two corresponding unit dual quaternions, by analogy to multiplying the two corresponding homogeneous transformation matrices.

Summing up, unit dual quaternions can be used to represent rigid 3D transformations in a similar compact way as unit quaternions can be

其中，对偶数 $\hat{\theta} = \theta + \varepsilon d$，用于表示螺旋的旋转角度和移动距离，对偶向量 $\hat{\boldsymbol{n}} = \boldsymbol{n} + \varepsilon \boldsymbol{o}$，用于表示螺旋轴的方向向量 $\boldsymbol{n}$ 以及矩向量 $\boldsymbol{o}$(其中 $\|\boldsymbol{n}\| = 1$ 且 $\boldsymbol{n} \cdot \boldsymbol{o} = 0$)。然后，一条直线 $\hat{\mathbf{l}}$ 可以通过刚性三维变换 $\hat{\mathbf{q}}$ 进行转换，详见文献（Rooney，1978；Daniilidis，1999）：

(3.230)

单位对偶四元数可以和用于表示刚性三维变换的齐次变换矩阵相互转换(Daniilidis, 1999)。对于直线的变换，对偶四元数表示更加高效。因此相比于齐次变换，在进行直线转换时，应优先考虑对偶四元数（Rooney, 1978）。与四元数中 $\mathbf{q}$ 和 $-\mathbf{q}$ 代表同一个旋转类似，$\hat{\mathbf{q}}$ 和 $-\hat{\mathbf{q}}$ 也表示同一个刚性三维变换。相比于齐次变换矩阵（12 个元素），单位对偶四元数（8 个元素）使得刚性三维变换的表示更加紧凑。值得注意的是，单位对偶四元数的 8 个参数对于刚性三维变换来说依然是过参数化的，因为这个变换只需要六个自由度即可。在对偶四元数的 8 个自由度中，有两个自由度由于使用了单位对偶四元数而受到了限制（见等式（3.227））。

两个连续的刚性三维变换可以表示为两个与它们对应的对偶四元数的乘法运算，就类似于齐次变换矩阵的乘运算。

综上所述，单位对偶四元数可用于表示刚性三维变换，就像单位四元数可以表示三维旋转一样。因此单位

used to represent 3D rotations. Therefore, unit dual quaternions are used in the linear hand–eye calibration approach that is introduced in the following section.

3.13.4 Linear Hand–Eye Calibration

In this section, we describe a linear approach for hand–eye calibration that is based on Daniilidis (1999). Because of the advantages mentioned above, it simultaneously solves for rotation and translation. This is accomplished by representing the poses as unit dual quaternions.

Equation (3.216) can be rewritten by using dual quaternions as $\hat{\mathbf{a}}\hat{\mathbf{x}} = \hat{\mathbf{x}}\hat{\mathbf{b}}$ or

$$\hat{\mathbf{a}} = \hat{\mathbf{x}}\hat{\mathbf{b}}\bar{\hat{\mathbf{x}}}$$

As we have seen before, the scalar part of the dual quaternion contains the angle and the translation of the corresponding screw. Because of Eq. (3.218), the scalar part Sc of a dual quaternion can be extracted by

$$\mathrm{Sc}(\hat{\mathbf{q}}) = (\hat{\mathbf{q}} + \bar{\hat{\mathbf{q}}})/2 \tag{3.232}$$

By extracting the scalar part of Eq. (3.231), we obtain

$$\mathrm{Sc}(\hat{\mathbf{a}}) = (\hat{\mathbf{a}} + \bar{\hat{\mathbf{a}}})/2 = (\hat{\mathbf{x}}\hat{\mathbf{b}}\bar{\hat{\mathbf{x}}} + \hat{\mathbf{x}}\bar{\hat{\mathbf{b}}}\bar{\hat{\mathbf{x}}})/2 = \\ (\hat{\mathbf{x}}(\hat{\mathbf{b}} + \bar{\hat{\mathbf{b}}})\bar{\hat{\mathbf{x}}})/2 = \hat{\mathbf{x}}\,\mathrm{Sc}(\hat{\mathbf{b}})\bar{\hat{\mathbf{x}}} = \mathrm{Sc}(\hat{\mathbf{b}})\hat{\mathbf{x}}\bar{\hat{\mathbf{x}}} = \mathrm{Sc}(\hat{\mathbf{b}}) \tag{3.233}$$

This is the fundamental result on which the linear approach to hand–eye calibration is based. It is also known as the screw congruence theorem (Chen, 1991), which says that the angle and the translation of the screw of the camera (for a moving camera) or of the calibration object (for a

对偶四元数可用于线性手眼标定算法中。下文中将为大家介绍这一算法。

3.13.4 线性手眼标定

在本章中，将为大家介绍，基于（Daniilidis，1999）论文中所提及的线性手眼标定算法原理。之所以选用这个算法，是因为它有着可以同时求解旋转和平移变换的优势。为了进行运算，需要将位姿表示为单位对偶四元数。

引入对偶四元数后，等式（3.126）可被写为 $\hat{\mathbf{a}}\hat{\mathbf{x}} = \hat{\mathbf{x}}\hat{\mathbf{b}}$ 或

$$\tag{3.231}$$

正如之前看到的，对偶四元数的标量部分包含了螺旋的角度和平移。由于存在等式（3.218），对偶四元数的标量部分 Sc 可通过下式获取

通过提取等式（3.231）的标量部分可得

这正是线性手眼标定方法的原理基础。它被称为螺旋一致定理。该定理指出，在两个机械手位姿之间，（对于运动摄像机）摄像机或（对于固定摄像机）标定板螺旋的角度和平移与工具螺旋的角度和平移是一致的。因此，

stationary camera) is identical to the angle and the translation of the screw of the tool between two robot poses. Consequently, only the vector part of Eq. (3.231) contributes to the determination of the unknown $\hat{\mathbf{x}}$:

$$\begin{pmatrix} 0 \\ \hat{\boldsymbol{a}} \sin \hat{\theta}_{\hat{\mathbf{a}}} \end{pmatrix} = \hat{\mathbf{x}} \begin{pmatrix} 0 \\ \hat{\boldsymbol{b}} \sin \hat{\theta}_{\hat{\mathbf{b}}} \end{pmatrix} \bar{\hat{\mathbf{x}}} \qquad (3.234)$$

where the dual vectors $\hat{\boldsymbol{a}}$ and $\hat{\boldsymbol{b}}$ represent the screw axes and the dual numbers $\hat{\theta}_{\hat{\mathbf{a}}}$ and $\hat{\theta}_{\hat{\mathbf{b}}}$ represent the angle and translation of the screws. Because of the screw congruence theorem, $\hat{\theta}_{\hat{\mathbf{a}}} = \hat{\theta}_{\hat{\mathbf{b}}}$. If the screw angles are not 0° or 360°, we obtain

$$\begin{pmatrix} 0 \\ \hat{\boldsymbol{a}} \end{pmatrix} = \hat{\mathbf{x}} \begin{pmatrix} 0 \\ \hat{\boldsymbol{b}} \end{pmatrix} \bar{\hat{\mathbf{x}}} \qquad (3.235)$$

Together with Eq. (3.230), this equation tells us that the hand–eye calibration is equivalent to finding the rigid 3D transformation $\hat{\mathbf{x}}$ that aligns the straight lines that represent the screw axes of the camera (for a moving camera) or of the calibration object (for a stationary camera) with the straight lines that represent the screw axes of the tool. Note that for each pair of robot poses, one pair of corresponding straight lines is obtained. To robustly solve the problem, the following requirements must be taken into account (Chen, 1991):

- At least two non-parallel screw axes are necessary to uniquely solve the hand–eye calibration.
- For only a single screw axis (i.e., two robot poses), the angle and the translation of the unknown screw $\hat{\mathbf{x}}$ cannot be determined.
- If all screw axes are parallel, the translation of the unknown screw $\hat{\mathbf{x}}$ cannot be determined.

只有等式（3.231）的矢量部分可以帮助我们确定未知项 $\hat{\mathbf{x}}$：

上式中，对偶向量 $\hat{\boldsymbol{a}}$ 和 $\hat{\boldsymbol{b}}$ 表示螺旋轴，对偶数 $\hat{\theta}_{\hat{\mathbf{a}}}$ 和 $\hat{\theta}_{\hat{\mathbf{b}}}$ 表示螺旋的角度和平移。由于螺旋一致定理，$\hat{\theta}_{\hat{\mathbf{a}}} = \hat{\theta}_{\hat{\mathbf{b}}}$。如果螺旋角度不等于 0° 或 360°，可得

结合等式（3.230），由上式可以看出，手眼标定等价于寻找一个刚性三维变换 $\hat{\mathbf{x}}$，使得表示摄像机（对于运动摄像机）或标定板（对于固定摄像机）的螺旋轴的直线与表示工具的螺旋轴直线对齐。值得注意的是，对于每一组机器人位姿都存在一一对应的直线。为了能够很好地解决这一问题，必须考虑如下情况（Chen，1991）：

- 至少需要两个不平行的螺旋轴才能确定手眼标定的唯一解。
- 对于仅已知一个螺旋轴的情况（例如，只有两个机器人位姿），无法确定未知螺旋 $\hat{\mathbf{x}}$ 中的平移和角度。
- 如果所有的螺旋轴都平行，无法确定未知螺旋 $\hat{\mathbf{x}}$ 中的平移参数。

- Further degenerate cases occur if the screw angles $\theta_{\hat{\mathbf{a}}} = \theta_{\hat{\mathbf{b}}} = 0°$, or if the screw angles are $\theta_{\hat{\mathbf{a}}} = \theta_{\hat{\mathbf{b}}} = 180°$ and the screw translations are $d_{\hat{\mathbf{a}}} = d_{\hat{\mathbf{b}}} = 0$.

- 此外，还需考虑参数发生退化的情况，$\theta_{\hat{\mathbf{a}}} = \theta_{\hat{\mathbf{b}}} = 0°$ 或 $\theta_{\hat{\mathbf{a}}} = \theta_{\hat{\mathbf{b}}} = 180°$ 或 $d_{\hat{\mathbf{a}}} = d_{\hat{\mathbf{b}}} = 0$。

To determine the unknown $\hat{\mathbf{x}}$, let us write a dual quaternion as $\hat{\mathbf{q}} = \mathbf{q} + \varepsilon \mathbf{q}'$ with the real part $\mathbf{q}$ and the dual part $\mathbf{q}'$. Then, we can rewrite Eq. (3.231) by separating the real and the dual part:

为了确定未知项 $\hat{\mathbf{x}}$，将对偶四元数写作 $\hat{\mathbf{q}} = \mathbf{q} + \varepsilon \mathbf{q}'$，其中 $\mathbf{q}$ 为实部，$\mathbf{q}'$ 为对偶部。随后，可以将等式（3.231）的实部和对偶部拆分：

$$\mathbf{a} = \mathbf{x}\mathbf{b}\bar{\mathbf{x}} \tag{3.236}$$

$$\mathbf{a}' = \mathbf{x}\mathbf{b}\bar{\mathbf{x}}' + \mathbf{x}\mathbf{b}'\bar{\mathbf{x}} + \mathbf{x}'\mathbf{b}\bar{\mathbf{x}} \tag{3.237}$$

Let $\mathbf{a} = (0, \boldsymbol{a})$, $\mathbf{a}' = (0, \boldsymbol{a}')$, $\mathbf{b} = (0, \boldsymbol{b})$, and $\mathbf{b}' = (0, \boldsymbol{b}')$. It can be shown that the two equations (3.236) and (3.237) can be written in matrix form as (Daniilidis, 1999):

令 $\mathbf{a} = (0, \boldsymbol{a})$, $\mathbf{a}' = (0, \boldsymbol{a}')$, $\mathbf{b} = (0, \boldsymbol{b})$, $\mathbf{b}' = (0, \boldsymbol{b}')$。则等式（3.236）和等式（3.237）可转换为矩阵形式（Daniilidis，1999）：

$$\begin{pmatrix} \boldsymbol{a} - \boldsymbol{b} & [\boldsymbol{a} + \boldsymbol{b}]_\times & 0 & 0 \\ \boldsymbol{a}' - \boldsymbol{b}' & [\boldsymbol{a}' + \boldsymbol{b}']_\times & \boldsymbol{a} - \boldsymbol{b} & [\boldsymbol{a} + \boldsymbol{b}]_\times \end{pmatrix} \begin{pmatrix} \mathbf{x} \\ \mathbf{x}' \end{pmatrix} = 0 \tag{3.238}$$

where $[\boldsymbol{a}]_\times$ denotes the antisymmetric 3×3 matrix of the vector $\boldsymbol{a}$, which corresponds to the cross product with $\boldsymbol{a}$, i.e., $[\boldsymbol{a}]_\times \boldsymbol{c} = \boldsymbol{a} \times \boldsymbol{c}$. Thus, for each pair of robot poses, i.e., for each robot movement i, we obtain a 6×8 matrix, which we will denote by S_i. For m robot movements, we obtain a $6m \times 8$ matrix $\boldsymbol{T}$ by stacking the matrices S_i:

其中，$[\boldsymbol{a}]_\times$ 为向量 $\boldsymbol{a}$ 的 3×3 反对称矩阵。该运算相当于 $\boldsymbol{a}$ 的叉乘，例如 $[\boldsymbol{a}]_\times \boldsymbol{c} = \boldsymbol{a} \times \boldsymbol{c}$。由此，对于每一对位姿，机器人的每个运动 i 都可以获得一个 6×8 矩阵。用 S_i 表示该矩阵，经过 m 次运动后，通过叠加矩阵 S_i，就得到了一个 $6m \times 8$ 矩阵 $\boldsymbol{T}$：

$$\boldsymbol{T} = (S_1^\top, S_2^\top, \cdots, S_m^\top)^\top \tag{3.239}$$

This corresponds to a homogeneous linear equation system with $6m$ equations and 8 unknowns. Although with at least two robot movements the number of equations would already exceed the number of unknowns, there would still be no unique solution. The reason is that the rank of

这相当于一个包含 $6m$ 个方程和 8 个未知数的线性齐次方程组。虽然存在两次以上的机器人运动可以使得方程的数目大于未知数的数量，但这里还是无法获得唯一解。因为 $\boldsymbol{T}$ 矩阵的秩为 6（在无噪声的情况下）。这个所谓

T is only 6 (in the noise-free case). This so-called datum deficiency or gauge freedom results in a 2D manifold of solutions. The manifold can be calculated by computing the singular value decomposition (SVD) of T = UDV$^\top$ (Press *et al.*, 2007). The diagonal elements in D contain the singular values σ_i, with $i = 1, \cdots, 8$, in descending order. Because of the datum deficiency, the last two singular values σ_7 and σ_8 are zero in the ideal noise-free case and close to zero in practical applications with noisy input data. The manifold of solutions is spanned by the last two columns $\boldsymbol{v}_7$ and $\boldsymbol{v}_8$ of the matrix V, which correspond to the vanishing singular values:

的数据不足或规范自由度导致了一个二维流形的解。可以通过 T = UDV$^\top$ 的奇异值分解（SVD）来计算这种多样性。D 矩阵中对角线上的元素包含有降序排列的奇异值 σ_i，其中 $i = 1, \cdots, 8$。在理想无噪声的状态下，或是在实际应用中输入数据噪声接近 0 的情况下，由于数据缺失，最后两个奇异值 σ_7 和 σ_8 为 0。多解的问题正是由于消失的两个奇异值在矩阵 V 中对应列 $\boldsymbol{v}_7$ 和 $\boldsymbol{v}_8$ 引起的：

$$\begin{pmatrix} \mathbf{x} \\ \mathbf{x}' \end{pmatrix} = \lambda_1 \boldsymbol{v}_7 + \lambda_2 \boldsymbol{v}_8 = \lambda_1 \begin{pmatrix} \boldsymbol{u}_1 \\ \boldsymbol{v}_1 \end{pmatrix} + \lambda_2 \begin{pmatrix} \boldsymbol{u}_2 \\ \boldsymbol{v}_2 \end{pmatrix} \quad (3.240)$$

To reduce the set of solutions to a single solution, we must apply two additional constraints. Since we know that the resulting dual quaternion $\hat{\mathbf{x}}$ represents a rigid 3D transformation, and hence must be a unit dual quaternion, we apply the two unity constraints of Eq. (3.227) to Eq. (3.240):

为了将解的数目变为一个，我们增加两个附加条件。由于我们知道所求对偶四元数结果 $\hat{\mathbf{x}}$ 是一个刚性三维变换，所以它一定是一个单位对偶四元数，可以将等式（3.227）中的两个约束条件应用于等式（3.240）中：

$$\lambda_1^2 \boldsymbol{u}_1^\top \boldsymbol{u}_1 + 2\lambda_1 \lambda_2 \boldsymbol{u}_1^\top \boldsymbol{u}_2 + \lambda_2^2 \boldsymbol{u}_2^\top \boldsymbol{u}_2 = 1 \quad (3.241)$$

$$\lambda_1^2 \boldsymbol{u}_1^\top \boldsymbol{v}_1 + \lambda_1 \lambda_2 (\boldsymbol{u}_1^\top \boldsymbol{v}_2 + \boldsymbol{u}_2^\top \boldsymbol{v}_1) + \lambda_2^2 \boldsymbol{u}_2^\top \boldsymbol{v}_2 = 0 \quad (3.242)$$

The two quadratic polynomials in the two unknowns λ_1 and λ_2 yield two solutions, from which the correct solution can be easily selected (Daniilidis, 1999). Note that because of Eqs. (3.241) and (3.242), the approach is actually not completely linear. Nevertheless, it provides a direct unique solution. Therefore, we call it linear in order to distinguish it from the nonlinear optimization that we will describe in Section 3.13.5.

这个二次方程关于未知数 λ_1 和 λ_2，并会产生两个解，这使得很容易对正确的解进行选择（Daniilidis, 1999）。值得注意的是由于等式（3.241）和等式（3.242）的引入，实际上该算法已经不是一个纯粹的线性方程了。但是，它却能够提供一个直接的唯一解。因此，称之为线性的主要原因是希望能够区别于 3.13.5 节中将要介绍的非线性算法。

To obtain the highest robustness and accuracy of the linear approach, the following should be considered (Tsai and Lenz, 1989):

- The robot's kinematics should be calibrated, i.e., the pose of the tool relative to the robot base should be known with high accuracy.
- The angle between the screw axes of different robot movements should be as large as possible.
- The rotation angle of the screw of a single robot movement should be as large as possible or should differ from 0° and 180° as much as possible for screws with a small translation.
- The distance between the projection center of the camera and the calibration object should be as small as possible.
- The distance between the tool center points of two robot poses should be as small as possible.
- Increasing the number of robot poses increases the accuracy.

Some of these criteria are difficult to consider for a non-expert user. Therefore, from the set of robot poses, the pose pairs (movements) that best fulfill the criteria can be automatically selected (Schmidt et al., 2003; Schmidt and Niemann, 2008). Nevertheless, the user must ensure that enough valid pose pairs are available that can be selected. In most practical applications, 10 to 15 valid robot poses are enough to obtain a sufficiently high accuracy.

Furthermore, providing incorrect input poses for hand–eye calibration is a frequent source of errors. Often, the robot controller returns the robot poses in a different format or in different units compared to the calibration object poses that are

为了获得更高的鲁棒性和精度还需要考虑如下几点（Tsai and Lenz, 1989）：

- 机器人需要经过运动学标定，例如，这里需要已知工具关于机器人基座的精确位置。
- 需要尽量增大不同的机器人运动所对应的螺旋轴的夹角。
- 在单次机器人运动中，平移较小时，需要确保螺旋的旋转角度足够大，或者说使得该角度与 0° 和 180° 差异足够大。
- 摄像机的投影中心和标定板的投影中心距离尽量小。
- 机械手运动中相邻的两个位姿工具中心点的距离尽量小。
- 增加机械手的位姿有助于提高精度。

对于普通用户而言，有些条件是很难实现的。因此，可以从一组机器人位姿中对满足条件的位姿（运动）进行自动选择（Schmidt et al., 2003; Schmidt and Niemann, 2008）。在大多数实际应用中，一般需要 10 到 15 个有效的机器人位姿就足以获得高精度的标定结果。

此外，在手眼标定过程中，常常会为算法提供错误的位姿。经常会碰到机器人控制器返回的位姿格式或者单位与算法不同的情况，我们希望通过标定算法进行识别并反馈给客户。

returned by the camera calibration. Also, providing the correct pairs of corresponding robot and calibration object poses is sometimes difficult. To support the user in this regard, the consistency of the input poses can be checked automatically. For this, the user must provide the approximate accuracy of the input poses, i.e., the approximate standard deviation of the rotation and the translation part. Then, for each pair of robot poses, the two dual quaternions that represent the movement of the robot tool and of the calibration object (for a stationary camera) or of the camera (for a moving camera) is computed. Because of the screw congruence theorem of Eq. (3.233), the scalar parts of both dual quaternions must be equal. Based on the provided standard deviations of the input poses, the standard deviation of the scalar parts can be derived through error propagation. From this, a confidence interval is computed by multiplying the standard deviation by an appropriate factor, e.g., 3.0. If the scalar parts differ by more than this value, a warning is issued.

同样，标定过程中，有时可能会误操作，导致获取的机器人位姿和拍摄到标定板图像的不匹配。为了解决上述的问题，需要算法有自动检测输入位姿一致性的功能。为此，用户需要输入近似精度，即旋转和平移部分的近似标准偏差。此后，对于每一对机器人位姿，算法将计算表示机器人工具运动的对偶四元数以及表示标定板（固定摄像机标定）或摄像机（运动摄像机标定）相对运动的对偶四元数。由等式（3.233）中的螺旋一致定理得出，上述两对偶四元数的标量部分相等。基于用户提供的标准输入位姿偏差，通过误差传递，可以推导出标量部分的标准偏差。自此，可以通过标准偏差乘以一个特定的系数（例如3.0）生成一个置信区间。如果标量部分超出这一数值则进行报警。

3.13.5 Nonlinear Hand–Eye Calibration

To further increase the accuracy, the resulting poses of the linear approach can be used as initial values in a nonlinear optimization framework. For this, we solve the fundamental equation (3.213) for B_i and compute the following error matrix

$$E_i = B_i - X^{-1} A_i^{-1} Y \qquad (3.243)$$

For the optimal solution, the norm of the elements in E_i must be minimized. Therefore, we minimize

3.13.5 非线性手眼标定

为了进一步提高精度，可以将线性算法的处理结果作为非线性框架的初始值进行非线性优化。为此，需要对关于 B_i 的理论等式（见等式（3.213））进行求解，并计算误差矩阵。

为了获得最优解，需要使得 E_i 中元素的范数最小。因此，对于 n 个机器人位姿不同的 X 和 Y 转换的 2×6 的未知位姿参数，需要最小化

$$e = \sum_{i=1}^{n} \mathrm{Tr}(\mathrm{E}_i \mathbf{W} \mathrm{E}_i^\top) \tag{3.244}$$

over the 2×6 unknown pose parameters of the transformations X and Y for all n robot poses. The matrix $\mathrm{W} = \mathrm{Diag}(1,1,1,9)$ balances the different number of entries in E for rotation (9 entries) and translation (3 entries). (The operator Diag constructs a diagonal matrix with the specified entries.) Additionally, the translation part of all input matrices is scaled by $1/d$, where d is the maximum extent of the workspace defined by the robot poses. This causes the results to be scale-invariant and the errors in translation and rotation to be weighted appropriately with respect to each other. A similar approach is proposed by Dornaika and Horaud (1998), who also apply different weights to the rotation and translation part of the error function.

Both approaches, the linear approach of Section 3.13.4 and the nonlinear approach minimize an algebraic error, which does not have a direct geometric meaning. An alternative approach is proposed by Strobl and Hirzinger (2006), which minimizes the weighted sum of the rotation and translation error. They close the chain of transformations of Eq. (3.213), yielding $A_i X B_i Y^{-1} = I$. Based on this equation, they compute a scalar value for the rotation and the translation error. The appealing advantage of the approach is that the weights that balance both error types are statistically computed. However, in our experience, this approach does not return more accurate results than the approach described above.

由于 E 包含旋转（9 项）和平移（3 项），矩阵 W = Diag(1,1,1,9) 用于平衡 E 中不同的项数。（算子 Diag 用于构造一个对角线为特定元素的对角矩阵。）此外，所有输入的平移部分都会使用 $1/d$ 进行缩放，其中 d 为空间中最大运动范围，由机器人位姿确定。这将使得处理结果具有尺度不变性，同时平移和旋转误差会相互进行加权。这类似于文献（Dornaika and Horaud，1998）提出的对旋转和平移误差函数采用不同权值的算法。

无论是 3.13.4 节中的线性算法还是本节中的非线性算法都是求一个最小化的代数误差，这使得这个误差并没有直接的几何意义。文献（Strobl and Hirzinger，2006）提出了一种替代算法，该算法可以最小化旋转和平移误差的加权和。他们提出，闭环等式（3.213）的转换约束，在运算过程中服从等式 $A_i X B_i Y^{-1} = I$。因此，基于这个等式，可以分别计算旋转和平移的标量误差。该算法引人注目的地方在于，引入平衡两种误差的权值进行统计计算。然而，在实际测试中，使用该算法并没有比这两节提到的算法获得更高的标定精度。

3.13.6 Hand–Eye Calibration of SCARA Robots

In many industrial applications, SCARA (Selective Compliant Arm for Robot Assembly) robots (ISO 8373:2012; ISO 9787:2013) are used instead of articulated robots. SCARA robots have at least two parallel rotary joints and one parallel prismatic joint. Figure 3.176 shows a model of a SCARA robot with three rotary joints. In contrast to articulated robots, which cover six degrees of freedom (three translations and three rotations), SCARA robots only cover four degrees of freedom (three translations and one rotation). Compared to articulated robots, they offer faster and more precise performance and typically have a more compact design. Therefore, they are best suited for pick-and-place, assembly, and packaging applications, and are preferred if only limited space is available.

3.13.6 SCARA 机器人手眼标定

在很多工业应用中，SCARA（Selective Compliant Arm for Robot Assembly）机器人（ISO 8373:2012；ISO 9787:2013）常常用于替代多关节机器人。SCARA 机器人拥有至少两个平行的旋转关节和一个柱状关节。图 3.176 为一个包含三个旋转关节的 SCARA 机器人模型。相比于覆盖 6 自由度（三个平移和三个旋转自由度）的多关节机器人，SCARA 机器人只覆盖了 4 个自由度（三个平移和一个旋转自由度）。相比于多关节机器人，它们在性能上更快、更精确，在结构上更加紧凑。因此它们非常适用于搬运、装配、包装这些应用场景，除此之外它们还适用于比较狭小的工作空间。

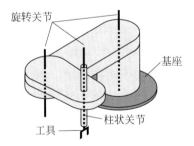

图 3.176　一个拥有三个平行旋转关节和一个平行柱状关节的 SCARA 机器人。图中旋转轴线用粗线标出

Tsai and Lenz (1989) showed that the error in hand–eye calibration is inversely proportional to the sine of the angle between the screw axes of the robot movement. For SCARA robots, all screw axes are parallel because all rotation axes are parallel, and hence the error would be infinitely large. Furthermore, Chen (1991) showed that one

文献（Tsai and Lenz, 1989）中提出，手眼标定的误差反比于机器人多次运动所形成的螺旋轴夹角的正弦。对于 SCARA 机器人，其运动过程中所有螺旋的轴都是平行的，因为从机械接口来看，所有的旋转轴都是平行的，这也就意味着标定误差将会无限

parameter cannot be determined by hand–eye calibration if all screw axes are parallel. Therefore, the linear approach of Section 3.13.4 for calibrating articulated robots does not work for SCARA robots. However, it can be extended for the calibration of SCARA robots as described in Ulrich and Steger (2016).

Because for SCARA robots all screw axes are parallel, the rank of the matrix T in Eq. (3.239) is further reduced by one, and hence is only five in the noise-free case. This results in three vanishing singular values. The three corresponding singular values v_6, v_7, and v_8, span the set of solutions:

$$\begin{pmatrix} \mathbf{x} \\ \mathbf{x}' \end{pmatrix} = \lambda_1 v_6 + \lambda_2 v_7 + \lambda_3 v_8 \qquad (3.245)$$

Without loss of generality, we assume that all joint axes of the SCARA robot are parallel to the z-axis of the robot's BCS and TCS. From Zhuang (1998) we know that the translation component t_z of the unknown transformation $\hat{\mathbf{x}}$ cannot be determined uniquely, and hence there exists a 1D manifold of equivalent solutions. From this manifold, we arbitrarily select one solution by temporarily requiring $t_z = 0$. Then, we can compute the unknown $\hat{\mathbf{x}}$ and finally determine the real t_z in an additional post-processing step. In the following, this approach is described in more detail.

By inserting Eq. (3.245) into the two unity constraints of Eq. (3.227), we obtain two quadratic equations in the three unknowns λ_1, λ_2, and λ_3 (Ulrich and Steger, 2016). The translation vector t of $\hat{\mathbf{x}}$ can be extracted as $(0, t) = 2\mathbf{x}\bar{\mathbf{x}}'$. Then, the third constraint $t_z = 0$ is formulated as

$$x_1 x_4' + x_2 x_3' - x_3 x_2' - x_4 x_1' = 0 \qquad (3.246)$$

大。此外，在标定过程中文献（Chen, 1991）指出，如果所有螺旋轴都平行，则无法求解单个参数。因此，3.13.4 节中所描述的针对多关节机器人的线性标定算法无法应用于 SCARA 机器人。然而这里可以对（Ulrich and Steger, 2016）所描述的方法进行扩展，并应用于 SCARA 机器人。

由于 SCARA 机器人所有螺旋轴都是平行的，因此在忽略误差的情况下，等式（3.239）中 T 矩阵的秩进一步下降，变为了 5。这就导致了有三个奇异值将会消失。这三个消失的奇异值将会使方程产生一组解。

一般情况下，假设 SCARA 机器人所有关节的轴都平行于机器人的 BCS 和 TCS 的 z 轴。从（Zhuang, 1998）的论文中可以得知，在这种情况下，未知转换对偶四元数 $\hat{\mathbf{x}}$ 的平移部分 t_z 是没有唯一解的，因此这里存在一个一维的等价解。从这个解中任意选择一项满足 $t_z = 0$ 的解，然后计算未知项 $\hat{\mathbf{x}}$，最终再经过后续算法确定真正的 t_z。在下文中会为大家描述这个算法的细节。

将等式（3.245）代入到两个单位约束等式（3.227）中，可得到两个包含三个未知数 λ_1、λ_2 和 λ_3 的二次方程。提取 $\hat{\mathbf{x}}$ 的平移向量 t 得 $(0, t) = 2\mathbf{x}\bar{\mathbf{x}}'$。随后，$t_z = 0$ 的约束条件可表示为如下方程

where $x_1, \cdots, x_4$ and $x'_1, \cdots, x'_4$ denote the elements of $\mathbf{x}$ and $\mathbf{x'}$, respectively. By inserting Eq. (3.245) into Eq. (3.246), we obtain a third quadratic equation in the three unknowns λ_1, λ_2, and λ_3 (Ulrich and Steger, 2016). One of the three equations represents an ellipsoid; the other two represent elliptic cones. The up to eight solutions correspond to the intersection points of the three quadrics. Because the exact computation of the solutions would be cumbersome and computationally expensive, Ulrich and Steger (2016) present a method that efficiently computes an approximation of the solution that is shown to be adequate for practical applications.

Also, the criteria for selecting suitable pose pairs, which were listed at the end of Section 3.13.4, must be slightly modified for SCARA robots. Because all screw axes are parallel, we cannot require that the angle between the screw axes should be as large as possible. Instead, we require that the difference of the scalar parts of the dual quaternions that represent two corresponding screws $\hat{\mathbf{a}}$ and $\hat{\mathbf{b}}$ should be as small as possible (Ulrich and Steger, 2016). Remember that the scalar parts of $\hat{\mathbf{a}}$ and $\hat{\mathbf{b}}$ represent the angle and the translation of the screws, and hence must be identical in the case of perfect noise-free data (see Eq. 3.233).

For further practical implementation issues like the handling of the sign ambiguity of screws and the treatment of antiparallel screw axes, the interested reader is referred to Ulrich and Steger (2016).

We can refine the solution that is obtained by the linear approach by a subsequent nonlinear optimization, similar to the approach for articulated

其中 $x_1, \cdots, x_4$ 和 $x'_1, \cdots, x'_4$ 表示 $\mathbf{x}$ 和 $\mathbf{x'}$ 的元素。将等式（3.245）代入等式（3.246），便可得到第三个包含三个未知数 λ_1、λ_2 和 λ_3 的二次方程（Ulrich and Steger, 2016）。其中一个方程表示椭球，另外两个表示椭圆锥。在这里会获得八个方程的解，他们分别对应了这三个形状的焦点。由于精确计算这些解是十分烦琐又耗时，（Ulrich and Steger，2016）提出了更适合实际应用的快速近似求解算法。

同样，在 3.13.4 节末尾所列举的位姿对筛选规则，在应用于 SCARA 机器人时也需要进行相应的调整。由于所有螺旋轴都是平行的，无法要求他们之间的角度尽量大。而是希望表示两个螺旋的对偶四元数 $\hat{\mathbf{a}}$ 和 $\hat{\mathbf{b}}$ 的标量部分相差尽量小 (Ulrich and Steger, 2016)。因为，$\hat{\mathbf{a}}$ 和 $\hat{\mathbf{b}}$ 的标量部分包含螺旋的角度和平移，因此在没有误差的情况下他们是一致的（见等式（3.233））。

在实际应用还会碰到更多的问题，诸如螺旋方向不一致的处理以及反向平行螺旋轴的处理等。对于这些问题感兴趣的读者可参见（Ulrich and Steger，2016）。

就像 3.13.5 节中对多关节机器人进行的处理一样，在获得线性解之后，还可以通过非线性算法对它进行优

robots that was described in Section 3.13.5. However, because t_z cannot be determined, the minimization only optimizes the remaining 11 pose parameters of the two unknown poses, while keeping t_z fixed.

In the last step, the real value of t_z must be determined. Let us first consider a scenario with a stationary camera. From the calibration process, $^t\mathrm{H}_o$ and $^b\mathrm{H}_c$ of Eq. (3.211) are known up to t_z of $^t\mathrm{H}_o$. To determine t_z, we detach the calibration object from the robot and place it at a position from which it is observable by the camera. Then, the pose $^c\tilde{\mathrm{H}}_o$ of the calibration object in the CCS is automatically determined. From this, we compute z_{calib}, which is the z component of the translation of $^b\tilde{\mathrm{H}}_o = {}^b\mathrm{H}_c{}^c\tilde{\mathrm{H}}_o$. Next, we manually move the tool of the robot to the origin of the calibration object and query the robot pose to obtain $^t\tilde{\mathrm{H}}_b$. The z component of the translation of $^b\tilde{\mathrm{H}}_t$ represents the true translation, which we denote z_{true}. The values of z_{true} and z_{calib} must be identical because they represent the same physical distance. We can achieve this by modifying the z component of $^t\mathrm{H}_o$ by $z_{\text{true}} - z_{\text{calib}}$. Finally, the desired matrix $^b\mathrm{H}_c$ is computed with Eq. (3.211).

The scenario of a moving camera can be treated similarly. However, for some setups, it is impossible for the camera to observe the calibration object if the tool is moved to the origin of the calibration object. Let us assume that $^b\mathrm{H}_o$ and $^c\mathrm{H}_t$ are known from the calibration up to the z component of the translation part of $^b\mathrm{H}_o$. In this case, the robot is manually moved to two poses. First, the tool is moved such that the camera can observe the calibration object. Now, an image of the calibration object is acquired and the tool pose

化。然而，由于 t_z 无法确定，在 t_z 不变的情况下，只能优化两个未知位姿中的 11 个参数。

在最后一步中，会计算真正的 t_z。首先，假设有一个固定摄像机的场景。在标定过程中，等式（3.211）中的 $^t\mathrm{H}_o$ 和 $^b\mathrm{H}_c$ 是已知的，包括 $^t\mathrm{H}_o$ 中的 t_z。为了确定 t_z，将标定板从机器人系统中分离出来，并将其置于一个摄像机可以拍摄到的位置。而后，就可以获得标定板在 CCS 坐标系下的位置关系 $^c\tilde{\mathrm{H}}_o$。通过该变量，可以计算平移中的 z 分量 $^b\tilde{\mathrm{H}}_o = {}^b\mathrm{H}_c{}^c\tilde{\mathrm{H}}_o$。在此之后，手动将机器人的工具移动到标定板的原点上并获取机器人位姿 $^t\tilde{\mathrm{H}}_b$。将这个实际测得的平移矩阵定义为 $^b\tilde{\mathrm{H}}_t$，并将其分量 z 定义为 z_{true}。理论上来说 z_{true} 和 z_{calib} 是相等的，因为它们代表同样的物理距离。这里可以通过 $z_{\text{true}} - z_{\text{calib}}$ 获得 $^t\mathrm{H}_o$ 在 z 分量上的差异。最后通过等式（3.211）计算矩阵 $^b\mathrm{H}_c$。

对于运动摄像机的场景也可以使用上述思路进行解决。然而，受限于安装条件，在某些情况下当工具移动到标定板原点时摄像机无法拍摄到标定板。假设 $^b\mathrm{H}_o$ 和 $^c\mathrm{H}_t$ 可以通过标定获取，包括 $^b\mathrm{H}_o$ 平移部分的 z 分量。在这种情况下，需要手动设置两个机器人位姿。首先，将机器人移动到摄像机可以拍摄到标定板的位置，记录下该位置标定板图像和机器人工具的位姿，由此可获得矩阵 $^c\tilde{\mathrm{H}}_o$ 和 $^b\tilde{\mathrm{H}}_t$。

is queried, which gives us $^c\tilde{H}_o$ and $^b\tilde{H}_t$. Second, the tool of the robot is moved to the origin of the calibration object, yielding $^b\check{H}_t$. The z component of the translation of $^t\tilde{H}_b{}^b\check{H}_t$ is given by z_{true}, while z_{calib} is the z component of the translation of $^t{H}_c{}^c\tilde{H}_o$. Again, $z_{\mathrm{true}} - z_{\mathrm{calib}}$ can be used to correct the z component of $^b{H}_o$.

Actually, to determine t_z, the tool does not need to be moved to the origin of the calibration object. Instead, it is sufficient to move the tool to a point that has the same z coordinate as the origin of the calibration object, where the z coordinate is measured in the robot BCS. Sometimes, however, the origin or even a point with the same z coordinate cannot be reached by the tool. In this case, the tool should be moved to a point with known height, i.e., vertical distance in the z direction of the BCS, above or below the origin. The z component of the transformation must additionally be corrected by this height.

3.14 Optical Character Recognition

In quite a few applications, we face the challenge of having to read characters on the object we are inspecting. For example, traceability requirements often lead to the fact that the objects to be inspected are labeled with a serial number and that we must read this serial number (see, for example, Figures 3.25–3.27). In other applications, reading a serial number might be necessary to control production flow.

OCR is the process of reading characters in images. It consists of two tasks: segmentation of the individual characters and classification of

接下来，将机器人的工具移动到标定板的原点并获得矩阵 $^b\check{H}_t$。可以通过 $^t\tilde{H}_b{}^b\check{H}_t$ 平移部分的 z 分量获取 z_{true}，通过 z_{calib} 平移部分的 z 分量获得 $^t{H}_c{}^c\tilde{H}_o$。最后，再次使用 $z_{\mathrm{true}} - z_{\mathrm{calib}}$ 计算矩阵 $^b{H}_o$。

实际上，获取 t_z 并不需要将机器人的工具移动到标定板的原点上。只要能够在 BCS 坐标系下，将工具移动到与标定板原点拥有相同的 z 坐标的位置即可。但有时候由于机构的限制，工具无法到达指定的位置。在这种情况下，工具需要移动到原点之上或之下的一个指定高度。即在 BCS 坐标系下一个指定的 z 坐标。如此，在计算平移 z 分量的同时还需要使用上述的高度进行校正。

3.14 光学字符识别 (OCR)

在非常多的应用中都需要将检测对象上印刷的字符识别出来。例如，产品的可追溯性经常需要在每个产品上贴上一个序列号，因此我们必须读取这个序列号（例如图 3.25～图 3.27）。在其他某些应用中，可能必须通过读取序列号来控制生产流程。

光学字符识别（OCR）是在图像中识别字符的过程。它包含两个任务：将图像中单个字符分割出来以及将分

the segmented characters, i.e., the assignment of a symbolic label to the segmented regions. We will examine the segmentation and feature extraction of the characters in this section. The classification will be described in Section 3.15.

3.14.1 Character Segmentation

The classification of the characters requires that we have segmented the text into individual characters, i.e., each character must correspond to exactly one region.

To segment the characters, we can use all the methods that we have discussed in Section 3.4: thresholding with fixed and automatically selected thresholds, dynamic thresholding, and the extraction of connected components.

Furthermore, we might have to use the morphological operations of Section 3.6 to connect separate parts of the same character, e.g., the dot of the character "i" to its main part (see Section 3.6.1.2) or parts of the same character that are disconnected, e.g., because of bad print quality. For characters on difficult surfaces, e.g., punched characters on a metal surface, gray value morphology may be necessary to segment the characters (see Section 3.6.2.4).

Additionally, in some applications it may be necessary to perform a geometric transformation of the image to transform the characters into a standard position, typically such that the text is horizontal. This process is called image rectification. For example, the text may have to be rotated (see Figure 3.19), perspectively rectified (see Figure 3.23), or rectified with a polar transformation (see Figure 3.24).

割得到的字符进行分类，也就是说为分割得到的区域分配一个符号标记。本节中将会介绍字符分割和特征提取，分类将会在 3.15 节讨论。

3.14.1 字符分割

将字符进行分类首先需要将文本分割成为单个的字符，也就是说每个字符必须对应于一个区域。

为了分割字符，可以使用在 3.4 节中讨论的所有方法：使用固定的阈值、自动选择的阈值、动态阈值以及提取图像中的连通区域。

另外，可能不得不使用 3.6 节中介绍的形态学方法将同一个字符分离的部分连接起来，例如将字符"i"的小圆点与下面的部分连接起来（见 3.6.1.2 节），也可以将可能由于印刷质量等原因造成同一个字符的多个部分连接起来。对于在复杂表面的字符来讲可能需要灰度值形态学技术来分割字符，如在金属表面的字符（见 3.6.2.4 节）。

另外，在一些应用中可能必须对图像进行几何变换将字符变换到一个标准位置，一般情况是将文本变换为水平的。这个过程称为图像校正。例如，文本可能不得不进行旋转（图 3.19）、透视校正（图 3.23）或者使用极性变换校正图像（图 3.24）。

Even though we have many segmentation strategies at our disposal, in some applications it may be difficult to segment the individual characters because the characters actually touch each other, either in reality or at the resolution at which we are looking at them in the image. Therefore, special methods to segment touching characters are sometimes required.

The simplest such strategy is to define a separate ROI for each character we are expecting in the image. This strategy sometimes can be used in industrial applications because the fonts typically have a fixed pitch (width) and we know a priori how many characters are present in the image, e.g., if we are trying to read serial numbers with a fixed length. The main problem with this approach is that the character ROIs must enclose the individual characters we are trying to separate. This is difficult if the position of the text can vary in the image. If this is the case, we first need to determine the pose of the text in the image based on another strategy, e.g., template matching, to find a distinct feature in the vicinity of the text we are trying to read, and to use the pose of the text either to rectify the text to a standard position or to move the character ROIs to the appropriate position.

While defining separate ROIs for each character works well in some applications, it is not very flexible. A better method can be derived by realizing that the characters typically touch only with a small number of pixels. An example of this is shown in Figures 3.177(a) and (b). To separate these characters, we can simply count the number of pixels per column in the segmented region. This is shown in Figure 3.177(c). Since the touch-

虽然有这么多的分割策略可供使用，但在一些应用中仍可能难以将单个字符分割出来，这是因为实际中字符之间可能互相粘连，这种粘连可能由于实际中确实粘连或者字符图像分辨率不够造成的。因此，有时需要一些特殊的方法来分割这种粘连的字符。

最简单的策略是为图像中每个预期的字符定义一个单独的感兴趣区域（ROI）。工业应用中有时可以使用这种策略，这是由于字体间距（宽度）固定，并且可以事先知道图像中多少个字符，例如试图在图像中读取一个固定长度的序列号时就可以使用这种策略。使用这种方法的主要问题就是字符的感兴趣区域（ROI）必须将需要分离的单个字符包含在内。如果图像中的文本位置可能发生变化，那么这种策略就很难达到效果。这种情况下，首先需要基于其他策略确定图像中文本的位姿，例如使用模板匹配找到正在尝试读取的文本附近的一个明显特征，并且使用这个特征的位姿将文本校正到标准位置或者将字符感兴趣区域（ROI）移动到适当的位置。

虽然在一些应用中为每个字符定义单独的感兴趣区域（ROI）可以达到目的，但是这种方法并不灵活。由于字符之间一般接触的部分像素数很少，因此可以由此得到一个更好的方法。例如图 3.177（a）和图 3.177（b）。为了将这些字符分离开，可以计算分割得到区域在每列的像素数量，结果如图 3.177（c）。由于字符之间连接部

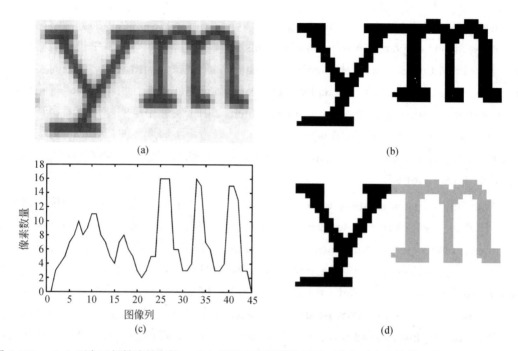

图 3.177 （a）两个互相粘连的字符；（b）阈值分割得到的区域，注意两个字符并没有分离开；（c）图（b）中每列像素数量的图表；（d）将两个字符在图（c）中像素数的最小值位置 21 处分离开

ing part is only a narrow bridge between the characters, the number of pixels in the region of the touching part only has a very small number of pixels per column. In fact, we can simply segment the characters by splitting them vertically at the position of the minimum in Figure 3.177(c). The result is shown in Figure 3.177(d). Note that in Figure 3.177(c) the optimal splitting point is the global minimum of the number of pixels per column. However, in general, this may not be the case. For example, if the strokes between the vertical bars of the letter "m" were slightly thinner, the letter "m" might be split erroneously. Therefore, to make this algorithm more robust, it is typically necessary to define a search space for the splitting of the characters based on the expected width of the characters. For example, in this application the characters are approximately 20 pixels wide. Therefore, we could restrict the search

分是一个非常狭窄的桥，也就是说粘连部分每列的像素数非常少。实际上，可以简单地使用图 3.177（c）中最小值的位置将字符分割开。结果显示在图 3.177（d）中。注意在图 3.177（c）中最佳的分割点是每列像素数的全局最小值。然而，一般情况下这个值可能不是最佳分割点。例如，如果字母 m 的竖笔画之间的部分稍微细一点的话，字母 m 可能被错误地分割为多个字符。因此，为了使这个算法更加可靠，一般情况下有必要为字符的分割处定义一个搜索区域，这个搜索区域可以基于字符预期的宽度确定。例如，在这个应用中字符的宽度大约为 20 个像素。因此可以约束最佳分割点的搜索空间为预期字符宽度左右 ±4 个像素（预期宽度的 20%）的区域。在实际应用中这个字符分割方法可以

space for the optimal splitting point to a range of ±4 pixels (20% of the expected width) around the expected width of the characters. This simple splitting method works very well in practice. Further approaches for segmenting characters are discussed by Casey and Lecolinet (1996).

3.14.2 Feature Extraction

As discussed in Section 3.15, classification can be regarded as a mapping f that maps a feature vector $\boldsymbol{x} \in \mathbb{R}^n$ to a class $\omega_i \in \Omega$. The vector $\boldsymbol{x}$ thus must have a fixed length n. In the following, we will discuss how $\boldsymbol{x}$ can be extracted from the segmented regions. This process is called feature extraction. Note that it is also possible to learn feature extraction (see Section 3.15.3.4).

For OCR, the features that are used for the classification are features that we extract from the segmented characters. Any of the region features described in Section 3.5.1 and the gray value features described in Section 3.5.2 can be used as features. The main requirement is that the features enable us to discern the different character classes. Figure 3.178 illustrates this point. The input image is shown in Figure 3.178(a). It contains examples of lowercase letters. Suppose that we want to classify the letters based on the region features anisometry and compactness. Figures 3.178(b) and (c) show that the letters "c" and "o" and "i" and "j" can be distinguished easily based on these two features. In fact, they can be distinguished solely based on their compactness. As Figures 3.178(d) and (e) show, however, these two features are not sufficient to distinguish between the classes "p" and "q" and "h" and "k"

达到非常好的效果。关于分割字符的更多方法可以参见文献（Casey and Lecolinet，1996）。

3.14.2 特征提取

如 3.15 节所讨论，分类可以看做是一个映射 f，把某个特征向量 $\boldsymbol{x} \in \mathbb{R}^n$ 映射到某个类别 $\omega_i \in \Omega$。特征向量 $\boldsymbol{x}$ 必须具有固定长度 n，下面将会讨论如何从分割得到的区域中提取 $\boldsymbol{x}$，这个过程叫做特征提取。注意，也可以在 3.15.3.4 节中学习特征提取。

对于 OCR 来讲，用来进行分类的特征是从单个字符中提取到的特征。这些特征可以使用在 3.5.1 节中介绍的所有区域特征和 3.5.2 节讨论的所有灰度值特征。主要需要这些特性能够帮助区分不同的字符类型，图 3.178 说明了这一点。图 3.178（a）是输入图像，图中包含很多小写字母的实例。假设希望基于区域特征 anisomery 和 compactness 来对这些字母进行分类。图 3.178（b）和图 3.178（c）中可以看出基于这两个特征可以非常轻松的区分出字母"c"和"o"以及字母"i"和"j"。实际上，只基于 compactness 特征就可以将这些字母区分开。然而，如图 3.178（d）和（e）所示，这两个特征不足以将类"p"和"q"区分开，同时也不能将类"h"和"k"区分开。

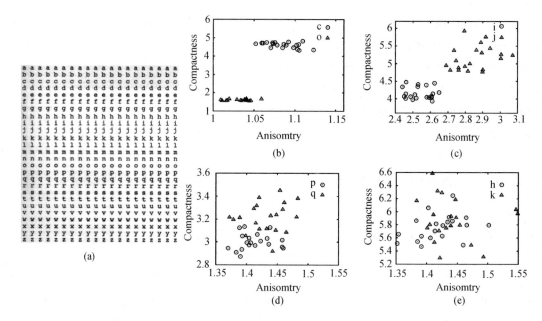

图 3.178　(a) 包含很多小写字母在内的图像；(b) 字母 "c" 和 "o"；(c) 字母 "i" 和 "j"；(d) 字母 "p" 和 "q"；(e) 字母 "h" 和 "k" 的区域特征 anisometry 和 compactness。注意在图 (b) 和图 (c) 中可以轻松将字母区分开，但在图 (d) 和图 (e) 中不能区分

From the above example, we can see that the features we use for the classification must be sufficiently powerful to enable us to classify all relevant classes correctly. The region and gray value features described in Sections 3.5.1 and 3.5.2, unfortunately, are often not powerful enough to achieve this. A set of features that is sufficiently powerful to distinguish all classes of characters is the gray values of the image themselves. Using the gray values directly, however, is not possible because the classifier requires a constant number of input features. To achieve this, we can use the smallest enclosing rectangle around the segmented character, enlarge it slightly to include a suitable amount of background of the character in the features (e.g., by one pixel in each direction), and then zoom the gray values within this rectangle to a standard size, e.g., 8×10 pixels. While transforming the image, we must take care to use the

从上面示例中，可以看出用来分类的特征必须足够有效，必须要能对所有相关的字符类型进行正确分类。不幸的是，要达到这个标准，3.5.1 节和 3.5.2 节中讨论的区域特征和灰度值特征通常都不能胜任。能够有效区分所有字符的一个特征集就是图像本身的灰度值。然而由于分类器的输入特征的数量必须是常数，因此不可能直接使用图像的灰度值作为输入特征。为了使图像灰度值能够用作分类器的输入特征，可以使用分割得到字符的最小外接矩形，将这个外接矩形稍微放大，(例如，沿边缘的每个方向放大一个像素) 使其包含适当的背景，然后将这个矩形中的灰度区域缩放到一个标准尺寸，如 8×10 像素中。在变换图像时必须注意使用 3.3.2 节中介绍的内插算法和平滑技术。然而

interpolation and smoothing techniques discussed in Section 3.3.2. Note, however, that by zooming the image to a standard size based on the surrounding rectangle of the segmented character, we lose the ability to distinguish characters like "−" (minus sign) and "I" (upper case I in fonts without serifs). The distinction can, however, easily be made based on a single additional feature: the ratio of the width to the height of the smallest surrounding rectangle of the segmented character.

Unfortunately, the gray value features defined above are not invariant to illumination changes in the image. This makes the classification very difficult. To achieve invariance to illumination changes, two options exist. The first option is to perform a robust contrast normalization of the character, as described in Section 3.2.1.3, before the character is zoomed to the standard size. The second option is to convert the segmented character into a binary image before the character is zoomed to the standard size. Since the gray values generally contain more information, the first strategy is preferable in most cases. The second strategy can be used whenever there is significant texture in the background of the segmented characters, which would make the classification more difficult.

Figure 3.179 displays two examples of the gray value feature extraction for OCR. Figures 3.179(a) and (d) display two instances of the letter "5," taken from images with different contrast (Figures 3.26(a) and (b)). Note that the characters have different sizes (14×21 and 13×20 pixels, respectively). The result of the robust contrast normalization is shown in Figures 3.179(b) and (e). Note that both characters now have full

需要注意的是，如果将分割得到的字符外接矩形内的图像缩放到一个标准尺寸，就不能将类似"−"（减号）和"I"（没有衬线的大写字母 I）的字符辨别开。但可以简单地基于单个额外的特征进行辨别：分割得到的字符的最小外接矩形的宽高比。

不幸的是，上面定义的灰度值特征会随着图像中光照变化而变化。这就使分类非常困难。为了使这些特征不随光照变化而变化，存在两种选择。第一种选择是在将字符缩放到标准尺寸前，按照 3.2.1.3 节中介绍的灰度值归一化算法，将字符的灰度值进行归一化。第二种选择是在将字符缩放到标准尺寸前，将分割得到的字符转换为二值图像。由于一般情况下灰度值包含更多的信息，因此第一种策略在大多数情况下更好一些。无论何时，如果分割得到的字符背景中有明显特征，那么可以使用第二种策略，因为背景中的特征会导致分类变得更困难。

图 3.179 中显示了 OCR 中灰度值特征提取的两个例子。图 3.179（a）和图 3.179（d）中显示了从图 3.26（a）和图 3.26（b）中提取得到的字符"5"对比度不同的两个实例。注意这两个字符的尺寸并不相同（分别为 14×21 和 13×20）。经过灰度值归一化操作后的结果显示在图 3.179（b）和图 3.179（e）中。注意此时对两种

contrast. Finally, the result of zooming the characters to a size of 8 × 10 pixels is shown in Figures 3.179(c) and (f). Note that this feature extraction automatically makes the OCR scale-invariant because of the zooming to a standard size.

情况都进行了充分对比。最后将字符尺寸缩放到 8 × 10 像素，结果如图 3.179（c）和图 3.179（f）。注意这种特征提取方式可以使 OCR 不受字符尺寸变化的影响，因为所有字符都被缩放到一个标准尺寸。

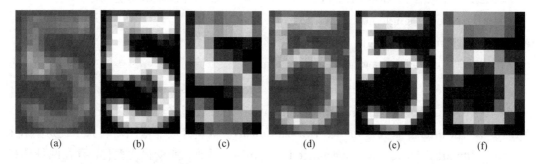

图 3.179　为 OCR 提取灰度值特征：(a) 从图 3.26（a）的第二列字符中提取得到的字符 "5"；(b) 对图（a）进行可靠的灰度值归一化操作；(c) 将图（b）缩放到 8 × 10 像素；(d) 图 3.26（b）中第二列字符中提取出字母 "5"；(e) 对图（d）进行可靠的灰度值归一化操作；(f) 将图（e）缩放到 8 × 10 像素

The preceding discussion has used a standard size of 8 × 10 for the normalized character. A large set of tests has shown that this size is appropriate for most industrial applications. If there are only a small number of classes to distinguish, e.g., only numbers, it may be possible to use slightly smaller sizes. For some applications involving a larger number of classes, e.g., numbers and uppercase and lowercase characters, a slightly larger size may be necessary (e.g., 10×12). On the other hand, using much larger sizes typically does not lead to better classification results because the features become progressively less robust against small segmentation errors if a large standard size is chosen. This happens because larger standard sizes imply that a segmentation error will lead to progressively larger position inaccuracies in the zoomed character as the standard size becomes larger. Therefore, it is best not to use

之前的讨论对于标准化字符均使用了标准尺寸 8 × 10，通过大的测试集表明这个尺寸对于大多数工业应用是合理的。如果在应用中只有很少类型需要辨别，例如只需要辨别数字，就可以使用稍微小一点的标准尺寸。对于一些有较多类型需要辨别的应用中，例如需要辨别数字、大写字母和小写字母，可能有必要使用稍微大一些的尺寸（如 10 × 12）。另一方面，一般情况下使用更大的尺寸不会导致更好的分类结果，因为使用大的标准尺寸时会放大分割误差的影响。这种情况发生是因为标准尺寸变大就意味着，分割的误差会导致缩放到标准尺寸时字符位置的误差变大。因此，使用的标准尺寸的大小最好不要比上面推荐的尺寸大太多。但存在一个例外的情况，就是在字符类型的数量非常

a standard size that is much larger than the above recommendations. One exception to this rule is the recognition of an extremely large set of classes, e.g., logographic characters like the Japanese kanji characters. Here, much larger standard sizes are necessary to distinguish the large number of different characters.

As we have seen in Section 3.11.5.6, gradient orientations provide a feature that has high discriminatory power and is invariant to illumination changes. Therefore, gradient orientations can also be used for feature extraction. The idea is to measure the distribution of gradient orientations in different regions within the smallest enclosing rectangle of the character (Liu *et al.*, 2004). To do so, the gray values of the character are scaled to a standard size, e.g., 35×35. This image is subdivided into 5×5 blocks, each of size 7×7. The gradient directions are computed within the 35×35 image using a Sobel filter (see Section 3.7.3.1). The normalized gradient direction vectors are then projected onto the eight canonical directions ($k\pi/4$, $k = 0, \cdots, 7$). This discretizes each gradient direction into its contribution to at most two of the eight direction channels that correspond to the canonical directions. Within each 7×7 block, the contents of each direction channel are weighted by a Gaussian that is centered within the block and are summed. This results in one feature per direction channel, i.e., in eight features per 7×7 block. Each of the eight features can be regarded as a weighted count of how many pixels in the block have a certain gradient direction. Since there are 5×5 blocks with eight features each, the gradient direction feature vector has 200 elements.

大的情况下，例如辨别日本 Kanji 文字这种标识字符时。此情况下必然需要更大的标准尺寸来辨别大量不同的字符类型。

如 3.11.5.6 节所看到的，梯度方向特征具有很高的区分能力，对于光照变化也是不变的。因此，梯度方向可以用来做特征提取。思路就是在不同的字符最小外接矩形区域里测量梯度方向分布（Liu *et al.*,2004）。为此，把字符的灰度值缩放到一个标准尺寸，例如 35×35。图像细分为 5×5 个区块，每个区块尺寸为 7×7。使用 Sobel 滤波器在 35×35 图像中计算梯度方向（见 3.7.3.1 节）。然后把归一化梯度方向向量投射到 8 个规范方向（$k\pi/4, k = 0, \cdots, 7$）。这就使每个梯度方向离散化为梯度方向的贡献度，这个贡献度是与规范方向相关的 8 个通道中的至多 2 个。在每个 7×7 区块中，每个方向通道就是通过块内的高斯函数加权求和。结果是每个方向通道对应一个特征，即每 7×7 区块对应 8 个特征。8 个特征中的每一个特征可以作为一个加权计数，这个加权计数就是指块内有多少个像素具有某一梯度方向。因为有 5×5 个区块，每个区块有 8 个特征，因此梯度方向特征向量有 200 个元素。

3.15 Classification

Classification is the task of assigning a class ω_i to a feature vector $\boldsymbol{x}$. For example, for OCR, $\boldsymbol{x}$ can be computed as described in Section 3.14.2. Furthermore, for OCR, the classes ω_i can be thought of as the interpretation of the character, i.e., the string that represents the character. If an application must read serial numbers, the classes $\{\omega_1, \cdots, \omega_{10}\}$ are simply the strings $\{0, \cdots, 9\}$. If numbers and uppercase letters must be read, the classes are $\{\omega_1, \cdots, \omega_{36}\} = \{0, \cdots, 9, A, \cdots, Z\}$. Hence, classification can be thought of as a function f that maps a feature vector $\boldsymbol{x}$ of fixed length n to the set of classes $\Omega = \{\omega_i, i = 1, \cdots, m\}$: $f: \mathbb{R}^n \mapsto \Omega$. In this section, we will take a closer look at how the mapping can be constructed.

3.15.1 Decision Theory

3.15.1.1 Bayes Decision Rule

First of all, we can note that the feature vector $\boldsymbol{x}$ that serves as the input to the mapping can be regarded as a random variable because of the variations that objects, e.g., the characters, exhibit. In the application, we are observing this random feature vector for each object we are trying to classify. It can be shown that, to minimize the probability of erroneously classifying the feature vector, we must maximize the probability that the class ω_i occurs under the condition that we observe the feature vector $\boldsymbol{x}$, i.e., we should maximize $P(\omega_i | \boldsymbol{x})$ over all classes ω_i, for $i = 1, \cdots, m$ (Theodoridis and Koutroumbas, 2009; Webb and Copsey, 2004). The probability $P(\omega_i | \boldsymbol{x})$ is also

3.15 分类

分类就是给一个特征向量 $\boldsymbol{x}$ 确定一个类 ω_i。例如，对 OCR 来说，$\boldsymbol{x}$ 可以通过 3.14.2 节描述的方法来计算。此外，对于 OCR，类 ω_i 可以理解为是代表一种字符，也就是说表示字符内容的字符串。如果一个应用必须读取一串数字，类 $\{\omega_1, \cdots, \omega_{10}\}$ 就是字符串 $\{0, \cdots, 9\}$。如果数字和大写字母必须读取，类 $\{\omega_1, \cdots, \omega_{36}\} = \{0, \cdots, 9, A, \cdots, Z\}$。因此，分类可以认为是一个把一个固定长度 n 的特征向量 $\boldsymbol{x}$ 映射到类集 $\Omega = \{\omega_i, i = 1, \cdots, m\}$: $f: \mathbb{R}^n \mapsto \Omega$ 的函数 f。本节将详细了解一下如何构造这个映射关系。

3.15.1 决策理论

3.15.1.1 贝叶斯决策规则

首先，作为映射输入的特征向量 $\boldsymbol{x}$ 可以看作为一个随机变量，这是因为不同的对象表现也不同，例如不同的字符。在这个应用中，将每一个待分类别看作一个随机特征向量。可以看出为了使特征向量错误分类的概率最小化，应该使特征向量为 $\boldsymbol{x}$ 的情况下类 ω_i 的可能性最大，也就是说应该使 $P(\omega_i | \boldsymbol{x})$ 与其他所有字符类型 ω_i, for $i = 1, \cdots, m$ (Theodoridis and Koutroumbas, 2009; Webb and Copsey, 2004) 出现的概率相比为最大。这个概率 $P(\omega_i | \boldsymbol{x})$ 也被称为后验概率，因为上面介绍过类 ω_i 的概率

called the a posteriori probability because of the above property that it describes the probability of class ω_i given that we have observed the feature vector $\boldsymbol{x}$. This decision rule is called the Bayes decision rule. It yields the best classifier if all errors have the same weight, which is a reasonable assumption for most classification problems (e.g., OCR).

We now face the problem of how to determine the a posteriori probability. Using Bayes' theorem, $P(\omega_i|\boldsymbol{x})$ can be computed as follows:

$$P(\omega_i|\boldsymbol{x}) = \frac{P(\boldsymbol{x}|\omega_i)P(\omega_i)}{P(\boldsymbol{x})} \tag{3.247}$$

where

$$P(\boldsymbol{x}) = \sum_{j=1}^{m} P(\boldsymbol{x}|\omega_i)P(\omega_i) \tag{3.248}$$

Hence, we can compute the a posteriori probability based on the a priori probability $P(\boldsymbol{x}|\omega_i)$ that the feature vector $\boldsymbol{x}$ occurs given that the class of the feature vector is ω_i, the probability $P(\omega_i)$ that the class ω_i occurs, and the probability $P(\boldsymbol{x})$ that the feature vector $\boldsymbol{x}$ occurs. To simplify the calculations, we can note that the Bayes decision rule only needs to maximize $P(\omega_i|\boldsymbol{x})$ and that $P(\boldsymbol{x})$ is a constant if $\boldsymbol{x}$ is given. Therefore, the Bayes decision rule can be written as

$$\boldsymbol{x} \in \omega_i \quad \Leftrightarrow \quad P(\boldsymbol{x}|\omega_i)P(\omega_i) > P(\boldsymbol{x}|\omega_j)P(\omega_j), \qquad j=1,\cdots,m, j \neq i \tag{3.249}$$

What do we gain by this transformation? As we will see below, the probabilities $P(\boldsymbol{x}|\omega_i)$ and $P(\omega_i)$ can, in principle, be determined from training samples. This enables us to evaluate $P(\omega_i|\boldsymbol{x})$, and hence to classify the feature vector $\boldsymbol{x}$.

是在观测特征向量后得到的。这种决策规则被称为贝叶斯 (Bayes) 决策规则。如果所有的误差权重相同的情况下, 这种决策规则可以得到最好的分类器。正好对于 OCR 来讲误差权重相同是非常合理的假设。

现在面对的问题是如何确定后验概率。使用贝叶斯定理, 可以按下式计算概率 $P(\omega_i|\boldsymbol{x})$:

式中

因此, 可以基于一个先验概率 $P(\boldsymbol{x}|\omega_i)$ 计算后验概率, 这个先验概率是指已知字符类型为 ω_i 时特征向量 $\boldsymbol{x}$ 出现的概率。概率 $P(\omega_i)$ 是指类 ω_i 出现的概率。概率 $P(\boldsymbol{x})$ 是指特征向量 $\boldsymbol{x}$ 出现的概率。为了使计算更加简单, 可以注意到贝叶斯决策规则只需要使概率 $P(\omega_i|\boldsymbol{x})$ 最大化, 在 $\boldsymbol{x}$ 给定的情况下 $P(\boldsymbol{x})$ 为常数。因此, 贝叶斯决策规则可以写为下式:

这种变换究竟得到了什么呢? 下面将看到, 概率 $P(\boldsymbol{x}|\omega_i)$ 和 $P(\omega_i)$ 从原理上讲可以通过训练样品得到。这样就可以计算出 $P(\omega_i|\boldsymbol{x})$, 并可以对特征向量 $\boldsymbol{x}$ 分类。然而, 在开始详细

Before we examine this point in detail, however, let us assume that the probabilities in Eq. (3.249) are known. For example, let us assume that the feature space is 1D ($n = 1$) and that there are two classes ($m = 2$). Furthermore, let us assume that $P(\omega_1) = 0.3$, $P(\omega_2) = 0.7$, and that the features of the two classes have a normal distribution $N(\mu, \sigma)$ such that $P(x|\omega_1) \sim N(-3, 1.5)$ and $P(x|\omega_2) \sim N(3, 2)$. The corresponding likelihoods $P(x|\omega_i)P(\omega_i)$ are shown in Figure 3.180. Note that features to the left of $x \approx -0.7122$ are classified as belonging to ω_1, while features to the right are classified as belonging to ω_2. Hence, there is a dividing point $x \approx -0.7122$ that separates the classes from each other.

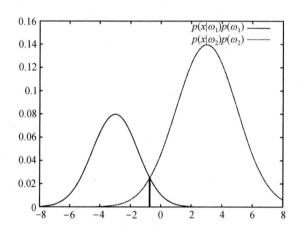

图 3.180　一维特征空间中两个类型分类示例，其中 $P(\omega_1) = 0.3$，$P(\omega_2) = 0.7$，$P(x|\omega_1) \sim N(-3, 1.5)$ 并且 $P(x|\omega_2) \sim N(3, 2)$。注意 $x \approx -0.7122$ 左边的特征向量都被分为类型 ω_1，$x \approx -0.7122$ 右边的特征向量都被分为类 ω_2

As a further example, consider a 2D feature space with three classes that have normal distributions with different means and covariances, as shown in Figure 3.181(a). Again, there are three regions in the 2D feature space in which the respective class has the highest probability, as shown in Figure 3.181(b). Note that now there are 1D curves that separate the regions in the feature space from each other.

中需要使用一维曲线将三个区域分割开。

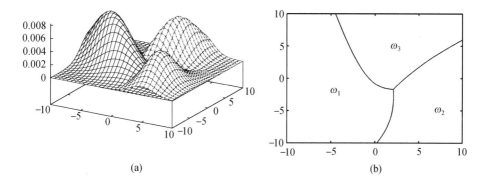

图 3.181　二维特征空间中对三个类型进行分类的示例，其中三个类型成正态分布，并且各自的均值和协方差不同：（a）出现三个类型的后验概率；（b）二维空间中相应类型概率最高的区域

As the above examples suggest, the Bayes decision rule partitions the feature space into mutually disjoint regions. This is obvious from the definition in Eq. (3.249): each region corresponds to the part of the feature space in which the class ω_i has the highest a posteriori probability. As also suggested by the above examples, the regions are separated by $(n-1)$-dimensional hypersurfaces (points for $n = 1$ and curves for $n = 2$, as in Figures 3.180 and 3.181). The hypersurfaces that separate the regions from each other are given by the points at which two classes are equally probable, i.e., by $P(\omega_i|\boldsymbol{x}) = P(\omega_j|\boldsymbol{x})$, for $i \neq j$.

从上面示例中可以看出，贝叶斯（Bayes）决策规则将特征空间分割为不相交的多个区域。这点可以从等式（3.249）的定义中看出：每个区域对应于特征空间中类型 ω_i 的后验概率最高的部分。通过上面的示例也可以看出，这些区域被 $(n-1)$ 维的超曲面分割开（$n = 1$ 时为点、$n = 2$ 时为曲线。如图 3.180 和图 3.181 所示）。将这些区域分割开的超曲面是由两个类型概率相等的点构成的，也就是 $P(\omega_i|\boldsymbol{x}) = P(\omega_j|\boldsymbol{x})$, for $i \neq j$。

3.15.1.2　Classifier Types

According to the above discussion, we can identify two different types of classifiers. The first type of classifier tries to estimate the a posteriori probabilities, typically via Bayes' theorem, from the a priori probabilities of the different classes. In contrast, the second type of classifier tries to construct the separating hypersurfaces between the

3.15.1.2　分类器类型

根据上面的讨论，可以确定两种不同类型的分类器。第一种类型的分类器一般使用贝叶斯定理，尝试通过不同类型的前验概率估计后验概率。与之不同，第二种类型的分类器尝试在类型之间构建分割的超曲面。在 3.15.2 节和 3.15.3 节，将分别讨论

classes. In Sections 3.15.2 and 3.15.3, we will discuss representatives for both types of classifier.

3.15.1.3 Training, Test, and Validation Sets

All classifiers require a method with which the probabilities or separating hypersurfaces are determined. To do this, a training set is required. The training set is a set of sample feature vectors x_k with corresponding class labels ω_k. For example, for OCR, the training set is a set of character samples from which the corresponding feature vectors can be calculated, along with the interpretation of the respective character. The training set should be representative of the data that can be expected in the application. For example, for OCR, the characters in the training set should contain the variations that will occur later, e.g., different character sets, stroke widths, noise, etc. Since it is often difficult to obtain a training set with all variations, the image processing system must provide means to extend the training set over time with samples that are collected in the field, and should optionally also provide means to artificially add variations to the training set.

Furthermore, to evaluate the classifier, in particular how well it has generalized the decision rule from the training samples, it is indispensable to have a test set that is independent of the training set. This test set is essential to determine the error rate that the classifier is likely to have in the application. Without an independent test set, no meaningful statement about the quality of the classifier can be made.

Finally, it is frequently necessary to adjust some hyper-parameters of a classifier to achieve

两种类型中典型的分类器。

3.15.1.3 训练集、测试集和验证集

所有分类器需要一个方法来得到概率或分割的超曲面。为了实现这个目的就需要一个训练集。这个训练集是一系列特征向量的样本 x_k 以及它们所对应的类别 ω_k。对 OCR 来说，训练集是一系列字符样本，从这些字符样本中计算它们的特征向量，同时也已知这些字符所属的类型。需要这个训练集能够代表实际应用中可能出现的情况。实际上，对于 OCR 来讲，训练集中的字符应该包含可能出现的一些变化，例如不同的字符集、笔画宽度、噪声等。由于一般情况下很难得到包含所有可能变化的训练集，因此图像处理系统必须提供一些手段基于收集到的字符样本扩展训练集，并且同时也可以人为地为训练集加入一些变化。

另外，为了评估一个分类器实际中是否很好地从训练集中概括了决策规则，必须提供一个与训练集不同的测试集。这个测试集实际上是用来确定分类器在实际应用中可能的错误率。如果没有这个独立的测试集，那么就不能对分类器的性能进行一个有效的评估。

最后，为了达到最佳性能，时常需要调整分类器的一些超参数（例

optimal performance (e.g., the number of hidden units in the classifier of Section 3.15.3.2). For this purpose, a third data set, the validation set, must be used. It must be independent of the training and test sets. To determine the optimal hyperparameters, the error rate on the validation set is optimized.

3.15.1.4 Novelty Detection

The Bayes decision rule in Eq. (3.249) assigns every feature vector $\boldsymbol{x}$ to exactly one class ω_i. It therefore makes a closed-world assumption: the training data is representative of all the classes that exist. Unfortunately, reality is often more complex in that we might encounter feature vectors that do not belong to any of the trained classes. For example, in an OCR application we might have trained a classifier to recognize only digits. Suppose the algorithm has segmented a character that actually is an "M" or a speck of dirt that actually is not a character at all. The Bayes decision rule will assign the character to one of the ten digits since these ten classes are all that it knows about. What is even worse, the probability $P(\omega_i|\boldsymbol{x})$ for the selected class will typically be very close to 1 in such cases (since, by Eq. (3.247), $P(\omega_i|\boldsymbol{x})$ will be close to 1 whenever $P(\boldsymbol{x}|\omega_i)P(\omega_i)$ is significantly larger than all the other $P(\boldsymbol{x}|\omega_j)P(\omega_j)$, no matter how small $P(\boldsymbol{x}|\omega_i)P(\omega_i)$ actually is).

In industrial applications, it is frequently essential that the classifier is able to recognize that a feature vector does not belong to any of the trained classes, i.e., that it does not sufficiently resemble any of the training data. In these cases, the feature vector should be rejected as not belonging

如，3.15.3.2 节分类器中隐层单元的数量）。为此，必须使用第三个数据集，即验证集。它必须独立于训练集和测试集。为了确定最优的超参数，验证集的错误率必须优化。

3.15.1.4 异常检测

等式（3.249）中的贝叶斯决策规则将每个特征向量 $\boldsymbol{x}$ 赋值给一个类 ω_i。因此，就有了一个封闭世界假设：训练数据代表了所有已有的类。不幸的是，实际情况往往更复杂，可能会遇到不属于任何训练过的类的特征向量。例如，在一个 OCR 应用中，可能已经训练了一个只能识别数字的分类器。假设算法已经分割出字符，该字符实际上是"M"或者根本不是字符而是一粒灰尘。贝叶斯决策规则将把这个字符分到 1 到 10 数字中的一个，因为它所知道的就这十个类别。更糟的是，对于这种情况，被选中类的概率 $P(\omega_i|\boldsymbol{x})$ 往往非常接近 1（因为，根据等式（3.247），当 $P(\boldsymbol{x}|\omega_i)P(\omega_i)$ 明显大于其他 $P(\boldsymbol{x}\omega_j)P(\omega_j)$ 时 $P(\omega_i|\boldsymbol{x})$ 将接近 1，无论 $P(\boldsymbol{x}|\omega_i)P(\omega_i)$ 实际上有多小）。

在工业应用中，有必要让分类器能够识别一个特征向量不属于任何一个已训练的类，也就是，该向量与任何训练数据都不完全相似。这种情况下，该特征向量因为不属于任何类，应该被拒识。这个问题被称为异常检

to any class. This problem is called novelty detection. In the above example, the segmented "M" or the speck of dirt should be rejected as not being a digit.

The capability for novelty detection is an important characteristic of a classifier. We will note below whether a particular classifier is capable of novelty detection or how it can be modified to have this capability.

3.15.2 Classifiers Based on Estimating Class Probabilities

The classifiers that are based on estimating probabilities, more precisely probability densities, are called Bayes classifiers because they try to implement the Bayes decision rule via the probability densities. The first problem they have to solve is how to obtain the probabilities $P(\omega_i)$ of the occurrence of the class ω_i. There are two basic strategies for this. The first strategy is to estimate $P(\omega_i)$ from the training set. For this, the training set must be representative not only in terms of the variations of the feature vectors but also in terms of the frequencies of the classes. Since this second requirement is often difficult to ensure, an alternative strategy for the estimation of $P(\omega_i)$ is to assume that each class is equally likely to occur, and hence to use $P(\omega_i) = 1/m$. In this case, the Bayes decision rule reduces to classification according to the a priori probabilities since $P(\omega_i|x) \sim P(\bm{x}|\omega_i)$ should now be maximized.

The remaining problem is how to estimate $P(\bm{x}|\omega_i)$. In principle, this could be done by determining the histogram of the feature vectors of the training set in the feature space. To do so,

测。在上面的例子中，"M"或是灰尘因为不是数字，应该被拒识。

异常检测能力是分类器的一个重要特征。下面会注意一个特定的分类器是否具有异常检测能力或者如何修改使之具备这种能力。

3.15.2 基于估计概率的分类器

基于估计概率 (准确地说应该是估计概率密度) 的分类器被称为贝叶斯分类器，因为它们尝试通过概率密度实现贝叶斯决策规则。它们需要解决的第一个问题就是如何得到所有类型 ω_i 出现的概率 $P(\omega_i)$。可以通过两个基本的策略解决这个问题。第一个策略就是从训练集中估计 $P(\omega_i)$。注意此时训练集绝对不能只表示特征向量的变化，而必须同时能够体现所有类型出现的频率。由于第二个需求经常难以得到保证，因此可以使用另一个估计概率 $P(\omega_i)$ 的策略，就是假设每个类型可能出现的概率相等，也就是说 $P(\omega_i) = 1/m$。注意这种情况下贝叶斯决策规则退化为取决于先验概率的分类，因为此时需要 $P(\omega_i|x) \sim P(\bm{x}|\omega_i)$ 最大化。

剩下的问题就是如何估计 $P(\bm{x}|\omega_i)$。原理上可以通过在特征空间上计算训练集中特征向量的直方图来估计 $P(\bm{x}|\omega_i)$。为了计算直方图，可

we could subdivide each dimension of the feature space into b bins. Hence, the feature space would be divided into b^n bins in total. Each bin would count the number of occurrences of the feature vectors in the training set that lie within this bin. If the training set and b are large enough, the histogram would be a good approximation to the probability density $P(\boldsymbol{x}|\omega_i)$. Unfortunately, this approach cannot be used in practice because of the so-called curse of dimensionality: the number of bins in the histogram is b^n, i.e., its size grows exponentially with the dimension of the feature space. For example, if we use the 81 features described in Section 3.14.2 and subdivide each dimension into a modest number of bins, e.g., $b = 10$, the histogram would have 10^{81} bins, which is much too large to fit into any computer memory.

3.15.2.1 k Nearest-Neighbor Classifiers

To obtain a classifier that can be used in practice, we can note that in the histogram approach, the size of the bin is kept constant while the number of samples in the bin varies. To get a different estimate for the probability of a feature vector, we can keep the number k of samples of class ω_i constant while varying the volume $v(\boldsymbol{x}, \omega_i)$ of the region in space around the feature vector $\boldsymbol{x}$ that contains the k samples. Then, if there are t feature vectors in the training set, the probability of occurrence of the class ω_i is approximately given by

$$P(\boldsymbol{x}|\omega_i) \approx \frac{k}{tv(\boldsymbol{x},\omega_i)} \qquad (3.250)$$

Since the volume $v(\boldsymbol{x}, \omega_i)$ depends on the k nearest neighbors of class ω_i, this type of density estimation is called k nearest-neighbor density estimation.

In practice, this approach is often modified as follows. Instead of determining the k nearest neighbors of a particular class and computing the volume $v(\boldsymbol{x},\omega_i)$, the k nearest neighbors in the training set of any class are determined. The feature vector $\boldsymbol{x}$ is then assigned to the class that has the largest number of samples among the k nearest neighbors. This classifier is called the k nearest-neighbor (kNN) classifier. For $k=1$, we obtain the nearest-neighbor (NN) classifier. It can be shown that the NN classifier has an error probability that is at most twice as large as the error probability of the optimal Bayes classifier that uses the correct probability densities (Theodoridis and Koutroumbas, 2009), i.e., $P_B \leq P_{NN} \leq 2P_B$. Furthermore, if P_B is small, we have $P_{NN} \approx 2P_B$ and $P_{3NN} \approx P_B + 3P_B^2$. Hence, the 3NN classifier is almost as good as the optimal Bayes classifier. Nevertheless, kNN classifiers are difficult to use in practice because they require that the entire training set (which can easily contain several hundred thousand samples) is stored with the classifier. Furthermore, the search for the k nearest neighbors is time consuming, even if optimized data structures are used to find exact (Friedman et al., 1977) or approximate (Arya et al., 1998; Muja and Lowe, 2014) nearest neighbors.

To provide the kNN classifier with the capability for novelty detection, one can, for example, use the distance of the feature vector $\boldsymbol{x}$ to the closest training sample. If this distance is too large, $\boldsymbol{x}$ is rejected. The distance threshold constitutes a hyper-parameter. Therefore, it should be determined using a validation set (see Section 3.15.1.3). For this to work, it is essential that the validation set includes representative samples of non-classes.

实际应用中，这个方法常被更改为下述方式：为了得到体积 $v(\boldsymbol{x},\omega_i)$，不需要求某个特定类型的 k 个最近邻域，而只是求训练集中所有类型的 k 个最近邻域。在 k 个最近邻域中属于哪个类型的训练样本最多，就将特征向量 $\boldsymbol{x}$ 分配给这个类型。这种分类器被称为 k 最近邻域分类器 (kNN 分类器)。当 $k=1$ 时，这个分类器就是最近邻域分类器 (NN 分类器)。NN 分类器的错误率是使用正确概率密度的最优贝叶斯（Bayes）分类器错误率的 2 倍以上（Theodoridis and Koutroumbas, 2009）：$P_B \leq P_{NN} \leq 2P_B$。另外，在 P_B 值很小的情况下，$P_{NN} \approx 2P_B$ 并且 $P_{3NN} \approx P_B + 3P_B^2$。因此，3NN 分类器基本和理想的贝叶斯分类器的效果差不多。不过，kNN 分类器很难在实际应用中使用，因为它们需要将整个训练集都保存在分类器中（很可能训练集中需要包含几十万个样本）。并且，即使使用优化的数据结构来寻找准确的（Friedman et al., 1977）或者近似的（Arya et al., 1998; Muja and Lowe, 2014）最近邻域，搜索 k 最近领域也是非常耗时的。

为了使 kNN 分类器有异常检测的能力，例如，可以使用特征向量 $\boldsymbol{x}$ 到最近的训练样本的距离。如果距离太大，$\boldsymbol{x}$ 就要排除掉。距离阈值组成一个超参数。因此，应该使用一个验证集来确定（参考 3.15.1.3 节）。为此，验证集必须包含非此类别的样本。这在实际应用中很难实现。如果特征向量包含不同单元的元素，会带来更

This is often difficult to achieve in practice. A further problem occurs if the feature vector contains elements in different units. In this case, the Euclidean distance is not semantically meaningful. For these reasons, the distance threshold is often set in an ad-hoc fashion or novelty detection is not used.

3.15.2.2 Gaussian Mixture Model Classifiers

As we have seen from the above discussion, direct estimation of the probability density function is not practicable, either because of the curse of dimensionality for histograms or because of efficiency considerations for the kNN classifier. To obtain an algorithm that can be used in practice, we can assume that $P(\boldsymbol{x}|\omega_i)$ follows a certain distribution, e.g., an n-dimensional normal distribution:

进一步的问题。这种情况下，欧几里得距离失去意义。由于这些原因，距离阈值常常通过特别方式设置，或者不使用异常检测。

3.15.2.2 高斯混合模型分类器

从上述讨论中可以看出，直接估计概率密度函数是不实际的，要么是由于直方图方法的维数灾难，要么是由于 kNN 分类器的效率问题。为了得到可以用在实际应用中的算法，可以假设 $P(\boldsymbol{x}|\omega_i)$ 的分布是某种确定的分布，如 n 维正态分布：

$$P(\boldsymbol{x}|\omega_i) = \frac{1}{(2\pi)^{n/2}|\boldsymbol{\Sigma}_i|^{1/2}} \exp\left(-\frac{1}{2}(\boldsymbol{x}-\boldsymbol{\mu}_i)^\top \boldsymbol{\Sigma}_i^{-1}(\boldsymbol{x}-\boldsymbol{\mu}_i)\right) \qquad (3.251)$$

With this, estimating the probability density function reduces to the estimation of the parameters of the probability density function. For the normal distribution, the parameters are the mean vector $\boldsymbol{\mu}_i$ and the covariance matrix $\boldsymbol{\Sigma}_i$ of each class. Since the covariance matrix is symmetric, the normal distribution has $(n^2 + 3n)/2$ parameters in total. They can, for example, be estimated via the standard maximum likelihood estimators

这样，对概率密度函数的估计就变成只需要估计概率密度函数的参数。对于正态分布来讲，函数的参数就是每个类型的均值向量 $\boldsymbol{\mu}_i$ 和协方差矩阵 $\boldsymbol{\Sigma}_i$。因为协方差矩阵是对称的，正态分布的概率密度函数共有 $(n^2+3n)/2$ 个参数。它们可以通过标准的最大似然估计方法进行估计。

$$\boldsymbol{\mu}_i = \frac{1}{n_i}\sum_{j=1}^{n_i} \boldsymbol{x}_{i,j} \quad \text{and} \quad \boldsymbol{\Sigma}_i = \frac{1}{n_i-1}\sum_{j=1}^{n_i}(\boldsymbol{x}_{i,j}-\boldsymbol{\mu}_i)(\boldsymbol{x}_{i,j}-\boldsymbol{\mu}_i)^\top \qquad (3.252)$$

Here, n_i is the number of samples for class ω_i, while $\boldsymbol{x}_{i,j}$ denotes the samples for class ω_i.

式中，n_i 是类型 ω_i 中的样本的数量，$\boldsymbol{x}_{i,j}$ 表示类型 ω_i 中的样本。

While the Bayes classifier based on the normal distribution can be quite powerful, often the assumption that the classes have a normal distribution does not hold in practice. In OCR applications, this happens frequently if characters in different fonts are to be recognized with the same classifier. One striking example of this is the shapes of the letters "a" and "g" in different fonts. For these letters, two basic shapes exist: *a* vs. a and g vs. g. It is clear that a single normal distribution is insufficient to capture these variations. In these cases, each font will typically lead to a different distribution. Hence, each class consists of a mixture of l_i different densities $P(\boldsymbol{x}|\omega_i,j)$, each of which occurs with probability $P_{i,j}$:

虽然基于正态分布的贝叶斯分类器非常强大，但通常实际应用中类型分布为正态分布的假设并不成立。在 OCR 应用中，如果不同字体的字符需要使用同一个分类器进行分类时，就不能假设各种类型为正态分布。一个突出的例子就是字母"a"和"g"使用不同字体时的形状。这两个字母存在两个基本的形状："*a*"和 a "*g*"和"g"。很明显一个单独的正态分布不能够将这些变化考虑在内。这种情况下，一般每种字体导致一个不同的分布。因此，每个类型由 l_i 个不同概率密度 $P(\boldsymbol{x}|\omega_i,j)$ 混合组成，每个概率密度出现的概率为 $P_{i,j}$：

$$P(\boldsymbol{x}|\omega_i) = \sum_{j=1}^{l_i} P(\boldsymbol{x}|\omega_i,j) P_{i,j} \qquad (3.253)$$

Typically, the mixture densities $P(\boldsymbol{x}|\omega_i,j)$ are assumed to be normally distributed. In this case, Eq. (3.253) is called a Gaussian mixture model (GMM). If we knew to which mixture density each sample belonged, we could easily estimate the parameters of the normal distribution with the above maximum likelihood estimators. Unfortunately, in real applications we typically do not have this knowledge, i.e., we do not know j in Eq. (3.253). Hence, determining the parameters of the mixture model requires the estimation of not only the parameters of the mixture densities but also the mixture density labels j. This is a much harder problem, which can be solved by the expectation maximization (EM) algorithm. The interested reader is referred to Nabney (2002); Theodoridis and Koutroumbas (2009) for details.

一般情况下，混合密度 $P(\boldsymbol{x}|\omega_i,j)$ 假设为正态分布。这种情况下，等式（3.253）被称为高斯混合模型（GMM）。如果已知每个样本属于混合密度中的哪个，就可以非常容易使用上面提到的标准的最大似然估计方法来估计正态分布函数的参数。不幸的是，在实际应用中一般无法事先知道样本属于哪个混合密度，也就是说不知道等式（3.253）中的 j。因此，求混合模型参数不单单需要估计混合密度的参数，同时还需要估计混合密度的标记 j，这是一个非常困难的问题，这个问题可以通过期望值最大化算法（EM 算法）解决。关于该算法的细节感兴趣的读者可以参见文献（Nabney，2002；Theodoridis and Koutroumbas，2009）。

Another problem in the mixture model approach is that we need to specify how many mixture densities there are in the mixture model, i.e., we need to specify l_i in Eq. (3.253). This is quite cumbersome to do manually. To solve this problem, algorithms that compute l_i automatically have been proposed. The interested reader is referred to (Figueiredo and Jain, 2002; Wang et al., 2004) for details.

The GMM classifier is, in principle, inherently capable of novelty detection by using the value of $P(\boldsymbol{x})$ in Eq. (3.248) to reject feature vectors $\boldsymbol{x}$ that are too unlikely. The problem with this approach is that $P(\boldsymbol{x})$ is a probability density. Consequently, the range of $P(\boldsymbol{x})$ is basically unrestricted and depends on the scaling of the feature space. Thus, it is difficult to select the threshold for rejecting feature vectors in practice.

A method for novelty detection that is easier to use can be obtained as follows. A $k\sigma$ error ellipsoid is defined as a locus of points for which

在混合模型方法中另一个问题是需要指定在混合模型中有多少个混合密度，也就是说需要指定等式（3.253）中的 l_i。手动设置这个值非常麻烦。最近已经有文献提出自动计算 l_i 值的方法，关于这个方法的细节感兴趣的读者可以参见（Figueiredo and Jain, 2002；Wang et al., 2004）。

原则上，通过使用等式（3.248）中 $P(\boldsymbol{x})$ 的值来排除不可能的特征向量 $\boldsymbol{x}$，GMM 具备异常检测的能力。这种方法的问题是 $P(\boldsymbol{x})$ 是一个概率密度，因此 $P(\boldsymbol{x})$ 的范围基本上是不受限制的，并且依赖于特征空间的缩放。这样，在实际中就很难选择阈值来剔除特征向量。

以下为获得一个比较易用的异常检测的方法。一个 $k\sigma$ 误差椭球定义为点的轨迹

$$(\boldsymbol{x}-\boldsymbol{\mu})^\top \boldsymbol{\Sigma}^{-1}(\boldsymbol{x}-\boldsymbol{\mu}) = k^2 \qquad (3.254)$$

In the one dimensional case, this is the interval $[\mu - k\sigma, \mu + k\sigma]$. For a 1D Gaussian distribution, $\approx 65\%$ of the occurrences of the random variable are within this range for $k = 1$, $\approx 95\%$ for $k = 2$, $\approx 99\%$ for $k = 3$, etc. Hence, the probability that a Gaussian distribution will generate a random variable outside this range is $\approx 35\%$, $\approx 5\%$, and $\approx 1\%$, respectively. This probability is called the $k\sigma$ probability and is denoted by $P(k)$. For Gaussian distributions, the value of $P(k)$ can be computed numerically. For GMMs, the $k\sigma$ probability of a single class ω_i is computed as

对于一维情况，存在一个区间 $[\mu - k\sigma, \mu + k\sigma]$。对于一维高斯分布，当 $k=1$ 时，大约 65% 的随机变量在这个范围内出现，$k=2$ 时，大约 95%，$k=3$ 时，大约 99%，等等。因此，高斯分布在这个范围之外产生一个随机变量的概率分别是大约 35%、5% 和 1%。这个概率称为 $k\sigma$ 概率，用 $P(k)$ 表示。对于高斯分布，$P(k)$ 的值可以通过数值计算。对于 GMM，单一类 ω_i 的 $k\sigma$ 概率可以通过下式计算

$$P_{k\sigma;i}(\boldsymbol{x}) = \sum_{j=1}^{l_i} P(k_{i,j}) P_{i,j} \qquad (3.255)$$

where $k_{i,j}^2 = (\boldsymbol{x} - \boldsymbol{\mu}_{i,j})^\top \boldsymbol{\Sigma}_{i,j}^{-1}(\boldsymbol{x} - \boldsymbol{\mu}_{i,j})$. With this, the $k\sigma$ probability over all classes can be computed as

$$P_{k\sigma}(\boldsymbol{x}) = \frac{\max_{i=1,\cdots,m} P(\omega_i) P_{k\sigma;i}(\boldsymbol{x})}{\max_{i=1,\cdots,m} P(\omega_i)} \tag{3.256}$$

The advantage of $P_{k\sigma}(\boldsymbol{x})$ over $P(\boldsymbol{x})$ is that $P_{k\sigma}(\boldsymbol{x})$ has a well-defined range that can be interpreted easily: $[0,1]$. Consequently, the rejection threshold can be selected in an intuitive manner.

Figure 3.182 displays an example GMM with one Gaussian per class. The feature space is 2D. Training samples were generated by uniformly sampling points from three ellipses, as shown in Figure 3.182(a). The likelihoods $P(\boldsymbol{x}|\omega_i)P(\omega_i)$, the probability density $P(\boldsymbol{x})$, and the a posteriori probabilities $P(\omega_i|\boldsymbol{x})$ are displayed in Figures 3.182(f)–(h). The range of $P(\boldsymbol{x})$ is $[0, 9.9 \times 10^{-5}]$. Note that for higher-dimensional feature spaces, it is impossible to perform an exhaustive plot like the one in Figure 3.182(e). Therefore, in practice it is difficult to determine the range of $P(\boldsymbol{x})$ and hence a suitable threshold for novelty detection based on $P(\boldsymbol{x})$. The range of $P(\omega_i|\boldsymbol{x})$ is $[0,1]$. Note that $P(\omega_i|\boldsymbol{x})$ has a value close to 1 for an entire sector of the feature space, even for feature values that are very far from the training samples. The classification result without rejection (i.e., without novelty detection) is shown in Figure 3.182(i). Every point in the feature space is assigned to a class, even if it lies arbitrarily far from the training samples. The value of $P_{k\sigma}(\boldsymbol{x})$ (range: $[0,1]$) is shown in Figure 3.182(j), while the classification result with a rejection threshold of $P = 0.01$ is shown in Figure 3.182(k). By using the $k\sigma$ probability, the selection of the rejection threshold is relatively straightforward.

这里 $k_{i,j}^2 = (\boldsymbol{x} - \boldsymbol{\mu}_{i,j})^\top \boldsymbol{\Sigma}_{i,j}^{-1}(\boldsymbol{x} - \boldsymbol{\mu}_{i,j})$。基于此，所有类的 $k\sigma$ 的概率可以计算如下

使用 $P_{k\sigma}(\boldsymbol{x})$ 替代 $P(\boldsymbol{x})$ 的好处是 $P_{k\sigma}(\boldsymbol{x})$ 有个容易理解的明确范围：$[0,1]$。因此，可以用直观的方式来选择剔除阈值。

图 3.182 所示为一个 GMM 的例子，每个类都有一个高斯分布。特征空间是二维的。训练样本通过对来自三个椭圆的点均匀采样来创建，如图 3.182（a）所示。似然概率 $P(\boldsymbol{x}|\omega_i)P(\omega_i)$、概率密度 $P(\boldsymbol{x})$ 和后验概率 $P(\omega_i|\boldsymbol{x})$ 分别如图 3.182(f)~(h)所示。$P(\boldsymbol{x})$ 的范围是 $[0, 9.9 \times 10^{-5}]$。注意，对于更高维度的特征空间来说，不可能像图 3.182(e) 中所示展现详细的细节。因此，在实际情况下，很难确定 $P(\boldsymbol{x})$ 的范围，因而很难基于 $P(\boldsymbol{x})$ 确定一个合适的异常检测阈值。$P(\omega_i|\boldsymbol{x})$ 的范围是 $[0,1]$。注意，即使特征值与训练样本相差甚远，对于一个完整的特征空间，$P(\omega_i|\boldsymbol{x})$ 都有接近于 1 的值。不带异常检测的分类结果如图 3.182(i) 所示。不管与训练样本相差多少，特征空间中的每个点都会归属于某一个类。$P_{k\sigma}(\boldsymbol{x})$ 的值（范围：$[0,1]$）如图 3.182（j），同时，使用剔除阈值 $P = 0.01$ 的分类结果如图 3.182（k）所示。通过使用 $k\sigma$ 概率，剔除阈值的选择相对简单一些。

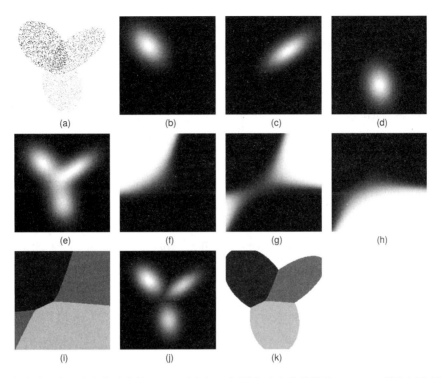

图 3.182 每个类别有一个高斯分布的 GMM 例子，不同的概率和分类结果：（a）二维特征空间中三个类的样本，用三个灰度级表示；（b）$P(\boldsymbol{x}|\omega_1)P(\omega_1)$；（c）$P(\boldsymbol{x}|\omega_2)P(\omega_2)$；（d）$P(\boldsymbol{x}|\omega_3)P(\omega_3)$；（e）$P(\boldsymbol{x})$，$P(\boldsymbol{x})$ 的范围是 $[0, 9.9 \times 10^{-5}]$；（f）$P(\omega_1|\boldsymbol{x})$；（g）$P(\omega_2|\boldsymbol{x})$；（h）$P(\omega_3|\boldsymbol{x})$，$P(\omega_i|\boldsymbol{x})$ 的范围是 $[0,1]$；（i）将特征空间分为三类，没有异常检测；（j）$P_{k\sigma}(\boldsymbol{x})$，$P(\boldsymbol{x})$ 的范围是 $[0,1]$；（k）将特征空间分为三类，异常检测阈值为 $P = 0.01$

3.15.3 Classifiers Based on Constructing Separating Hypersurfaces

3.15.3.1 Single-Layer Perceptrons

Let us now turn our attention to classifiers that construct separating hypersurfaces between the classes. Of all possible surfaces, the simplest ones are planes. Therefore, it is instructive to consider this special case first. Planes in the n-dimensional feature space are given by

$$\boldsymbol{w}^\top \boldsymbol{x} + b = 0 \tag{3.257}$$

Here, $\boldsymbol{x}$ is an n-dimensional vector that describes a point, while $\boldsymbol{w}$ is an n-dimensional vector that

3.15.3 基于构造分离超曲面的分类器

3.15.3.1 单层感知器

下面将注意力转移到在类型之间构建分离超曲面的分类器上。对于所有可能存在的表面来说，最简单的就是平面。因此，首先考虑这种特殊情况是具有一定启发性的。在 n 维特征空间中的平面可以使用下式表示：

式中，$\boldsymbol{x}$ 是表示一个点的 n 维向量，$\boldsymbol{w}$ 是表示平面法向矢量的 n 维向量。

describes the normal vector to the plane. This equation is linear. Because of this, classifiers based on separating hyperplanes are called linear classifiers.

Let us first consider the problem of classifying two classes with the plane. We can assign a feature vector to the first class ω_1 if $\boldsymbol{x}$ lies on one side of the plane, while we can assign it to the second class ω_2 if it lies on the other side of the plane. Mathematically, the test on which side of the plane a point lies is performed by looking at the sign of $\boldsymbol{w}^\top \boldsymbol{x} + b$. Without loss of generality, we can assign $\boldsymbol{x}$ to ω_1 if $\boldsymbol{w}^\top \boldsymbol{x} + b > 0$, while we assign $\boldsymbol{x}$ to ω_2 if $\boldsymbol{w}^\top \boldsymbol{x} + b < 0$.

For classification problems with more than two classes, we construct m separating planes $(\boldsymbol{w}_i, b_i)$ and use the following classification rule (see Theodoridis and Koutroumbas, 2009):

$$\boldsymbol{x} \in \omega_i \quad \Leftrightarrow \quad \boldsymbol{w}_i^\top \boldsymbol{x} + b_i > \boldsymbol{w}_j^\top \boldsymbol{x} + b_j, \qquad j = 1, \cdots, m, j \neq i \qquad (3.258)$$

Note that, in this case, the separating planes do not have the same meaning as in the two-class case, where the plane actually separates the data. The interpretation of Eq. (3.258) is that the plane is chosen such that the feature vectors of the correct class have the largest positive distance from the plane of all feature vectors.

Linear classifiers can also be regarded as neural networks, as shown in Figure 3.183 for the two-class and n-class cases. The neural network has processing units (neurons) that are visualized by circles. They first compute the linear combination of the feature vector $\boldsymbol{x}$ and the weights $\boldsymbol{w}$: $\boldsymbol{w}^\top \boldsymbol{x} + b$. Then, a nonlinear activation function f is applied. For the two-class case, the activation function is simply $\mathrm{sgn}(\boldsymbol{w}^\top \boldsymbol{x} + b)$, i.e., the side of the hyperplane on which the feature vector lies.

这个等式是线性的。因此,基于分离超曲面的分类器也被称为线性分类器。

首先考虑使用平面分类两个类型的问题。将位于平面一边的特征向量 $\boldsymbol{x}$ 分为类型 ω_1,同时将位于平面另一边的特征向量分为类型 ω_2。从数学角度看,测试一个点位于平面的哪边可通过求解式 $\boldsymbol{w}^\top \boldsymbol{x} + b$ 的正负号来得到。不失一般性,如果 $\boldsymbol{w}^\top \boldsymbol{x} + b > 0$,就将 $\boldsymbol{x}$ 分为类型 ω_1,同理如果 $\boldsymbol{w}^\top \boldsymbol{x} + b < 0$,就将 $\boldsymbol{x}$ 分为类型 ω_2。

对于多于两个类型的分类问题,构建 m 个分离平面 $(\boldsymbol{w}_i, b_i)$ 并且使用下面的分类规则(参见 Theodoridis and Koutroumbas,2009):

注意这种情况下,分离平面与存在两个类型情况下的意义不相同,在只有两个类型的情况下,分离平面确实将数据分离开。等式(3.258)的解释是选择的分离平面需要使正确类型的特征向量在所有特征向量中与分离平面的正向距离最大。

线性分类器还可以看作为神经网络,图 3.183 中分别显示两个类型和 n 个类型的神经网络结构。神经网络中有一些使用圆圈表示的处理单元(神经元)。它们首先计算特征向量 $\boldsymbol{x}$ 和权值 $\boldsymbol{w}$ 的线性组合:$\boldsymbol{w}^\top \boldsymbol{x} + b$。然后应用一个非线性的激活函数 f。对于两个类型的情况,激活函数为 $\mathrm{sgn}(\boldsymbol{w}^\top \boldsymbol{x} + b)$,也就是判断特征向量位于超平面的哪一侧。因此,函数的

Hence, the output is mapped to its essence: −1 or +1. This type of activation function essentially thresholds the input value. For the n-class case, the activation function f is typically chosen such that input values < 0 are mapped to 0, while input values $\geqslant 0$ are mapped to 1. The goal in this approach is that a single processing unit returns the value 1, while all other units return the value 0. The index of the unit that returns 1 indicates the class of the feature vector. Note that the plane in Eq. (3.258) needs to be modified for this activation function to work since the plane is chosen such that the feature vectors have the largest distance from the plane. Therefore, $\boldsymbol{w}_j^\top \boldsymbol{x} + b_j$ is not necessarily < 0 for all values that do not belong to the class. Nevertheless, the two definitions are equivalent. Because the neural network has one layer of processing units, this type of neural network is also called a single-layer perceptron.

输出就是：1 或 −1。这个类型的激活函数本质上是将输入值进行阈值分割。对于有 n 个类型的情况，激活函数 f 一般情况下在输入值小于 0 时映射为 0，在输入值大于等于 0 时映射为 1。这种方法的目的就是使其中单个处理单元返回 1，同时其他的处理单元返回 0。返回值是 1 的处理单元的序号就表示这个特征向量所属的类型。注意等式（3.258）中表示的平面需要为这个激活函数进行修改，这是因为此平面选择是需要特征向量与平面的正向距离最大。因此，对于不属于这个类型的所有值，$\boldsymbol{w}_j^\top \boldsymbol{x} + b_j$ 不一定 < 0。然而，这两种定义是等价的。由于这个神经网络有一层处理单元，这种类型的神经网络也被称为单层感知器。

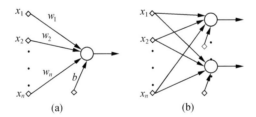

图 3.183 使用神经网络（单层感知器）表示的线性分类器的结构：（a）两个类型的神经网络；（b）n 个类型的神经网络。在这两种情况中，神经网络都只包括一层使用圆圈表示的处理单元。它们首先计算特征向量与权值的线性组合。然后应用一个非线性的激活函数，这个激活函数将输出映射到 −1 或 1（两个类型的神经网络），或映射为 0 或 1（n 个类型的神经网络）

While linear classifiers are simple and easy to understand, they have very limited classification capabilities. By construction, the classes must be linearly separable, i.e., separable by a hyperplane, for the classifier to produce the correct output. Unfortunately, this is rarely the case in practice. In fact, linear classifiers are unable to represent

虽然线性分类器非常简单并且容易理解，但它们的分类能力非常有限。从线性分类器的构造上看，分类器将不同类型使用一个超平面线性分离开来试图得到正确的输出。不幸的是实际应用中很少存在这种情况。实际上，线性分类器甚至都不能够表现一个类似

a simple function like the XOR function, as illustrated in Figure 3.184, because there is no line that can separate the two classes. Furthermore, for n-class linear classifiers, there is often no separating hyperplane for each class against all the other classes, although each pair of classes can be separated by a hyperplane. This happens, for example, if the samples of one class lie completely within the convex hull of all the other classes.

XOR 功能的简单函数，如图 3.184 所示，这里没有一条线可以将两个类型分离开。另外，对于 n 个类型的线性分类器，就算每对类型之间可以使用一个超平面分开，但通常没有一个分离超平面可以将每个类型与其他所有类型分开。如果有一个类型的样本完全位于所有其他类型构成的凸壳中，那么就不可能找到一个分离超平面将这个类型与其他所有类型分离开。

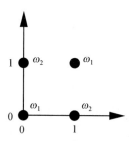

图 3.184　一个线性分类器不能够表示 XOR 函数，因为对应于 XOR 函数两个输出的两种类型不能够被任意一条线分离开

3.15.3.2　Multilayer Perceptrons

To obtain a classifier that is able to construct more general separating hypersurfaces, one approach is simply to add more layers to the neural network, as shown in Figure 3.185. This architecture is called a multilayer perceptron (MLP). Each layer first computes the linear combination of the feature vector or the results from the previous layer:

3.15.3.2　多层感知器

为了能够得到一个可以构建的普遍分离超曲面分类器，一种简单的方法就是在神经网络中添加更多的层，如图 3.185 所示。这种结构称为多层感知器（MLP），在每层中首先计算特征向量或上层结果的线性组合

$$a_j^{(l)} = \sum_{i=1}^{n_l} w_{ji}^{(l)} x_i^{(l-1)} + b_j^{(l)} \tag{3.259}$$

Here, $x_i^{(0)}$ is the feature vector, while $x_i^{(l)}$, with $l \geq 1$, is the result vector of layer l. The coefficients $w_{ji}^{(l)}$ and $b_j^{(l)}$ are the weights of layer l. Then, the results are passed through a nonlinear activation function.

式中，$x_i^{(0)}$ 是特征向量，同时，$x_i^{(l)}(l \geq 1)$ 是第 l 层的结果向量。式中的系数 $w_{ji}^{(l)}$ 和 $b_j^{(l)}$ 是 l 层的权值。然后，结果 $x_i^{(l)}$ 将传递给一个非线性的激活函数。

$$x_j^{(l)} = f(a_j^{(l)}) \tag{3.260}$$

Let us assume for the moment that the activation function in each processing unit is the threshold function that is also used in the single-layer perceptron, i.e., the function that maps input values < 0 to 0 while mapping input values ⩾ 0 to 1. Then, it can be seen that the first layer of processing units maps the feature space to the corners of the hypercube $\{0,1\}^p$, where p is the number of processing units in the first layer. Hence, the feature space is subdivided by hyperplanes into half-spaces (Theodoridis and Koutroumbas, 2009).

现在假设每个处理单元的激活函数都是与单层感知器相同的阈值分割函数，也就是说这个激活函数将输入值小于 0 的值映射为 0，同时将输入值大于等于 0 的值映射为 1。此时可以看出，第一层的处理单元将特征空间投影到超立方体的角点 $\{0,1\}^p$ 处，这里 p 指第一层中处理单元的数量。因此，特征空间就被超平面分成了半空间（Theodoridis and Koutroumbas，2009）。

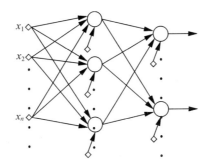

图 3.185　多层感知器的结构。神经网络有多层处理单元，用圆圈表示。它们计算前一层结果以及网络权值的线性组合，然后将结果传递给一个非线性激活函数

The second layer of processing units separates the points on the hypercube by hyperplanes. This corresponds to intersections of half-spaces, i.e., convex polyhedra. Hence, the second layer is capable of constructing the boundaries of convex polyhedra as the separating hypersurfaces (Theodoridis and Koutroumbas, 2009). This is still not general enough, however, since the separating hypersurfaces might need to be more complex than this. If a third layer is added, the network can compute unions of the convex polyhedra (Theodoridis and Koutroumbas, 2009). Hence,

第二层处理单元使用超平面将超立方体上的点分离开。这就相当于得到与半空间的交集，也就是一个凸多面体。因此，第二层可以与分离超曲面相似的构建凸多面体的边界（Theodoridis and Koutroumbas，2009）。然而这种分割仍然不够普遍，因为分离超曲面可能需要比这种凸多面体边界更加复杂。如果加入第三层，这个神经网络可以得到多个凸多面体的并集（Theodoridis and Koutroumbas，2009）。因此，如果使用的激活

three layers are sufficient to approximate any separating hypersurface arbitrarily closely if the threshold function is used as the activation function.

In practice, the above threshold function is rarely used because it has a discontinuity at $x = 0$, which is detrimental for the determination of the network weights by numerical optimization. Instead, often a sigmoid activation function is used, e.g., the logistic function (see Figure 3.186(a)):

$$f(x) = \frac{1}{1 + e^{-x}} \qquad (3.261)$$

Similar to the hard threshold function, it maps its input to a value between 0 and 1. However, it is continuous and differentiable, which is a requirement for most numerical optimization algorithms. Another choice for the activation functions is to use the hyperbolic tangent function (see Figure 3.186(b)):

$$f(x) = \tanh(x) = \frac{e^x - e^{-x}}{e^x + e^{-x}} \qquad (3.262)$$

in all layers except the output layer (Bishop, 1995). In the output layer, the softmax activation function is used (see Figure 3.186(c)):

函数是阈值分割函数的话，三层的神经网络大约可以近似满足任意情况的分离超曲面。

在实际应用中，上面的阈值分割函数很少用来做激活函数，因为它在 $x = 0$ 处有一个不连续点，这对于通过数值最优化方法计算网络权值非常不利。因此，通常使用一个 sigmoid 激活函数来替代阈值函数，例如使用逻辑函数（见图 3.186（a））：

与硬阈值函数相似，它将输入值映射为一个 0 到 1 之间的值。只不过它是连续的并且是可微的，在大部分数值最优化算法中都需要连续并可微。激活函数的另一个选择是在除输出层之外的其他所有层中使用双曲正切函数（见图 3.186（b））：

而在输出层中使用 softmax 激活函数（Bishop，1995）（见图 3.186（c））：

$$f(x) = \frac{e^{x_i}}{\sum_{j=1}^{m} e^{x_j}} \qquad (3.263)$$

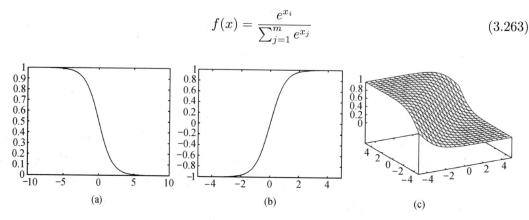

图 3.186 （a）逻辑激活函数（见等式（3.261））；（b）双曲线正切激活函数（见等式（3.262））；（c）softmax 激活函数（见等式（3.263））

The hyperbolic tangent function behaves similarly to the logistic function. The major difference is that it maps its input to values between -1 and $+1$. There is experimental evidence that the hyperbolic tangent function leads to faster training of the network than the logistic function (Bishop, 1995). In the output layer, the softmax function maps the input to the range $[0, 1]$, as desired. Furthermore, it ensures that the output values sum to 1, and hence have the same properties as a probability distribution (Bishop, 1995). With any of these choices of the activation function, it can be shown that two layers are sufficient to approximate any separating hypersurface and, in fact, any output function with values in $[0, 1]$, arbitrarily closely (Bishop, 1995). The only requirement for this is that there is a sufficient number of processing units in the first layer (the hidden layer).

After having discussed the architecture of the MLP, we can now examine how the network is trained. Training the network means that the weights $w_{ji}^{(l)}$ and $b_j^{(l)}$ ($l \in \{1, 2\}$) of the network must be determined. Let us denote the number of input features by n_i, the number of hidden units (first layer units) by n_h, and the number of output units (second layer units) by n_o. Note that n_i is the dimensionality of the feature vector, while n_o is the number of classes in the classifier. Hence, the only free hyper-parameter is the number n_h of units in the hidden layer. There are

$$n_w = (n_i + 1)n_h + (n_h + 1)n_o \tag{3.264}$$

weights in total. For example, if $n_i = 81$, $n_h = 40$, and $n_o = 10$, there are 3690 weights that must be determined. It is clear that this is a very complex

双曲正切函数的表现与逻辑函数相似。这两者之间的主要区别是双曲线正切函数将输入值映射为一个 -1 到 1 之间的值。实验证明使用双曲线正切激活函数可以使网络的训练速度更快（Bishop, 1995）。然后在输出层上，使用 softmax 函数将输入值映射为一个 0 到 1 之间的值。另外，它保证所有输出值总和为 1，因此它和概率密度有同样的属性（Bishop, 1995）。选择这些激活函数中的任意一个，都可以看出两层的神经网络已足够接近任意的分离超曲面，实际上选择值在 0 到 1 之间的任意输出函数均可任意接近（Bishop, 1995）。唯一需要的就是在第一层 (隐藏层) 中有足够多的处理单元。

在讨论过多层感知器的结构后，现在可以讨论如何训练神经网络。训练神经网络就意味着要计算得到网络中的所有权值 $w_{ji}^{(l)}, b_j^{(l)}$ ($l \in \{1, 2\}$)。我们使用 n_i 表示输入特征的数量，n_h 表示隐藏层单元（第一层单元）的数量，并使用 n_o 表示输出单元（第二层单元）的数量。注意 n_i 表示特征向量的维度，而 n_o 是分类器中类别的数量。因此，唯一不固定的超参数就是在隐藏层中处理单元的数量 n_h。在整个神经网络中共有个

权值。例如，如果 $n_i = 81$, $n_h = 40$ 并且 $n_o = 10$，那么神经网络中共有 3690 个权值。可以看出这是一个非常

problem and that we can hope to determine the weights uniquely only if the number of training samples is of the same order of magnitude as the number of weights.

As described above, the training of the network is performed based on a training set, which consists of sample feature vectors $\boldsymbol{x}_k$ with corresponding class labels ω_k, for $k = 1,\cdots,l$. The sample feature vectors can be used as they are. The class labels, however, must be transformed into a representation that can be used in an optimization procedure. As described above, ideally we would like to have the MLP return a 1 in the output unit that corresponds to the class of the sample. Hence, a suitable representation of the classes is a target vector $\boldsymbol{y}_k \in \{0,1\}^{n_o}$, chosen such that there is a 1 at the index that corresponds to the class of the sample and a 0 in all other positions. With this, we can train the network by minimizing the cross-entropy error of the outputs of the network on all the training samples (Bishop, 1995). In the notation of Eq. (3.260), the training minimizes

复杂的问题,如果训练样本和权值的数量是一个数量级时能够得到权值的唯一解。

与上面讨论的一样,神经网络的训练是基于一个训练集进行的,这个训练集中包括一些样本的特征向量 $\boldsymbol{x}_k$ 以及它们对应的类型 ω_k,$k = 1,\cdots,l$。这些样本的特征向量可以直接使用。然而类型必须转换为一种可以用在一个最优化过程中的表示方法。上面已经讨论过,理想上希望多层感知器在样本类型相应的输出单元上返回 1。因此,这些类别的一个合适的表示法就是一个标签向量 $\boldsymbol{y}_k \in \{0,1\}^{n_o}$,这个标签向量中样本的类型所对应序号的值为 1,并且其他所有的值都为 0。这时,就可以通过神经网络输出的交叉熵最小化来训练网络(Bishop,1995)。在等式(3.260)中,最小化

$$\varepsilon = -\sum_{k=1}^{l}\sum_{j=1}^{n_o} y_{k,j}\ln x_{k,j}^{(2)} \qquad (3.265)$$

Here, $x_{k,j}^{(2)}$ and $y_{k,j}$ denote the jth element of the output vector $\boldsymbol{x}_k^{(2)}$ and target vector $\boldsymbol{y}_k$, respectively. Note that $\boldsymbol{x}_k^{(2)}$ implicitly depends on all the weights $w_{ji}^{(l)}$ and $b_j^{(l)}$ of the MLP. Hence, minimization of Eq. (3.265) determines the optimum weights. Numerical minimization algorithms, such as the conjugate gradient algorithm (Press et al., 2007; Bishop, 1995) or the scaled conjugate gradient algorithm (Bishop, 1995), can be used to minimize Eq. (3.265).

式中,$x_{k,j}^{(2)}$ 和 $y_{k,j}$ 分别表示输出向量 $\boldsymbol{x}_k^{(2)}$ 和目标向量 $\boldsymbol{y}_k$ 的第 j 个元素。注意,其实 $\boldsymbol{x}_k^{(2)}$ 取决于网络中所有的权值 $w_{ji}^{(l)}$ 和 $b_j^{(l)}$。因此,通过最小化等式(3.265)可以得到最优权值。数值最小化算法,如共轭梯度算法(Press et al.,2007;Bishop,1995)或者尺度共轭梯度算法(Bishop,1995),可以用来最小化等式(3.265)。

Figures 3.187(a)–(e) show an example MLP with $n_i = 2$, $n_h = 5$, and $n_o = 3$ applied to the same problem as in Figure 3.182 (see Section 3.15.2.2). Training samples were generated by uniformly sampling points from three ellipses, as shown in Figure 3.187(a). The resulting output activations for classes 1–3 are displayed in Figures 3.187(b)–(d). Note that the output activations show a qualitatively similar behavior as the a posteriori probabilities $P(\omega_i|\boldsymbol{x})$ of the GMM classifier (see Figures 3.182(f)–(h)): they have a value close to 1 for an entire sector of the feature space, even for feature values that are very far from the training samples. In fact, the only places where they do not have a value of 0 or 1 are the regions of the feature space where the training samples overlap. Consequently, the value of the output activations cannot be used for rejecting feature vectors. The classification result is shown in Figure 3.187(e). Every point in the feature space is assigned to a class, even if it lies arbitrarily far from the training samples. This shows that MLPs, as described so far, are incapable of novelty detection.

Another property that is striking about Figures 3.187(b)–(d) is that the output activations change their values abruptly, in distinct contrast to the outputs of the GMM classifier (see Figure 3.182). This can result in unintuitive behavior. Let us illustrate the problem by way of an OCR example. The letter "B" and the number "8" can look relatively similar in some fonts. Now suppose the MLP must classify a slightly distorted "B" that lies somewhere between the training samples for the classes "B" and "8" in the feature space, i.e., in the range of the feature space that was not covered by any training

图 3.187（a）～（e）所示为一个 MLP 的例子，示例中 MLP 应用于图 3.182（见 3.15.2.2 节）中提到的同样的问题，其中 $n_i = 2$, $n_h = 5$, $n_o = 3$。训练样本通过对来自三个椭圆的点均匀采样产生，如图 3.187（a）所示。类 1～3 输出激励结果如图 3.187（b）～（d）所示。注意此输出激励表现出和 GMM 分类器后验概率 $P(\omega_i|\boldsymbol{x})$ 性质相似的行为（见图 3.182（f）～（h））：注意，即使特征值与训练样本相差甚远，对于一个完整的特征空间，$P(\omega_i|\boldsymbol{x})$ 都有接近于 1 的值。实际上，唯一取值不为 0 或者 1 的位置是特征空间中训练样本重叠的区域。因此，输出激励结果不能用来剔除特征向量。分类结果如图 3.187（e）所示。不管与训练样本相差多大，特征空间中的每个点都归属于某一个类。如到目前为止所描述的，MLP 没有异常检测的能力。

如图 3.187（b）～（d）所示，另一个显著的属性是输出激励值的变化非常突然，这与 GMM 分类器的输出形成鲜明对比（如图 3.182 所示）。这会产生一些不符合直觉的行为。通过 OCR 例子来说明该问题。字母 "B" 和数字 "8" 在某些字体中看起来非常相似。现在假设 MLP 必须分类一个有些扭曲的 "B"，在特征空间中，这个 "B" 位于训练样本 "B" 和 "8" 之间，即在不包含任何训练样本的特征空间中（你可以想象成图 3.187（b）中的类 1 对应 "B"，图 3.187（c）

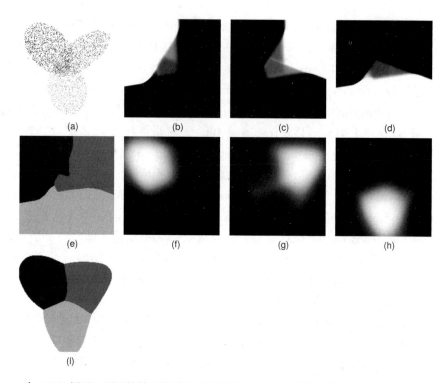

图 3.187　一个 MLP 例子，不同的输出激励和分类结果：（a）二维特征空间中三个类的样本，分别用三个灰度级表示；（b）~(d) 为类 1~3 的输出激励；（e）把特征空间分为三类，没有异常检测；（f）~(h) 使用归一化和拒绝类样本来训练 MLP，为类 1~3 的输出激励。后面的步骤中 MLP 使用了异常检测；（i）使用了异常检测把特征空间分为 3 类

samples. (You might want to imagine that class 1 in Figure 3.187(b) corresponds to "B," class 2 in Figure 3.187(c) corresponds to "8," and the "B" in question lies somewhere above the two respective ellipses in Figure 3.187(a) in the vicinity of the class boundary.) Because of the abrupt transitions between the classes, moving the feature vector slightly might change the "B" (with an output activation of 1) to an "8" (also with an output activation of 1). Since the feature space at the "B" in question was not covered in the training, the class boundary might not be correct. This, in itself is not a problem. What is a problem, however, is that the output activations will indicate that the MLP is highly confident of its classifica-

中的类 2 对应"8"，有问题的"B"位于图 3.187（a）中两个椭圆上方类的边界附近的某个位置）。由于两个类之间突然转换，轻微的移动特征向量可能把"B"（输出激励为 1）变成"8"（输出激励也为 1）。由于训练时，没有覆盖到特征空间中有问题的"B"，类的边界可能不准确。这本身并不是问题。然而，问题是输出激励表明 MLP 高度信任这两种情况下的分类结果。如果 MLP 返回了一个"8"，信任度为 1，这就非常不符合直觉。考虑到有问题的"B"是一个轻度变形版本的"B"，我们期望的是 MLP 返回一个较低信任度的"B"（如信任度

tion results in both cases. If the MLP returns an "8" with a confidence of 1, this is very unintuitive. Given that the "B" in question is a slightly distorted version of a "B," we would expect that the MLP returns a "B" with a lower confidence (e.g., 0.7) and, an "8" with a confidence different from 0 (e.g., 0.3).

The problem of the abrupt transitions between the classes is caused by the fact that the MLP weights are unrestricted in magnitude. The larger the weights are, the steeper the transitions between the classes become. To create smoother class transitions, the training of the MLP can be regularized by adding a weight decay penalty to the error function (see Nabney, 2002):

为 0.7）和一个信任度不为 0 的"8"（如 0.3）。

MLP 权值大小不受限制导致了两个类之间的突然转换。权值越大，两个类之间的变换就越陡峭。为了使变换更平缓，可以在误差函数上增加一个权值衰减惩罚项，使 MLP 的训练正则化（参见（Nabney，2002））：

$$\varepsilon_{\mathrm{r}} = \varepsilon + \frac{\alpha}{2}\|\boldsymbol{w}\|_2^2 \qquad (3.266)$$

where ε is given by Eq. (3.265) and the vector $\boldsymbol{w}$ denotes the union of all MLP weights $w_{ji}^{(l)}$ and $b_j^{(l)}$ ($l \in \{1,2\}$). As described in Section 3.15.1.3, α constitutes a hyper-parameter that can, for example, be determined based on a validation set. We will describe a different method to determine α below.

这里 ε 由等式（3.265）给定，向量 $\boldsymbol{w}$ 表示所有 MLP 权值 $w_{ji}^{(l)}$ 和 $b_j^{(l)}$ ($l \in \{1,2\}$) 的合集。如 3.15.1.3 节所述，α 构成一个超参数，它可以由验证集来确定。下面将会描述一个不同的方法来确定 α。

One problem with a single weight decay parameter is that the resulting MLP will not be invariant to a linear rescaling of the training data (Bishop, 1995; Nabney, 2002). To achieve invariance, each logical subset of weights must be weighted with a separate weight decay penalty, resulting in the following error function (see Nabney, 2002):

使用单个权值衰减参数带来的问题是，所产生的 MLP 将无法对训练数据的线性缩放保持不变（Bishop，1995；Nabney，2002）。为实现不变，每个权值的逻辑子集必须增加一个独立的权值衰减惩罚权值，会产生如下误差函数（参见（Nabney，2002））：

$$\varepsilon_{\mathrm{r}} = \varepsilon + \frac{1}{2}(\alpha_{w1}\|\boldsymbol{w}^{(1)}\|_2^2 + \alpha_{b1}\|\boldsymbol{b}^{(1)}\|_2^2 + \alpha_{w2}\|\boldsymbol{w}^{(2)}\|_2^2 + \alpha_{b2}\|\boldsymbol{b}^{(2)}\|_2^2) \qquad (3.267)$$

where $\boldsymbol{w}^{(l)}$ and $\boldsymbol{b}^{(l)}$ denote the union of all the weights $w_{ji}^{(l)}$ and $b_j^{(l)}$ for the respective layer ($l \in$

这里 $\boldsymbol{w}^{(l)}$ 和 $\boldsymbol{b}^{(l)}$ 分别表示对各层 ($l \in \{1,2\}$) 所有权值 $w_{ji}^{(l)}$ 和 $b_j^{(l)}$ 的合集。

{1, 2}). This leaves us with four hyper-parameters that must be determined. Doing a 4D optimization manually using a validation set is infeasible. Fortunately, the hyper-parameters can be determined automatically using a Bayesian framework called the evidence procedure. The details of the evidence procedure are beyond the scope of this book. The interested reader is referred to MacKay (1992a,b,c); Nabney (2002) for details.

As a practical matter, the evidence procedure must compute and invert the Hessian matrix (the second derivatives) of Eqs. (3.266) or (3.267). Storing the Hessian requires $O(n_w^2)$ of memory and inverting it requires $O(n_w^3)$ of computation time, where n_w is given by Eq. (3.264). Therefore, the evidence procedure can only be used for MLPs with a moderate number of weights (up to about ten thousand) to keep the memory consumption and run time within reasonable bounds. Furthermore, it requires an outer iteration around the optimization of Eqs. (3.266) or (3.267) that adjusts the hyper-parameters. Therefore, the evidence procedure is significantly more time consuming than simply optimizing Eqs. (3.266) or (3.267) with fixed hyper-parameters (which has memory and runtime complexity $O(n_w)$). Hence, for the single hyper-parameter in Eq. (3.266), it might be faster to determine α manually based on a validation set.

We now turn to the problem of novelty detection. As mentioned above, the architecture of the MLPs that we have discussed so far is incapable of novelty detection. The reason for this is the softmax function: if one of the x_i has a value that is somewhat larger than the remaining components of the input vector to the softmax function, the

必须确定 4 个超参数。使用验证集人工进行四维优化是不可行的。幸运的是，通过被证明过的贝叶斯框架，超参数可以自动确定。证明过程的细节不在本书的讨论范围之内。有兴趣的读者可以参见（MacKay，1992a,c,b；Nabney，2002）来了解更多细节。

事实上，证明过程必须对等式（3.266）或等式（3.267）的海森矩阵（二阶求导）求逆。存储海森矩阵的内存复杂度为 $O(n_w^2)$，求逆的计算时间复杂度为 $O(n_w^3)$，这里 n_w 由等式（3.264）给出。因此，证明过程只能用于适度权值数（最多是万）的 MLP，以确保内存和计算时间在可接受范围内。此外，围绕等式（3.266）或等式（3.267）（调整超参数）的优化，它需要一个外部迭代。因此，证明过程显然比等式（3.266）或等式（3.267）用固定超参数做简单优化（固定超参数优化算法的内存和时间复杂度为 $O(n_w)$）更耗时。因此，对于等式（3.266）中的单个超参数，基于验证集人工确定 α 可能更快。

现在来讨论一下异常检测问题。如上所述，到目前为止所讨论的 MLP 结构不具备异常检测能力。理由是 softmax 函数：如果 x_i 某个值比 softmax 函数输入向量的剩余元素略大，元素 i 将主导等式（3.263）中分母的和，并且 softmax 函数的输出将会是

component i will dominate the sum in the denominator of Eq. (3.263) and the output of the softmax function will be (close to) 1. This leads to the fact that at least one component of the output vector has a value of 1 in an entire region of the feature space that extends to infinity. For novelty detection, ideally we would like the output vector to behave in a fashion similar to $P(\boldsymbol{x})$ or $P_{k\sigma}(\boldsymbol{x})$ for GMMs (see Section 3.15.2.2). Hence, the output vector of an MLP should consist of all zeros for feature vectors that are too far from the training data. Then, we could simply threshold the output activations and reject feature vectors for which the maximum activation is too small. However, the above property of the softmax function prevents this strategy. Instead, we must add an additional class to the MLP for the sole purpose of novelty detection. We will call this class the "rejection class." If the activation of the rejection class is larger than the largest activation of all the regular classes, the feature vector is novel and is rejected.

To train the rejection class, a strategy similar to the one proposed by Singh and Markou (2004) can be used. During the training of the MLP, random training vectors for the rejection class are generated. First, the bounding hyper-boxes around the training samples of each class are computed. This defines an inner shell, in which no training samples for the rejection class are generated. The hyper-boxes are then enlarged by a suitable amount in each direction. This defines an outer shell beyond which no training samples are generated. The region of the feature space between the inner and outer shell constitutes the region in which training samples for the rejection

（接近）1。这导致在整个可扩展到无穷的特征空间中至少有一个输出向量的值是 1。对于异常检测，理想情况下，希望输出向量的行为方式类似于 GMM 的 $P(\boldsymbol{x})$ 或 $P_{k\sigma}(\boldsymbol{x})$（见 3.15.2.2 节）。因此，MLP 的输出向量应该包含全部为 0 的特征向量（与训练数据相差太远）。然后，可以对输出激励做简单的阈值分割，并剔除那些最大激励都很小的特征向量。然而 softmax 的上述特点阻止了这个策略。相反，必须给 MLP 增加一个额外的类，其唯一目的就是为了异常检测。称该类为"拒绝类"。如果拒绝类的激励比普通类的最大的激励还大，则异常的特征向量被剔除。

为了训练拒绝类，可以使用与（Singh and Markou, 2004）所提出策略类似的策略。在 MLP 训练过程中，生成拒绝类的随机训练向量。首先计算每个类训练样本的边界 hyper-boxes。这定义了一个内壳，在内壳中不会生成拒绝类的训练样本。然后 hyper-boxes 在每个方向上扩大一定范围。这又定义了一个外壳，超出外壳范围，不会生成训练样本。位于内壳和外壳之间的特征空间组成一个区域，在这个区域中，会生成拒绝类的训练样本。在二维空间，该区域看起来像一个矩形的"环"。因此，把这些

class are potentially generated. In 2D, this region looks like a rectangular "ring." Therefore, we will refer to these regions as hyper-box rings. Each class possesses a hyper-box ring. Figure 3.187(a) makes clear that we cannot simply generate training samples for all hyper-box rings since the ring of one class may overlap with the samples of another class. Therefore, training samples are created only within the region of each hyper-box ring that does not overlap with the inner shell (the bounding hyper-box) of any other class.

The above strategy is required for low-dimensional feature spaces. For high-dimensional feature spaces, an even simpler strategy can be used: the training samples for the rejection class are simply computed within the outer shell of each class. This works in practice since the data for a class is typically distributed relatively compactly within each class. For simplicity, we may assume that the data is distributed roughly spherically. In an n-dimensional space, the volume of a hyper-sphere of radius r is given by $(r^n \pi^{n/2})/\Gamma(n/2+1)$, whereas the hyper-cube around the hyper-sphere has a volume of $(2r)^n$ (here, $\Gamma(x)$ denotes the gamma function). For large n, the ratio of the volume of the hyper-sphere to the volume of the hyper-cube is very small. For example, for $n = 81$, the ratio is approximately 10^{-53}. Therefore, it is extremely unlikely that a random training sample for the rejection class falls within any of the regions of the regular classes.

Figures 3.187(f)–(h) show the output activations of an MLP with $n_i = 2$, $n_h = 5$, and $n_o = 4$ that was trained on the training data of Figure 3.187(a) with regularization ($\alpha = 10$) and with a rejection class (class 4, output acti-

区域作为 hyper-boxes 环。每个类处理一个 hyper-boxes 环。从图 3.187（a）可以很清楚地看到不能简单地对所有的 hper-boxes 环创建训练样本，因为某个类的环可能和其他类的环的训练样本有重叠。因此，只有在每个 hyper-boxes 环与其他类的内壳（byper-boxes 边界）没有交叠的地方才可以创建训练样本。

低维特征空间需要用上述策略。对于高维特征空间，可以使用更简单的策略：拒绝类的训练样本可以在每个类的外壳中简单计算。这在实际中是有作用的，因为类的数据通常在每个类中分布相对紧凑。为简单起见，假设数据大体上是以球状分布的。在一个 n 维空间中，半径为 r 的超球体的体积由 $(r^n \pi^{n/2})/\Gamma(n/2+1)$ 确定，而在超球体周围的超立方体的体积为 $(2r)^n$（这里，$\Gamma(x)$ 表示伽马函数）。对于 n 比较大的情况，超球体体积与超立方体体积的比例是非常小的。例如，对于 $n = 81$，比例接近 10^{-53}。因此，拒绝类的随机训练样本落在任何一个普通类的区域是极不可能的。

图 3.187（f）～（h）所示为一个 MLP 在 $n_i = 2$、$n_h = 5$ 和 $n_o = 4$ 时的输出激励，MLP 基于图 3.187（a）中正则化（$\alpha = 10$）的方法训练数据，并使用了拒绝类（类 4，输出激励未

vations not shown) with random samples drawn from hyper-box rings around each class, as described above. Figure 3.187(i) shows the classification result of this MLP on the entire feature space. As desired, feature vectors that lie far from the training samples are rejected as novel. One noteworthy aspect is that the activations and classes roughly follow the data distribution of the regular training samples even though the rejection samples were generated in rectangular rings around each class.

3.15.3.3 Support Vector Machines

Another approach to obtaining a classifier that is able to construct arbitrary separating hypersurfaces is to transform the feature vector into a space of higher dimension, in which the features are linearly separable, and to use a linear classifier in the higher dimensional space. Classifiers of this type have been known for a long time as generalized linear classifiers (Theodoridis and Koutroumbas, 2009). One instance of this approach is the polynomial classifier, which transforms the feature vector by a polynomial of degree $\leqslant d$. For example, for $d = 2$ the transformation is

$$\Phi(x_1,\cdots,x_n) = (x_1,\cdots,x_n,x_1^2,\cdots,x_1x_n,x_2^2\cdots,x_2x_n,\cdots,x_n^2) \quad (3.268)$$

To show the potential impact of a nonlinear transformation, Figure 3.188(a) displays a 2D feature space with samples from two classes that are not linearly separable. The samples can be transformed into a 3D space with a quadratic transformation $\Phi(x_1, x_2) = (x_1^2, x_1x_2, x_2^2)$. As shown in Figure 3.188(b), the classes can be separated in the transformed feature space by a hyperplane, i.e., they have become linearly separable.

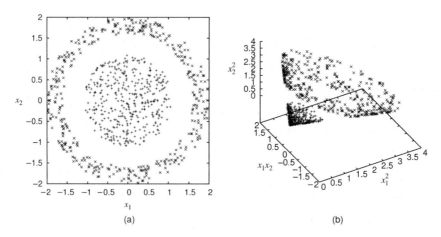

图 3.188 （a）一个二维特征空间，样本来自非线性可分的两个类；（b）使用二次变换转换到三维特征空间，使两个类线性可分

The problem with this approach is again the curse of dimensionality: the dimension of the feature space grows exponentially with the degree d of the polynomial. In fact, there are $\binom{d+n-1}{d}$ monomials of degree $=d$ alone. Hence, the dimension of the transformed feature space is

这种方法的问题也是维数灾难问题：特征空间的维数随多项式次数 d 成指数增长。实际上，在多项式中存在 $\binom{d+n-1}{d}$ 个次数为 d 的单项式。因此，变换后特征空间的维数就是

$$n' = \sum_{i=1}^{d} \binom{i+n-1}{i} = \binom{d+n}{d} - 1 \qquad (3.269)$$

For example, if $n = 81$ and $d = 5$, the dimension is 34 826 301. Even for $d = 2$ the dimension already is 3402. Hence, transforming the features into the larger feature space seems to be infeasible, at least from an efficiency point of view. Fortunately, however, there is an elegant way to perform the classification with generalized linear classifiers that avoids the curse of dimensionality. This is achieved by support vector machine (SVM) classifiers (Schölkopf and Smola, 2002; Christianini and ShaweTaylor, 2000).

Before we can examine how SVMs avoid the curse of dimensionality, we must take a closer look

例如，如果 $n = 81$ 并且 $d = 5$，得到的维数是 34 826 301。甚至在 $d = 2$ 时维数仍是 3402。因此，将特征变换到更大的特征空间看起来不可行，至少从方法效率的角度看是不可行的。然而，幸运的是存在一种非常好的方式，使用一般线性分类器的同时避免维数灾难。这可以通过支持向量机（SVM）分类器（Schölkopf and Smola, 2002; Christianini and ShaweTaylor, 2000）实现。

在关注支持向量机是如何避免维数灾难之前，首先深入了解一下如何

at how the optimal separating hyperplane can be constructed. Let us consider the two-class case. As described in Eq. (3.257) for linear classifiers, the separating hyperplane is given by $\boldsymbol{w}^\top\boldsymbol{x}+b=0$. As noted in Section 3.15.3.1, the classification is performed based on the sign of $\boldsymbol{w}^\top\boldsymbol{x}+b$. Hence, the classification function is

构建最优的分离超平面。首先考虑两个类型的情况，如等式（3.257）中所示的线性分类器，分离超平面为 $\boldsymbol{w}^\top\boldsymbol{x}+b=0$。上面已经提到过，这个分类器是基于判断 $\boldsymbol{w}^\top\boldsymbol{x}+b$ 的正、负号实现的。因此，分类器函数就是

$$f(x) = \mathrm{sgn}(\boldsymbol{w}^\top\boldsymbol{x}+b) \qquad (3.270)$$

Let the training samples be denoted by $\boldsymbol{x}_i$ and their corresponding class labels by $y_i = \pm 1$. Then, a feature is classified correctly if $y_i(\boldsymbol{w}^\top\boldsymbol{x}_i + b) > 0$. However, this restriction is not sufficient to determine the hyperplane uniquely. This can be achieved by requiring that the margin between the two classes be as large as possible. The margin is defined as the closest distance of any training sample to the separating hyperplane.

使用 $\boldsymbol{x}_i$ 表示训练样本，使用 $y_i = \pm 1$ 表示这些训练样本相应的类型标签。此时，在 $y_i(\boldsymbol{w}^\top\boldsymbol{x}_i + b) > 0$ 时这个特征就被正确分类。然而，这个约束不足以得到唯一的超平面。通过下面的约束可以保证得到唯一解，这个约束就是需要两个类型之间的间隔应该尽可能地大。这个两类型之间的间隔定义为所有训练样本中与分割超平面最近的距离。

Let us look at a small example of the optimal separating hyperplane, shown in Figure 3.189. If we want to maximize the margin (shown as a dotted line), there will be samples from both classes

在图 3.189 中可以看到一个最优分离超平面的小示例。注意如果要使两个类别之间间隔（如点线所示）最大，那么两个类别中都将有训练样本

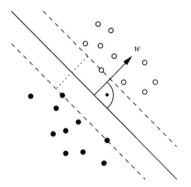

图 3.189　两个类型之间的最优分离超平面。两个类型的样本分别使用实心圆和空心圆表示，分离超平面如图实线所示。图中两条虚线表示两个超平面，在这两个超平面上有一些样本，这两个超平面之间的间隔表示为点线，这也就是两个类型之间的最短距离。在两个超平面上的样本定义了分离超平面，因为它们"支持"分离超平面，所以它们被称为支持向量

that attain the minimum distance to the separating hyperplane defined by the margin. These samples "support" the two hyperplanes that have the margin as the distance to the separating hyperplane (shown as the dashed lines). Hence, the samples are called the support vectors. In fact, the optimal separating hyperplane is defined entirely by the support vectors, i.e., a subset of the training samples:

与这个间隔定义的分离超平面之间的距离最小。这些样本"支持"两个超平面,这两个超平面之间的间隔被看作分离超平面(如虚线所示)的距离。因此,这些样本被称为支持向量。实际上,最优分离超平面完全通过支持向量定义,也就是训练样本的一个子集:

$$\boldsymbol{w} = \sum_{i=1}^{l} \alpha_i y_i \boldsymbol{x}_i \qquad \alpha_i \geqslant 0 \qquad (3.271)$$

where $\alpha_i > 0$ if and only if the training sample is a support vector (Schölkopf and Smola, 2002). With this, the classification function can be written as

只有在训练样本是支持向量时 $\alpha_i > 0$ (Schölkopf and Smola, 2002)。此时,分类函数可以重写为:

$$f(\boldsymbol{x}) = \text{sgn}(\boldsymbol{w}^\top \boldsymbol{x} + b) = \text{sgn}\left(\sum_{i=1}^{l} \alpha_i y_i \boldsymbol{x}_i^\top \boldsymbol{x} + b\right) \qquad (3.272)$$

Hence, to determine the optimal hyperplane, the coefficients α_i of the support vectors must be determined. This can be achieved by solving the following quadratic programming problem (Schölkopf and Smola, 2002): maximize

因此,为了确定最优超平面就必须得到支持向量的系数 α_i。这可以通过解决下面的二次规划问题来实现(Schökopf and Smola, 2002):最大化

$$\sum_{i=1}^{l} \alpha_i - \frac{1}{2} \sum_{i=1}^{l} \sum_{j=1}^{l} \alpha_i \alpha_j y_i y_j \boldsymbol{x}_i^\top \boldsymbol{x}_j \qquad (3.273)$$

subject to

并要满足

$$\alpha_i \geqslant 0, \quad i = 1, \cdots, l \qquad (3.274)$$

$$\sum_{i=1}^{l} \alpha_i y_i = 0 \qquad (3.275)$$

An efficient algorithm for solving this optimization problem was proposed by Platt (1999). Further speed improvements were described by Fan et al. (2005).

文献(Platt, 1999)提出了一个解决这个优化问题的有效算法。文献(Fan et al., 2005)则描述了如何进一步的提升速度。

Note that in both the classification function (3.272) and the optimization function (3.273), the feature vectors $\boldsymbol{x}$, $\boldsymbol{x}_i$, and $\boldsymbol{x}_j$ only are present in the dot product.

We now turn our attention back to the case in which the feature vector x is first transformed into a higher dimensional space by a function $\Phi(\boldsymbol{x})$, e.g., by the polynomial function in Eq. (3.268). Then, the only change in the above discussion is that we substitute the feature vectors $\boldsymbol{x}$, $\boldsymbol{x}_i$, and $\boldsymbol{x}_j$ by their transformations $\Phi(\boldsymbol{x})$, $\Phi(\boldsymbol{x}_i)$, and $\Phi(\boldsymbol{x}_j)$. Hence, the dot products are simply computed in the higher dimensional space. The dot products become functions of two input feature vectors: $\Phi(\boldsymbol{x})^\top \Phi(\boldsymbol{x}')$. These dot products of transformed feature vectors are called kernels in the SVM literature and are denoted by $k(\boldsymbol{x}, \boldsymbol{x}') = \Phi(\boldsymbol{x})^\top \Phi(\boldsymbol{x}')$. Hence, the decision function becomes a function of the kernel $k(\boldsymbol{x}, \boldsymbol{x}')$:

需要注意的是在等式（3.272）的分类函数和等式（3.273）最优化函数中，特征向量 $\boldsymbol{x}$, $\boldsymbol{x}_i$ 和 $\boldsymbol{x}_j$ 之间都是点乘。

现在将注意力转向特征向量 x，首先通过函数 $\Phi(\boldsymbol{x})$ 变换到一个更高维数空间中的情况，例如可以通过等式（3.268）中多项式函数变换到更高维的空间。然后使用变换结果 $\boldsymbol{x}$, $\boldsymbol{x}_i$, $\boldsymbol{x}_j$ 来替换式中的特征向量 $\Phi(\boldsymbol{x})$, $\Phi(\boldsymbol{x}_i)$ 和 $\Phi(\boldsymbol{x}_j)$。因此，只需要在更高维空间中简单计算点积。点积变为两个输入特征向量的函数: $\Phi(\boldsymbol{x})^\top \Phi(\boldsymbol{x}')$。这些变换后特征向量的点积在支持向量机（SVM）的文献中被称为核（Kernels），并且表示为 $k(\boldsymbol{x}, \boldsymbol{x}') = \Phi(\boldsymbol{x})^\top \Phi(\boldsymbol{x}')$。因此，判定函数变成核 $k(\boldsymbol{x}, \boldsymbol{x}')$ 的函数：

$$f(\boldsymbol{x}) = \text{sgn}\left(\sum_{i=1}^{l} \alpha_i y_i k(\boldsymbol{x}_i, \boldsymbol{x}) + b\right) \tag{3.276}$$

The same happens with the optimization function Eq. (3.273).

So far, it seems that we do not gain anything from the kernel because we still must transform the data into a feature space of a prohibitively large dimension. The ingenious trick of SVM classification is that, for a large class of kernels, the kernel can be evaluated efficiently without explicitly transforming the features into the higher dimensional space, thus making the evaluation of the classification function (3.276) feasible. For example, if we transform the features by a polynomial of degree d, it easily can be shown that

该函数的作用与等式（3.273）中最优化函数相同。

目前为止，看起来这个核并不能带来什么好处，因为仍然不得不将数据变换到一个特别高维数的特征空间中。SVM 分类器的一个非常奇妙的技巧就是对于一大类的核而言，不需要将特征明确地变换到高维空间中，也可以高效地计算出核的值，因此使分类函数（见等式（3.276））的计算可行。例如，如果使用一个 d 次多项式变换特征，非常简单就可以得到：

$$k(\boldsymbol{x}, \boldsymbol{x}') = (\boldsymbol{x}^\top \boldsymbol{x}')^d \tag{3.277}$$

Hence, the kernel can be evaluated solely based on the input features without going to the higher dimensional space. This kernel is called a homogeneous polynomial kernel. As another example, the transformation by a polynomial of degree $\leq d$ can simply be evaluated as

因此，可以基于输入特征单独对核进行计算，而不需要将特征变换到更高维空间。这个核被称为一个齐次多项式核。作为另一个例子，通过次数小于等于 d 的多项式变换可以按下式简单计算

$$k(\boldsymbol{x}, \boldsymbol{x}') = (\boldsymbol{x}^\top \boldsymbol{x}' + 1)^d \tag{3.278}$$

This kernel is called an inhomogeneous polynomial kernel. Further examples of possible kernels include the Gaussian radial basis function kernel

这个核被称为一个非齐次多项式核。其他可能使用到的核包括高斯径向基函数核

$$k(\boldsymbol{x}, \boldsymbol{x}') = \exp\left(-\frac{\|\boldsymbol{x} - \boldsymbol{x}'\|^2}{2\sigma^2}\right) \tag{3.279}$$

and the sigmoid kernel

和 sigmoid 核

$$k(\boldsymbol{x}, \boldsymbol{x}') = \tanh(\kappa \boldsymbol{x}^\top \boldsymbol{x}' + \vartheta) \tag{3.280}$$

This is the same function that is also used in the hidden layer of the MLP. With any of the above four kernels, SVMs can approximate any separating hypersurface arbitrarily closely.

这个函数与多层感知器隐藏层中使用的函数相同。使用上面四个核中的任意一个，支持向量机可以近似任意分离超曲面。

The above training algorithm that determines the support vectors still assumes that the classes can be separated by a hyperplane in the higher dimensional transformed feature space. This may not always be achievable. For these cases, the training algorithm can be extended to handle a certain number of training samples that are classified incorrectly or lie between the margins (so-called margin errors). Thus, a margin error does not necessarily imply that a training sample is classified incorrectly. It may still lie on the correct side of the separating hyperplane. To handle margin errors, a parameter ν is introduced (ν is the Greek letter nu). The value of ν is an upper bound on the fraction of margin errors. It is also a lower bound on the fraction of support vectors

上面介绍的训练算法仍然假设在变换到更高维特征空间中可以使用超平面将类别分离开。这个假设在实际应用中并不总是可以满足。针对这些情况，训练算法可以扩展为处理一定数量的训练样本，这些样本或者分类错误，或者位于两个边界之间（所谓的边际误差）。因此边际误差并不一定意味着训练样本被错误分类。也许样本仍然在分离超平面的正确的一边。为了处理边际误差，引入参数 ν（ν 是希腊字母 nu）。ν 的值是边际误差范围的上界，也是训练数据中支持向量的下界。SVM 的扩展叫做 ν-SVM（Schölkopf and Smola, 2002）。为了训练 ν-SVM，必须解决下面的二次

in the training data. This extension of SVMs is called a ν-SVM (Schölkopf and Smola, 2002). To train a ν-SVM, the following quadratic programming problem must be solved: maximize

规划问题：最大化

$$-\frac{1}{2}\sum_{i=1}^{l}\sum_{j=1}^{l}\alpha_i\alpha_j y_i y_j k(\boldsymbol{x}_i,\boldsymbol{x}_j) \tag{3.281}$$

subject to

满足

$$0\leqslant \alpha_i \leqslant \frac{1}{l},\quad i=1,\cdots,l \tag{3.282}$$

$$\sum_{i=1}^{l}\alpha_i y_i = 0 \tag{3.283}$$

$$\sum_{i=1}^{l}\alpha_i \geqslant \nu \tag{3.284}$$

The decision function for ν-SVMs is still given by Eq. (3.276).

By its nature, SVM classification only can handle two-class problems. To extend SVMs to multiclass problems, two basic approaches are possible. The first strategy is to perform a pairwise classification of the feature vector against all pairs of classes and to use the class that obtains the most votes, i.e., is selected most often as the result of the pairwise classification. Note that this implies that $m(m-1)/2$ classifications must be performed if there are m classes. This strategy is called "one-versus-one." The second strategy is to perform m classifications of one class against the union of the rest of the classes. This strategy is called "one-versus-all." From an efficiency point of view, the one-versus-all strategy may seem to be preferable since it depends linearly on the number of classes. However, in the one-versus-all strategy, typically there will be a larger number of sup-

ν-SVMs 的决策函数仍然由等式（3.276）给出。

SVM 分类器本身只能处理两个类型的问题。为了将 SVM 分类器扩展为可处理多类型的问题，可以使用两种基本方法。第一个策略是将特征向量进行两两分类，也就是对所有可能的类别进行分类，最后使用得到最多选票的类别，也就是挑选最多的作为两两分类的结果。注意这就意味着如果有 m 个类型的话就需要执行 $m(m-1)/2$ 次分类。这个策略称为"一对一"策略。第二种策略是执行 m 次分类，每次分类用来分离一种类型与其他所有剩余类型，称为"一对所有"策略。从效率的角度看，一对所有策略可能更好一些，因为它与类型数量的关系是线性的。然而需要注意的是在一对所有策略中一般情况下会比两两分类时需要的支持向量更多。

port vectors than in the one-versus-one strategy. Since the run time depends linearly on the number of support vectors, this number must grow less than quadratically for the one-versus-all strategy to be faster.

Figures 3.190(a) and (b) show an example ν-SVM with a Gaussian radial basis function kernel applied to the same problem as in Figures 3.182 and 3.187 (see Sections 3.15.2.2 and 3.15.3.2). The training samples are shown in Figure 3.190(a). Because of the large overlap of the training samples, ν was set to 0.25. To classify the three classes, the one-versus-one strategy was used. The resulting classes are displayed in Figure 3.190(b). Every point in the feature space is assigned to a class, even if it lies arbitrarily far from the training samples. This shows that SVMs, as discussed so far, are incapable of novelty detection.

由于运行时间与支持向量数量成线性关系，因此只有支持向量数量增长比二次方程慢时一对所有策略才更快一些。

图 3.190（a）和图 3.190（b）所示为一个 ν-SVM 示例，示例中 ν-SVM 使用高斯径向基函数核分别应用到图 3.182 和 3.187 的相同问题上（见 3.15.2.2 节和 3.15.3.2 节）。训练样本如图 3.190（a）所示。因为训练样本重叠度很高，ν 设置为 0.25。为了分类这三类，则使用一对多策略。产生的类如图 3.190（b）所示。特征空间中的每一个点都被归属于某一个类，无论该点与训练样本相差多远。到目前为止所讨论的，表明 SVM 没有异常检测能力。

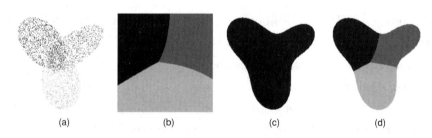

图 3.190　两个 SVM 例子的分类结果：(a) 二维特征空间中三个类型的样本，三个类型的样本分别使用不同灰度值表示；(b) 通过 ν-SVM 把特征空间分为三个类，没有异常检测；(c) 带异常检测的 SVM 分类结果，非异常的特征向量用深灰色区域表示；(d) 在 (b) 和 (c) 中 SVM 分类结果的组合

To construct an SVM that is capable of novelty detection, the architecture of the SVM must be modified since in novelty detection there are no class labels y_i. The basic idea is to construct a separating hypersurface around the training data that separates the training samples from the rest of the feature space. This effectively is a single-class classification problem: separate the known class from everything that is novel.

为构建一个具有异常检测能力的 SVM，SVM 的结构必须修改，因为在异常检测中没有类标签 y_i。基本思路是围绕训练数据构建一个分离超曲面，这个分离超曲面可以把训练样本从特征空间中剩余部分分离出来。这实际上是个单一类群分类问题：把已知类和任何异常类分开。

As for regular SVMs, the feature vectors are conceptually transformed into a higher-dimensional space. To determine the class boundary, a hyperplane is constructed in the higher-dimensional space that maximally separates the training samples from the origin. Not every kernel function can be used for this purpose. Gaussian radial basis functions can be used since they always return values > 0. Therefore, all transformed feature vectors lie in the same "octant" in the higher-dimensional space and are separated from the origin.

The above ideas are shown in Figure 3.191. The hyperplane is parameterized by its normal vector $\boldsymbol{w}$ and offset ρ. It is chosen such that the margin (the distance $\rho/\|\boldsymbol{w}\|$ of the hyperplane to the origin) is maximal. As for the other SVMs, the decision function is specified solely in terms of the support vectors via the kernel:

对于常规的 SVM，特征向量在概念上转换到更高维度空间。为了确定类的边界，在更高纬度空间中构建一个超平面，该超平面最大程度上从原点分离训练样本。不是每个核函数都可以用作此目的。高斯径向基函数可以使用，因为它们的返回值总大于 0。因此，在更高纬度空间中，所有转换后的特征向量都位于相同的"八分圆"（octant）并与原点分离。

以上思路如图 3.191 所示。超平面基于法向量 $\boldsymbol{w}$ 和偏移量 ρ 来参数化。挑选此分离超平面可以使得间隔最大（超平面到原点的距离 $\rho/\|\boldsymbol{w}\|$）。对于其他 SVM，决策函数仅由支持向量通过内核指定：

$$f(x) = \operatorname{sgn}\left(\sum_{i=1}^{l} \alpha_i k(\boldsymbol{x}_i, \boldsymbol{x}) - \rho\right) \qquad (3.285)$$

A feature vector $\boldsymbol{x}$ is classified as novel if $f(\boldsymbol{x}) = -1$.

如果 $f(\boldsymbol{x}) = -1$，特征向量 $\boldsymbol{x}$ 将被归类为异常。

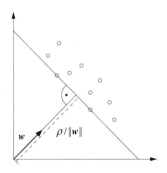

图 3.191 SVM 中的异常检测通过把特征向量转换到更高维度的空间中来执行，并且通过超平面把训练样本和原点分离开，超平面基于法向量 $\boldsymbol{w}$ 和偏移量 ρ 进行参数化，挑选这个分离超平面可以使得间隔最大（超平面到原点的距离 $\rho/\|\boldsymbol{w}\|$）

As for ν-SVMs, it is useful to add a provision to the SVM that allows for some margin errors on the training data, e.g., to account for training data that may contain some outliers. For this purpose, the parameter ν is used. It is an upper bound for the fraction of outliers in the training data and a lower bound for the fraction of support vectors (Schölkopf and Smola, 2002). To train the SVM for novelty detection, the following quadratic programming problem must be solved: maximize

对于 ν-SVM，给 SVM 增加如下规定是有用的：允许训练数据存在一些边际误差，例如，对可能包含一些异常点的训练数据进行解释。为此，会使用参数 ν。它是训练数据中异常值部分的上界，是支持向量部分的下界（Schölkopf and Smola, 2002）。为训练有异常检测能力的 SVM，如下二次规划问题必须解决：最大化

$$\frac{1}{2}\sum_{i=1}^{l}\sum_{j=1}^{l}\alpha_i\alpha_j k(\boldsymbol{x}_i,\boldsymbol{x}_j) \qquad (3.286)$$

subject to 满足

$$0 \leqslant \alpha_i \leqslant \frac{1}{\nu l}, \quad i=1,\cdots,l \qquad (3.287)$$

$$\sum_{i=1}^{l}\alpha_i = 1 \qquad (3.288)$$

Figure 3.190(c) shows the results of novelty detection with an SVM that was trained on the union of the training samples in Figure 3.190(a) with $\nu = 0.001$. The known class is displayed in dark gray, while all novel feature vectors are displayed in white. The class region tightly encloses the training samples. The results of combining novelty detection with the regular classification in Figure 3.190(b) is shown in Figure 3.190(d). Note that, in contrast to GMMs and MLPs, two separate SVMs are required for this purpose.

图 3.190（c）所示为一个 SVM 的异常检测结果，SVM 基于图 3.190（a）训练样本的合集，其中 $\nu = 0.001$。已知类用深灰色表示，同时所有的异常特征向量用白色表示。类的区域严格包含训练样本。异常检测结果和图 3.190（b）的普通分类结果相结合，如图 3.190（d）所示。注意，与 GMM 和 MLP 相比，需要两个独立的 SVM 来实现此目的。

3.15.3.4 Convolutional Neural Networks

The classifiers we have discussed so far rely on the availability of features that are able to separate the classes. How these features are computed must be specified by a person who has knowledge

3.15.3.4 卷积神经网络

到目前为止所讨论的分类器都依赖于能够把类分离开的特征的有效性。如何获取这些特征需要某个人来指定，这个人要具备所用分类器的相

of the application domain for which the classifier is used. For example, for OCR applications, features that were designed by domain experts are described in Section 3.14.2. Features that work well in one application domain do not necessarily work well in a different domain. Designing special features for each new application domain manually is very time-consuming and requires expert knowledge. Thus, it would be convenient if the classifier would not only learn, i.e., optimize, the actual classification but also the feature extraction. Ideally, it should be possible to train the classifier by merely using a set of images (without any feature extraction and preprocessing except for, potentially, zooming the images to a standard size and contrast-normalizing them) and the corresponding class labels.

As described in Section 3.15.3.2, MLPs with one hidden layer are universal approximators, i.e., they can, in theory, solve any classification problem. Therefore, we could simply try to train an MLP with a sufficient number of hidden units on the training images. This should result in an implicit feature representation being built up within the MLP. Unfortunately, the number of hidden units that are necessary for this kind of approach is typically too large to be used in practice (Goodfellow et al., 2016). To solve this problem, more hidden layers can be added to the MLP. With more layers, the MLP may achieve the same ability to solve a problem (the same capacity) with fewer parameters. Since MLPs can be regarded as becoming deeper the more layers are added, these networks are called "deep neural networks" and training them is called "deep learning" (Goodfellow et al., 2016).

关应用领域的知识。例如，对于 OCR 应用，由相关领域专家设计的特征如 3.14.2 节所描述。在某一应用领域适用的特征不一定适用于其他不同的领域。为每个新领域人工设计特定的特征非常耗时且需要专业知识。因此，如果分类器不仅可以学习，即优化实际的分类，而且可以学习特征提取，将会非常方便。理想情况下，仅仅通过一组图像（除了把图片缩放到标准尺寸，对比度归一化等操作外，不用任何特征提取和处理）和对应的类标签是有可能训练分类器的。

如 3.15.3.2 节所描述，有一个隐含层的 MLP 是通用逼近器，即，理论上它们可以解决任何分类问题。因此，可以简单地尝试训练一个 MLP，该 MLP 对于训练的图像有足够数量的隐含单元。这相当于在 MLP 内部构建一个隐式的特征表示。不幸的是，这类方法所必需的隐含单元的数量通常都会太大而不能在实际中应用 (Goodfellow et al., 2016)。为了解决这个问题，需要为 MLP 增加更多的隐含层。使用更多的层数，MLP 可以用更少的参数来获得解决某个问题的相同的能力。MLP 层数增加的越多，可以认为变得越深，所以这类网络被称为"深度神经网络"，并且训练它们称为"深度学习"(Goodfellow et al., 2016)。

While adding more layers helps to increase the capacity of an MLP and to learn the feature extraction, the architecture is not optimal in practice. Suppose we scale the input images of the classifier to 200×200 pixels. Thus, the input layer of the MLP has 40 000 features. As shown in Figure 3.185, all of the layers in an MLP are fully connected with each other. In particular, by Eq. (3.259), the first layer of the MLP has $40\,000 \times n_1$ weights and must compute a multiplication of this weight matrix with the input feature vector. The memory required to store the weight matrix and the time required to compute the matrix–vector product are too large for practical purposes. Therefore, it is essential to have a layer architecture that results in a more economical representation. One method to achieve this is to replace the matrix multiplication with a convolution operation (see Eq. (3.20)). Often, the operation that is actually used is a correlation (see Eq. (3.33); Goodfellow *et al.*, 2016). However, this distinction is immaterial for the purposes of this section. Therefore, as is common practice, we will refer to these operations as convolutions throughout this section.

Equations (3.20) and (3.33) are only defined for single-channel images. Often, multichannel images are used as input. It is straightforward to extend Eqs. (3.20) and (3.33) for multichannel images by adding a sum over the channels. A filter of size $s \times s$ for an image with c channels therefore has cs^2 filter parameters. Furthermore, as in MLPs, a bias parameter b is added after the convolution, resulting in a total number of $cs^2 + 1$ weights.

尽管 MLP 添加更多的层数可以帮助其增加能力并且学会特征提取，实际上其架构并没有被优化。假设把分类器的输入图像缩放到 200×200 像素。因此，MLP 的输入层有 40 000 个特征。如图 3.185 所示，MLP 中的所有层都是彼此完全连接的。尤其通过等式（3.259），MLP 的第一层有 $40\,000 \times n_1$ 个权值，并且必须计算权值矩阵与输入特征向量的乘积。对于实际用途来说，用来存储权值矩阵的内存和用来计算矩阵向量乘积的时间都太大。因此，必须有一层的架构能有更经济的表示方式。一种方法是用卷积操作代替矩阵相乘（见等式（3.20））。实际上经常使用的卷积操作是相关性（见等式（3.33）；Goodfellow *et al.*, 2016）。然而，这个差别对于本节来说并不重要。因此，按照惯例，在本节中把这些操作称之为卷积。

等式（3.20）和等式（3.33）只是对单通道图像的定义。多通道的图像也经常用来作为输入。对于多通道图像，通过增加通道求和来扩展等式（3.20）和等式（3.33）是比较直接的方式。一个滤波器尺寸为 $s \times s$，有 c 个通道的图像因此会有 cs^2 个参数。此外，在 MLP 中，卷积常常加上一个偏差参数 b，使总的权值数为 $cs^2 + 1$。

To obtain a useful feature representation, a single convolution filter is insufficient. For this purpose, more filters, say n_f, are required. Training a layer that performs convolutions therefore consists of optimizing the $n_f(cs^2+1)$ filter coefficients.

When a single filter is applied to an image, the filter output is called a feature map in the context of deep learning. Applying the n_f filters therefore results in n_f feature maps. These feature maps collectively can be regarded as an n_f-channel image. The feature maps can serve as the input of a subsequent layer of convolutions. Therefore, it is natural to replace the first few layers of a deep neural network by layers of convolutions. Such a network is called a convolutional neural network (CNN).

As described in Section 3.2.3.3, border treatment is an important aspect of any filter. In CNNs, one popular strategy is to compute the feature map only for the points for which the filter lies completely within the input feature map. This effectively reduces the size of the output feature map by $s-1$ pixels horizontally and vertically. If this is not desired, a popular strategy is to use zero-padding (Goodfellow et al., 2016). For simplicity, we will assume in the following that the first strategy is used.

So far, we have assumed that the filters are applied to each position in the input feature map. To reduce the computational burden, it is also possible to apply the filter only at every rth position horizontally and vertically. The parameter r is called the stride of the filter. Using a filter stride $r > 1$ has the same effect as subsampling the output feature maps by a factor of $r \times r$. For simplicity, we will assume $r = 1$ in the following.

为了得到一个有用的特征表示方式，单一卷积滤波器是不够的。为此，需要更多的滤波器，称为 n_f。因此，训练执行卷积操作的层由优化 $n_f(cs^2+1)$ 个滤波器系数组成。

当单一滤波器应用于图像时，在深度学习的背景下，滤波器的输出称为特征映射。因此 n_f 个滤波器表示 n_f 个特征映射。这些特征映射可以看作是一幅 n_f 通道的图片。特征映射可以作为卷积网络后续层的输入。因此，很自然的可以用卷积层替换深度神经网络中的前几层。这样的网络称为卷积神经网络（CNN）。

如 3.2.3.3 节所描述，对于任何一个滤波器来说，边界处理都是很重要的问题。在 CNN 中，一个流行的策略是对于滤波器完全位于输入特征映射中的那些点计算特征映射。这有效的减少了输出特征映射的尺寸，在垂直和水平方向上各 $s-1$ 个像素。如果这还不是所希望的，另外一个流行的策略是使用补零（Goodfellow et al., 2016）。为简单起见，假设下文中会使用第一种策略。

至此，假设滤波器已经应用到输入特征映射的各个位置。为减少计算负担，也有可能会只在水平和垂直方向上的每第 r 个位置使用滤波器。参数 r 称作滤波器的步幅。使用一个步幅大于 1 的滤波器和对输出特征映射进行一个因子为 $r \times r$ 的二次抽样一个效果。为简单起见，下面假设 $r=1$。

As for MLPs (cf. Section 3.15.3.2), it is important that a nonlinear activation function is applied to the convolutions (which are linear). Instead of using the logistic or hyperbolic tangent activation functions, CNNs typically use the following simpler function (see Goodfellow *et al.*, 2016)

$$f(x) = \max(0, x) \tag{3.289}$$

This function is nonlinear since negative inputs are clipped to 0. Units with this activation function are called rectified linear units (ReLUs). Like MLPs, the output layer of a CNN uses the softmax activation function. Neural networks with ReLUs are universal approximators (Goodfellow *et al.*, 2016).

CNNs with multiple convolution layers aggregate information. To see this, let us assume a CNN with two convolution layers with filters of size $s \times s$ and $t \times t$. We can see that the output feature map of the second convolution layer effectively computes a nonlinear filter of size $(s + t - 1) \times (s + t - 1)$ on the input image. The area of the input image over which a unit in a CNN effectively computes its output is called the receptive field of the unit. Each convolution layer in a CNN increases the receptive field and therefore aggregates information over a larger area of the input image.

To increase the aggregation capabilities of a CNN, it is often useful to add an explicit aggregation step to the architecture of the CNN. In the aggregation, the information of $p \times p$ pixels of a feature map is reduced to a single value. This step is called pooling. The most commonly used pooling function is to compute the maximum value of

与 MLP（见 3.15.3.2 节）一样，一个非线性激励函数应用于卷积（卷积为线性）是很重要的。代替使用逻辑（logistic）或者双曲正切激励函数，CNN 通常使用如下更简单的函数（参见（Goodfellow *et al.*，2016））

因为负的输入被截取为 0，这个函数是非线性。使用此激励函数的单元被称为修正线性单元（ReLUs）。像 MLP 那样，CNN 的输出层使用 softmax 激励函数。基于 ReLUs 的神经网络是通用逼近器（Goodfellow *et al.*, 2016）。

具有多个卷积层的 CNN 汇聚合信息。为了说明这一点，假设一个有两个卷积层，滤波器尺寸为 $s \times s$ 和 $t \times t$ 的 CNN。可以看到第二卷积层的输出特征映射在输入图像上有效地计算了一个尺寸为 $(s+t-1) \times (s+t-1)$ 的非线性滤波器。一个 CNN 单元基于输入图像可以有效计算其输出，则该输入图像的面积被称为单元的感受野。CNN 上每个卷积层都增加感受野，因此在输入图像的更大面积上聚合信息。

为增加 CNN 的聚合能力，在 CNN 架构上增加一个显式聚合步骤通常是有用的。在聚合过程中，特征映射中 $p \times p$ 像素信息减少到一个值。这个步骤称为池化。最常用的池化函数是计算一个 $p \times p$ 像素的最大值。这被称为最大池化（Goodfellow *et al.*,

the $p \times p$ pixels. This is called max pooling (Goodfellow et al., 2016). Since the goal is to aggregate information, max pooling is typically performed with a filter stride of p. This is what we will assume in the following.

The architecture of CNNs typically consists of fixed building blocks that consist of a convolution stage, a nonlinear activation stage, and a pooling stage. There are two sets of terminology that are commonly used (Goodfellow et al., 2016). In one terminology, the combination of convolution, nonlinear activation, and pooling is called a "convolutional layer." An alternative terminology is to call each of the above stages a layer of the CNN.

We now have the building blocks to construct an example CNN for OCR, shown in Figure 3.192. The architecture of this CNN is similar to that of the CNN proposed by LeCun et al. (1998). The input for the CNN is an image of the character that has been contrast normalized robustly and scaled to 28×28 pixels (see Section 3.14.2). In the first convolutional layer, 20 convolutions of size 5×5 are used, resulting in 20 feature maps of size 24×24. After this, ReLUs and 2×2 max pooling are applied, resulting in 20 feature maps of size 12×12. In the second convolutional layer, 50 convolutions of size $5 \times 5 \times 20$, ReLUs, and 2×2 max pooling are used, resulting in 50 feature maps of size 4×4. These 800 features serve as the input of an MLP that consists of two fully connected layers: a hidden layer with 500 units and an output layer with 82 units, which enables the CNN to read digits, uppercase and lowercase characters, and several special characters like parentheses, commas, colons, etc.

CNNs have a very large number of parameters. Even the simple CNN in Figure 3.192 already

2016）。因为目标是聚合信息，最大池化通常以滤波器步长为 p 来执行。这就是接下来要假设的。

CNN 的结构通常由固定的构件组成：卷积阶段、非线性激活阶段和池化阶段。通常使用两套术语（Goodfellow et al., 2016）。其中一套术语中，卷积组合、非线性激活和池化一起称为"卷积层"。另外一套术语把以上每个阶段称为 CNN 的一层。

现在来构建一个用于 OCR 的 CNN 例子，如图 3.192 所示。CNN 架构与（LeCun et al., 1998）提出的 CNN 架构相似。此 CNN 的输入是一幅图像，图像已经做了可靠的对比度归一化且缩放到 28×28 像素（见 3.14.2 节）。在第一卷积层，使用了 20 个 5×5 的卷积，产生了 20 个 24×24 的特征映射。然后，使用 ReLUs 和 2×2 池化，产生 20 个 12×12 的特征映射。在第二个卷积层，使用了 50 个 $5 \times 5 \times 20$ 的卷积、ReLUs，并使用了 2×2 最大池化，产生了 50 个 4×4 的特征映射。这 800 个特征作为一个由两层全连接层组成的 MLP 的输入：一个 500 个单元的隐含层和一个 82 个单元的输出层，该 CNN 能读取数字、大写小写字母和一些特殊字符。如括号、逗号、冒号等。

CNN 有大量的参数。即使图 3.192 中简单的 CNN 也有 467 152 个

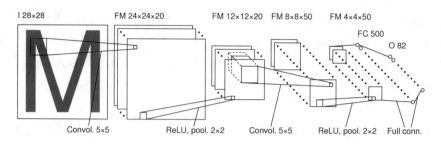

图 3.192　一个用于 OCR 的 CNN 架构的例子。一个字符缩放到 28×28，作为输入图像（I），该图像使用 20 个 5×5 的滤波器来卷积，产生 20 个尺寸为 24×24 的特征映射（FM）。然后，使用 ReLUs，并且特征映射使用 2×2 的最大池化来二次抽样，产生 20 个 12×12 特征映射。这些特征映射和 50 个 5×5×20 的滤波器做卷积，然后是 ReLUs 和最大池化，产生 50 个 4×4 的特征映射。这 800 个特征作为一个有两个全连接（FC）层的 MLP 的输入：一个含有 500 个单元的隐含层和一个含有 82 个单元的输出层

has 467 152 parameters (520 for the first convolutional layer, 25 050 for the second convolutional layer, and 441 582 for the two fully connected layers). The ability of a CNN to solve complex problems increases with the depth of the CNN (Goodfellow et al., 2016). There are CNNs with tens or even hundreds of layers. These CNNs can have tens of millions of parameters. It is obvious that hundreds of thousands or even millions of training images are necessary to train a CNN from scratch. This creates a problem in practice since collecting and labeling data sets of this size can be very costly and therefore sometimes economically infeasible.

Experience has shown that, relatively independent of the application domain, training CNNs on large training sets results in the fact that the first few convolutional layers learn low-level feature extractors for edges, lines, points, etc. Therefore, to train a CNN with fewer training samples (albeit still tens of thousands), the following strategy can be used. First, the CNN is trained with a large training set, which must be available from a different application. This initializes the parameters of the convolutional layers. Then,

参数（第一个卷积层有 520 个，第二个卷积层有 25 050 个，两个全连接层有 441 582 个）。CNN 解决复杂问题的能力随着 CNN 的深度而增加（Goodfellow et al., 2016）。有的 CNN 有 10 层或者数百层。这些 CNN 有数以千万计的参数。很明显训练一个用于划伤检测的 CNN 需要成千上万或者甚至是数百万的训练图片。这在实际应用中会产生一些问题，因为收集和标注这个规模的数据集会非常昂贵，因此有时在经济上是不可行的。

经验表明，相对独立的应用领域，在大的训练集上训练 CNN 事实上会导致前面几个卷积层学习边缘、直线和点等低级特征提取。因此，使用更少的训练样本训练 CNN（尽管还是要成千上百），将使用如下策略。首先，CNN 使用一个大的训练集训练，该训练集必须在不同应用中都可获得。这会初始化卷积层参数。然后，使用来自实际应用的训练集对 CNN 进行微调。视情况，前面几个卷积层可以不

the CNN is fine-tuned using the training set from the actual application. Optionally, the first few convolutional layers can be excluded from the training to preserve the low-level filters that were learned on the large training set.

To train CNNs, as for MLPs, the cross-entropy error in Eq. (3.265) is minimized. However, the fact that a large training set must be used to train CNNs leads to different training algorithms being used than for MLPs. For MLPs, typically the entire training set is used to compute the gradient of the cross-entropy error, which is then used in numerical minimization algorithms like the conjugate gradient or the scaled conjugate gradient algorithms (cf. Section 3.15.3.2). For CNNs, computing the gradient on the entire training set of millions of images is computationally too demanding. Furthermore, it is also wasteful since the precision with which the gradient can be determined is inversely proportional to the square root of the number of training images. Hence, using more training images offers an increasingly small benefit. Therefore, to train CNNs, an algorithm called stochastic gradient descent (SGD) is typically used (Goodfellow et al., 2016). This is a variant of the gradient descent algorithm that computes the gradient on a random subset of the training data at each iteration (optimization step). Each random subset of the training data is called a "minibatch." In practice, the minibatch is sometimes not chosen randomly at each iteration. Instead, the training data is shuffled randomly and the gradient descent algorithm simply uses consecutive minibatches of the randomly shuffled training data, cycling back to the beginning of the data once it reaches the end of the data.

训练，从而保护基于大量训练集学习过的低级滤波器。

为了与训练 MLP 一样训练 CNN，等式（3.265）中的交叉熵误差要最小化。然而，大量的训练集必须用来训练 CNN，这导致和 MLP 相比 CNN 需要使用不同的训练算法。对 MLP 来说，通常全部训练集用来计算交叉熵误差梯度，然后交叉熵误差又会用在数值最小化算法，如共轭梯度或缩放共轭梯度算法（参考 3.15.3.2 节）。对于 CNN，计算整个数百万图像训练集的梯度在计算方面来讲要求太高。而且，这也很浪费，因为梯度精度与训练图像数量的平方根成正比。因此，使用更多的训练图像提供的好处越来越小。为了训练 CNN，通常使用一个称为随机梯度下降（SGD）的算法（Goodfellow et al., 2016）。这是梯度下降算法的一个变形，该算法在每次迭代（优化步骤）时计算训练数据中一个随机子集的梯度。每个训练数据的随机子集被称为一个"小批量"。实际上，有时候小批量是不能在每次迭代中自由选择的。相反，训练数据被随机打乱，梯度下降算法仅仅连续使用随机打乱的训练数据的小批量数据，当到达数据末尾时，循环回到数据的开始。

To minimize the error function, SGD takes a step of fixed size in the negtive gradient direction at each iteration. The step size is called the "learning rate" in the context of CNNs. It is an important hyper-parameter that must be set correctly for the training to be successful and fast. In practice, the learning rate must be decreased as the optimization progresses (Goodfellow et al., 2016).

Even if the learning rate is set appropriately, the training may still be too slow in practice because SGD has a tendency to take small zigzag steps towards the minimum of the error function. To prevent zigzagging, the method of momentum can be used. It modifies the current gradient by a weighted average of previous gradients to tweak the current descent direction into the direction of the previous descent directions. Effectively, this results in an adaptive learning rate and causes the minimum of the error function to be found faster (Goodfellow et al., 2016). There are a few other algorithms that determine the learning rate adaptively, e.g., AdaGrad or Adam. The interested reader is referred to Goodfellow et al. (2016) for details.

Because of the large number of parameters, there is a danger that CNNs overfit the training data, even if a large amount of training data is used. Therefore, as for MLPs (cf. Section 3.15.3.2), it is essential that CNNs are regularized during training. One popular approach is to use weight decay with a single hyper-parameter, as in Eq. (3.266). Furthermore, stopping the SGD algorithm early, i.e., before it has converged to the minimum, also can have a regularizing effect (Goodfellow et al., 2016). Another important

为了最小化误差函数，随机梯度下降算法（SGD）在每次迭代中都在负梯度方向上采取固定大小的步长。该步长在 CNN 中称为"学习率"。这是一个重要的超参数，这个参数需要正确设置来保证训练能够快速成功。实际上，随着优化的进行必须降低学习率。

即使设置了恰当的学习率，实际训练也可能仍然很慢，因为 SGD 对于最小化误差函数倾向采取小的 z 字形步长。为了阻止 z 字形运动，可以使用动量方法。通过之前梯度的加权平均来修正当前梯度，从而把当前梯度下降方向调整到之前梯度下降方向。实际上，这产生了一个自适应学习率并使误差函数的最小值能够更快速的找到（Goodfellow et al., 2016）。还有一些其他的算法来确定自适应学习率，如 AdaGrad 算法和 Adam 算法。有兴趣的读者可以参见文献（Goodfellow et al., 2016）。

因为大量的参数，即使使用大量的训练数据，存在的危险就是 CNN 过度适应训练数据。因此，和 MLP 一样（见 3.15.3.2 节），有必要让 CNN 在训练时正则化。一个流行的方法是使用一个单一超参的权值衰退，如等式（3.266）所述。此外，早一些停止 SGD 算法，即在它收敛到最小值之前停止，也能有正则化效果（Goodfellow et al., 2016）。另外一个重要的策略是用训练样本的系统变化

strategy is to extend the training data set with systematic variations of the training samples. This is called "dataset augmentation." For example, additional training samples can be generated by adding random noise to the existing training samples. Furthermore, the existing training samples can be altered by random or systematic geometric distortions. For example, for OCR, the training samples can be subjected to small random rotations. In addition, gray value dilations and erosions can be used to alter the stroke width of the characters. Further regularization strategies are described in Goodfellow *et al.* (2016).

Computing the output of a CNN on an image (either a training image or an image to be classified) is called "inference" or "forward pass." Computing the gradient of the error function is called "backpropagation" or "backward pass." Both operations are computationally very demanding, especially if a large number of convolutional layers is used. This leads to the fact that CNNs are often too slow for practical applications when used on CPUs. To achieve run times that are adequate for practical applications, currently graphics processing units (GPUs) are commonly used because they offer a significantly greater computing power than CPUs for this kind of algorithm. Having to use GPUs in machines creates its own set of problems, however. GPUs have a high power consumption and therefore must be cooled actively using fans. This makes it difficult to use them in harsh factory environments. Furthermore, the availability of a certain GPU model is typically very short, often only a few months to a few years. This is problematic for machine builders, who must ensure the availability and maintenance of their machines for many years, sometimes for decades. The few GPU

来扩展训练数据集。这被称为"数据增强"。例如，给已有的训练样本增加随机噪声来增加训练样本。此外，已有的训练样本也可以通过随机或者系统的几何畸变来改变。例如，对于OCR，训练样本可以进行小的随机旋转。另外，灰度值膨胀和腐蚀可以用来改变字符的笔画宽度。更多的正规化策略可以参见（Goodfellow *et al.*，2016）。

计算一幅图像的CNN输出（无论是训练图像或者是待分类图片）被称作"推断"或"前向传播"。计算误差函数的梯度被称为"反向传播"或者"反向迭代"。两种操作从计算量上来说都要求非常高，特别是使用大量的卷积层的时候。这导致在实际应用中，如果用CPU，计算会非常慢。对于实际应用为了实现合适的运行时间，现在通常使用图形处理单元（GPU），因为对于这种算法，GPU计算能力明显大于CPU。但是，在机器中使用GPU也有自己的问题。GPU功率较大，因此需要使用风扇降温。这使得它很难在严苛的工业环境中使用。此外，特定型号的GPU使用寿命通常很短，常常就是几个月到几年。这对于机器制造商来说是有问题的，他们必须确保机器的有效性和维护性能够持续很多年，也可能几十年。少数能用这么久的GPU都非常贵，因此，大幅提高了机器的成本。

models that are available over a long period are very expensive and therefore can raise machine costs substantially.

Like MLPs, CNNs are inherently incapable of novelty detection. Automatic novelty detection for CNNs is essentially an open research problem. The approach that was described in Section 3.15.3.2 cannot be used for CNNs since the input features (the images) are typically not distributed compactly in the feature space. Therefore, the bounding boxes around the features would be too loose to be useful. Furthermore, due to the high dimension of the feature space, the number of rejection samples would have to be extremely large, which would slow down the training too much. To equip CNNs with the capability of novelty detection, the most popular option is to train an explicit rejection class with manually or automatically collected training samples. The automatic collection is typically performed by cropping random sub-images from images that do not show any of the classes of interest. Alternatively, sub-images of images that show the classes of interest can be randomly cropped, while ensuring that the objects of interest are not contained in the sub-images.

像 MLP 一样，CNN 本身无法进行异常检测。CNN 的自动异常检测本质上是一个开放的研究问题。3.15.3.2 节中描述的方法不能用于 CNN，因为输入特征（图像）在特征空间通常不是紧凑分布的。因此，在特征周围的边界框太松散而无法使用。此外，因为高维度的特征空间，剔除样本的数量将会特别大，使得训练速度大幅度下降。为了给 CNN 提供异常检测的能力，最流行的选择是人工或者自动收集训练样本来训练一个显式拒绝类。自动收集通常通过裁减对任何类没有表明兴趣的图像的随机子图像。或者，显示出感兴趣类的图像可以自由裁减成子图像，同时，要确保感兴趣的对象没有包含在子图像中。

3.15.4 Example of Using Classifiers for OCR

We conclude this section with an example that uses the ICs that we have already used in Section 3.4. In this application, the goal is to read the characters in the last two lines of the print in the ICs. Figures 3.193(a) and (b) show images of two sample ICs. The segmentation is performed with a threshold that is selected automatically based on

3.15.4 使用分类器用于 OCR 的例子

使用 3.4 节中已经使用过的 IC 作为示例来结束本节。在这个应用中，目的是读取 IC 表面上印刷的最后两行字符。图 3.193（a）和图 3.193（b）中显示两个 IC 样本的图像。首先利用图像灰度值直方图（图 3.26）自动选择阈值进行阈值分割。分割后，基

the gray value histogram (see Figure 3.26). After this, the last two lines of characters are selected based on the smallest surrounding rectangle of the characters. For this, the character with the largest row coordinate is determined, and characters lying within an interval above this character are selected. Furthermore, irrelevant characters like the "–" are suppressed based on the height of the characters. The characters are classified with an MLP that has been trained with several tens of thousands of samples of characters on electronic components, which do not include the characters on the ICs in Figure 3.193. The result of the segmentation and classification is shown in Figures 3.193(c) and (d). Note that all characters have been read correctly.

于字符的最小外接矩形选择最后两行字符，为此，首先确定行坐标最大的字符，然后选择这些字符上面一定间隔的字符作为倒数第二行的字符。另外，类似"–"这种无关的字符通过字符高度进行剔除。这个字符分类使用一个多层感知器（MLP），这个分类器是通过电子元器件上几万个字符作为样本进行训练得到的，这个训练集中不包括图 3.193 上 IC 上的字符。图 3.193（c）和图 3.193（d）中分别显示了图像分割和分类的结果。可以看到所有的字符都被正确读出。

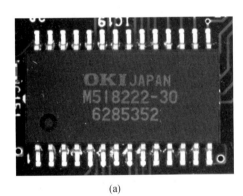

(a)

(b)

(c)

(d)

图 3.193　(a)、(b) IC 上印刷体的图像；(c)、(d) 字符分割结果（浅灰）和 OCR 结果（黑）。基于图像灰度值直方图（图 3.26）自动选择阈值进行阈值分割。而且，只选择印刷体最后两行字符。另外类似"–"这种无关的字符通过字符高度进行剔除

4 Machine Vision Applications

To emphasize the engineering aspects of machine vision, this chapter contains a wealth of examples and exercises that show how the machine vision algorithms discussed in Chapter 3 can be combined in non-trivial ways to solve typical machine vision problems. The examples are based on the machine vision software HALCON. As described in the preface, you can download all the applications presented in this chapter to get a hands-on experience with machine vision. Throughout this chapter, we will mention the location where you can find the example programs. We use "..." to indicate the directory into which you have installed the downloaded applications.

4.1 Wafer Dicing

Semiconductor wafers generally contain multiple dies that are arranged in a rectangular grid. To obtain single dies, the wafer must be diced at the gaps between the dies. Because these gaps are typically very narrow ($<100\,\mu m$), they must be located with a very high accuracy in order not to damage the dies during the cutting process.

The algorithms used in this application are:
- Fast Fourier transform (Section 3.2.4.2)
- Correlation (Section 3.2.4.1)
- Shape-based matching (Section 3.11.5.6)

The corresponding example program is:
···/machine_vision_book/wafer_dicing/wafer_dicing.hdev

4. 机器视觉应用

为了强调机器视觉的工程应用，本章用大量的例子和练习来介绍第3章中所讲的机器视觉算法如何综合应用，展示算法在解决典型机器视觉应用问题的强大威力。所有的例子都是基于HALCON软件。如在前言中所讲，本章的所有应用都可以下载，每个人都可以亲自感受一下机器视觉。本章会告诉大家例子程序所在位置，"···"表示下载目录。

4.1 半导体晶片切割

半导体晶片通常含有多个矩形栅格排列的芯片。为了得到每一片芯片，需要在芯片间的空隙处切割。由于空隙通常小于$100\,\mu m$，为了保证在切割过程中不损坏芯片，需要在切割过程中精确定位。

本应用使用的算法：
- 快速Fourier变换（3.2.4.2节）。
- 相关（3.2.4.1节）。
- 基于形状的模板匹配（3.11.5.6节）。

相应的示例程序位于：
···/machine_vision_book/wafer_dicing/wafer_dicing.hdev

In this application, we locate the center lines of the gaps on a wafer (see Figure 4.1). In the first step, we determine the width and height of the rectangular dies. The exact position of the dies, and hence the position of the gaps, can then be extracted in the second step.

本应用要在图 4.1 中找到晶片缝隙的中线。第一步要测定矩形半导体晶片的宽和高。这样才能在第二步中提取出每个芯片的精确位置及缝隙的位置。

图 4.1　晶片图像。芯片水平对齐排列在矩形栅格上

We assume that all dies have a rectangular shape and have the same size in the image, i.e., the camera must be perpendicular to the wafer. If this camera setup cannot be realized or if the lens produces heavy distortions, the camera must be calibrated and the images must be rectified (see Section 3.9). Furthermore, in wafer dicing applications, in general the wafer is horizontally aligned in the image. Consequently, we do not need to worry about the orientation of the dies in the image, which reduces the complexity of the solution.

首先假设所有的芯片在图像上都是矩形，尺寸大小一样，因此摄像机必须与晶片垂直。如果摄像机做不到这一点或者镜头有较大的畸变，就必须对摄像机进行标定并对图像进行校正，见 3.9 节。在晶片切割应用中图像中晶片是水平的，因此不需要担心芯片在图像上的方向，这大大降低了解决方案的难度。

4.1.1　Determining the Width and Height of the Dies

4.1.1　确定芯片的宽度和高度

In the first step, the width and the height of the rectangular dies is determined by using autocorrelation. The autocorrelation of an image g is the

第一步，使用自相关的算法确定矩形芯片的宽度和高度。一幅图像的自相关 g 是图像本身的相关性 $(g \star g)$，

correlation of the image with itself ($g \star g$), and hence can be used to find repeating patterns in the image, which in our case is the rectangular structure of the die. As described in Section 3.2.4.1, the correlation can be computed in the frequency domain by a simple multiplication. This can be performed by the following operations:

因此可以用于在图像中找到相同的样品，在这个例子中就是矩形的芯片结构。在 3.2.4.1 节中讲过，相关性可以在频域中通过简单的乘法计算得出。通过如下运算可以得到：

```
rft_generic (WaferDies, ImageFFT, 'to_freq', 'none', 'complex', Width)
correlation_fft (ImageFFT, ImageFFT, CorrelationFFT)
rft_generic (CorrelationFFT, Correlation, 'from_freq', 'n',
             'real', Width)
```

First, the input image is transformed into the frequency domain with the fast Fourier transform. Because we know that the image is real-valued, we do not need to compute the complete Fourier transform, but only one half, by using the real-valued Fourier transform (see Section 3.2.4.2). Then, the correlation is computed by multiplying the Fourier-transformed image with its complex conjugate. Finally, the resulting correlation, which is represented in the frequency domain, is transformed back into the spatial domain by using the inverse Fourier transform.

首先利用快速傅里叶变换将图像转换到频域。因为图像的数据一定是实数，所以不需要完全傅里叶变换，仅用实数傅里叶变换做一半即可，见 3.2.4.2 节。然后，通过傅里叶变换后图像与其复共轭相乘计算出相关。最后将在频域中得到的相关结果利用反傅里叶变换变换回时域。

Figure 4.2(a) shows the result of the autocorrelation. The gray value of the pixel $(r, c)^\top$ in the autocorrelation image corresponds to the correlation value that is obtained when shifting the image by $(r, c)^\top$ and correlating it with the unshifted original image. Consequently, the pixel in the upper left corner (origin) of the image has a high gray value because it represents the correlation of the (unshifted) image with itself. Furthermore, if periodic rectangular structures of width and height (w, h) are present in the image, we also can expect a high correlation value at the position

图 4.2(a) 显示了自相关结果。自相关图像上 $(r, c)^\top$ 点的灰度值是将图像水平移动 r、垂直移动 c 后与原始图像做相关运算得到的相关值。因此，图像左上角（原点）的灰度值很高，因为这点代表没有移动的图像与其自身的相关性。如果图像上有周期性宽度 w 和高度 h 的矩形结构，在自相关图像的 (w, h) 位置上也会得到高的相关值。通过在相关图像上找到与左上角最近的局部极大值，直接得到周期结构的尺寸。亚像素精度局部

(w, h) in the autocorrelation image. Thus, by extracting the local maximum in the correlation image that is closest to the upper left corner, we directly obtain the size of the periodic structure from the position of the maximum. The local maxima can be extracted with subpixel precision by using the following lines of code:

极大值可以通过如下代码得到：

```
gen_rectangle1 (Rectangle, 1, 1, Height/2, Width/2)
reduce_domain (Correlation, Rectangle, CorrelationReduced)
local_max_sub_pix (CorrelationReduced, 'gauss', 2, 5000000, Row, Col)
```

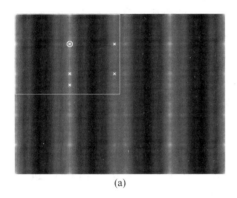

(a)

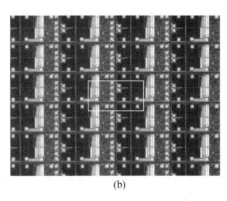

(b)

图 4.2　（a）晶片图像的自相关。较亮的灰度值代表较高的相关值。用于提取极大值的感兴趣区域为白色矩形。局部极大值以白色叉表示。白圈表示代表芯片大小的局部极大值；（b）矩形代表了提取出的芯片大小。为了便于观看，矩形人为被放置到图像中心

Note that, with the first two operators, the ROI is restricted to the inner part of the correlation image, excluding a one-pixel-wide border. This prevents local maxima from being extracted at the image border. On the one hand, local maxima at the image border can only be extracted with a limited subpixel precision because not all gray values in the neighborhood can be used for subpixel interpolation. On the other hand, we would have to identify the two corresponding local maxima at the image border to determine the width and the height of the dies instead of identifying only one maximum in the

前两个操作使得感兴趣区域限定在边缘为一个像素宽的相关图像内部。这避免了在图像边缘提取局部极大值。一方面，图像边缘局部极大值提取只能做到有限的亚像素精度，因为在边缘不是每个灰度值都可以用于亚像素插值。另一方面，在图像边缘必须得到两个局域极大值才能确定芯片相应的高度和宽度，而在其他部位仅需要得到一个极大值就可以。另外，可以将感兴趣区域限定在自相关图像的左上 1/4 部分，因为当假定晶片在图像是水平的，自相关是关于两个坐

diagonal direction. Additionally, the ROI can be restricted to the upper left quarter of the autocorrelation image. This is possible because the autocorrelation is an even function with respect to both coordinate axes when assuming that the wafer is horizontally aligned in the image. The ROI is visualized in Figure 4.2(a) by the white rectangle.

The resulting maxima are shown in Figure 4.2(a) as white crosses. From these maxima, the one that is closest to the origin can easily be selected by computing the distances of all maxima to the origin. It is indicated by the white circle in Figure 4.2(a). In our example, the size of the dies is $(w, h) = (159.55, 83.86)$ pixels. For visualization purposes, a rectangle of the corresponding size is displayed in Figure 4.2(b).

标轴的偶函数。在图 4.2（a）中感兴趣区域为白色矩形。

得到的极大值在图 4.2（a）中为白色叉。从得到的极大值中，通过计算所有极大值与原点的距离可以轻易地找出与原点最近的极大值，在图 4.2（a）中用白圈表示。在本例中芯片大小为 $(w, h) = (159.55, 83.86)$ 像素。为看起来方便，在图 4.2（b）中用矩形表示了对应尺寸。

4.1.2 Determining the Position of the Dies

The position of the dies can be determined by using HALCON's shape-based matching (see Section 3.11.5.6). Since we have calculated the width and height of the dies, we can create an artificial template image that contains a representation of four adjacent dies:

4.1.2 确定芯片的位置

利用 HALCON 基于形状的匹配算法可以得到芯片的位置，见 3.11.5.6 节。因为已经计算出芯片的宽度和高度，可以人为制作一个表示含有 4 个相邻芯片的模板图像：

```
LineWidth := 7
RefRow := round(0.5*Height)
RefCol := round(0.5*Width)
for Row := -0.5 to 0.5 by 1
    for Col := -0.5 to 0.5 by 1
        gen_rectangle2_contour_xld (Rectangle,
                                    RefRow+Row*DieHeight,
                                    RefCol+Col*DieWidth, 0,
                                    0.5*DieWidth-0.5*LineWidth,
```

```
                         0.5*DieHeight-0.5*LineWidth)
        paint_xld (Rectangle, Template, Template, 0)
    endfor
endfor
```

To obtain a correct representation, we additionally have to know the line width `LineWidth` between the dies. The line width is assumed to be constant for our application because the distance between the camera and the wafer does not change. Otherwise, we would have to introduce a scaling factor that can be used to compute the line width based on the determined die size. First, four black rectangles of the appropriate size are painted into an image that previously has been initialized with a gray value of 128. The center of the four rectangles is arbitrarily set to the image center.

为了得到正确的芯片尺寸，还需要知道每个芯片间缝隙线的宽度 LineWidth。由于摄像机与晶片间的距离不变，本例中假设线宽为常数。否则，需要引入比例因子来通过已经得到的芯片尺寸计算出线宽。首先在图像上画出 4 个初始化值为 128 灰度、尺寸大致为芯片大小的黑色矩形。4 个矩形的中心人为地设定在图像中心。

```
LineWidthFraction := 0.6
gen_rectangle2_contour_xld (Rectangle, RefRow, RefCol, 0,
                            0.5*LineWidthFraction*LineWidth, DieHeight)
paint_xld (Rectangle, Template, Template, 0)
gen_rectangle2_contour_xld (Rectangle, RefRow, RefCol, 0, DieWidth,
                            0.5*LineWidthFraction*LineWidth)
paint_xld (Rectangle, Template, Template, 0)
gen_rectangle2 (ROI, RefRow, RefCol, 0, DieWidth+5, DieHeight+5)
reduce_domain (Template, ROI, TemplateReduced)
```

After this, two additional black rectangles that reflect the non-uniform gray value of the lines are added, each of which covers 60% of the line width. Note that `paint_xld` paints the contour onto the background using anti-aliasing. Consequently, the gray value of a pixel depends on the fraction by which the pixel is covered by the rectangle. For example, if only half of the pixel is covered, the gray value is set to the mean gray value of the background and the rectangle. The

再加上两个黑色矩形，每个矩形覆盖 60% 的线宽，用于表示线的灰度值的不一致性。Paint_xld 无锯齿地将轮廓画于背景上。因此一个像素的灰度值取决于这个像素被这个矩形覆盖的比例。举例说明，如果像素的一半被覆盖，灰度值将为背景和矩形值的平均值。最后的矩形感兴趣区域代表为匹配算法产生形状模型的感兴趣区域。得到的模型图像和感兴趣区域

last rectangle `ROI` represents the ROI that is used to create the shape model that can be used for the matching. The resulting template image together with its ROI is shown in Figure 4.3(a). Finally, the shape model is created by using the following operation:

示于图 4.3（a）。最后使用下面操作产生形状模型：

```
create_shape_model (TemplateReduced, 'auto', 0, 0, 'auto', 'auto',
                    'ignore_local_polarity', 'auto', 5, ModelID)
```

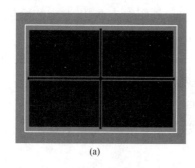 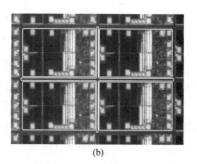

图 4.3　（a）匹配使用的模板，白色显示的是建立形状模型所用的感兴趣区域；
（b）基于形状匹配得到的最好匹配结果

The model can be used to find the position of the best match of the four adjacent dies in the original image that was shown in Figure 4.1:

此模型用于找到图 4.1 原始图中 4 个相邻芯片最匹配的位置：

```
NumMatches := 1
find_shape_model (WaferDies, ModelID, 0, 0, MinScore, NumMatches,
                  0.5, 'least_squares', 0, Greediness, MatchRow,
                  MatchColumn, MatchAngle, MatchScore)
```

The match obtained is visualized in Figure 4.3(b). Note that, although more than one instance of our model is present in the image, we are only interested in the best-fitting match, and hence the parameter `NumMatches` is set to 1. Also note that the coordinates `MatchRow` and `MatchColumn` are obtained with subpixel accuracy.

图 4.3（b）显示了匹配的结果。尽管图像中有多个模型，但是仅对最匹配的感兴趣，因此参数 NumMatches 置为 1。坐标 MatchRow 和 MatchColumn 为亚像素精度。

In the last step, the cutting lines can be computed based on the position of the found match, e.g., by using the following lines of code:

最后根据匹配的位置计算得出切割线：

```
NumRowMax := ceil(Height/DieHeight)
NumColMax := ceil(Width/DieWidth)
for RowIndex := -NumRowMax to NumRowMax by 1
    RowCurrent := MatchRow+RowIndex*DieHeight
    gen_contour_polygon_xld (CuttingLine, [RowCurrent,RowCurrent],
                             [0,Width-1])
    dev_display (CuttingLine)
endfor
for ColIndex := -NumColMax to NumColMax by 1
    ColCurrent := MatchColumn+ColIndex*DieWidth
    gen_contour_polygon_xld (CuttingLine, [0,Height-1],
                             [ColCurrent,ColCurrent])
    dev_display (CuttingLine)
endfor
```

First, the maximum number of cutting lines that might be present in the image in the horizontal and vertical directions is estimated based on the size of the image and of the dies. Then, starting from the obtained match coordinates MatchRow and MatchCol, parallel horizontal and vertical lines are computed using a step width of DieHeight and DieWidth, respectively. The result is shown in Figure 4.4.

首先根据图像和芯片大小估计出水平和垂直方向最多可能的切割线数量，然后从匹配得出的坐标 MatchRow 和 MatchCol 点开始计算出相距 DieHeight 相互平行的水平线，和相距 DieWidth 相互平行的垂直线。计算结果如图 4.4 所示。

To conclude this example, it should be noted that the proposed algorithm can also be used for wafers for which the dies have a different size or even a completely different appearance compared to the dies shown in this example. However, the algorithm requires that the size and the appearance of the gaps between the dies are known and constant. Therefore, for wafers with gaps that differ from the gaps in this example, a new appropriate template image would have to be created.

这个例子中建议使用的算法可以用于与本例不同大小甚至完全不同外观的晶片。算法要求芯片间缝隙的大小和外观是已知的并且是不变的。因此，如果芯片间的缝隙与本例中的不同，就需要创建新的合适的模板图像。

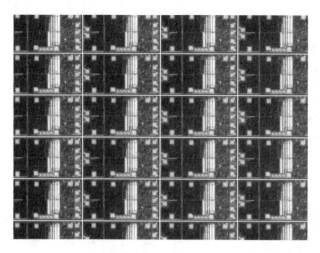

图 4.4　用于切割晶片的切割线

4.1.3　exercises

1. In the above application, we used the matching to find only the best-fitting match. Alternatively, the matching could be used to find all instances of the model in the image. Modify the program such that the cutting lines are computed based on all found matches.

2. The above program assumes that the wafer is horizontally aligned in the image. However, for some applications this assumption is not valid. Modify the program so that it can also be used for images in which the wafer appears slightly rotated, e.g., by up to ±20°. Tip 1: By extracting two maxima in the correlation image, the size as well as the orientation of the dies can be computed. For this, it is easiest to rearrange the correlation image such that the origin is in the center of the image before extracting the maxima (for example, by using the operators `crop_part` and `tile_images_offset`). Tip 2: The model that is used for the matching must be generated in the corresponding orientation.

4.1.3　练习

1. 在上例中使用匹配算法仅需要找到最匹配的一个。匹配算法还可以找到图像中所有相匹配的对象。修改程序使切割线基于所有匹配而计算得出。

2. 上面程序假设晶片在图像中水平放置。然而，对于有些应用中这种假设不成立。修改程序对于图像中晶片有一些旋转，比如小于 ±20° 时仍然有效。要点 1：在相关图像中提取两个极大值可以计算出芯片大小及方向。对于本问题最简单的方法是在提取极大值之前将相关图像进行重新排列使得原点位于图像中心。比如使用 `crop_part` 及 `tile_images_offset`。要点 2：匹配所用模型必须在相对应的方向生成。

4.2 Reading of Serial Numbers

In this application, we read the serial number that is printed on a CD by using OCR. The serial number is printed along a circle that is concentric with the CD's center. First, we rectify the image using a polar transformation to transform the characters into a standard position. Then, we segment the individual characters in the rectified image. In the last step, the characters are read by using a neural network classifier.

The algorithms used in this application are:

- Mean filter (Section 3.2.3.2).
- Dynamic thresholding (Section 3.4.1.3).
- Robust circle fitting (Section 3.8.2.2).
- Polar transformation (Section 3.3.4).
- Extraction of connected components (Section 3.4.2).
- Region features (Section 3.5.1).
- OCR (Section 3.14).
- Classification (Section 3.15).

The corresponding example program is:
···/machine_vision_book/reading_of_serial_numbers/reading_of_serial_numbers.hdev

Figure 4.5 shows the image of the center part of the CD. The image was acquired using diffuse bright-field front light illumination (see Section 2.1.5.1). The characters of the serial number, which is printed in the outermost annulus, appear dark on a bright background.

图 4.5 CD 中心部分图像。序列号印于最外环面，为了便于观看，用白色边界突出

4.2.1 Rectifying the Image Using a Polar Transformation

The first task is to determine the center of the CD and to compute the radius of the outer border of the outermost annulus. Based on this information, we will later rectify the image using a polar transformation. Because the outer border is darker than its local neighborhood, it can be segmented with a dynamic thresholding operation. To estimate the gray value of the local background, a 51×51 mean filter is applied. The filter size is chosen relatively large in order to suppress noise in the segmentation better. Because no neighboring objects can occur outside the annulus, we do not need to worry about the maximum filter size (see Section 3.4.1.3).

4.2.1 使用极坐标变换对图像进行校正

首要任务是确定 CD 的中心并计算出外环最外边界的直径。基于这些信息稍后对图像做极坐标变换。由于最外端边界比其邻近的像素要暗，可以通过动态阈值操作分割。使用 51×51 的均值滤波来估计局部背景的灰度值。可以选择相对较大的滤波器大小，这样分割时可更好地抑制噪声。由于环面外不可能有其他被测物，不必担心最大滤波器大小（见 3.4.1.3 节）。

```
mean_image (Image, ImageMean, 51, 51)
dyn_threshold (Image, ImageMean, RegionDynThresh, 15, 'dark')
```

```
fill_up (RegionDynThresh, RegionFillUp)
gen_contour_region_xld (RegionFillUp, ContourBorder, 'border')
fit_circle_contour_xld (ContourBorder, 'ahuber', -1, 0, 0, 3, 2,
                        CenterRow, CenterColumn, Radius,
                        StartPhi, EndPhi, PointOrder)
```

All image structures that are darker than the estimated background by 15 gray values are segmented. The result of the mean filter and the segmented region are shown in Figure 4.6(a). Note that the border of the outer annulus is completely contained in the result. To find the outer border, we fill up the holes in the segmentation result. Then, the center of the CD and the radius of the outer border can be obtained by fitting a circle to the outer border of the border pixels, which can be obtained with `gen_contour_region_xld`. The circle that best fits the resulting contour points is shown in Figure 4.6(b).

所有比估计的背景低 15 个灰度值的图像结构被分割出。中值滤波和区域分割的结果显示于图 4.6（a）。结果中最外环面的边界完全得到。为了找到外边界，将分割结果中的空洞填充。这样就可以使用 gen_contour_region_xld 得到外边界像素的拟合圆，从而得到 CD 的中心和外边界的直径。与轮廓拟合最好的圆如图 4.6（b）所示。

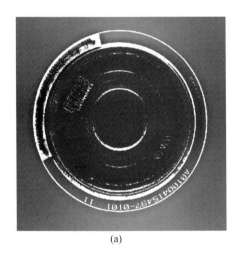 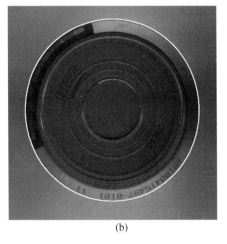

(a)   (b)

图 4.6　（a）中值滤波图像并叠加了白色来表示动态阈值操作结果；（b）原始图像并叠加了白色来表示边界分割结果的拟合圆。在做拟合圆之前分割结果中的空洞已被填充

Based on the circle center and the radius of the outer border, we can compute the parameters of the polar transformation that rectifies the annulus containing the characters:

基于圆心及外边界半径，可以计算出极坐标变换的参数来调整含有字符的环面：

```
AnnulusInner := 0.90
AnnulusOuter := 0.99
```

The inner and outer radii of the annulus to be rectified are assumed to be constant fractions (0.90 and 0.99, respectively) of the radius of the outer border. With this, we can achieve scale invariance in our application. In Figure 4.7, a zoomed part of the original image is shown. The fitted circle (solid line) and the inner and outer borders of the annulus (dashed lines) are overlaid in white. Note that the outer border of the annulus is chosen in such a way that the black line is excluded from the polar transformation, which simplifies the subsequent segmentation of the serial number.

假设环面的内径和外径与外边界半径之比是恒定的，在此是 0.90 和 0.99。这样可以做到尺度变换不变性。图 4.7 是放大了的原始图像，并叠加了白实线表示拟合圆，白虚线表示环面最内部和最外部边界。注意的是环面外边界选择不含黑线，这样将简化序列号的分割。

图 4.7　部分放大的原始图像。白实线表示拟合圆，白虚线表示确定极坐标变换的环面内、外边界叠加在图像上

The width and height of the transformed image are chosen such that no information is lost, i.e., the image resolution is kept constant:

变换图像的宽度和高度要保障无信息丢失，也就是图像分辨率不变：

```
WidthPolar  := 2*Pi*Radius*AnnulusOuter
HeightPolar := Radius*(AnnulusOuter-AnnulusInner)
```

Finally, the annulus arc that is to be transformed can be specified by its radius and angle interval:

最后由环面的径向范围和角度大小决定了需要做变换的环面弧：

```
RadiusStart := Radius*AnnulusOuter
RadiusEnd := Radius*AnnulusInner
AngleStart := 2*Pi-2*Pi/WidthPolar
AngleEnd := 0
```

Note that the start radius corresponds to the outer border of the annulus, while the end radius corresponds to its inner border. This ensures that the characters appear upright in the transformed image. Also note that the end angle is smaller than the start angle. This ensures that the characters can be read from left to right, since the polar transformation is performed in the mathematically positive orientation (counterclockwise). Finally, to avoid the first column of the transformed image being identical to the last column, we exclude the last discrete angle from the transformation. Now, we are ready to perform the polar transformation:

起始半径对应环面的外边界,而结束半径对应环面的内边界,这样保证变换后图像上的字符向上。极坐标变换是逆时针进行的,所以还需要使结束的角度小于开始的角度,以保证字符从左向右。最后为了避免变换后图像第一列与最后一列一样,变换时不含最后一个离散角。现在可以进行极坐标变换了:

```
polar_trans_image_ext (Image, PolarTransImage,
                       CenterRow, CenterColumn,
                       AngleStart, AngleEnd,
                       RadiusStart, RadiusEnd,
                       WidthPolar, HeightPolar, 'bilinear')
```

The center point of the polar transformation is set to the center of the CD. The transformed image is shown in Figure 4.8. Note that the characters of the serial number have been rectified successfully.

极坐标变换的中心点为 CD 的中心。变换后的图像如图 4.8 所示,序列号字符已成功调整。

图 4.8 变换到极坐标后含序列号的环面。为了方便观看,将变换后的图像分为三部分。图中从上至下对应的角度范围为 $[0, 2\pi/3]$、$[2\pi/3, 4\pi/3]$ 和 $[4\pi/3, 2\pi]$

4.2.2 Segmenting the Characters

In the next step, OCR can be applied. As described in Section 3.14, OCR consists of two tasks: segmentation and classification. Consequently, we first segment the single characters in the transformed image. For a robust segmentation, we must compute the approximate size of the characters in the image:

4.2.2 字符分割

下一步可进行 OCR。3.14 节中讲过 OCR 有分割和分类两大任务。首先在变换后的图像上将每个字符分割出来。为了确保分割的鲁棒性，必须计算出图像中字符的大致尺寸：

```
CharWidthFraction := 0.01
CharWidth := WidthPolar*CharWidthFraction
CharHeight := CharWidth
```

To maintain scale invariance, we compute the width of the characters as a constant fraction of the full circle, i.e., the width of the polar transformation. In this application, the width is approximately 1% of the full circle. Furthermore, the height of the characters is approximately the same as their width.

为保证尺度变换不变性，计算字符宽度时把它算为整个圆周也就是极坐标变换宽度的比。在这个应用中字符宽约为整个圆的 1%。字符的高度与宽度大致相等。

Because the characters are darker than their local neighborhood, we can use the dynamic thresholding again for the segmentation. In this case, the background is estimated with a mean filter that is twice the size of the characters:

由于字符比邻近的其他部分要暗，可以再次使用动态阈值进行分割。本例中背景值使用大小为字符尺寸两倍均值滤波器来估计：

```
mean_image (PolarTransImage, ImageMean, 2*CharWidth, 2*CharHeight)
dyn_threshold (PolarTransImage, ImageMean, RegionThreshold, 10, 'dark')
```

图 4.9 图 4.8 中三部分图像动态阈值分割的结果。三部分加边框是为了方便观看

The result of the thresholding operation is shown in Figure 4.9. Note that, in addition to the characters, several unwanted image regions are segmented. The latter can be eliminated by checking the values of appropriate region features of the connected components:

图 4.9 为阈值分割操作的结果。除了字符外，还有一些不需要的部分被分割出来。后者通过检测连通区域的区域特征值可以剔除掉：

```
connection (RegionThreshold, ConnectedRegions)
select_shape (ConnectedRegions, RegionChar,
          ['height','width','row'], 'and',
          [CharHeight*0.1, CharWidth*0.3, HeightPolar*0.25],
          [CharHeight*1.1, CharWidth*1.1, HeightPolar*0.75])
sort_region (RegionChar, RegionCharSort, 'character', 'true', 'row')
```

In this application, three region features are sufficient to single out the characters: the region height, the region width, and the row coordinate of the center of gravity of the region. The minimum value for the height must be set to a relatively small value because we want to segment the hyphen in the serial number. Furthermore, the minimum width must be set to a small value to include characters like "I" or "1" in the segmentation. The respective maximum values can be set to the character size enlarged by a small tolerance factor. Finally, the center of gravity must be around the center row of the transformed image. Note that the latter restriction would not be reasonable if punctuation marks like "." or "," had to be read. To obtain the characters in the correct order, the connected components are sorted from left to right. The resulting character regions are displayed in Figure 4.10.

本应用中三个区域特征就足够将字符挑出了。这三个特征就是区域高度、区域宽度和区域重心纵向坐标。高度的最小值必须置得相对很小，因为需要分割出序列号中的连字符；最小宽度也需要很小，因为分割中可能包含"I"和"1"这样的字符。最大值可以比字符尺寸稍大。最后，重心需要在变换后的图像纵向中心左右。最后一项在需要读出像"."或","这样标点符号时就没有意义了。从左到右从连通区域中挑选，以保证正确的字符顺序。挑选出的字符区域示于图 4.10。

图 4.10 图 4.8 中图像三个部分分割出的字符。字符由三种不同灰度显示。三部分加了边框是为了方便观看

4.2.3 Reading the Characters

In the last step, the segmented characters are classified, i.e., a symbolic label is assigned to each segmented region. HALCON provides several pretrained fonts that can be used for OCR. In this application, we use the pretrained font `Industrial`, which is similar to the font of the serial number:

```
read_ocr_class_mlp ('Industrial_Rej.omc', OCRHandle)
do_ocr_multi_class_mlp (RegionCharSort, PolarTransImage, OCRHandle,
                        Class, Confidence)
SNString := sum(Class)
```

First, the pretrained OCR classifier is read from file. The classifier is based on an MLP (see Section 3.15.3.2). In addition to gray value and gradient direction features (see Section 3.14.2), a few region features were used to train the classifier (see Section 3.5.1). The result of the classification is a tuple of class labels that are returned in `Class`, one for each input region. For convenience reasons, the class labels are converted into a single string.

For visualization purposes, the character regions can be transformed back from polar coordinates into the original image:

```
polar_trans_region_inv (RegionCharSort, XYTransRegion,
                        CenterRow, CenterColumn,
                        AngleStart, AngleEnd,
                        RadiusStart, RadiusEnd,
                        WidthPolar, HeightPolar,
                        Width, Height, 'nearest_neighbor')
```

In Figure 4.11, the original image, the transformed character region, and the result of the OCR are shown.

4.2.3 读取字符

最后一步是把分割出的字符分类,也就是把每个分割出的区域赋予一个符号标记。HALCON 提供了一些可用于 OCR 的训练过的字体,这个应用中我们使用与序列号相类似的 Industrial 字体:

图 4.11 中显示了原始图像、变换后的字符区域及 OCR 结果。

图 4.11 OCR 结果以白色文字叠加在图像的左上角。分割出的字符区域经反变换以白色叠加在原始图像上

4.2.4　exercises

1. The described program assumes that the serial number does not cross the 0° border in the polar transformation. Extend the program so that this restriction can be discarded. Tip: One possibility is to determine the orientation of the CD in the image. Another possibility is to generate an image that holds two copies of the polar transformation in a row.

2. In many applications, the segmentation of the characters is more difficult than in the described example. Some of the problems that might occur are discussed in Section 3.14. For such cases, HALCON provides the operator `find_text`, which eases the segmentation considerably. Modify the above program by replacing the segmentation and reading of the characters with this operator.

4.2.4　练习

1. 上述程序假设序列号不过极坐标变换 0° 边界，修改程序以取消此限制。要点：一种可能是确定 CD 的方向。另外一种可能是生成一幅图像并排放两幅相同的极坐标变换后图像。

2. 在许多应用中，字符分割比这个例子要难许多。3.14 节中描述了一些可能出现的情况。对此 HALCON 提供了算子 `find_text` 使得分割简单了许多。修改程序使用上述算子替代原有的字符分割算子并读出字符。

4.3 Inspection of Saw Blades

During the production process of saw blades, it is important to inspect the individual saw teeth to ensure that the shape of each tooth is within predefined limits. In such applications, high inspection speed with high inspection accuracy are the primary concerns.

Because in this application the essential information can be derived from the contour of the saw blades and the depth of the saw blade is small, we can take the images by using diffuse back lighting (see Section 2.1.5.4). Figure 4.12 shows an image of a saw blade that was taken with this kind of illumination. This simplifies the segmentation process significantly. First, the contour of the saw blade is extracted in the image. Then, the individual teeth are obtained by splitting the contour appropriately. Finally, for each tooth, the included angle between its two sides, i.e., the tooth face and the tooth back, is computed. This angle can then easily be compared to a reference value.

4.3 锯片检测

在锯片生产过程中检测锯片的每个锯齿以保证锯齿的外形在预先确定的范围内是非常重要的。在这个应用中检测速度和检测准确度是关注要点。

由于本应用从锯片轮廓中可以得到基本信息，锯片厚度很薄，可以使用漫反射背光照明（见 2.1.5.4 节）。图 4.12 显示的就是通过这种照明采集的锯片图像。这样可以大大简化分割的过程。首先，从图像中提取锯片的轮廓，然后通过适当的分离轮廓得到每个锯齿，最后每个锯齿两侧的坡口角度就可以计算出来并与参考值进行比较。

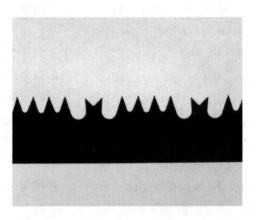

图 4.12　用于锯齿检测的背光照明锯片图像

The algorithms used in this application are:
- Subpixel-precise thresholding (Section 3.4.3).
- Segmentation of contours into lines and circles (Section 3.8.4.2).
- Contour length (Section 3.5.3.1).
- Robust line fitting (Section 3.8.1.2).

The corresponding example program is

···/machine_vision_book/saw_blade_inspection/saw_blade_inspection.hdev

4.3.1 Extracting the Saw Blade Contour

Because we want to extract the included angles of the tooth sides with high accuracy, it is advisable to use subpixel-precise algorithms in this application. First, we want to obtain the subpixel-precise contour of the saw blade in the image. In Chapter 3 we have seen that two possible ways to extract subpixel-precise contours are subpixel thresholding (see Section 3.4.3) and subpixel edge extraction (see Section 3.7.3.5). Unfortunately, the threshold that must be specified when using subpixel-precise thresholding influences the position of the contour. Nevertheless, in our case applying subpixel-precise thresholding is preferable because it is considerably faster than the computationally expensive edge extraction. Because of the back-lit saw blades, the background appears white while the saw blade appears black in the image. Therefore, it is not difficult to find an appropriate threshold, for example by simply using a mean gray value, e.g., 128. Moreover, the dependence of the contour position on the threshold value can be neglected in this application, as we will see later. Hence, the contour, which is visualized in Figure 4.13, is obtained by the following operator:

本应用使用的算法：
- 亚像素精度阈值分割（3.4.3 节）。
- 将轮廓分割为线和圆（3.8.4.2 节）。
- 轮廓长度（3.5.3.1 节）。
- 鲁棒的线段拟合（3.8.1.2 节）。

相应的例子程序位于：

4.3.1 提取锯片的轮廓

由于需要高准确度提取锯齿两侧的坡口角度，本应用建议使用亚像素精度算法。首先需要提取锯片图像亚像素精度的轮廓。第 3 章讲过有两种亚像素精度轮廓提取的方法。一种是亚像素阈值分割（见 3.4.3 节），另外一种是亚像素边缘提取（见 3.7.3.5 节）。不幸的是，使用亚像素阈值分割方法必须指定的阈值会影响轮廓的位置。然而，这个例子中使用亚像素阈值非常合适，因为比边缘提取运算快许多。由于使用背光，图像上背景为白，锯片为黑色，不难找到一个合适的阈值，比如均值 128。后面将看到，本应用中阈值对轮廓位置的影响可以忽略不计。图 4.13 中的轮廓可以通过如下操作得到：

```
threshold_sub_pix (Image, Border, 128)
```

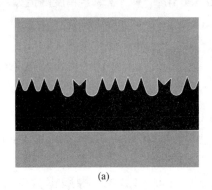
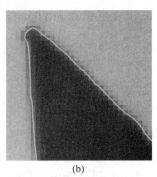

图 4.13　(a) 亚像素精度阈值分割得到的轮廓。为了便于观看提取出的轮廓，图像的对比度减小了；(b)（a) 图的细节（原始对比度）

4.3.2 Extracting the Teeth of the Saw Blade

4.3.2 提取锯片上的锯齿

In the next step, we get rid of all contour parts that are not part of the tooth sides:

下一步去除轮廓上不是锯齿的部分：

```
segment_contours_xld (Border, ContoursSplit, 'lines_circles', 5, 6, 4)
select_shape_xld (ContoursSplit, SelectedContours,
                  'contlength', 'and' 30, 200)
```

First, the contour is split into line segments and circular arcs by using the algorithm described in Section 3.8.4.2. On the one hand, this enables us to separate the linear contour parts of the tooth sides from the circularly shaped gullets between adjacent teeth. On the other hand, because polygons in the Ramer algorithm are approximated by line segments, we can also separate the face and the back of each tooth from each other. The resulting contour parts are shown in Figure 4.14(a). In the second step, the length of each contour part so obtained is calculated as described in Section 3.5.3.1. Because we know the approximate size of the tooth sides, we can discard all contour parts that are shorter than 30 and longer than 200 pixels by using `select_contours_xld`

首先，通过使用 3.8.4.2 节描述的算法将轮廓分为线段和圆弧。这样一方面可以将轮廓中锯齿直线部分与相邻锯齿间空隙的圆弧形分开，另一方面 Ramer 算法中的多边形是通过线段逼近的，可以分开每个锯齿的正面和后面。图 4.14（a）为得到的轮廓结果。第二步如 3.5.3.1 节所述计算出得到的轮廓的长度。由于大致知道锯齿每面的尺寸，可以使用 select_contours_xld 操作去掉比 30 个像素短或比 200 像素长的轮廓部分。

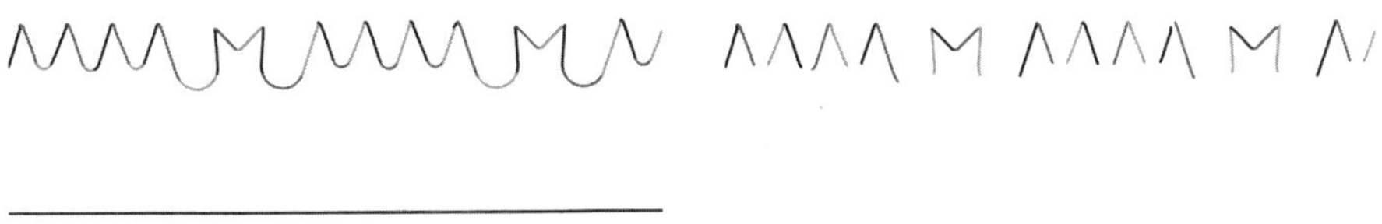

图 4.14 （a）将原始轮廓分为圆弧和线段后的轮廓，轮廓以三种不同的灰度显示；（b）去掉过长和过短及圆弧后剩下的锯齿

Finally, we can exclude all circular arcs, which represent the gullets, from further processing because we are only interested in the line segments, which represent the tooth sides:

最后因为仅对代表锯齿的线段感兴趣，所以通过进一步处理去掉所有代表锯齿间空隙的圆弧：

```
count_obj (SelectedContours, Number)
gen_empty_obj (ToothSides)
for Index2 := 1 to Number by 1
    select_obj (SelectedContours, SelectedContour, Index2)
    get_contour_global_attrib_xld (SelectedContour, 'cont_approx', Attrib)
    if (Attrib == -1)
        concat_obj (ToothSides, SelectedContour, ToothSides)
    endif
endfor
sort_contours_xld (ToothSides, ToothSidesSorted, 'upper_left',
                   'true', 'column')
```

This can be achieved by querying the attribute `cont_approx` for each contour part, which was set in `segment_contours_xld`. For line segments, the attribute returned in `Attrib` is `-1`, while for circular arcs it is `1`. The line segments correspond to the remaining tooth sides and are collected in the array `ToothSides`. They are shown in Figure 4.14(b). Finally, the tooth sides are sorted with respect to the column coor-

可以通过查询设置在 segment_contours_xld 中的每个轮廓部分的 "cont_approx" 属性来实现。对于线段 Attrib 返回值为 −1，对于圆弧返回值为 1。代表锯齿两侧的线段集中于数组 ToothSides，显示于图 4.14（b）。最后锯齿边缘按其外接矩形左上角坐标升序排序，这样的结果是把图像上的锯齿边缘按从左至右的

dinate of the upper left corner of their surrounding rectangles. The sorting is performed in ascending order. Consequently, the resulting `ToothSidesSorted` are sorted from left to right in the image, which later enables us to easily group the tooth sides into pairs, and hence to obtain the face and the back of each saw tooth.

顺序排序放于 ToothSidesSorted，据此可以轻易地将锯齿边缘分为一对一对的，就得到了每个锯齿的前面与后面。

4.3.3 Measuring the Angles of the Teeth of the Saw Blade

4.3.3 测量锯片锯齿的角度

In the next step, the orientation of the teeth's sides is computed. The straightforward way to do this would be to take the first and the last contour point of the contour part, which represents a line segment, and compute the orientation of the line running through both points. Unfortunately, this method would not be robust against outliers. As can be seen from Figure 4.15(a), which shows a zoomed part of a saw tooth, the tooth tip is not necessarily a perfect peak. Obviously, the last point of the tooth side lies far from the ideal tooth back, which would falsify the computation of the orientation. Figure 4.15(b) shows a line that has the same start and end points as the contour of the tooth back shown in Figure 4.15(a). A better way to compute the orientation is to use all contour points of the line segment instead of only the end points and to identify outliers. This can be achieved by robustly fitting a line through the contour points as described in Section 3.8.1.2. We use the Tukey weight function with a clipping factor of $2\sigma_\delta$ and five iterations:

下一步计算锯齿每侧的方向。最直接的办法就是找到轮廓的开始点和结束点，这两点代表分割出的一条线，利用这两点连成一条直线，计算这条直线的角度。但是对于轮廓线这种方法不是很可靠。从图 4.15（a）放大的锯齿图像中可以看出锯齿顶部不一定是非常尖的。显然锯齿结束点与理想的锯齿有很大差异，这就会造成锯齿方向的计算错误。图 4.15（b）显示了与图 4.15（a）锯齿轮廓开始点和结束点一样的直线。因此计算方向较好的方法是利用轮廓线段上所有的点，而不是仅使用开始和结束两点来代表外边界。可以通过使用 3.8.1.2 节办法得到拟合很好的直线。这里使用 5 次迭代，削波因数为 $2\sigma_\delta$ 的 Tukey 加权函数。

```
fit_line_contour_xld (ToothSidesSorted, 'tukey', -1, 0, 5, 2,
        Rows1, Columns1, Rows2, Columns2, Nr, Nc, Dist)
```

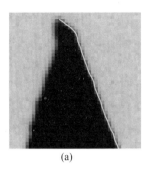 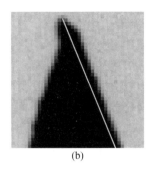 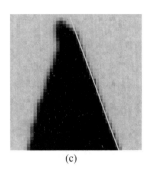

(a)　　　　　　　　　(b)　　　　　　　　　(c)

图 4.15　(a) 放大后的锯齿，代表锯齿后面的轮廓以白线显示；(b) 与 (a) 中轮廓线同起点、同终点的直线；(c) 与 (a) 中轮廓很好拟合的直线

The orientation of each fitted line can be computed with the following operation:

使用如下运算计算得出每条拟合出的直线方向：

```
line_orientation (Rows1, Columns1, Rows2, Columns2, Orientations)
```

At this point we can see why the threshold dependence of the contour position, which is introduced by the subpixel-precise thresholding, can be neglected. If the threshold were changed, the contour points of the tooth side would be shifted in a direction perpendicular to the tooth side. Because this shift would be approximately constant for all contour points, the direction of the fitted line would not change.

可以看出为什么亚像素阈值分割过程中，与阈值大小相关的轮廓位置的变化可以忽略不计。当阈值大小变化时，锯齿边轮廓点会沿与锯齿边垂直的方向移动。由于这种移动对于轮廓上各点大致都一样，所以拟合直线方向不变。

In the last step, for each tooth, the included angle between its face and its back can be computed with the following operations:

最后一步中每个锯齿前面与后面的夹角通过下面操作得到：

```
for Index2 := Start to |Orientations|-1 by 2
    Angle := abs(Orientations[Index2-1]-Orientations[Index2])
    if (Angle > PI_2)
        Angle := PI-Angle
    endif
endfor
```

Because we have sorted the tooth sides from left to right, we can simply step through the sorted

由于已经把锯齿边缘按从左到右排列，可以简单地每次在数组中使用

array by using a step size of 2 and select two consecutive tooth sides each time. This ensures that both tooth sides belong to the same tooth, of which the absolute angle difference can be computed. Finally, the obtained **Angle** is reduced to the interval $[0, \pi/2]$. In Figure 4.16, three examples of the computed tooth angles are displayed by showing the fitted lines from which the angle is computed. For visualization purposes, the two lines are extended up to their intersection point.

步长为 2，就选择出了连续的 2 个锯齿边缘，并保障了这两个边缘属于同一个锯齿，从而计算出锯齿角度。最后得到的 Angle 限定到 $[0, \pi/2]$ 之间。图 4.16 显示了三个例子，含有计算角度所用的拟合直线。为了视觉效果，直线延长到交点。

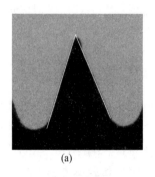
(a)
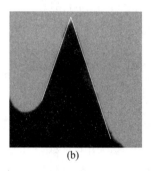
(b)
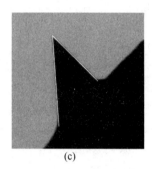
(c)

图 4.16 三个计算出角度的锯齿例子：(a) 41.35°；(b) 38.08°；(c) 41.64°。为了视觉效果对比度降低了

4.3.4 exercises

In the above program, only the angles of the tooth sides are computed. Often, it is also important to ensure that the teeth are of a perfect triangular shape, i.e., they do not show any indentations that exceed a certain size tolerance. Extend the program so that these irregularities of the tooth shape can be identified as well.

4.3.4 练习

上述程序仅计算出锯齿的角度。确保每个锯齿是完美的三角形常常也非常重要，也就是不可以有超出一定范围的缺口。修改程序检测出锯齿形状的缺陷。

4.4 Print Inspection

Print inspection is used to guarantee the quality of prints on arbitrary objects. Depending on the type of material and the printing technology, different image acquisition setups are used. For

4.4 印刷检测

印刷检测用于保障不同材质印刷时的质量。由于印刷材料和技术的不同需要不同的图像采集设置。对于透明物体，需要使用漫反射、明场背光

transparent objects, diffuse bright-field back light illumination may be used, while for other objects, e.g., for embossed characters or braille print, a directed dark-field front light illumination is necessary.

In this application, we inspect the textual information as well as the wiring diagram on the surface of a relay. We check for smears and splashes as well as for missing or misaligned parts of the print.

The algorithms used in this application are:
- 2D edge extraction (Section 3.7.3).
- Gray value morphology (Section 3.6.2).
- Shape-based matching (Section 3.11.5.6).
- Image transformation (Section 3.3.2).
- Variation model (Section 3.4.1.4).

The corresponding example program is:
···/machine_vision_book/print_inspection/print_inspection.hdev

本应用检测继电器上的文字和电路图。检查是否有模糊、飞墨以及印刷中的缺失或移位。

本应用使用的算法：
- 二维边缘提取（3.7.3 节）。
- 灰度形态学（3.6.2 节）。
- 基于形状的模板匹配（3.11.5.6 节）。
- 图像变换（3.3.2 节）。
- 变差模型（3.4.1.4 节）。

相应的例子位于：

4.4.1 Creating the Model of the Correct Print on the Relay

4.4.1 创建继电器上正确印刷信息的模型

The image of a relay shown in Figure 4.17(a) was acquired with diffuse bright-field front light illumination (see Section 2.1.5.1). The printed textual information and the wiring diagram appear dark on the bright surface of the relay. Because we only want to check the print and not the boundary of the relay or the position of the print on the relay, a predefined, manually generated, ROI is used to reduce the domain of the image.

图 4.17（a）为使用漫射、明场正面照明（见 2.1.5.1 节）采集的继电器图像。印刷的文字信息及电路图为黑色，继电器背景为白色。为了减少图像范围，人为限定感兴趣区域，因为仅需检查印刷的信息，并不需要检查继电器的边界或印刷信息在继电器上的位置。

```
reduce_domain (Image, ROI, ImageReduced)
```

The reduced image (see Figure 4.17(b)) is used as the reference image. Because only small size and position tolerances are allowed, the com-

图 4.17（b）为简化后的图像，作为参考图像。由于允许的大小和位置误差较小，质量合格的继电器印刷信

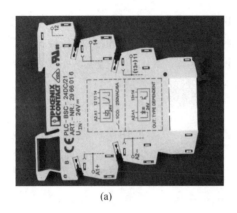 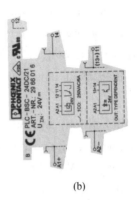

(a) (b)

图 4.17 （a）使用漫反射、明场正面照明采集的继电器图像；（b）仅显示感兴趣区域的参考图像

parison of images of correctly printed relays with the reference image will result in larger differences only near gray value edges of the reference image. For this reason, the allowed variations can be approximated by computing the edge amplitude from the reference image:

息与参考图像仅在边缘处灰度有较大的不同。为此，可以通过计算参考图像边缘振幅来估计允许的误差：

```
Sigma := 0.5
edges_image (ImageReduced, ImaAmp, ImaDir, 'canny', Sigma, 'none', 20, 40)
gray_dilation_rect (ImaAmp, VariationImage, 3, 3)
```

The reference image contains edges that lie close to each other. To minimize the influence of neighboring edges on the calculated edge amplitude, a small filter mask is used. For the Canny edge filter, $\sigma = 0.5$ corresponds to a mask size of 3×3 pixels. To broaden the area of allowed differences between the test image and the reference image, a gray value dilation is applied to the amplitude image. The resulting variation image is shown in Figure 4.18.

参考图像含有相互靠近的边缘。为了使相邻边缘对边缘振幅计算造成的影响最小化，使用一个小的滤波器掩模。对于 Canny 边缘滤波器，$\sigma = 0.5$ 对应的掩模大小为 3×3 像素。振幅图像使用灰度膨胀以使被检测图像与参考图像间的允许差异增加。图 4.18 为得到的偏差图像。

With the reference image and the variation image, we can create the variation model directly:

有了参考图像和偏差图像，可以直接生成变差模型：

```
AbsThreshold := 15
VarThreshold := 1
```

```
create_variation_model (Width, Height, 'byte', 'direct', VarModelID)
prepare_direct_variation_model (ImageReduced, VariationImage,
                    VarModelID, AbsThreshold, VarThreshold)
```

(a)

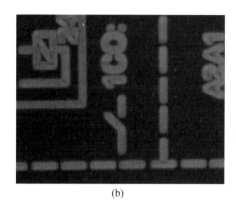
(b)

图 4.18　（a）参考图 4.17（b）的偏差图像；（b）偏差图像的细节

4.4.2 Creating the Model to Align the Relays

For the comparison of the test image with the reference image, it is essential that the two images are perfectly aligned. To determine the transformation between the two images, robust template matching is used:

4.4.2 创建一个用于对齐继电器的模型

为了能够将被测图像与参考图像进行比较，两个图像必须很好地对齐。使用鲁棒的模板匹配来确定这两幅图像间的偏差量。

```
reduce_domain (Image, MatchingROI, Template)
create_shape_model (Template, 5, -rad(5), rad(10),
                    'auto', 'auto', 'use_polarity',
                    'auto', 'auto', ShapeModelID)
area_center (MatchingROI, ModelArea, ModelRow, ModelColumn)
```

To speed up the matching, we use only selected parts of the print as template. These parts surround the area to be inspected to ensure that test images can be aligned with high accuracy. The shape model is created such that the test images may be rotated by ±5 degrees with respect to the reference image.

为了提高匹配速度，仅把印刷信息选择出一部分作为模板。这些部分包围需要检测的区域以保障被测图像可以高精度对准。形状模型使得被测图像可以相对参考图像旋转 ±5°。

4.4.3 Performing the Print Inspection

With the above preparations, we are ready to check the test images. For each test image, we must determine the transformation parameters that align it with the reference image:

```
find_shape_model (Image, ShapeModelID, -rad(5), rad(10),
                  0.5, 1, 0.5, 'least_squares', 0, 0.9,
                  Row, Column, Angle, Score)
vector_angle_to_rigid (Row, Column, Angle,
                  ModelRow, ModelColumn, 0, HomMat2D)
affine_trans_image (Image, ImageAligned, HomMat2D,
                  'constant', 'false')
```

First, we determine the position and orientation of the template in the test image. Using these values, we can calculate the transformation parameters of a rigid transformation that aligns the test image with the reference image. Then, we transform the test image using the determined transformation parameters. Now, we compare the transformed test image with the reference image:

```
reduce_domain (ImageAligned, ROI, ImageAlignedReduced)
compare_variation_model (ImageAlignedReduced, RegionDiff, VarModelID)
```

The resulting region may contain some clutter, and the parts of the regions that indicate defects may be disconnected. To separate the defect regions from the clutter, we assume that the latter is equally distributed while defect regions that belong to the same defect lie close to each other. Therefore, we can close small gaps in the defect regions by performing a dilation:

```
MinComponentSize := 5
dilation_circle (RegionDiff, RegionDilation, 3.5)
connection (RegionDilation, ConnectedRegions)
```

4.4.3 印刷检测

做好上述准备后，就可以检测被测图像了。对于每个被测图像，需要确定与参考图像重叠的转换参数：

首先，确定模板在被测图像上的位置和方向，计算出使被测图像与参考图像对准的刚性变换的变换参数。然后，使用这些变换参数将被测图像进行转换，将转换后的被测图像与参考图像进行比较：

得到的区域中可能含有杂乱的东西，部分有缺陷的区域可能不连通。为了分辨有缺陷的区域与杂乱的东西，假设杂乱的东西是均匀分布的，而同一个缺陷区域彼此靠得很近。所以可以通过膨胀运算将有缺陷区域的缝隙闭合：

```
intersection (ConnectedRegions, RegionDiff, RegionIntersection)
select_shape (RegionIntersection, SelectedRegions,
              'area', 'and', MinComponentSize, ModelArea)
```

After the dilation, previously disjoint defect regions are merged. From the merged regions, the connected components are computed. To obtain the original shape of the defects, the connected components are intersected with the original non-dilated region. Note that the connectivity of the components is preserved during the intersection computation. Thus, with the dilation, we can simply extend the neighborhood definition that is used in the computation of the connected components. Finally, we select all components with an area exceeding a predefined minimum size.

The results are shown in Figure 4.19(a) and (b). The detected errors are indicated by black ellipses along with the expected edges from the reference image, which are outlined in white.

膨胀运算后,先前断开的缺陷区域闭合了。从闭合的区域计算连通区域。为了得到缺陷的原始形状,取膨胀后区域与原始未膨胀区域交集,这一过程连通性不变。因此膨胀仅扩大了用于计算连通区域的边界。最后选出所有面积超过预定最小值的区域。

结果显示于图 4.19(a) 和图 4.19(b) 中。发现的缺陷用黑色椭圆表示,参考图像中正确的边缘用白色表示。

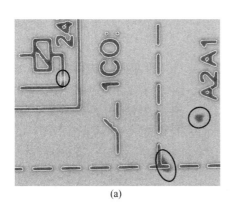 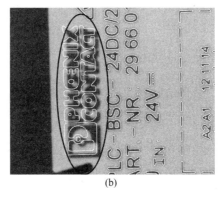

(a)   (b)

图 4.19 印刷检测结果。发现的缺陷用黑色椭圆表示,参考图像中正确的边缘用白色表示,为了视觉效果,图像对比度减小了:(a) 印刷中漏印、模糊和飞墨;(b) 套印偏差

4.4.4 exercises

1. In the above application, the variation image has been created by computing the edge amplitude image of the reference image. Another

4.4.4 练习

1. 上述应用中偏差图像是通过计算参考图像边缘振幅得到的。另外一种可能是生成人造变差模型。修改

possibility for the creation of the variation image from one reference image is to create artificial variations of the model. Modify the example program such that the variation image is created from multiple artificially created reference images.

2. In the example program described above, the application of the variation model approach is only possible if non-varying illumination of the object can be assured. Extend the program such that it works even for illumination conditions that vary over time.

4.5 Inspection of Ball Grid Arrays

A BGA is a chip package having solder balls on the underside for mounting on a PCB. To ensure a proper connection to the PCB, it is important that all individual balls are at the correct position, have the correct size, and are not damaged in any way.

In this application, we inspect BGAs. First, we check the size and shape of the balls. Then, we test the BGA for missing or extraneous balls as well as for wrongly positioned balls.

The algorithms used in this application are:
- Thresholding (Section 3.4.1.1).
- Extraction of connected components (Section 3.4.2).
- Region features (Section 3.5.1).
- Subpixel-precise thresholding (Section 3.4.3).
- Contour moments (Section 3.5.3.2).
- Geometric transformations (Section 3.3).

The corresponding example program is:
··· /machin_vision_book/bga_inspection/bga_inspection.hdev

例子程序利用多个人造参考图像生成变差图像。

2. 上述例子程序中，必须保证被测物的照明条件不变才可以使用此方法得到变差模型，扩充程序使得即使照明条件不断变化也可以工作。

4.5 BGA 封装检查

BGA 是一种芯片封装，其底部有球形焊点，用于安装在印刷电路板上。为了保障与印刷电路板良好连接，每个球的正确位置、正确尺寸以及不能有任何损坏都是非常重要的。

本应用检测 BGA。首先，检查焊锡球的大小和形状。然后检查是否有缺少的或多余焊锡球以及是否有位置不对的。

本应用使用的算法：
- 阈值分割（3.4.1.1 节）。
- 连通区域提取（3.4.2 节）。
- 区域特征（3.5.1 节）。
- 亚像素阈值分割（3.4.3 节）。
- 轮廓矩（3.5.3.2 节）。
- 几何变换（3.3 节）。

相应的例子程序位于：

For BGA inspection, the images are typically acquired with directed dark-field front light illumination (see Section 2.1.5.3). With this kind of illumination, the balls appear as doughnut-like structures, while the surroundings of the balls are dark (see Figure 4.20). To ensure that all correct balls appear in the same size and that they form a rectangular grid in the image, the image plane of the camera must be parallel to the BGA. If this camera setup cannot be realized for the image acquisition or if the lenses produce heavy distortions, the camera must be calibrated and the images must be rectified (see Section 3.9).

对于 BGA 检测通常使用暗场直接正面照明（见 2.1.5.3 节）。使用这种照明，焊锡球为多纳圈状结构，而球的周围是暗的（如图 4.20 所示）。为保障所有正确的焊锡球在图像上呈现同样的大小并形成矩形网格，摄像机像平面必须与 BGA 平行。如果摄像机安装不能保证这一点或者镜头产生很大的径向畸变，摄像机必须进行标定，图像必须经过校正（见 3.9 节）。

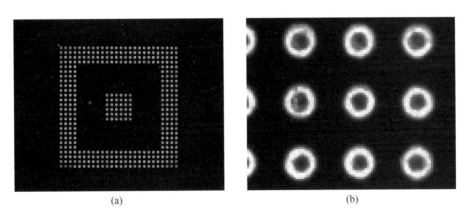

图 4.20　BGA 图像。图像采集时使用暗场直接正面照明。焊锡球为多纳圈状结构，而球的周围是暗的。（a）整个图像；（b）BGA 左上部分

4.5.1 Finding Balls with Shape Defects

First, we will determine wrongly sized and damaged balls. For this, we have to segment the balls in the image. Then, we calculate features of the segmented regions that describe their size and shape. If the feature values of a region are within a given range, the respective ball is assumed to be correct. Otherwise, it will be marked as wrongly sized or damaged.

4.5.1 找出有形状缺陷的焊锡球

首先确定尺寸不对和已损坏的焊锡球。为此要在图像中将焊锡球分割出来。然后计算分割出的区域中表征其尺寸和形状的特征。如果区域特征值在一定范围内，相对应的焊锡球就认为是正确无误的。否则就标记为尺寸不对或有损坏。

There are different ways to perform this check. One possibility is to carry out a pixel-precise segmentation and to derive the features either from the segmented regions (see Section 3.5.1) or from the gray values that lie inside the segmented regions (see Section 3.5.2). Another possibility is to use a subpixel-precise segmentation (see Section 3.4.3) and to derive the features from the contours (see Section 3.5.3). In this application, we use pixel-precise segmentation with region-based features as well as subpixel-precise segmentation.

有多种方法可完成这个检查。一种是进行像素精度分割，然后从分割出的区域中提取特征（见 3.5.1 节）或从分割出的区域内灰度值中提取特征（见 3.5.2 节）。另外一种可能性就是使用亚像素分割（见 3.4.3 节），从轮廓中提取特征（见 3.5.3 节）。本应用使用像素精度分割、区域特征提取及亚像素分割。

The pixel-precise segmentation of the balls can be performed as follows:

通过如下操作完成像素分割：

```
threshold (Image, Region, BrighterThan, 255)
connection (Region, ConnectedRegions)
fill_up (ConnectedRegions, RegionFillUp)
select_shape (RegionFillUp, Balls, ['area','circularity'], 'and',
              [0.8*MinArea,0.75], [1.2*MaxArea,1.0])
```

First, the region that is brighter than a given threshold is determined. Then, the connected components of this region are calculated and all holes in these connected components are filled. Finally, the components that have an area in a given range and that have a more or less circular shape are selected. To be able to detect wrongly sized balls, we select regions even if they are slightly smaller or larger than expected. A part of a test image that contains wrongly sized and shaped balls along with the extracted balls is displayed in Figure 4.21(a).

首先，确定比给定阈值更亮的区域。然后计算出区域中相连的成分，其中的空洞要填充上。最后选择出一定面积范围内、近似于圆的成分。为了能够检出尺寸不对的焊锡球，那些稍大或稍小的区域也要选择上。图 4.21(a) 显示了含有错误尺寸和形状的焊锡球以及提取结果的部分测试图像。

Now we select the regions that have an incorrect size or shape:

现在选择含有错误尺寸或形状的区域：

```
select_shape (Balls, WrongAreaBalls, ['area','area'],
              'or', [0,MaxArea], [MinArea,10000])
```

```
select_shape (Balls, WrongAnisometryBalls, 'anisometry',
              'and', MaxAnisometry, 100)
```

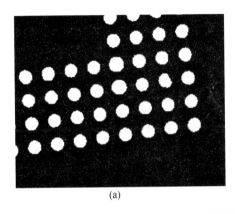 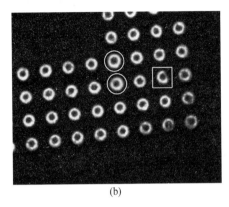

图 4.21　含有错误尺寸和形状的焊锡球的部分图像：（a）分割结果；（b）用白色圆圈标示的尺寸不对的焊锡球和用白色方框标示的错误形状的焊锡球

The first operator selects all regions that have an area less than `MinArea` pixels or larger than `MaxArea` pixels. These regions represent the wrongly sized balls. The second operator selects the regions that have an anisometry larger than a given value, e.g., `MaxAnisometry = 1.2`. The detected defective balls are shown in Figure 4.21(b). Wrongly sized balls are indicated by white circles around them and wrongly shaped balls by white squares.

The segmentation and feature extraction can also be done with subpixel-precise segmentation methods and contour-based features. First, we use the balls extracted above to define an ROI for the subpixel-precise extraction of the boundaries of the balls. In the image that is reduced to this ROI, we perform the following operations:

第一个操作选择了所有面积比 MinArea 个像素少或面积比 MaxArea 个像素多的区域。这些区域代表尺寸不对的焊锡球。第二个操作选出了异向性大于给定值的区域比如 MaxAnisometry=1.2。检测出的不合格焊锡球如图 4.21（b）所示。白色圆圈标示的为尺寸不对的焊锡球，用白色方框圈起来的为错误形状的焊锡球。

分割和特征提取还可以通过亚像素分割和基于轮廓特征。首先，利用上面提取出的焊锡球定义亚像素精度轮廓提取的感兴趣区域。对于图像中减小后的感兴趣区域进行如下操作：

```
threshold_sub_pix (ImageReduced, Boundary, BrighterThan)
select_shape_xld (Boundary, Balls, 'area', 'or', 0.8*MinArea, 1.2*MaxArea)
```

The boundaries of the balls are extracted with subpixel-precise thresholding. To eliminate

焊锡球的边界以亚像素精度被提取出来。为了消除杂点，仅选择包围

clutter, only the boundaries that enclose a suitable area are selected.

As above, we can now select the boundaries that have an incorrect size or shape:

一定面积的边界。

同上可以选择出尺寸及形状不正确的那些边界：

```
select_shape_xld (Balls, WrongAreaBalls, ['area','area'],
                 'or', [0,MaxArea], [MinArea,10000])
select_shape_xld (Balls, WrongAnisometryBalls, 'anisometry',
                 'or', MaxAnisometry, 100)
```

The above two methods—pixel-precise and subpixel-precise segmentation and feature extraction—yield similar results if the regions to be extracted are large enough. In this case, the pixel-precise approximation of the regions is sufficient. However, this does not hold for small regions, where the subpixel-precise approach provides better results.

如果需要提取的区域足够大，上述的像素精度和亚像素精度分割和特征提取两种方法产生的结果类似。对于这种情况像素精度就够了。然而对于很小的区域，像素精度就不够了，此时需要使用亚像素精度才能得到较好的结果。

4.5.2 Constructing a Geometric Model of a Correct BGA

4.5.2 构造一个正确的 BGA 几何模型

Up to now, we have not detected falsely positioned and missing balls. To be able to detect such balls, we need a model of a correct BGA. This model can be stored in matrix representation. In this matrix, a value of −1 indicates that there is no ball at the respective position of the BGA. If there is a ball, the respective entry of the matrix contains its (non-negative) index. The index points to two other arrays that hold the exact reference coordinates of the ball given, for example, in millimeters.

至此还不能找到位置不对和缺少的焊锡球。为了检测出这类缺陷，需要一个正确的 BGA 模型。这个模型可以存储于矩阵中。在矩阵中 −1 表示 BGA 相应的位置没有焊锡球。如有焊锡球，相应的矩阵登记了其索引。索引指向另外两个数组，数组中含有焊锡球对应的以毫米为单位的实际参考坐标。

For BGAs with concentric squares of balls, we can define the BGA layout in a tuple that contains a 1 if the respective square contains balls and 0 if not. The first entry of this tuple belongs to the

对于有些 BGA 其焊锡球分布为同心的正方形，对这些 BGA 可以将其布局放于一个元组中，含有焊锡球的正方形值为 1，没有焊锡球的为 0。

innermost square, the last entry to the outermost square of balls. Furthermore, we need to define the distance between neighboring balls.

元组的第一个记录为最内部的正方形，最后一个记录为最外端的正方形。进一步需要定义相邻焊锡球的距离。

```
BgaLayout := [1,1,1,0,0,0,0,0,0,1,1,1,1]
BallDistRowRef := 0.05*25.4
BallDistColRef := 0.05*25.4
```

Using this information, we can create the BGA model. We store the matrix in linearized form.

利用这些信息可以生成 BGA 模型。把矩阵存成线性的形式。

```
BallsPerRow := 2*|BgaLayout|
BallsPerCol := 2*|BgaLayout|
BallMatrixRef := gen_tuple_const(BallsPerRow*BallsPerCol,-1)
BallsRowsRef := []
BallsColsRef := []
CenterRow := (BallsPerRow-1)*0.5
CenterCol := (BallsPerCol-1)*0.5
I := 0
for R := 0 to BallsPerRow-1 by 1
    for C := 0 to BallsPerCol-1 by 1
        Dist := max(int(fabs([R-CenterRow,C-CenterCol])))
        if (BgaLayout[Dist])
            BallMatrixRef[R*BallsPerCol+C] := I
            BallsRowsRef := [BallsRowsRef,R*BallDistRowRef]
            BallsColsRef := [BallsColsRef,C*BallDistColRef]
            I := I+1
        endif
    endfor
endfor
```

First, the numbers of balls per row and column are determined from the given BGA layout tuple. Then, the whole BGA model matrix is initialized with -1, and empty fields for the x and y coordinates are created. For any position on the BGA, the index of the respective square in the BGA layout tuple is given by the chessboard distance (see Section 3.6.1.8) between the ball

首先，通过给定的 BGA 布局元组确定每行和每列焊锡球数量。然后，整个 BGA 模型矩阵初始化为 -1，x 和 y 坐标也清空。对于 BGA 上任意位置，相应方块在 BGA 布局元组中的索引为此焊锡球位置距 BGA 中心点的 8 邻域距离（见 3.6.1.8 节）。如果目前位置上有焊锡球，将 I 存于

position and the center of the BGA. If a ball exists at the current position, its index I stored in the linearized model matrix and its coordinates are appended to the fields of row and column coordinates.

We can display the BGA matrix with the following lines of code (see Figure 4.22):

线性化模型矩阵，相应的坐标加于行和列坐标中。

用下列代码可以简单显示 BGA 矩阵，如图 4.22 所示。

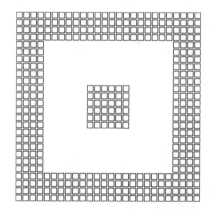

图 4.22　图 4.20 所示 BGA 的模型。矩形为焊锡球必须所在的位置

```
Scale := RectangleSize/(0.8*min([BallDistanceRow,BallDistanceCol]))
gen_rectangle2_contour_xld (Matrix,
                            RectangleSize/2.0+BallsRows*Scale,
                            RectangleSize/2.0+BallsCols*Scale,
                            gen_tuple_const(|BallsRows|,0),
                            gen_tuple_const(|BallsRows|,
                                            RectangleSize/2.0),
                            gen_tuple_const(|BallsRows|,
                                            RectangleSize/2.0))
dev_display (Matrix)
```

4.5.3 Finding Missing and Extraneous Balls

To detect erroneously placed and missing balls, we must create a similar matrix for the image to be investigated. This BGA matrix contains the actual presence and exact position of the balls.

4.5.3　检测缺失或多余的焊锡球

为了检测位置不正确及缺失的焊锡球，必须为需要检测的图像建立类似的矩阵。这个 BGA 矩阵包含实际存在的焊锡球的位置。

To build this matrix, we must relate the position of the balls in the image to the matrix positions, i.e., we need to know the relative position of the balls in the BGA. For this, we must know the size of the BGA in the current image:

```
area_center_xld (BallsSubPix, Area, BallsRows, BallsCols, PointOrder)
gen_region_points (RegionBallCenters, BallsRows, BallsCols)
smallest_rectangle2 (RegionBallCenters, RowBGARect, ColumnBGARect,
            PhiBGARect, Length1BGARect, Length2BGARect)
```

We create a region that contains all extracted ball positions and determine the smallest surrounding rectangle of this region. With this, we can determine the transformation of the ball positions into indices of the BGA matrix. Note that this only works if no extraneous balls lie outside the BGA.

```
hom_mat2d_identity (HomMat2DIdentity)
hom_mat2d_rotate (HomMat2DIdentity, -PhiBGARect,
            RowBGARect, ColumnBGARect, HomMat2DRotate)
hom_mat2d_translate (HomMat2DRotate,
            -RowBGARect+Length2BGARect,
            -ColumnBGARect+Length1BGARect,
            HomMat2DTranslate)
BallDistCol := 2*Length1BGARect/(BallsPerCol-1)
BallDistRow := 2*Length2BGARect/(BallsPerRow-1)
hom_mat2d_scale (HomMat2DTranslate, 1/BallDistRow, 1/BallDistCol,
            0, 0, HomMat2DScale)
```

The first part of this transformation is a rotation around the center of the BGA to align the BGA with the rows and columns of the model matrix. Then, we translate the rotated ball positions such that the upper left ball lies at the origin of the BGA model. Finally, the distance between the balls in the image is calculated in the row and column directions. The inverse distances are used

建立这样一个矩阵需要知道焊锡球在图像上的位置与矩阵位置的关系，也就是焊锡球在 BGA 上的相对位置。为此需要知道 BGA 在当前图像上的尺寸：

创建包含所有提取出焊锡球位置的区域，确定包含此区域的最小外接矩形。这样可以把焊锡球位置转换为 BGA 矩阵索引。此方法仅在没有任何位于 BGA 之外的焊锡球存在时有效。

变换的第一步是绕 BGA 中心点旋转使 BGA 行和列与模型矩阵重合。然后做坐标变换使左上角焊锡球位于 BGA 模型起点。最后计算出图像上焊锡球间横向和纵向的距离。距离的倒数用作焊锡球坐标的尺度，这样焊锡球横向和纵向的距离将为 1。

to scale the ball coordinates such that the distance between them in the row and column directions will be one.

Now, we transform the ball positions with the transformation matrix derived above and round them to get the index in the BGA model matrix for each extracted ball.

现在使用上面推导出的转换矩阵来转换焊锡球的位置，逐个得到每个焊锡球在 BGA 模型矩阵中的索引。

```
affine_trans_point_2d (HomMat2DScale, BallsRows, BallsCols,
                      RowNormalized, ColNormalized)
BallRowIndex := round(RowNormalized)
BallColIndex := round(ColNormalized)
```

Finally, we set the indices of the balls in the BGA matrix:

最后在 BGA 矩阵中放入焊锡球的索引：

```
BallMatrix := gen_tuple_const(BallsPerRow*BallsPerCol,-1)
for I := 0 to NumBalls-1 by 1
    BallMatrix[BallRowIndex[I]*BallsPerCol+BallColIndex[I]] := I
endfor
```

With this representation of the extracted balls, it is easy to detect missing or additional balls. Missing balls have a non-negative index in the BGA model matrix and a negative index in the BGA matrix of the current image, while extraneous balls have a negative index in the BGA model matrix and a non-negative index in the BGA matrix derived from the image.

有了代表提取出的焊锡球的索引，非常容易检测出是否有缺少或多余的焊锡球。缺少焊锡球在 BGA 模型中为非负索引而在图像的 BGA 矩阵中为负索引。多余的焊锡球在 BGA 模型中为负索引，在图像的 BGA 矩阵索引中为非负索引。

```
for I := 0 to BallsPerRow*BallsPerCol-1 by 1
    if (BallMatrixRef[I] >= 0 and BallMatrix[I] < 0)
        * Missing ball.
    endif
    if (BallMatrixRef[I] < 0 and BallMatrix[I] >= 0)
        * Extraneous ball.
    endif
endfor
```

The missing and extraneous balls are displayed in Figure 4.23(a). Missing balls are indicated by white diamonds while extraneous balls are indicated by white crosses.

图 4.23（a）显示了缺少和多余焊锡球。缺少的焊锡球以白色菱形标示，多余的焊锡球以白色叉标示。

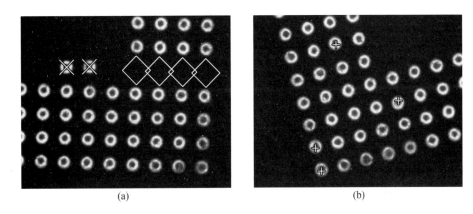

图 4.23　（a）缺少焊锡球（以白色菱形标示）和多余焊锡球（以白色叉标示）；
　　　　　（b）位置不对的焊锡球（正确的位置用白 + 号标示）

4.5.4　Finding Displaced Balls

Finally, we check the position of the balls. For this, we must transform the pixel coordinates of the extracted balls into the WCS in which the reference coordinates are given.

4.5.4　检测位置错误的焊锡球

最后检查焊锡球的位置。为此需要将提取出的焊锡球像素坐标转换到含有参考坐标的世界坐标系。

```
BallsMatchedRowsImage := []
BallsMatchedColsImage := []
IndexImage := []
BallsMatchedRowsRef := []
BallsMatchedColsRef := []
IndexRef := []
K := 0
for I := 0 to BallsPerRow*BallsPerCol-1 by 1
    if (BallMatrixRef[I] >= 0 and BallMatrix[I] >= 0)
        BallsMatchedRowsImage := [BallsMatchedRowsImage,
                        BallsRows[BallMatrix[I]]]
        BallsMatchedColsImage := [BallsMatchedColsImage,
                        BallsCols[BallMatrix[I]]]
        IndexImage := [IndexImage,BallMatrix[I]]
        BallsMatchedRowsRef := [BallsMatchedRowsRef,
```

```
                            BallsRowsRef[BallMatrixRef[I]]]
        BallsMatchedColsRef := [BallsMatchedColsRef,
                            BallsColsRef[BallMatrixRef[I]]]
        IndexRef := [IndexRef,BallMatrixRef[I]]
        K := K+1
    endif
endfor
```

To calculate the transformation from the pixel coordinate system to the WCS, we must establish a set of corresponding points in the two coordinate systems. Because we already know the mapping from the extracted ball positions to the positions of the balls in the BGA model, this can be done by using all points that are defined in both matrices. With this, we can determine the transformation and transform the extracted ball positions into the reference coordinate system.

为了能够将像素坐标系转换至世界坐标系，必须在两个坐标系中建立一系列对应点。因为已经知道了提取出的焊锡球位置到 BGA 模型焊锡球位置的映射关系，所以利用这两个矩阵中定义的所有点可以轻易地完成这一工作。利用这些对应点，确定转换参数并将提取出的焊锡球位置转换到参考坐标系。

```
vector_to_similarity (BallsMatchedRowsImage, BallsMatchedColsImage,
                BallsMatchedRowsRef, BallsMatchedColsRef, HomMat2D)
affine_trans_point_2d (HomMat2D,
                BallsMatchedRowsImage, BallsMatchedColsImage,
                BallsMatchedRowsWorld, BallsMatchedColsWorld)
```

Note that, if the camera was calibrated, the ball positions could be transformed into the WCS. Then, a rigid transformation could be used instead of the similarity transformation.

如果摄像机标定过，焊锡球的位置可以转换到世界坐标系，也就可以使用刚体变换代替相似变换。

The deviation of the ball positions from their reference positions can be determined simply by calculating the distance between the reference coordinates and the transformed pixel coordinates extracted from the image.

通过计算从图像中提取出的转换后像素坐标与参考坐标的距离就可以确定焊锡球与参考位置的偏离。

```
distance_pp (BallsMatchedRowsRef, BallsMatchedColsRef,
        BallsMatchedRowsWorld, BallsMatchedColsWorld,
        Distances)
for I := 0 to |Distances|-1 by 1
    if (Distances[I] > MaxDistance)
```

```
        * Wrongly placed ball.
    endif
endfor
```

Each ball for which the distance is larger than a given tolerance can be marked as wrongly placed. Figure 4.23(b) shows the wrongly placed balls indicated by white plus signs at the correct position.

4.5.5 exercises

1. In the above application, we calculated the features in two ways: from pixel-precise regions, and from subpixel-precise contours. A third possibility is to derive the features from the gray values that lie inside the segmented regions (see Section 3.5.2). Modify the above program so that the positions of the balls are determined in this way. Compare the results with the results achieved above. What conditions must be fulfilled so that the gray value features can be applied successfully to determine the center of segmented objects?

2. The creation of the BGA model described above is restricted to BGAs with balls located in concentric squares. To be more flexible, write a procedure that creates the BGA model from an image of a correct BGA and the reference distances between the balls in the row and column directions.

4.6 Surface Inspection

A typical machine vision application is to inspect the surface of an object. Often, it is essential to identify defects like scratches or ridges.

每个与参考位置偏差大于规定公差的焊锡球都作出位置错误的标记。图 4.23（b）显示了位置不对的焊锡球并以白色加号显示了正确的位置。

4.5.5 练习

1. 上述应用中使用两种方法进行特征提取：像素精度区域和亚像素轮廓的方法。第三种方法是从分割出的区域内灰度值提取特征（见 3.5.2 节）。修改程序使用这种方法确定焊锡球的位置。将结果与上述方法得到的结果进行比较。为了使灰度特征提取的方法有效，确定分割物体的中心必须满足的条件是什么？

2. 上述 BGA 模型的建立要求 BGA 上焊锡球分布为同心正方形。为了灵活，写一段程序从正确的 BGA 图像建立模型，含有焊锡球横向和纵向间距。

4.6 表面检测

物体表面检测是一种典型的机器视觉应用。识别诸如划痕或褶皱类的缺陷通常是最基本的。

In this application, we inspect the surface of doorknobs. Other industry sectors in which surface inspection is important are, for example, optics, the automobile industry, and the metal-working industry. An example image of a doorknob with a typical scratch is shown in Figure 4.24(a). First, we must select an appropriate illumination to highlight scratches in the surface. Then we can segment the doorknob in the image. After that, we create an ROI that contains the planar surface of the doorknob. In the last step, we search for scratches in the surface within the ROI.

本例检测门把手的表面。在其他一些像光学加工、汽车生产和金属加工等工业领域，表面检测都是很重要的。图 4.24（a）是门把手的图像，上面有常见的划痕。首先需要选择合适的照明来加强表面的划痕，然后，在图像上分割出门把手，创建含有门把手平面的感兴趣区域。最后，在感兴趣区域中寻找划痕。

 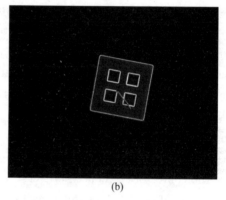

(a) (b)

图 4.24 （a）使用明场直接正面照明得到的门把手图像；（b）使用 LED 环形光暗场直接正面照明采集的同一个门把手图像，上面的划痕清晰可见

The algorithms used in this application are:
- Mean filter (Section 3.2.3.2).
- Dynamic thresholding (Section 3.4.1.3).
- Region features (Section 3.5.1).
- Region morphology (Section 3.6.1).
- Extraction of connected components (Section 3.4.2).
- Affine transformations (Section 3.3.1).

本应用使用的算法：
- 均值滤波（3.2.3.2 节）。
- 动态阈值分割（3.4.1.3 节）。
- 区域特征（3.5.1 节）。
- 区域形态学（3.6.1 节）。
- 连通区域提取（3.4.2 节）。
- 仿射变换（3.3.1 节）。

The corresponding example program is:
··· /machine_vision_book/surface_inspection/surface_inspection.hdev

相应的例子程序位于：

For surface inspection, the images typically are acquired with directed dark-field front light illumination (see Section 2.1.5.3). Figure 4.24(b) shows an image of the doorknob of Figure 4.24(a) acquired with directed dark-field front light illumination by using an LED ring light. The edges of the doorknobs appear bright, while its planar surface appears dark. This simplifies the segmentation process significantly. Note that the illumination also makes the scratch in the surface clearly visible.

4.6.1 Segmenting the Doorknob

Scratches appear bright in the dark regions. Unfortunately, the border of the doorknob and the borders of the four inner squares also appear bright. To distinguish the bright borders from the scratches, we first segment the bright border regions. We then subtract the segmented regions from the doorknob region and reduce the ROI for the scratch detection to the difference region.

Because the edges are locally brighter than the background, we can segment the doorknob in the image with dynamic thresholding (see Section 3.4.1.3):

```
KnobSizeMin := 100
mean_image (Image, ImageMean, KnobSizeMin, KnobSizeMin)
dyn_threshold (Image, ImageMean, RegionBrightBorder, 50, 'light')
fill_up (RegionBrightBorder, RegionDoorknob)
```

To estimate the gray value of the local background, we use a mean filter. We set the size of the mean filter to the minimum size at which the doorknob appears in the image to ensure that all local structures are eliminated by the smoothing, and hence to better suppress noise in the segmentation. Then, all image structures that are brighter than the background by 50 gray values are segmented, yielding the region of the bright border. The result of applying the dynamic thresholding to the image of Figure 4.24(b) is shown in Figure 4.25(a). To obtain the complete region of the doorknob, holes in the border region are filled (see Figure 4.25(b)).

出 50 个灰度级的所有图像结构被分割出来，产生高亮的边界区域。对图 4.24（b）应用动态阈值运算的结果如图 4.25（a）所示。为得到完整的门把手区域，需要将边界区域内的空洞填充，如图 4.25（b）所示。

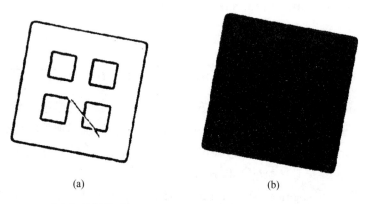

(a) (b)

图 4.25 （a）动态阈值操作结果，为了便于观看，仅显示了门把手放大后的一部分；
（b）（a）图中的空洞填充后得到的门把手区域

4.6.2 Finding the Surface to Inspect

4.6.2 找到需要检测的平面

In the next step, the planar surface that we would like to inspect is determined. For this, we must eliminate the bright outer border of the doorknob as well as the bright borders of the four small inner squares from the segmentation result. To do so, we first must determine the orientation and the size of the doorknob in the image:

下一步来确定需要检测的平面。为此需要从分割结果中去掉门把手的亮边界和中间 4 个小的正方形的亮边界。为此首先必须知道门把手在图像中的方向和大小：

```
erosion_circle (RegionDoorknob, RegionErosion, KnobSizeMin/4)
smallest_rectangle2 (RegionErosion, Row, Column, Phi, KnobLen1, KnobLen2)
KnobSize := KnobLen1+KnobLen2+KnobSizeMin/2
```

Because the doorknob has a square shape, its orientation corresponds to the orientation of the smallest enclosing rectangle of the segmented

因为门把手是正方形，其方向与分割区域最小外接矩形的方向是一致的（见 3.5.1.4 节）。由于分割区域

region (see Section 3.5.1.4). Since clutter or small protrusions of the segmented region would falsify the computation, they are eliminated in advance by applying an erosion to the doorknob region shown in Figure 4.25(b) (see Section 3.6.1.3). The radius of the circle that is used as the structuring element is set to a quarter of the minimum doorknob size. On the one hand, this allows relatively large protrusions to be eliminated. On the other hand, it ensures that the orientation can still be determined with sufficient accuracy. The size of the doorknob can be computed from the size of the smallest enclosing rectangle by adding the diameter of the circle that was used for the erosion.

Based on the orientation and the size of the doorknob, we segment the four inner squares by again using region morphology (see Section 3.6.1.6). First, small gaps in the previously segmented border of the inner squares are closed by performing two closing operations:

中杂点和小的突出物会导致计算错误，需要对图 4.25（b）门把手区域使用腐蚀运算将这些干扰去掉（见 3.6.1.3 节）。作为结构元素的圆的半径取为门把手最小尺寸的 1/4。这样一方面可以去除较大的突起，另一方面可以确保方向计算上有足够的精度。门把手的尺寸可以通过最小外接矩形尺寸加上腐蚀所用圆的直径得到。

基于门把手的方向和尺寸，再次使用区域形态学（见 3.6.1.6 节）分割出内部的 4 个正方形。首先使用 2 次闭运算填充前面分割出的内部正方形边缘上的小空洞：

```
ScratchWidthMax := 11
gen_rectangle2 (StructElement1, 0, 0, Phi, ScratchWidthMax/2, 1)
gen_rectangle2 (StructElement2, 0, 0, Phi+rad(90), ScratchWidthMax/2, 1)
closing (Region, StructElement1, RegionClosing)
closing (RegionClosing, StructElement2, RegionClosing)
```

For this, two perpendicular rectangular structuring elements are generated in the appropriate orientation. The size of the rectangles is chosen such that gaps with a maximum width of `ScratchWidthMax` can be closed. Figure 4.26(a) shows a detailed view of the segmented region of Figure 4.25(a). Note the small gaps in the border of the inner square that are caused by the crossing scratch. The result of the closing operation is shown in Figure 4.26(b): the gaps have been successfully closed.

为此要在合适的方向产生两个正交的矩形结构元素。矩形大小的选择要使宽度小于 ScratchWidthMax 的缝隙均可以闭合。图 4.26（a）显示了图 4.25（a）分割区域的细节。由于划痕，内部正方形边界上有缝隙。闭运算的结果如图 4.26（b）所示：缝隙成功地闭合。

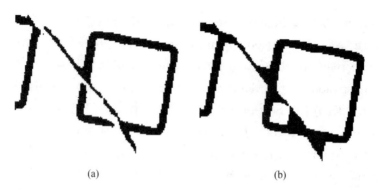

图 4.26 （a）图 4.25（a）分割区域的细节，正方形边缘区域有缝隙；
（b）(a) 图经过闭运算的结果，缝隙成功闭合

Up to now, all scratches in the surface are still contained in our segmentation result of the bright border region. To be able to detect the scratches, we must separate the scratches from the segmentation result. Because we know the appearance of the border region of the inner squares, we can get rid of the scratches by using opening operations with appropriate structuring elements. For this, we generate a structuring element that consists of two axis-parallel rectangles that represent two opposite borders of the inner squares:

至此，划痕仍在分割出的亮的边界区域中。为了能够检测出划痕，需要将划痕从分割结果中分离出来。由于已知内部正方形的边界区域的形状，可以使用合适的结构元素开运算去除划痕。为此生成一个结构元素，由 2 个轴平行的矩形组成，代表内部正方形的两个对边：

```
InnerSquareSizeFraction = 0.205
InnerSquareSize := KnobSize*InnerSquareSizeFraction
gen_rectangle2 (Rectangle1, -InnerSquareSize/2.0, 0, 0,
                InnerSquareSize/4.0, 0)
gen_rectangle2 (Rectangle2, InnerSquareSize/2.0, 0, 0,
                InnerSquareSize/4.0, 0)
union2 (Rectangle1, Rectangle2, StructElementRef)
```

The size of the inner squares is computed as a predefined fraction of the size of the doorknob. The distance between the rectangles is set to the size of the inner squares. The structuring element that represents the upper and lower borders of the inner squares is obtained by rotating the generated region according to the determined orienta-

内部正方形尺寸与整个门把手尺寸成一定比例，其大小可以通过计算得出。矩形间的距离置为内部正方形的尺寸。通过按照已经确定的门把手方向来旋转产生的区域得到代表内部正方形上、下边界的结构元素（仿射变换的细节见 3.3.1 节）。代表内部

tion of the doorknob (see Section 3.3.1 for details about affine transformations). The structuring element that represents the left and right borders of the inner squares is obtained accordingly by adding a rotation angle of 90°:

正方形左、右边界的结构元素由代表内部正方形上、下边界的结构元素再旋转 90° 而得到：

```
hom_mat2d_rotate (HomMat2DIdentity, Phi, 0, 0, HomMat2DRotate)
affine_trans_region (StructElementRef, StructElement1,
                     HomMat2DRotate, 'false')
hom_mat2d_rotate (HomMat2DIdentity, Phi + rad(90), 0, 0, HomMat2DRotate)
affine_trans_region (StructElementRef, StructElement2,
                     HomMat2DRotate, 'false')
```

Figures 4.27(a) and (b) show the two generated structuring elements. Note that the rotation of the structuring elements could be omitted when creating rectangles that already are in the appropriate orientations. However, this requires the centers of the rectangles to be transformed according to the orientation.

图 4.27（a）和图 4.27（b）为产生的两个结构元素。当在合适的方向生成矩形时，结构元素可以不作旋转。但是需要根据方向变换矩形中心。

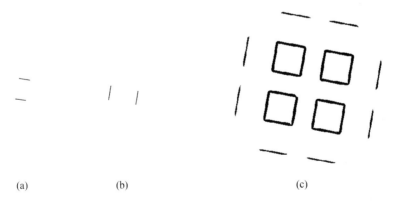

(a)　　　　　(b)　　　　　(c)

图 4.27　（a）对应内部正方形边界上、下的结构元素；（b）对应内部正方形边界左、右的结构元素；（c）使用结构元素（a）和（b）进行开运算的结果

As we have seen in Section 3.6.1.6, the opening operation can be used as a template matching operation returning all points of the input region into which the structuring element fits:

从 3.6.1.6 节中可以看出，开运算可以用作模板匹配，会返回输入区域内所有与结构元素相匹配的点：

```
opening (RegionClosing, StructElement1, RegionOpening1)
opening (RegionClosing, StructElement2, RegionOpening2)
union2 (RegionOpening1, RegionOpening2, RegionOpening)
```

Figure 4.27(c) shows the union of the results of both opening operations. As expected, the result contains the border of the inner squares. However, the result also contains parts of the outer border of the doorknob because the distance from the inner squares to the border of the doorknob is the same as the size of the inner squares. To exclude the part of the outer border, we intersect the result of the opening with an eroded version of the doorknob region:

图 4.27（c）显示了两次开运算的合并后结果。正如所期待的，结果含有内部正方形边界。然而结果仍含有门把手部分外边界，这是因为内正方形到门把手边界的距离与内正方形的大小一样。为去掉外边界部分，取开运算的结果和腐蚀后的门把手区域交集：

```
erosion_circle (RegionDoorknob, RegionInner, InnerSquareSize/2)
intersection (RegionInner, RegionOpening, RegionSquares)
```

The obtained region `RegionSquares` now only contains the border of the four inner squares. Finally, the region of the planar surface is the difference of the doorknob region and the border of the inner squares:

这样得到仅含有 4 个内部正方形边界的区域 RegionSquares。最后要检查的表面就是门把手区域与内正方形边界的差：

```
BorderWidth := 7
BorderTolerance := 3
erosion_circle (RegionDoorknob, RegionInner, BorderWidth+BorderTolerance)
dilation_circle (RegionSquares, RegionSquaresDilation, BorderTolerance)
difference (RegionInner, RegionSquaresDilation, RegionSurface)
```

Before computing the difference, the doorknob region is eroded using a circular structuring element to exclude the border from the surface inspection. The radius of the circle is the sum of `BorderWidth` and `BorderTolerance`, which are both predefined values. By adding `BorderTolerance` to the radius, pixels in the close vicinity of the border are also excluded from the inspection because their gray values are still influ-

在计算差值之前，使用圆形结构元素对门把手区域进行腐蚀以去除边界。圆的半径为 BorderWidth 与 BorderTolerance 的和，这两个值都是事先定义的。半径加上 BorderTolerance 是为了检测时去掉与边界非常靠近的像素，这些像素灰度会受到边界的影响，可能被错误地判作缺陷。同理，代表内正方形边界区域也要膨胀一些。

enced by the border region, and hence would be erroneously interpreted as defects. For the same reason, the region that represents the border of the inner squares is slightly dilated accordingly. The resulting ROI `RegionSurface` that contains the planar surface of the doorknob is shown in Figure 4.28. Note that neither the white border of the doorknob nor the white borders of the inner squares are contained in the region.

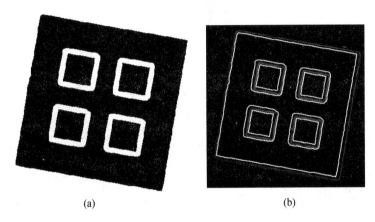

(a)   (b)

图 4.28 （a）含有门把手平面的感兴趣区域（黑色）；（b）感兴趣区域边界用白色叠加到原始图像上，不包含门把手白色边界及内正方形白边。为了视觉效果降低了图像的对比度

4.6.3 Detecting Defects

Within the ROI, we can now perform the defect detection:

```
ScratchGrayDiffMin := 15
reduce_domain (Image, RegionSurface, ImageReduced)
median_image (ImageReduced, ImageMedian, 'circle', ScratchWidthMax,
              'mirrored')
dyn_threshold (ImageReduced, ImageMedian, RegionDeviation,
               ScratchGrayDiffMin, 'light')
```

The defects are detected by dynamic thresholding. However, now we can use the median filter (see Section 3.2.3.9) to estimate the background. Based on the known maximum scratch width `ScratchWidthMax`, we can eliminate all scratches by passing `ScratchWidthMax` for the radius of the median filter. Because of the dark-field front light illumination, scratches appear as bright regions, which can easily be segmented by using the predefined threshold `ScratchGrayDiffMin`. In Figure 4.29(a), the result of the dynamic thresholding is shown. As can be seen, noise is included in the resulting region, which must be eliminated in a post-processing step:

器半径去除所有划痕。由于采用暗场正面照明，划痕在图像中为亮的区域，可以容易地使用预先定义的 ScratchGrayDiffMin 作为阈值进行分割。图 4.29（a）为动态阈值分割的结果，如图所示，结果中含有的噪声，需要在后处理中去掉：

```
connection (RegionDeviation, ConnectedRegions)
select_shape (ConnectedRegions, RegionDeviationNoNoise, 'area',
              'and', 4, 10000000)
```

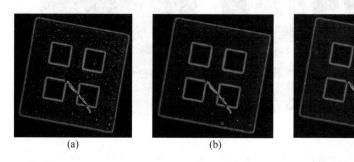

图 4.29 （a）动态阈值分割结果以白色叠加到原始图像上，对比度已减小，区域中有噪声；（b）去除（a）中小于 4 个像素的连通区域后的结果，并不是所有噪声都去掉了；（c）表面检测结果，检测出的划痕以白色显示

First, connected components (see Section 3.4.2) that are smaller than 4 pixels are assumed to be noise, and hence are eliminated. Unfortunately, this does not completely eliminate the noise from the segmentation, as can be seen from Figure 4.29(b). However, further increasing the threshold would also eliminate parts of interrupted defect regions, which is not desirable. To separate the interrupted defect regions from the noise, we assume that noise is evenly distributed while defect regions that belong to the same scratch lie close to each other. Therefore, we can close small gaps in the defect regions by performing a dilation:

在这种情况下，所有少于 4 个像素的连通区域（见 3.4.2 节）被看作噪声并被去除。但是从图 4.29（b）中可以看出，并不是所有噪声都完全被去除了，进一步提高阈值可能会同时去除部分不连续的缺陷区域，这是不可取的。为了区分噪声和缺陷，假设噪声是均匀分布的，而同属一个划痕的缺陷是彼此靠近的，因此，可以通过膨胀将缺陷区域中小的缝隙闭合：

```
ScratchAreaMin := 20
dilation_circle (RegionDeviationNoNoise, RegionDilation, 2)
connection (RegionDilation, ConnectedRegions)
intersection (ConnectedRegions, RegionDeviationNoNoise,
              RegionIntersection)
select_shape (RegionIntersection, RegionErrors, 'area',
              'and', ScratchAreaMin, 10000000)
```

After the dilation, previously interrupted defect regions are merged. The connected components are recomputed from the merged regions. To obtain the original shape of the defects, the connected components are intersected with the original non-dilated region. Note that the connectivity of the components is preserved during the intersection computation. Thus, with the dilation, we can simply extend the neighborhood definition that is used in the computation of the connected components. Finally, we select all components with an area exceeding a predefined threshold for the minimum scratch area. The final result is shown in Figure 4.29(c).

In Figure 4.30(a), the zoomed part of a second example image is shown. Note the low contrast of the small scratch at the lower right corner

原来断开的缺陷经过膨胀后连在一起了，对膨胀后的区域重新计算连通区域。为了得到缺陷的原始形状，取未膨胀前的原始区域与连通区域的交集。注意交集运算不影响各成分的连通性，于是，通过膨胀仅增加了连通区域的轮廓。最后选出所有比预定最小划痕大的区域。最终结果如图 4.29（c）所示。

图 4.30（a）为第二个示例图像的放大部分。请注意门把手右下方对比度很差的小的划痕。4.30（b）为使用

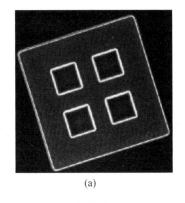
(a)

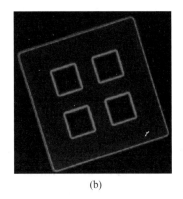
(b)

图 4.30　（a）第二个示例图像的放大部分；（b）图像（a）的检测结果。检测出的缺陷边缘以白色叠加在原始图像上（对比度已降低了）

of the doorknob. The surface inspection detects this defect, as shown in Figure 4.30(b).

4.6.4 exercises

1. In the above application, defects can only be identified on the planar surface of the doorknob. Modify the program such that it is also possible to detect defects in the bright border region of the doorknob and in the bright border regions of the inner squares.

2. The program described above uses region morphology to locate the doorknob and to segment its surface. Alternatively, the doorknob could be located by using shape-based matching (see Section 3.11.5.6). Furthermore, the inspection could be performed by using an appropriate variation model (see Section 3.4.1.4). Use these two methods to perform the inspection of the surface and of the bright border regions simultaneously, without using region morphology.

4.7 Measurement of Spark Plugs

In shape and dimensional inspection applications, distances between objects or object parts often must be measured to decide whether they lie within the required tolerances.

In this application, we measure the size of spark plug gaps. The size of the gap between the center electrode and the side or ground electrode must lie within a certain tolerance range to ensure that the spark plug can be used within a specific engine type. We use a camera with a telecentric lens as well as a telecentric back light illumination

表面检测检测到这个缺陷。

4.6.4 练习

1. 上述应用中仅能检测门把手表面的缺陷。修改程序，使得门把手亮的边界和内部正方形亮的边界上的缺陷也可以检测。

2. 上述程序使用区域形态学来定位门把手及表面分割。另外一种方法是使用基于形状的匹配算法来定位门把手（见 3.11.5.6 节）。另外检测也可以使用变差模型（见 3.4.1.4 节）。不使用形态学，而使用以上两种方法来同时实现表面和亮边界区域的检测。

4.7 火花塞测量

在形状和尺寸检测应用中，通常需要测量被测物间距离或被测物上部件间距离以确定是否在允许公差范围内。

本应用旨在测量火花塞间隙的尺寸。火花塞中心电极与侧边或地电极间的距离必须在一定范围内才能保证火花塞可以用于特定的发动机类型。使用远心镜头和平行背光源采集图像。首先标定摄像机，然后使用鲁棒的匹配算法在图像中找到火花塞。根

for image acquisition. First, we calibrate the camera setup. Then, a robust matching algorithm is used to find the spark plug in an image. Based on the matching pose, the ROI in which the measurement is performed is aligned. The actual measuring is performed with 1D edge extraction in a gray value profile across the spark gap.

The algorithms used in this application are:
- Camera calibration (Section 3.9).
- Shape-based matching (Section 3.11.5.6).
- Region features (Section 3.5.1).
- Affine transformations (Section 3.3.1).
- 1D edge extraction (Section 3.7.2).

The corresponding example program is:
···/machine_vision_book/spark_plug_measuring/spark_plug_measuring.hdev

Figures 4.31(a) and (b) show an image of a spark plug with a correct gap size and with a gap size that is too large, respectively. We use a telecentric lens to avoid perspective distortions in the image that would make an exact measurement difficult (see Section 2.2.4). With the telecentric illumination, very sharp edges are obtained in the silhouette of the spark plug (see Section 2.1.5.5). Furthermore, reflections at the spark plug on the camera side are avoided.

据匹配的姿态，对用于测量的感兴趣区域进行对齐。测量是对火花塞间隙进行一维边缘提取得到其灰度轮廓来实现的。

本应用使用的算法：
- 摄像机标定（3.9 节）。
- 基于形状的模板匹配（3.11.5.6 节）。
- 区域特征（3.5.1 节）。
- 仿射变换（3.3.1 节）。
- 一维边缘提取（3.7.2 节）。

相应的例子程序位于：

图 4.31（a）和图 4.31（b）分别对应正确间隙的火花塞图像和间隙过大的火花塞图像。使用远心镜头可以避免由于图像的透视变形而对精确测量带来的困难（见 2.2.4 节），使用平行光照明可以得到非常锐利的火花塞边缘轮廓的图像（见 2.1.5.5 节），同时还能避免火花塞在摄像机侧的反射。

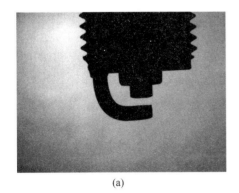
(a)

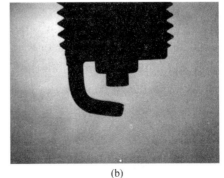
(b)

图 4.31 两个火花塞图像：（a）间隙正常的火花塞；（b）间隙过大的火花塞

4.7.1 Calibrating the Camera

In the first step, we calibrate the camera. In this application, camera calibration is essential for two reasons. First, we want to measure the spark gap accurately, and hence must consider possible lens distortions. Second, the size of the gap should be determined in metric units. To calibrate the camera, we acquire multiple images of a transparent calibration target in various poses. For historical reasons (this example was developed for the first edition of this book), we use an old calibration target (cf. Section 3.9.4.1). In new applications, we would use the new calibration target, of course. Figure 4.32 shows two of the 14 images used for the calibration.

4.7.1 标定摄像机

第一步标定摄像机。在这个应用中有两个原因必须标定摄像机。首先如果想准确测量火花塞，镜头有一定畸变是必须加以考虑的。其次，间隙需要以公制单位表示。为了标定摄像机，需要对不同位姿的、清晰的标定目标采集多幅图像。由于历史原因（该示例程序开发于本书的第一版），我们采用了旧版的标定板（见 3.9.4.1 节），在新的应用中，我们会采用新版的标定板。图 4.32 为用于标定的 14 幅图像中的 2 幅。

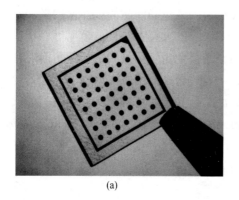
(a)

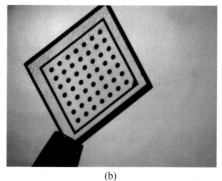
(b)

图 4.32　14 幅不同姿态的标定板图像中的 2 幅。这些图像用于标定摄像机

First, we specify the calibration target used, define initial values for the interior orientation of the camera, and initialize the handle that will be used to collect the image coordinates of the calibration marks and the initial poses of the calibration targets:

首先指定使用的标定目标，定义摄像机内参初始值，初始化用于存储标定标志的图像坐标和标定目标初始姿态的句柄：

```
CaltabName := 'caltab_10mm.descr'
gen_cam_par_area_scan_telecentric_division (0.17, 0, 6.6e-06, 6.6e-06,
```

```
                                    320, 240, 640, 480,
                                    StartCamParam)
create_calib_data ('calibration_object', 1, 1, CalibDataID)
set_calib_data_calib_object (CalibDataID, 0, CaltabName)
set_calib_data_cam_param (CalibDataID, 0, [], StartCamParam)
```

The telecentric lens that was used has a nominal magnification of 0.17. The initial value of the distortion coefficient is set to zero, the pixel size is 6.6 µm, and the principal point is assumed to be in the center of the image. These values are just initial values. They will be improved during the calibration process.

所使用的远心镜头的标称放大率为 0.17,畸变系数的初始值也设置为 0,像素尺寸为 6.6 µm,并假设主点位于图像的中心。这些值仅是初始值,在标定过程中会对这些值进行修正。

Then, within a loop over all calibration images, the subpixel-accurate positions of the circular calibration marks and initial estimates for the poses of the calibration target with respect to the camera are determined:

在对所有标定图像的一轮标定中,确定了具有亚像素精度的圆形标定标志的位置并初步估计标定板相对摄像机的位姿:

```
find_calib_object (Image, CalibDataID, 0, 0, Index, [], [])
```

Now, we determine the exact interior orientation of the camera:

现在来精确确定摄像机内参:

```
calibrate_cameras (CalibDataID, Error)
get_calib_data (CalibDataID, 'camera', 0, 'params', CamParam)
```

The camera is calibrated using the operator `calibrate_cameras`. After the calibration has been performed, we can read out the interior orientation of the camera. In the online phase, the interior orientation is used to transform the image measurements into world units. Note that, although the exterior orientation of the camera is also determined by `calibrate_cameras`, it is not used below because we do not need to know the position of the spark plug with respect to a WCS.

使用算子 calibrate_cameras 来对摄像机进行标定,在完成标定之后,可以读出摄像机的内参。在线测量过程中,摄像机内参用于将图像测量用世界单位表示。注意,尽管摄像机的外参已经由算子 calibrate_cameras 确定了,但是后面不会用到,因为不需要知道火花塞在世界坐标系中的位置。

4.7.2 Determining the Position of the Spark Plug

The position of the spark plug in the image can be determined by using shape-based matching, which is described in Section 3.11.5.6. For this, we create a model representation of the spark plug from a template image:

```
gen_rectangle1 (ModelRegion, 120, 230, 220, 445)
reduce_domain (ModelImage, ModelRegion, TemplateImage)
create_shape_model (TemplateImage, 'auto', rad(-30), rad(60), 'auto',
                    'none', 'use_polarity', 'auto', 'auto', ModelID)
area_center (ModelRegion, Area, RefRow, RefCol)
```

The template image is shown in Figure 4.33(a). Because the thread rolling may be different for different spark plugs, we must not include it in the model representation. Furthermore, the side electrode may bend as the gap size changes, and hence also must not be used. Therefore, the model is created only from a small part of the spark plug that always appears in the same way, and hence permits robust matching. In Figure 4.33(a), the corresponding image region is indicated by the gray rectangle. It only contains the center electrode, the insulator, and the top part of the thread. The contour representation of the resulting model is shown in Figure 4.33(b). Note that, because the orientation of the spark plugs is similar in all images, the model is created only within an angle range of ±30°. Finally, to be able to correctly align the measurement in the online phase, we must know the reference point of the model, which is the center of gravity of the model region. In Figure 4.33(a), the reference point is displayed as a gray cross.

4.7.2 确定火花塞的位置

火花塞在图像上的位置可通过使用鲁棒的匹配算法来确定。这里使用基于形状的匹配，见 3.11.5.6 节。从模板图像中建立代表火花塞的模型：

图 4.33（a）显示了模板图像。由于不同种类的火花塞的滚丝是不同的，所以模型中不能包含滚丝。此外，侧电极弯曲度随缝隙变化，也不能用作模型。模型仅能从火花塞保持不变的一小部分中创建以保障可靠的匹配。图 4.33（a）中灰色矩形为相应的图像区域，区域中仅包括中心电极、绝缘体和螺纹上面的部分。图 4.33（b）为模型的轮廓。由于所有图像中火花塞的方向大致相同，模型创建时仅考虑 ±30° 范围。最后为了能够在线测量时正确地匹配，必须知道模型的参考点，也就是模型区域的重心。图 4.33（a）中参考点为灰十字。相对模型参考点，定义实际缝隙测量的矩形：

Relative to the reference point of the model, we define a rectangle in which the actual gap measurement is to be performed:

```
RectRelRow := 80
RectRelCol := -26
RectRelPhi := rad(90)
RectLen1 := 50
RectLen2 := 15
```

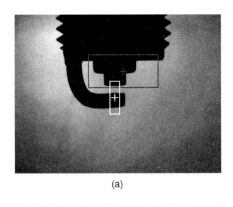 

(a) (b)

图 4.33 （a）用于创建模型的模板图像。模型区域以灰色矩形表示，模型的参考点为灰色十字。白色矩形及中心为进行缝隙大小测量的范围；（b）创建的模型的轮廓

The center of the rectangle is 80 pixels below and 26 pixels to the left of the model reference point. The rectangle's orientation is chosen such that its major axis points in the direction in which the measurement, i.e., the 1D edge extraction, is performed. Thus, edges that are perpendicular to the major axis can be measured. Here, the major axis is oriented vertically in the model image. During measuring in the online phase, the gray values are averaged perpendicular to the major axis to obtain a 1D edge profile. The range in which the averaging is performed is defined by the width of the rectangle. In Figure 4.33(a), the measurement rectangle and its center point are displayed in white.

矩形中心位于模型参考点下方 80 像素，左侧 26 像素。矩形的方向选择是使其主轴上的点与测量方向一致，这里的测量是指一维边缘提取。这样与主轴相垂直的边缘就可以测量了。这里主轴的方向在模板图像中垂直。在线检测过程中，与主轴垂直方向上灰度进行平均以得到一维边缘轮廓。平均的范围定义为矩形的宽度。在图 4.33（a）中，测量用矩形和中心点用白色显示。

4.7.3 Performing the Measurement

All the steps described so far are performed offline. In the online phase, we use the shape model to find the spark plug in an image:

```
find_shape_model (Image, ModelID, rad(-30), rad(60), 0.7, 1, 0.5,
                 'least_squares', 0, 0.9, Row, Column, Angle, Score)
```

Based on the pose of the found spark plug, we create the measurement rectangle:

```
vector_angle_to_rigid (0, 0, 0, Row, Column, Angle, HomMat2D)
affine_trans_pixel (HomMat2D, RectRelRow, RectRelCol,
                   TransRow, TransCol)
gen_measure_rectangle2 (TransRow, TransCol, Angle+RectRelPhi,
                       RectLen1, RectLen2, Width, Height,
                       'bilinear', MeasureHandle)
```

First, we transform the matching pose into a 2D rigid transformation, which aligns the model with the spark plug in the image. Then, with the obtained transformation, the relative position of the center of the measurement rectangle is transformed. Finally, the orientation of the rectangle is adapted according to the orientation of the spark plug. Figure 4.34(a) shows an

4.7.3 测量

至此为止，以上所有描述的步骤均是离线的。在线检测过程中使用形状模型在图像上找火花塞：

根据找到的火花塞位姿，创建测量用的矩形：

首先对匹配的位姿进行二维刚性变换，使得模型与图像中的火花塞重合。然后，利用得到的变换对测量矩形中心的相对位置进行变换，最后使矩形的方向与火花塞的方向一致。图 4.34（a）为待测的火花塞图像。火花塞轮廓以粗白线叠加在上面，矩形测量区域用细白线显示。使用双线性插

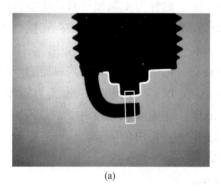

(a)

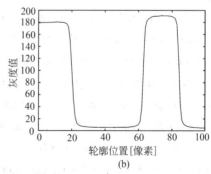

(b)

图 4.34 （a）待测的火花塞图像。找到的火花塞的位姿变换后的模型轮廓为粗白线，变换后的矩形测量区域为细白线；（b）测量矩形的灰度轮廓。边缘提取就是计算灰度轮廓的一阶导数的局部极值

image of the spark plug to be measured. The contour of the found spark plug is overlaid as a bold white line. The transformed measurement rectangle is displayed as a thin white line. With `gen_measure_rectangle2`, the gray value profile is computed by using bilinear interpolation. It is shown in Figure 4.34(b). The actual measurement is performed with the following operation:

值，利用 gen_measure_rectangle2 计算出灰度值轮廓，见图 4.34（b）。实际测量按如下操作进行：

```
measure_pairs (Image, MeasureHandle, 1, 30, 'positive_strongest',
               'all', Row1, Col1, Amplitude1, Row2, Col2,
               Amplitude2, IntraDistance, InterDistance)
```

The 1D edges are extracted by computing the local extrema of the first derivative of the gray value profile. Because we are only interested in the edge pair that corresponds to the gap, we use `measure_pairs`, which automatically groups the edges into pairs. To ensure that the correct edge pair is returned, we demand that the first edge of the pair has a positive transition. With this, we can avoid obtaining the edge pair that corresponds to the lower and upper borders of the side electrode. Finally, the position of the resulting 1D edge pair is transformed back into the image and returned in (`Row1, Col1`) and (`Row2, Col2`).

通过计算灰度值轮廓一阶导数的局部极值得到一维边界。因为仅对代表缝隙的一对边缘感兴趣，使用 measure_pairs 将边缘自动配对。为保证返回正确的一对边缘，要求第一对边缘有正的转换，这样我们就可以避免得到侧电极上、下两个边缘。最后一维边缘对的位置变换回图像并从（Row1，Col1）和（Row2，Col2）中返回。

To eliminate the effect of lens distortions and to obtain the result in metric units, we must transform the two edge points into the CCS. This can be done by constructing the two lines of sight that pass through the two edge points in the image:

为了消除镜头畸变的影响并得到以公制为单位的结果，必须将两个边缘点变换到摄像机坐标系统。这通过构建图像上通过边缘点的两条视线来完成：

```
get_line_of_sight ([Row1,Row2], [Col1,Col2], CamParam, X, Y, Z,
                   XH, YH, ZH)
DX := X[1]-X[0]
DY := Y[1]-Y[0]
GapSize := sqrt(DX*DX+DY*DY)
```

For both edge points, the line of sight is returned as two points on the line, given in world units. Because of the telecentric camera, all lines of sight are parallel, and hence the two points that define the line have the same X and Y coordinates. The size of the gap corresponds to the distance of the parallel lines that pass through the two edge points. In Figures 4.35(a)–(c), three examples are shown. The measured edge points are visualized by two lines that are perpendicular to the measurement direction. The measured gap size is overlaid as white text. Finally, based on a predefined range of valid gap sizes, we can decide whether the gap size is within the tolerances:

对于两个边缘点，视线以线上两点的世界单位返回。由于使用远心镜头，所有的视线是平行的，确定线的两个点的 X 和 Y 坐标是相同的。缝隙的尺寸也就与通过这两个边缘点的平行线间的距离一致。图 4.35（a）～（c）为三个例子，被测量的边界点用与测量方向垂直的两条白线表示，测量出的缝隙尺寸以白色文字叠加显示。最后根据事先定义的有效缝隙大小，确定缝隙是否在公差范围内：

```
GapSizeMin := 0.78e-3
GapSizeMax := 0.88e-3
if (GapSize < GapSizeMin)
    * Gap size is too small.
elseif (GapSize > GapSizeMax)
    * Gap size is too large.
else
    * Gap size is within tolerances.
endif
```

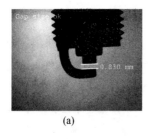 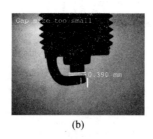 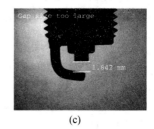

(a)　　　　　　　　　(b)　　　　　　　　　(c)

图 4.35　火花塞测量结果的三个例子。测量的边界点用与测量方向垂直的白线表示：（a）缝隙大小合适的火花塞；（b）缝隙太小的火花塞；（c）缝隙太大的火花塞

According to the above decision, an appropriate action can be performed, e.g., by rejecting the current spark plug.

根据测量结果，采取如剔除当前的火花塞等适当的动作。

4.7.4 exercises

1. As an alternative to shape-based matching, NCC could be used for template matching (see Section 3.11.1.2). Adapt the above program such that the pose of the spark plug is determined using NCC.

2. Overheating, oil leakage, or bad fuel quality may lead to the accumulation of deposits on the center electrode and on the insulator during the use of the spark plug. Therefore, in the automobile industry, especially in racing applications, used spark plugs are analyzed because they indicate conditions within the running engine. Extend the above program so that irregularities in the shape of the center electrode and of the insulator can be detected. Furthermore, measure the diameter of both components to detect heavy deposits and erosion.

4.8 Molding Flash Detection

Especially in the metal- or plastic-working industry, a frequent task is to detect flashes on castings or molded plastic parts. Flashes often cannot be avoided because the molds do not fit together tightly. Therefore, it is important to detect the flashes and either to remove them in a secondary operation or to reject the object.

In this application, we detect molding flashes on a circular plastic part that has been manufactured by injection molding. After the flashes are detected, their size and position on the circular border are determined. For illustration purposes, this task is solved with two alternative approaches.

While the first approach uses region morphology, the second approach is based on the processing of subpixel-precise contours.

The algorithms used in this application are:
- Thresholding (Section 3.4.1.1).
- Region morphology (Section 3.6.1).
- Robust circle fitting (Section 3.8.2.2).
- Contour features (Section 3.5.3).
- Subpixel-precise 2D edge extraction (Section 3.7.3.5).

本应用使用的算法：
- 阈值分割（3.4.1.1 节）。
- 区域形态学（3.6.1 节）。
- 鲁棒的圆拟合（3.8.2.2 节）。
- 轮廓特征（3.5.3 节）。
- 亚像素精度的二维边缘提取（3.7.3.5 节）。

The corresponding example program is:

···/machine_vision_book/molding_flash_detection/molding_flash_detection.hdev

相应的例子程序位于：

Figure 4.36(a) shows an image of the molded plastic part we want to inspect. Note the flash at the upper right part of its border. The image was acquired with diffuse bright-field back light illumination (see Section 2.1.5.4). With this kind of illumination, the object appears dark while the background appears bright, which simplifies the segmentation process significantly.

图 4.36（a）为需要检测的模压塑料工件，工件边缘的右上侧有毛边。图像采集使用漫反射、明场背光照明（见 2.1.5.4 节），这种照明方式采集的图像被测物为黑，背景为亮，大大简化了分割处理。

 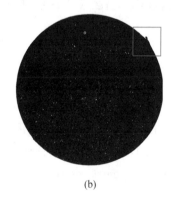

(a)　　　　　　　　　(b)

图 4.36　（a）边缘右上有毛边的圆形模压塑料工件；（b）阈值分割结果。为了视觉效果，下面仅显示以灰色矩形标示部件的一部分并经过放大

In the first step, which is identical for both presented approaches, the dark object is segmented by a thresholding operation:

第一步通过阈值分割操作将黑色物体分割出来，这对于两种方法都一样：

```
threshold (Image, Object, 0, 180)
```

The result is shown in Figure 4.36(b).

结果如图 4.36（b）所示。

4.8.1 Molding Flash Detection Using Region Morphology

4.8.1 区域形态学方法检测模制品毛边

The first approach, which uses region morphology, is described in the following. It is similar to the approach that is described in Section 3.6.1.6. However, an extended version is presented here.

下面介绍第一种方法：基于区域形态学。这种方法与 3.6.1.6 节介绍的类似，但有些扩充。

The flash appears as a protrusion of the object region (see Figure 4.36(b)). Therefore, the flash can be segmented by performing an opening on the object region and subtracting the opened region from the original segmentation:

毛边呈现为被测物区域的凸起（见图 4.36（b））。因此通过对被测区域开运算并从原始图像中减去开运算结果从而分割出毛边：

```
opening_circle (Object, OpenedObject, 400.5)
difference (Object, OpenedObject, RegionDifference)
```

For the opening, a circle is used as the structuring element. The circle is chosen almost as large as the object to ensure that even large flashes can be eliminated and the circular shape of the molded plastic can be recovered. Note that the radius must be smaller than the radius of the molded plastic part. Otherwise, the opening would remove the object entirely. Figures 4.37(a) and (b) show the result of the opening and the difference operation, respectively, for a zoomed image part. Small components of the difference region that are caused by minor irregularities at the border must be eliminated. For this, the difference region is opened with a rectangular structuring element of size 5 × 5:

开运算采用圆作为结构元素。圆的大小与被测物体近似，这样可以保证非常大的毛边也可以去除，塑料盘的圆形可以复原。但是圆的半径要小于塑料圆的半径，否则开运算将会除去整个被测物。图 4.37（a）和图 4.37（b）分别为放大的部分图像开运算结果及原始图像与开运算结果的差。边界上非常小的无规律的成分必须要去掉。为此使用 5 × 5 矩形结构元素的开运算：

```
MinFlashSize := 5
opening_rectangle1 (RegionDifference, FlashRegion, MinFlashSize,
                    MinFlashSize)
```

The resulting region, which is shown in Figure 4.37(c), represents the molding flash.

图 4.37（c）为运算结果，代表模制品的毛边。

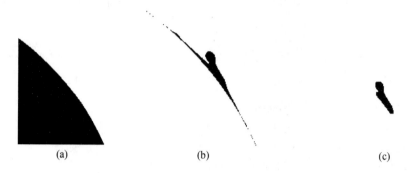

图 4.37 （a）图 4.36（b）放大图像开运算区域；（b）原始图像与开运算结果的差；（c）对（b）图像利用 5×5 矩形结构元开运算得到的分割出的毛边区域

An important quality feature in this application is the maximum distance of the flashes from the ideal circular shape of the object, which we will call the reference shape below. Here, the reference shape is represented by the opened object region. To determine the maximum distance, we compute the distance transform of the opened object region:

本应用中一个重要的质量特征就是毛边与被测物理想圆形边界的最大距离，后续章节把被测物理想圆形边界称为参考形状，在这里参考形状为开运算后的被测区域。为得到最大距离，对开运算区域进行距离变换：

```
distance_transform (OpenedObject, DistanceImage, 'euclidean',
                   'false', Width, Height)
min_max_gray (FlashRegion, DistanceImage, 0, DistanceMin,
              DistanceMax, DistanceRange)
```

The resulting distance image, shown in Figure 4.38, contains the shortest distance to the reference shape for each point in the background region. The latter is the complement of the reference shape. Thus, the maximum distance `DistanceMax` is obtained by searching for the maximum value in the distance image within the flash region.

图 4.38 为得到的距离图像，含有背景中各点与参考形状最近距离，后者与参考形状互补。最大距离 DistanceMax 为毛边区域内距离图像的局部最大值。

To compute the angle range of the segmented flashes with respect to the object center, we need

为了计算分割出来的毛边相对被测物中心的角度，需要知道在注塑工

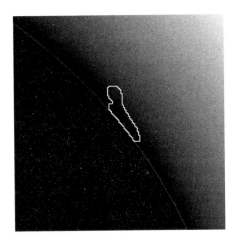

图 4.38 对开运算区域进行距离变换后得到的距离图像。被测物的边缘以灰色显示。分割出的毛边区域边缘以白色显示。在距离图像中亮的地方代表较远的距离，为了视觉效果，距离图像通过平方根查找表后显示

to know the start and end point of each flash on the border of the molded plastic part. These points can also be obtained with region morphology:

件边缘每个毛边的起点和终点。这些可以通过使用区域形态学得到：

```
boundary (OpenedObject, RegionBoundary, 'outer')
connection (FlashRegion, FlashRegions)
intersection (FlashRegions, RegionBoundary, RegionCircleSeg)
junctions_skeleton (RegionCircleSeg, EndPoints, JuncPoints)
```

First, the one-pixel-wide boundary of the opened object region is computed (see Figure 4.39(a)). Note that the boundary obtained lies one pixel outside of the original region, and hence inside the flash region. Then, each connected component of the flash region is intersected with the boundary. Thus, for each flash, a one-pixel-wide region at the boundary of the object is obtained. Finally, the region that only contains the end points of the intersected regions can be obtained with the operator `junctions_skeleton`. The intersected regions and the corresponding end points are visualized in Figure 4.39(b).

首先计算出经过开运算后被测物一个像素宽度的边界，如图 4.39（a）所示。注意，这个边界是在原有区域向外扩张一个像素的位置上，即位于毛边内部。毛边区域的每个连通部分与边界相交，于是，对于每个毛边，在被测物边界处，得到一个像素宽度的区域。最后使用 junctions_skeleton 算子，可以得到仅含有相交区域端点的区域。图 4.39（b）显示相交部分及相应的端点。

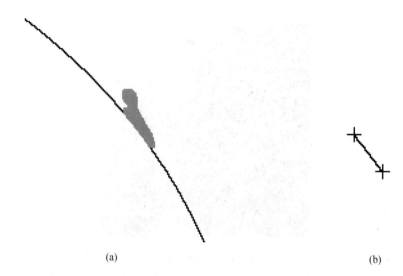

图 4.39　（a）图 4.37（a）被测物开运算后的一个像素宽度的边界（黑色）区域及毛边区域（灰色）；
　　　　　（b）边界与毛边相交的区域，端点以十字表示，可用来计算模塑毛边的角度

From the two end points of each segmented molding flash, the corresponding angle interval can be computed. For this, we must know the center of the circular plastic part. The circle center is the center of gravity of the opened object region:

从每个分割出来的模塑毛边的两个端点，可以计算出对应的间隔角。为此需要知道注塑工件的中心。这个圆的中心就是开运算后被测物区域的重心：

```
area_center (OpenedObject, Area, CenterRow, CenterCol)
```

Now, we are able to compute the angle range of each flash:

现在可以计算每个毛边的角度了：

```
count_obj (EndPoints, NumFlash)
for Index := 1 to NumFlash by 1
    select_obj (EndPoints, EndPointsSelected, Index)
    get_region_points (EndPointsSelected, SegRow, SegCol)
    Angle1 := atan2(CenterRow-SegRow[0],SegCol[0]-CenterCol)
    Angle2 := atan2(CenterRow-SegRow[1],SegCol[1]-CenterCol)
    AngleRange := Angle2-Angle1
endfor
```

Here get_region_points returns the row and column coordinates of the two end points in the two tuples SegRow and SegCol, respectively, each of which contains two elements, one for each

get_region_points 操作将两个端点的行、列坐标返回到两个元组 SegRow 和 SegCol 中，每个元组含两个元素，分别对应两个端点。最后

end point. Finally, the corresponding angles are computed using the arc tangent function. The result is shown in Figure 4.40.

利用反正切函数计算出角度，结果如图 4.40 所示。

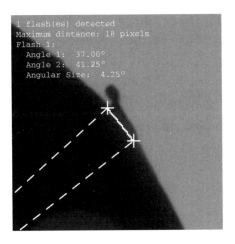

图 4.40 使用区域形态学检测模制品毛边的结果。毛边最大距离、起点和终点的角度以及毛边的角度以白色文字叠加在图像左上角。对应的扇形（点线）、分割出的边界（实线）及端点（十字）也以白色显示。为了更好的视觉效果，图像的对比度降低了

4.8.2 Molding Flash Detection with Subpixel-Precise Contours

In the following, the second approach is described, which is based on the processing of subpixel-precise contours. The idea is to compute the distance from the contour that represents the boundary of the plastic part to the circular reference shape. The contour representation can be obtained by using subpixel-precise edge extraction (see Section 3.7.3.5). Because edge extraction is computationally expensive, we determine an ROI first:

4.8.2 使用亚像素精度轮廓检测模制品毛边

下面讲述基于亚像素轮廓处理的方法。这个方法就是计算代表注塑件边缘的轮廓到圆形参考形状的距离。轮廓可以通过使用亚像素边缘提取（见 3.7.3.5 节）得到。由于边缘提取的计算量非常大，首先要选出感兴趣区域：

```
boundary (Object, RegionBorder, 'outer')
dilation_circle (RegionBorder, RegionDilation, 5.5)
reduce_domain (Image, RegionDilation, ImageReduced)
```

As in the first approach, we start with the segmented object region, which is shown in Figure 4.36(b). First, the one-pixel-wide boundary of the

与第一种方法一样，从分割被测物区域开始，如图 4.36（b）所示。首先计算出一个像素宽度的被测物边

object region is computed. We broaden this border using morphology to get a band-shaped ROI for the subpixel-precise extraction of the edges:

界。利用形态学，扩宽这个边界，得到用于亚像素精度边界提取的带状感兴趣区域：

```
edges_sub_pix (ImageReduced, Edges, 'canny', 2.0, 20, 40)
union_adjacent_contours_xld (Edges, BorderContour, 20, 1, 'attr_keep')
```

Because of small irregularities of the object border, the edges obtained may be interrupted. To obtain one connected contour, adjacent edge segments are merged. The resulting contour is shown in Figure 4.41(a).

由于被测物边界上小的不规则点，得到的边界可能不连续。为了得到连续的轮廓，相邻的边界段合并在一起得到如图 4.41（a）所示的轮廓。

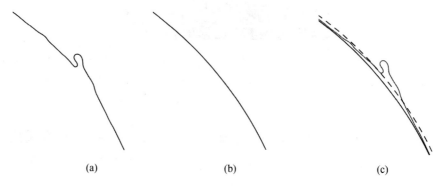

图 4.41 （a）图 4.36 中放大的图像部分边界提取的结果。相邻的边界已被合并；（b）与图（a）鲁棒拟合的圆弧；（c）灰线为提取出的轮廓，实黑线为拟合圆的轮廓，黑虚线为半径扩展 MinFlashSize 后圆的轮廓。提取出的轮廓与扩展圆的交点表示毛边的起点和终点

In the next step, the reference shape must be extracted. This is done by fitting a circle to the extracted contour. The contour points that lie on the flash would falsify the result of the fitting, and hence should be treated as outliers. Therefore, we use a robust fitting algorithm as described in Section 3.8.1.2. Because we want to completely disregard the outliers from the computation, we use the Tukey weight function:

下一步需要提取参考形状。参考形状是提取出轮廓的拟合圆。毛边上的轮廓点可能会导致虚假的拟合结果，应该被丢弃。使用 3.8.1.2 节讲述的鲁棒的拟合算法。由于需要将多余的点完全摒弃，这里使用 Tukey 加权函数：

```
fit_circle_contour_xld (BorderContour, 'geotukey', -1, 0, 0, 3, 2,
                       CenterRow, CenterCol, Radius,
                       StartPhi, EndPhi, PointOrder)
```

The contour of the circle obtained is shown in Figure 4.41(b). Now, we can compute the distances of all contours points to the fitted circle:

图 4.41（b）为得到的圆的轮廓。现在计算所有轮廓点与拟合圆的距离：

```
dist_ellipse_contour_points_xld (BorderContour, 'unsigned', 0,
                      CenterRow, CenterCol, 0,
                      Radius, Radius, Distances)
```

In this approach, the maximum distance of the flashes from the ideal circular shape is obtained easily:

在这个方法中，很容易得到毛边到理想圆的距离：

```
DistanceMax := max(Distances)
```

In the next step, we compute the start and end points of each contour interval whose points all have a distance exceeding the predefined threshold MinFlashSize. For this, we define a function whose argument is the index of the contour points. The function values are the distance values corrected by the threshold value. The function is shown in Figure 4.42. Thus, the indices of the start and end points correspond to the zero crossings of this function:

下一步计算出所有距离超出预定义阈值 MinFlashSize 轮廓间隔的起点和终点。对此定义一个函数，其自变量为轮廓点的索引。函数值为被阈值修正后的距离值。函数如图 4.42 所示。这样要找的起点和终点就是对应的函数零相交点：

```
DistancesOffset := Distances-MinFlashSize
create_funct_1d_array (DistancesOffset, Function)
zero_crossings_funct_1d (Function, ZeroCrossings)
```

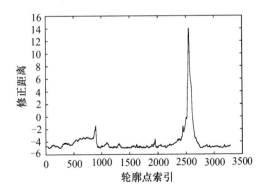

图 4.42 所有轮廓点的修正距离 DistanceOffset 的函数。函数值为正数的区域为毛边所在

The principle is illustrated in Figure 4.41(c). Effectively, we enlarge the radius of the fitted circle by the threshold `MinFlashSize` and compute the intersection points of the contour with the enlarged circle. The points obtained are the start and end points of the intervals that represent the detected flashes. If the flash crosses the end of the contour, we must take the 360° wrap-around into account. Note that, because of the closed contour, the first and last function values are identical. The flash crosses the end of the contour if the first (or the last) function value is positive. In this case, we move the last zero crossing to the first position in the tuple:

上述方法的原理如图 4.41（c）所示。将拟合圆的半径扩大 MinFlashSize，计算出轮廓与扩大后的圆的交点，即为检测到的毛边区域的起点和终点。如果毛边过轮廓的终点，需要考虑毛边过 360° 的点。由于轮廓是闭合的，第一个和最后一个函数值是一样的。如果第一个函数值是正的，毛边过轮廓的终点，此时需要将元组中最后一个零相交点移至首位：

```
if (DistancesOffset[0] > 0)
    Num := |ZeroCrossings|
    ZeroCrossings := [ZeroCrossings[Num-1],ZeroCrossings[0:Num-2]]
endif
```

Now, we can group two consecutive zero crossings to one interval. The corresponding angle range can be computed from the circle center and the two contour points at the position of the zero crossings:

现在可以将两个连续的零相交点作为一组。可以由圆心和在零相交点位置的二个轮廓点计算出对应的角度范围：

```
get_contour_xld (BorderContour, ContRow, ContCol)
for Index := 0 to |ZeroCrossings|-1 by 2
    Start := round(ZeroCrossings[Index])
    End := round(ZeroCrossings[Index+1])
    Angle1 := atan2(CenterRow-ContRow[Start],ContCol[Start]-CenterCol)
    Angle2 := atan2(CenterRow-ContRow[End],ContCol[End]-CenterCol)
    AngleRange := Angle2-Angle1
endfor
```

Note that the zero crossings are extracted with subpixel precision. Therefore, we must round them to the nearest integer value before using them as indices to the array of contour points.

注意由于零相交点是以亚像素精度提取的，因此必须取整后用作轮廓点数组的索引。

The result of the region-based molding flash detection is shown in Figure 4.43.

基于亚像素精度轮廓的注塑件毛边检测的结果如图 4.43 所示。

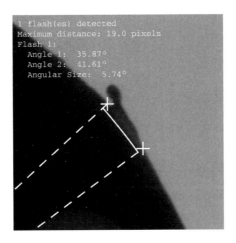

图 4.43 利用亚像素精度轮廓检测注塑件毛边的结果。检测出的毛边的最大距离、起点和终点的角度以及相应的角度范围以白色文字叠加在图像的左上角。拟合圆对应的区域（点线）、对应圆上线段（实线）和毛边的起点、终点（十字）以白色显示。为了更好的视觉效果，图像的对比度降低了

4.8.3 exercises

The example program described is able to detect flashes, i.e., protrusions, on the boundary of the molded part. Another molding defect that frequently must be detected are voids, i.e., indentations at the object boundary. Modify the program so that indentations as well as protrusions can be detected.

4.8.3 练习

上述例子程序可以检测出毛边也就是注塑件边缘的凸起。另外一种必须检测的缺陷就是被测物边缘的缺口。修改程序使其可以检测出毛边及缺口。

4.9 Inspection of Punched Sheets

4.9 冲孔板检查

A punched sheet is sheet metal that has been cut by using a punch. The punched sheets must be inspected to guarantee predefined tolerances.

In this application, we check the size of the holes in a punched sheet. First, we extract the boundaries of the holes in the sheet metal. Then the size of the holes is calculated from the extracted contours.

冲孔板是利用冲床切割的金属薄板。冲孔板必须检测以保证预先确定的公差。

在这个应用中检查冲压打孔板上孔的尺寸。先提取金属板上孔洞的边界，然后通过提取轮廓计算孔洞的尺寸。

The algorithms used in this application are:
- Thresholding (Section 3.4.1.1).
- Region boundaries (Section 3.6.1.4).
- Subpixel-precise 2D edge extraction (Section 3.7.3.5).
- Segmentation of contours into lines and circles (Section 3.8.4.2).
- Robust circle fitting (Section 3.8.2.2).

The corresponding example program is:

···/machine_vision_book/punching_sheet_inspection/punching_sheet_inspection.hdev

To inspect the outline of thin planar objects, the images are acquired with diffuse bright-field back light illumination (see Section 2.1.5.4). With this kind of illumination, the object appears dark while the surrounding areas are bright (see Figure 4.44(a)). This simplifies the segmentation process significantly. To avoid perspective distortions, the image plane of the camera must be parallel to the sheet metal. If this camera setup cannot be realized for the image acquisition, or if the lenses produce heavy distortions, the camera must be calibrated and the images must be rectified (see Section 3.9). If the lens distortions can be neglected, the rectification can also be done with a projective transformation (see Section 3.3.1.1).

本应用使用的算法：
- 阈值分割（3.4.1.1 节）。
- 区域边界（3.6.1.4 节）。
- 亚像素精度二维边缘提取（3.7.3.5 节）。
- 轮廓的线、圆分割（3.8.4.2 节）。
- 鲁棒的圆拟合（3.8.2.2 节）。

相应的示例程序位于：

为了检测薄平面物体的轮廓，在漫射、明场背光照明环境下采集图像，见 2.1.5.4 节。采用这种照明方式，被测物为黑色，周围区域为亮（如图 4.44（a）所示），使分割处理大大简化。为避免畸变，摄像机像平面必须与金属板平面平行。如果摄像机无法满足这一要求或者镜头有很大的径向畸变，摄像机必须标定，图像需要调整，见 3.9 节。如果径向畸变可以忽略不计，这个调整也可通过投影变换来完成，见 3.3.1.1 节。

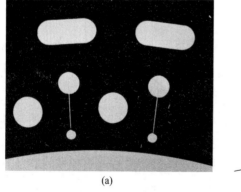
(a)

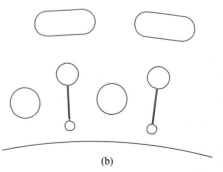
(b)

图 4.44　(a) 含有圆和椭圆孔的冲孔板图像；(b) 从 (a) 中提取的边缘

First, the boundaries of the punched sheet are extracted in the image. The resulting contours are then segmented into lines and circular arcs. Finally, circles are fitted to the circular arcs to obtain the radii of the holes in the sheet metal.

4.9.1 Extracting the Boundaries of the Punched Sheets

We use subpixel-precise edge extraction (see Section 3.7.3.5) for the determination of the boundary. Because edge extraction is computationally expensive, we determine an ROI first:

```
threshold (Image, Region, 128, 255)
boundary (Region, RegionBorder, 'inner')
dilation_circle (RegionBorder, RegionBorderDilation, 3.5)
reduce_domain (Image, RegionBorderDilation, ImageReduced)
```

Because of the diffuse bright-field back light illumination, it is easy to find an appropriate threshold range for the segmentation of the bright background, e.g., 128 to 255. With this, we can be sure that the edges to be extracted lie in the vicinity of the border of the segmented region. We broaden this border using morphology to get a band-shaped ROI for the subpixel-precise extraction of the edges.

```
edges_sub_pix (ImageReduced, Boundary, 'canny', 2.0, 20, 40)
```

The extracted edges are shown in Figure 4.44(b). In the next step, these edges are segmented into lines and circular arcs:

```
segment_contours_xld (Boundary, ContoursSplit, 'lines_circles', 5, 10, 5)
```

Now, we collect all circular arcs. The type of the split contours can be determined by reading out the global contour attribute `'cont_approx'`,

4.9.1 提取冲孔板的边界

用亚像素边缘提取来确定边界（见 3.7.3.5 节）。由于边缘提取计算量很大，先确定感兴趣区域：

由于使用漫射、明场背光照明，很容易找到合适的阈值范围来分割亮的背景，比如，从 128~255。由此，可以确保提取出来的边界在分割出的区域边界附近。利用形态学，拓宽这个边界得到带状感兴趣区域来进行亚像素精度边缘提取。

图 4.44（b）为提取出的边界。下一步将这些边界分割为线段和圆弧：

现在选出所有的圆弧。通过读取全局轮廓属性 cont_approx 可以确定分离轮廓的类型。属性为 1 的轮廓近

which has the value 1 for contours that are best approximated by circular arcs. For contours that can be approximated by lines, the attribute has the value −1.

似为圆弧，属性为 −1 的轮廓近似为线段。

```
gen_empty_obj (CircularArcs)
Number := |ContoursSplit|
for i := 1 to Number by 1
    select_obj (ContoursSplit, ObjectSelected, i)
    get_contour_global_attrib_xld (ObjectSelected, 'cont_approx', Attrib)
    if (Attrib == 1)
        concat_obj (CircularArcs, ObjectSelected, CircularArcs)
    endif
endfor
```

The contours may have been oversegmented during the segmentation into lines and circular arcs. To obtain one contour for each circular border of the punched sheet, we merge cocircular arcs:

在将轮廓分割成线段和圆弧过程中可能会发生过度分割。为使冲孔板每一个圆形边界得到一个轮廓，我们合并共圆弧：

```
union_cocircular_contours_xld (CircularArcs, UnionContours,
                               0.5, 0.5, 0.2, 30, 10, 20, 'true', 1)
```

The resulting circular arcs are displayed in Figure 4.45(a). Figure 4.45(b) shows a detail of the image from Figure 4.44(a) along with the circular arc.

图 4.45（a）为得到的圆弧。图 4.45（b）为图 4.44（a）图像的细节及圆弧。

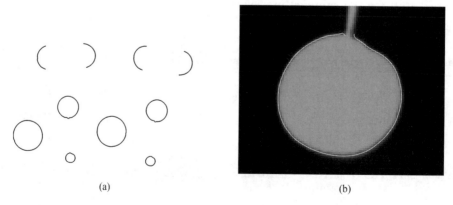

图 4.45 (a) 图 4.44(b) 边界中选出的圆弧；(b) 4.44(a) 的细节及选出的边界。为了更好的视觉效果，图像的对比度降低了

4.9.2 Performing the Inspection

Now we determine the parameters of the best-fitting circle for each circular arc and select those circles that have a radius smaller than a relatively large value, e.g., 500 pixels, to get rid of the large arc at the lower image border.

4.9.2 边缘检测

现在针对每个圆弧确定最佳拟合圆的参数并选出那些半径小于一个相对较大的数值的圆，比如 500 像素，这样是为了去除图像下部的大圆弧：

```
fit_circle_contour_xld (UnionContours, 'geotukey', -1, 0, 0, 3, 1,
                        Row, Column, Radius, StartPhi, EndPhi, PointOrder)
CircleIds := find(sgn(Radius-500),-1)
select_obj (UnionContours, Circles, CircleIds+1)
```

To determine the indices of the circles that have a radius below 500 pixels, we use the following operations. First, we subtract 500 from the tuple that holds the radii and determine the sign of the resulting values. If a particular radius is smaller than 500, `sgn(Radius-500)` will be `-1`. Using the function `find`, we collect all indices of circles with a radius smaller than 500. The radii of these circles are displayed in Figure 4.46(a). Figure 4.46(b) shows a detail of the image from Figure 4.44(a) along with the fitted circle.

为了确定半径小于 500 像素的圆的索引，使用下面的操作。首先在保存半径的元组中减去 500，判断结果的正、负符号。如果半径小于 500，sgn(Radius-500) 为 −1。使用函数 find 得到所有半径小于 500 的圆的索引。图 4.46（a）显示了这些圆的半径。图 4.46（b）显示了图 4.44（a）的细节及拟合圆。

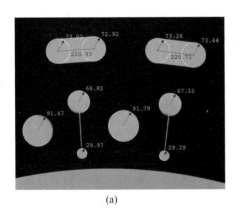

(a)   (b)

图 4.46 （a）检测结果。拟合圆及其半径以白色显示。对于椭圆孔，还给出了两个圆圆心间的距离；（b）图 4.44（a）的细节及拟合圆，可以看出孔的形状不是完美的圆。为了视觉效果图像的对比度降低了

Finally, we determine the distance between the centers of the two circles of the two oval holes at the top of the image. For simplicity, we assume that the oval holes are the uppermost holes in the image. Hence we can select the four respective circle centers simply by sorting them appropriately:

最后确定图像上方两个椭圆孔的圆心间的距离。为了简化，假设椭圆孔是位于图像最上方的孔，这样只是通过适当地分类就可以选出这四个圆心：

```
IndicesUppermost := sort_index(Row)[0:3]
RowOval := subset(Row,IndicesUppermost)
ColOval := subset(Column,IndicesUppermost)
```

First, we sort the row coordinates in ascending order and take the indices of the first four circle centers. By using these indices, we can select the four uppermost circles.

首先，按递增顺序对行坐标排序并且取出前四个圆心的索引，根据这些索引，得到最上面的四个圆。

Now, we sort these circles according to their column position:

然后，再根据它们所在列的位置对这些圆排序：

```
Indices := sort_index(ColOval)
RowOvalSorted := subset(RowOval,Indices)
ColOvalSorted := subset(ColOval,Indices)
```

The sorted indices are used to arrange the centers of the four circles from left to right. With this, it is easy to determine the distance between the two circle centers of each oval hole:

得到的索引用于从左到右得到四个圆的圆心。这样可以容易地确定每个椭圆孔的两个圆的圆心间的距离：

```
for CircleNo := 0 to |Indices|-1 by 2
    distance_pp (RowOvalSorted[CircleNo], ColOvalSorted[CircleNo],
                 RowOvalSorted[CircleNo+1], ColOvalSorted[CircleNo+1],
                 Distance)
endfor
```

Because we have sorted the circle centers from left to right, we can simply step through the sorted array by using a step width of 2 and select two consecutive circle centers each time. This ensures that both circle centers belong to the same oval hole. The resulting distances are displayed in Figure 4.46(a).

由于已经将圆心从左到右排序，每次在排序后的矩阵中选择两个连续的圆心就可以保证这两个圆心属于同一个椭圆。距离结果如图 4.46（a）所示。

4.9.3 exercises

1. For the detection of the two oval holes, we made the assumption that the punched sheet appears in roughly the same orientation in each image. Write a procedure that determines the pairs of circles that belong to the same oval hole even if the punched sheet is arbitrarily oriented. Tip: You can test whether the connection between two circle centers lies completely within a hole by means of set operations on regions (see Section 3.6.1.1).

2. As can be seen in Figure 4.46(b), the actual boundary of the hole in the sheet metal deviates from a perfect circle. Extend the above program such that the deviations of the extracted boundary from the fitted circle are determined. Visually highlight the parts of the boundary that deviate from the circle by more than some given threshold.

4.10 3D Plane Reconstruction with Stereo

A binocular stereo system consists of two cameras looking at the same object. If the interior orientations and the relative orientation of the two cameras are known, the 3D surface of the object can be reconstructed from the images of the two cameras.

In this application, we determine the angle between two planes on an intake manifold using a stereo camera setup. First, we calibrate the stereo camera setup. Then, from a stereo image pair of the intake manifold, a distance map is determined that describes the 3D surface of the object. We determine the two planes that form the indentation in the cylindrical surface. Finally, we calculate the indentation angle between the two planar faces of the indentation.

The algorithms used in this application are:
- Camera calibration (Section 3.9).
- Stereo reconstruction (Section 3.10.1).
- 2D edge extraction (Section 3.7.3).
- Region morphology (Section 3.6.1).

The corresponding example program is:

···/machine_vision_book/3d_plane_reconstruction_with_stereo/3d_plane_reconstruction_with_stereo.hdev

系统进行标定，然后通过进气歧管的立体图像对求出一个距离图，这个距离图描述了物体的三维表面。下一步从距离图像中提取出圆柱表面上 V 形缺口的两个平面。最后计算 V 形缺口两个平面之间的角度。

本应用使用的算法：
- 摄像机标定（3.9 节）。
- 立体重构（3.10.1 节）。
- 二维边缘提取（3.7.3 节）。
- 区域形态学（3.6.1 节）。

相应的例子程序位于：

4.10.1 Calibrating the Stereo Setup

To be able to reconstruct the 3D surface of the object, we must calibrate the stereo system. For this, we acquire stereo images of a calibration target. Figure 4.47 shows two stereo image pairs of a calibration target. In Figures 4.47(a) and (c), the images of the first camera are shown, while Figures 4.47(b) and (d) show the respective images of the second camera. For historical reasons (this example was developed for the first edition of this book), we use an old calibration target (cf. Section 3.9.4.1). In new applications, we would of course use the new calibration target.

First, we specify the calibration target used, define initial values for the interior orientation of both cameras, and initialize the handle that will be used to collect the image coordinates of the calibration marks and the initial poses of the calibration targets:

4.10.1 标定立体视觉系统

为了能够重构物体的三维表面，我们必须首先标定立体视觉系统。为此，我们需要拍摄得到标定板的立体图像对。图 4.47 中显示了标定板的两个立体图像对。在图 4.47（a）和图 4.47（c）中显示的为第一个摄像机拍摄得到的图像，图 4.47（b）和（d）中显示的是第二个摄像机拍摄得到的图像。由于历史原因（该示例程序开发于本书的第一版），我们采用了旧版的标定板（见 3.9.4.1 节），在新的应用中，我们会采用新版的标定板。

首先，指定使用的标定板，定义两个摄像机内参的初始值并且初始化用来保存标定点的图像坐标以及标定板初始位姿的句柄：

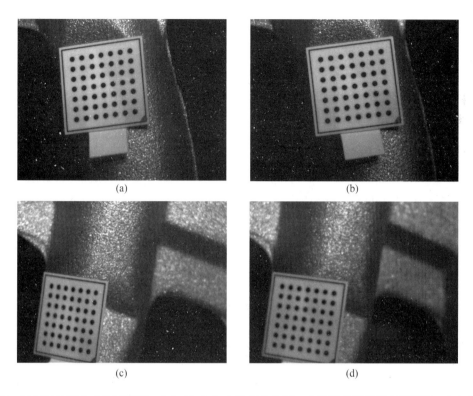

图 4.47 标定板的两个立体图像对。(a) 和 (c) 中显示的为第一个摄像机拍摄得到的图像，(b) 和 (d) 中显示的是第二个摄像机拍摄得到的图像

```
CaltabName := 'caltab_30mm.descr'
gen_cam_par_area_scan_division (0.025, 0, 7.5e-6, 7.5e-6, Width1/2.0,
                                Height1/2.0, Width1, Height1,
                                StartCamParam1)
gen_cam_par_area_scan_division (0.025, 0, 7.5e-6, 7.5e-6, Width2/2.0,
                                Height2/2.0, Width2, Height2,
                                StartCamParam2)
create_calib_data ('calibration_object', 2, 1, CalibDataID)
set_calib_data_calib_object (CalibDataID, 0, CaltabName)
set_calib_data_cam_param (CalibDataID, 0, [], StartCamParam1)
set_calib_data_cam_param (CalibDataID, 1, [], StartCamParam2)
```

Both cameras have a focal length of approximately 25 mm and the pixel size is 7.5 μm. The distortion coefficient is set to zero and the principal point is assumed to be in the center of the image. These values are just initial values. They will be improved during the calibration process.

两个摄像机的焦距约为 25 mm，像元尺寸为 7.5 μm。首先将畸变系数设置为 0，并假设主点的位置在图像的中心。这些值都只是初始值，在标定过程中要对这些值进行校准。

Then, within a loop over all available calibration image pairs, the subpixel-accurate positions of the calibration marks and initial estimates for the poses of the calibration target with respect to the two cameras are determined:

然后，在一个遍历所有可用的标定图像对的循环中，确定标定点的亚像素精度的坐标及确定对标定板相对于两个摄像机的位姿的初始估算：

```
find_calib_object (Image1, CalibDataID, 0, 0, Index, 'alpha', 0.5)
find_calib_object (Image2, CalibDataID, 1, 0, Index, 'alpha', 0.5)
```

Now, we determine the exact interior orientations as well as the relative pose of the two cameras:

此时可以得到两个摄像机准确的内参、外参以及两个摄像机之间的相对姿态：

```
calibrate_cameras (CalibDataID, Error)
get_calib_data (CalibDataID, 'camera', 0, 'params', CamParam1)
get_calib_data (CalibDataID, 'camera', 1, 'params', CamParam2)
get_calib_data (CalibDataID, 'camera', 1, 'pose', RelPose)
```

For the calibrated stereo system, we determine two rectification maps that are used to rectify the stereo image pairs to the epipolar standard geometry:

对于标定后的双目立体视觉系统，可以计算得到两个校正映射图，这两幅映射图可以用来将立体图像对校正为标准极线几何结构。

```
gen_binocular_rectification_map (Map1, Map2, CamParam1, CamParam2,
                RelPose, 1, 'geometric', 'bilinear',
                CamParamRect1, CamParamRect2,
                CamPoseRect1, CamPoseRect2, RelPoseRect)
```

The rectification maps define a mapping from the original stereo images to the rectified stereo images in which corresponding points, i.e., points that belong to the same object point, have identical row coordinates.

校正映射图定义了一个从原始立体图像到校正后的立体图像的映射关系，在校正后的立体图像对中，相同空间点在两幅图像中的对应点在同一行。

4.10.2 Performing the 3D Reconstruction and Inspection

4.10.2 进行三维重构及检测

All steps described so far are performed offline. In the online phase, we use the calibrated stereo

目前为止介绍的所有步骤都是离线执行的。在线阶段，使用校正后的

system to reconstruct the 3D surface of the intake manifold. For this, we need a stereo image pair of the object. Figure 4.48 shows one stereo image pair of an intake manifold acquired with a calibrated stereo system.

立体视觉系统重构进气歧管的三维表面。为此需要得到被测物体的立体图像对。图 4.48 中显示使用标定后的立体视觉系统拍摄得到的进气歧管的立体图像对。

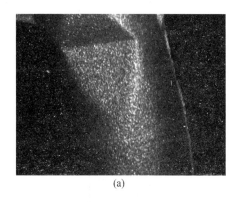 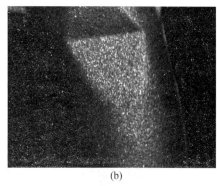

(a) (b)

图 4.48 使用标定后的立体视觉系统拍摄得到的进气歧管的立体图像对：（a）第一个摄像机拍摄得到的图像；（b）第二个摄像机拍摄得到的图像

First, the images are rectified to the epipolar standard geometry with the rectification maps created above:

首先，使用上面创建的校正映射图将立体图像对校正为标准极线几何结构：

```
map_image (Image1, Map1, ImageRect1)
map_image (Image2, Map2, ImageRect2)
```

The rectified images of the two cameras are shown in Figure 4.49.

两个摄像机拍摄得到的图像校正后的图像如图 4.49 所示。

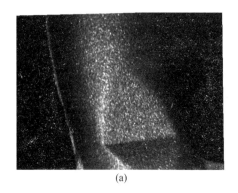 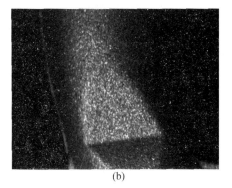

(a) (b)

图 4.49 校正为标准极线几何结构后第一个摄像机图像（a）和第二个摄像机图像（b）

Now, we determine the distance of each point on the object surface from the stereo system:

现在就可以计算被测物体表面上每个点与立体视觉系统之间的距离。

```
binocular_distance (ImageRect1, ImageRect2, Distance, Score,
                   CamParamRect1, CamParamRect2, RelPoseRect,
                   'ncc', 21, 21, 5, MinDisparity, MaxDisparity,
                   5, 0.1, 'left_right_check', 'interpolation')
```

The distance of the object surface from the stereo system is shown in Figure 4.50(a). Figure 4.50(b) shows the quality of the matches between the two rectified images. Bright values indicate good matches.

图 4.50（a）显示被测物体表面与立体视觉系统的距离。图 4.50（b）中显示的是校正后两幅图像之间匹配的质量，亮的部分表示匹配分值较高。

(a)

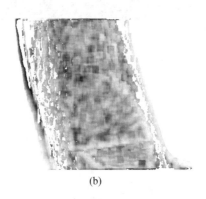
(b)

图 4.50　（a）被测物体表面与双目立体视觉系统之间的距离。亮的区域比暗的区域距离更远。为了使图像看起来更清晰，显示的值为 $d_{r,c}^{1/2}$；（b）校正后图像之间匹配的质量，亮的部分表示匹配分值较高

Now, we determine the two planes that form the indentation. First we determine the direction of the gradients in the distance image. Then, we segment the image into regions that have homogeneous gradient directions. Finally, we test how well these regions represent planes. We select the best two regions and determine the normal vectors of the respective planes.

此时需要确定构成 V 形缺口的两个平面。首先计算得到距离图像中的梯度方向；然后将图像中梯度方向相似的区域分割出来；最后测试这些区域能否很好地表示平面。选择最符合平面特征的两个区域并确定各自的法向量。

We determine the gradient direction of the distance image as follows:

为了确定距离图像的梯度方向，可以使用下面的操作：

```
get_domain (Distance, Domain)
min_max_gray (Domain, Distance, 0, Min, Max, Range)
scale_image (Distance, DistanceScaled,
             pow(2,16)/Range, -Min*pow(2,16)/Range)
convert_image_type (DistanceScaled, DistanceUInt2, 'uint2')
edges_image (DistanceUInt2, ImaAmp, ImaDir, 'canny', 1.5, 'none', 20, 40)
```

Because the distance image is a real-valued image and this data type is not supported by the edge extraction operator, we must convert the image such that the distance values are represented by integer values. To preserve the accuracy of the distance values, the distance image is converted into an image with a gray value depth of 16 bits. The original (real) gray values are scaled such that they fully exploit the gray value range of the 16-bit image. Then, the gradient directions are determined with the operator `edges_image`. In the resulting direction image, regions that have homogeneous directions are determined.

由于距离图像中每个像素的值为实数，而边缘提取算子不支持这种数据类型的图像，因此不得不将图像进行变换，使距离值用整数表示。为了保证距离值的准确度，将距离图像转换为 16 位的灰度值的图像。将原始的灰度值（实数）进行缩放使得到的灰度值能够完全覆盖 16 位灰度图像的灰度值范围。然后调用 edges_image 算子计算梯度方向。在得到的方向图像中，确定方向相似的区域。

In the direction image, the gradient directions are stored in 2-degree steps, i.e., a direction of x degrees with respect to the horizontal axis is stored as $x/2$ in the direction image. Points with edge amplitude 0 are assigned the edge direction 255, which indicates an undefined direction. Hence, only gray values in the range $[0, 179]$ are of interest for the following analysis. Because we are searching for planes, we must find regions in the image that have a homogeneous gradient direction. For this reason, the histogram of the direction image should contain peaks that represent the mean directions of the regions with homogeneous gradient direction. The histogram is obtained with the following operation:

在方向图像中，梯度方向是按 2 度间隔保存，也就是说与水平轴方向成角度为 x 的方向在方向图像中保存为 $x/2$。将边缘振幅（幅角）为 0 的点的边缘方向设置为 255，这个值表示未定义的方向。因此下面分析中，只有灰度值区域为 $[0, 179]$ 是感兴趣的。由于搜索的是平面，因此图像中一定会存在梯度方向相似的区域。因此，方向图像的灰度值直方图中一定包含波峰，这些波峰表示了梯度方向相似区域的平均方向。可以通过下面算子得到灰度值直方图：

```
gray_histo (Domain, ImaDir, AbsoluteHisto, RelativeHisto)
```

Figure 4.51(a) shows the histogram of the direction image. It has four major peaks, one of them at the wrap-around at 360 degrees. Two of the peaks belong to the planes we want to extract. The two other peaks represent areas on the cylindrical surface of the object. To illustrate this, Figure 4.51(b) displays an image of the intake manifold along with the mean gradient directions in the regions that correspond to the four major peaks of the histogram of the direction image.

图 4.51（a）中显示了方向图像的灰度值直方图。直方图中含有四个主要的波峰，其中一个波峰在 360° 的附近。这些波峰中有两个波峰表示了希望提取的平面。另外两个波峰表示的区域是被测物体的圆柱表面。图 4.51（b）中显示了进气歧管的图像，同时显示了方向图像的直方图中四个波峰相应的区域的平均梯度方向。

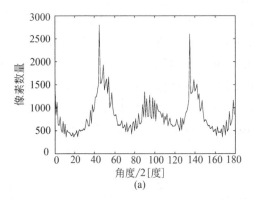
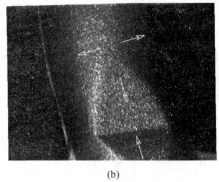

图 4.51　（a）方向图像的灰度值直方图。注意梯度方向是按 2 度间隔保存；（b）进气歧管的图像，图上显示的四个箭头表示图（a）灰度值直方图中四个波峰相应区域的平均梯度方向

The appropriate thresholds for the segmentation of the image into regions with homogeneous gradient directions are the minima in the histogram of the direction image. To determine the minima in a stable manner, the histogram must be smoothed. Because of the cyclic nature of the direction image, we must take care to ensure correct smoothing around the position 0 degrees. This is done by creating a function that consists of two consecutive copies of the histogram. We then smooth this function. From the local minima of the smoothed function, we select a sequence that covers a range of 360 degrees; however, we should

使用方向图像灰度值直方图中最小值作为阈值进行图像分割得到具有相似梯度方向的区域。为了能够以可靠的方式确定最小值，必须首先对灰度值直方图进行平滑处理。由于方向图像的周期特性，必须注意在 0° 的位置进行正确的平滑处理，这可以通过创建一个包含两个直方图连续拷贝的函数来实现。然后对该函数进行平滑处理。从平滑后的函数局部的最小值中，必须选择一个覆盖整个 360° 的序列，此时不应该使用第一个或最后一个最小值，因为它们很可能

not use the first or the last minimum because their positions might be disturbed if they lie close to 0 degrees or 360 degrees, respectively.

由于与 0° 或 360° 过近而导致位置被干扰。

```
create_funct_1d_array ([AbsoluteHisto[0:179],AbsoluteHisto[0:179]],
                       HistoFunction)
smooth_funct_1d_gauss (HistoFunction, 4, SmoothedHistoFunction)
local_min_max_funct_1d (SmoothedHistoFunction, 'strict_min_max', 'true',
                        Minima, Maxima)
MinThreshShifted := Minima[1:(find(sgn(Minima-(Minima[1]+179)),1))[0]-1]
MinThresh := sort(fmod(MinThreshShifted,180))
```

We can now segment the distance image using the thresholds determined above. Because of the cyclic nature of the direction image from which the thresholds were determined, we must merge the region that starts at 0 degrees with the one that ends at 360 degrees if there is no minimum in the direction histogram at 0 degrees:

现在可以使用上面确定的阈值对距离图像进行分割。由于用于确定阈值的方向图像具有周期属性，如果在方向直方图 0° 处没有最小值，就必须将从 0° 开始的区域和在 360° 结束的区域合并起来：

```
threshold (ImaDir, Region, [0,int(MinThresh)+1], [int(MinThresh),179])
count_obj (Region, Number)
select_obj (Region, FirstRegion, 1)
select_obj (Region, LastRegion, Number)
union2 (FirstRegion, LastRegion, RegionUnion)
copy_obj (Region, ObjectsSelected, 2, Number-2)
concat_obj (RegionUnion, ObjectsSelected, Regions)
```

The resulting regions all have a homogeneous gradient direction. To eliminate clutter, we select the largest connected component from each region and apply an opening operation to it. Then, we select the two regions that represent planes:

得到的区域中都有一个相似的梯度方向。为了消除一些干扰区域，从每个区域中选择最大的连通区域并且在选中的区域上应用一个开运算操作。然后，选择出表示平面的两个区域：

```
count_obj (Regions, Number)
for Index := 1 to Number by 1
    select_obj (Regions, ObjectSelected, Index)
    connection (ObjectSelected, ConnectedRegions)
    area_center (ConnectedRegions, Area, RowConnected, ColumnConnected)
    select_obj (ConnectedRegions, LargestRegion,
                sort_index(-AreaConnected)[0]+1)
```

```
    opening_circle (LargestRegion, PlaneRegion, 3.5)
    area_center (PlaneRegion, AreaPlane, RowPlane, ColumnPlane)
    fit_surface_first_order (PlaneRegion, Distance, 'regression', 5, 2,
                             Alpha, Beta, Gamma)
    gen_image_surface_first_order (ImageSurface, 'real', Alpha, Beta,
                                   Gamma, Row, Col, Width, Height)
    reduce_domain (ImageSurface, PlaneRegion, ImageReduced)
    sub_image (Distance, ImageReduced, DeviationFromPlane, 1, 0)
    intensity (PlaneRegion, DeviationFromPlane, Mean, Deviation)
endfor
```

Similarly to the fitting of lines described in Section 3.8.1, we fit planes into the regions of the distance image that were described above. Then, we determine the deviation of the respective parts of the distance image from these planes. The two regions for which the distance image deviates least from the respective planes are selected.

与 3.8.1 节介绍的线段拟合相似，用平面拟合上面得到的距离图像的区域。然后，求出从距离图像中得到的所有区域与它们拟合得到的平面之间的偏差。将偏差最小的两个区域选为平面。

We determine the normal vectors for these regions from the plane parameters and the pixel size of the rectified image:

然后从平面参数以及校正后图像的像素尺寸计算出这些区域的法向量：

```
Nx := -Sy*Alpha
Ny := -Sx*Beta
Nz := Sx*Sy
Length := sqrt(Nx*Nx+Ny*Ny+Nz*Nz)
Nx := Nx/Length
Ny := Ny/Length
Nz := Nz/Length
```

From the two normal vectors, the angle between the two planes easily can be determined.

有了这两个法向量，就可以轻松计算出两个平面之间的夹角。

Figure 4.52 shows the final results. In Figure 4.52(a), the rectified image of the intake manifold is shown along with the outlines of the two planes. Figure 4.52(b) shows a 3D plot of the surface of the intake manifold.

图 4.52 显示了最终的计算结果。在图 4.52（a）中，显示了校正后的进气歧管的图像以及最终得到两个平面的轮廓。图 4.52（b）中显示了进气歧管表面的三维图。

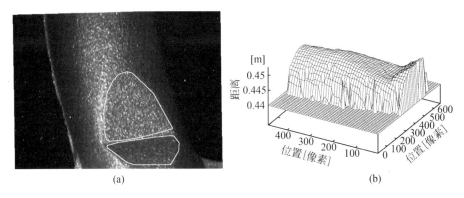

图 4.52 （a）校正后的进气歧管的图像以及得到的两个平面的轮廓；（b）进气歧管表面的三维图

4.10.3 exercises

We have used the operator `binocular_distance` to perform the 3D reconstruction. Modify the program to use the operator `reconstruct_surface_stereo`.

4.11 Pose Verification of Resistors

Typically, electronic components are mounted on a PCB by a pick-and-place machine. Before the completed boards are tested, they are visually inspected for missing or misaligned components. Furthermore, in some applications it is necessary to verify whether the correct type of electronic component has been mounted at the intended place.

In this application, we verify the pose and the type of different resistors that have been mounted on a PCB. First, the pose of the resistor on the board is determined and missing resistors are detected. Then, the type of the resistor (if present) is extracted and compared to the known reference type.

4.10.3 练习

我们已经使用了算子 binocular_distance 来进行 3D 重构。请修改程序使用算子 reconstruct_surface_stereo 来实现。

4.11 电阻姿态检验

一般情况下，电子元器件通过贴片机安放到印刷电路板（PCB）上。在整个电路板被检测之前，首先使用视觉手段检测电路板上是否缺少某些电子元器件并且检测电子元器件的位置是否有偏差。另外，在一些应用中还有必要检测在指定的位置上安装的电子元器件的类型是否正确。

在这个应用中，我们检测安装在一个印刷电路板上的不同电阻的位姿和类型。首先，检测电路板上电阻的姿态以及检测是否缺失电阻。然后，如果电阻上印有型号，提取电阻的型号并与已知的参考型号做比较。

The algorithms used in this application are:
- Shape-based matching (Section 3.11.5.6).
- Affine transformations (Section 3.3.1).
- Image transformations (Section 3.3.2).

The corresponding example program is:
··· /machine_vision_book/resistor_verification/resistor_verification.hdev

To prevent specular reflections in the images, we use diffuse bright-field front light illumination (see Section 2.1.5.1). Figures 4.53(a) and (b) show the two types of resistors that we want to verify: 33 Ω and 1.1 Ω, respectively.

图 4.53 必须进行检测的两种电阻：(a) 33Ω；(b) 1.1Ω

4.11.1 Creating Models of the Resistors

The first task is to determine the pose of the resistor in the image. All resistors have a rectangular shape. However, the size of the resistors and the aspect ratio of their sides are not identical. Consequently, determining the pose of the resistor implies determining its position, its orientation, and two scaling factors, which represent the size and the aspect ratio of the resistor's sides. We determine the pose of the resistor by using shape-based matching (see Section 3.11.5.6). First, we create an artificial template image of a generic resistor with average size and aspect ratio:

姿态。首先，创建一个普通电阻的人工模板图像，它的尺寸和形状比为平均值：

```
MeanModelHeight := 185
MeanModelWidth := 100
gen_image_const (Image, 'byte', Width, Height)
gen_rectangle2_contour_xld (Rectangle, Height/2, Width/2, 0,
                            MeanModelWidth/2.0, MeanModelHeight/2.0)
paint_xld (Rectangle, Image, ModelImageGeneric, 128)
```

For this, we generate a rectangular contour that represents the boundary of an average resistor. Then, the contour is painted into an empty image. Note that `paint_xld` paints the contour onto the background using anti-aliasing. Consequently, the gray value of a pixel depends on the fraction by which the pixel is covered by the rectangle. For example, if only half of the pixel is covered, the gray value is set to the mean gray value of the background and the rectangle. The resulting template image is shown in Figure 4.54(a). After this, a shape model is created from the template image by using the following operations:

为了创建模板图像，创建一个矩形轮廓来表示普通电阻的边界。然后，将这个轮廓绘制到一个空的图像中，注意使用 paint_xld 算子可以使用反锯齿方式在图像背景中绘制这个轮廓。因此，一个像素的灰度值取决于矩形覆盖了这个像素的多大部分。例如，如果矩形区域覆盖了这个像素的一半，那么这个像素的灰度值就等于矩形背景灰度值的中值。最终得到的模板图像如图 4.54（a）所示。然后就可以通过下面的操作从这个模板图像中创建基于形状的模板：

```
AngleTol := rad(5)
ScaleTol := 0.1
create_aniso_shape_model (ModelImageGeneric, 3,
                          -AngleTol, 2.0*AngleTol, 'auto',
                          1.0-ScaleTol, 1.0+ScaleTol, 'auto',
                          1.0-ScaleTol, 1.0+ScaleTol, 'auto',
                          'auto', 'ignore_local_polarity', 'auto',
                          10, ModelIDGeneric)
```

In this application, it can be assumed that the PCB is aligned horizontally and that the resistors are mounted on the board with an angle tolerance of ±5°. Furthermore, the length of the resistor's sides may vary by ±10% with respect to the average values MeanModelHeight

在这个应用中，假设 PCB 板水平放置，并且安装在电路板上的电阻的角度公差是 ±5°。另外，电阻的边长相对于平均值 MeanModelHeight 和 MeanModelWidth 可能存在 10% 的偏差。因此，要依据相应的公差值

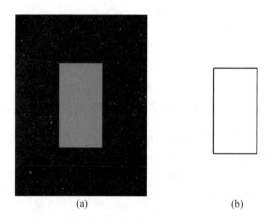

图 4.54 （a）普通电阻的人工模板图像；（b）用于匹配的模板的轮廓表示

and `MeanModelWidth`. The shape model of the generic resistor is created accordingly by passing the corresponding tolerance values. The model can be used to determine the pose of the resistor in the online phase even if it appears rotated and anisotropically scaled.

The second task is to determine the type of the resistor. This task can be solved by means of the printed characters on top of the resistors. For this, we create two additional model representations, one for the print "330" on the 33 Ω resistor and one for the print "1R1" on the 1.1 Ω resistor. In the online phase, the best matching model is assumed to represent the present resistor type. If neither of the two models can be found in the online phase, the resistor type is set to "unknown." In the following, the model creation is described for the 33 Ω resistor. The model creation for the 1.1 Ω resistor is performed in the same manner.

The model is created from the print on the resistor shown in Figure 4.53(a). To ease the future use of new resistor types, the model generation process is done automatically. First, the pose of the resistor in the model image is determined by using the generic resistor model generated above:

创建普通电阻基于形状的模板。使用这个模板就可以在在线阶段确定电阻的姿态，即使电阻存在一定的角度旋转或者它的尺寸存在不同比例的变化。

其次就是确定电阻的类型。这可以依靠电阻上印刷的字符来解决。为了解决这个问题，另外再创建两个模板，一个模板表示在 33 Ω 的电阻上印刷的 "330"，另一个模板表示在 1.1 Ω 电阻上印刷的 "1R1"。在线阶段中，认为这两个模板中最佳的匹配者表示当前电阻的类型。如果在线阶段中，两个模板都不能被找到，那么当前电阻的类型被设置为 "未知"。下面介绍如何为 33 Ω 电阻上的印刷字符创建模板。可以使用同样的方法为 1.1 Ω 电阻上印刷字符创建模板。

使用图 4.53（a）中电阻上的印刷字符创建模板。为了使将来对于新类型的电阻创建模板更加方便，整个创建模板的过程都是自动的。首先，使用上面创建的普通电阻的模板确定模板图像中电阻的姿态：

```
find_aniso_shape_model (Image, ModelIDGeneric,
                        -AngleTol, 2.0*AngleTol,
                        1.0-ScaleTol, 1.0+ScaleTol,
                        1.0-ScaleTol, 1.0+ScaleTol,
                        0.7, 1, 0.5, 'least_squares', 0, 0.7,
                        Row, Column, Angle, ScaleR, ScaleC, Score)
```

The resulting pose parameters refer to the reference point of the generic model. The reference point is the center of gravity of the domain of the model image. Since the domain of the model image comprises the whole image, the reference point is simply the center of the model image, and hence the center of the resistor. Consequently, the pose of the generic model can be used to generate a rectangular ROI that contains the print on the resistor. For this, we assume that the print is contained within a rectangle that has the same position and orientation as the resistor but only half of its side lengths:

所得到的姿态参数与普通模板的参考点有关。参考点是模板图像的区域的重心。由于模板图像的区域包含整幅图像,因此参考点就是整幅模板图像的中心点,也就是电阻的中心点。因此,普通模板的姿态可以用来生成一个矩形感兴趣区域,在这个感兴趣区域中包含电阻上的印刷字符。在此,假设包含印刷字符在内的矩形区域的位置和方向与整个电阻相同,只是边长是电阻边长的一半:

```
PrintFraction := 0.5
gen_rectangle2 (Rectangle, Row, Column, Angle,
                PrintFraction*ScaleC*0.5*MeanModelWidth,
                PrintFraction*ScaleR*0.5*MeanModelHeight)
reduce_domain (Image, Rectangle, ModelImage)
create_shape_model (ModelImage, 'auto', rad(-90), rad(360),
                    'auto', 'auto', 'use_polarity',
                    ['auto_contrast_hyst',20], 'auto', ModelID330)
```

The domain of the model image is reduced to the generated rectangle. Figure 4.55(a) shows the model image, the border of the found resistor, and the generated rectangle. Because the size and the aspect ratio of the print are constant in all images, this time it is sufficient to create a model that allows only rotated instances of the print to be found. Note, however, that the model is created within the full angle range

模板图像的范围缩小为上面程序生成的矩形感兴趣区域。图 4.55(a) 中显示了模板图像,模板图像中搜索得到电阻的边界以及生成的矩形感兴趣区域的边界。由于在所有图像中印刷字符的尺寸和形状比都不变,因此这次只创建一个在存在一定旋转的情况下能够找到印刷字符的模板就够了。然而,需要注意的是,模板是在

of 360°. The reason for this is that the resistors might be mounted not only in the reference orientation shown in Figure 4.55 but also rotated by 180°. The contour representation of the resulting shape model is shown in Figure 4.55(b). The model image and the resulting shape model of the print "1R1" are shown in Figures 4.55(c) and (d), respectively. Note that the print on the resistor in both model images should appear at an average orientation. If this cannot be guaranteed, either the model image must be rectified before creating the model, or the angle range that is used to search the print in the online phase must be adapted to the orientation difference of the print with respect to the resistor.

360°角度范围内创建的。这主要是因为电阻的安装不一定与图 4.55 的方向一致，也可能旋转 180°。图 4.55（b）显示了印刷字符形状模板的轮廓。印刷字符"1R1"的模板图像与模板轮廓分别如图 4.55（c）和图 4.55（d）所示。注意两个模板图像中电阻上印刷字符的方向都应该是通常的方向，如果不能保证这一点，就必须在创建模板前将模板图像进行校正，或者在线阶段用来搜索印刷字符的角度范围必须适应印刷字符相对于电阻的角度差。

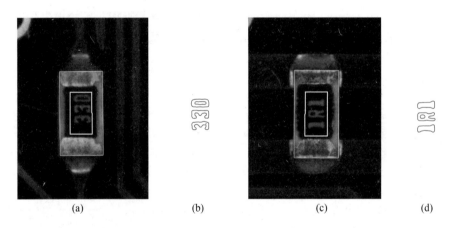

图 4.55　(a)(c) 用来创建印刷体"330"和"1R1"模板的图像，找到的电阻用外接白色矩形表示，创建模板的感兴趣区域用里面的白色矩形表示；(b)(d) 创建得到模板的轮廓

4.11.2 Verifying the Pose and Type of the Resistors

4.11.2 检测电阻的位姿和类型

All the steps described so far can be performed offline. In the online phase, we use the generated models to verify the pose and the type of the resistors. To verify the pose, the generic resistor model

目前为止介绍的所有步骤都可以在离线状态下执行。在线阶段中，使用上面创建得到的模板来检测电阻的姿态和类型。使用普通电阻模板来检

is used. It is searched in the same pose range in which it was created:

```
find_aniso_shape_model (ImageOnline, ModelIDGeneric,
                -AngleTol, 2.0*AngleTol,
                1.0-ScaleTol, 1.0+ScaleTol,
                1.0-ScaleTol, 1.0+ScaleTol,
                0.7, 1, 0.5, 'least_squares', 0, 0.7,
                Row, Column, Angle, ScaleR, ScaleC, Score)
```

If the model cannot be found, we assume that the resistor is missing on the board. Otherwise, we proceed with the type verification by using the two models that represent the prints on the resistors. Since we know the pose of the resistor, we can restrict the search for the print to an appropriate image region. This region must contain the reference point of the model. Because the position of the print does not vary much with respect to the outline of the resistor, a relatively small region is sufficient. In this example, the region is set to a rectangle the side lengths of which are a quarter of the resistor's side lengths:

验电阻的姿态。搜索电阻的姿态的范围与创建模板的范围相同：

如果不能在图像中找到模板，我们认为电路板上缺少了这个电阻。否则，继续用上面创建的代表两种印刷字符的模板来检测电阻的类型。由于已知电阻的位姿，可以将搜索印刷字符的区域限定到一个合适的区域中。这个区域中必须包含模板的参考点。由于印刷字符与电阻轮廓的相对位置不会发生很大的变化，一个相对较小的区域就足够了。在这个示例中，将这个区域设置为一个边长为电阻边长 1/4 的矩形：

```
PrintRefFraction := 0.25
gen_rectangle2 (Rectangle, Row, Column, Angle,
                PrintRefFraction*ScaleC*0.5*MeanModelWidth,
                PrintRefFraction*ScaleR*0.5*MeanModelHeight)
reduce_domain (ImageOnline, Rectangle, ImageReduced)
```

Finally, the models of the prints are searched in the reduced image domain:

最后，在这个缩小的图像区域中搜索印刷字符的模板：

```
find_shape_models (ImageReduced,
                [ModelID330,ModelID330,ModelID1R1,ModelID1R1],
                [0,rad(180),0,rad(180)]-AngleTol, 2*AngleTol,
                0.5, 1, 0.5, 'least_squares', 0, 0.9,
                PrintRow, PrintColumn, PrintAngle,
                PrintScore, PrintModel)
```

As mentioned above, the resistors may be mounted in the two orientations 0° and 180° (plus the tolerances). Consequently, we do not need to search the prints in the full range of 360°. Instead, each model is only searched in the two angle ranges [−AngleTol, AngleTol] and [π−AngleTol, π+AngleTol]. With `find_shape_models`, multiple models can be searched simultaneously. Therefore, altogether four model handles are passed to `find_shape_models`: each of the two models is passed twice, once for each angle range. Alternatively, `find_shape_model` could be called four times, once for each model and each angle range. However, this would be computationally more expensive because some computations must be performed multiple times. Furthermore, the best match would have to be determined in an additional post-processing step. The index of the model that yielded the best match is returned in `PrintModel`. This index refers to the tuple of shape model handles that was passed as input, and hence is in the range [0...3]. If no match was found, we assume that the resistor type is "unknown." Otherwise, the resistor type can be computed based on the returned index:

正如上面提到过的，电阻可能在两个方向上安装：0°和180°（加上公差值）。因此，不需要在整个360°的范围内搜索印刷字符模板。相反，只需要在两个角度范围内搜索每个印刷字符模板：[−AngleTol, AngleTol]和[π−AngleTol, π+AngleTol]。使用算子 find_shape_models 可以同时搜索多个模板。因此，可以将4个模板句柄同时传递给算子 find_shape_models：每个印刷字符模板传递两次，每次对应不同的角度范围。或者调用四次 find_shape_models，每次对应一个模板和一个角度范围。然而后面这种方式需要耗费更长的计算时间，因为这种方式下某些计算必须执行多次。此外，必须在一个附加的后处理步骤中确定哪个是最佳匹配对象。变量 PrintModel 返回最佳匹配对象的模板序号。这个序号是作为输入的形状模板句柄数组的序号，因此它的范围是 [0...3]。如果没有找到匹配对象，就假设电阻类型为"未知"。否则，电阻的类型就可以基于返回的序号计算得到：

```
ResistorTypes := ['330','1R1']
Model := PrintModel/2
ResistorType := ResistorTypes[Model]
```

Finally, if the determined resistor type does not correspond to the expected one, an appropriate action can be triggered. In Figures 4.56(a) and (b), for each of the two resistor types, an example of the verification result is shown.

最后，如果得到的电阻类型与期望的类型不同，就需要触发一个适当的动作。在图 4.56（a）和图 4.56（b）中显示了对两个电阻类型进行检测的示例图像。

图 4.56 一个 33Ω（a）和一个 1.1Ω（b）电阻示例图像的检测结果。找到的电阻的边界显示为一个白色矩形框。找到的印刷体的姿态用白色轮廓线表示。电阻的姿态和它的类型显示在图像的左上角

4.11.3 exercises

1. In some applications, the resistors additionally appear in an orientation of ±90°. Modify the program appropriately so that resistors with orientations 0°, 90°, 180°, and 270° can be verified.

2. As an alternative to shape-based matching, NCC could be used for the template matching (see Section 3.11.1.2). Adapt the above program so that the pose and type of the resistor are determined with NCC. Tip: The generic model should be extended to include more details of the resistor. It should contain the dark rectangular center part, the gray border regions at the sides, and the bright regions of the contacts at the upper and lower parts of the resistor. Note that NCC is more robust against small deformations than shape-based matching. Therefore, no (anisotropic) scaling of the resistor model needs to be taken into account.

4.11.3 练习

1. 在一些应用中，电阻的方向还可能是 ±90°。将程序进行适当的修改，使其可以检测安装方向为 0°、90°、180° 和 270° 的电阻。提示：为了避免在整个 360° 空间内进行搜索，可以为每种印刷字符创建 4 个模板。

2. 除了基于形状的模板匹配外，归一化互相关系数方法同样可以用于模板匹配（见 3.11.1.2 节）。修改上面的程序，使用归一化互相关系数匹配方法确定电阻的姿态和类型。提示：上面使用的普通电阻模板图像应该进行扩展，使其包含更多的细节。它应该包含中间的黑色矩形、边缘部分灰色区域和电阻上下两个亮的焊接区域。注意归一化互相关系数的模板匹配方法在搜索对象出现小的变形的情况下比基于形状的模板匹配更加稳定。因此不需要考虑电阻模板在纵横方向出现不同缩放比例的情况。

3. In the above application, the type of the resistor is determined by using template matching. Alternatively, the print could be segmented by using region morphology (see Section 3.6.1) and read using OCR (see Sections 3.14 and 3.15). Modify the above program accordingly.

3. 上面的应用中，电阻的类型是通过模板匹配的方式进行检测。另一种方法是首先使用区域形态学的方法分割得到印刷字符区域（见 3.6.1 节），然后使用光学字符识别技术（见 3.14 节和 3.15 节）将字符读出。请按照这种方法相应地修改上面的程序。

4.12 Classification of Non-Woven Fabrics

4.12 非织造布分类

Through classification, a sample can be assigned to one of a set of predefined categories. The goal of texture classification is to assign an unknown image to one of a set of known texture classes.

通过分类，可以确定一个样本属于预先定义的类型集合中的哪一类。纹理分类的目的就是确定一个未知的图像属于已知的纹理类型中的哪一种类型。

In this application, we classify images of different types of non-woven fabrics using an MLP.

在这个应用中，使用多层感知器 (MLP) 形式的神经网络分类器来分类不同类型非织造布的图像。

The algorithms used in this application are:
- 2D edge extraction (Section 3.7.3).
- Gray value features (Section 3.5.2).
- Classification (Section 3.15).

本应用使用的算法：
- 二维边缘提取（3.7.3 节）。
- 灰度值特征（3.5.2 节）。
- 分类器（3.15 节）。

The corresponding example program is:

相应的示例程序位于：

···/machine_vision_book/classification_of_nonwoven_fabrics/classification_of_nonwoven_fabrics.hdev

4.12.1 Training the Classifier

4.12.1 训练分类器

Texture classification involves two phases: the training phase and the recognition phase. In the training phase, a classifier is created for the texture content of each texture class that is present in the training data. The training data consists of images with known class labels. The texture con-

特征分类的过程分为两个阶段：训练阶段和识别阶段。在训练阶段，为出现在训练样本中每个纹理类型的纹理内容创建分类器。训练样本由很多类型已知的图像组成。这些图像中包含的纹理内容可以通过一系列纹理

tent of the images is captured by a set of texture features. These features characterize the texture properties of the images.

It is advantageous to separate the training data into a set of training images and a set of independent test images. The former is used to train the classifier as well as to carry out a first test of the classifier by reclassifying the training images. The latter is used to test the classifier and to improve it by adding those test images to the set of training images that could not be classified correctly. In real applications, to rate the performance of the final classifier, another set of independent test images must be available.

In the recognition phase, the texture content of the images is described by the same texture features that were used in the training phase. Then, the classifier assigns each image to the best-matching class based on the texture features.

Figure 4.57 shows three samples out of a set of 22 different types of non-woven fabrics. The images are taken with diffuse bright-field front light illumination (see Section 2.1.5.1). In the current application, only six training images and four independent test images are available for each of the 22 classes. For real applications, much more training data must be available to achieve a reliable

特征得到。这些特征表现出图像的纹理属性。

将训练数据分成一个训练图像集和一个独立的测试图像集是非常有益的。训练图像集用来训练分类器，并且通过对这个图像集进行重新分类进行分类器第一次测试。测试图像集可以用来测试分类器并且可以通过将那些不能正确分类的测试图像加入训练图像集中重新训练来改进分类器。在实际应用中，为了评价最终分类器的性能必须提供另一个独立的测试图像集。

在识别阶段中，图像纹理内容由训练阶段中的相同的纹理特征描述。然后，分类器将基于纹理特征为每幅图像指定最佳匹配类型。

图 4.57 中显示了 22 种不同类型的非织造布中 3 种类型的样本。拍摄图像时使用的是明场漫射正面照明（见 2.1.5.1 节）。在当前这个应用中，为 22 种不同类型中的每种类型都只提供了 6 幅训练图像和 4 幅单独的测试图像。需要注意的是，在实际应用中为了创建一个更可靠的分类

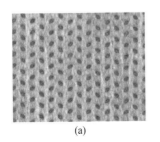 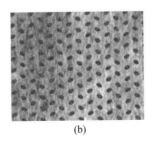 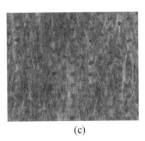

(a)　　　　　　　　　(b)　　　　　　　　　(c)

图 4.57　22 种不同类型的非织造布的三种类型的样本

classifier. It is not unusual that several thousands of training samples are used for each class.

器，需要提供更多的训练样本。为每种类型提供几千个训练样本是非常常见的。

First, we initialize the classifier:

首先，初始化分类器：

```
NumHidden := 10
create_class_mlp (NumFeatures, NumHidden, NumClasses, 'softmax',
                 'normalization', -1, 42, MLPHandle)
```

The number of hidden neurons defines the size of the MLP. Generally, it should be in the range of the number of features and the number of classes. If too few hidden neurons are chosen, the reclassification of the training images will give a large number of misclassified images. If the number of hidden neurons is chosen too large, the MLP may overfit the training data, which typically leads to bad generalization properties, i.e., the MLP learns the training data very well, but does not return very good results on unknown data.

神经网络中隐含神经元的数量定义了神经网络的大小。一般情况下，这个数量应该在特征数量和类型数量的范围内。如果选择的隐含神经元的数量太小，重新分类训练图像时将出现很多错误。如果隐含神经元的数量太大，那么，MLP 可能会过于符合训练数据，从而导致分类器的普遍性很差，也就是说 MLP 非常好地学习了训练数据，但是对于未知的数据不能返回比较好的结果。

We now add the texture features of the training images to the classifier. We use features that characterize the number and strength of edges in the image as well as features that measure the distribution of the gray values across the image:

下面将训练图像的纹理特征添加到分类器中。使用图像中边缘数量和强度特征以及评估图像上灰度值分布的特征：

```
Features := []
gen_gauss_pyramid (Image, ImagePyramid, 'constant', 0.5)
for Index := 1 to 3 by 1
    select_obj (ImagePyramid, ImageRR, Index)
    sobel_amp (ImageRR, EdgeAmplitude, 'sum_abs', 3)
    gray_histo_abs (EdgeAmplitude, EdgeAmplitude, 8, AbsoluteHisto)
    SobelFeatures := real(AbsoluteHisto)/sum(AbsoluteHisto)
    Features := [Features,SobelFeatures]
endfor
```

First, edges are extracted from the image. Then, the relative histogram of the edge ampli-

首先，从图像中提取边缘。然后得到边缘幅度的相关直方图。需要在

tudes is derived. These features are calculated for the original image as well as for reduced-resolution copies of the image.

原始图像中计算这些特征，也需要在降低分辨率的图像拷贝中计算这些特征。

The distribution of the gray values across the image is measured by the entropy and the anisotropy, again calculated for the original image as well as for reduced-resolution copies of the image:

图像上灰度值分布可以通过一致性（entropy）和各向异性（anisotropy）来评估。同样这种计算也需要同时在原始图像和降低分辨率后图像拷贝中计算。

```
for Index := 1 to 3 by 1
    select_obj (ImagePyramid, ImageRR, Index)
    entropy_gray (ImageRR, ImageRR, Entropy, Anisotropy)
    Features := [Features,Entropy,Anisotropy]
endfor
```

All features are calculated within the procedure gen_features. It is called for each training image, and the resulting features are added to the classifier:

上面提到的所有特征都在函数 gen_features 中计算得到。对每个训练图像都调用这个函数来计算图像特征，然后将这些特征添加到分类器中：

```
gen_features (Image, Features)
add_sample_class_mlp (MLPHandle, Features, Class)
```

The internal weights of the classifier are determined based on the training data:

分类器的内部权重基于训练数据来决定：

```
train_class_mlp (MLPHandle, 200, 0.1, 0.001, Error, ErrorLog)
```

This is an iterative process that terminates when both the internal weights and the error of the MLP on the training samples become stable. To judge the training phase, the progression of the error can be plotted as a function against the number of iterations (see Figure 4.58). The error should drop off steeply at first, leveling out to almost flat at the end.

这是一个迭代的过程，它将在 MLP 的内部权重和误差趋于稳定时停止。为了判定训练阶段，将误差的变化等级作为迭代次数的函数绘制如图 4.58。在开始阶段，误差随着迭代次数增加急剧下降，而在最后的部分变化非常平缓。

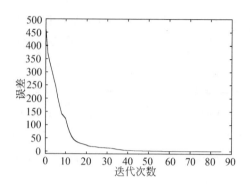

图 4.58　在分类器训练过程中误差的变化。注意在开始阶段，误差随迭代次数增加急剧下降，而在最后部分变化非常平缓

4.12.2 Performing the Texture Classification

Using the classifier created above, we can reclassify all the training images. If all the images can be classified correctly, we can test the classifier by classifying the independent test images. Typically, some of the test images are classified incorrectly. Figure 4.59 shows two pairs of images. Misclassified test images are displayed in the upper row, while the lower row shows one image of the class to which the image displayed above has been assigned erroneously.

If the number of misclassifications is very large, either the number of training samples must be increased significantly or the number of hidden neurons must be reduced. Furthermore, it should be checked whether the features used are suitable for separating the different classes. If the number of misclassifications is moderate, the classifier is already reasonably good. The wrongly classified images should be added to the training images and the classifier should be trained anew using the extended training data. This creates an improved classifier.

4.12.2 进行纹理分类

使用上面创建的分类器，可以将所有的训练图像重新分类。如果得到所有图像的类型都正确，那么就可以通过分类独立的测试图像集来对分类器进行测试。一般情况下，会有一些测试图像不能被正确分类。图 4.59 中显示了两对图像。在上面一行中显示的是不能正确分类的测试图像，分类器将这两幅图像分到了错误的类型中，下面一行显示的就是相应的错误类型的样本。

如果分类错误的数量太多，那么就必须增加更多的训练样本，或者减少隐含层神经元的数量。此外，还应该考虑使用的特征是否适合用来区分不同类型。如果分类错误的数量不是很多，分类器已经比较好了。应该把错误分类的图像添加到分类器的训练样本中，然后使用这个扩展的训练样本集重新训练，这将最终得到一个改进的分类器。

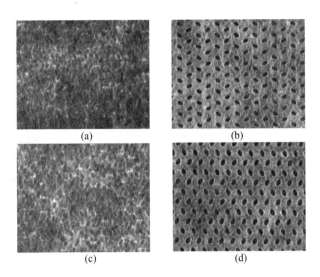

图 4.59 分类结果错误的图像。在上面一行中显示的是分类错误的测试图像，分类器将这两幅图像分到了错误的类型中，下面一行显示的就是相应的错误类型的样本

If much training data is available, the classification of independent test images and the adding of the misclassified test images to the training images can be repeated several times. This procedure has two main advantages over feeding all the available training data into the classifier in one step. First, the training is faster because fewer training images are used to train the classifier. Second, the performance of the classifier can be judged already in an early stage of the training phase.

Once the training phase is completed, the classifier can be used to classify images of unknown content:

如果有很多训练数据可用，可以将对独立测试图像进行分类和将错误分类的测试图像添加到训练图像集中这个过程重复多次。这种方式与直接将所有的图像放入训练图像样本中相比有两个主要的优点：第一，由于训练分类器的训练图像的数量少，因此训练的速度更快。第二，分类器的性能可以在训练阶段的前期就得到评判。

一旦训练阶段完成，那么分类器可以用来将未知类型的图像进行分类：

```
gen_features (Image, Features)
classify_class_mlp (MLPHandleImproved, Features, 1,
                    ClassifiedClass, Confidence)
```

Figure 4.60 shows one sample out of each of the 22 classes of non-woven fabrics. The improved classifier is able to classify all images correctly.

图 4.60 中显示了所有 22 种非织造布样本。改进后的分类器可以正确分类所有图像。

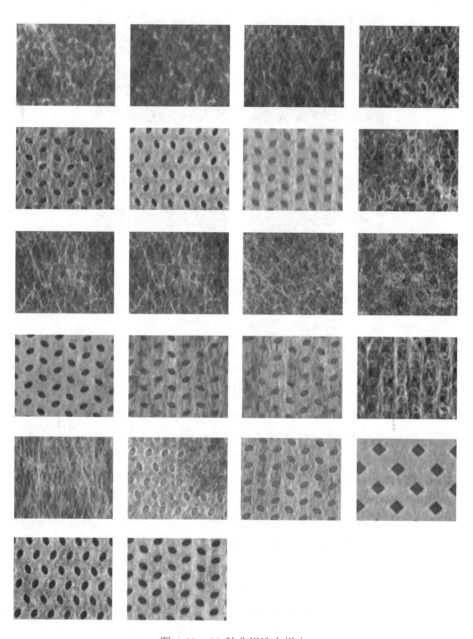

图 4.60 22 种非织造布样本

4.12.3 exercises

Explore the behavior of the classifier for different texture features and different sizes of the MLP.

4.12.3 练习

考察使用不同纹理特征或不同大小的神经网络创建的分类器的效果。

4.13 Surface Comparison

Surface comparison is used to ensure that the surface of produced objects (test objects) is similar to the surface of a reference object.

In this application, we reconstruct the surface of injection molds with a calibrated sheet-of-light system, which produces metric 3D coordinates. We then check for the presence and correctness of holes in the injection molds.

In the first part of the example program, the reference object model is created, together with a model for surface-based 3D matching. In the second part, the test objects are aligned with the reference object. Then, for each surface point, the distance from the test object to the reference object and the distance from the reference object to the test object are calculated. The results of these two distance measurements are analyzed for indications of

- additional holes,
- missing holes,
- holes with a wrong size, and
- holes at a slightly wrong position.

The result of this analysis is then visualized.

The algorithms used in this application are:
- Sheet of light reconstruction (Section 3.10.2).
- Surface-based 3D matching (Section 3.12.3).
- Region features (Section 3.5.1).

The corresponding example program is:
···/machine_vision_book/surface_comparison/surface_comparison.hdev

4.13.1 Creating the Reference Model

First we create a sheet of light model. In this example, the calibration information, i.e., the camera parameters, the camera pose, the pose of the laser plane, and the movement pose is already known. We configure the sheet of light model to use this calibration information.

就是摄像机内参、摄像机位姿、激光平面位姿和运动姿态是已知的。我们用这些已知信息来配置（片光）激光三角测量模型。

```
create_sheet_of_light_model (Rectangle, [], [], SheetOfLightModelID)
set_sheet_of_light_param (SheetOfLightModelID, 'calibration', 'xyz')
set_sheet_of_light_param (SheetOfLightModelID, 'scale', 'm')
set_sheet_of_light_param (SheetOfLightModelID, 'camera_parameter',
                          CameraParam)
set_sheet_of_light_param (SheetOfLightModelID, 'camera_pose',
                          CameraPose)
set_sheet_of_light_param (SheetOfLightModelID, 'lightplane_pose',
                          LightplanePose)
set_sheet_of_light_param (SheetOfLightModelID, 'movement_pose',
                          MovementPose)
```

To create the reference model, we reconstruct the surface of the reference object with calibrated sheet of light. Figure 4.61 shows the disparity image returned directly by the sensor.

为了创建参考模型，使用标定的激光三角测量系统来重构参考物表面。图 4.61 所示为传感器直接返回的视差图。

图 4.61　激光三角测量传感器返回的视差图像

Using the parameters of the calibrated setup, the disparity measurements are transformed into a 3D object model.

通过已经标定好的参数，将视差测量转换为一个三维对象模型。

```
set_profile_sheet_of_light (Disparity, SheetOfLightModelID, [])
get_sheet_of_light_result_object_model_3d (SheetOfLightModelID,
                          ReferenceOrig)
```

Figure 4.62(a) shows the reference model as seen from the top. Here, the gray values are proportional to the z value of the respective surface point, i.e., higher parts of the object are displayed brighter.

图 4.62（a）所示为参考模型的顶视图。这里，各个表面点的灰度值和 z 值是成比例的，也就是说物体上越高的部位，图中显示得越亮。

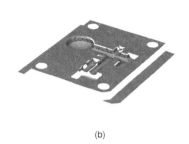

图 4.62　参考物的三维重构模型。(a) 灰度图顶视图，物体越高的部分显示越亮；(b) 透视图

To align the reconstruction of the test objects with the reconstruction of the reference object, we create a model for surface-based 3D matching. This model should only contain points on the object. In particular, the ground plane, on which the object was placed, must be eliminated. This can be done by selecting only those points that have a z coordinate in a suitable range. In our example, a z range from 0.01 m to 1 m is suitable. This range can easily be determined by inspecting the histogram of the z coordinates of all reconstructed points.

为了对齐被测物和参考物的重构结果，我们创建了基于表面的三维匹配模板。此模板应该只包括物体上的点，尤其是物体所在的参考平面必须被去除。通过选取 z 值在某个合适范围内那些点可以做到这点。在例子中，z 值范围从 0.01 m 到 1 m 是合适的。通过检测所有重构点的 z 值直方图可以很容易确定这个范围。

```
select_points_object_model_3d (ReferenceOrig, 'point_coord_z',
                               0.01, 1, Reference)
```

Figure 4.62(b) shows a perspective view of the reference object model without the ground plane. This reference object model is used to create the model for surface-based 3D matching.

图 4.62（b）所示为参考物的透视图，没有参考平面。参考物模型用来创建基于表面的三维匹配模板。

```
create_surface_model (Reference, 0.02, [], [], SurfaceModelID)
```

4.13.2 Reconstructing and Aligning Objects

4.13.2 重构和对齐物体

In the first step of the online phase, the objects to be compared must be reconstructed.

在线阶段的第一步,被比对的物体必须先重构。

```
reset_sheet_of_light_model (SheetOfLightModelID)
set_profile_sheet_of_light (Disparity, SheetOfLightModelID, [])
get_sheet_of_light_result_object_model_3d (SheetOfLightModelID,
                                    TestObjectOrig)
select_points_object_model_3d (TestObjectOrig, 'point_coord_z',
                          0.01, 1, TestObject)
```

The reconstructed surface of the produced object is typically not aligned with the reference model (see Figure 4.63(a)). To align the produced object with the reference part, we determine the transformation between the two parts with surface-based 3D matching and transform the object using this transformation (see Figure 4.63(b)).

所重构的待测物体表面通常情况下是没有和参考模型对齐的(见图 4.63(a))。为了和参考模型对齐,使用基于表面的三维匹配方法可以确定两个模型之间的转换关系,然后利用转换关系对物体进行转换(见图 4.63(b))。

```
find_surface_model (SurfaceModelID, TestObject, 0.02, 0.5, 0, 'false',
                 'pose_ref_sub_sampling', 1, Pose, Score, NotUsed)
pose_invert (Pose, PosesInvert)
rigid_trans_object_model_3d (TestObject, PosesInvert, TestObjectTrans)
```

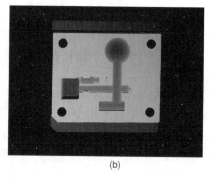

(a)　　　　　　　　　　(b)

图 4.63　被测物所重构的三维模型:(a)被测物方向与参考物方向没有对应;(b)所重构的模型已经和参考模型对齐

4.13.3 Comparing Objects and Classifying Errors

Surface comparison can be performed by calculating the distance between the surfaces of the reference object and the produced object. If this distance is below a predefined threshold for each surface point, the produced object is correct and passes the test. If there are larger deviations, we must analyze them. As mentioned above, we are looking for the following errors:

- additional holes,
- missing holes,
- holes with a wrong size, and
- holes at slightly wrong positions.

These errors become apparent through the following effects in the calculated distances:

- Additional holes result in circular blobs of larger distances when calculating the distances from each surface point of the reference object to the surface of the produced object.
- Missing holes result in circular blobs of larger distances when calculating the distances from each surface point of the produced object to the reference object.
- Too small holes result in annular blobs of larger distances when calculating the distance from each surface point of the produced object to the reference object.
- Too large holes result in annular blobs of larger distances when calculating the distances from each surface point of the reference object to the surface of the produced object.

4.13.3 对比物体并且对错误进行分类

通过计算参考物表面和被测物表面之间的距离来执行表面比对，对于表面上每个点如果距离小于定义好的阈值，则被测物是好的并且通过测试。如果存在较大偏差，必须进行分析。如上面所提到的，要寻找以下错误：

- 多余的孔洞。
- 缺失的孔洞。
- 错误尺寸的孔洞。
- 位置稍微偏差的孔洞。

在计算出的距离中通过以下效果，可以使这些错误看起来更明显：

- 当计算参考物表面每个点到被测物表面的距离时，多余的孔洞会导致较大距离的圆形斑点。
- 当计算被测物表面每个点到参考物表面的距离时，缺失的孔洞会导致较大距离的圆形斑点。
- 当计算被测物表面每个点到参考物表面的距离时，太小的孔洞会导致较大距离的环形斑点。
- 当计算参考物表面每个点到被测物表面的距离时，太大的孔洞会导致较大距离的环形斑点。

- Holes at slightly wrong positions result in pairs of crescent-shaped blobs of larger distances, where one blob appears in the distances from the reference object to the produced object and the other blob appears in the distances from the produced object to the reference object.

We calculate the distance between the two aligned surfaces and select all surface points that have a distance larger than a predefined threshold value (here, `MaxDist` has been set to 1 mm). As we have seen above, we must calculate the distances in both directions separately to be able to classify the various defects correctly.

First, we calculate the distances from all points of the reference object to the test object and select the points that have a distance above the predefined threshold, i.e., where the respective part of the surface is missing from the test object (`ErrorSource 'missing in test'`).

- 稍微出现位置偏差的孔洞会导致较大距离的成对的新月形斑点，其中一个斑点出现在从参考物到被测物的距离上，另外一个斑点出现在从被测物到参考物的距离上。

我们计算两个对齐表面之间的距离并且选择所有距离值大于预定义阈值的所有的点（这里，`MaxDist` 设置为 1 mm）。如上面所看到的那样，必须单独计算两个方向上的距离值才能正确分类各种不同的缺陷。

首先，计算参考物上所有点到被测物表面的距离并且选择距离大于预定义阈值的那些点，也就是，被测物表面上缺失的那些部分（`ErrorSource 'missing in test'`）。

```
distance_object_model_3d (Reference, TestObjectTrans, [], 0.0, [], [])
select_points_object_model_3d (Reference, '&distance', MaxDist, 1,
                                ObjectModel3DThresholded)
```

If there are such points, we compute the connected components of these points based on the 3D distance to their neighboring points and remove all small components, which typically are caused by noise. Note that the minimum size of the components is given by the minimum number of points and must therefore be chosen according to the sampling distance and the minimum size of the defects.

如果存在这样的点，我们会基于这些点到临近点的三维距离来计算与这些点的连通部分，并且去掉所有尺寸较小的连通部分，这部分是由噪声导致的。注意，这些连通部分的最小尺寸由这些最小数量的点给出，因此必须按照采样距离和缺陷的最小尺寸来选择。

```
get_object_model_3d_params (ObjectModel3DThresholded,
                            'num_points', NumPoints)
if (NumPoints > 0)
    connection_object_model_3d (ObjectModel3DThresholded, 'distance_3d',
                                0.001, ObjectModel3DConnected)
    select_object_model_3d (ObjectModel3DConnected, 'num_points',
                            'and', 200, 1000000,
                            SurfacePointsMissingInTestObject)
endif
```

Similarly, we calculate the distances from all points of the test object to the reference object and determine the surface points of the test object that are missing in the reference object (`ErrorSource 'missing in reference'`).

The components resulting from the above steps represent potential errors.

In our example, we want to ensure the correctness of the holes. Because the surroundings of the holes are planar, we can perform the classification of the potential errors in 2D. Therefore, we create a region from each of the above components by fitting a plane through the points of the component, projecting the points into this plane, creating a region from these points, and dilating it slightly to fill gaps between the projected points.

For each region, we calculate several features to classify the potential errors.

First, we create two auxiliary regions, one where all holes are filled and one that is the difference between the filled region and the original region, i.e., that represents only the holes in the original region:

同样，计算被测物上所有点到参考物表面的距离并且确定在参考物表面缺失的被测物的所有点（`ErrorSource 'missing in reference'`）。

由上述步骤得出的连通部分表示潜在的错误（缺陷）。

在例子中，希望确保这些孔洞的正确性。因为孔洞周围是平面的，可以在二维空间中对这些潜在的缺陷进行分类。因此，依靠拟合一个通过这些点的平面，从上面的连通部件中创建一个区域，把这些点投影到这个平面，从这些点中创建一个区域，并做轻微的膨胀操作来填充这些投影点之间的空隙。

对于每个区域，通过计算一些特征来分类潜在的缺陷。

首先，创建两个辅助区域，其中一个区域所有的孔洞被填充，另外一个区域是填充区域和原始区域之间的差异，也就是仅表示原始区域的孔洞:

```
fill_up (Region, RegionFillUp)
difference (RegionFillUp, Region, RegionDifference)
```

We calculate the circularity and the area for

对于这三个区域的每一个，分别

each of the three regions:

计算其似圆度和面积：

```
circularity (Region, CircularityReg)
circularity (RegionFillUp, CircularityRegFillUp)
circularity (RegionDifference, CircularityRegDifference)
area_center (Region, AreaRegDilation, Row1, Column1)
area_center (RegionFillUp, AreaRegFillUp, Row1, Column1)
area_center (RegionDifference, AreaRegDifference, Row1, Column1)
```

Based on these features, we can classify the potential errors as follows.

基于这些特征，我们能够如下分类潜在的缺陷。

Approximately circular regions with only small holes indicate either additional (`ErrorType 1`) or missing (`ErrorType 2`) holes. Approximately circular regions with an approximately circular hole indicate too small (`ErrorType 3`) or too large (`ErrorType 4`) holes. We assign the error type 0, which means "unknown," to all potential errors that cannot be classified with the above feature set.

只有小孔洞的圆形区域可以近似地表示多余孔洞缺陷（`ErrorType 1`）或者缺失孔洞（`ErrorType 2`）。有近似圆形孔洞的圆形区域可以近似地表示孔洞过小（`ErrorType 3`）或者孔洞过大（`ErrorType 4`）。对于在以上特征集里面不能分类的所有缺陷，我们分配了错误类型为 0 的错误码，表示"未知的缺陷"。

```
MinCircularity := 0.7
if (CircularityReg > MinCircularity and
    real(AreaRegDifference)/AreaReg < 0.1)
    if (ErrorSource == 'missing in test')
        ErrorType := 1
    else
        ErrorType := 2
    endif
elseif (CircularityRegFillUp > MinCircularity and
        CircularityRegDifference > MinCircularity and
        real(AreaRegDifference)/AreaReg > 0.1)
    if (ErrorSource == 'missing in test')
        ErrorType := 3
    else
        ErrorType := 4
    endif
else
    ErrorType := 0
endif
```

We will now inspect the unknown error types for whether there are clues for holes that have just a slightly wrong position. This kind of error would result in symmetric blobs of larger distances, because some part of the hole would be missing in the test object and some part would be missing in the reference object at the opposite side of the hole.

对于位置有轻微偏差的孔洞无论是否有线索，现在要检查未知的缺陷类型。这种缺陷会导致较大距离的对称斑点，因为在被测物上孔洞的某些部分会缺失，在参考物上孔洞的另一边某些部分会缺失。

This analysis can be done very easily in 3D. For each pair of potentially symmetric blobs, we calculate a hypothesis for the symmetry plane based on the centers of gravity of the two blobs. The plane is represented in the Hesse normal form $\vec{N}_{\text{symm}} \cdot \vec{X} - C_{\text{symm}} = 0$.

这个分析在三维空间中很容易进行，对于每一对潜在的对称斑点（连通区域），基于两个斑点的重心计算对称平面的一个假设，平面用海塞正规式 $\vec{N}_{\text{symm}} \cdot \vec{X} - C_{\text{symm}} = 0$ 来表示。

```
get_object_model_3d_params (OM1, 'point_coord_x', X1)
get_object_model_3d_params (OM1, 'point_coord_y', Y1)
get_object_model_3d_params (OM1, 'point_coord_z', Z1)
get_object_model_3d_params (OM2, 'point_coord_x', X2)
get_object_model_3d_params (OM2, 'point_coord_y', Y2)
get_object_model_3d_params (OM2, 'point_coord_z', Z2)
Center1 := [mean(X1),mean(Y1),mean(Z1)]
Center2 := [mean(X2),mean(Y2),mean(Z2)]
NSymm := Center2-Center1
NSymm := NSymm/sqrt(sum(NSymm*NSymm))
PSymm := Center1+0.5*(Center2-Center1)
CSymm := sum(NSymm*PSymm)
if (CSymm < 0)
    NSymm := -NSymm
    CSymm := -CSymm
endif
```

Now, we determine a 3D transformation that maps the symmetry plane to the plane $z = 0$.

现在，来确定一个三维变换把对称平面映射到平面 $z = 0$。

```
hnf_to_hom_mat3d (NSymm[0], NSymm[1], NSymm[2], CSymm, HomMat3D)
```

Using this 3D transformation, we transform the points of the two blobs. For the transformed points, the symmetry plane is the plane $z = 0$.

使用此三维变换，来转换两个斑点上的点。对于变换后的点，对称平面是 $z = 0$。这就简化了后续的对称

This simplifies the subsequent symmetry analysis, because mirroring at the symmetry plane reduces to switching the sign of the z coordinate.

分析，因为对称平面的镜像分析已经减少到只需改变 z 坐标的符号。

```
hom_mat3d_invert (HomMat3D, HomMat3DInvert)
affine_trans_point_3d (HomMat3DInvert, X1, Y1, Z1, X1T, Y1T, Z1T)
affine_trans_point_3d (HomMat3DInvert, X2, Y2, Z2, X2T, Y2T, Z2T)
```

To measure the symmetry of the two point clouds, we can mirror the first point cloud at the symmetry plane and determine the overlap of the mirrored point cloud with the second one. For this, the distances of points to the closest point of the other point cloud must be determined, which can be done, for example, with the help of a kNN classifier (see Section 3.15.2.1).

为了测量两个点云的对称性，可以以对称平面来镜像第一个点云，然后确定镜像后的点云和第二个点云的交叠部分。为此，必须先确定点到另一个点云上最近点的距离，例如，可以借助 KNN 分类器来实现（见 3.15.2.1 节）。

If we find a pair of symmetric blobs, they indicate a hole at a slightly wrong position (ErrorType 5). We merge these two blobs and remove the two individual blobs from the list of potential errors.

如果发现一对对称的斑点，则表明孔洞位置有些轻微偏差（ErrorType 5）。合并这两个斑点，同时在潜在错误列表中删除这两个独立的斑点。

In the corresponding example program, the above procedures are applied to a set of produced objects, some of which show defects. Figures 4.64, 4.65, and 4.66 show some results of the surface comparison for produced objects with defects.

在对应的例程中，以上过程应用于一批被测物，其中一些显示了缺陷。图 4.64、图 4.65 和图 4.66 显示了存在缺陷的被测物表面比对结果。

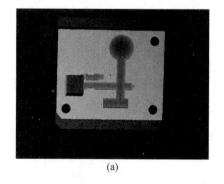
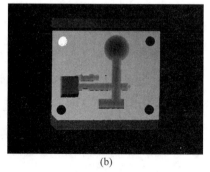

(a)   (b)

图 4.64 被测工件表面比对结果，其缺失一个孔洞：（a）重构表面；（b）表面比对结果，缺失孔洞已标识

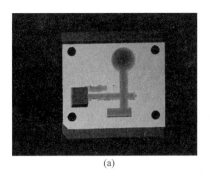
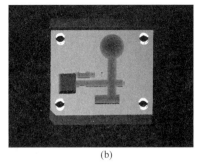

图 4.65 被测工件表面比对结果,其所有孔洞位置有轻微偏移:(a)重构表面;(b)表面比对结果,所有位置偏移的孔洞已被标识出

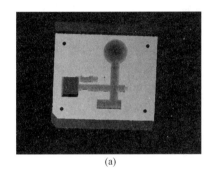

图 4.66 被测工件表面比对结果,其所有孔洞尺寸偏小:(a)重构表面;(b)表面比对结果,所有尺寸有错误的孔洞均被标识出

Figure 4.64(a) shows the reconstructed (unaligned) surface of a produced object where the upper left hole is missing. Figure 4.64(b) shows the result of the comparison with the surface of the reference object (see Figure 4.62). The missing hole was detected and is indicated in white.

Figure 4.65(a) shows the reconstructed surface of a produced object where all holes are in a slightly wrong position. Figure 4.65(b) shows the result of the surface comparison. The wrong position of the holes was detected and is indicated in white.

Figure 4.66(a) shows the reconstructed surface of a produced object where all holes are too small. Figure 4.66(b) shows the result of the surface comparison. The wrong size of the holes was detected and is indicated in white.

图 4.64(a)所示为一个被测物体的重构(未对齐)表面,其左上角孔洞是缺失的。图 4.64(b)所示为与参考物比对的结果(见图 4.62),缺失的孔洞已经检测出并标识为白色。

图 4.65(a)所示为一个被测工件的重构表面,其所有孔洞的位置都有轻微偏移。图 4.65(b)所示为表面比对结果,位置偏移的孔洞已经检测出并标识为白色。

图 4.66(a)所示为一个被测工件的重构表面,其所有孔洞尺寸偏小。图 4.66(b)所示为表面比对结果,错误尺寸的孔洞已经检测出并标识为白色。

4.13.4 exercises

The example program only classifies errors with respect to holes. Extend the program to be able to find dents and bulges in the surfaces of the produced objects.

4.14 3D Pick-and-Place

Pick-and-place robots are very popular systems for material handling. In the past, vision-based pick-and-place systems were restricted to 2D. Typically, the objects had to lie on a planar workspace or conveyor belt and they had to be separated from each other so that the vision system was able to determine their positions.

3D image acquisition devices (see Section 2.5), 3D object recognition technologies like surface-based 3D matching (see Section 3.12.3), and hand–eye calibration (see Section 3.13) have made 3D pick-and-place systems feasible. Surface-based 3D matching allows determining the 3D pose of arbitrary objects, even if they are not lying on a plane in the workspace or on the conveyor belt but on top of other objects. Hand–eye calibration links the coordinate system of the robot with that of the vision system.

In this application, we show how to implement a 3D pick-and-place application using a stationary 3D image acquisition device. The application consists of three major steps: hand–eye calibration, definition of the grasping point, and picking and placing of objects.

Figure 4.67(a) shows the hardware setup of our pick-and-place application: the articulated six-axis robot, the 3D image acquisition device (a

4.13.4 练习

示例程序仅对孔的缺陷进行了分类，可以扩展程序使其能够找到被测物表面的凹陷和凸起。

4.14 三维取放

取放机器人是非常流行的物料搬运系统。过去，视觉引导的取放系统被限制在二维空间。通常情况下，物体必须放置在一个平面工作空间或者传送带上，而且相互之间必须分离以便视觉系统能够确定它们的位置。

三维图像采集设备（见 2.5 节），三维物体识别技术，例如基于表面的三维匹配（见 3.12.3 节），还有手眼标定（见 3.13 节）这些技术使得三维取放系统变得可行。基于表面的三维匹配技术可以确定任意物体的三维姿态，即使物体不在平面工作空间或者传送带上，甚至是叠放在其他物体上。手眼标定则把机器人坐标系统和视觉坐标系统关联到一起。

在本节的应用中，演示了如何使用静态三维图像采集设备来实现一个三维取放应用。此应用由三个主要步骤组成：手眼标定，定义抓取点，取放物体。

图 4.67（a）所示为取放应用的硬件配置：关节型六轴机器人，三维图像采集设备（集成了图案投影仪的

stereo sensor with integrated pattern projector), and a pile of objects to be handled. Figure 4.67(b) shows one of the objects, a screwdriver, that is to be handled by the robot.

立体传感器），和一堆待抓取的物体。图 4.67（b）所示为其中一个待抓取物体，一把螺丝刀。

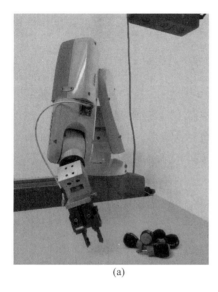
(a)

(b)

图 4.67　（a）使用静态三维图像采集设备的取放应用硬件配置；（b）待抓取的螺丝刀

The algorithms used in this application are:
- Hand–eye calibration (Section 3.13).
- Surface-based 3D matching (Section 3.12.3).

本应用使用的算法：
- 手眼标定（3.13 节）。
- 基于表面的三维匹配（3.12.3 节）。

The corresponding example program are:
··· /machine_vision_book/3d_pick_and_place/hand_eye_calibration.hdev
··· /machine_vision_book/3d_pick_and_place/grasping_point_definition.hdev
··· /machine_vision_book/3d_pick_and_place/3d_pick_and_place.hdev

对应的示例程序位于：

4.14.1 Performing the Hand–Eye Calibration

In the first step, we must establish the relationship between the coordinate systems of the robot and the stationary vision system to allow the robot to move to positions determined by the vision system (see Section 3.13).

4.14.1 手眼标定

第一步，必须建立机器人坐标系和静态视觉系统坐标系之间的关系，使机器人能够移动到视觉系统所确定的位置（见 3.13 节）。

In principle, we can use an arbitrary object as calibration object. The only condition is that the 3D object recognition must be able to uniquely determine the pose of the calibration object. In practice, it is often reasonable to use the object to be picked and placed in the final application also as the calibration object for the hand–eye calibration.

Because our application is intended to handle screwdrivers, we also use a screwdriver as calibration object. Note, however, that the pose of the screwdriver cannot be determined uniquely because the screwdriver is rotationally symmetric. To overcome this, we extend our calibration object to also include a part of the robot's gripper. Figure 4.68(a) shows an image of the screwdriver held by the robot.

In the example program, we assume that we already have a surface model (i.e., a global model description as introduced in Section 3.12.3.1) that can be used to determine the pose of the calibration object with surface-based 3D matching. In practice, such a surface model can be provided easily: acquire a 3D image of the robot gripper

原则上，可以使用任意物体作为标定物体。唯一需要的条件是三维物体识别系统必须能够唯一地确定标定物体的位姿。实际上，经常把最终应用中取放的物体来作为手眼标定的标定物体来使用，这也是合理的。

因为本节应用要抓取螺丝刀，使用螺丝刀作为标定物。然而需要注意的是，因为螺丝刀的外形是旋转对称的，因此螺丝刀的位姿不能唯一确定。为了解决这个问题，把机器人的夹具部分也包含在内和螺丝刀一起作为标定物。图 4.68（a）所示为被机器人夹持的螺丝刀。

在示例程序中，假设已经有一个表面模型（也就是 3.12.3.1 节中介绍的全局模型描述），这个表面模型可以用来确定标定物体的位姿。实际上，很容易提供这样一个表面模型：获取一张夹持物体的机器人夹具三维图像，去掉所有不属于标定物体的点，

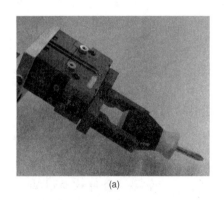
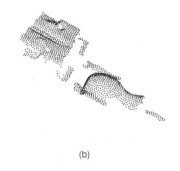

(a)　(b)

图 4.68　用于手眼标定的标定物体：（a）被机器人夹具夹持的螺丝刀；（b）标定物体的表面模型。注意必须把机器人夹具部分包含进表面模型，也就是说，标定物体由螺丝刀和机器人夹具的一部分组成，以便能够唯一确定标定物体的位姿

holding the object, remove all points that do not belong to the calibration object, and create a surface model from that point cloud. Figure 4.68(b) shows the surface points of the calibration object.

Before we start the calibration, we must provide the input data for the calibration:

然后基于点云创建一个表面模型。图4.68（b）所示为标定物体的表面的点。

开始标定之前，必须给标定过程提供输入数据：

```
create_calib_data ('hand_eye_stationary_cam', 0, 0, CalibDataID)
set_calib_data (CalibDataID, 'model', 'general', 'optimization_method',
                'nonlinear')
```

To perform the hand–eye calibration, the robot moves the calibration object in front of the 3D image acquisition device. We must then provide the 3D poses of the calibration object in the coordinate system of the 3D image acquisition device together with the respective robot poses, i.e., the poses of the mechanical interface in the robot's base coordinate system, for multiple positions and orientations of the calibration object.

为了执行手眼标定，首先机器人在三维图像采集设备前移动标定物体。然后，对于标定物体的多个位置和方向，必须提供在三维图像采集设备坐标系下的标定物体的三维位姿和各自对应的机器人位姿，也就是在机器人基坐标系下的机械接口的位姿。

In the example program, we assume that we have already defined the robot poses to which the robot moves the calibration object and that we can read these poses from files.

在示例程序中，假设已经定义了机器人移动标定物体所到位置的机器人位姿，并且可以从文件读取这些位姿数据。

```
NumCalibPoses := 23
for I := 0 to NumCalibPoses-1 by 1
    read_pose (PathPoses+'robot_pose_calib_'+(I+1)$'02d'+'.dat',
               RobotPose)
```

Then, the robot moves to the current robot pose

然后，移动机器人到当前机器人位姿采集三维图像。

```
move_robot (RobotPose)
```

and we acquire the 3D image. Depending on the 3D image acquisition device, it might be necessary to convert the 3D image into meters, which can simply be done by scaling the 3D image by the respective scaling factor.

根据所使用的三维图像采集设备，也许有必要把三维图像转化为米制，通过单独的比例因子来缩放三维图像很容易实现转换。

```
grab_data (ImageData, Region, Contours, AcqHandle, Data)
scale_image (ImageData, ImageDataM, 0.001, 0)
```

We can eliminate some outliers from the data by thresholding the image that holds the z coordinates. Here, `ThresholdZOutlier` was set to `[0.2, 1.0]`, which means that we eliminate all parts of the 3D image for which the measured distance from the 3D sensor is less than 0.2 m or greater then 1.0 m.

通过在 z 坐标分量图像中使用阈值分割算法，可以去掉一些异常值。这里，`ThresholdZOutlier` 设置为 `[0.2, 1.0]`，意味着在三维图像中距离三维传感器小于 0.2 m 或者大于 1.0 m 的部分会被去掉。

```
decompose3 (ImageDataM, Xm, Ym, Zm)
threshold (Zm, Region, ThresholdZOutlier[0], ThresholdZOutlier[1])
reduce_domain (Xm, Region, ImageX)
reduce_domain (Ym, Region, ImageY)
reduce_domain (Zm, Region, ImageZTmp)
```

Finally, we perform an edge preserving smoothing to reduce noise in the 3D image.

最后，使用边缘保持平滑算法来减少三维图像中的噪声。

```
median_image (ImageZTmp, ImageZ, 'circle', 3, 'mirrored')
```

In the next step, we create two 3D object models from the 3D image data. The first one is only used for visualization and contains the complete 3D information.

下一步，用三维图像数据来创建两个三维物体模型，第一个模型仅用来可视化，并且包含完整的三维信息。

```
xyz_to_object_model_3d (ImageX, ImageY, ImageZ, OM3DSceneVis)
```

The second 3D object model is used for the 3D object recognition. To speed up the search, we can eliminate the background plane from the 3D data. Since we use a stationary 3D image acquisition device that is mounted such that it looks approximately perpendicularly at the ground plane, we can eliminate this plane very easily from the 3D image by applying a suitable threshold to the z coordinates. Here, `ThresholdZWorkingArea` was set to `[0.2, 0.5]`, which means that—in addition to the above outlier elimination—we eliminate

第二个模型用来进行三维物体识别。为了提高搜索速度，可以在三维数据中去掉背景平面。由于使用了静态的三维图像采集设备，其固定方式近似于垂直地平面，因此在 z 坐标图中通过设置一个合适的阈值，可以很容易地去掉三维图像中的背景平面。这里，`ThresholdZWorkingArea` 设置为 `[0.2, 0.5]`，意味着除了可以去掉上面所提到的异常值外，还可以去掉三维图像中距离三维传感器大于

all parts of the 3D image for which the measured distance from the 3D sensor is greater than 0.5 m. Since the 3D image acquisition device was mounted approximately 0.55 m above the ground plane, this eliminates the complete background plane in the 3D image.

0.5 m 的部分。因为三维图像采集设备安装在距离地平面大约 0.55 m 的位置，因此可以在三维图像中去掉整个背景平面。

```
threshold (ImageZ, Roi, ThresholdZWorkingArea[0],
          ThresholdZWorkingArea[1])
reduce_domain (ImageZ, Roi, ImageZWorkingArea)
xyz_to_object_model_3d (ImageX, ImageY, ImageZWorkingArea, OM3DMatch)
```

Now we can perform the 3D object recognition using surface-based 3D matching:

现在可以使用基于表面的三维匹配技术来进行三维物体识别：

```
find_surface_model (SurfaceModelID, OM3DMatch, 0.05, 0.2, 0.2, 'false',
                   ['dense_pose_refinement','pose_ref_sub_sampling'],
                   ['true',1], Pose, Score, SurfaceMatchingResultID)
```

Figure 4.69(a) shows the 3D object model that has been created from one 3D calibration image. Figure 4.69(b) shows the recognized calibration object.

图 4.69（a）所示为一幅三维标定图像生成的三维物体模型。图 4.69（b）所示为识别的标定物体。

(a)　　　　　　　　　(b)

图 4.69　由三维标定图像生成的三维物体模型：（a）三维图像的灰度编码表示方式，越高的部分越暗；（b）识别到的三维标定物体（用白色表示）

For each calibration object we find, we collect its pose as determined by the vision system as well as the respective robot pose. Note that, depending on the manufacturer of the robot, the robot pose

对于每一个找到的标定物体，收集其位姿和各自的机器人位姿，标定物位姿是由视觉系统确定的。注意，取决于机器人制造商，机器人的位姿

might be provided in a form that is different from the one used in HALCON. In particular, the order of rotations and the units of the translations might differ. Therefore, typically, the robot pose must first be converted to the HALCON format.

形式也许和 HALCON 中所用的是不同的。尤其旋转顺序和平移单位也许是不同的,因此,通常情况下,机器人位姿首先要转换为 HALCON 格式。

```
    if (|Pose| and Score > 0.2)
        set_calib_data_observ_pose (CalibDataID, 0, 0, I, Pose)
        * Convert the robot pose into HALCON pose format.
        RobotPoseTmp := RobotPose
        RobotPoseTmp[0:2] := RobotPoseTmp[0:2]*0.001
        create_pose (RobotPoseTmp[0], RobotPoseTmp[1], RobotPoseTmp[2],
                     RobotPoseTmp[3], RobotPoseTmp[4], RobotPoseTmp[5],
                     'Rp+T', 'abg', 'point', RobotPoseHALCON)
        set_calib_data (CalibDataID, 'tool', I, 'tool_in_base_pose',
                        RobotPoseHALCON)
    endif
endfor
```

Finally, we can perform the hand–eye calibration and access the pose ^cH_b of the robot base in the coordinate system of the 3D image acquisition device.

最后,可以完成手眼标定得到三维图像采集设备坐标系下的机器人基座位姿 ^cH_b。

```
calibrate_hand_eye (CalibDataID, Errors)
get_calib_data (CalibDataID, 'camera', 0, 'base_in_cam_pose',
                BaseInCamPose)
```

4.14.2 Defining the Grasping Point

4.14.2 定义抓取点

In this step, we interactively define the grasping point by providing the pose of the object relative to the camera and the pose of the robot when it is in the right position and orientation to grasp the object.

在此步骤中,当机器人在合适的位置和方向来抓取物体时,通过提供物体相对于摄像机的位姿和机器人位姿来交互式地确定抓取点。

First, we place the object to be grasped in an arbitrary position in the working area, i.e., in a position where it can be seen from the 3D image acquisition device and where it can be grasped by

首先,把待抓取的物体放置在工作区域内的任意位置,也就是说,这个位置是可以被三维图像采集设备拍摄到而且能被机器人抓取到,

the robot. Figure 4.70(a) shows the screwdriver placed on the working area.

图 4.70（a）所示为放置在工作区域内的螺丝刀。

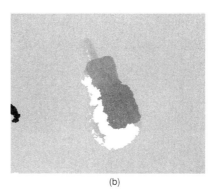

图 4.70 （a）放置在工作区域内的待抓取物体；（b）待抓取物体的三维图像

Then, we acquire a 3D image with the 3D image acquisition device (see Figure 4.70(b), which shows a part of this 3D image) and determine the pose $^{c}H_{o}$ of the object with 3D surface-based matching.

然后，获取三维图像采集设备中的一幅三维图像（见图 4.70（b），为三维图像的一部分）并使用三维表面匹配技术确定物体的位姿 $^{c}H_{o}$。

```
find_surface_model (SurfaceModelID, OM3DScene, 0.05, 0.2, 0.2,
                    'false', 'num_matches', 1, ObjInCamPose,
                    Score, SurfaceMatchingResultID)
```

The surface model can be provided similarly to the surface model of the calibration object (see Section 4.14.1). Note, however, that here, only points must be used that lie on the surface of the actual object to be grasped. In the example program, we assume that we already have a surface model of the object.

物体的表面模型和标定物体的表面模型是类似的（见 4.14.1 节）。然而，需要注意的是，只有那些位于被抓取物体表面上的点会用到。在示例程序中，假设已经有物体的表面模型。

Figure 4.71(a) shows the recognized object to be grasped overlaid in white on the 3D image.

图 4.71（a）所示为识别到的待抓取物体，在三维图像中标示为白色。

Now, we manually move the robot to the grasping pose. Once the robot is in the correct position and orientation for grasping the object (see Figure 4.71(b)), we obtain the robot pose $^{b}H_{t}$ from the robot controller.

现在，手动操作机器人移动到抓取位置，一旦机器人移动到正确的位置和方向（见图 4.71（b）），通过机器人控制器可以得到机器人的位姿 $^{b}H_{t}$。

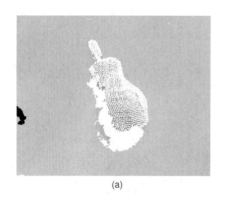 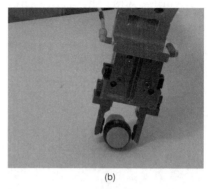

图 4.71 （a）识别到的待抓取物体（标示为白色）；（b）移动到抓取位置的机器人夹具

The pose ^tH_o of the object relative to the tool can now be calculated with Eq. (3.212) ($^tH_o = {^tH_b}{^bH_c}{^cH_o}$). For this, we must invert ^bH_t to compute the required pose ^tH_b.

物体相对于夹具的位姿 ^tH_o 可以通过等式（3.212）($^tH_o = {^tH_b}{^bH_c}{^cH_o}$）计算得到，为此，必须转置 ^bH_t 来计算所需的位姿 ^tH_b。

```
pose_invert (ToolInBasePose, BaseInToolPose)
```

Furthermore, we must invert the pose ^cH_b, which has been determined in the previous step (see Section 4.14.1), to obtain the pose ^bH_c.

此外，必须转置前面步骤所确定的 ^cH_b 来得到位姿 ^bH_c（见 4.14.1 节）。

```
pose_invert (BaseInCamPose, CamInBasePose)
```

Finally, we calculate the pose ^tH_o based on Eq. (3.212).

最后，基于等式（3.212）来计算位姿 ^tH_o。

```
pose_compose (BaseInToolPose, CamInBasePose, CamInToolPose)
pose_compose (CamInToolPose, ObjInCamPose, ObjInToolPose)
```

Note that if we determine the pose ^tH_o as described above, i.e., by manually moving the robot to the grasping pose, it is unnecessary to explicitly add the pose of the tool with respect to the mechanical interface because it is already included in ^tH_o.

注意，如果按照上面描述的方式通过手动移动机器人到抓取位置来确定位姿 ^tH_o，那么不需要显式添加夹具相对于机械接口的位姿，因为已经包含在位姿 ^tH_o 内了。

4.14.3 Picking and Placing Objects

Using the above determined poses ^tH_o and ^bH_c, we are now able to run the pick-and-place application.

We first acquire a 3D image of the pile of objects to be grasped (see Figure 4.72(a) and (b)) and use surface-based matching to find instances of the screwdrivers.

```
find_surface_model (SurfaceModelID, OM3DScene, 0.05, 0.2, 0.2,
            'false', 'num_matches', 4, ObjsInCamPose,
            Score, SurfaceMatchingResultID)
```

4.14.3 取放物体

使用上面得到的位姿 ^tH_o 和 ^bH_c，现在可以执行取放应用了。

首先获取一堆待抓取物体的三维图像（见图 4.72（a）和（b）），使用表面匹配技术找到螺丝刀的实例。

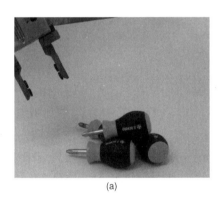
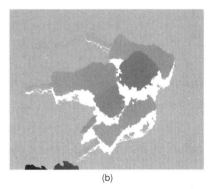

图 4.72 一堆待抓取物体：（a）有四个螺丝刀的场景图片；（b）物体三维图像的灰度表示，越高的部分显示越暗

To ensure that the robot can grasp the object, we search for more than one (here, up to 4) instances of the object and select the pose of the top-most instance. In our application, the 3D image acquisition device looks perpendicularly at the ground plane. Therefore, this selection can be done based on the z translation component of the objects' poses.

为了确保机器人能够抓取到物体，我们会搜索到多个实例（这里是四个），然后位于最顶层实例的位姿。在示例应用中，三维图像采集设备安装位置垂直于地面，因此，可以基于物体位姿的 z 平移分量来选择最顶层物体。

```
ZValuesMatches := ObjsInCamPose[[2:7:|ObjsInCamPose|]]
IdxTopMatch := sort_index(ZValuesMatches)[0]
ObjInCamPose := ObjsInCamPose[IdxTopMatch*7:IdxTopMatch*7+6]
```

Using the pose ^cH_o of the top-most object relative to the camera, we can determine the robot pose ^bH_t by solving Eq. (3.212) for ^bH_t, which yields

基于最顶层物体相对于摄像机的位姿 ^cH_o，通过解等式（3.212）来确定机器人位姿 ^bH_t：

$$^bH_t = {^bH_c}\,{^cH_o}\,{^oH_t} \tag{4.1}$$

The equivalent code is:

相应的程序代码是：

```
pose_compose (CamInBasePose, ObjInCamPose, ObjInBasePose)
pose_compose (ObjInBasePose, ToolInObjPose, ToolInBasePose)
```

Typically, we must transform the pose from the HALCON format to the format of the robot:

通常情况下，需要把位姿从 HALCON 格式转换成机器人支持的格式：

```
convert_pose_type (ToolInBasePose, 'Rp+T', 'abg', 'point', RobotPose)
RobotPose[0:2] := RobotPose[0:2]*1000
```

With this pose, we move the robot to the position and orientation to grasp the top-most object (see Figure 4.73(a)). Note that, in general, suitable path planning is required to avoid collisions with surrounding objects.

基于这个位姿，移动机器人到相应的位置和方向来抓取最顶层物体（见图 4.73（a））。需要注意的是，一般需要合适的路径规划来避免机器人和周围物体发生碰撞。

Finally, we can move the object to a predefined position (see Figure 4.73(b)).

最后，移动物体到预先确定的位置（见图 4.73（b））。

(a)

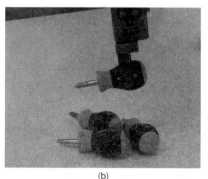
(b)

图 4.73 （a）抓取；（b）移动最顶层物体

We can now repeat the steps described above—3D image acquisition, recognition of the

现在可以对剩余物体重复以上步骤，三维图像采集，最顶层物体识别，

top-most object, calculation of the robot pose, and picking of the object—for all remaining objects (see Figure 4.74).

计算机器人位姿，最后抓取物体（见图 4.74）。

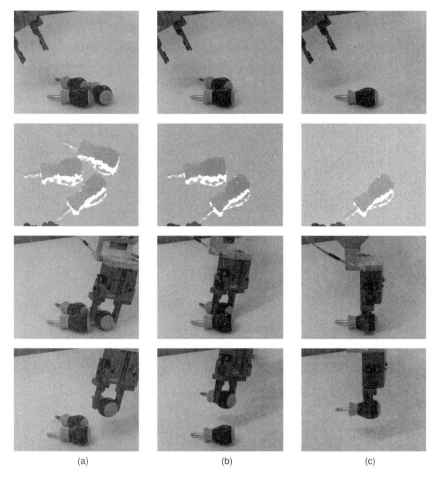

(a)　　　　　　　　(b)　　　　　　　　(c)

图 4.74 抓取剩余物体。从顶部到底部：剩余物体的图片，物体的三维图像，位于抓取位置的机器人，抓取的物体

4.14.4　exercises

1. In the example, the background was removed from the 3D images by simply thresholding the z coordinate image. This was possible because the 3D image acquisition device was oriented perpendicular to the ground plane. In practice, this cannot always be realized. Describe

4.14.4　练习

1. 在例子中，通过在 z 坐标分量图像中使用简单的阈值分割可以去除三维图像中的背景。因为例子中三维图像采集设备安装位置是垂直于地平面的，所以这是可能的。但实际应用中，情况并不总是这样，

a possible way to remove the background if the 3D image acquisition device is at an angle to the ground plane.

2. The screwdrivers handled in the example have an imprint. Describe the necessary extensions of the example (hardware and software) to ensure that the screwdrivers are placed with the imprint facing upwards.

如果三维图像采集设备与地平面成某一角度，请描述一个可以去除背景的可行方案。

2. 在例子中螺丝刀上面是有印记的。请描述对示例应用进行何种必要的扩展（硬件和软件），能确保螺丝刀放置时印记朝上。

References 参考文献

1394 Trade Association (2008) IIDC 1394-based digital camera specification, version 1.32.

Ahn, S.J., Rauh, W., and Warnecke, H.J. (2001) Least-squares orthogonal distances fitting of circle, sphere, ellipse, hyperbola, and parabola. *Pattern Recognition*, **34** (12), 2283–2303.

Ahn, S.J., Warnecke, H.J., and Kotowski, R. (1999) Systematic geometric image measurement errors of circular object targets: Mathematical formulation and correction. *Photogrammetric Record*, **16** (93), 485–502.

AIA (2012a) Camera Link HS: Specifications of the Camera Link HS interface standard for digital cameras and frame grabbers, version 1.0.

AIA (2012b) Camera Link: Specifications of the Camera Link interface standard for digital cameras and frame grabbers, version 2.0.

AIA (2013) GigE Vision: Video streaming and device control over ethernet standard, version 2.0.03.

AIA (2015) USB3 Vision, version 1.0.1.

Alexander, S. and Droms, R. (1997) DHCP options and BOOTP vendor extensions, RFC 2132.

Alleysson, D., Süsstrunk, S., and Hérault, J. (2005) Linear demosaicing inspired by the human visual system. *IEEE Transactions on Image Processing*, **14** (4), 439–449.

Ando, S. (2000) Consistent gradient operators. *IEEE Transactions on Pattern Analysis and Machine Intelligence*, **22** (3), 252–265.

Anisimov, V.A. and Gorsky, N.D. (1993) Fast hierarchical matching of an arbitrarily oriented template. *Pattern Recognition Letters*, **14** (2), 95–101.

Arya, S., Mount, D.M., Netanyahu, N.S., Silverman, R., and Wu, A.Y. (1998) An optimal algorithm for approximate nearest neighbor searching in fixed dimensions. *Journal of the ACM*, **45** (6), 891–923.

Aubert, G. and Kornprobst, P. (2006) *Mathematical Problems in Image Processing: Partial Differential Equations and the Calculus of Variations*, Springer-Verlag, New York, NY, 2nd edn.

Ayache, N. and Faugeras, O.D. (1986) HYPER: A new approach for the recognition and positioning of two-dimensional objects. *IEEE Transactions on Pattern Analysis and Machine Intelligence*, **8** (1), 44–54.

Babaud, J., Witkin, A.P., Baudin, M., and Duda, R.O. (1986) Uniqueness of the Gaussian kernel for scale-space filtering. *IEEE Transactions on Pattern Analysis and Machine Intelligence*, **8** (1), 26–33.

Baillot, Y., Julier, S.J., Brown, D., and Livingston, M.A. (2003) A tracker alignment framework for augmented reality, in *International Symposium on Mixed and Augmented Reality*, pp. 142–150.

Ballard, D.H. (1981) Generalizing the Hough transform to detect arbitrary shapes. *Pattern Recognition*, **13** (2), 111–122.

Bayer, B.E. (1976) Color imaging array, US Patent 3 971 065.

Bell, T., Li, B., and Zhang, S. (2016) Structured light techniques and applications, in *Wiley Encyclopedia of Electrical and Electronics Engineering* (ed. J.G. Webster), John Wiley & Sons, doi:10.1002/047134608X.W8298.

Bentley, J.L. (1975) Multidimensional binary search trees used for associative searching. *Communications of the ACM*, **18** (9), 509–517.

Berge, J. (2004) *Fieldbuses for Process Control: Engineering, Operation, and Maintenance*, ISA — The Instrumentation, Systems, and Automation Society, Research Triangle Park, NC.

Berzins, V. (1984) Accuracy of Laplacian edge detectors. *Computer Vision, Graphics, and Image Processing*, **27** (2), 195–210.

Besl, P.J. (1988) Active, optical range imaging sensors. *Machine Vision and Applications*, **1** (2), 127–152.

Besl, P.J. and McKay, N.D. (1992) A method for registration of 3-D shapes. *IEEE Transactions on Pattern Analysis and Machine Intelligence*, **14** (2), 239–256.

Bishop, C.M. (1995) *Neural Networks for Pattern Recognition*, Oxford University Press, Oxford.

Blais, F. (2004) Review of 20 years of range sensor development. *Journal of Electronic Imaging*, **13** (1), 231–243.

Bloomberg, D.S. (2002) Implementation efficiency of binary morphology, in *International Symposium on Mathematical Morphology VI*, pp. 209–218.

Bookstein, F.L. (1989) Principal warps: Thin-plate splines and the decomposition of deformations. *IEEE Transactions on Pattern Analysis and Machine Intelligence*, **11** (6), 567–585.

Borgefors, G. (1984) Distance transformation in arbitrary dimensions. *Computer Vision, Graphics, and Image Processing*, **27** (3), 321–345.

Borgefors, G. (1988) Hierarchical chamfer matching: A parametric edge matching algorithm. *IEEE Transactions on Pattern Analysis and Machine Intelligence*, **10** (6), 849–865.

Born, M. and Wolf, E. (1999) *Principles of Optics*, Cambridge University Press, Cambridge, 7th edn.

Bray, T., Paoli, J., Sperberg-McQueen, C.M., Maler, E., and Yergeau, F. (2008) Extensible markup language (XML) 1.0 (fifth edition), W3C Recommendation.

Brenner, C., Böhm, J., and Gühring, J. (1998) Photogrammetric calibration and accuracy evaluation of a cross-pattern stripe projector, in *Videometrics VI* (eds S.F. El-Hakim and A. Gruen), Proc. SPIE 3641, pp. 164–172.

Brigham, E.O. (1988) *The Fast Fourier Transform and its Applications*, Prentice-Hall, Upper Saddle River, NJ.

Brown, D.C. (1966) Decentering distortion of lenses. *Photogrammetric Engineering*, **32** (3), 444–462.

Brown, D.C. (1971) Close-range camera calibration. *Photogrammetric Engineering*, **37** (8), 855–866.

Brown, L.G. (1992) A survey of image registration techniques. *ACM Computing Surveys*, **24** (4), 325–376.

Brown, M.Z., Burschka, D., and Hager, G.D. (2003) Advances in computational stereo. *IEEE Transactions on Pattern Analysis and Machine Intelligence*, **25** (8), 993–1008.

Calow, R., Ilchev, T., Lilienblum, E., Schnitzlein, M., and Michaelis, B. (2010) Schnelles Zeilensensorsystem zur gleichzeitigen Erfassung von Farbe und 3D-Form, in *Forum Bildverarbeitung* (eds F. Puente Léon and M. Heinzmann), KIT Scientific Publishing, Karlsruhe, pp. 181–192.

Canny, J. (1986) A computational approach to edge detection. *IEEE Transactions on Pattern Analysis and Machine Intelligence*, **8** (6), 679–698.

Caro, D. (2003) *Automation Network Selection*, ISA — The Instrumentation, Systems, and Automation Society, Research Triangle Park, NC.

Casey, R.G. and Lecolinet, E. (1996) A survey of methods and strategies in character segmentation. *IEEE Transactions on Pattern Analysis and Machine Intelligence*, **18** (7), 690–706.

Chen, H.H. (1991) A screw motion approach to uniqueness analysis of head–eye geometry, in *Computer Vision and Pattern Recognition*, pp. 145–151.

Chen, J.M., Ventura, J.A., and Wu, C.H. (1996) Segmentation of planar curves into circular arcs and line segments. *Image and Vision Computing*, **14** (1), 71–83.

Chen, X., Xi, J., Jin, Y., and Sun, J. (2009) Accurate calibration for a camera–projector measurement system based on structured light projection. *Optics and Lasers in Engineering*, **47** (3), 310–319.

Chen, Y. and Medioni, G. (1992) Object modelling by registration of multiple range images. *Image and Vision Computing*, **10** (3), 145–155.

Cheshire, S., Aboba, B., and Guttman, E. (2005) Dynamic configuration of IPv4 link-local addresses, RFC 3927.

Chou, J.C.K. and Kamel, M. (1991) Finding the position and orientation of a sensor on a robot manipulator using quaternions. *International Journal of Robotics Research*, **10** (3), 240–254.

Christianini, N. and Shawe-Taylor, J. (2000) *An Introduction to Support Vector Machines and Other Kernel-Based Learning Methods*, Cambridge University Press, Cambridge.

CIE 15:2004 (2004) Colorimetry, 3rd edition.

Clark, J.J. (1989) Authenticating edges produced by zero-crossing algorithms. *IEEE Transactions on Pattern Analysis and Machine Intelligence*, **11** (1), 43–57.

Connolly, C. (2010) Smart cameras tackle 3D industrial inspection. *Sensor Review*, **30** (2), 107–110.

Costa, M.S. and Shapiro, L.G. (2000) 3D object recognition and pose with relational indexing. *Computer Vision and Image Understanding*, **79** (3), 364–407.

Danielsson, P.E. (1980) Euclidean distance mapping. *Computer Graphics and Image Processing*, **14** (3), 227–248.

Daniilidis, K. (1999) Hand–eye calibration using dual quaternions. *International Journal of Robotics Research*, **18** (3), 286–298.

Davis, J., Diego, N., Ramamoorthi, R., and Rusinkiewicz, S. (2005) Spacetime stereo: A unifying framework for depth from triangulation. *IEEE Transactions on Pattern Analysis and Machine Intelligence*, **27** (2), 296–302.

de Berg, M., Cheong, O., van Kreveld, M., and Overmars, M. (2010) *Computational Geometry: Algorithms and Applications*, Springer-Verlag, Berlin, 3rd edn.

Deering, S. and Hinden, R. (1998) Internet protocol, version 6 (IPv6), RFC 2460.

Deriche, R. (1987) Using Canny's criteria to derive a recursively implemented optimal edge detector. *International Journal of Computer Vision*, **1** (2), 167–187.

Deriche, R. (1990) Fast algorithms for low-level vision. *IEEE Transactions on Pattern Analysis and Machine Intelligence*, **12** (1), 78–87.

Deriche, R. (1993) Recursively implementing the Gaussian and its derivatives, *Rapport de Recherche 1893*, INRIA, Sophia Antipolis.

Di Stefano, L., Mattoccia, S., and Mola, M. (2003) An efficient algorithm for exhaustive template matching based on normalized cross correlation, in *12th International Conference on Image Analysis and Processing*, pp. 322–327.

Di Zenzo, S. (1986) A note on the gradient of a multi-image. *Computer Vision, Graphics, and Image Processing*, **33** (1), 116–125.

DIN 1335:2003-12 (2003) Geometrische Optik — Bezeichnungen und Definitionen.

Dornaika, F. and Horaud, R. (1998) Simultaneous robot–world and hand–eye calibration. *IEEE Transactions on Robotics and Automation*, **14** (4), 617–622.

Dorsch, R.G., Häusler, G., and Herrmann, J.M. (1994) Laser triangulation: fundamental uncertainty in distance measurement. *Applied Optics*, **33** (7), 1306–1314.

Driewer, A., Hosticka, B.J., Spickermann, A., and Vogt, H. (2016) Modeling of the charge transfer in a lateral drift field photo detector. *Solid-State Electronics*, **126**, 51–58.

Droms, R. (1997) Dynamic host configuration protocol, RFC 2131.

Drost, B. (2017) *Point Cloud Computing for Rigid and Deformable 3D Object Recognition*, Dissertation, Technische Universität München.

Drost, B. and Ilic, S. (2012) 3D object detection and localization using multimodal point pair features, in *Second International Conference on 3D Imaging, Modeling, Processing, Visualization & Transmission*, pp. 9–16.

Drost, B. and Ilic, S. (2013) A hierarchical voxel hash for fast 3D nearest neighbor lookup, in *Pattern Recognition, Lecture Notes in Computer Science*, vol. 8142 (eds J. Weickert, M. Hein, and B. Schiele), Springer-Verlag, Berlin, pp. 302–312.

Drost, B. and Ilic, S. (2015) Graph-based deformable 3D object matching, in *Pattern Recognition, Lecture Notes in Computer Science*, vol. 9358 (eds J. Gall, P. Gehler, and B. Leibe), Springer-Verlag, Cham, pp. 222–233.

Drost, B. and Ulrich, M. (2012) Recognition and pose determination of 3D objects in 3D scenes, European Patent 2 385 483.

Drost, B. and Ulrich, M. (2014) Recognition and pose determination of 3D objects in 3D scenes, US Patent 8 830 229.

Drost, B. and Ulrich, M. (2015a) Recognition and pose determination of 3D objects in 3D scenes, Japanese Patent 5 677 798.

Drost, B. and Ulrich, M. (2015b) Recognition and pose determination of 3D objects in 3D scenes, Chinese Patent 102 236 794.

Drost, B. and Ulrich, M. (2015c) Recognition and pose determination of 3D objects in multimodal scenes, European Patent 2 720 171.

Drost, B. and Ulrich, M. (2015d) Recognition and pose determination of 3D objects in multimodal scenes, US Patent 8 994 723.

Drost, B., Ulrich, M., Navab, N., and Ilic, S. (2010) Model globally, match locally: Efficient and robust 3d object recognition, in *Computer Vision and Pattern Recognition*, pp. 998–1005.

Duda, R.O. and Hart, P.E. (1972) Use of the Hough transformation to detect lines and curves in pictures. *Communications of the ACM*, **15** (1), 11–15.

Eckhard, T. (2015) *Design considerations for line-scan multi-spectral imaging systems: Application in reflectance and color measurements*, Dissertation, Universidad de Granada.

Eckhardt, U. and Maderlechner, G. (1993) Invariant thinning. *International Journal of Pattern Recognition and Artificial Intelligence*, **7** (5), 1115–1144.

El Gamal, A. and Eltoukhy, H. (2005) CMOS image sensors. *IEEE Circuits and Devices Magazine*, **21** (3), 6–20.

EMVA (2011) GenICam CLProtocol module: Using GenApi with CameraLink, version 1.1.

EMVA (2015) GenICam GenTL standard, version 1.5.

EMVA (2016a) EMVA standard 1288: Standard for characterization of image sensors and cameras, release 3.1.

EMVA (2016b) GenICam GenCP: Generic control protocol, version 1.2.

EMVA (2016c) GenICam pixel format naming convention, version 2.1.

EMVA (2016d) GenICam standard features naming convention, version 2.3.

EMVA (2016e) GenICam standard: Generic interface for cameras, version 2.1.1.

EMVA (2017a) GenICam GenApi reference implementation, version 3.0.2. URL http://www.emva.org/standards-technology/genicam/genicam-downloads/, accessed 3 May 2017.

EMVA (2017b) GenICam GenTL standard features naming convention, version 1.1.1.

Evens, L. (2008) View camera geometry. URL http://www.math.northwestern.edu/len/photos/pages/vc.pdf, accessed 2 January 2017.

Fan, R.E., Chen, P.H., and Lin, C.J. (2005) Working set selection using second order information for training support vector machines. *Journal of Machine Learning Research*, **6**, 1889–1918.

Faugeras, O. (1993) *Three-Dimensional Computer Vision: A Geometric Viewpoint*, MIT Press, Cambridge, MA.

Faugeras, O., Hotz, B., Mathieu, H., Viéville, T., Zhang, Z., Fua, P., Théron, E., Moll, L., Berry, G., Vuillemin, J., Bertin, P., and Proy, C. (1993) Real time correlation-based stereo: algorithm, implementations and applications, *Rapport de Recherche 2013*, INRIA, Sophia-Antipolis.

Faugeras, O. and Luong, Q.T. (2001) *The Geometry of Multiple Images: The Laws That Govern the Formation of Multiple Images of a Scene and Some of Their Applications*, MIT Press, Cambridge, MA.

Fielding, R.T., Gettys, J., Mogul, J.C., Nielsen, H.F., Masinter, L., Leach, P.J., and Berners-Lee, T. (1999) Hypertext transfer protocol — HTTP/1.1, RFC 2616.

Figueiredo, M.A.T. and Jain, A.K. (2002) Unsupervised learning of finite mixture models. *IEEE Transactions on Pattern Analysis and Machine Intelligence*, **24** (3), 381–396.

Fischler, M.A. and Bolles, R.C. (1981) Random sample consensus: A paradigm for model fitting with applications to image analysis and automated cartography. *Communications of the ACM*, **24** (6), 381–395.

Fitzgibbon, A., Pilu, M., and Fisher, R.B. (1999) Direct least square fitting of ellipses. *IEEE*

Transactions on Pattern Analysis and Machine Intelligence, **21** (5), 476–480.

Florack, L.M.J., ter Haar Romeny, B.M., Koenderink, J.J., and Viergever, M.A. (1992) Scale and the differential structure of images. *Image and Vision Computing*, **10** (6), 376–388.

Flusser, J. and Suk, T. (1993) Pattern recognition by affine moment invariants. *Pattern Recognition*, **26** (1), 167–174.

Foix, S., Alenyà, G., and Torras, C. (2011) Lock-in time-of-flight (ToF) cameras: A survey. *IEEE Sensors Journal*, **11** (9), 1917–1926.

Förstner, W. (1994) A framework for low level feature extraction, in *Third European Conference on Computer Vision, Lecture Notes in Computer Science*, vol. 801 (ed. J.O. Eklundh), Springer-Verlag, Berlin, pp. 383–394.

Fraunhofer Allianz Vision (2003) *Guideline for Industrial Image Processing*, Fraunhofer-Gesellschaft zur Förderung der angewandten Forschung, Erlangen.

Friedman, J.H., Bentley, J.L., and Finkel, R.A. (1977) An algorithm for finding best matches in logarithmic expected time. *ACM Transactions on Mathematical Software*, **3** (3), 209–226.

Frigo, M. and Johnson, S.G. (2005) The design and implementation of FFTW3. *Proceedings of the IEEE*, **93** (2), 216–231.

Fürsattel, P., Placht, S., Balda, M., Schaller, C., Hofmann, H., Maier, A., and Riess, C. (2016) A comparative error analysis of current time-of-flight sensors. *IEEE Transactions on Computational Imaging*, **2** (1), 27–41.

Geiger, A., Lenz, P., Stiller, C., and Urtasun, R. (2017) The KITTI vision benchmark suite. URL http://www.cvlibs.net/datasets/kitti/, accessed 28 March 2017.

Geng, J. (2011) Structured-light 3D surface imaging: a tutorial. *Advances in Optics and Photonics*, **3** (2), 128–160.

Gharavi-Alkhansari, M. (2001) A fast globally optimal algorithm for template matching using low-resolution pruning. *IEEE Transactions on Image Processing*, **10** (4), 526–533.

Gil, J. and Kimmel, R. (2002) Efficient dilation, erosion, opening, and closing algorithms. *IEEE Transactions on Pattern Analysis and Machine Intelligence*, **24** (12), 1606–1617.

Glazer, F., Reynolds, G., and Anandan, P. (1983) Scene matching by hierarchical correlation, in *Computer Vision and Pattern Recognition*, pp. 432–441.

Goodfellow, I., Bengio, Y., and Courville, A. (2016) *Deep Learning*, MIT Press, Cambridge, MA.

Gorthi, S.S. and Rastog, P. (2010) Fringe projection techniques: Whither we are? *Optics and Lasers in Engineering*, **48** (2), 133–140.

Grimson, W.E.L. and Lozano-Pérez, T. (1987) Localizing overlapping parts by searching the interpretation tree. *IEEE Transactions on Pattern Analysis and Machine Intelligence*, **9** (4), 469–482.

Gruen, A. and Huang, T.S. (eds) (2001) *Calibration and Orientation of Cameras in Computer Vision*, Springer-Verlag, Berlin.

Gupta, R. and Hartley, R.I. (1997) Linear pushbroom cameras. *IEEE Transactions on Pattern Analysis and Machine Intelligence*, **19** (9), 963–975.

Hagen, N. and Kudenov, M.W. (2013) Review of snapshot spectral imaging technologies. *Optical Engineering*, **52** (9), 090901-1–090901-23.

Hanson, A.J. (2006) *Visualizing Quaternions*, Morgan Kaufmann Publishers, San Francisco, CA.

Haralick, R.M. and Shapiro, L.G. (1992) *Computer and Robot Vision*, vol. I, Addison-Wesley, Reading, MA.

Haralick, R.M. and Shapiro, L.G. (1993) *Computer and Robot Vision*, vol. II, Addison-Wesley, Reading, MA.

Haralick, R.M., Watson, L.T., and Laffey, T.J. (1983) The topographic primal sketch. *International Journal of Robotics Research*, **2** (1), 50–72.

Hartley, R. and Zisserman, A. (2003) *Multiple View Geometry in Computer Vision*, Cambridge University Press, Cambridge, 2nd edn.

Heikkilä, J. (2000) Geometric camera calibration using circular control points. *IEEE Transactions on Pattern Analysis and Machine Intelligence*, **22** (10), 1066–1077.

Hel-Or, Y. and Hel-Or, H. (2003) Real time pattern matching using projection kernels, in *9th International Conference on Computer Vision*, vol. 2, pp. 1486–1493.

Hirakawa, K. and Parks, T.W. (2005) Adaptive homogeneity-directed demosaicing algorithm. *IEEE Transactions on Image Processing*, **14** (3), 360–369.

Hirschmüller, H., Innocent, P.R., and Garibaldi, J. (2002) Real-time correlation-based stereo vision with reduced border errors. *International Journal of Computer Vision*, **47** (1–3), 229–246.

Hirschmüller, H. and Scharstein, D. (2009) Evaluation of stereo matching costs on images with radiometric differences. *IEEE Transactions on Pattern Analysis and Machine Intelligence*, **31** (9), 1582–1599.

Hofhauser, A. and Steger, C. (2010) System and method for deformable object recognition, European Patent 2 081 133.

Hofhauser, A. and Steger, C. (2011) System and method for deformable object recognition, Japanese Patent 4 825 253.

Hofhauser, A. and Steger, C. (2012) System and method for deformable object recognition, US Patent 8 260 059.

Hofhauser, A. and Steger, C. (2013) System and method for deformable object recognition, Chinese Patent 101 488 187.

Hofhauser, A., Steger, C., and Navab, N. (2008) Edge-based template matching and tracking for perspectively distorted planar objects, in *Advances in Visual Computing, Lecture Notes in Computer Science*, vol. 5358 (eds G. Bebis, R. Boyle, B. Parvin, D. Koracin, P. Remagnino, F. Porikli, J. Peters, J. Klosowski, L. Arns, Y.K. Chun, T.M. Rhyne, and L. Monroe), Springer-Verlag, Berlin, pp. 35–44.

Hofhauser, A., Steger, C., and Navab, N. (2009a) Edge-based template matching with a harmonic deformation model, in *Computer Vision and Computer Graphics: Theory and Applications — International Conference VISIGRAPP 2008, Revised Selected Papers, Communications in Computer and Information Science*, vol. 24 (eds A.K. Ranchordas, H.J. Araújo, J.M. Pereira, and J. Braz), Springer-Verlag, Berlin, pp. 176–187.

Hofhauser, A., Steger, C., and Navab, N. (2009b) Perspective planar shape matching, in *Image Processing: Machine Vision Applications II* (eds K.S. Niel and D. Fofi), Proc. SPIE 7251.

Holland, P.W. and Welsch, R.E. (1977) Robust regression using iteratively reweighted least-squares. *Communications in Statistics — Theory and Methods*, **6** (9), 813–827.

Holst, G.C. and Lomheim, T.S. (2011) *CMOS/CCD Sensors and Camera Systems*, SPIE Press, Bellingham, WA, 2nd edn.

Horaud, R. and Dornaika, F. (1995) Hand–eye calibration. *International Journal of Robotics Research*, **14** (3), 195–210.

Horaud, R., Hansard, M., Evangelidis, G., and Ménier, C. (2016) An overview of depth cameras and range scanners based on time-of-flight technologies. *Machine Vision and Applications*, **27** (7), 1005–1020.

Hough, P.V.C. (1962) Method and means for recognizing complex patterns, US Patent 3 069 654.

Hu, M.K. (1962) Visual pattern recognition by moment invariants. *IRE Transactions on Information Theory*, **8** (2), 179–187.

Hu, X. and Mordohai, P. (2012) A quantitative evaluation of confidence measures for stereo vision. *IEEE Transactions on Pattern Analysis and Machine Intelligence*, **34** (11), 2121–2133.

Huang, P.S., Zhang, C., and Chiang, F.P. (2003) High-speed 3-D shape measurement based on digital fringe projection. *Optical Engineering*, **42** (1), 163–168.

Huang, T.S., Yang, G.J., and Tang, G.Y. (1979) A fast two-dimensional median filtering algorithm. *IEEE Transactions on Acoustics, Speech, and Signal Processing*, **27** (1), 13–18.

Huber, P.J. (1981) *Robust Statistics*, John Wiley & Sons, New York, NY.

IEC 61883-1:1998 (1998) Consumer audio/video equipment — Digital interface — Part 1: General.

IEC 61883-1:2008 (2008) Consumer audio/video equipment — Digital interface — Part 1: General, 3rd edn.

IEC 61883-2:1998 (1998) Consumer audio/video equipment — Digital interface — Part 2: SD-DVCR data transmission.

IEC 61883-2:2004 (2004) Consumer audio/video equipment — Digital interface — Part 2: SD-DVCR data transmission, 2nd edn.

IEC 61883-3:1998 (1998) Consumer audio/video equipment — Digital interface — Part 3: HD-DVCR data transmission.

IEC 61883-3:1998 (2004) Consumer audio/video equipment — Digital interface — Part 3: HD-DVCR data transmission, 2nd edn.

IEC 61883-4:1998 (1998) Consumer audio/video equipment — Digital interface — Part 4: MPEG2-TS data transmission.

IEC 61883-4:2004 (2004) Consumer audio/video equipment — Digital interface — Part 4: MPEG2-TS data transmission, 2nd edn.

IEC 61883-5:1998 (1998) Consumer audio/video equipment — Digital interface — Part 5: SDL-DVCR data transmission.

IEC 61883-5:2004 (2004) Consumer audio/video equipment — Digital interface — Part 5: SDL-DVCR data transmission, 2nd edn.

IEC 61883-8:2008 (2008) Consumer audio/video equipment — Digital interface — Part 8: Transmission of ITU-R BT.601 style digital video data.

IEC 62680-2-1:2015 (2015) Universal serial bus interfaces for data and power — Part 2-1: Universal Serial Bus Specification, Revision 2.0.

IEEE Std 1394-1995 (1995) IEEE standard for a high performance serial bus.

IEEE Std 1394-2008 (2008) IEEE standard for a high performance serial bus.

IEEE Std 1394a-2000 (2000) IEEE standard for a high performance serial bus — Amendment 1.

IEEE Std 1394b-2002 (2002) IEEE standard for a high performance serial bus — Amendment 2.

IEEE Std 1588-2008 (2008) IEEE standard for a precision clock synchronization protocol for networked measurement and control systems.

IEEE Std 802.3-2015 (2015) IEEE standard for Ethernet.

Ilchev, T., Lilienblum, E., Joedicke, B., Michaelis, B., and Schnitzlein, M. (2012) A stereo line sensor system to high speed capturing of surfaces in color and 3D shape, in *International Conference on Computer Graphics Theory and Applications* (eds P. Richard, M. Kraus, R.S. Laramee, and J. Braz), pp. 809–812.

ISO 14524:2009 (2009) Photography — Electronic still-picture cameras — Methods for measuring opto-electronic conversion functions (OECFs).

ISO 15739:2013 (2013) Photography — Electronic still-picture imaging — Noise measurements.

ISO 15795:2002 (2002) Optics and optical instruments — Quality evaluation of optical systems — Assessing the image quality degradation due to chromatic aberrations.

ISO 20473:2007 (2007) Optics and photonics — Spectral bands.

ISO 517:2008 (2008) Photography — Apertures and related properties pertaining to photographic lenses — Designations and measurements.

ISO 8373:2012 (2012) Robots and robotic devices — Vocabulary.

ISO 9039:2008 (2008) Optics and optical instruments — Quality evaluation of optical systems — Determination of distortion.

ISO 9787:2013 (2013) Robots and robotic devices — Coordinate systems and motion nomenclatures.

ITU-R BT.470-6 (1998) Conventional television systems.

Jain, R., Kasturi, R., and Schunck, B.G. (1995) *Machine Vision*, McGraw-Hill, New York, NY.

JCGM 200:2012 (2012) International vocabulary of metrology — basic and general concepts and associated terms (VIM), 3rd edition.

JIIA CXP-001-2015 (2015) CoaXPress standard, version 1.1.1.

Joseph, S.H. (1994) Unbiased least squares fitting of circular arcs. *Computer Vision, Graphics, and Image Processing: Graphical Models and Image Processing*, **56** (5), 424–432.

Joseph, S.H. (1999) Analysing and reducing the cost of exhaustive correspondence search. *Image and Vision Computing*, **17** (11), 815–830.

Kaiser, B., Tauro, R.A., and Wörn, H. (2008) Extrinsic calibration of a robot mounted 3D imaging sensor. *International Journal of Intelligent Systems Technologies and Applications*, **5** (3/4), 374–379.

Keys, R.G. (1981) Cubic convolution interpolation for digital image processing. *IEEE Transactions on Acoustics, Speech, and Signal Processing*, **29** (6), 1153–1160.

Koch, M.W. and Kashyap, R.L. (1987) Using polygons to recognize and locate partially occluded objects. *IEEE Transactions on Pattern Analysis and Machine Intelligence*, **9** (4), 483–494.

Kuipers, J.B. (1999) *Quaternions and Rotation Sequences: A Primer with Applications to Orbits, Aerospace, and Virtual Reality*, Princeton University Press, Princeton, NJ.

Kwon, O.K., Sim, D.G., and Park, R.H. (2001) Robust Hausdorff distance matching algorithms using pyramidal structures. *Pattern Recognition*, **34** (10), 2005–2013.

Lai, S.H. and Fang, M. (1999) Accurate and fast pattern localization algorithm for automated visual inspection. *Real-Time Imaging*, **5** (1), 3–14.

Lam, L., Lee, S.W., and Suen, C.Y. (1992) Thinning methodologies—a comprehensive survey. *IEEE Transactions on Pattern Analysis and Machine Intelligence*, **14** (9), 869–885.

Lamdan, Y., Schwartz, J.T., and Wolfson, H.J. (1990) Affine invariant model-based object recognition. *IEEE Transactions on Robotics and Automation*, **6** (5), 578–589.

Lanser, S. (1997) *Modellbasierte Lokalisation gestützt auf monokulare Videobilder*, PhD thesis, Forschungs- und Lehreinheit Informatik IX, Technische Universität München. Shaker Verlag, Aachen.

Lanser, S. and Eckstein, W. (1992) A modification of Deriche's approach to edge detection, in *11th International Conference on Pattern Recognition*, vol. III, pp. 633–637.

Lanser, S., Zierl, C., and Beutlhauser, R. (1995) Multibildkalibrierung einer CCD-Kamera, in *Mustererkennung* (eds G. Sagerer, S. Posch, and F. Kummert), Springer-Verlag, Berlin, Informatik aktuell, pp. 481–491.

Lapray, P.J., Wang, X., Thomas, J.B., and Gouton, P. (2014) Multispectral filter arrays: Recent advances and practical implementation. *Sensors*, **14** (11), 21 626–21 659.

LeCun, Y., Bottou, L., Bengio, Y., and Haffner, P. (1998) Gradient-based learning applied to document recognition. *Proceedings of the IEEE*, **86** (11), 2278–2324.

Lee, H.C. (2005) *Introduction to Color Imaging Science*, Cambridge University Press, Cambridge.

Lenhardt, K. (2017) Optical systems in machine vision, in *Handbook of Machine and Computer Vision* (ed. A. Hornberg), Wiley-VCH, Weinheim, pp. 179–290, 2nd edn.

Lenz, R. (1988) *Viedeometrie mit CCD-Sensoren und ihre Anwendung in der Robotik*, Habilitationsschrift, Lehrstuhl für Nachrichtentechnik der Technischen Universität München.

Lenz, R. and Fritsch, D. (1990) Accuracy of videometry with CCD sensors. *ISPRS Journal of Photogrammetry and Remote Sensing*, **45** (2), 90–110.

Lindeberg, T. (1994) *Scale-Space Theory in Computer Vision*, Kluwer Academic, Dordrecht.

Liu, C.L., Nakashima, K., Sako, H., and Fujisawa, H. (2004) Handwritten digit recognition: investigation of normalization and feature extraction techniques. *Pattern Recognition*, **37** (2), 265–279.

MacKay, D.J.C. (1992a) Bayesian interpolation. *Neural Computation*, **4** (3), 415–447.

MacKay, D.J.C. (1992b) The evidence framework applied to classification networks. *Neural Computation*, **4** (5), 720–736.

MacKay, D.J.C. (1992c) A practical Bayesian framework for backpropagation networks. *Neural Computation*, **4** (3), 448–472.

Mahajan, V.N. (1998) *Optical Imaging and Aberrations — Part I: Ray Geometrical Optics*, SPIE Press, Bellingham, WA.

Mahalik, N.P. (ed.) (2003) *Fieldbus Technology: Industrial Network Standards for Real-Time Distributed Control*, Springer-Verlag, Berlin.

Mamistvalov, A.G. (1998) n-dimensional moment invariants and conceptual mathematical theory of recognition n-dimensional solids. *IEEE Transactions on Pattern Analysis and Machine Intelligence*, **20** (8), 819–831.

Mann, S. and Mann, R. (2001) Quantigraphic imaging: Estimating the camera response and exposures from differently exposed images, in *Computer Vision and Pattern Recognition*, vol. I, pp. 842–849.

Mendel, J.M. (1995) Fuzzy logic systems for engineering: A tutorial. *Proceedings of the IEEE*, **83** (3), 345–377.

Merklinger, H.M. (2010) Focusing the view camera: A scientific way to focus the view camera and estimate depth of field, version 1.6.1. URL http://www.trenholm.org/hmmerk/FVC161.pdf, accessed 2 January 2017.

Metcalfe, R.M. and Boggs, D.R. (1976) Ethernet: Distributed packet switching for local computer networks. *Communications of the ACM*, **19** (7), 395–404.

Mitschke, M.M. and Navab, N. (2000) Recovering projection geometry: How a cheap camera can outperform an expensive stereo system, in *Computer Vision and Pattern Recognition*, vol. 1, pp. 193–200.

Mitsunaga, T. and Nayar, S.K. (1999) Radiometric self calibration, in *Computer Vision and Pattern Recognition*, vol. I, pp. 374–380.

Moganti, M., Ercal, F., Dagli, C.H., and Tsunekawa, S. (1996) Automatic PCB inspection algorithms: A survey. *Computer Vision and Image Understanding*, **63** (2), 287–313.

Moreno-Noguer, F., Lepetit, V., and Fua, P. (2007) Accurate non-iterative $o(n)$ solution to the PnP problem, in *11th International Conference on Computer Vision*.

Mosteller, F. and Tukey, J.W. (1977) *Data Analysis and Regression*, Addison-Wesley, Reading, MA.

Mühlmann, K., Maier, D., Hesser, J., and Männer, R. (2002) Calculating dense disparity maps from color stereo images, an efficient implementation. *International Journal of Computer Vision*, **47** (1–3), 79–88.

Muja, M. and Lowe, D.G. (2014) Scalable nearest neighbor algorithms for high dimensional data. *IEEE Transactions on Pattern Analysis and Machine Intelligence*, **36** (11), 2227–2240.

Myronenko, A. and Song, X. (2010) Point set registration: Coherent point drift. *IEEE Transactions on Pattern Analysis and Machine Intelligence*, **32** (12), 2262–2275.

Nabney, I.T. (2002) *NETLAB: Algorithms for Pattern Recognition*, Springer-Verlag, London.

Olson, C.F. and Huttenlocher, D.P. (1997) Automatic target recognition by matching oriented edge pixels. *IEEE Transactions on Image Processing*, **6** (1), 103–113.

O'Rourke, J. (1998) *Computational Geometry in C*, Cambridge University Press, Cambridge, 2nd edn.

Otsu, N. (1979) A threshold selection method from gray-level histograms. *IEEE Transactions on Systems, Man, and Cybernetics*, **9** (1), 62–66.

Papoulis, A. and Pillai, S.U. (2002) *Probability, Random Variables, and Stochastic Processes*, McGraw-Hill, New York, NY, 4th edn.

Perreault, S. and Hébert, P. (2007) Median filtering in constant time. *IEEE Transactions on Image Processing*, **16** (9), 2389–2394.

Planck, M. (1901) Ueber das Gesetz der Energieverteilung im Normalspectrum. *Annalen der Physik*, **309** (3), 553–563.

Platt, J.C. (1999) Fast training of support vector machines using sequential minimal optimization, in *Advances in Kernel Methods — Support Vector Learning* (eds B. Schölkopf, C.J.C. Burges, and A.J. Smola), MIT Press, Cambridge, MA, pp. 185–208.

Postel, J. (1980) User datagram protocol, RFC 768.

Postel, J. (1981a) Internet protocol, RFC 791.

Postel, J. (1981b) Transmission control protocol, RFC 793.

Powell, I. (1989) Linear diverging lens, US Patent 4 826 299.

Press, W.H., Teukolsky, S.A., Vetterling, W.T., and Flannery, B.P. (2007) *Numerical Recipes: The Art of Scientific Computing*, Cambridge University Press, Cambridge, 3rd edn.

Proll, K.P., Nivet, J.M., Körner, K., and Tiziani, H.J. (2003) Microscopic three-dimensional topometry with ferroelectric liquid-crystal-on-silicon displays. *Applied Optics*, **42** (10), 1773–1778.

Ramer, U. (1972) An iterative procedure for the polygonal approximation of plane curves. *Computer Graphics and Image Processing*, **1** (3), 244–256.

Remondino, F. and Stoppa, D. (eds) (2013) *TOF Range-Imaging Cameras*, Springer-Verlag, Berlin.

Rooney, J. (1977) A survey of representations of spatial rotation about a fixed point. *Environment and Planning B*, **4** (2), 185–210.

Rooney, J. (1978) A comparison of representations of general spatial screw displacement. *Environment and Planning B*, **5** (1), 45–88.

Rosin, P.L. (1997) Techniques for assessing polygonal approximations of curves. *IEEE Transactions on Pattern Analysis and Machine Intelligence*, **19** (6), 659–666.

Rosin, P.L. (2003) Assessing the behaviour of polygonal approximation algorithms. *Pattern Recognition*, **36** (2), 508–518.

Rosin, P.L. and West, G.A.W. (1995) Nonparametric segmentation of curves into various representations. *IEEE Transactions on Pattern Analysis and Machine Intelligence*, **17** (12), 1140–1153.

Rucklidge, W.J. (1997) Efficiently locating objects using the Hausdorff distance. *International Journal of Computer Vision*, **24** (3), 251–270.

Salvi, J., Fernandez, S., Pribanic, T., and Llado, X. (2010) A state of the art in structured light patterns for surface profilometry. *Pattern Recognition*, **43** (8), 2666–2680.

Sansoni, G., Carocci, M., and Rodella, R. (1999) Three-dimensional vision based on a combination of gray-code and phase-shift light projection: analysis and compensation of the systematic errors. *Applied Optics*, **38** (31), 6565–6573.

Sansoni, G., Corini, S., Lazzari, S., Rodella, R., and Docchio, F. (1997) Three-dimensional imaging based on Gray-code light projection: characterization of the measuring algorithm and development of a measuring system for industrial applications. *Applied Optics*, **36** (19), 4463–4472.

Schaffer, M., Grosse, M., and Kowarschik, R. (2010) High-speed pattern projection for three-dimensional shape measurement using laser speckles. *Applied Optics*, **49** (18), 3622–3629.

Scharstein, D. and Szeliski, R. (2002) A taxonomy and evaluation of dense two-frame stereo correspondence algorithms. *International Journal of Computer Vision*, **47** (1–3), 7–42.

Scharstein, D., Szeliski, R., and Hirschmüller, H. (2017) Middlebury stereo vision page. URL http://vision.middlebury.edu/stereo/, accessed 28 March 2017.

Schmid, C., Mohr, R., and Bauckhage, C. (2000) Evaluation of interest point detectors. *International Journal of Computer Vision*, **37** (2), 151–172.

Schmidt, J. and Niemann, H. (2008) Data selection for hand–eye calibration: A vector quantization approach. *International Journal of Robotics Research*, **27** (9), 1027–1053.

Schmidt, J., Vogt, F., and Niemann, H. (2003) Robust hand–eye calibration of an endoscopic surgery robot using dual quaternions, in *Pattern Recognition, Lecture Notes in Computer Science*, vol. 2781 (eds B. Michaelis and G. Krell), Springer-Verlag, Berlin, pp. 548–556.

Schölkopf, B. and Smola, A.J. (2002) *Learning with Kernels — Support Vector Machines, Regularization, Optimization, and Beyond*, MIT Press, Cambridge, MA.

Sedgewick, R. (1990) *Algorithms in C*, Addison-Wesley, Reading, MA.

Seitz, S.M., Curless, B., Diebel, J., Scharstein, D., and Szeliski, R. (2006) A comparison and evaluation of multi-view stereo reconstruction algorithms, in *Computer Vision and Pattern Recognition*, vol. 1, pp. 519–528.

Seitz, S.M., Curless, B., Diebel, J., Scharstein, D., and Szeliski, R. (2017) Middlebury multi-view stereo vision page. URL http://vision.middlebury.edu/mview/, accessed 28 March 2017.

Serra, J. (1982) *Image Analysis and Mathematical Morphology*, vol. 1, Academic Press, London.

Shapiro, L.G. and Stockman, G.C. (2001) *Computer Vision*, Prentice-Hall, Upper Saddle River, NJ.

Sheu, H.T. and Hu, W.C. (1999) Multiprimitive segmentation of planar curves—a two-level breakpoint classification and tuning approach. *IEEE Transactions on Pattern Analysis and Machine Intelligence*, **21** (8), 791–797.

Singh, S. and Markou, M. (2004) An approach to novelty detection applied to the classification of image regions. *IEEE Transactions on Knowledge and Data Engineering*, **16** (4), 396–407.

Soille, P. (2003) *Morphological Image Analysis*, Springer-Verlag, Berlin, 2nd edn.

Spickermann, A., Durini, D., Süss, A., Ulfig, W., Brockherde, W., Hosticka, B.J.H., Schwope, S., and Grabmaier, A. (2011) CMOS 3D image sensor based on pulse modulated time-of-flight principle and intrinsic lateral drift-field photodiode pixels, in *37th European Solid State Circuits Conference (ESSCIRC)*, pp. 111–114.

Steger, C. (1996) On the calculation of arbitrary moments of polygons, *Technical Report FGBV–96–05*, Forschungsgruppe Bildverstehen (FG BV), Informatik IX, Technische Universität München.

Steger, C. (1998a) Analytical and empirical performance evaluation of subpixel line and edge detection, in *Empirical Evaluation Methods in Computer Vision* (eds K.J. Bowyer and P.J. Phillips), pp. 188–210.

Steger, C. (1998b) *Unbiased Extraction of Curvilinear Structures from 2D and 3D Images*, PhD thesis, Fakultät für Informatik, Technische Universität München. Herbert Utz Verlag, München.

Steger, C. (2000) Subpixel-precise extraction of lines and edges, in *International Archives of*

Photogrammetry and Remote Sensing, vol. XXXIII, part B3, pp. 141–156.

Steger, C. (2001) Similarity measures for occlusion, clutter, and illumination invariant object recognition, in *Pattern Recognition, Lecture Notes in Computer Science*, vol. 2191 (eds B. Radig and S. Florczyk), Springer-Verlag, Berlin, pp. 148–154.

Steger, C. (2002) Occlusion, clutter, and illumination invariant object recognition, in *International Archives of Photogrammetry and Remote Sensing*, vol. XXXIV, part 3A, pp. 345–350.

Steger, C. (2005) System and method for object recognition, European Patent 1 193 642.

Steger, C. (2006a) System and method for object recognition, Japanese Patent 3 776 340.

Steger, C. (2006b) System and method for object recognition, US Patent 7 062 093.

Steger, C. (2017) A comprehensive and versatile camera model for cameras with tilt lenses. *International Journal of Computer Vision*, **123** (2), 121–159.

Stewart, C.V. (1999) Robust parameter estimation in computer vision. *SIAM Review*, **41** (3), 513–537.

Strobl, K.H. and Hirzinger, G. (2006) Optimal hand–eye calibration, in *International Conference on Intelligent Robots and Systems*, pp. 4647–4653.

Sturm, P., Ramalingam, S., Tardif, J.P., Gasparini, S., and Barreto, J. (2010) Camera models and fundamental concepts used in geometric computer vision. *Foundations and Trends in Computer Graphics and Vision*, **6** (1–2), 1–183.

Sturm, P.F. and Maybank, S.J. (1999) On plane-based camera calibration: A general algorithm, singularities, applications, in *Computer Vision and Pattern Recognition*, vol. I, pp. 432–437.

Szeliski, R. (2011) *Computer Vision: Algorithms and Applications*, Springer-Verlag, London.

Tanimoto, S.L. (1981) Template matching in pyramids. *Computer Graphics and Image Processing*, **16** (4), 356–369.

Theodoridis, S. and Koutroumbas, K. (2009) *Pattern Recognition*, Academic Press, Burlington, MA, 4th edn.

Tian, Q. and Huhns, M.N. (1986) Algorithms for subpixel registration. *Computer Vision, Graphics, and Image Processing*, **35** (2), 220–233.

Toussaint, G. (1983) Solving geometric problems with the rotating calipers, in *Proceedings of IEEE MELECON '83*, IEEE Press, Los Alamitos, CA, pp. A10.02/1–4.

Tsai, R.Y. and Lenz, R.K. (1989) A new technique for fully autonomous and efficient 3D robotics hand/eye calibration. *IEEE Transactions on Robotics and Automation*, **5** (3), 345–358.

Ulrich, M. (2003) *Hierarchical Real-Time Recognition of Compound Objects in Images*, Reihe C, vol. 569, Deutsche Geodätische Kommission bei der Bayerischen Akademie der Wissenschaften, München.

Ulrich, M., Baumgartner, A., and Steger, C. (2002) Automatic hierarchical object decomposition for object recognition, in *International Archives of Photogrammetry and Remote Sensing*, vol. XXXIV, part 5, pp. 99–104.

Ulrich, M. and Steger, C. (2001) Empirical performance evaluation of object recognition methods, in *Empirical Evaluation Methods in Computer Vision* (eds H.I. Christensen and P.J. Phillips), IEEE Computer Society Press, Los Alamitos, CA, pp. 62–76.

Ulrich, M. and Steger, C. (2002) Performance comparison of 2d object recognition techniques, in *International Archives of Photogrammetry* and *Remote Sensing*, vol. XXXIV, part 3A, pp. 368–374.

Ulrich, M. and Steger, C. (2007) Hierarchical component based object recognition, US Patent 7 239 929.

Ulrich, M. and Steger, C. (2009) Hierarchical component based object recognition, Japanese Patent 4 334 301.

Ulrich, M. and Steger, C. (2011) Hierarchical component based object recognition, European Patent 1 394 727.

Ulrich, M. and Steger, C. (2013a) Hierarchical component based object recognition, Japanese Patent 5 330 579.

Ulrich, M. and Steger, C. (2013b) Hierarchical component based object recognition, Japanese Patent 5 329 254.

Ulrich, M. and Steger, C. (2016) Hand–eye calibration of SCARA robots using dual quaternions. *Pattern Recognition and Image Analysis*, **26** (1), 231–239.

Ulrich, M., Steger, C., and Baumgartner, A. (2003) Real-time object recognition using a modified generalized Hough transform. *Pattern Recognition*, **36** (11), 2557–2570.

Ulrich, M., Steger, C., Baumgartner, A., and Ebner, H. (2004) Erkennung von zusammengesetzten Objekten in Bildern unter Echtzeit-Anforderungen. *Zeitschrift für Geodäsie, Geoinformation und Landmanagement*, **129** (3), 184–194.

Ulrich, M., Wiedemann, C., and Steger, C. (2009) CAD-based recognition of 3D objects in monocular images, in *International Conference on Robotics and Automation*, pp. 1191–1198.

Ulrich, M., Wiedemann, C., and Steger, C. (2012) Combining scale-space and similarity-based aspect graphs for fast 3D object recognition. *IEEE Transactions on Pattern Analysis and Machine Intelligence*, **34** (10), 1902–1914.

USB Implementers Forum (2000) Universal Serial Bus specification, revision 2.0.

USB Implementers Forum (2012) Universal Serial Bus device class definition for video devices, revision 1.5.

USB Implementers Forum (2013) Universal Serial Bus 3.1 specification.

Van der Jeught, S. and Dirckx, J.J.J. (2016) Real-time structured light profilometry: a review. *Optics and Lasers in Engineering*, **87**, 18–31.

Van Droogenbroeck, M. and Talbot, H. (1996) Fast computation of morphological operations with arbitrary structuring elements. *Pattern Recognition Letters*, **17** (14), 1451–1460.

Ventura, J.A. and Wan, W. (1997) Accurate matching of two-dimensional shapes using the minimal tolerance error zone. *Image and Vision Computing*, **15** (12), 889–899.

Wang, H.X., Luo, B., Zhang, Q.B., and Wei, S. (2004) Estimation for the number of components in a mixture model using stepwise split-and-merge EM algorithm. *Pattern Recognition Letters*, **25** (16), 1799–1809.

Wang, Z., Nguyen, D.A., and Barnes, J.C. (2010) Some practical considerations in fringe projection profilometry. *Optics and Lasers in Engineering*, **48** (2), 218–225.

Wäny, M. and Israel, G.P. (2003) CMOS image sensor with NMOS-only global shutter and enhanced responsivity. *IEEE Transactions on Electron Devices*, **50** (1), 57–62.

Webb, A. and Copsey, K.D. (2004) *Statistical Pattern Recognition*, John Wiley & Sons, Chichester, 3rd edn.

Welzl, E. (1991) Smallest enclosing disks (balls and ellipsoids), in *New Results and Trends in Computer Science, Lecture Notes in Computer Science*, vol. 555 (ed. H. Maurer), Springer-Verlag, Berlin, pp. 359–370.

Wiedemann, C., Ulrich, M., and Steger, C. (2008) Recognition and tracking of 3D objects, in *Pattern Recognition, Lecture Notes in Computer Science*, vol. 5096 (ed. G. Rigoll), Springer-Verlag, Berlin, pp. 132–141.

Wiedemann, C., Ulrich, M., and Steger, C. (2009) System and method for 3D object recognition, European Patent 2 048 599.

Wiedemann, C., Ulrich, M., and Steger, C. (2011) System and method for 3D object recognition, Japanese Patent 4 785 880.

Wiedemann, C., Ulrich, M., and Steger, C. (2013a) System and method for 3D object recognition, US Patent 8 379 014.

Wiedemann, C., Ulrich, M., and Steger, C. (2013b) System and method for 3D object recognition, Chinese Patent 101 408 931.

Wiegmann, A., Wagner, H., and Kowarschik, R. (2006) Human face measurement by projecting bandlimited random patterns. *Optics Express*, **14** (17), 7692–7698.

Witkin, A.P. (1983) Scale-space filtering, in *Eigth International Joint Conference on Artificial Intelligence*, vol. 2, pp. 1019–1022.

Wuescher, D.M. and Boyer, K.L. (1991) Robust contour decomposition using a constant curvature criterion. *IEEE Transactions on Pattern Analysis and Machine Intelligence*, **13** (1), 41–51.

Wyszecki, G. and Stiles, W.S. (1982) *Color Science: Concepts and Methods, Quantitative Data and Formulae*, John Wiley & Sons, New York, NY, 2nd edn.

Yadid-Pecht, O. and Etienne-Cummings, R. (eds) (2004) *CMOS Imagers: From Phototransduction to Image Processing*, Kluwer Academic, Dordrecht.

Yan, P. and Bowyer, K.W. (2007) A fast algorithm for ICP-based 3D shape biometrics. *Computer Vision and Image Understanding*, **107** (3), 195–202.

Young, I.T. and van Vliet, L.J. (1995) Recursive implementation of the Gaussian filter. *Signal Processing*, **44** (2), 139–151.

Zhang, L., Curless, B., and Seitz, S.M. (2003) Spacetime stereo: Shape recovery for dynamic scenes, in *Computer Vision and Pattern Recognition*, vol. 2, pp. 367–374.

Zhang, S. (2010) Recent progresses on real-time 3D shape measurement using digital fringe projection techniques. *Optics and Lasers in Engineering*, **48** (2), 149–158.

Zhang, Z. (2000) A flexible new technique for camera calibration. *IEEE Transactions on Pattern Analysis and Machine Intelligence*, **22** (11), 1330–1334.

Zhuang, H. (1998) Hand/eye calibration for electronic assembly robots. *IEEE Transactions on Robotics and Automation*, **14** (4), 612–616.

Index/索引

3D image acquisition device/三维图像采集设备, 134–156, 477, 511, 514, 524
    sheet of light sensor/激光三角测量法 (片光) 传感器, 139–142
    stereo sensor/立体视觉传感器, 135–138
    structured light sensor/结构光传感器, 142–150
3D object recognition/三维物体识别, 476–526
    3D data/三维数据, 478, 510–526
    3D object/三维物体, 477, 478, 493–526
    deformable matching/变形匹配, 477–493
        3D pose/三维位姿, 486–489
        calibrated/标定的, 486–489
        circular structure/圆形结构, 481–483
        clustering/聚类, 481–482
        deformable object/可变形物体, 488–493
        hierarchical search/分层搜索, 484–485
        linear structure/线性结构, 481–483
        model generation/生成模板, 481–482
        model parts/模板 parts, 481–482
        pose refinement/位姿优化, 485–486
        principle/原理, 479–481
        similarity measure/相似度量, 482–484
        uncalibrated/未标定的, 478, 485–486
    deformable object/可变形物体, 477, 488–493
        deformation model/形变模型, 488–492
    depth data/深度数据, 510–526
    image data/图像数据, 478–510
    planar object/平面物体, 477–489
        calibrated/标定的, 478, 486–489
        uncalibrated/未标定的, 478, 485–486
    shape-based 3D matching/基于形状的三维匹配, 478, 493–510
        2D matching pose/二维匹配位姿, 496–498
        2D model generation/生成二维模板, 503–506
        3D pose/三维位姿, 496–498
        accuracy/准确度, 498–499, 508–509
        applications/应用, 509–510
        CAD model/CAD 模型, 503–506
        degrees of freedom/自由度, 495, 497–498
        examples/案例, 509–510
        hierarchical model/分层模型, 500–503
        hierarchical search/分层搜索, 502–503
        image pyramid/图像金字塔, 501–502, 504, 507–508
        minimum face angle/最小面角, 506
        model image/模板图像, 504–506
        multi-channel edge tensor/多通道边缘张量, 505
        perspective correction/投影校正, 506, 508
        perspective distortions/透视畸变, 498–499, 506–508
        pose range/位姿范围, 499–500
        pose refinement/位姿优化, 508–509
        principle/原理, 495–499
        robustness/鲁棒性，健壮, 498–500, 504, 506–507, 524
        run time/运行时间, 498–500, 503
        spherical coordinates/球面坐标系, 499–500
        spherical projection/球面投影, 508
        view/视图, 497–498, 500–503
        view sphere/球面视图, 495–500
        view-based approach/基于视图的方法, 495–499
        virtual camera/虚拟摄像机, 495–496, 500–503
    surface-based 3D matching/基于表面的三维匹配, 478, 510–526, 711–734
        accumulator array/累加器数组, 517–519
        accuracy/准确度, 519–522
        corresponding points/对应点, 520–522
        deformable object/可变形物体, 522–524

global model description/全局模型描述, 512–515
hash table/哈希表, 514–515, 518–519, 526
Hough transform/霍夫变换, 517–519
iterative closest point/迭代最近点, 520–522
local parameters/局部参数, 515–518
multimodal data/多模态数据, 524–526
non-maximum suppression/非极大值抑制, 519
normal vector/法向量, 514, 516
planar object/平面物体, 526
point pair/点对, 512–515
point pair feature/点对特征, 512–515, 518, 525–526
point sampling/点采样, 514
pose refinement/位姿优化, 519–522, 526
reference point/参考点, 516–519, 525
robustness/鲁棒性，健壮, 506
score/分值, 519
search scene/搜索场景, 512, 514, 518
voting/投票, 517–519, 526

3D reconstruction/三维重构, 477, 511, 514
sheet of light/片光, 49–142, 412–416
calibration/标定, 414–416
extraction of laser line/激光线提取, 413–414
stereo/立体, 48, 135–138, 390–412, 685–695
structured light/结构光, 50, 52–53, 142–150, 416–424
binary code patterns/二值编码模式, 145–146
camera calibration/摄像机标定, 420–423
fringe projection/光栅投影, 147–150
Gray code decoding/格雷码解码, 417–420
Gray code patterns/格雷码图案, 146, 150
phase decoding/相位解码, 418–420
phase shift/移相, 147–150
projector calibration/投影仪标定, 420–423
radiometric calibration/辐射标定, 423–424
stripe decoding/条纹解码, 416–420

a

a posteriori probability/后验概率, 561
a priori probability/先验概率, 561, 566
aberrations see lens aberrations/像差
absolute phase/绝对相位, 149
absolute sensitivity threshold/绝对灵敏度阈值, 81
absolute sum of normalized dot products/归一化点积绝对总和, 471
accumulator array/累加器数组, 451, 454, 455, 518
accuracy/准确度, 317–318
camera parameters/摄像机参数, 386–390
contour moments/轮廓矩, 256
edges/边缘, 171, 320–327
gray value features/灰度值特征, 171
gray value moments/灰度值矩, 250–252, 256
hardware requirements/硬件需求, 327
region moments/区域矩, 250–252
subpixel-precise threshold/亚像素阈值, 171
achromatic lens/消色差镜, 58
active pixel sensor/主动像素传感器, 70
ADC, 63, 70, 89
affine transformation/仿射变换, 206–207, 426, 245, 457, 469, 660–669, 695–704
Airy disk/艾里斑, 40
algebraic distance/代数距离, 339
algebraic error/代数误差, 339
aliasing/混淆, 65, 67, 70, 201–202, 216, 434
alignment/对准, 206, 228, 229
amplifier noise/放大器噪声, 79
analog-to-digital converter see camera, ADC/模数转换器
Ando filter/Ando 滤波器, 306
anisometry/各向异性, 244, 250, 254, 555, 617–625
anti-bloom drain/溢流沟道, 68
anti-extensive operation/非外延性运算
erosion/腐蚀, 266, 284
Minkowski subtraction/闵可夫斯基减法, 265, 284
opening/开运算, 272, 285
antisymmetric matrix/反对称矩阵, 542

aperture stop/孔径光阑, 33, 40, 42–43, 45–46, 48, 54–56, 58, 173
apochromatic lens/复消色差镜头, 58
APS see camera, active pixel sensor/主动像素传感器
area/面积, 241, 249, 251, 254, 255, 617–625, 649–660, 669–679
area of interest/感兴趣区域, 70, 121, 123
area sensor/面阵传感器, 64–71, 348
articulated robot/关节型机器人, 527–528
aspherical lens/非球面镜头, 54
astigmatism/像散, 55, 324
asynchronous reset/异步重置, 68, 72, 132

b

back light/背光, 19
back porch/后肩, 87
Bayes decision rule/贝叶斯决策规则, 561, 563
Bayes theorem/贝叶斯定理, 561
bilateral telecentric lens/双远心镜头, 45–47, 348, 357, 360
binary image/二值图像, 161, 236, 241, 257, 258, 262, 268, 277
binocular stereo reconstruction/双目立体重构, 48, 135–138, 390–412, 685–695
black body/黑体, 7–9
blooming/高光溢出, 67
boundary/边界, 237, 247, 269–270
bounding box/边界框, 245–246, 255, 556, 617–632, 649–660
bright-field illumination/明场照明, 19

c

CAD model/CAD 模型, 486, 494, 496, 503–506, 509, 512–513
calibration/标定
    geometric/几何, 325, 326, 347–391, 420–423, 660–669
        accuracy of interior orientation/内方位准确度, 386–390
        binocular stereo calibration/双目立体视觉标定, 393–394, 421–423, 685–695
        calibration target/标定物, 370–373, 422
        camera constant see calibration/摄像机常量
        geometric, principal distance/几何, 像主距
        camera coordinate system/摄像机坐标系, 351–352, 364–365, 392
        camera motion vector/摄像机运动向量, 363–364
        distortion coefficient (division model)/畸变系数 (除法模型), 352–354, 357–358, 366, 386–390, 392, 397
        distortion coefficients (polynomial model)/畸变系数 (多项式模型), 354–358, 392
        exterior orientation/外方位参数, 326, 373–380, 414, 685–695
        focal length see calibration, geometric/焦距
    principal distance/主距
        image coordinate system/图像坐标系, 356, 366
        image plane coordinate system/成像面坐标系, 352, 365–366
        image plane distance/成像面距离, 360–362, 392
        interior orientation/内方位参数, 137, 140, 143, 326, 373–380, 391, 414, 685–695
        magnification/放大率, 352, 357
        pixel size/像素尺寸, 356, 366, 392
        principal distance/主距, 26, 36, 349–350, 352, 357, 365–366, 386–390, 392
        principal point/主点, 356, 366, 386–390, 392
        projection center/投影中心, 26, 35, 137, 349, 392, 393, 396, 400, 421
        relative orientation/相对位姿, 137, 143, 391, 392, 423
        world coordinate system/世界坐标系, 351–352, 365
    hand-eye see hand-eye calibration/手眼标定
    radiometric/辐射, 170–180, 324, 423–424
        calibration target/标定物, 171
        chart-based/基于图表, 171–173
        chartless/无图表, 173–180

defining equation for chartless calibration/无图表标定的定义公式, 174
discretization of inverse response function/逆响应函数的离散化, 175–176
gamma response function/伽马响应函数, 171, 179
inverse response function/逆响应函数, 174
normalization of inverse response function/逆响应函数的标准化, 177
polynomial inverse response function/逆响应函数多项式, 178
response function/响应函数, 171, 174, 179
smoothness constraint/平滑约束, 177–178

camera/摄像机, 158
absolute sensitivity threshold/绝对灵敏度阈值, 81
active pixel sensor/主动像素传感器, 70
ADC/模数转换器, 63, 70, 89
analog-to-digital converter see camera/模数转换器

ADC
area scan/面扫描, 64–71
asynchronous reset/异步复位, 68, 72, 132
calibration/标定
image rectification/图像校正, 385
world coordinates from single image/通过单幅图像获取世界坐标, 414–415, 326–327, 380–385, 486–487
world coordinates from stereo reconstruction/通过立体重构获取世界坐标, 390–412
CCD, 62–69
anti-bloom drain/溢流沟道, 68
blooming/高光溢出, 67
frame transfer sensor/帧转移图像传感器, 65–66
full frame sensor/全帧图像传感器, 64–65
interlaced scan/隔行扫描, 69, 86
interline transfer sensor/行间转移图像传感器, 66–68
lateral overflow drain/侧溢流沟道, 68
line sensor/线阵传感器, 62–64
progressive scan/逐行扫描, 69, 89
vertical overflow drain/垂直溢流沟道, 68

CMOS, 69–72
area of interest 感兴趣区域, 70
global shutter/全局快门, 71, 136
line sensor/线阵传感器, 71
rolling shutter/卷帘快门，行曝光, 71
color/彩色, 73–75
single-chip/单芯片, 73–74
three-chip/三芯片, 74–75
color filter array/彩色滤镜阵列, 73
Bayer, 74
demosaicking/去马赛克 (颜色插值), 74
configuration/架构, 95, 97, 99, 100, 102, 104, 110
GenICam, 95, 99, 102, 110, 116–126, 130
GenICam CLProtocol, 95, 118
GenICam GenApi, 99, 102, 110, 116–126, 130
GenICam GenCP, 99, 110, 116–126
GenICam PFNC see camera, configuration, GenICam pixel format naming convention/GenICam PFNC
GenICam pixel format naming convention/GenICam 像素格式命名约定, 115, 116–126
GenICam SFNC see camera, configuration, GenICam standard features naming convention/GenICam SFNC
GenICam standard features naming convention/GenICam SFNC, 99, 102, 110, 114, 116–126
IIDC, 104
control see camera, configuration/控制
dark signal nonuniformity/暗信号不均匀, 82
digital pixel sensor/数字像素传感器, 71
dynamic range/动态范围, 81
exposure time/曝光时间, 39, 64, 68, 173
fill factor/填充因子, 65, 67, 70, 72, 253, 323, 327
gamma response function/伽马响应函数, 171, 179
gray value response/灰度值响应, 170–171, 423–424
linear/线性, 170, 324, 327, 423
nonlinear/非线性, 171, 324, 423

inverse response function/逆响应函数, 174
line scan/线扫描, 62–64, 71, 348, 363–370, 384–385
    encoder/编码器, 64
noise/噪声
    amplifier noise/放大器噪声, 79
    dark current noise/暗电流噪声, 79
    dark noise/暗噪声, 79
    noise floor/本底噪声, 79
    overall system gain/总系统增益, 80
    pattern noise/模式噪声, 82
    photon noise/光子噪声, 78
    quantization noise/量化噪声, 80
    reset noise/复位噪声, 79
    signal-to-noise ratio/信噪比, 80, 149, 152, 156
    spatial noise/空间噪声, 82
    temporal noise/随机噪声, 80
performance/性能, 77–84
perspective/透视, 349–357, 382–384
photoresponse nonuniformity/光响应不均匀性, 82
pinhole/针孔, 26, 35–37, 350
quantum efficiency/量子效率, 78, 79
response function/响应函数, 171, 174, 179
saturation capacity/饱和量, 81
sensor size/传感器尺寸, 75–77
spectral response/光谱响应, 72–73, 75
telecentric/远心, 348–357, 381–382, 660–669
time-of-flight/飞行时间, 151–156
    continuous-wave-modulated/连续波调制信号, 151–153
    distance computation/距离计算, 153, 156
    distance range/距离范围, 152, 154
    phase demodulation/相位解调, 152
    pulse-modulated/脉冲调制, 153–156
    random errors/随机误差, 153, 156
    resolution/分辨率, 153, 156
    scene intensity/场景灰度值, 152
    systematic errors/系统误差, 153, 156
    time-of-flight computation/飞行时间计算, 154–156
    trigger/触发, 64, 68, 132–133
Camera Link, 93–96
Camera Link HS, 96–99
Canny filter/Canny 滤波, 300–301, 308, 319, 632–638, 669–679
    edge accuracy/边缘准确度, 320, 322
    edge precision/边缘精度, 319
cardinal elements/基本要素, 31
CCD see camera, CCD
CCIR, 85
center of gravity/重心, 242–244, 250–251, 255, 330, 332, 414, 617–625, 660–669
central moments/中心矩, 242–244, 250, 255, 330
CFA see camera, color filter array/彩色滤镜阵列
chamfer-3-4 distance/chamfer 3-4 距离, 280
characteristic function/特征函数, 161, 249, 250, 282
charge-coupled device see camera, CCD/电气耦合器件
chessboard distance/棋盘距离, 279, 280
chief ray see principal ray/主光线
chromatic aberration/色差, 57–58, 324
circle fitting/圆拟合, 336–338, 617–625, 669–685
    outlier suppression/抑制离群值, 337
    robust/鲁棒性, 337
circle of confusion/弥散圆, 37, 54
city-block distance/4-连通距离, 279, 280
classification/分类, 560–607, 617–625, 704–710
    a posteriori probability/后验概率, 561
    a priori probability/先验概率, 561, 566
    Bayes classifier/贝叶斯分类器, 566–572
    classifier types/分类器类型, 563–564
    curse of dimensionality/维度灾难, 567, 588
    decision theory/决策理论, 560–563
        Bayes decision rule/贝叶斯决策规则, 561, 563
        Bayes theorem/贝叶斯定理, 561
    error rate/误差率, 564–565
    expectation maximization algorithm/最大期望值算法, 570
    features/特征, 555, 560, 617–625
    Gaussian mixture model classifier/高斯混合模型分类器, 569–572
        k-probability/k-最近邻域分类器, 571–572
        novelty detection/异常检测, 571–572

generalized linear classifier/广义线性分类器, 587
hyper-parameter/超参数, 564, 568
k-nearest-neighbor classifier/k-最近邻域分类器, 567–568
    novelty detection/异常检测, 568–569
linear classifier/线性分类器, 573–576
neural network/神经网络, 574–587, 617–625, 704–710
    convolutional neural network/卷积神经网络, 598–606
    cross-entropy error/交叉熵误差, 580, 603
    dataset augmentation/数据增强, 605
    evidence procedure/证明过程, 584
    hyperbolic tangent activation function/双曲线正切激活函数, 578
    logistic activation function/逻辑激活函数, 578
    multilayer perceptron/多层感知器, 576–587, 597, 617–625, 704–710
    multilayer perceptron training/多层感知器训练, 579–586
    novelty detection/异常检测, 584–586, 606
    rectified linear unit activation function/修正线性单元激活函数, 600
    regularization/正则化, 583–584, 604–605
    sigmoid activation function/sigmoid 激活函数, 578–579
    single-layer perceptron/单层感知器, 574–576
    softmax activation function/softmax 激活函数, 578, 600
    threshold activation function/阈值激活函数, 575, 577
    universal approximator/泛逼近器, 578, 579, 597, 600
    weight decay/权重衰减, 583–584, 604
nonlinear classifier/非线性分类器, 576–606
novelty detection/异常检测, 565–566
    Gaussian mixture model classifier/高斯混合模型分类器, 571–572
    k-nearest-neighbor classifier/k-最近邻域分类器, 568–569
polynomial classifier/多项式分类器, 587
rejection/拒绝类, 565–566
support vector machine/支持向量机, 587–596
    Gaussian radial basis function kernel/高斯径向基函数核, 592
    homogeneous polynomial kernel/齐次多项式核, 592
    inhomogeneous polynomial kernel/非齐次多项式核, 592
    kernel/核, 591
    margin/边缘, 589
    margin errors/边缘误差, 592, 596
    novelty detection/异常检测, 594–596
    ν-SVM/支持向量机, 592–593
    one-versus-all/一对所有, 593–594
    one-versus-one/一对一, 593–594
    separating hyperplane/分离超平面, 588–591
    sigmoid kernel/sigmoid 核, 592
    universal approximator/泛逼近器, 592
test set/测试集, 564
training set/训练集, 564, 580
validation set/验证集, 564–565, 568, 583, 584
closing/闭运算, 274–276, 285–287, 649–660
clutter/混乱, 杂乱, 443, 448, 456, 468
CMOS see camera, CMOS
CNN see convolutional neural network
CoaXPress, 99–102
color filter/彩色滤镜, 16, 75
color filter array see camera, color filter array/彩色滤镜阵列
color temperature see correlated color/色温 temperature/温度
coma/彗差, 54, 324
compactness/紧凑度, 247, 555, 617–625
complement/补集, 258–259
complementary metal-oxide-semiconductor see camera, CMOS
completeness checking/完整性检测, 1
component labeling/区域标记, 236
composite video/复合视频, 88
connected components/连通区域, 233–237, 413, 552, 617–625, 638–660

connectivity/连通性, 234–236, 259, 269–270, 278
contour/轮廓, 164–165
contour feature see features, contour/轮廓特征
contour length/轮廓长度, 247
contour segmentation/轮廓分割, 341–347, 444, 626–632, 679–685
    lines/直线, 341–344
    lines and circles/直线和圆, 344–347, 626–632, 679–685
    lines and ellipses/直线和椭圆, 344–347
contrast enhancement/对比度增强, 166–170
contrast normalization/对比度归一化, 167–170
    robust/鲁棒性, 168–170, 557, 601
convex hull/凸包, 246–247, 255
convexity/凸性, 246–247, 617–625
convolution/卷积, 188–189, 199–200, 598
    kernel/卷积核, 188
convolutional neural network/卷积神经网络, 598–606
    convolution/卷积, 598–599
    convolutional layer/卷积层, 601
    filter stride/滤波器步幅, 599, 601
    fine-tuning/参数调整, 602–603
    learning rate/学习率, 604
    minibatch/小批量, 603
    momentum/动量, 604
    novelty detection/异常检测, 606
    pooling/池化, 600–601
    receptive filed/感受野, 600
    rectified linear unit/修正线形单元, 600
    softmax activation function/softmax 激活函数, 600
    stochastic gradient descent/随机梯度下降, 603–604
    training/训练
        cross-entropy error/交叉熵误差, 603
        dataset augmentation/数据增强, 605
        regularization/正则化, 604–605
        weight decay/权重衰减, 604
    universal approximator/泛逼近器, 600
coordinates/坐标系
    homogeneous/齐次, 207, 208
    inhomogeneous/非齐次, 207, 208
    polar/极, 218

correlated color temperature/相关色温, 9, 10
correlation/相关系数, 200–201, 598, 608–616
see also normalized cross-correlation/归一化相关系数
cross-entropy error/交叉熵误差, 580, 603
cumulative histogram/累加直方图, 169, 249
curvature of field/场曲率, 55–56
cylindrical lens/柱面透镜, 139

d

dark current noise/暗电流噪声, 79
dark noise/暗噪声, 79
dark signal nonuniformity/暗信号非均匀性, 82
dark-field illumination/暗场照明, 19
data structures/数据结构
    images/图像, 158–159
    regions/区域, 160–164
    subpixel-precise contours/亚像素精度轮廓, 164–165
dataset augmentation/数据增强, 605
datum deficiency/数据不足, 543
DCS see distributed control system/分布式控制系统
decision theory/决策理论, 560–563
    a posteriori probability/后验概率, 561
    a priori probability/先验概率, 561, 566
    Bayes decision rule/贝叶斯决策规则, 561, 563
    Bayes theorem/贝叶斯原理, 561
deep learning/深度学习, 597
deep neural network/深度神经网络, 597
deformable matching see 3D object recognition, deformable matching/变形匹配
demosaicking/去马赛克 (颜色插值), 74
depth of field/景深, 37–41, 44, 45, 48–53
depth-first search/深度优先搜索, 235
Deriche filter/Deriche 滤波器, 301–302, 309, 319
    edge accuracy/边缘准确性, 320, 324
    edge precision/边缘精度, 319
derivative/导数
    directional/方向导数, 292
    first/一阶导数, 290, 295
    gradient/梯度, 292
    Laplacian/拉普拉斯算子, 293

partial/偏导数, 292, 293, 305, 308
second/二阶导数, 291, 295
DFT see discrete Fourier transform/不连续傅里叶变换
DHCP see Dynamic Host Configuration/动态主机配置 Protocol/协议
diaphragm/光圈, 33, 38, 54
difference/差异, 258, 649–660, 669–685
diffraction/衍射, 40
diffuse bright-field back light illumination/明场背光漫射照明, 23–24
diffuse bright-field front light illumination/明场漫射正面照明, 19–21
diffuse illumination/漫射照明, 18
digital input/output/数字输入/输出, 4
digital light processing/数字光处理, 144
digital micromirror device/数字微镜器件, 143–145
    diamond pixel array layout/菱形像素阵列布局, 144–145, 147, 419
    regular pixel array layout/规则像素阵列布局, 144–145
digital pixel sensor/数字像素传感器, 71
digital signal processor/数字信号处理器, 3
dilation/膨胀, 261–264, 268–270, 282–283, 314, 632–638, 649–660, 679–685
dimensional inspection/尺寸检测, 2
direct memory access/直接内存读取 (DMA), 89, 132
directed bright-field front light illumination/明场直接正面照明, 21
directed dark-field front light illumination/暗场直接正面照明, 22
directed illumination/直接照明, 18
discrete Fourier transform/不连续傅里叶变换, 201–205
    see also Fourier transform/傅里叶变换
disparity/视差, 402–404
dispersion/色散, 28
distance/距离
    chamfer-3-4, 280
    chessboard/棋盘距离, 279, 280
    city-block/4-连通距离, 279, 280

    Euclidean/欧几里得距离, 279, 280, 669–679
distance transform/距离变换, 278–281, 445, 448, 669–679
distortion/畸变, 56–57, 325, 352–358, 366, 495
    barrel/桶形畸变, 56–57, 352–356, 366
    division model/除法模型, 352–354, 357–358, 423
    pincushion/枕形畸变, 56–57, 352–356, 366
    polynomial model/多项式模型, 354–358, 423
distributed control system/分布式控制系统, 4
DLP see digital light processing/数字光处理器 (DLP)
DMA see direct memory access/直接内存读取 (DMA)
DMD see digital micromirror device/数字微镜器件 (DMD)
DSNU see dark signal nonuniformity/暗信号不均匀
DSP see digital signal processor/数字信号处理 (DSP)
dual number/对偶数, 537–538
    dual part/对偶部, 537
    dual unit/对偶单位, 537
    Plücker coordinates/Plücker 坐标, 537
    real part/实部, 537
dual quaternion/对偶四元数, 533, 537–540
    advantages/特点, 539
    ambiguity/歧义, 539
    conjugation/共轭, 538
    dual part/对偶部, 538
    inversion/反演变换, 538
    line transformation/线性变换, 539
    multiplication/乘法, 538, 539
    overparameterization/过参数化, 539
    Plücker coordinates/Plucker 坐标, 538
    pure/纯, 538
    real part/实部, 538
    scalar part/标量部分, 540, 549
    screw/螺旋, 538–540
    unit/单位, 538–540, 543
    vector part/向量部分, 541
dual vector/对偶向量, 537, 538
duality/二元性
    dilation–erosion/膨胀-腐蚀, 268, 284

hit-or-miss transform/击中 -击不中变换, 271
opening–closing/开 -闭运算, 274, 285
Dynamic Host Configuration Protocol/动态主机配置 (DHCP), 113
dynamic range/动态范围, 81
dynamic thresholding/动态阈值法, 224–228, 287, 417, 552, 617–625, 649–660

e

edge/边缘
 amplitude/振幅, 233, 293, 307, 505, 506
 definition/定义
  1D/一维, 290–291
  2D/二维, 292–295
 gradient magnitude/梯度幅度, 233, 293, 307
 gradient vector/梯度向量, 292
 Laplacian/拉普拉斯算子, 293, 315–317
 non-maximum suppression/非最大抑制, 291, 303, 310–311
 polarity/极性, 291
edge extraction/边缘提取, 288–327, 444, 679–685
 1D/一维, 295–305, 660–669
  Canny filter/Canny 滤波器, 300–301, 319
  Deriche filter/Deriche 滤波器, 301–302, 319
  derivative/导数, 290, 291, 295–296
  gray value profile/灰度值轮廓, 297–298, 660–669
  non-maximum suppression/非最大抑制, 303
  subpixel-accurate/亚像素精度, 303–304
 2D/二维, 233, 305–317, 632–638, 669–695, 704–710
  Ando filter/Ando 滤波器, 306
  Canny filter/Canny 滤波器, 308
  Deriche filter/Deriche 滤波器, 309
  Frei filter/Frei 滤波器, 306
  gradient/梯度, 292
  hysteresis thresholding/滞后阈值, 311–313
  Lanser filter/Lanser 滤波器, 309, 319
  Laplacian/拉普拉斯算子, 293, 315–317
  non-maximum suppression/非最大抑制, 310–311
  Prewitt filter/Prewitt 滤波器, 306
  Sobel filter/Sobel 滤波器, 306, 559
  subpixel-accurate/亚像素精度, 313–317
edge filter/边缘滤波器, 233, 300–302
 Ando, 306
 Canny, 300–301, 308, 319, 632–638, 669–679
  edge accuracy/边缘准确度, 320, 322
  edge precision/边缘精度, 319
 Deriche, 301–302, 309, 319
  edge accuracy/边缘准确度, 320, 324
  edge precision/边缘精度, 319
 Frei, 306
 Lanser, 309, 319
  edge accuracy/边缘准确度, 320
  edge precision/边缘精度, 319
 optimal/最优，最佳, 300–302, 308–309
 Prewitt, 306
 Sobel, 306, 559
edge-spread function/边缘扩散函数, 58–59
EIA-170, 85
electromagnetic radiation/电磁辐射, 6–9
 black body/黑体, 7–9
 infrared/近红外, 7, 72, 151, 153
 spectrum/光谱, 6–7
 ultraviolet/紫外光, 7, 72
 visible/可见光, 6
ellipse fitting/椭圆拟合, 338–340
 algebraic error/代数误差, 339
 geometric error/几何误差, 339
 outlier suppression/抑制离群值, 339–340
 robust/鲁棒性, 339–340
ellipse parameters/椭圆参数, 242–246, 250, 254, 256, 330, 332, 339
enclosing circle/外接圆, 245–246, 255
enclosing rectangl/外接矩形 e, 245–246, 255–556, 617–632, 649–660
encoder/编码器, 64
entocentric lens/近心镜头 (普通透视投影镜头), 42, 46–47, 348, 357, 360–362
entrance pupil/入射光瞳, 33, 35, 43, 137, 349, 421
epipolar image rectification/极线图像校正, 399–402, 685–695
epipolar line/核线, 395
epipolar plane/核面, 395

epipolar standard geometry/极标准几何, 398–399, 685–695
epipole/核点, 396
erosion/腐蚀, 265–270, 283–284, 649–660, 679–685
Ethernet/以太网, 111–116
Euclidean distance/欧氏距离, 279, 280, 669–679
evidence procedure/证明过程, 584
exit pupil/出瞳, 34, 35, 44, 45, 360, 361, 421
exposure time/曝光时间, 39, 64, 68, 173
extensible markup language see XML/可扩展标志语言 (XML)
extensive operation/粗放算子
    closing/闭运算, 275, 285
    dilation/膨胀, 262, 283
    Minkowski addition/闵可夫斯基加法, 260, 283
exterior orientation/外方位参数, 326, 351–352, 365, 373–380, 414, 422, 530, 660–669, 685–695
    world coordinate system/世界坐标系, 351–352, 365

f

f-number/光圈值, 38, 44, 48, 54–56, 58, 174
facet model/facet 模型, 313
fast Fourier transform/快速傅里叶变换, 203, 608–616
    see also Fourier transform/傅里叶变换
feature extraction/特征提取, 240–256, 638–649
features/特征
    contour/轮廓, 240, 254–256, 626–632, 638–649
        area/面积, 255
        center of gravity/重心, 255
        central moments/中心矩, 255
        contour length/轮廓线长度, 254, 626–632
        ellipse parameters/椭圆参数, 256
        major axis/主轴, 256
        minor axis/短轴, 256
        moments/矩, 255
        normalized moments/归一化的矩, 155, 255
        orientation/方向, 256
        smallest enclosing circle/最小外接圆, 255
        smallest enclosing rectangle/最小外接矩形, 255, 626–632
    gray value/灰度值, 240, 248–254, 704–710
        α-quantile/α-分位数, 249
        anisometry/各向异性, 250, 254
        area/面积, 254
        center of gravity/重心, 250, 414
        central moments/中心矩, 250
        ellipse parameters/椭圆参数, 250
        major axis/主轴, 254
        maximum/最大值, 168
        mean/平均值, 414
        median/中值, 249
        minimum/最小值, 168, 248
        minor axis/短轴, 250, 254
        moments/矩, 249–254
        orientation/方位, 250
        standard deviation/标准差, 248
        variance/方差, 248
    region/区域, 240–247, 617–625, 638–660, 669–679, 711–722
        anisometry/各向异性, 244, 555, 617–625
        area/面积, 241, 251, 617–625, 649–660, 669–679
        center of gravity/重心, 242–244, 251, 617–625
        central moments/中心矩, 242–244
        compactness/紧凑度, 247, 555, 617–625
        contour length/轮廓线长, 247
        convexity/凸包, 246–247, 617–625
        ellipse parameters/椭圆参数, 242–246
        major axis/主轴, 242–244
        minor axis/短轴, 242–244
        moments/矩, 242–245, 617–625
        normalized moments/归一化的矩, 242
        orientation/方位, 242–246
        smallest enclosing circle/最小外接圆, 245–246
        smallest enclosing rectangle/最小外接矩形, 245–246, 556, 617–625, 649–660
FFT see fast Fourier transform/快速傅里叶变换
field angle/视场角, 35

field-programmable gate array/现场可编程门阵列 (FPGA), 3
fieldbus/现场总线, 4
fill factor/填充因子, 65, 67, 70, 72, 253, 323, 327
filter/滤波器
    anisotropic/各向异性, 191, 309
    border treatment/边界处理, 185–186
    convolution/卷积, 188–189, 199–200, 598
        kernel/卷积核, 188
    definition/定义, 188
    edge/边缘, 300–302
        Ando, 306
        Canny, 300–301, 308, 319, 632–638, 669–679
        Deriche, 301–302, 309, 319
        Frei, 306
        Lanser, 309, 319
        optimal/最优, 300–302, 308–309
        Prewitt, 306
        Sobel, 306, 559
    Gaussian/高斯, 191–195, 216, 223, 225, 226, 301, 308, 435
        frequency response/频率响应, 194, 200
    isotropic/各向同性, 192, 193, 309
    linear/线性, 188–189, 296
    mask/掩码, 掩模
    maximum see morphology, gray value, dilation/最大值
    mean/均值, 183–188, 195, 216, 225, 226, 297, 435, 617–625, 649–660
        frequency response/频率响应, 190–191, 200, 436
    median/中值, 196–198, 225, 226, 649–660
    minimum see morphology, gray value, erosion/最小值
    nonlinear/非线性, 196–198, 284
    optical/光学的
        anti-aliasing/抗锯齿, 74
        color/颜色, 16, 75
        infrared cut/红外截止, 16, 75
        infrared pass/红外通过, 16, 73
        polarizing/偏振, 17
    rank/排列, 198, 284
    recursive/递归, 187, 189, 193, 285
    runtime complexity/时间复杂度, 186–187
    separable/可分离的, 187, 189, 193
    smoothing/平滑, 181–198
        optimal/最优, 191–193, 301, 308
    spatial averaging/空间平均, 183–187
    temporal averaging/时间平均, 182–183, 230
FireWire see IEEE 1394
fitting/拟合
    circles/圆, 336–338, 617–625, 669–685
        outlier suppression/抑制离群值, 337
        robust/鲁棒性, 337
    ellipses/椭圆, 338–340
        algebraic error/代数误差, 339
        geometric error/几何误差, 339
        outlier suppression/抑制离群值, 339–340
        robust/鲁棒性, 339–340
    lines/直线, 329–336, 626–632
        outlier suppression/抑制离群值, 331–336
        robust/鲁棒性, 331–336
fluorescent lamp/荧光灯, 10–11
focal length/焦距, 30
see also principal distance
focal point/焦点, 29
focusing plane/焦平面, 37, 48, 50
Fourier transform/傅里叶变换, 190, 198–205, 608–616
    1D/一维, 198
        inverse/逆, 199
    2D/二维, 199, 608–616
        inverse/逆, 199, 608–616
    continuous/连续, 198–201
    convolution/卷积, 199–200
    discrete/离散, 201–205
        inverse/逆, 202
    fast/快速, 203, 608–616
    frequency domain/频域, 198
    Nyquist frequency/奈奎斯特频率, 201
    real-valued/实值, 203, 608–616
    spatial domain/空域, 198
    texture removal/纹理去除, 204, 205
FPGA see field-programmable gate array
frame grabber/图像采集卡, 84
    analog/模拟的, 89–92

line jitter/行抖动 (像素抖动), 90–92, 320
pixel clock/像素时钟, 89, 92
frame transfer sensor/帧转移传感器, 65–66
Frei filter/Frei 滤波器, 306
frequency domain/频域, 198
fringe projection/光栅投影, 147–150, 418–420
front light/正面光, 18
front porch/前肩, 87
full frame sensor/全帧转移传感器, 64–65
fuzzy membership/模糊类属度, 250–254
fuzzy set/模糊集, 250–254

g

gamma response function/伽马响应函数, 171, 179
gauge freedom/规范自由度, 543
Gaussian filter/高斯滤波器, 191–195, 216, 223, 225–226, 301, 308, 435
  frequency response/频率响应, 194, 200
Gaussian mixture model/高斯混合模型, 569–572
  k-probability/k-最近邻域分类器, 571–572
  novelty detection/异常检测, 571–572
Gaussian optics/高斯光学, 27–46
GenApi see GenICam, GenApi
GenCP see GenICam, GenCP
generalized Hough transform/广义霍夫变换, 450–456, 517
  accumulator array/累加器数组, 451, 454, 455
  R-table, 454
GenICam, 95, 99, 102, 110, 116–126, 130
  CLProtocol, 95, 118
  GenApi, 99, 102, 110, 116–126, 130
  GenCP, 99, 110, 116–126
  GenTL, 102, 126–131
    GenTL consumer, 127
    GenTL producer, 127
    SFNC see GenICam, GenTL, standard features naming convention/特征命名约定
    standard features naming convention/标准特征命名约定, 126–131
PFNC see GenICam, pixel format naming convention/像素格式命名约定
  pixel format naming convention/像素格式命名约定, 115–126

SFNC see GenICam, standard features naming convention/标准特征命名约定
  standard features naming convention/标准特征命名约定, 99, 102, 110, 114, 116–126
  transport layer/传输层, 119, 126–131
GenTL see GenICam, GenTL
geometric camera calibration/几何摄像机标定, 325, 326, 347–391, 420–423, 660–669
  binocular stereo calibration/双目立体视觉标定, 393–394, 421–423, 685–695
  calibration target/标定物, 370–373, 422
  exterior orientation/外方位参数, 326, 351–352, 365, 373–380, 414, 422, 660–669, 685–695
    world coordinate system/世界坐标系, 351–352, 365
  interior orientation/内方位参数, 137, 140, 143, 326, 349–357, 373–380, 391, 414, 422–423, 486, 495, 660–669, 685–695
    accuracy/准确度, 386–390
    camera constant see geometric camera/摄像机常量
    calibration, interior orientation, principal distance
    camera coordinate system/摄像机坐标系, 351–352, 364–365, 392
    camera motion vector/摄像机运动向量, 363–364
    distortion coefficient (division model)/畸变系数 (除法模型), 352–354, 357–358, 366, 386–390, 392, 397
    distortion coefficients (polynomial model)/畸变系数 (多项式模型), 354–358, 392
    focal length see geometric camera calibration, interior orientation, principal distance/焦距
    image coordinate system/图像坐标系, 356, 366
    image plane coordinate system/成像面坐标系, 352, 365–366
    image plane distance/成像面距离, 360–362, 392

magnification/放大率, 352, 357
pixel size/像素尺寸, 356, 366, 392
principal distance/主距, 26, 36, 349–350, 352, 357, 365–366, 386–390, 392
principal point/主点, 356, 366, 386–390, 392
projection center/投影中心, 26, 35, 137, 349, 392–393, 396, 400, 421
tilt angle/倾角, 358–362, 392
tilt axis angle/倾斜轴角度, 358–362, 392
relative orientation/相对位姿, 137, 143, 391, 392, 423
 base/基, 137, 393
 base line/基线, 396
geometric error/几何误差, 339
geometric hashing/几何哈希, 457–460
geometric matching/几何匹配, 456–475
GEV see GigE Vision
Gigabit Ethernet see GigE Vision/千兆以太网
GigE Vision, 113–116
 control channel/控制信道, 114
 GigE Vision Control Protocol/GigE Vision 控制协议, 114
 GigE Vision Streaming Protocol/GigE Vision 数据流协议, 115
 GVCP see GigE Vision, GigE Vision GVCP Control Protocol/控制协议
 GVSP see GigE Vision, GigE Vision Streaming Protocol/数据流协议
  message channel/消息信道, 114
  stream channel/数据流信道, 114
global shutter/全局快门, 71, 136
GMM see Gaussian mixture model/高斯混合模型 (GMM)
gradient/梯度, 292
 amplitude/振幅, 233, 293, 307
 angle/角度, 293, 453
 direction/方向, 293, 450, 453
 length/长度, 292
 magnitude/量级, 233, 293, 307
 morphological/形态学的, 287–288
Gray code/格雷码, 146, 150, 417–420
gray value/灰度值, 158
 1D histogram/一维直方图, 168–170, 221–224, 249
  cumulative/累积的, 169, 249
  maximum/最大值, 221–224
  minimum/最小值, 221–224
  peak/峰值, 221–224
 2D histogram/二维直方图, 176
 α-quantile/α-分位数, 249
 camera response/摄像机响应, 170–171, 423–424
  linear/线性, 170, 324, 327, 423
  nonlinear/非线性, 171, 324, 423
 feature see features, gray value/特征
 maximum/最大值, 168, 248
 mean/均值, 248, 414
 median/中值, 249
 minimum/最小值, 168, 248
 normalization/归一化, 167–170
  robust/鲁棒性, 168–170, 557
 profile/轮廓, 297
 robust normalization/鲁棒性归一化, 557, 601
 scaling/缩放, 166
 standard deviation/标准偏差, 248
 transformation/变换, 166–170, 248, 253
 variance/方差, 248
GVCP see GigE Vision, GigE Vision Control Protocol
GVSP see GigE Vision, GigE Vision Streaming Protocol

h

Hamming distance/汉明 (Hamming) 距离, 146
hand–eye calibration/手眼标定, 526–551, 722–734
 algebraic error/代数误差, 546
 articulated robot/多关节型机器人, 540–546, 722–734
  linear/线性, 532–533, 540–546
  nonlinear/非线性, 533, 545–546
 base coordinate system/基坐标系, 528–529
 calibration object/标定物, 529–532
 camera coordinate system/摄像机坐标系, 528–529
 coordinate systems/坐标系, 528–529
 input poses/输入位姿, 544–545

moving camera/移动（运动）摄像机, 528, 529, 531
poses/位姿, 530–531
practical advice/实用性建议, 544, 549
requirements/需求, 541
robot pose/机器人姿态, 529
SCARA robot/水平关节型机器人, 547–551
    ambiguity/歧义, 548, 550–551
    linear/线性, 548–549
    nonlinear/非线性, 550
screw congruence theorem/螺旋同余定理, 540–541, 545
stationary camera/固定摄像机, 528–529, 531
tool coordinate system/工具坐标系, 528–529
transformations/变换, 530–531
unity constraint/一致性约束, 543, 548
world coordinate system/世界坐标系, 528–529

Hausdorff distance/Hausdorff 距离, 447–450
Hessian normal form/Hessian 范式, 329
hinge line/铰合线, 51
histogram/直方图
    1D/一维, 168–170, 221–224, 249
        cumulative/累积, 169, 249
        maximum/最大值, 221–224
        minimum/最小值, 221–224
        peak/峰值, 221–224
    2D/二维, 176
hit-or-miss opening/击中-击不中开操作, 273
hit-or-miss transform/击中-击不中变换, 270–271, 277
homogeneous coordinates/齐次坐标系, 207, 208
horizontal blanking interval/行消隐时间, 87
horizontal synchronization pulse/行同步脉冲, 87
Huber weight function/Huber 权函数, 333
hypothesize-and-test paradigm/假设与测试范例, 456
hysteresis Thresholding/滞后阈值, 311–313

i

ICP see iterative closest point
IDE see integrated development environment
idempotent operation/幂等运算
    closing/闭运算, 274, 285
    opening/开运算, 272, 285
identification/识别, 1
IEEE 1394, 102–105
    asynchronous data transfer/异步数据传输, 104
    IIDC, 103–105
    isochronous data transfer/同步数据传输, 104
IIDC, 103–105
illumination/照明, 6–25
    back light/背光, 19
    bright-field/明场, 19
    dark-field/暗场, 19
    diffuse/漫射, 18
    diffuse bright-field back light illumination/明场背光漫射照明, 23–24
    diffuse bright-field front light illumination/暗场漫射正面照明, 19–21
    directed/直接照明, 18
    directed bright-field front light illumination/明场直接正面照明, 21
    directed dark-field front light illumination/暗场直接正面照明, 22
    front light/正面光, 18
    light sources/光源, 9–12
        fluorescent lamp/荧光灯, 10–11
        incandescent lamp, 9–10
    LED see illumination, light sources, light-emitting diode
        light-emitting diode/白炽灯, 11–12
        xenon lamp/氙灯, 10
    telecentric/远心, 18
    telecentric bright-field back light illumination/明场背光平行照明, 24–25
image/图像, 158–159
    binary/二值, 161, 236, 241, 257, 258, 262, 268, 277
    bit depth/位深, 159
    complement/补集, 284
    domain see region of interest/区域
    enhancement/增强, 165–198
    function/函数, 158–160
    gray value/灰度值, 158
    gray value normalization/灰度值归一化, 167–170

robust/鲁棒性, 168–170, 557, 601
gray value scaling/灰度值缩放, 166
gray value transformation/灰度值变换, 166–170
label/标记, 161, 236
multichannel/多通道, 158
noise see noise/噪声
pyramid/金字塔, 434–441, 472, 479–481, 484–485, 489–490, 500–504, 506–508
rectification/校正, 212, 216–217, 385, 399–402, 476, 478, 490, 495, 552, 685–695
RGB, 158
segmentation see segmentation/分割
single-channel/单通道, 158
smoothing/平滑, 181–198
spatial averaging/空间平均, 183–187
temporal averaging/时间平均, 182–183, 230
transformation/变换, 209–219, 228, 632–638, 695–704
image acquisition modes/图像采集模式, 131–134
asynchronous acquisition/异步采集, 132
continuous acquisition/连续采集, 134
queued acquisition/排队采集, 134
synchronous acquisition/同步采集, 131–132
triggered acquisition/触发采集, 132–133
image distance/像距, 30, 37
image plane/像平面, 26, 31, 37, 48, 347, 357, 392, 398, 400, 402
tilted/倾斜, 50–52, 136, 140, 142, 357–362
image-side telecentric lens/像方远心镜头, 46–47, 348, 357, 360
incandescent lamp/白炽灯, 9–10
increasing operation/加运算, 相加运算
closing/闭运算, 275, 285
dilation/膨胀, 262, 283
erosion/腐蚀, 266, 284
Minkowski addition/闵可夫斯基加法, 261, 283
Minkowski subtraction/闵可夫斯基减法, 265, 284
opening/开运算, 272, 285
infrared cut filter/红外截止滤光片, 16, 75
infrared pass filter/红外通过滤光片, 16, 73
inhomogeneous coordinates/非齐次坐标, 207, 208

integrated development environment/集成开发环境, 121
interior orientation/内方位参数, 137, 140, 143, 326, 349–357, 363–370, 373–380, 391, 414, 422, 423, 486, 495, 530, 660–669, 685–695
accuracy/准确度, 386–390
camera constant see interior orientation, camera constant/摄像机常量
camera coordinate system/摄像机坐标系, 351–352, 364–365, 392
camera motion vector/摄像机运动向量, 363–364
distortion coefficient (division model)/畸变系数 (除法模型), 352–354, 357–358, 366, 386–390, 392, 397
distortion coefficients (polynomial model)/畸变系数 (多项式模型), 354–358, 392
focal length see interior orientation, principal distance/焦距
image plane coordinate system/成像面坐标系, 352, 356, 365–366
image plane distance/成像面距离, 360–362, 392
magnification/放大率, 352, 357
pixel size/像素尺寸, 356–366, 392
principal distance/主距, 26, 36, 349–350, 352, 357, 365–366, 386–390
principal point/主点, 356, 366, 386–390, 392
projection center/投影中心, 26, 35, 137, 349, 392–393, 396, 400, 421
tilt angle/倾角, 358–362, 392
tilt axis angle/倾斜轴角度, 358–362, 392
interlaced scan/隔行扫描, 69, 86
interline transfer sensor/行间转移传感器, 66–68
Internet Protocol/互联网协议 (IP), 112
interpolation/插值
bicubic/双三次插值, 213–215
bilinear/双线性插值, 212–213, 237, 298, 423
nearest-neighbor/最近邻域插值, 211–212, 298
intersection/交集, 258, 649–660, 669–679
invariant moments/不变矩, 245

IP see Internet Protocol
iterative closest point/迭代最近点, 520–522
iteratively reweighted least-squares/迭代再加权最小二乘, 333

j

junction/交叉点, 165

k

kernel see convolution, kernel and support/内核 vector machine, kernel/向量机

l

label image/标记图像, 161, 236
labeling/标记, 236
Lanser filter/Lanser 滤波器, 309, 319
    edge accuracy/边缘准确度, 320
    edge precision/边缘精度, 319
Laplacian/拉普拉斯算子, 293, 315–317
laser projector/激光投影仪, 139, 140
    cylindrical lens/柱面透镜, 139
    Powell lens/Powell 透镜, 139
    raster lens/光栅透镜, 140
laser triangulation/激光三角测量, 139–142, 412–416
    calibration/标定, 414–416
    extraction of laser line/激光线提取, 413–414
lateral overflow drain/侧溢流沟道, 68
law of refraction/折射定律, 27
LCD see liquid-crystal display
LCOS see liquid crystal on silicon
LED see light-emitting diode
lens/镜头, 27–46
    achromatic/消色差, 58
    Airy disk/艾里斑, 40
    aperture stop/孔径光阑, 33, 40, 42, 43, 45, 46, 48, 54–56, 58, 173
    apochromatic/复消色差, 58
    aspherical/非球面, 54
    cardinal elements/基本要素, 31
    chief ray see lens, principal ray/主光线
    circle of confusion/弥散圆, 37, 54
    cylindrical/柱面, 139
    depth of field/景深, 37–41, 44, 45, 48–53

diaphragm/光圈, 33, 38, 54
diffraction/衍射, 40
entocentric/近心, 42, 46–47, 348, 357, 360–362
entrance pupil/入射光瞳, 33, 35, 43, 137, 349, 421
exit pupil/出射光瞳, 34–35, 44, 45, 360, 361, 421
f-number/光圈值, 38, 44, 48, 54–56, 58, 174
field angle/视场角, 35
focal length/焦距, 30
focal point/焦点, 29
focusing plane/焦平面, 37, 48, 50
image distance/像距, 30, 37
image plane/像平面, 26, 31, 37, 48, 347, 357
    tilted/倾斜, 50–52, 136, 140, 142, 357–362
magnification/放大率, 31, 40, 48, 49, 136, 142, 352
nodal point/节点, 30
numerical aperture/数值孔径, 44
object distance/物距, 30, 37
optical axis/光轴, 30, 357
perspective/透视, 47, 348
Powell lens/Powell 镜头, 139
principal plane/主平面, 30
principal ray/主光线, 34, 35
pupil magnification factor/透光因子, 35, 362
raster lens/光栅透镜, 140
sagittal focal surface/弧矢焦平面, 56
sagittal image/矢形图像, 55
Scheimpflug lens/沙姆镜头, 53, 136, 140, 142, 419
Scheimpflug optics/沙姆光学, 53
Scheimpflug principle/沙姆定律, 48–53
surface vertex/表面顶点, 30
system of lenses/透镜系统, 32
tangential focal surface/切向焦面, 56
tangential image/切向图像, 55
telecentric/远心, 42–47, 348
thick/厚, 29–32
tilt lens/倾斜镜头, 48–53, 136, 140, 142, 357–362, 419
    hinge line/铰合线, 51
    Scheimpflug line/沙姆直线, 50

vignetting/光晕, 60–61, 172
lens aberrations/像差, 29, 53–59
    astigmatism/散光, 55, 324
    chromatic aberration/色差, 57–58, 324
    coma/彗差, 54, 324
    curvature of field/场曲率, 55–56
    distortion/畸变, 56–57, 325, 352–358, 366, 495
        barrel/桶形失真, 56–57, 352–356, 366
        pincushion/枕形失真, 56–57, 352–356, 366
    spherical aberration/球面像差, 53–54
light/光, 6, 72
    absorption/吸收, 14
    polarized/偏振, 13
    reflection/反射, 12
    refraction/折射, 13, 27
    spectrum/光谱, 6–7
        black body/黑体, 7–9
light sources/光源, 9–12
    fluorescent lamp/荧光灯, 10–11
    incandescent lamp/白炽灯, 9–10
    LED see light sources, light-emitting diode
    light-emitting diode/LED, 11–12
    xenon lamp/氙灯, 10
light-emitting diode/ LED, 11–12
line/直线
    Hessian normal form/海塞正规式, 329
    Plücker coordinates/Pluker 坐标, 537, 538
line fitting/直线拟合, 329–336, 626–632
    outlier suppression/抑制离群值, 331–336
    robust/鲁棒性, 331–336
line jitter/行抖动 (像素抖动), 90–92, 320
line scan camera/线扫摄像机, 62–64, 71, 348, 363–370, 384–385
line sensor/线阵传感器, 62–64, 71, 348
Link-Local Address/本地链路地址, 113
liquid crystal on silicon/LCoS 硅基液晶, 144
liquid-crystal display/液晶显示器, 144
LLA see Link-Local Address
local deformation/局部畸变, 477
look-up table/查找表, 166, 174
low-voltage differential signaling/低电压差分信号 (LVDS), 94
LUT see look-up table/查找表

LVDS see low-voltage differential signaling/低电压查分信号

m

magnification/放大率, 31, 48, 49, 136, 142, 352
major axis/主轴, 242–244, 250, 254, 256, 330, 332
maximum filter see morphology, gray value/最大值滤波器
dilation/膨胀
maximum likelihood estimator/最大似然估计值, 569
mean filter/均值滤波器, 183–187, 195, 216, 225, 226, 297, 435, 617–625, 649–660
    frequency response/频率响应, 190–191, 200, 436
mean squared edge distance/均方边缘距离 (SED), 444–447
median filter/中值滤波器, 196–198, 225, 226, 649–660
minimum filter see morphology, gray value/最小值滤波器
erosion/腐蚀
Minkowski addition/闵可夫斯基加法, 260–263, 268, 282
Minkowski subtraction/闵可夫斯基减法, 264–266, 268, 283
minor axis/短轴, 242–244, 250, 254, 256, 330
MLP see multilayer perceptron/多层感知器
moments/矩, 242–245, 249–254, 617–625
    invariant/恒定不变, 245
morphology/形态学, 256–288, 552, 649–660, 669–685
    anti-extensive operation/非外延性运算
        erosion/腐蚀, 266, 284
        Minkowski subtraction/闵可夫斯基减法, 265, 284
        opening/开运算, 272, 285
    duality/二元性
        dilation-erosion/膨胀 -腐蚀, 268, 284
        hit-or-miss transform/击中击不中变换, 271
        opening-closing/开 -闭运算, 274, 285
    extensive operation/外延性运算
        closing/闭运算, 275, 285

dilation/膨胀, 262, 283
Minkowski addition/闵可夫斯基加法, 260, 283
gray value/灰度值
　　closing/闭运算, 285–287
　　complement/补集, 284
　　dilation/膨胀, 282–283, 632–638
　　erosion/腐蚀, 283–284
　　gradient/梯度, 287–288
　　Minkowski addition/闵可夫斯基加法, 282
　　Minkowski subtraction/闵可夫斯基减法, 283
　　opening/开运算, 285–287
　　range/范围, 287–288
idempotent operation/幂运算
　　closing/闭运算, 274, 285
　　opening/开运算, 272, 285
increasing operation/增量运算
　　closing/闭运算, 275, 285
　　dilation/膨胀, 262, 283
　　erosion/腐蚀, 266, 284
　　Minkowski addition/闵可夫斯基加法, 261, 283
　　Minkowski subtraction/闵可夫斯基减法, 265, 284
　　opening/开运算, 272, 285
region/区域, 257–281, 649–660, 669–695
　　boundary/边界, 269–270
　　closing/闭运算, 274–276, 649–660
　　complement/补集, 258–259
　　difference/差, 258, 649–660, 669–685
　　dilation/膨胀, 261–264, 314, 649–660, 679–685
　　distance transform/距离变换, 278–281, 445, 448, 669–679
　　erosion/腐蚀, 265–270, 649–660, 679–685
　　hit-or-miss opening/击中击不中开运算, 273
　　hit-or-miss transform/击中击不中变换, 270–271, 277
　　intersection/交集, 258, 649–660, 669–679
　　Minkowski addition/闵可夫斯基加法, 260–263, 268
　　Minkowski subtraction/闵可夫斯基减法, 264–266, 268

　　opening/开运算, 271–273, 649–660, 669–679
　　skeleton/骨架, 276–278, 669–679
　　translation/平移, 259
　　transposition/变换, 259
　　union/并集，合并, 257–258, 649–660
　　structuring element/结构元素, 259, 269, 282, 283, 649–660
　　translation-invariant operation/平移不变运算
　　　　closing/闭运算, 274
　　　　opening/开运算, 272
multi-channel edge tensor/多通道边缘张量, 505
multilayer perceptron/多层感知器, 576–587, 597
　　evidence procedure/证明过程, 584
　　novelty detection/异常检测, 584–586
　　training/训练, 579–586
　　　　cross-entropy error/交叉熵误差, 580
　　　　regularization/正则化, 583–584
　　　　weight decay/权重衰减, 583–584
　　universal approximator/泛逼近器, 578, 579, 597

n

neighborhood/邻域, 234–236, 269–270, 278
neural network/神经网络, 574–587, 617–625, 704–710
　　activation function/激活函数
　　　　hyperbolic tangent/双曲正切, 578
　　　　logistic/逻辑, 578
　　　　rectified linear unit/修正线性单元, 600
　　　　sigmoid, 578–579
　　　　softmax, 578, 600
　　　　threshold/阈值, 575, 577
　　convolutional neural network/卷积神经网络, 598–606
　　　　convolution/卷积, 598–599
　　　　convolutional layer/卷积层, 601
　　　　cross-entropy error/交叉熵误差, 603
　　　　dataset augmentation/数据增强, 605
　　　　filter stride/滤波器步长, 599, 601
　　　　fine-tuning/微调, 602–603
　　　　learning rate/学习率, 604
　　　　minibatch/小批量, 603

momentum/动量, 604
novelty detection/异常检测, 606
pooling/池化, 600–601
receptive field/感受野, 600
regularization/正则化, 604–605
stochastic gradient descent/随机梯度下降, 603–604
universal approximator/泛逼近器, 600
weight decay/权重衰减, 604
multilayer perceptron/多层感知器, 576–587, 597, 617–625, 704–710
cross-entropy error/交叉熵误差, 580
evidence procedure/证明过程, 584
novelty detection/异常检测, 584–586
regularization/正则化, 583–584
training/训练, 579–586
universal approximator/泛逼近器, 578, 579, 597
weight decay/权重衰减, 583–584
single-layer perceptron/单层感知器, 574–576
nodal point/节点, 30
noise/噪声, 181–182
amplifier noise/放大器噪声, 79
dark current noise/暗电流噪声, 79
dark noise/暗噪声, 79
noise floor/本底噪声, 79
pattern noise/模式噪声, 82
photon noise/光子噪声, 78
quantization noise/量化噪声, 80
reset noise/复位噪声, 79
signal-to-noise ratio/信噪比, 80, 149, 152, 156
spatial noise/空间噪声, 82
speckle/散斑, 141
suppression/抑制, 182–198
temporal noise/随机噪声, 80
variance/方差, 181, 183, 184, 194, 297, 319
non-maximum suppression/非最大抑制, 291, 303, 310–311
normal distribution/正态分布, 569
normalized cross-correlation/归一化互相关, 406, 601, 430–433, 695–704
normalized moments/归一化的矩, 242, 250, 255, 330, 332

novelty detection see classification, novelty detection/异常检测
NTSC, 85
numerical aperture/数值孔径, 44
Nyquist frequency/奈奎斯特频率, 201

O

object distance/物距, 30, 37
object identification/物体识别, 1
object recognition see 3D object recognition/物体识别
object-side telecentric lens/物方远心镜头, 42–44, 46–47, 348, 357, 360–362
occlusion/闭塞，遮挡, 137, 141, 143–468
OCR see optical character recognition opening
opening/开运算, 271–273, 285–287, 649–660, 669–679
optical anti-aliasing filter/光学抗失真滤光片, 74
optical axis/光轴, 30, 357
optical character recognition/光学字符识别 OCR, 170, 206, 207, 240–241, 551–559, 617–625, 695–704
character segmentation/字符分割, 552–555, 617–625
touching characters/粘连字符, 553–555
classification see classification/分类
features/特征, 555–559, 617–625
image rectification/图像校正, 212, 216–217, 552
orientation/方位, 242–246, 250, 256
exterior/外部, 326, 351–352, 365, 373–380, 414, 422, 660–669, 685–695
world coordinate system/世界坐标系, 351–352, 365
interior/内部, 137, 140, 143, 326, 349–357, 363–370, 373–380, 391, 414, 422, 423, 486, 495, 660–669, 685–695
accuracy/准确度, 386–390
camera constant see orientation, interior,principal distance/摄像机常量
camera coordinate system/摄像机坐标系, 351–352, 392, 364–365
camera motion vector/摄像机运动向量, 363–364

distortion coefficient (division model)/畸变系数 (除法模型), 352–354, 357–358, 366, 386–390, 392, 397

distortion coefficients (polynomial model)/畸变系数 (多项式模型), 354–358, 392

focal length see orientation, interior, principal distance/景深

image plane coordinate system/成像面坐标系, 352, 356, 365–366

image plane distance/成像面距离, 360–362, 392

magnification/放大率, 352, 357

pixel size/像素尺寸, 356, 366, 392

principal distance/主距, 26, 36, 349–350, 352, 357, 365–366, 386–390

principal point/主点, 356, 366, 386–390, 392

projection center/投影中心, 26, 35, 137, 349, 392–393, 396, 400, 421

tilt angle/倾角, 358–362, 392

tilt axis angle/倾斜轴角度, 358–362, 392

relative/相对, 137, 143, 391, 392, 423

base/基, 137, 393

base line/基线, 396

outlier/离群值，异常值, 328, 331

outlier suppression/抑制离群值, 331–337, 339–340

Huber weight function/Huber 权函数, 333

iteratively reweighted least-squares/迭代再加权最小二乘, 333

random sample consensus/随机样本一致性, 335

RANSAC, 335

Tukey weight function/Tukey 权函数, 333

overall system gain/总系统增益, 80

p

PAL, 85

parallax/视差, 136, 140

paraxial approximation/旁轴近似, 28

pattern noise/模式噪声, 82

perspective camera/透视摄像机, 349–357, 382–384, 506

projection center/投影中心, 137, 349, 392–393, 396, 400, 421, 507–508

perspective lens/透视镜头, 47, 348, 506

perspective transformation/透视变换, 208–209, 216–217, 361, 476–479, 552

PFNC see GenICam, pixel format naming/convention PFNC

phase shift/相移, 147–150, 418–420

phase unwrapping/相位展开, 149, 152

photon noise/光子噪声, 78

photoresponse nonuniformity/光响应不均匀性, 82

pinhole camera/针孔摄像机, 26, 35–37, 350

projection center/投影中心, 26, 35

pixel/像素, 158

pixel clock/像素时钟, 89, 92

pixel vignetting/像素渐晕, 46

PLC see programmable logic controller

Plücker coordinates/Plucker 坐标系, 537, 538

line representation/线表示法, 537

polar coordinates/极坐标, 218

polar transformation/极坐标变换, 218–219, 552, 617–625

polarization/偏振, 13

polarizing filter/偏振滤光片, 17

polygonal approximation/多边形近似, 341–344

Ramer algorithm/Ramer 算法, 341–344

pose/位姿, 205–206, 351–352, 365, 425, 476, 478, 479, 486, 488, 496–498, 511, 516, 518, 527–528, 531, 536, 660–669

position detection/定位, 1

Powell lens/Powell 透镜, 139

precision/精度, 317–318

edge angle/棱角, 455

edges/边缘, 318–320

hardware requirements/硬件需求, 320

phase/相位, 148–149

Precision Time Protocol/精确时间协议, 116

Prewitt filter/Prewitt 滤波器, 306

principal distance/主距, 26, 36, 349–350

principal plane/主平面, 30

principal point/主点, 356, 366, 386–390, 392

principal ray/主光线, 34, 35

print inspection/印刷检测, 227–233
PRNU see photoresponse nonuniformity programmable logic controller 2
progressive scan/逐行扫描, 69, 89
projection center/投影中心, 26, 35, 137, 349, 392–393, 396, 400, 421
projective transformation/投影变换, 208–209, 216–217, 361, 426, 478, 479, 483, 485, 487, 507, 552
projector/投影仪
    laser/激光, 139, 140
        cylindrical lens/柱面透镜, 139
        Powell lens/Powell 透镜, 139
        raster lens/光栅透镜, 140
    random texture/随机纹理, 137–138
    structured light/结构光, 142
        digital light processing/数字光处理, 144
        digital micromirror device/数字微镜器件, 143–145
        liquid crystal on silicon display/硅基液晶显示器, 144
        liquid-crystal/液晶, 144
PTP see Precision Time Protocol/精确时间协议
pupil magnification factor/透光因子, 35, 362
pure dual quaternion/纯对偶四元数, 538
pure quaternion/纯四元数, 534

q

quadratic programming/二次规划, 590, 593, 596
quantization noise/量化噪声, 80
quantum efficiency/量子效应, 78, 79
quaternion/四元数, 533–536, 538
    advantages/优点, 535–536
    ambiguity/歧义, 535
    basis elements/基础要素, 534
    conjugation/共轭, 533
    exponential form/指数形式, 535
    interpolation/插值, 536
    inversion/反演, 534
    multiplication/乘法, 533, 535
    norm/标准，规范, 533
    overparameterization/过参数化, 535
    pure/纯, 534, 538
    rotation/翻转, 534–536

scalar part/标量部分, 533
unit/单元, 533–535, 538
vector part/矢量部分, 533

r

radiometric camera calibration/辐射摄像机标定, 170–180, 324, 423–424
    calibration target/标定板, 171
    chart-based/基于图表, 171–173
    chartless/无图表, 173–180
        defining equation/定义方程, 174
    gamma response function/伽马响应函数, 171, 179
    inverse response function/逆响应函数, 174
        discretization/离散, 175–176
        normalization/标准化，归一化, 177
        polynomial/多项式, 178
        smoothness constraint/平滑约束, 177–178
    response function/响应函数, 171, 174, 179
Ramer algorithm/Ramer 算法, 341–344
rank filter/排序滤波器, 198, 284
raster lens/光栅透镜, 140
reflection/反射, 12
reflectivity/反射率, 13
refraction/折射, 13, 27
refractive index/折射率, 27
region/区域, 160–164
    as binary image/二值图像表示 161, 236, 241, 257–258, 262, 268, 277
    boundary/边界, 247, 269–270
    characteristic function/特征函数, 161, 249, 250, 282
    complement/补集, 258–259
    connected components/连通部分，连通区域, 233–237, 413, 552, 617–625, 638–660
    convex hull/凸包, 246–247
    definition/定义, 160
    difference/差异, 258, 649–660, 669–685
    feature see features, region/特征
    intersection/交集, 258, 649–660, 669–679
    run-length representation/行程编码, 162–164, 235, 241, 242, 257, 258, 262, 277
    translation/平移, 259
    transposition/变换, 259

union/并集，合并, 257–258, 649–660
region of interest/感兴趣区域, 160, 206, 210, 264, 281, 314, 426
regularization/正则化, 583–584, 604–605
relative orientation/相对位姿, 137, 143, 391, 392, 423
    base/基, 137, 393
    base line/基线, 396
reset noise/复位噪声, 79
RGB video/RGB 视频, 88
rigid 3D transformation/刚性三维变换, 352, 531, 539, 536–539, 541, 543
rigid transformation/刚性变换, 351, 425, 442, 458, 475, 660–669
robot/机器人, 526–551
    articulated/关节 (关节机器人), 527–528, 547
    base/基, 527, 531
    controller/控制器, 531
    end effector/末端执行器, 528
    joints/关节, 527–528, 547
    kinematics/运动学, 528
    mechanical interface/机械接口, 528, 529
    movement/运动, 532, 542
    SCARA, 547
    tool/工具, 528, 529, 531
    tool center point/工具中心点, 529
ROI see region of interest/感兴趣区域
rolling shutter/卷帘快门，行曝光, 71
rotation/旋转, 207, 212, 245, 425, 442, 552, 649–660
R-table/R 表, 454
run-length encoding/行程编码, 162–164, 235, 241, 242, 257, 258, 262, 277

S

sagittal focal surface/弧矢焦平面, 56
sagittal image/矢形图像, 55
saturation capacity/饱和量, 81
scaling/缩放, 207, 216, 245, 425, 442
SCARA robot/平面关节型机器人, 547
Scheimpflug lens/沙姆镜头, 59, 136, 140, 142, 419
Scheimpflug line/沙姆直线, 50
Scheimpflug optics/沙姆光学, 53
Scheimpflug principle/沙姆定律, 48–53

screw/螺旋, 536–538, 540, 549
    angle/角度, 536, 537, 539–541
    axis/轴, 536, 537, 541, 548, 549
    Chasles' theorem/Chasles 定理, 536
    direction/方向, 536, 537, 539
    dual quaternion/对偶四元数, 538–539
    moment/矩, 536, 537, 539
    Plücker coordinates/Plucker 坐标系, 537
    rotation/旋转, 537
    translation/平移, 536, 537, 539–541
screw theory/螺旋理论, 532, 533, 536–537
segmentation/分割, 220–239, 638–649, 669–685
    connected components/连通区域, 233–237, 413, 552, 617–625, 638–660
    dynamic thresholding/动态阈值法, 224–228, 287, 417, 552, 617–625, 649–660
    hysteresis thresholding/滞后阈值分割, 311–313
    subpixel-precise thresholding/亚像素精度阈值分割, 237–239, 315, 626–632, 638–649
    thresholding/阈值分割, 220–224, 413, 552, 638–649, 669–685
    automatic threshold selection/动态阈值选择, 221–224, 552
    variation model/变差模型, 228–233, 632–638, 649–660
sensor/传感器, 158
serial interface/串口, 4
SFNC see GenICam, standard features naming convention and GenICam, GenTL, standard features naming onvention
shape inspection/形状检测, 2
shape-based 3D matching see 3D object recognition, shape-based 3D matching/基于形状的三维匹配
shape-based matching/基于形状的匹配, 467–475, 477, 479, 482, 493–494, 497, 506, 608–616, 660–669, 695–704
sheet of light reconstruction/基于激光三角测量法 (片光技术) 重构, 52, 53, 140–142, 412–416, 711–722
    calibration/标定, 414–416
    extraction of laser line/激光线提取, 413–414
    occlusion/闭塞，遮挡, 141, 413

shutter/快门
    electronic/电子的, 68
    global/全局, 71, 136
    mechanical/机械, 65, 67
    rolling/旋转, 71
signal-to-noise ratio/信噪比, 79–80, 149, 152, 156, 300, 319, 320, 455
similarity measure/相似度量, 426–434, 469–472
    absolute sum of normalized dot products/归一化点积绝对总和, 471
    normalized cross-correlation/归一化互相关, 406, 430–433, 695–704
    sum of absolute gray value differences/绝对灰度值插值和, 405, 427–429, 432–433
    sum of absolute normalized dot products/绝对归一化点积总和, 471, 501
    sum of normalized dot products/归一化点积总和, 470–471
    sum of squared gray value differences/灰度值平方差总和, 405, 427–429
    sum of unnormalized dot products/非归一化点积总和, 469–470
similarity transformation/相似变换, 426, 442, 458, 476, 482, 484, 497, 638–649
singular value/奇异值, 543, 548
singular value decomposition/奇异值分解, 543
skeleton/骨架, 276–278, 669–679
skew/不对称，倾斜, 207
slant/不对称，倾斜, 207
smallest enclosing circle/最小外接圆, 245–246, 255
smallest enclosing rectangle/最小外接矩形, 245–246, 255, 556, 617–632, 649–660
smart camera/智能摄像机, 3
smoothing filter/平滑滤波器, 181–198
    Gaussian/高斯滤波器, 191–195, 216, 223, 225–226, 301, 308, 435
    frequency response/频率响应, 194, 200
    mean/均值, 183–187, 195, 216, 225–226, 297, 435, 617–625, 649–660
    frequency response/频率响应, 190–191, 200, 436
    median/中值, 196–198, 225, 226, 649–660
    optimal/最优, 191–193, 301, 308

spatial averaging/空间平均, 183–187
temporal averaging/时间平均, 182–183, 230
SNR see signal-to-noise ratio/SNR 信噪比
Sobel filter/Sobel 滤波器, 306, 559
spacetime stereo/时空立体视觉, 411–412
spatial averaging/空间平均, 183–187
spatial domain/空域, 198
spatial noise/空间噪声, 82
speckle noise/斑点噪声, 141
spectral response/光谱响应
    Gaussian filter/高斯滤波器, 194, 200
    human visual system/人类视觉系统, 72
    mean filter/均值滤波器, 190–191, 200, 436
    sensor/传感器, 72–73, 75, 158
speed of light/光速, 8, 152
spherical aberration/球面像差, 53–54
stereo geometry/立体几何, 135–137, 391–404
    corresponding points/对应点, 137–138, 395
    disparity/视差, 402–404
    epipolar line/极线, 395
    epipolar plane/极平面, 395
    epipolar standard geometry/外极线标准几何, 398–399, 685–695
    epipole/核点, 396
    image rectification/图像校正, 399–402, 685–695
    parallax/视差, 136, 140
stereo matching/立体匹配, 404–412, 685–695
    occlusion/闭塞，遮挡, 137, 407, 410
    robust/鲁棒性, 409–411
    disparity consistency check/视差一致性检验, 410–411
    excluding weakly textured areas/排除弱纹理区域, 410
    similarity measure/相似度量
    normalized cross-correlation/归一化互相关, 406
    sum of absolute gray value differences/绝对灰度值插值和, 405
    sum of squared gray value differences/灰度值平方差求和, 405
    spacetime stereo/时空立体视觉, 411–412
    subpixel-accurate/亚像素精度, 408
    window size/窗口尺寸, 138, 408–409

stereo reconstruction/立体重构, 48, 135–138, 390–412, 512, 524, 685–695
stochastic process/随机过程, 181, 183
　　ergodic/遍历, 184
　　stationary/静止, 181
structured light reconstruction/结构光重构, 50, 52–53, 142–150, 416–424
　　binary code patterns/二值编码模式, 145–146
　　camera calibration/摄像机标定, 420–423
　　fringe projection/光栅投影, 147–150
　　Gray code decoding/格雷码解码, 417–420
　　Gray code patterns/格雷码图案, 146, 150
　　occlusion/闭塞, 遮挡, 143, 418
　　phase decoding/解相, 418–420
　　phase shift/相移, 147–150
　　projector calibration/投影仪标定, 420–423
　　radiometric calibration/辐射标定, 423–424
　　stripe decoding/条纹解码, 416–420
structuring element/结构元素, 259, 269, 282, 283, 649–660
subpixel-precise contour/亚像素精度轮廓, 164–165
　　convex hull/凸包, 255
　　features see features, contour/特征
subpixel-precise thresholding/亚像素精度阈值分割, 237–239, 315, 626–632, 638–649
sum of absolute gray value differences/绝对灰度值插值总和, 427–429, 405, 432–433
sum of absolute normalized dot products/绝对归一化点积总和, 471, 501
sum of normalized dot products/归一化点积总和, 470–471
sum of squared gray value differences/灰度值平方差总和, 405, 427–429
sum of unnormalized dot products/非归一化点积总和, 469–470
support vector machine/支持向量机, 587–596
　　kernel/核, 591
　　　　Gaussian radial basis function/高斯径向基函数, 592
　　　　homogeneous polynomial/齐次多项式, 592
　　　　inhomogeneous polynomial/非齐次多项式, 592

　　　　sigmoid, 592
　　margin/边缘, 589
　　margin errors/边缘误差, 592, 596
　　novelty detection/异常检测, 594–596
　　ν-SVM/支持向量机, 592–593
　　one-versus-all/一对所有, 593–594
　　one-versus-one/一对一, 593–594
　　separating hyperplane/分离超平面, 588–591
　　universal approximator/泛逼近器, 592
surface inspection/表面检测, 2
surface vertex/表面顶点, 30
surface-based 3D matching/基于表面的三维匹配, 478, 510–526
see 3D object recognition, surface-based 3D matching/基于表面的三维匹配, 711–734
SVD, 543
S-Video, 88
SVM see support vector machine

t

tangential focal surface/切向焦面, 56
tangential image/切向图像, 55
TCP see Transmission Control Protocol/TCP 传输控制协议
telecentric bright-field back light illumination/明场背光平行照明, 24–25
telecentric camera/远心摄像机, 348–357, 381–382, 660–669
telecentric illumination/平行光照明, 18
telecentric lens/远心镜头, 42–47, 348
　　bilateral/双远心, 45–47, 348, 357, 360
　　image-side/像方, 46–47, 348, 357, 360
　　object-side/物方, 42–44, 46–47, 348, 357, 360–362
template matching/模板匹配, 228, 424–477, 553
　　clutter/混乱, 443, 448, 456, 468
　　erosion/腐蚀, 267
　　generalized Hough transform/广义霍夫变换, 450–456
　　　　accumulator array/累加器数组, 451, 454-455
　　　　R-table/R 表, 454
　　geometric hashing/几何哈希法, 457–460

geometric matching/几何匹配, 456–475
Hausdorff distance/Hausdorff 距离, 447–450
hierarchical search/分层搜索, 438–441, 473
hit-or-miss transform/击中击不中变换, 270
hypothesize-and-test paradigm/假设与测试范例, 456
image pyramid/图像金字塔, 435–441, 472–473, 501, 502
linear illumination changes/线性照明变化, 430
matching geometric primitives/匹配几何基元, 461–466
mean squared edge distance/均方边缘距离, 444–447
nonlinear illumination changes/非线性照明变化, 443, 456, 468, 470–471
occlusion/闭塞，遮挡, 443, 448, 456, 468
opening/开运算, 272, 649–660
robust/鲁棒性, 443–476, 608–616, 632–638, 649–669, 695–704
rotation/旋转, 441–443
scaling/缩放, 441–443
shape-based matching/基于形状的匹配, 467–475, 477, 479, 482, 501, 608–616, 660–669, 695–704
similarity measure/相似度量, 426–434, 469–472
    absolute sum of normalized dot products/归一化点积绝对总和, 471
    normalized cross-correlation/归一化互相关, 406, 430–433, 695, 704
    sum of absolute gray value differences/绝对灰度值插值和, 405, 427–429, 432–433
    sum of absolute normalized dot products/绝对归一化点积总和, 471, 501
    sum of normalized dot products/归一化点积总和, 470–471
    sum of squared gray value differences/灰度值平方差总和, 405, 427–429
    sum of unnormalized dot products/非归一化点积总和, 469–470
stopping criterion/停止标准, 432–434
    normalized cross-correlation/归一化互相关, 433
    sum of absolute gray value differences/绝对灰度值插值总和, 432–433
    sum of normalized dot products/归一化点积总和, 472
    subpixel-accurate/亚像素精度, 441, 473–474
    translation/平移, 426–441
temporal averaging/时间平均, 182–183, 230
temporal noise/随机噪声, 80
texture/纹理
    removal/移除, 204, 205
thick lens/厚透镜, 29–32
    cardinal elements/基本元素, 31
    focal length/焦距, 30
    focal point/焦点, 29
    image distance/像距, 30, 37
    magnification/放大率, 31, 40, 48, 49, 136, 142, 352
    nodal point/节点, 30
    object distance/物距, 30, 37
    optical axis/光轴, 30, 357
    principal plane/主平面, 30
    surface vertex/表面顶点, 30
thresholding/阈值分割, 220–224, 413, 552, 638–649, 669–685
    automatic threshold selection/自动阈值选择, 221–224, 552
    subpixel-precise/亚像素精度, 237–239, 315, 626–632, 638–649
tilt lens/倾斜镜头, 48–53, 136, 140, 142, 357–362, 419
    hinge line/铰合线, 51
    Scheimpflug line/沙姆直线, 50
tilted image plane/倾斜像平面, 50-52, 136, 140, 142, 357–362
time-of-flight camera/飞行时间摄像机 (TOF 摄像机), 151–156
    continuous-wave-modulated/连续波调制信号, 151–153
    distance computation/距离计算, 153
    distance range/距离范围, 152
    phase demodulation/相位解调, 152
    random errors/随机误差, 153

resolution/分辨率, 153
scene intensity/场景灰度值, 152
systematic errors/系统误差, 153
pulse-modulated/脉冲调制, 153–156
distance computation/距离计算, 156
distance range/距离范围, 154
random errors/随机误差, 156
resolution/分辨率, 156
systematic errors/系统误差, 156
time-of-flight computation/飞行时间计算, 154–156
TOF camera see time-of-flight camera/飞行时间摄像机
transformation/变换
affine/仿射, 206–207, 245, 426, 457, 469, 660–669, 695–704
geometric/几何的, 205–219, 552, 617–625, 638–669, 695–704
gray value/灰度值, 166–170
image/图像, 209–219, 228, 617–625, 632–638, 660–669, 695–704
local deformation/局部畸变, 477
perspective/透视, 208–209, 216–217, 361, 476–479, 552
polar/极, 218–219, 552, 617–625
projective/投影, 208–209, 216–217, 361, 426, 478, 479, 483, 485, 487, 507, 552
rigid/刚性, 351, 425, 442, 458, 475, 660–669
rigid 3D/刚性三维, 352, 531, 536–539, 541, 543
rotation/旋转, 207, 212, 245, 425, 442, 552, 649–660
scaling/缩放, 207, 216, 245, 425, 442
similarity/相似性, 426, 442, 458, 476, 482, 484, 497, 638–649
skew/不对称, 207
slant/不对称, 207
translation/平移, 207, 245, 259, 425, 442
translation/平移, 207, 245, 259, 425, 442
translation-invariant operation/平移不变运算
closing/闭运算, 274
opening/开运算, 272

Transmission Control Protocol/传输控制协议, 112
transmittance/透光率, 14
transposition/变换, 259
triangulation/三角测量, 135
sheet of light sensor/激光三角测量法 (片光) 传感器, 139
stereo sensor/立体视觉传感器, 137
structured light sensor/结构光传感器, 142
trigger/触发, 64, 68, 132–133
Tukey weight function/Tukey 权函数, 333

u

U3V see USB3 Vision
U3VCP see USB3 Vision, USB3 Vision Control Protocol/控制协议
U3VSP see USB3 Vision, USB3 Vision Streaming Protocol/数据流协议
UDP see User Datagram Protocol
union/合并, 257–258, 649–660
unit dual quaternion/单位对偶四元数, 538–539, 540, 543
transformation matrix/变换矩阵, 539
unit quaternion/单位四元数, 533–535, 538
universal approximator/泛逼近器, 578, 579, 592, 597, 600
universal serial bus see USB/USB 通用串行总线
USB
bulk data transfers/批量数据传输, 107, 110
control transfers/控制传输, 107, 110
interrupt data transfers/中断数据传输, 107, 110
isochronous data transfers/同步数据传输, 107, 110
USB 2.0, 105–108
USB3 Vision, 108–111
U3VCP see USB3 Vision, USB3 Vision Control Protocol/控制协议
U3VSP see USB3 Vision, USB3 Streaming Control Protocol/控制协议
USB3 Vision Streaming Protocol/USB3 Vision 控制协议, 111

USB3 Vision Control Protocol/USB3 Vision 数据流协议, 110
User Datagram Protocol/用户数据报协议 UDP, 112

v

variation model/变差模型, 228–233, 632–638, 649–660
vertical blanking interval/场消隐间隔, 87
vertical overflow drain/垂直溢流沟道, 68
vertical synchronization pulse/帧同步脉冲, 87
video signal/视频信号
    analog/模拟
        back porch/后沿, 87
        CCIR, 85
        EIA-170, 85
        front porch/前沿, 87
        horizontal blanking interval/行消隐间隔, 87
        horizontal synchronization pulse/行同步脉冲, 87
        interlaced scan/隔行扫描, 86
        NTSC, 85
        PAL, 85
        progressive scan/逐行扫描, 89
        vertical blanking interval/场消隐间隔, 87
        vertical synchronization pulse/帧同步脉冲, 87
    color/彩色
        composite video/复合视频, 88
        RGB, 88
        S-Video, 88
        Y/C, 88
    digital/数字, 92–116, 126–131
        Camera Link, 93–96
        Camera Link HS, 96–99
        CoaXPress, 99–102
        FireWire see video signal, digital, IEEE 1394
        frame valid/帧有效, 92
        GenICam GenTL, 102, 119, 126–131
see video signal, digital, GenICam GenTL standard features naming convention
        GenICam GenTL standard features naming convention, 126–131
        Gigabit Ethernet see video signal, digital, GigE Vision/千兆以太网
    GigE Vision
        GigE Vision, 113–116
        IEEE 1394, 103–105
        IIDC, 103–105
        line valid/行有效, 92
        low-voltage differential signaling/低电压差分信号 (LVDS), 94
        LVDS see video signal, digital, low-voltage differential signaling LVDS
        pixel clock/像素时钟, 92
        USB 2.0, 105–108
        USB3 Vision, 108–111
vignetting/光晕, 60–61, 172

w

weight decay/权重衰减, 583–584, 604
weight function/权函数
    Huber, 333
    Tukey, 333
world coordinates/世界坐标系
    from sheet of light reconstruction/使用激光三角测量法 (片光技术) 重构, 52, 53, 140–142, 412–416
        calibration/标定, 414–416
        extraction of laser line/提取激光线, 413–414
    from single image/使用单幅图像, 326–327, 380–385, 414, 415, 486–487
        line scan camera/线阵摄像机, 384–385
        perspective camera/透视摄像机, 382–384
        telecentric camera/远心摄像机, 381–382
    from stereo reconstruction/使用立体视觉重构, 135–138, 390–412
    from structured light reconstruction/使用结构光重构, 50, 52–53, 142–150, 416–424
        camera calibration/摄像机标定, 420–423
        Gray code decoding/格雷码解码, 417–420
        phase decoding/解相, 418–420

projector calibration/投影仪标定, 420–423

radiometric calibration/辐射标定, 423–424

stripe decoding/条纹解码, 416–420

x

xenon lamp/氙灯, 10

XML, 99, 102, 110, 114, 129

y

Y/C video / Y/C 视频, 88

z

zero-crossing/零交点, 291, 293, 315